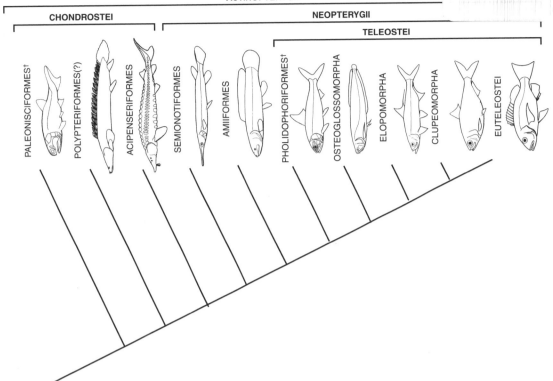

Phylogenetic relationships among actinopterygian fishes (see Fig. 11.13).

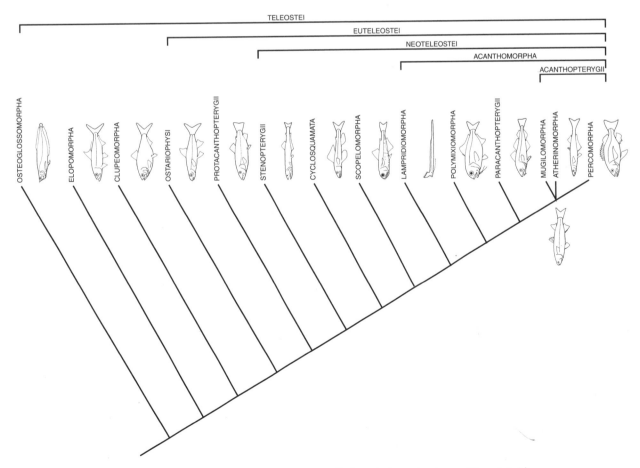

Phylogenetic relationships among living teleosts (see Fig. 14.1).

THE DIVERSITY
OF FISHES

THE DIVERSITY OF FISHES

GENE S. HELFMAN

Institute of Ecology
University of Georgia
Athens, Georgia

BRUCE B. COLLETTE

National Marine Fisheries Service
National Museum of Natural History
Washington, District of Columbia

DOUGLAS E. FACEY

Department of Biology
St. Michael's College
Colchester, Vermont

b

**Blackwell
Science**

Blackwell Science Editorial offices:

Commerce Place, 350 Main Street,
    Malden, Massachusetts 02148, USA

Osney Mead, Oxford OX2 0El, England

25 John Street, London WC1N 2BL, England

23 Ainslie Place, Edinburgh EH3 6AJ, Scotland

54 University Street, Carlton, Victoria 3053, Australia

Other Editorial offices:

Blackwell Wissenschafts-Verlag GmbH Kurfürstendamm 57,
    10707 Berlin, Germany

Zehetnergasse 6, A-1140 Vienna, Austria

Distributors:

USA

Blackwell Science, Inc.
Commerce Place
350 Main Street
Malden, Massachusetts 02148
    (Telephone orders: 800-215-1000 or 617-388-8250;
    fax orders: 617-388-8270)

CANADA

Copp Clark Professional
200 Adelaide Street, West, 3rd Floor
Toronto, Ontario
Canada, M5H 1W7
    (Telephone orders: 416-597-1616 or 1-800-815-9417;
    fax orders: 416-597-1617)

AUSTRALIA

Blackwell Science Pty, Ltd.
54 University Street
Carlton, Victoria 3053
    (Telephone orders: 03-9347-0300;
    fax orders: 03-9349-3016)

OUTSIDE NORTH AMERICA AND AUSTRALIA

Blackwell Science, Ltd.
c/o Marston Book Services, Ltd.
P.O. Box 269
Abingdon
Oxon OX14 4YN
England
    (Telephone orders: 44-01235-465500;
    fax orders: 44-01235-465555)

Acquisitions: Jane Humphreys
Development: Kathleen Broderick
Production: Ellen Samia
Manufacturing: Lisa Flanagan
Typeset by Northeastern Graphic Services, Inc.
Printed and bound by Braun-Brumfield, Inc.

Front cover illustrations, clockwise from upper right:

1. Orangethroat darter, *Etheostoma spectabile*, by S. F. Denton, from the Smithsonian poster, *Drawn from the Sea*.
2. Lionfish, *Pterois antennata*, by K. Ito, from *Drawn from the Sea*.
3. Razorback sucker, *Xyrauchen texanus*, by J. R. Tomelleri, used with permission of the artist.
4. Stonefish, *Erosa erosa*, by K. Morita, from *Drawn from the Sea*.
5. Holiday darter, *Etheostoma brevirostrum*, by J. R. Tomelleri, used with permission of the artist.
6. Yellowtail coris, *Coris gaimard*, by K. Morita, from *Drawn from the Sea*.
7. Saddleback butterflyfish, *Chaetodon ephippium*, by K. Ito, from *Drawn from the Sea*.
8. Rainbow trout, *Oncorhynchus mykiss*, by C. B. Hudson, from *Drawn from the Sea*.

Back cover, clockwise from upper left:

1. American eel, *Anguilla rostrata*, by C. B. Hudson, from *Drawn from the Sea*.
2. Southern redbelly dace, *Phoxinus erythrogaster*, by J. R. Tomelleri, used with permission of the artist.
3. Spotted gar, *Lepisosteus oculatus*, by J. R. Tomelleri, used with permission of the artist.

Library of Congress Cataloging-in-Publication Data

Helfman, Gene S.
    The diversity of fishes / Gene S. Helfman. Bruce B.
Collette, Douglas E. Facey.
        p.   cm.
    Includes bibliographical references and index.
    ISBN 0-86542-256-7
    1. Fishes. 2. Fishes—Adaptation. I. Collette, Bruce B.
II. Facey, Douglas E. III. Title.
QL615.H44 1997
597—dc21   96-51651
                                                              CIP

DEDICATION

To our parents, for their encouragement of our nascent interest in things biological;

To our wives—Judy, Sara, and Janice—for their patience and understanding during the production of this volume;

And to the diversity of fishes, with the hope that students and lovers of fishes will do what is needed to preserve that diversity for future generations.

Contents

Preface

Two types of people are likely to pick up this book: those with an interest in fishes and those with a fascination for fishes. This book is written by the latter, directed at the former, with the intent of turning interest into fascination.

Our two major themes are **adaptation** and **diversity**. These themes recur throughout the chapters. Wherever possible, we have attempted to understand the adaptive significance of an anatomical, physiological, ecological, or behavioral trait, pointing out how the trait affects an individual's probability of surviving and reproducing. Our focus on diversity has prompted us to provide numerous lists of species that display particular traits, emphasizing the parallel evolution that has occurred repeatedly in the history of fishes, as different lineages exposed to similar selection pressures have converged on similar adaptations.

The intended audience of this book is the senior undergraduate or graduate student taking an introductory course in ichthyology, although we also hope that the more seasoned professional will find it a useful review and reference for many topics. We have written this book assuming that the student has had an introductory course in comparative anatomy of the vertebrates, with at least background knowledge in the workings of evolution. To understand ichthyology, or any natural science, a person should have a solid foundation in evolutionary theory. This book is not the place to review much more than some basic ideas about how evolutionary processes operate and their application to fishes, and we strongly encourage all students to take a course in evolution.

Although a good comparative anatomy or evolution course will have treated fish anatomy and systematics at some length, we go into considerable detail in our introductory chapters on the anatomy and systematics of fishes. The nomenclature introduced in these early chapters is critical to understanding much of the information presented later in the book. Extra care spent reading those chapters will reduce confusion about terminology used in most other chapters.

More than 25,000 species of fishes are alive at present. Students at the introductory level are likely to be overwhelmed by the diversity of taxa and of unfamiliar names. To facilitate this introduction, we have been selectively inconsistent in our use of scientific versus common names. Some common names are likely to be familiar to most readers, such as salmons, minnows, tunas, and freshwater sunfishes; for these and many others, we have used the common family designation freely. For other, less familiar groups (e.g., Sundaland noodlefishes, trahiras, morwongs), we are as likely to use scientific as common names. Many fish families have no common English name, and for these we use the anglicized scientific designation (e.g., cichlids, galaxiids, labrisomids). In all cases, the first time a family is encountered in a chapter we give the scientific family name in parentheses after the common name. Both scientific and common designations for families are also listed in the index.

According to an accepted convention, where lists of families occur, taxa are listed in phylogenetic order. We follow Robins et al. (1991) on names of North American fishes and Nelson (1994) on classification and names of families and of higher taxa. In the few instances where we disagree with these sources, we have tried to explain our rationale.

Any textbook is a compilation of facts. Every statement of fact results from the research efforts of usually several people, often over several years. Students often lose sight of the origins of this information, namely the effort that has gone into verifying an observation, repeating an experiment, or making the countless measurements necessary to establish the validity of a fact. An entire dissertation, representing 3 to 5 or more years of intensive work, may be distilled down to a single sentence in a textbook. It is our hope that as you read through the chapters in this book, you will not only appreciate the diversity of adaptation in fishes but also consider the many ichthyologists who have put their fascination to practical use to obtain the facts and ideas we have compiled here. To acknowledge these efforts, and because it is just good scientific practice, we have gone to considerable lengths to cite the sources of our information in the text, which correspond to the entries in the lengthy bibliography at the end of the book. This will make it possible for the reader to go to a cited work and learn the details of a study that we can treat only superficially. Additionally, the end of each chapter contains a list of supplemental readings, including books or longer review articles that can provide an interested reader with a much greater understanding of the subjects covered in the chapter.

This book is not designed as a text for a course in fisheries science. It contains relatively little material directly relevant to such applied aspects of ichthyology as commercial or sport fisheries or aquaculture; several good texts and reference books deal specifically with those topics (for starters, see the edited volumes by Lackey and

Nielsen 1980; Nielsen and Johnson 1983; Schreck and Moyle 1990; and Kohler and Hubert 1993). We recognize, however, that many students in a college-level ichthyology class are training to become professionals in those or related disciplines. Our objectives here are to provide such readers with enough information on the general aspects of ichthyology to make informed, biologically sound judgments and decisions, and to gain a larger appreciation of the diversity of fishes beyond the relatively small number of species with which fisheries professionals often deal.

Adaptations Versus Adaptationists

Our emphasis throughout this text on evolved traits and the selection pressures responsible for them does not mean that we view every characteristic of a fish as an adaptation. It is important to realize that a living animal is the result of past evolutionary events and that animals will be adapted to current environmental forces only if those forces are similar to what has happened to the individual's ancestors in the past. Such **phylogenetic constraints** arise from the long-term history of a species.

For example, tunas are masters of the open sea as a result of a streamlined morphology, large locomotory muscle mass connected via efficient tendons to fused tail bones, and highly efficient respiratory and circulatory systems. However, they rely on water flowing passively into their mouths and over their gills to breath and have reduced the branchiostegal bones in the throat region that help pump water over their gills. Therefore, tunas are constrained phylogenetically from using habitats or foraging modes that require them to stop and hover, because by ceasing swimming they would also cease breathing.

Animals are also imperfect because characteristics that have evolved in response to one set of selective pressures often create problems with respect to other pressures. Everything in life involves a **trade-off**, another recurring theme in this text. The elongate pectoral fins ("wings") of a flyingfish allow the animal to glide over the water's surface faster than it can swim through the much denser water medium. However, the added surface area of the enlarged fins creates drag when the fish is swimming. This drag increases costs in terms of a need for larger muscles to push the body through the water, requiring greater food intake, time spent feeding, and so forth. The final mix of traits evolved in a species represents a compromise involving often-conflicting demands placed on an organism. Because of phylogenetic constraints, trade-offs, and other factors, some fishes and some characteristics of fishes appear to be and are poorly adapted. Our emphasis in this book is on traits for which function has been adequately demonstrated or appears obvious. Skepticism about apparent adaptations can only lead to greater understanding of the complexities of the evolutionary process. We encourage and try to practice such skepticism.

Acknowledgments

This book results from effort expended and information acquired over most of our professional lives. Each of us has been tutored, coaxed, aided, and instructed by many fellow scientists. A few people have been particularly instrumental in facilitating our careers as ichthyologists and deserve special thanks: George Barlow, John Heiser, Bill McFarland, and Jack Randall for GSH; Ed Raney, Bob Gibbs, Ernie Lachner, and Dan Cohen for BBC; Gary Grossman and George LaBar for DEF. The help of many others is acknowledged and deeply appreciated, although they go unmentioned here.

Specific aid in the production of this book has come from an additional host of colleagues. Students in our ichthyology classes have written term papers that served as literature surveys for many of the topics treated here; they have also critiqued drafts of chapters. Many colleagues have answered questions, commented on chapters and chapter sections, loaned photographs, and sent us reprints, both requested and volunteered. Singling out a few who have been particularly helpful, we thank C. Barbour, T. Berra, J. Beets, W. Bemis, J. Briggs, E. Brothers, S. Concelman, J. Crim, D. Evans, S. Hales, B. Hall, D. Helfmeyer, C. Jeffrey, D. Johnson, G. Lauder, C. Lowe, D. Mann, D. Martin, A. McCune, J. Meyer, J. Miller, J. Moore, L. Parenti, L. Privitera, T. Targett, B. Thompson, P. Wainwright, J. Webb, S. Weitzman, D. Winkelman, J. Willis, and G. Wippelhauser. Joe Nelson provided us logistic aid and an early draft of the classification incorporated into the third edition of his indispensable *Fishes of the World*. Often animated and frequently heated discussions with ichthyological colleagues at annual meetings of the American Society of Ichthyologists and Herpetologists have been invaluable for separating fact from conventional wisdom. Academic departmental administrators gave us encouragement and made funds and personnel available at several crucial junctures. At the University of Georgia we thank J. Willis (Zoology), R. Damian (Cell Biology), and G. Barrett, R. Carroll, and R. Pulliam (Ecology) for their support. At St. Michael's College, we thank D. Bean (Biology). Gretchen Hummelman and Natasha Rajack labored long and hard over copyright permissions and many other details. Cecile Duray-Bito impressed us with her diligence in producing accurate illustrations. The personnel of Blackwell Science, especially Jane Humphreys and Ellen Samia, assisted us during all stages of production.

Finally, a note on the accuracy of the information contained in this text. As Nelson Hairston Sr. has so aptly pointed out, "Statements in textbooks develop a life independent of their validity." We have gone to considerable lengths to get our facts straight or to admit where uncertainties lie. We accept full responsibility for the inevitable errors that do appear, and we welcome hearing about them. Please write directly to us with any corrections or comments. Chief responsibilities fell on GSH for Chapters 1, 8 through 15, and 17 through 25; on BBC for Chapters 2 through 4 and 16; and on DEF for Chapters 5 through 7.

PART 1

Introduction

The Science of Ichthyology

Fishes make up more than half of the 48,000 species of living vertebrates. Along with this remarkable taxonomic diversity comes an equally impressive habitat diversity. Today, and in the past, fishes have occupied nearly all major aquatic habitats, from lakes and polar oceans that are ice-covered through much of the year, to tropical swamps, temporary ponds, intertidal pools, ocean depths, and all the more benign environments that lie within these various extremes.

Fishes have been ecological dominants in aquatic habitats through much of the history of complex life. To colonize and thrive in such a variety of environments, fishes have evolved obvious and striking anatomic, physiological, behavioral, and ecological adaptations. Students of evolution in general and of fish evolution in particular are aided by an extensive fossil record dating back 500 million years. All told, fishes are excellent showcases of the evolutionary process, exemplifying the intimate relationship between form and function, between habitat and adaptation. Adaptation and diversity are interwoven throughout the evolutionary history of fishes and are a recurring theme throughout this book.

WHAT IS A FISH?

It may in fact be unrealistic to attempt to define a *fish*, given the diversity of adaptation that characterizes the thousands of species alive today, each with a unique evolutionary history going back millions of years and including many more species. Recognizing this diversity, one can define a fish as "a poikilothermic, aquatic chordate with appendages (when present) developed as fins, whose chief respiratory organs are gills and whose body is usually covered with scales" (Berra 1981, xxi), or more simply, a fish is an aquatic vertebrate with gills and with limbs in the shape of fins (Nelson 1994). To most biologists, the term *fish* is not so much a taxonomic ranking as a convenient description for aquatic organisms as diverse as hagfishes, lampreys, sharks, rays, lungfishes, sturgeons, gars, and advanced ray-finned fishes.

Definitions are dangerous, since exceptions are often viewed as falsifications of the statement. Exceptions to the definitions above do not negate them but instead give clues to adaptations arising from particularly powerful selection pressures. Hence loss of scales and fins in many eel-shaped fishes tells us something about the normal function of these structures and their inappropriateness in benthic fishes with an elongate body.

Similarly, homeothermy in tunas and lamnid sharks instructs us about the metabolic requirements of fast-moving predators in open-sea environments, and lungs or other accessory breathing structures in lungfishes, gars, African catfishes, and gouramis indicate periodic environmental conditions where gills are inefficient for transferring water-dissolved oxygen to the blood. Deviation from "normal" in these and other exceptions are part of the lesson that fishes have to teach us about evolutionary processes.

The Diversity of Fishes

Numerically, valid scientific descriptions exist for approximately 24,600 living species of fishes in 482 families and 57 orders (Nelson 1994; Table 1.1). (Note: *Fish* is singular and plural for a single species; *fishes* refers to more than one species.) (See Fig. 1.1.) Of these, 85 are jawless fishes (hagfishes and lampreys); 850 are cartilaginous sharks, skates, rays, and chimaeras; and the remaining 23,000 or more species are bony fishes. Many others remain to be formally described. When broken down by major habitats, 41% of species live in freshwater, 58% live in seawater, and 1% move between freshwater and the sea during their life cycles (Cohen 1970).

Geographically, the highest diversities are found in the tropics. The Indo-West Pacific area that includes the western Pacific and Indian Oceans and the Red Sea have the highest diversity for a marine area, whereas Southeast Asia, South America, and Africa have the most freshwater fishes. Fishes occupy essentially all aquatic habitats that have liquid water throughout the year, including thermal and alkaline springs, hypersaline lakes, sunless caves, anoxic swamps, temporary ponds, torrential rivers, wave-swept coasts, and high-altitude and high-latitude environments. The altitudinal record is set by some nemacheiline river loaches that inhabit Tibetan hot springs at elevations of 5200 m. The record for unheated waters is Lake Titicaca in northern South America, where pupfishes live at an altitude of 3812 m. The deepest living fishes are cusk-eels, which occur 8000 m down in the deep sea.

Diversity in body size ranges from *Trimmatom nanus*, an Indian Ocean goby that matures at 8 mm long, to the world's largest fish, the 12-m-long (or longer) whale

Table 1.1.   The diversity of fishes. Below is a brief listing of higher taxonomic categories of living fishes, in phylogenetic order. This list is meant as an introduction to major groups of living fishes as they will be discussed in the initial two sections of this book. Many intermediate taxonomic levels, such as infraclasses, subdivisions, and series, are not presented here; they will be detailed when the actual groups are discussed in Part III. Only a few representatives of interesting or diverse groups are listed.

Superclass **Agnatha**: jawless fishes

    Order **Myxiniformes**: hagfishes

    Order **Petromyzontiformes**: lampreys

Superclass **Gnathostomata**: jawed fishes

  Class **Chondrichthyes**: cartilaginous fishes

   Subclass **Holocephali**: chimaeras

   Subclass **Elasmobranchii**: sharks, skates, rays

Grade **Teleostomi**: bony fishes

  Class **Sarcopterygii**: lobe-finned fishes

    Order **Coelacanthiformes**: coelacanth

   Infraclass **Dipnoi**: lungfishes

Class **Actinopterygii**: ray-finned fishes

  ??Subclass **Brachiopterygii** (Order **Polypteriformes**): bichirs

  Subclass **Chondrostei**

    Order **Acipenseriformes**: paddlefishes, sturgeons

  Subclass **Neopterygii**

    Order **Semionotiformes**: gars

    Order **Amiiformes**: bowfin

  Division **Teleostei:** modern bony fishes

   Subdivision **Osteoglossomorpha**: bonytongues

   Subdivision **Elopomorpha**: tarpons, bonefishes, eels

   Subdivision **Clupeomorpha**: herrings

   Subdivision **Euteleostei**: advanced bony fishes

    Superorder **Ostariophysi**: minnows, suckers, characins, loaches, catfishes

    Superorder **Protacanthopterygii**: pickerels, smelts, salmons

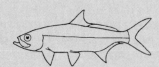

Table 1.1. (Continued)

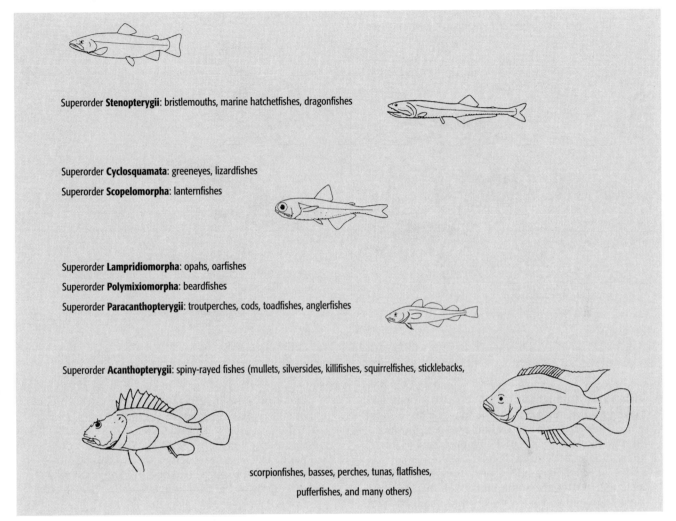

Superorder **Stenopterygii**: bristlemouths, marine hatchetfishes, dragonfishes

Superorder **Cyclosquamata**: greeneyes, lizardfishes
Superorder **Scopelomorpha**: lanternfishes

Superorder **Lampridiomorpha**: opahs, oarfishes
Superorder **Polymixiomorpha**: beardfishes
Superorder **Paracanthopterygii**: troutperches, cods, toadfishes, anglerfishes

Superorder **Acanthopterygii**: spiny-rayed fishes (mullets, silversides, killifishes, squirrelfishes, sticklebacks,

scorpionfishes, basses, perches, tunas, flatfishes,

pufferfishes, and many others)

Taxa and illustrations largely from Nelson 1994.

shark. Diversity in form includes relatively fishlike shapes such as minnows, trouts, perches, basses, and tunas but also such unexpected shapes as boxlike trunkfishes, elongate eels and catfishes, globose lumpsuckers and frogfishes, rectangular ocean sunfishes, question-mark-shaped seahorses, and flattened and circular flatfishes and batfishes, ignoring the exceptionally bizarre fishes of the deep sea (see cover illustrations).

SUPERLATIVE FISHES

A large part of ichthyology's fascination is the spectacular and unusual nature of the subject matter. As a few examples:

- Coelacanths, an offshoot of the lineage that gave rise to the amphibians, were thought to have died out with the dinosaurs at the end of the Cretaceous, 65 million years ago. However, in 1938, fishermen in

South Africa trawled up a very live coelacanth. This fortuitous capture of a living fossil not only rekindled debates about the evolution of higher vertebrates but also underscored the international and political nature of conservation efforts (see Chap. 13).
- Lungfishes can live in a state of dry "suspended animation" for up to 4 years, becoming dormant when their ponds dry up and reviving quickly when immersed in water (see Chaps. 5, 13).
- Antarctic fishes live in water that is colder than the freezing point of their blood. The fishes keep from freezing by avoiding free ice and because their blood contains antifreeze proteins that depress their blood's freezing point to $-2°C$. Some antarctic fishes have no hemoglobin (see Chap. 17).
- Deep-sea fishes include many forms that can swallow prey larger than themselves. Some deep-sea anglerfishes are characterized by females that are 10 times larger than males, the males existing as small parasites permanently fused to the side of the female, living off her bloodstream (see Chap. 17).

FIGURE 1.1.   *Fish versus fishes. By convention, "fish" refers to one or more individuals of a single species. "Fishes" is used when discussing more than one species, regardless of the number of individuals involved.*

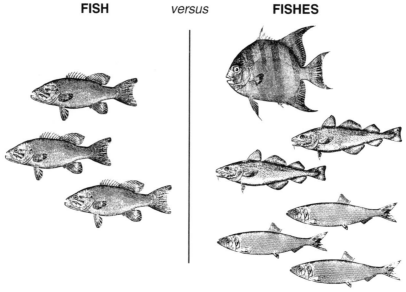

Drawings from Jordan 1905.

- Fishes grow throughout their lives, changing their ecological role several times. In some fishes, differences between larvae and adults are so pronounced that many larvae were originally described as entirely different taxa (see Chap. 9).
- Fishes have maximum lifespans of less than 1 year to as long as 150 years. Some short-lived species are annuals, surviving drought as eggs that hatch with the advent of rains. Longer-lived species may not begin reproducing until they are 20 years old, and then only at 5-year (or longer) intervals (see Chap. 10).
- Gender change is common among fishes. Some species are simultaneously male and female, whereas others change from male to female or from female to male (see Chaps. 10, 20).
- Fishes engage in parental care that ranges from simple nest guarding to mouth brooding to the production of external or internal body substances upon which young feed. Many sharks have a placental structure as complex as any found in mammals. Egg-laying fishes may construct nests by themselves, whereas some species deposit eggs in the siphon of living clams, on the undersides of leaves of terrestrial plants, or in the nests of other fishes (see Chaps. 12, 20).
- Fishes are unique among organisms with respect to the use of bioelectricity. Many fishes can detect biologically meaningful, minute quantities of electricity, which they use to find prey, competitors, or predators and for navigation. Some groups have converged on the ability to produce an electrical field and obtain information about their surroundings from disturbances to the field, whereas others produce large

amounts of high-voltage electricity to deter predators or stun prey (see Chaps. 6, 18, 19).
- Fishes are unique among vertebrates in their ability to produce light; this ability has evolved independently in different lineages and can be either autogenic (produced by the fish itself) or symbiotic (produced by bacteria living on or in the fish) (see Chap. 17).
- Although classically thought of as cold-blooded, some pelagic sharks, billfishes, and tunas maintain body temperatures warmer than their surroundings and have circulatory systems specifically designed for such temperature maintenance (see Chap. 7).
- Predatory tactics include attracting prey with modified body parts disguised as lures or by feigning death. Fishes include specialists that feed on ectoparasites, feces, blood, fins, scales, young, and eyes of other fishes (see Chaps. 18, 19).
- Fishes can significantly change the depth of their bodies by erecting their fins or by filling themselves with water, an effective technique for deterring many predators. In turn, the ligamentous and levering arrangement of mouth bones in some fishes allows them to increase mouth volume when open by as much as 40-fold (see Chaps. 8, 19).

A BRIEF HISTORY OF ICHTHYOLOGY

Fishes would be just as diverse and successful without ichthyologists studying them, but what we know about their diversity is the product of the efforts of workers worldwide over several centuries. Students in an introductory course often have difficulty appreciating historical treatments of the subject; the names are strange, the people are dead (sometimes as a result of their scientific efforts), and the relevance is elusive. However, science is a human endeavor, and knowing something about early ichthyologists, their activities, and their contributions to the storehouse of knowledge that we possess today should help give a sense of the dynamics and continuity of this long-established science.

Although natural historians in most cultures have studied fishes for millennia, modern science generally places its roots in the works of Carl Linne (Linnaeus). Linnaeus produced the first real attempt at an organized system of classification. Zoologists have agreed to use the tenth edition of his *Systema Naturae* (1758) as the starting point for our formal nomenclature. The genius of Linnaeus's system is what we refer to as **binomial nomenclature**, naming every organism with a two-part name based on **genus** (plural **genera**) and **species** (singular and plural, abbreviated **sp.** and **spp.**, respectively). Linnaeus did not care much for fishes, so his ichthyological classification is actually based largely on the efforts of Peter Artedi, the acknowledged "father of ichthyology." Artedi reportedly drowned one night after falling into a canal in Amsterdam while drunk.

In the mid-1800s, the great French anatomist Georges Cuvier joined forces with Achille Valenciennes to produce the first complete list of the fishes of the world. During those times, French explorers were active throughout much of the world, and many of their expeditions included naturalists who collected and saved material. Thus Cuvier and Valenciennes's *Histoire Naturelle de Poissons* includes descriptions of many previously undescribed species of fishes in its 24 volumes. This major reference is still of great importance to systematic ichthyologists today, as are the specimens upon which it is based, many of which are housed in the Museum National d'Histoire Naturelle in Paris.

A few years later, Albert Günther produced a multivolume *Catalogue of Fishes in the British Museum*. Although initially designed simply to list all the specimens in the British collections, Günther included all the species of which he was aware, making this catalog the second attempt at listing the known fishes of the world.

The efforts of Linnaeus, Artedi, Cuvier and Valenciennes, and Günther all placed species in genera and genera in families based on overall resemblance. A modern philosophical background to classification was first developed by Charles Darwin with the publication of his *On the Origin of Species* in 1859. His theory of evolution meant that species placed together in a genus were assumed to have had a common origin, a concept that underlies all important subsequent classifications of fishes and other organisms.

The major force in American ichthyology was David Starr Jordan. Jordan moved from Cornell University to the University of Indiana and then to the presidency of Stanford University. He and his students and colleagues were involved in describing the fishes collected during explorations of the United States and elsewhere in the late 1800s and early 1900s. In addition to a long list of papers, Jordan

and his coworkers, including B. W. Evermann, produced several publications that form the basis of our present knowledge of North American fishes. This includes the four-volume *Fishes of North and Middle America* (1896–1898), which described all the freshwater and marine fishes known from the Americas north of the Isthmus of Panama. In 1923, Jordan and Evermann published a list of all the genera of fishes that had ever been described, which served as the standard reference until recently, when it was updated and replaced by Eschmeyer (1990).

Overlapping with Jordan was the distinguished British ichthyologist, C. Tate Regan, based at the British Museum of Natural History. Regan revised many groups, and his work formed the basis of most recent classifications. Unfortunately, this classification was never published in one place; the best summary of it is in the individual sections on fishes in the fourth edition of the *Encyclopaedia Britannica* (1929).

A Russian ichthyologist, Leo S. Berg, first integrated paleoichthyology into the study of living fishes in his 1947 monograph *Classification of Fishes, Recent and Fossil*, published originally in Russian and English. He was also the first ichthyologist to apply the **-iformes** uniform endings to orders of fishes, replacing the classic and often confusing group names.

In 1966, three young ichthyologists—P. Humphry Greenwood at the British Museum, Donn Eric Rosen at the American Museum of Natural History, and Stanley H. Weitzman at the U.S. National Museum of Natural History—joined with an old-school ichthyologist, George S. Myers of Stanford University, to produce the first modern classification of the majority of present-day fishes, the Teleostei. This classification was updated in Greenwood's third edition of J. R. Norman's classic *A History of Fishes* (1975) and is the framework, with modifications based on more recent findings, of the classification used by Nelson (1994) and followed in this book.

Details of the early history of ichthyology are available in D. S. Jordan's classic *A Guide to the Study of Fishes*, volume I (1905). For a more thorough treatment of the history of North American ichthyology, we recommend Myers (1964) and Hubbs (1964). An excellent historical synopsis of European and North American ichthyologists can also be found in the Introduction of Pietsch and Grobecker (1987); a compilation focusing on the contributions of women ichthyologists appears in Balon et al. (1994).

ADDITIONAL SOURCES OF INFORMATION

This book is one view of ichthyology, with an emphasis on diversity and adaptation (please read the Preface). It is neither the final word nor the only perspective available. As undergraduates, we learned about fishes from other textbooks, some of which are in updated editions from which we have taught our own classes. All of these books are valuable. We have read or reread them during the production of this book to check on topics deserving coverage, and we frequently turn to them for alternative approaches and additional information. Among the most useful are Lagler et al. (1977), Bone et al. (1994), Bond (1996), and Moyle and Cech (1996); for laboratory purposes, Cailliet et al. (1996) is very helpful. From a historical perspective, books by Jordan (1905, 1922), Nikolsky (1961), and Norman and Greenwood (1975) are informative and enjoyable.

Three references have proven indispensable during the production of this book, and their ready access is recommended to anyone desiring additional information and particularly for anyone contemplating a career in ichthyology or fisheries science. Most valuable is Nelson's *Fishes of the World,* third edition (1994). For North American workers, the fifth edition of Robins et al.'s *Common and Scientific Names of Fishes from the United States and Canada* (1991) is especially useful. Finally, of a specialized but no less valuable nature is Eschmeyer's *Catalog of the Genera of Recent Fishes* (1990).

The first two books, although primarily taxonomic lists, are organized in such a way that they provide information on currently accepted phylogenies, characters, and nomenclature; Nelson (1994) is remarkably helpful with anatomic, ecological, evolutionary, and zoogeographic information on most families. Eschmeyer's volume is invaluable when reading older or international literature, because it gives other names that have been used for a fish (**synonymies**) and indicates the family to which a genus belongs.

Of a less technical but useful nature are fish encyclopedias, such as Wheeler's *Fishes of the World* (1975), also published as *The World Encyclopedia of Fishes* (1985); *McClane's New Standard Fishing Encyclopedia* (1974); or Paxton and Eschmeyer's (1994) *Encyclopedia of Fishes* (the latter is fact-filled and lavishly illustrated). Species guides exist for most states and provinces in North America, most countries in Europe (including current and former British Commonwealth nations), and some tropical nations and regions. These are too numerous and too variable in quality for listing here; a good source for titles is Berra (1981). Two of our favorite geographic treatments of fishes are as much anthropological as they are ichthyological, namely Johannes's (1981) *Words of the Lagoon* and Goulding's (1980) *The Fishes and the Forest.*

A stroll through the shelves of any decent public or academic library is potentially fascinating, with their collections of ichthyology texts dating back a century, geographic and taxonomic guides to fishes, specialty texts and edited volumes, and works in or translated from many languages. Among the better-known, established journals that specialize in or often focus on fish research are *Copeia, Transactions of the American Fisheries Society, Environmental Biology of Fishes, North American Journal of Fisheries Management, U.S. Fishery Bulletin, Canadian Journal of Fisheries and Aquatic Sciences, Canadian Journal of Zoology, Journal of Fish Biology, Journal of Ichthyology* (the translation of the Russian journal *Voprosy Ikhtiologii*), *Australian and New Zealand Journals of Marine and Freshwater Research, Bulletin of Marine Science,* and *Japanese Journal of Ichthyology.*

Although diving does not in itself constitute a biological science any more than does casual bird watching, snorkeling and scuba diving are essential methods for acquiring detailed information on fish biology. Two of us (Helfman, Collette) credit the thousands of hours we have spent underwater as formative and essential to our understanding of fishes. A full appreciation for the wonders of adaptation in fishes requires that they be viewed in their natural habitat, as they would be viewed by their conspecifics, competitors, predators, and neighbors (it is fun to try to think like a fish). We strongly urge anyone seriously interested in any aspect of fish biology to acquire basic diving skills, including the patience necessary to watch fishes going about their daily lives.

Public and commercial aquaria are almost as valuable, particularly because they expose an interested person to a wide zoogeographic range of species or to an intense selection of local fishes that are otherwise only seen dying in a bait bucket or at the end of a fishing line. Our complaint about such facilities is that, perhaps because of space constraints or an anticipated short attention span on the part of viewers, large aquaria seldom provide details about the fascinating lives of the animals they hold in captivity. Home aquaria are an additional source for inspiration and fascination, although we are deeply ambivalent about their value because so many tropical fishes are killed or habitats destroyed in the process of providing animals for the commercial aquarium trade, particularly for marine tropicals.

SUMMARY

1. Fishes account for more than half of all living vertebrates and are the most successful vertebrates in aquatic habitats worldwide. There are about 25,000 living species of fishes, of which approximately 850 are cartilaginous (sharks, skates, and rays), 85 are jawless (hagfishes, lampreys), and the remaining 24,000 are bony fishes.

2. A fish can be defined as an aquatic vertebrate with gills and with limbs in the shape of fins. Included in this definition is a tremendous diversity of sizes (from an 8-mm goby to a 12-m or longer whale shark), shapes, ecological functions, life history scenarios, anatomic specializations, and evolutionary histories.

3. Most (about 60%) of living fishes are primarily marine, and the remainder live in freshwater; about 1% move between salt and freshwater as a normal part of their life cycle. The greatest diversity of fishes is found in the tropics, particularly the Indo-West Pacific region for marine fishes and tropical South America, Africa, and Southeast Asia for freshwater species.

4. Unusual adaptations among fishes include African lungfishes that can live in dry mud for up to 4 years, supercooled antarctic fishes that live in water colder than the freezing point of their blood, deep-sea fishes that can swallow prey larger than themselves (some deep-sea fishes exist as small males fused to and entirely parasitic on larger females), annual species that

live less than a year and other species that may live 150 years, fishes that change sex from female to male or vice versa, sharks that provide nutrition for developing young via a complex placenta, fishes that create an electrical field around themselves and detect biologically significant disturbances of the field, light-emitting fishes, warm-blooded fishes, and at least one species (the coelacanth) that was thought to have gone extinct with the dinosaurs.

5. Historically important contributions to ichthyology have been made by Linnaeus, Peter Artedi, Georges Cuvier, Achille Valenciennes, Albert Günther, David Starr Jordan, B. W. Evermann, C. Tate Regan, and Leo S. Berg, among many others.

6. The literature on fishes is voluminous, including a diversity of college-level textbooks; popular and technical books on particular geographic regions, taxonomic groups, or species sought by anglers or best suited for aquarium keeping or aquaculture; and scientific journals in many countries. Another valuable source of knowledge is public aquaria. Observing fishes by snorkel or scuba diving will provide anyone interested in fishes with indispensable first-hand knowledge and appreciation.

2 Systematic Procedures

The basis of a taxonomically oriented discipline such as ichthyology is an organized, hierarchical system of names and evolutionary hypotheses associated with those names. This underlying structure provides a basis for identifying and discriminating among species and for understanding relationships among species and higher taxa. It also provides the common language that allows communication and discussion among practitioners of the discipline. This structure is generally known as **systematics**. In this chapter, we discuss the need, value, functions, and goals of systematic procedures, different philosophies of classifying organisms, and how systematic procedures lead to an increase in our understanding of fishes.

Why do we need a system of classification? Things must be divided into categories before we can talk about them. This includes cars, athletes, books, plants, and animals. We cannot deal with all the members of a class (such as the 25,000 species of fishes) individually, so we must put them into some sort of classification. Different types of classifications are designed for different functions. For example, one can classify automobiles by function (sedan, van, pickup, etc.) or by manufacturer (Ford, General Motors, Toyota, etc.). Baseball players can be classified by position (catcher, pitcher, first baseman, etc.) or by team (Braves, Orioles, etc.). Books may be shelved in a library by subject or by author. Similarly, plants and animals can be classified ecologically as grazers, detritivores, carnivores, and so forth or phylogenetically, on the basis of their evolutionary relationships.

Good reasons exist for ecologists to classify organisms ecologically, but this is a special classification for special purposes. The most general classification is considered to be the most **natural classification**, defined as the classification that best represents the phylogenetic (5 evolutionary) history of an organism and its relatives. A phylogenetic classification of taxonomic groups (**taxa**) holds extra information because the categories are predictive. Just as experience with one bad Ford automobile may lead an owner to generalize about other Fords, phylogenetic classification can also be predictive. If one species of fish in a genus builds a nest, it is likely that other species in that genus also do so.

SPECIES

Species are the fundamental unit of classification schemes. What is a species, and how should species be arranged in a phylogenetic classification? The British ichthyologist C. Tate Regan defined a species as "a group of organisms with distinctive enough morphological charac-

ters that, in the opinion of a competent systematist, are sufficiently definite to entitle them to a specific name" (Norman 1948, 365). This practical, but somewhat circular, definition of a species, now termed a **morphospecies**, does not depend on evolutionary concepts.

In the late 1930s and early 1940s, the first major attempts were made to integrate classification with evolution. Julian Huxley integrated genetics into evolution in his book *The New Systematics* in 1940. In *Systematics and the Evolution of Species*, Ernst Mayr (1942, 120) introduced the **biological species concept**. To Mayr, species are "groups of actually or potentially interbreeding populations which are reproductively isolated from other such groups." This was an important effort to move away from defining species strictly on the basis of morphological characters. This definition has been modified to better fit current concepts of evolution. "An **evolutionary species** is a single lineage of ancestor-descendant populations which maintains its identity from other such lineages and which has its own evolutionary tendencies and historical fate" (Wiley 1981, 25).

TAXONOMY VERSUS SYSTEMATICS

These two words are not exact synonyms but rather describe somewhat overlapping fields. Taxonomy deals with the theory and practice of describing biodiversity (including naming undescribed species), arranging this diversity into a system of classification, and devising identification keys. It includes the rules of nomenclature that govern use of taxonomic names. **Systematics** emphasizes the study of relationships postulated to exist among species or higher taxa, such as families and orders. Lundberg and McCade (1990) have presented a good summary of systematics oriented toward those interested in fishes. The two primary journals dealing with systematics of animals are *Systematic Biology* (formerly *Systematic Zoology*), published by the Society of Systematic Biologists, and *Cladistics*, published by the Willi Hennig Society.

APPROACHES TO CLASSIFICATION

Three recent general philosophies of classification have dominated scientific thought in the area of systematics: cladistics, or phylogenetic systematics; phenetics, or numerical taxonomy; and evolutionary systematics.

A revolution in systematic methodology was begun by a German entomologist, Willi Hennig. He introduced what has become known as **cladistics**, or **phylogenetic systematics**, following publication of the 1966 English translation of his 1950 German monograph. His fundamental principle was to divide characters into two groups: **apomorphies** (more recently evolved, derived, or advanced characters) and **plesiomorphies** (more ancestral, primitive, or generalized characters). The goal is to find **synapomorphies** (shared derived characters) that define **monophyletic groups**, or **clades** (groups containing an ancestor and all its descendant taxa). **Symplesiomorphies** (shared primitive characters) do not provide data useful for constructing phylogenetic classifications because primitive characters may be retained in a wide variety of taxa; advanced as well as primitive taxa may possess symplesiomorphies. **Autapomorphies**, specialized characters that are present in only a single taxon, are important in defining that taxon but are also not useful in constructing a phylogenetic tree.

All three major systematic approaches produce some sort of graphic illustration that depicts the different taxa, arranged in a manner that reflects their hypothesized relationships. In cladistics, taxa are arranged on a branching diagram called a **cladogram** (see Box 2.1, Fig. 2.1). Monophyletic groups are defined by at least one synapomorphy at a **node**, or branching point, on the cladogram. Deciding whether a character is plesiomorphic or apomorphic is based largely on outgroup analysis, that is, finding out what characters are present in **outgroups**, closely related groups outside the taxon under study, which is designated the **ingroup**. More than one outgroup should be used to protect against the problem of occasional apomorphies in one outgroup. The ancestral-derived dichotomy is referred to as the **polarity** of a character. **Sister groups** are the most closely related clades at the nodes of a cladogram. Problems arise when there are **homoplasies**, shared independently derived similarities such as parellelisms, convergences, or secondary losses. These do not reflect the evolutionary history of a taxon.

A primary goal of phylogenetic systematics is the definition of monophyletic groups. Current researchers agree on the necessity of avoiding **polyphyletic groups**, groups containing the descendants of different ancestors. Most researchers are equally adamant that **monophyletic** should be equal to **holophyletic**, groups containing all the descendants of a single ancestor and avoiding **paraphyletic groups**, groups that do not contain all the descendants of a single ancestor. **Grades** are groups that are defined by their morphological or ecological distinctness and not necessarily by synapomorphies.

Ideally, when constructing a classification, a taxon can be defined by a number of synapomorphies. However, conflicting evidence frequently exists. Some characters show the relationships of group A to group B, but other characters may show relationships of group A to group C. The principle used to sort out the confusion is that of **parsimony**. In other words, select the hypothesis that explains the data in the simplest or most economical manner (Box 2.1).

With large numbers of characters and large numbers of taxa, it frequently becomes necessary to utilize computer programs to identify the most parsimonious hypotheses, which are usually defined as the hypotheses requiring the fewest number of steps to progress from the outgroup to the terminal taxa on a cladogram. The most often used programs are PAUP (Phylogenetic Analysis Using Parsimony) by David Swofford (1993) and Hennig86 by James Farris (1988). A thorough explanation of cladistic methodology is presented by Wiley (1981), and cogent, brief summaries can be found in Lundberg and McDade (1990) and Funk (1995).

BOX 2.1

Cladistic Success: The Louvar

An ideal example of how cladistics should work concerns the oceanic fish known as the louvar (*Luvarus imperialis*). Most ichthyologists have classified the louvar as a strange sort of scombroid fish (Scombroidei), the perciform suborder that contains the tunas, billfishes, and snake mackerels. However, a comprehensive morphological and osteological study (Tyler et al. 1989) showed clearly that the louvar is actually an aberrant pelagic relative of the surgeonfishes (Acanthuroidei). This example is instructive because the study utilized 60 characters from adults and 30 more from juveniles (Fig. 2.1). Homoplasies—characters postulated to be **reversals** (return to original condition) or **independent acquisitions** (independently evolved)—were minimal. With the cladistic approach, synapomorphies show that the relationships of the louvar are with the acanthuroids, whereas noncladistic analysis overemphasized caudal skeletal characters, leading to placement among the scombroids.

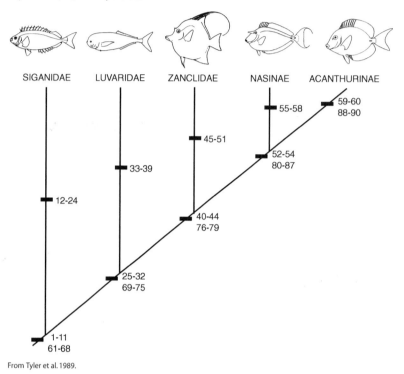

FIGURE 2.1. *Cladogram of hypothesized relationships of the louvar (Luvarus) and other Acanthuroidei. Arabic numerals show synapomorphies: numbers 1 to 60 represent characters from adults, 61 to 90 from juveniles. Some sample synapomorphies: 2) branchiostegal rays reduced to four or five; 6) premaxillae and maxillae (upper jawbones) bound together; 25) vertebrae reduced to 9 precaudal plus 13 caudal; 32) single postcleithrum behind the pectoral girdle; 54) spine or plate on caudal peduncle; 59) teeth spatulate.*

SIGANIDAE   LUVARIDAE   ZANCLIDAE   NASINAE   ACANTHURINAE

59-60
88-90

55-58

45-51

52-54
80-87

33-39

12-24

40-44
76-79

25-32
69-75

1-11
61-68

From Tyler et al. 1989.

Cladistic techniques and good classifications based on these techniques have proved particularly useful in analyzing the geographic distribution of plants and animals in a process called vicariance biogeography (see Chap. 16, Box 16.2).

Phenetics, or **numerical taxonomy**, is a second approach to systematics. Phenetics starts with species or other taxa as **operational taxonomic units** (OTUs) and then clusters the OTUs on the basis of overall similarity, using an array of numerical techniques. Advocates of this school believe that the more characters used, the better and more natural the classification should be. The last synthesis of the field was by Sneath and Sokal (1973). Some of the numerical techniques devised by this school are useful in dealing with masses of data and have been incorporated into cladistics. However, few modern systematists subscribe to the view that using a host of characters, without distinguishing between plesiomorphies and apomorphies, will provide a natural classification. Some molecular systematists still use phenetic methods to treat their data. Graphic representations in phenetics, known as **phenograms**, look like tennis ladders, with OTUs in place of the competitors. Relatedness is determined by comparing measured linear distances between OTUs; the closer two units are, the more closely related they are.

Evolutionary systematics, as summarized by Mayr (1974), holds that **anagenesis**, the amount of time and differentiation that have taken place since groups divided, must also be taken into consideration along with **cladogenesis**, the process of branch or lineage splitting between sister groups. Evolutionary relationships are expressed on a tree called a **phylogram**. The contrast between the cladistic and evolutionary schools can be demonstrated by considering how to classify birds. Cladists emphasize the fact that crocodiles and birds belong to the same evolutionary line by insisting they must be included within a monophyletic group, Archosauria, in a phylogenetic classification. Evolutionary systematists emphasize the long time gap between fossil crocodilians and modern birds and believe that birds and crocodiles must be treated as separate evolutionary units.

An attempt has been made recently (Carpenter et al. 1995) to try to combine the benefits of a cladistic analysis with anagenetic information in an approach termed **quantitative evolutionary systematics**. This approach begins with a cladogram but then adds back in additional information not useful in deciding on the location of the branching point. This procedure has the advantage of maintaining a stable classification while allowing retrieval of a cladistic hypothesis from an annotated form of the classification.

Most leading ichthyological theorists favor the cladistic school and tend to consider any problems resulting from strictly following cladistic theory as minor. On the other hand, many practical ichthyologists, working at the species level, ignore the controversy so they can get on with the business of describing and cataloging ichthyological diversity before humans exterminate large segments of it.

TAXONOMIC CHARACTERS

Whichever system of classification is employed, characters are needed to differentiate taxa and assess their interrelationships. Characters, as Stanford ichthyologist George Myers once said, are like gold—they are where you find them. Characters are variations of a homologous structure, and to be useful, they must show some variation in the taxon under study. Useful definitions of a wide variety of characters were presented by Strauss and Bond (1990). Characters can be divided, somewhat arbitrarily, into different categories.

Meristic characters originally referred to characters that correspond to body segments (myomeres), such as numbers of vertebrae and fin rays. Now, meristic is used for almost any countable structure, including numbers of scales, gill rakers, cephalic pores, and so on. These characters are useful because they are clearly definable, and usually other investigators will produce the same counts. In most cases, they are stable over a wide range of body size. Also, meristic characters are easier to treat statistically, so comparisons can be made between populations or species with a minimum of computational effort.

Morphometric characters refer to measurable structures such as fin lengths, head length, eye diameter, or ratios between such measurements. Some morphometric characters are harder to define exactly, and being continuous variables, they can be measured to different levels of precision and so are less easily repeated. Furthermore, there is the problem of **allometry**, whereby lengths of different body parts change at different rates with growth (see Chap. 10). Thus analysis of differences is more complex than with meristic characters. Size factors have to be compensated for through use of such techniques as regression analysis, analysis of variance (ANOVA), and analysis of covariance (ANCOVA) so that comparisons can be made between actual differences in characters and not differences due to body size. Principal components analysis (PCA) also adjusts for size, particularly if size components are removed by shear coefficients, as recommended by Humphries et al. (1981).

Widely used definitions of most meristic and morphometric characters were presented by Hubbs and Lagler (1964); some of these are illustrated in Figure 2.2.

Anatomical characters include characters of the skeleton (osteology) and characters of the soft anatomy, such as position of the viscera, divisions of muscles, and branches of blood vessels. Some investigators favor osteological characters because such characters have been thought to vary less than other characters. In some cases, this supposition has been due to the use of much smaller sample sizes than with the analysis of meristic or morphometric characters.

Other characters can include almost any fixed, describable differences among taxa. For example, color can include such characters as the presence of stripes, bars, spots, or specific colors. Photophores are light-producing structures that vary in number and position among different taxa. Sexually dimorphic ("two forms") structures can be of functional value, including copulatory organs used by males to inseminate females, like the gonopodium of

FIGURE 2.2. *Some meristic and morphometric characters shown on a hypothetical scombrid fish.*

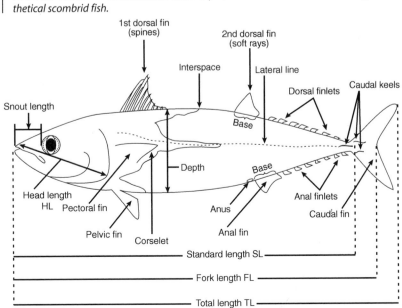

a guppy (modified anal fin), or the claspers of chondrichthyans (modified pelvic fins). Cytological (including karyological), electrophoretic, serological, and molecular techniques analyzing nuclear DNA or mitochondrial DNA (mtDNA) are becoming increasingly important at all levels of classification (Avise 1994). Behavioral and physiological characters are useful in some groups.

STEPS IN CLASSIFICATION

Ernst Mayr divided classification into three stages. He characterized the **alpha** level as that of species descriptions. The **beta** level is the arrangement of species into a natural system of classification. The **gamma** level is the analysis of infraspecific variation and the study of evolution.

VERTEBRATE CLASSES

Most textbooks list five classes of vertebrates: Pisces (25,000 species), Amphibia (4300), Reptilia (6000), Aves (9000), and Mammalia (4800). But as Nelson (1969) clearly demonstrated, this five-class system is anthropomorphic, with bird and mammal groups overemphasized by the mammal doing the classification, that is, us. The morphological and evolutionary gap between the Agnatha, the jawless vertebrates (lampreys and hagfishes), and other groups of fishes is much greater than between the classes of jawed fishes on the one hand and the tetrapods on the other hand. Thus fishes (or Pisces) is not a monophyletic group but a grouping used for convenience for the Agnatha, Chondrichthyes, bony fishes, and the fossil Acanthodii and ostracoderms.

UNITS OF CLASSIFICATION

Systematists use a large number of units to show relationships at different levels. Most of these units are not necessary except to the specialist in a particular group.

Ray-finned fishes fall into the following units: kingdom—Animalia; phylum—Chordata (chordates); subphylum—Vertebrata (vertebrates); superclass—Gnathostomata (jawed vertebrates); grade—Teleostomi or Osteichthyes (bony fishes); and class—Actinopterygii (ray-finned fishes). Classification of three representative fishes is shown in Table 2.1.

Note the uniform endings for order (*-iformes*), suborder (*-oidei*), family (*-idae*), subfamily (*-inae*), and tribe (*-ini*). Also, note that the group name is formed from a stem plus the ending. This means that if you learn that the yellow perch is *Perca flavescens*, you can construct much of the rest of classification by adding the endings. Percidae is the family including the perch, Percoidei is the suborder of perchlike fishes, and Perciformes is the order containing the perchlike fishes and their relatives.

It is conventional to italicize the generic and specific names of animals and plants to indicate their origin from Latin (or latinized Greek or other language). Generic names are always capitalized, but species names are always in lowercase (unlike the situation for some plant species names). The names of higher taxonomic units such as families and orders are never italicized but are always capitalized because they are proper nouns. Sometimes it is convenient to convert the name of a family or order into English (e.g., Percidae into percid, Scombridae into scombrid), in which case the name is no longer capitalized. Common names of fishes are not capitalized (unlike the situation with birds), unless part of the name is a proper name.

It is also conventional to list higher taxa down to orders in phylogenetic sequence, beginning with the most primitive and working up to the most advanced, reflecting the course of evolution. This procedure has the additional advantage that closely related species are listed near each other, facilitating comparisons. As knowledge about the relationships of organisms increases, changes need to be made in their classification. An instructive example of justification for changing the order in classification was presented by Smith (1988) in a paper entitled "Minnows first, then trout." Smith explained that he placed the minnows and relatives (Cypriniformes) before

Table 2.1.   Classification of Atlantic herring, yellow perch, and Atlantic mackerel.

Taxonomic Unit	Herring	Perch	Mackerel
Division	Teleostei	→	→
Subdivision	Clupeomorpha	Euteleostei	→
Order	Clupeiformes	Perciformes	→
Suborder	Clupeoidei	Percoidei	Scombroidei
Family	Clupeidae	Percidae	Scombridae
Subfamily	Clupeinae	Percinae	Scombrinae
Tribe	Clupeini	Percini	Scombrini
Genus	*Clupea*	*Perca flavescens*	*Scomber*
species	*harengus*		*scombrus*
subspecies	*harengus*		
Author	Linnaeus	Mitchill	Linnaeus

the trouts and salmons (Salmoniformes) in his book on the fishes of New York State to reflect the more primitive phylogenetic position of the Cypriniformes.

A contrasting example shows problems sometimes found with phylogenetic systematics. Analyses of similar data sets for the Scombroidei by Collette et al. (1984) and by Johnson (1986) produced different cladograms that resulted in very different classifications. For example, in a computer-generated cladogram (WAGNER 78 by Farris 1970), Collette et al. (1984) postulated a sister-group relationship of the wahoo (*Acanthocybium*) and Spanish mackerels (*Scomberomorus*) within the family Scombridae. In contrast, Johnson (1986) placed the wahoo as the sister group of the billfishes within a greatly expanded Scombridae that includes billfishes as a tribe, instead of being in the separate families Xiphiidae and Istiophoridae.

In part, the different authors reached different conclusions because they analyzed the data sets differently. Another part of the differences in classification centers around the large amount of homoplasy present. No matter which classification is employed, a large number of characters must be postulated to show reversal or independent acquisition. Either more data or a different method of analysis seems needed to resolve the conflict. Recent use of molecular data, DNA sequences from the mitochondrial gene cytyochrome *b* (Finnerty and Block 1995), supports the view that the wahoo is a scombrid but strongly refutes a close relationship between billfishes and scombroids.

INTERNATIONAL CODE OF ZOOLOGICAL NOMENCLATURE

The International Code of Zoological Nomenclature is a system of rules designed to foster stability of scientific names for animals. Rules deal with such topics as the definition of publication, authorship of new scientific names, and types of taxa. Much of the Code is based on the **Principle of Priority**, which states that the first validly described name for a taxon is the name to be used. Most of the rules deal with groups at the family level and below. Interpretations of the Code and exceptions to it are controlled by the International Commission of Zoological Nomenclature, members of which are distinguished systematists who specialize in different taxonomic groups.

Species and subspecies are based on **type specimens**, the specimens used by an author in describing new taxa at this level. Type specimens should be placed in permanent archival collections (see below) where they can be examined by future researchers. **Primary types** include the **holotype**, the single specimen upon which the description of a new species is based; **lectotype**, a specimen subsequently selected to be the primary type from a number of **syntypes**, a series of specimens upon which the description of a new species was based before the Code was changed to disallow this practice; and **neotype**, a replacement primary type specimen that is permitted only when there is strong evidence that the original primary type specimen was lost or destroyed and when a complex nomenclatorial problem exists that can only be solved by selection of a neotype.

Secondary types include **paratypes**, additional specimens used in the description of a new species, and **paralectotypes**, the remainder of a series of syntypes when a lectotype has been selected from the syntypes. Among the many other kinds of types, mention should also be made of **topotype**, a specimen taken from the same locality as the primary type and, therefore, useful in understanding variation of the population that included the specimen upon which the description was based.

Taxa above the species level are based on type taxa. For example, the **type species** of a genus is not a specimen but a particular species. Similarly, a family is based on a particular genus.

NAME CHANGES

Why do the scientific names of fishes sometimes change? There are four primary reasons that systematists change names of organisms: 1) **splitting** what was considered to be a single species into two (or more); 2) **lumping** two species that were considered distinct into one; 3) changes in classification (e.g., a species is shown to belong in a different genus); and 4) an earlier name is discovered and becomes the valid name by the Principle of Priority. Frequently, name changes involve more than one of these reasons, as shown in the following examples.

An example of the first situation is the Spanish mackerel of the western Atlantic (*Scomberomorus maculatus*), which was considered to extend from Cape Cod, Massachusetts, south to Brazil. However, populations referred to this species from Central and South America have 47 to 49 vertebrae, whereas *S. maculatus* from the Atlantic and Gulf of Mexico coasts of North America have 50 to 53 vertebrae. This difference, plus other morphometric and anatomic differences, was the basis for recognizing the southern populations as a separate species, *S. brasiliensis* (Collette et al. 1978).

An example of the second situation concerns tunas of the genus *Thunnus*. Many researchers believed that the species of tunas occurring off their coasts must be different from species in other parts of the world. Throughout the years, 10 generic and 37 specific names were applied to the seven species of *Thunnus* recognized by Gibbs and Collette (1967). Fishery workers in Japan and Hawaii recorded information on their yellowfin tuna as *Neothunnus macropterus*, those in the western Atlantic as *Thunnus albacares*, and those in the eastern Atlantic as *Neothunnus albacora*. Large, long-finned individuals, the so-called Allison tuna, were known as *Thunnus* or *Neothunnus allisoni*. Based on a lack of morphological differences among the nominal species, Gibbs and Collette postulated that the yellowfin tuna is a single worldwide species. Gene exchange among the yellowfin populations was subsequently confirmed using molecular techniques (Scoles and Graves 1993), further justifying lumping the different nominal species. Following the Principle of Priority, the correct name is the **senior synonym**, the earliest species name for a yellowfin tuna, which is *albacares* Bonnaterre 1788. Other, later names are **junior synonyms**.

Tunas also illustrate the other two kinds of name changes. Some researchers placed the bluefin tuna in the genus *Thunnus*, the albacore in *Germo*, the bigeye in *Parathunnus*, the yellowfin tuna in *Neothunnus*, and the longtail in *Kishinoella*, almost a genus for each species. Gibbs and Collette (1967) showed that the differences are really among species rather than among genera, so all seven species should be grouped together in one genus. But which genus? Under the Principle of Priority, *Thunnus* South 1845 is the senior synonym, and the other, later names are junior synonyms—*Germo* Jordan 1888, *Parathunnus* Kishinouye 1923, *Neothunnus* Kishinouye 1923, and *Kishinoella* Jordan and Hubbs 1925.

The name of the rainbow trout was changed from *Salmo gairdnerii* to *Oncorhynchus mykiss* in 1988 (Smith and Stearley 1989), affecting many fishery biologists and experimental biologists as well as ichthyologists (see Box 14.1). As with the tunas, this change involved a new generic classification as well as lumping of species previously considered distinct.

COLLECTIONS

Important scientific specimens are generally stored in **collections** where they serve as **vouchers** to document identification in published scientific research. Collections are similar to libraries in many respects. Specimens are filed in an orderly and retrievable fashion. Curators care for their collections and conduct research on certain segments of them, much as librarians care for their collections. Qualified investigators can borrow material from collections or libraries for their scholarly study.

Collections may be housed in national museums, state or city museums, private museums, or university museums. The eight major fish collections in the United States (and their acronyms) include the National Museum of Natural History (USNM), Washington, D.C.; University of Michigan Museum of Zoology (UMMZ), Ann Arbor; California Academy of Sciences (CAS), San Francisco; American Museum of Natural History (AMNH), New York; Academy of Natural Sciences (ANSP), Philadelphia; Museum of Comparative Zoology (MCZ), Harvard University, Cambridge, Massachusetts; Field Museum of Natural History (FMNH), Chicago; and Natural History Museum of Los Angeles County (LACM). These eight collections contain more than 24.2 million fishes (Poss and Collette 1995). An additional 118 fish collections in the United States and Canada hold 63.7 million more specimens.

The most significant fish collections in other countries are located in major cities of nations that played important roles in the exploration of the world in earlier times (Berra and Berra 1977). These include the Natural History Museum (formerly British Museum [Natural History]) (BMNH), London; Museum national d'Histoire naturelle (MNHN), Paris; Naturhistorisches Museum (NHMV), Vi-

enna; Rijksmuseum van Natuurlijke Historie (RMNH), Leiden, The Netherlands; Zoological Museum, University of Copenhagen (ZMUC); and the Australian Museum (AMS), Sydney. Leviton et al. (1985) list most of the major fish collections of the world.

Use of museum specimens has been primarily by systematists in the past. This will continue to be an important role of collections in the future, but other uses will become increasingly important. Examples include surveys of parasites (Cressey and Collette 1970) and breeding tubercles (Wiley and Collette 1970); comparison of heavy metal levels in fish flesh today with material up to 100 years old (Gibbs et al. 1974); long-term changes in biodiversity at specific sites (Gunning and Suttkus 1991); and pre- and postimpoundment surveys that could show the effects of dam construction. Many major collections are now computerized (Poss and Collette 1994), and more and more data are becoming accessible as computerized databases via online services such as Gopher and the Internet. A good example is NEODAT, a database of South American fishes held by a number of North and South American fish collections.

SUMMARY

1. The best classification is the most natural one, that which best represents the phylogenetic (= evolutionary) history of an organism and its relatives.

2. Species are the fundamental unit of classification and can be defined as actually or potentially interbreeding populations that are reproductively isolated.

3. Taxonomy deals with describing biodiversity (including naming undescribed species), arranging biodiversity into a system of classification, and devising identification keys. Rules of nomenclature govern use of taxonomic names. Systematics focuses on relationships among species or higher taxa.

4. Cladistics, or phylogenetic systematics, is a widely used system of classification in which characters are divided into apomorphies (derived or advanced) and plesiomorphies (primitive or generalized). The goal is to find synapomorphies (shared derived characters) that define monophyletic groups, or clades (groups containing an ancestor and all its descendant taxa).

5. Taxonomic characters can be meristic (countable), morphometric (measurable), morphological (including color), cytologic, behavioral, electrophoretic, or molecular.

6. Ray-finned fishes are generally classified as: kingdom—Animalia; phylum—Chordata (chordates); subphylum—Vertebrata (vertebrates); superclass—Gnathostomata (jawed vertebrates); grade—Teleostomi or Osteichthyes (bony fishes); and class—Actinopterygii (ray-finned fishes).

7. The International Code of Zoological Nomenclature promotes stability of scientific names for animals. These rules deal with such matters as the definition of publication, authorship of new scientific names, and types of taxa.

8. Species and subspecies are based on type specimens, higher taxa on type taxa. Primary types include the holotype, the single specimen upon which the description of a new species is based. Secondary types include paratypes, which are additional specimens used in the description of a new species.

SUPPLEMENTAL READING

Collette and Vecchione 1995; Mayr and Ashlock 1991; Smith 1988; Smith and Stearley 1989.

PART 2

Form, Function, and Ontogeny

3

Skeleton, Skin, and Scales

Fundamental to appreciating the biology of any group of organisms is knowledge of basic anatomy. We present here an outline of fish anatomy in four sections: osteology and the integumentary skeleton (skin and scales) in this chapter, soft anatomy and the nervous system in the next chapter. The skeleton provides much of the framework and support for the remainder of the body, and the skin and scales insulate and protect the organism from the surrounding environment. The general osteological description given here and many of the figures are based on members of a family of advanced perciform fishes, the tunas (Scombridae). Comparative notes on other bony fishes are added where needed. For a more comprehensive treatment of fish anatomy, see Harder (1975).

SKELETON

The **osteology** (study of bones) of fishes is more complicated than in higher vertebrates because fish skulls, girdles, and caudal skeletons are made up of many more bones. For example, humans have 28 skull bones, a primitive reptile has 72, and a fossil chondrostean fish more than 150 skull bones (Harder 1975). The general evolutionary trend from primitive bony fishes to more advanced fishes and higher vertebrates has been toward fusion and reduction in number of bony elements.

The **skull**, or cranium (Fig. 3.1), is the part of the axial endoskeleton that encloses and protects the brain and most of the sense organs. It is a complex structure because it is derived from several sources. Homologies of some fish skull bones are still in doubt (e.g., the vomer in the roof of the mouth). The skull has two major components: the neurocranium and the branchiocranium. The **neurocranium** is composed of the chondrocranium and the dermatocranium. The **chondrocranium** is the original cartilaginous braincase. Its bones ossify during ontogeny as cartilage is replaced by bone. **Cartilage replacement bones** and **dermal bones** have similar histological structure and differ in that cartilage bones are preformed in cartilage before they ossify. Some bones, however, are of complex origin coming from both sources. The **dermatocranium** consists of dermal bones. It is believed that the bones of the dermatocranium evolved from scales that became attached to the chondrocranium.

The **branchiocranium**, or visceral cranium, consists of a series of endoskeletal arches that originally formed as gill arch supports. The branchiocranium is also known as the splanchnocranium because it is derived from splanchnic mesoderm. The circumorbital bones, opercular bones, and branchiostegals overlie the branchiocranium, which overlies the neurocranium and pectoral girdle.

Skulls differ among the three basic groups of fishes. Hagfishes and lampreys (Agnatha) lack true biting jaws.

Toothlike structures are present, but these are horny rasps, not true teeth (see Chap. 13, Jawless Fishes). The round mouth has some internal cartilaginous support, hence the alternative name Cyclostomata. It was once thought that lamprey jaws had been lost in association with parasitism. However, the probable fossil ancestors of the lampreys, the primitive cephalaspidomorphs (see Chap. 11), also lacked jaws, so lack of jaws is now thought to be a primitive character. The neurocranium of the Chondrichthyes is a single cartilaginous structure, the jaws and branchial arches consisting of a series of cartilages.

Neurocranium

The neurocranium of bony fishes is derived from cartilaginous capsules that formed around the sense organs. To clarify spatial relationships among the large number of bones in the skull, it helps to divide the skull into four regions associated with major centers of ossification. From anterior to posterior, these regions are ethmoid, orbital, otic, and basicranial. For each region, the cartilage bones will be discussed first, followed by the dermal bones, which tend to roof over the underlying cartilage bones. Consult Harder (1975, pl. 1A–C) for a three-part plate of overlays that greatly helps visualize how the teleost skull bones fit together.

Ethmoid region.   The ethmoid region remains variably cartilaginous and bony even in adults of most teleosts (see Table 1.1). Two sets of cartilage bones form the ethmoid region. Paired **lateral ethmoids** (or parethmoids) form the posterolateral wall of the ethmoid region and the anterior wall of the orbit (Figs. 3.2 through 3.4). The median **ethmoid** is the most anterodorsal skull bone. It may have a dermal element fused to it, in which case it is sometimes termed dermethmoid. There are also two sets of dermal bones in this region. The median dentigerous (tooth-bearing) **vomer**, absent in a few teleosts, lies ven-

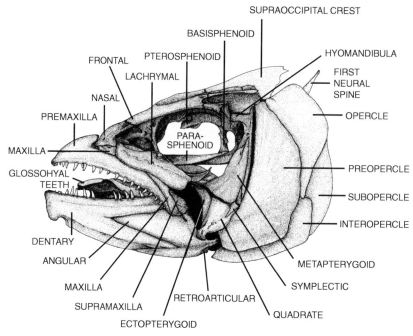

FIGURE 3.1. *Lateral view of the skull of the dogtooth tuna (Gymnosarda unicolor).*

From Collette and Chao 1975.

tral to the ethmoid, whereas the paired **nasals** are lateral to the ethmoid region, associated with the nasal capsule.

Orbital region.   The region that surrounds the orbit is composed of three sets of cartilage bones and two sets of dermal bones. Cartilage bone components include paired **pterosphenoids** (alisphenoids in earlier literature), which

meet along the ventral median line of the skull. The median **basisphenoid** extends from the pterosphenoids down to the parasphenoid and may divide the orbit into left and right halves. The **sclerotic** cartilages or bones protect and support the eye itself. The two sets of dermal bones are the paired **frontals**, which cover most of the dorsal surface of the cranium, and the **infraorbitals**. The

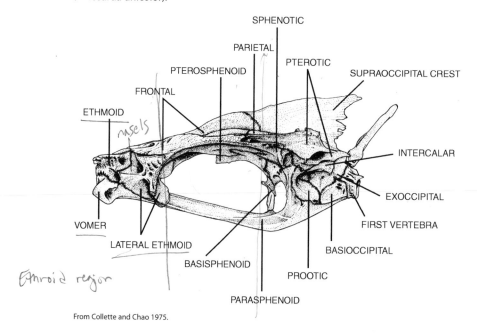

FIGURE 3.2. *Lateral view of the neurocranium of the dogtooth tuna (Gymnosarda unicolor).*

From Collette and Chao 1975.

FIGURE 3.3.   *Dorsal view of the neurocranium of the dogtooth tuna (Gymnosarda unicolor).*

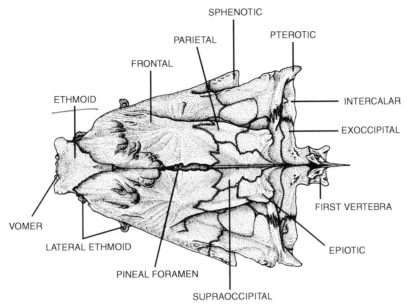

From Collette and Chao 1975.

infraorbitals, also known as circumorbital or suborbital bones, form a ring around the eye in primitive bony fishes. However, this ring is reduced to a chain of small bones under and behind the eye in advanced bony fishes. Advanced teleosts usually have **infraorbital 1**; the **lachrymal**, or preorbital; IO₂, or jugal; IO₃, or true suborbital, which may bear a suborbital shelf that supports the eye;

and the **dermosphenotic bones**, or postorbitals, which bear the infraorbital or suborbital lateral line canal (Fig. 3.5). Many primitive teleosts also have an antorbital and a supraorbital.

Otic region.   Five cartilage bones enclose each bilateral otic (ear) chamber inside the skull (see Figs. 3.2 through

FIGURE 3.4.   *Ventral view of the neurocranium of the dogtooth tuna (Gymnosarda unicolor).*

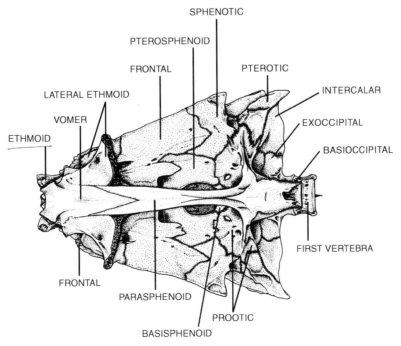

From Collette and Chao 1975.

FIGURE 3.5. *Left infraorbital bones in lateral view of a Spanish mackerel* (Scomberomorus maculatus).

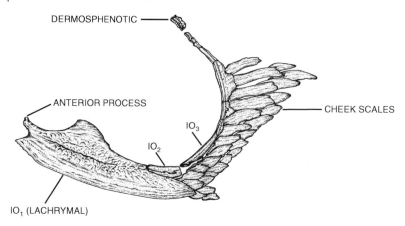

From Collette and Russo 1985b.

3.4). Paired **sphenotics** form the most posterior dorsolateral part of the orbit roof. Paired **pterotics** form the posterior outer corners of the neurocranium and enclose the horizontal semicircular canal. Paired **prootics** form the floor of the neurocranium and enclose the utriculus of the inner ear. Paired **epiotics**, more recently called epioccipitals, lie posterior to the parietals and lateral to the supraoccipital and contain the posterior vertical semicircular canal. The median process of the **posttemporal**, by which the pectoral girdle attaches to the posterior region of the skull, attaches to the epiotics. The epiotics enclose part of the posterior semicircular canal. Paired **intercalars** (or opisthotics) fit between the pterotics and exoccipitals and articulate with the lateral process of the posttempo-

ral. There is only one pair of dermal bones in the otic region, the paired **parietals**, which roof part of the otic region and articulate with the frontals anteriorly, supraoccipital medially, and the epiotics posteriorly.

Basicranial region. Three sets of cartilage bones, one pair plus two median bones, form the cranial base. Paired lateral **exoccipitals** form the sides of the foramen magnum (Fig. 3.6), which is the passageway for the spinal cord. The median **basioccipital** is the most posteroventral neurocranium bone and articulates with the first vertebra. The dorsal median **supraoccipital** bone usually bears a posteriorly directed supraoccipital crest that varies among teleosts from a slight ridge to a prominent crest.

FIGURE 3.6. *Rear view of the skull of a bonito (Sarda chiliensis). Xs indicate points of attachment of epineural bones.*

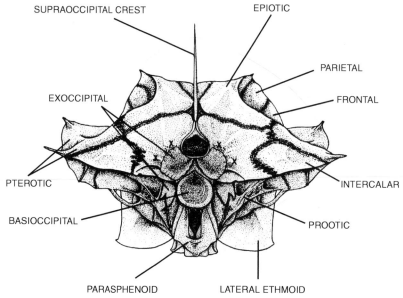

From Collette and Chao 1975.

The only dermal bone in the basicranial region is the median **parasphenoid**, a long cross-shaped bone that articulates with the vomer anteriorly and forms the posteroventral base of the skull.

Branchiocranium

The branchiocranium is divisible into five arches: mandibular, palatine, hyoid, opercular, and branchial.

The mandibular arch. The **mandibular arch** forms the upper jaw and is known as the palatoquadrate cartilage in Chondrichthyes. It is composed entirely of dermal bones in bony fishes. The mandibular arch may have three sets of bones. The dentigerous **premaxillae** are the anteriormost elements. The **maxillae** are dentigerous in some soft-rayed (malacopterygian grade) fishes, but the maxilla is excluded from the gape in more advanced spiny-rayed fishes (acanthopterygian grade). The third bone that may be present is the **supramaxilla**. It is a small bone on the posterodorsal margin of the maxilla. Some herringlike fishes (Clupeoidei) have multiple supramaxillae.

The lower jaw consists of Meckel's cartilage in Chondrichthyes. In bony fishes, the dermal, dentigerous **dentary bone** (Fig. 3.7) covers Meckel's cartilage, which is reduced to a thin rod extending posteriorly along the inner face of the dentary to the angular. The **angular** (sometimes called articular) is a large posterior cartilage bone that fits into the V of the dentary. The dermal **retroarticular** (sometimes called angular) is a small bone attached to the posteroventral corner of the angular. The lower jaw forms a single functional unit in most bony fishes, but in the kissing gourami (*Helostoma temmincki*), African characins of the genus *Distochodus*, and some parrotfishes of the genus *Scarus*, there is a mobile joint between the dentary and the angular (Liem 1967; Vari 1979; Bellwood 1994).

Teeth. The jaws and pharyngeal bones may bear teeth. Many different terms have been applied to the different sizes and shapes of teeth (see also Chap. 8, Dentition). Although the different kinds form a continuum, they can be divided into several types.

1. Canine: large conical teeth frequently located at the corners of the mouth; for example, snappers (*Lutjanus*).
2. Villiform: small, fine teeth.
3. Molariform: pavementlike crushing teeth, as in cownose rays (Rhinopterinae).
4. Cardiform: fine, pointed teeth arranged as in a wool card; for example, the pharyngeal teeth in pickerels (*Esox*).
5. Incisor: large teeth with flattened cutting surfaces adapted for feeding on mollusks and crustaceans; for example, chimaeras (Holocephali).
6. Teeth fused into beaks for scraping algae off corals, as in parrotfishes (Scaridae), or for biting crustaceans or echinoderms, as in blowfishes (Tetraodontiformes).
7. Flattened triangular cutting teeth, as in sharks and piranhas. Sharp, cutting teeth are uncommon in bony fishes with the exception of some characins (Characoidei) such as *Myleus* that feed on plants, and *Serrasalmus*, one genus of the infamous carnivorous piranhas.
8. Pharyngeal teeth. Minnows (Cypriniformes), which lack teeth in their jaws, have well-developed pharyngeal teeth.

Palatine and hyoid arches. The **palatine arch** consists of four pairs of bones in the roof of the mouth (see Fig. 3.7). The **palatines** are cartilage bones that are frequently dentigerous. They have been called "plowshare" bones because of their characteristic shape. The dermal **ectopterygoids** are

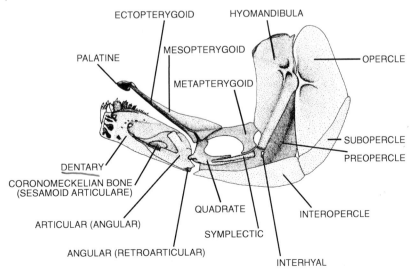

FIGURE 3.7. *Lateral bones of face and lower jaw suspension of a generalized characin (Brycon meeki).*

From Weitzman 1962

narrow bones, sometimes T-shaped, sometimes dentiger-ous. The dermal **entopterygoids** (or mesopterygoids) are thin bones that roof the mouth. The **metapterygoids** are cartilage bones, quadrangular-shaped and articulating with the quadrate and hyomandibula.

The **hyoid arch**, or **suspensorium**, consists of a chain of primarily cartilage bones that attach the lower jaw and opercular apparatus to the skull (see Fig. 3.7). The **hyomandibula** is an inverted L-shaped bone that con-nects the lower jaw and opercular bones to the neurocra-nium. The **symplectic** is a small bone that fits into the groove of the quadrate. The **quadrate** is a triangular bone with a groove for the symplectic; it has an articulating facet to which the lower jaw is attached.

The **hyoid complex** is a series of five pairs of bones (Fig. 3.8) that lie medial to the lower jaw and opercular bones and lateral to the branchiostegal rays that attach to them. The anteriormost bones are the dorsal and ventral **hypo-hyals** (or basihyals). They are followed by the **ceratohyal**, a long flat bone that interdigitates with the epihyal posteri-orly and to which some of the branchiostegal rays attach. The **epihyal** (or posterior ceratohyal) is a triangular bone to which some of the branchiostegal rays attach. The **in-terhyal** is a small rod-shaped bone that attaches the hyoid complex to the neurocranium and opercular apparatus. The **glossohyal** is an unpaired spatulate bone that lies over the anterior basibranchial and supports the tongue.

The dermal bones of the hyoid arch are the bran-chiostegal rays and the median urohyal bone. The **bran-chiostegal rays** are elongate, flattened, riblike structures (see Fig. 3.8) that attach to the ceratohyal and epihyal. They are important in respiration, particularly in bottom-dwelling species. Their number and arrangement are use-ful in tracing phylogenies (see McAllister 1968). The **urohyal** is a flattened, elongate, unpaired bone that lies inside the rami of the lower jaw.

Opercular and branchial series.   The **opercular apparatus** con-sists of four pairs of wide, flat dermal bones that form the gill covers that protect the underlying gill arches. The **opercle** is usually more or less rectangular and is usually the largest and heaviest of the opercular bones (see Figs. 3.1, 3.7). It has an anterior articulation facet connecting with the hyomandibula. The **subopercle** is the innermost

and most posterior element. The **preopercle** is the anteri-ormost element. It overlies parts of the other three oper-cular bones. The **interopercle** is the most ventral bone.

The **branchial arch** consists of the four pairs of gill arches, gill rakers, pharyngeal tooth patches, and support-ing bones (Fig. 3.10). All elements of the branchial arch are cartilage bones. The three **basibranchials** form a chain from anterior to posterior. The first basibranchial is cov-ered by the glossohyal; the second and third serve as at-tachments for the hypobranchials and ceratobranchials. Three pairs of **hypobranchials** connect the basibranchials with the ceratobranchials. **Ceratobranchials** are the long-est bones in the branchial arch, and they support most of the gill filaments and gill rakers. The anterior three cerato-branchials are unmodified and connect with their respec-tive hypobranchials. The fourth is more irregular. The fifth bears a tooth plate and is sometimes called the **lower pha-ryngeal bone**. Four pairs of **epibranchials** attach basally to the ceratobranchials. They vary from being long and slender (like a short ceratobranchial) to short and stubby. Four pairs of **pharyngobranchials** attach to the epibran-chials. The third and fourth may have dermal tooth patches attached to them and are then termed **upper pha-ryngeal bones**. For a detailed account of the gill arches and their importance in fish phylogeny, see Nelson (1969).

For a general reference to the osteology of the skull of many species, consult Gregory (1933); for more complete treatment of ostariophysan fishes, Harrington (1955) and Weitzman (1962) are excellent sources. A complete review of the braincase of actinopterygian fishes and their fossil ancestors, pholidophorids and leptolepids, was presented by Patterson (1975).

Postcranial Skeleton

The **notochord** is the most primitive supporting structure in chordates. It is a simple longitudinal rod composed of a group of cells that, when viewed in cross section, appear to be arranged as concentric circles. The most primitive chordate to possess a notochord is the "tadpole" larva of tunicates. The notochord provides support for an elon-gate body while swimming. Notochordal cells inside the notochord are few in number and contain large vacuoles. Turgor of the notochordal cells provides rigidity. The no-tochord is found during embryonic development in all

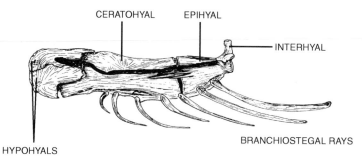

FIGURE 3.8.   *Left hyoid complex in lateral view in a Spanish mackerel* (Scomberomorus commerson).

CERATOHYAL   EPIHYAL

INTERHYAL

HYPOHYALS

BRANCHIOSTEGAL RAYS

From Collette and Russo 1985b.

Jaw Suspension

Much interest and controversy has arisen over which of the gill arches of the agnathan ancestors of gnathostomes gave rise to the jaws. Zoologists are not certain whether the jaw-forming arch was the first in the series, or whether it is posterior to a premandibular arch that has been lost (Walker and Liem 1994). Classically, four principal types of jaw attachment have been recognized.

1. **Amphistylic** jaw suspension is found in primitive sharks (Fig. 3.9). The upper jaw is attached to the cranium by ligaments at orbital and otic processes of the palatoquadrate. The hyoid arch is attached to the chondrocranium and lower jaw and is involved in suspension of both jaws.
2. **Hyostylic** suspension is found in most chondrichthyans and all actinopterygians (but Maisey 1980 found no dividing line between amphistyly and hyostyly in living sharks). The otic contact of the palatoquadrate has been lost, so that both jaws are suspended from the chondrocranium by way of ligamentous attachments to the hyomandibula, which is attached to the otic region of the neurocranium. **Methyostylic** suspension is a variety of hyostyly present in Actinopterygii. Remnants of the second gill arch (palatine and pterygoid bones) connect in the roof of the mouth. Dermal bones, the premaxilla and maxilla, form a new upper jaw. A new dermal anterior lower jaw element, the dentary, is connected with the angular, which is suspended from the otic capsule by hyoid derivatives.
3. **Autostylic** suspension is present in most non-fish vertebrates, lungfishes, and tetrapods. The processes of the palatoquadrate articulate to or fuse with the chondrocranium. The hyoid arch is no longer involved with jaw suspension. The hyomandibula becomes the columella of the inner ear in tetrapods.
4. **Holostylic** suspension is a variety of autostyly found only in the Holocephali, the chimaeras. The palatoquadrate is fused to the chondrocranium and supports the lower jaw in the quadrate region. The name Holocephali means "whole head," in reference to the upper jaw being a part of the cranium.

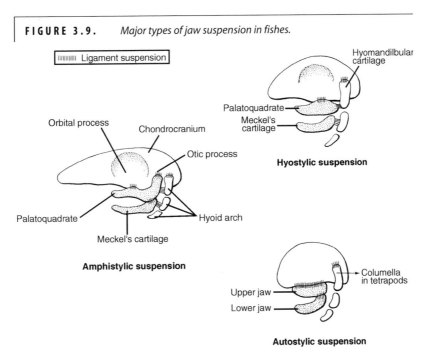

FIGURE 3.9. *Major types of jaw suspension in fishes.*

From Walker and Liem 1994.

FIGURE 3.10. *Branchial arch of a Spanish mackerel (Scomberomorus semifasciatus). Dorsal view of gill arches with the dorsal region folded back to show their ventral aspect. Epidermis removed from right-hand side to reveal underlying bones.*

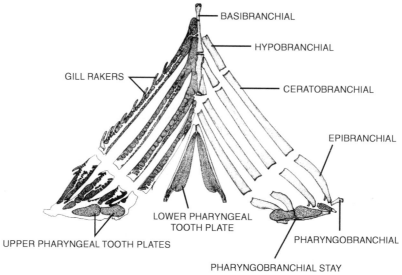

From Collette and Russo 1985b.

chordates, but intervertebral disks are all that remain of the notochord in most adults. However, it is present in adult lancelets, Chondrichthyes, Dipnoi, sturgeons (Acipenseridae), paddlefishes (Polyodontidae), and the coelacanth. A 1-m-long sturgeon may have a notochord nearly as long and about 12 mm in diameter.

Vertebral column. Vertebrae arise and form around the notochord where muscular myosepta intersect with dorsal, ventral, and horizontal septa. Vertebrae form from cartilaginous blocks called **arcualia**. Typically, there is one vertebra per body segment, the **monospondylous** condition. The basidorsal, interdorsal, basiventral, and interventral arcualia all fuse together to form a single vertebra. In the **diplospondylous** condition, the basidorsal fuses to the basiventral and the interdorsal fuses to the interventral, producing two vertebrae per body segment. Diplospondyly is present in the tail region of sharks and rays, in lungfishes, and in the caudal vertebrae of the bowfin (*Amia*). Diplospondyly permits increased body flexibility.

Vertebrae are usually divided into **precaudal** (anterior vertebrae extending posteriorly to the end of the body cavity and bearing ribs) and **caudal** vertebrae (posterior vertebrae beginning with the first vertebra bearing an elongate **haemal spine** extending ventral to the vertebral centrum) (Fig. 3.11).

Vertebrae may have various bony elements projecting from them. Dorsally, there is an elongate **neural spine** housing a **neural arch** through which the spinal cord passes (Fig. 3.12A). Ventrally, there may be **parapophyses** that extend ventrolaterally and to which the ribs usually attach (see Fig. 3.12A). The main artery of the body, the dorsal aorta, passes ventral to the precaudal vertebrae and enters the closed **haemal canal** (Fig. 3.12B) toward the

end of the abdominal cavity as the caudal artery. Other projections include **neural prezygapophyses** and **postzygapophyses** on the dorsolateral margins of the vertebrae and **haemal prezygapophyses** and **postzygapophyses** on the ventrolateral margins (Fig. 3.12D).

Ribs and intermuscular bones. **Ribs** (pleural ribs) form in the peritoneal membrane and attach to the vertebrae, usually from the third vertebra to the last precaudal vertebra. They are distinct from intermuscular bones and serve to protect the viscera.

Intermuscular bones occur only in teleosts and are segmental, serially homologous ossifications in the myosepta.

FIGURE 3.11. *Junction of precaudal and caudal vertebrae in left lateral view of the king mackerel (Scomberomorus cavalla). Middle vertebra, with elongate haemal spine, is first caudal vertebra. Vertebrae numbered from anterior.*

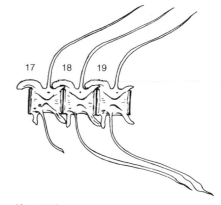

From Collette and Russo 1985b.

FIGURE 3.12. *Representative precaudal and caudal vertebrae of a generalized charancin* (Brycon meeki). *(A) Anterior view of twentieth precaudal vertebra. (B) Anterior view of twenty-fourth precaudal vertebra. (C) Anterior view of second caudal vertebra. (D) Lateral view of twentieth precaudal through second caudal vertebrae.*

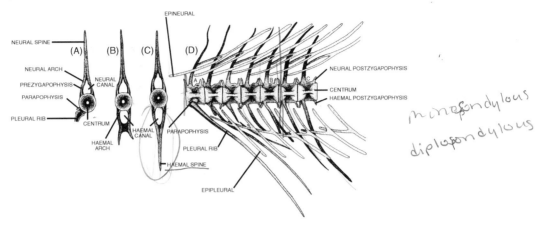

From Weitzman 1962.

Terminology used for these bones, and for ribs, was confused until Patterson and Johnson (1995) clarified the situation. Previously, the several series of bones projecting laterally from the vertebral column had been named topographically as dorsal and ventral ribs, or epineural, epicentral, and epipleural bones. Patterson and Johnson recognized three series of intermuscular bones: epineurals, epicentrals, and epipleurals. Primitively, ossified **epineurals** may be fused with the neural arches. Some are free (unfused) and may develop an anteroventral branch as in charancins (see Fig. 3.12D). Epineurals usually start on the first vertebra (sometimes on the back of the skull; see Fig. 3.6) and continue along the vertebrae well posterior to the ribs. **Epicentrals** lie in the horizontal septum and are primitively ligamentous. **Epipleurals** lie below the horizontal septum and are posteroventrally directed. Epicentrals and epipleurals have been lost in many advanced teleosts, leaving a series of short, straight epineurals lateral to the vertebral column and dorsal to the ribs.

Caudal complex. The tail of a fish is a complex of vertebrae, vertebral accessories, and fin rays that have been modified during evolution to propel the fish forward in a linear fashion. The functional morphology of the fish tail and the history of its progressive change are discussed in Chapters 8 (Locomotory Types) and 11 (Division Teleostei). The teleostean caudal skeleton was largely neglected as a source of systematic characters until Monod (1968) surveyed the caudal skeleton of a broad range of teleosts and established a coherent and homogeneous terminology. In primitive teleosts, a number of **hypurals** (enlarged haemal spines) support most of the branched **principal caudal fin rays** that form the caudal fin (Fig. 3.13). **Epurals** (enlarged neural spines) and the last haemal spine support the small spinelike **procurrent** caudal fin rays. In many advanced teleosts, the number of hypurals has been reduced to five. In some groups, such as atherinomorphs, sticklebacks, sculpins, the louvar, tunas

and mackerels, and flatfishes, the posterior vertebrae have been shortened and some of the hypurals fuse to form a **hypural plate**. In scombrids, hypurals 3 and 4 are united into the upper part of the plate and hypurals 1 and 2 into the lower part (Fig. 3.14).

Caudal fin types. Caudal fins of fishes vary in both external shape and internal anatomy. The different types of caudal fins provide useful information about swimming as well as about phylogeny. There are three basic types of fish tails, with an additional three types recognized for special groups of fishes.

The **protocercal** tail is the primitive undifferentiated caudal fin that extends around the posterior end in adult lancelets, agnathans, and larvae of more advanced fishes.

In the **heterocercal**, or unequal-lobed, tail, the vertebral column extends out into the upper lobe of the tail. This type of tail is found in Chondrichthyes and primitive bony fishes such as sturgeons (Acipenseridae) and is still recognizable in gars (Lepisosteidae). *Amia*, the bowfin, has what has been termed a hemihomocercal tail (Harder 1975), intermediate between heterocercal and homocercal, with external but not internal symmetry.

Most advanced bony fishes have a **homocercal**, or equal-lobed, tail (see Figs. 3.13, 3.14). In this type of tail, the caudal fin rays are arranged symmetrically but attach to a series of hypural bones posterior to the last vertebra that supports the caudal fin rays. These plates are ventral to the upward-directed **urostyle**, so this type of tail could be considered an abbreviated heterocercal tail.

The **leptocercal** (or diphycercal) tail is similar to the protocercal in having the dorsal and anal rays joined with the caudal around the posterior part of the fish, but this is considered to have been secondarily derived, not primitive. This type of tail is found in lungfishes (Dipnoi), the coelacanth, and rattails (Macrouridae).

The last vertebra of the **isocercal** tail, not the original urostyle, has been secondarily modified into a small flat-

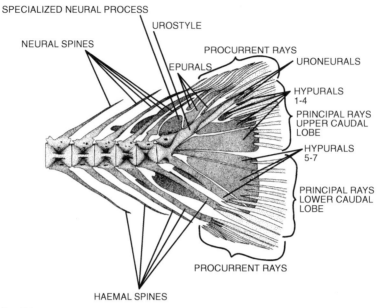

FIGURE 3.13. *Posterior vertebrae and caudal complex in a generalized charactin* (Brycon meeki).

SPECIALIZED NEURAL PROCESS

UROSTYLE

NEURAL SPINES

PROCURRENT RAYS

EPURALS

URONEURALS

HYPURALS 1-4

PRINCIPAL RAYS UPPER CAUDAL LOBE

HYPURALS 5-7

PRINCIPAL RAYS LOWER CAUDAL LOBE

PROCURRENT RAYS

HAEMAL SPINES

From Weitzman 1962.

tened plate to which the caudal fin rays attach in the cods (Gadidae).

Ocean sunfishes (Molidae) have lost the posterior end of the vertebral column, including the hypural plate. The dorsal and anal fins have grown around the posterior end of the fish to form what has been termed a **gephyrocercal** (or bridge) tail.

Appendicular Skeleton

Pectoral and pelvic girdles are absent in the agnathans, presumably because they had not yet evolved. Sharks have a coracoscapular cartilage that hangs more or less freely inside the body wall and has no attachment to the vertebral column. In rays, the pectoral girdle is attached to the fused anterior section of the vertebral column (synarchial condition) and also, by way of the propterygium of the pectoral girdle and antorbital cartilage, to the nasal capsules of the skull.

Pectoral girdle. Unlike the condition in tetrapods, the pectoral girdle in bony fishes has no attachment to the vertebral column. It attaches to the back of the skull by the posttemporal bone. Rather than dividing the bones into cartilage and dermal, as done for the skull, it seems more practical to present the bones in sequence from the skull to the girdle bones themselves.

Three dermal bones are involved in the suspension of the pectoral girdle from the skull. The **posttemporal** usually has two anterior projections that attach to the epiotic and intercalar bones on the back of the skull. The **supratemporal** (or extrascapular) is a very thin bone that

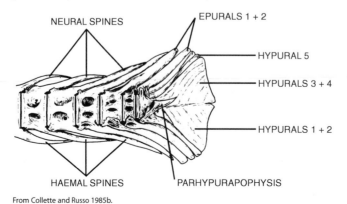

FIGURE 3.14. *Caudal complex in left lateral view of a Spanish mackerel* (Scomberomorus semifasciatus).

NEURAL SPINES

EPURALS 1 + 2

HYPURAL 5

HYPURALS 3 + 4

HYPURALS 1 + 2

HAEMAL SPINES

PARHYPURAPOPHYSIS

From Collette and Russo 1985b.

bears part of the lateral line canal. It usually lies right under the skin dorsal to the posttemporal (Fig. 3.15). The **supracleithrum** is a heavy bone that lies between the posttemporal and the pectoral girdle.

The pectoral girdle is composed of three cartilage bones in acanthopterygians. The **cleithrum** is the largest, dorsal-most, and anteriormost element of the pectoral girdle. The **scapula** is a small bone, usually with a round scapular foramen, lying between the cleithrum and the radials. The **coracoid** is a long thin bone that makes up the posterior part of the pectoral girdle and may support some of the pectoral fin radials. There is an additional element found between the coracoid and cleithrum in soft-rayed mala-copterygian grade fishes, the **mesocoracoid**. This bone is lost in acanthopterygians as the pectoral fin moves up and assumes a vertical instead of oblique position.

The **radials** (or actinosts) are hourglass-shaped cartilage bones that support the pectoral fin rays. There are typically four in teleosts, attached to the coracoid and scapula.

Posterior and internal to the pectoral girdle are the dermal **postcleithra**. Soft-rayed teleosts typically have three; two are elongate and scalelike, and one is rodlike. Spiny-rayed teleosts typically have two, one scalelike, the other more riblike.

Pelvic girdle. The pelvic girdle, like the pectoral girdle, is usually not attached to the vertebral column in fishes as it is in tetrapods. In sharks, the pelvic girdle consists of the **ischiopubic cartilages**, which float freely in the muscles of the posterior region of the body. In primitive bony fishes, there is a separate pelvic bone, a **basipterygium**, and radials to which the pelvic fin rays attach. In higher bony fishes, both the pelvic bone itself and the radials are lost or fused so that the fin rays attach directly to the single remaining element, the basipterygium.

In soft-rayed teleosts, the pelvic fins are **abdominal** in position, ventrally located, slightly anterior to the anal fin. The pelvic fins move forward to a **thoracic** position, directly below the pectoral fins, in spiny-rayed fishes. In some fishes (i.e., ophidiiforms and gadiforms), the pelvic fins lie anterior to the pectoral fins, a condition known as **jugular** pelvic fins. Jugular pelvic girdles may have attachments to the pectoral girdle.

Pelvic fin rays are frequently lost, and in some cases, such as eels (Anguilliformes), the neotenic South American needlefish *Belonion apodion*, and puffers, the pelvic girdle has also been lost.

Median fins. The median or unpaired fins consist of the dorsal, anal, and adipose fins along the dorsal and ventral profiles of the fish. In the Agnatha, the median fins are supported by cartilaginous rods. In Chondrichthyes, the median fins are supported by **ceratotrichia**, horny fin rays composed of elastin and supported by dermal cells. Below the ceratotrichia are three layers of **radials**, rodlike cartilages that support the fin rays and extend inward

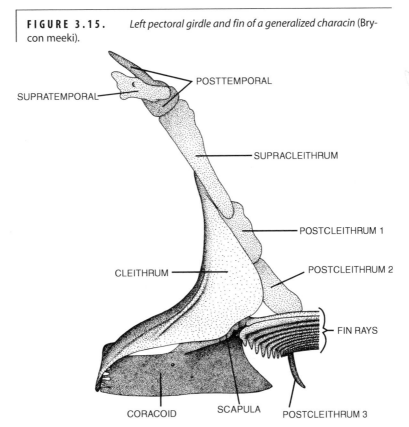

FIGURE 3.15. *Left pectoral girdle and fin of a generalized characin (Brycon meeki).*

From Weitzman 1962.

toward the vertebral column. If a spine is present at the anterior end of a median fin in a chondrichthyan such as the spiny dogfish (*Squalus acanthias*), it is not a true spine such as is found in spiny-rayed fishes (Acanthopterygii) but is rather a fusion of radials.

In bony fishes, the ceratotrichia are replaced during ontogeny by **lepidotrichia**, bony supporting elements that are derived from scales. Ceratotrichia are present in lungfishes and larval bony fishes. Primitive bony fishes such as the bowfin (*Amia*) still have three radials supporting each median fin ray, but these are reduced to two and then one in higher teleosts. The remaining element is then known as an **interneural bone** if it is under the dorsal fins or **interhaemal bone** if it is above the anal fin.

Primitive soft-rayed teleosts have a single dorsal fin that is composed entirely of soft rays. Advanced teleosts (Paracanthopterygii and Acanthopterygii) usually have two dorsal fins, with the anterior one (**first dorsal fin**) composed of spines and the posterior one (**second dorsal fin**) composed largely of soft rays, although there may be one spine at the anterior margin of the fin. Some soft-rayed fishes may have a single spine at the anterior end of the dorsal fin, but this is a bundle of fused rays, not a true spine.

True **spines** differ from **soft rays** in several characters.

Spines	Soft Rays
usually hard and pointed	usually soft and not pointed
unsegmented	segmented
unbranched	usually branched
solid	bilateral, with left and right halves

Some fast-swimming fishes such as the mackerels and tunas may have a series of **dorsal finlets**, small fins with one soft ray each, following the second dorsal fin.

Several groups of soft-rayed fishes have an additional fin posterior to the dorsal fin, the **adipose** fin. This is not a good term for this fin because it is rarely fatty. The adipose fin usually lacks lepidotrichia and is supported only by ceratotrichia, although some catfishes have secondarily developed a spine, composed of fused rays, at its anterior margin. The function or functions of adipose fins remain something of a mystery, but their presence is useful in identifying members of four groups that usually have them: trouts and salmons (Salmoniformes), characins (Characiformes), catfishes (Siluriformes), and lanternfishes and relatives (Myctophiformes).

The original function of the dorsal fin was as a stabilizer during swimming, but it has been modified in many different ways. It has been reduced or lost in rays (Batoidei) and South American knifefishes (Gymnotiformes). The dorsal and anal fins become **confluent**, joined with the caudal fin, around the posterior part of the body in many eels (Anguilliformes). The individual spines in the first dorsal fin have become shortened in fishes such as the bluefish (*Pomatomus saltatrix*) and the cobia (*Rachycentron canadum*). The first dorsal fin has been converted into a suction disk in the remoras (Echeneidae). The membranes between the spines have lost their attachment to each other in the bichirs (Polypteridae) and sticklebacks (Gasterosteidae). Venom glands have become

associated with dorsal fin spines, and other spines, in fishes such as the stonefish (*Synanceja*), the weeverfishes (Trachinidae), and venomous toadfishes (Thalassophryninae). The spiny dorsal fin has been converted into a locking mechanism in the triggerfishes (Balistidae). It is depressible into a groove during fast swimming in the tunas (Scombridae). Perhaps the most extreme modification of a dorsal fin is the conversion of the first dorsal spine into an **ilicium**, or fishing rod, with an **esca**, or bait, at its tip in the anglerfishes (Lophiiformes).

The **anal fin** usually lies just posterior to the anus. In soft-rayed fishes, it is composed entirely of soft rays, as is the single dorsal fin of these fishes. In spiny-rayed fishes, the anal fin usually contains one or several anterior spines, followed by soft rays. Fast-swimming fishes that have dorsal finlets usually also have **anal finlets**, small individual fins following the anal fin.

The anal fin shows the least variation among fishes. It has been lost in the ribbonfishes (Trachipteridae). It is very long and serves as the primary locomotory fin in the South American knifefishes (Gymnotoiformes). The anterior part of the anal fin has been modified into a **gonopodium** for spermatophore transfer in male livebearers (Poeciliidae). It is also variously modified into what has been called an **andropodium** in males of *Zenarchopterus* and several related viviparous genera of halfbeaks.

INTEGUMENTARY SKELETON

The **integument** is composed of the skin and skin derivatives, and includes scales in fishes and feathers and hair in birds and mammals. The integument forms an external protective structure parallel to the internal endoskeleton and serves as the boundary between the fish and the external environment. The structure of the skin in fishes is similar to that of other vertebrates, with two layers: an outer epidermis and an inner dermis.

Epidermis

The epidermis is ectodermal in origin. In agnathans and higher vertebrates, the epidermis is stratified. The lowest layer is the **stratum germinativum**, composed of columnar cells (Fig. 3.16). It is the generating layer that gives rise to new cells. In agnathans and bony fishes, there is an outer thin film of noncellular dead cuticle (Whitear 1970). The outer part of the epidermis in terrestrial vertebrates is the **stratum corneum**, which is composed of dead, horny, keratinized squamous cells that form hair and feathers. Breeding tubercles in fishes may also contain keratin.

Dermis

The inner dermis contains blood vessels, nerves, sense organs, and connective tissue. It is derived from embryonic mesenchyme of mesodermal origin. It is composed of fibroelastic and nonelastic collagenous connective tissue with relatively few cells. Dermal layers include an upper, relatively thin layer of loose cells, the **stratum** laxum (or stratum spongiosum) and a lower, compact thick layer, the **stratum compactum** (see Fig. 3.16). In adult fishes, the dermis is much thicker than the epidermis. The thicknesss of the integument depends on the thickness of the dermis. Scaleless species, such as catfishes of the genus *Ictalurus*, have relatively thick, leathery skin. The ocean sunfish (*Mola*) has the skin reinforced by a hard cartilage layer, 50 to 75 mm thick. Snailfishes (*Liparis*, Liparidae) have a transparent jellylike substance up to 25 mm thick in their dermis.

The chemical composition of fish skin is poorly studied, but some generalizations can be made. There is less water in fish skin than in fish flesh, a higher ash content, and similar amounts of protein. The main protein is collagen, which is why fish skin has been used to manufacture glue. The chief minerals in fish skin are phosphorus, potassium, and calcium (Van Oosten 1957). The ash composition of the skin of the coho salmon (*Oncorhynchus kisutch*) is

P_2O_5	33%	CaO	14%
Cl	21	Na_2O	9
K_2O	17	MgO	2

Among the functions of the skin are mechanical protection and production of mucus by epidermal mucous cells. **Mucin** is a glycoprotein, made up largely of albumin. Threads of mucin hold a large amount of water. It is possible to wring the water out of mucus, leaving threads of mucin. Among the first multicellular glands to evolve were the mucous glands of hagfishes (Myxinidae), called **thread cells** (Fernholm 1981). The oft-told story is that a hagfish 1 a bucket of water 5 a bucket of slime.

Other structures in the skin of fishes include epidermal venom glands associated with spines on fins (weeverfish, Trachinidae; madtom catfish, *Noturus*), opercles (venomous toadfishes, Thalassophryninae), and tail (stingrays, Dasyatidae). **Photophore**s, which produce bioluminescence, develop from the germinative layer of the epidermis. Color is due to **chromatophores**, which are modified dermal cells containing pigment. The skin also contains important receptors of physical and chemical stimuli.

Scales

Scales are the characteristic external covering of fishes. There are four basic types of scales.

1. **Placoid** scales are characteristic of all of the Chondrichthyes, although they have a more restricted distribution in rays and chimaeras than in sharks. This type of scale is sometimes called a "dermal denticle," but this is not accurate terminology because there are both epidermal and dermal portions, as in mammalian teeth. Each placoid scale consists of a flattened rectangular basal plate in the upper part of the dermis, from which a protruding spine projects posteriorly on the surface. The outer layer of the placoid scale is hard enamel-like **vitrodentine**, derived from ectoderm. Vitrodentine is noncellular and has a very low organic content. The scale has a cup or cone of **dentine** with a pulp cavity richly supplied with blood capillaries, just as in mammalian teeth. Placoid scales do not increase in size with growth; instead, new scales are added between older scales. Placoid scales are homologous with teeth in all vertebrates.
2. **Cosmoid** scales were present in fossil crossopterygians and fossil lungfishes. The scales of Recent lungfishes are highly modified by loss of the dentine layer. Cosmoid scales are similar to placoid scales and probably arose from fusion of placoid scales. Cosmoid scales are composed of two basal layers of bone: **isopedine**, which is the basal layer of dense lamellar bone, and cancellous (or spongy) bone, which is supplied with canals for blood vessels. Over the bone layers is a layer of **cosmine**, a noncellular dentinelike substance.

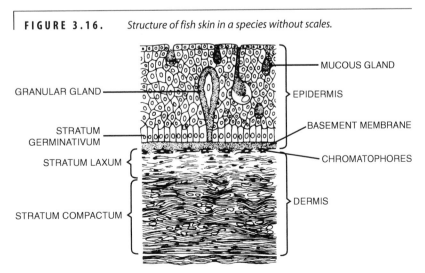

FIGURE 3.16. *Structure of fish skin in a species without scales.*

MUCOUS GLAND

GRANULAR GLAND

EPIDERMIS

STRATUM GERMINATIVUM

BASEMENT MEMBRANE

STRATUM LAXUM

CHROMATOPHORES

STRATUM COMPACTUM

DERMIS

From Walker and Liem 1994.

Over the cosmoid layer is a thin superficial layer of vitrodentine. Growth is by addition of new lamellar bone underneath, not over the upper surface.

3. **Ganoid** scales were present in fossil paleoniscoids and in Chondrostei. They are modified cosmoid scales, with the cosmine replaced by dentine and the surface vitrodentine replaced by **ganoine**, an inorganic bone salt secreted by the dermis. Ganoine is a calcified noncellular material without canals. Ganoid scales are usually rhomboidal in shape and have articulating peg and socket joints between them. The fossil paleoniscoid scale is least modified in the bichirs, Polypteridae (three layers: ganoine, dentine, and isopedine). Ganoid scales are more modified in the sturgeons (Acipenseridae) and paddlefishes (Polyodontidae), in which lamellae of ganoine lay above a layer of isopedine. Sturgeon scales are modified into large plates, with most of the rest of the body naked.

The scales of gars (Lepisosteidae) are similar to Polypteridae in external appearance but are more similar to those of the Acipenseridae and Polyodontidae in structure. In the bowfin (*Amia*), the scale is greatly reduced in thickness to merely a collagenous plate with bony particles, very similar to the cycloid scales of Teleostei.

4. **Cycloid** and **ctenoid** scales are almost completely dermal. There is no enamel-like layer except perhaps the **ctenii** (teeth on posterior border) and the most posterior and superficial ridges of the scale. These types of scales were derived evolutionarily from the ganoid scale by loss of the ganoin and thinning of the bony dermal plate. Two major portions make up these scales: 1) a surface "bony" layer, an organic framework impregnated with salts, mainly calcium phosphate (as hydroxyapatite) and calcium carbonate; and 2) a deeper fibrous layer, or **fibrillary plate**, composed largely of collagen.

Cycloid or ctenoid scales are present in the Teleostei, the vast majority of bony fishes. They have the advantage of being imbricate, overlapping like shingles on a roof, which gives great flexibility compared with cosmoid and ganoid scales. Small muscles pull unequally on the dermis, causing the anterior portion of the scale to become depressed in the dermis and covered over by the posterior margin of the preceding scale. Cycloid scales lack ctenii. Breeding tubercles and contact organs (see Chap. 20, Fig. 20.2) are present in many groups of fishes that lack ctenoid scales.

Including all scales with spines on their posterior margins under the term ctenoid is an oversimplification of the situation (Johnson 1984; Roberts 1993). Three different, general types of spined scales exist: **crenate,** with simple marginal indentations and projections; **spinoid,** with spines continuous with the main body of the scales; and ctenoid, with ctenii formed as separate ossifications distinct from the main body of the scale

(Roberts 1993). Crenate scales occur widely in the Elopomorpha and Clupeomorpha; spinoid scales occur widely in the Euteleostei; peripheral ctenoid scales (whole ctenii in one row) occur, probably independently, in the Ostariophysi, Paracanthopterygii, and Percomorpha; and transforming ctenoid scales (ctenii arising in two or three rows and transforming into truncated spines) are a synapomorphy diagnosing the Percomorpha.

As with fish skin, the chemical composition of scales is poorly known. About 41% to 84% is organic protein, mostly albuminoids such as collagen (24%) and ichthylepidin (76%). Up to 59% is bone, mostly $Ca_3(PO_4)_2$ and $CaCO_3$.

Phylogenetic significance of scale types. Scales have been used as a taxonomic tool since the beginnings of systematic ichthyology (Roberts 1993). For example, Louis Agassiz divided fishes into four groups based on their scale type. More recent classifications are based on more characters but are similar to the system used by Agassiz.

Agassiz System	*Recent Classification*
a. Placodermi ——————→	a. Chondrichthyes
b. Ganoidei ⎡——————→	b. Chondrostei
⎣——————→	c. Holostei
	d. Teleostei
c. Cycloidei ——————→	malacopterygian grade
d. Ctenoidei ——————→	acanthopterygian grade

Whereas most groups of advanced acanthopterygian teleosts have ctenoid scales, some "ctenoid" groups may also have cycloid scales. In the flatfishes, Pleuronectiformes, some species have ctenoid scales on the eyed side and cycloid scales on the blind side. Some flatfishes are sexually dimorphic, males having ctenoid scales and females having cycloid scales.

Scale size varies greatly in fishes. Scales may be microscopic and embedded as in freshwater eels (Anguillidae), which led to their being classified as nonkosher because of the supposed absence of scales. Scales are small in mackerel (*Scomber*), "normal" in perch (*Perca*), large enough to be used for junk jewelry in tarpon (*Megalops*), and huge (the size of the palm of a human hand) in the Indian mahseer (*Tor tor*, a cyprinid gamefish reaching 43 kg in weight).

Development pattern of scales. In actinopterygian (ray-finned) fishes, scales usually develop first along the lateral line on the caudal peduncle, then in rows dorsal and ventral to the lateral line, and then spread anterior (see Fig. 9.8). The last regions to develop scales in ontogeny are the first to lose scales in phylogeny. Once the full complement of scales is attained in ontogeny, the number remains fixed. Therefore, the number of scales is an important taxonomic character. Most scales remain in place for the life of the fish, which makes scales valuable in recording events in the life history of an individual fish, such as reduced growth that generally occurs during the winter or during the breeding season. Scales disappear with age in the swordfish, *Xiphias*, causing a question as to whether swordfish are kosher, because kosher dietary laws require that a fish have both fins and scales.

Geographic variation can occur in the relative development of ctenoid scales in some species. For example, in the swamp darter (*Etheostoma fusiforme*) of the Atlantic Coastal Plain of the United States, the number of scales in the interorbital area increases from north to south (Collette 1962). In the northern part of the range, the few scales present are embedded and cycloid. Moving further south, there is an increase in number and in relative "ctenoidy" of the scales; more scales have the posterior surface of the scale projecting through the epidermis, and these scales have more ctenii on them.

Lateral line scales form pores on scales from head to tail. Most fishes have **complete** lateral lines, that is, the pored scales extend from behind the opercular region all the way to the base of the caudal fin. Some species, such as the swamp darter, have **incomplete** lateral lines, with the pores extending only part way to the caudal base.

Modifications of scales. Some fishes have scales that are **deciduous**, that is, easily shed. This is true of many species of herrings (Clupeidae) and anchovies (Engraulidae). It may be true of one species in a genus but not of another. For example, of two species of common Australian halfbeaks or garfishes (Hemiramphidae), scales remain in the river garfish, *Hyporhamphus regularis*, but are easily lost in the sea garfish, *H. australis*.

Male darters of the genus *Percina* have **caducous scales**, a single row of enlarged scales along the ventral surface between the pelvic fins and the anus.

Several structures in chondrichthyans may have arisen from fusions of modified placoid scales. These include the "spines" at the beginning of the first and second dorsal fins of the spiny dogfish, the prominent dorsal "spine" in some chimaeras (Holocephali), the caudal fin spine in stingrays (Dasyatidae), and the teeth on the rostrum of the sawfish (*Pristis*).

The structure of placoid scales in Chondrichthyes is the same as the structure of teeth in vertebrates, leading to the question, Which came first? Did some primitive chondrichthyan ancestor develop teeth that then spread over the body? Or did the ancestor first develop scales that then spread into the mouth and became modified into teeth? Apparently, the dermal armor of the earliest known vertebrates, the ostracoderms, broke up into smaller units, and some of these scales in the mouth evolved into teeth (Walker and Liem 1994).

In many teleosts, there is an external dermal skeleton in addition to the internal supporting skeleton. This is composed of segmented bony plates in pipefishes (Syngnathidae) and poachers (Agonidae) and bony shields similar to placoid scales with vitrodentine in several South American armored catfish families such as the Loricariidae. The body is enclosed in a bony cuirass (armor) in the shrimpfishes (Centriscidae) and is completely enclosed in a rigid bony box in the trunkfishes (Ostraciidae).

Many fishes have protective scutes or spines. The ventral row of scales is modified into **scutes** with sharp, posteriorly directed points in herrings, such as the river herrings (*Alosa*) and the threadfins (*Harengula*). Some

jacks (Carangidae) have lateral scutes along the posterior part of the lateral line. Sticklebacks (Gasterosteidae) have bony lateral plates. These plates vary in number and size in *Gasterosteus aculeatus*, roughly correlated with the salinity of the habitat and presence or absence of predators. Sharp erectable spines derived from scales are present in the porcupinefishes (Diodontidae). Large bony "warts" characterize the lumpfish (*Cyclopterus*). Surgeonfishes (Acanthuridae) are so named because of the pair of sharp, anteriorly directed spines on the caudal peduncle.

Three other modifications of scales are discussed elsewhere. Lateral line scales bear sensory structures (see Chap. 6). Lepidotrichia, fin rays supporting the fins, probably originated from scales (see above). The superficial bones of the skull originated as scales and have become modified into dermal bones (see above).

Scale morphology in taxonomy and life history. For studying taxonomy and life history, various parts of the scale are distinguished. Cycloid and ctenoid scales can be divided into four fields (Fig. 3.17): anterior (which is frequently embedded under the preceding scale), posterior, dorsal, and ventral. The focus is the area where scale growth begins. The position and shape of the focus may vary, being oval, circular, rectangular, or triangular. Radially arranged straight lines called radii may extend across any of the fields. A **primary radius** extends from the focus to the margin of the scale. A **secondary radius** does not extend all the way out to the margin of the scale. Radii may be present in different fields: only anterior, as in pickerels (*Esox*); only posterior, as in shiners (*Notropis*); anterior and posterior, as in suckers (Catostomidae); or even in all four fields, as in barbs (*Barbus*). Ctenii may occur in a single marginal row or in two or more rows located on the posterior field.

Circuli are growth rings around the scale. Life history studies, particularly those dealing with age and growth, utilize such growth rings. This is especially useful in temperate waters where pronounced retardation of growth of body and scales occurs in fall and winter, causing the spacing between the circuli to decrease and thus leaving a dark band on the scales called an **annulus**. However, interpreting such marks as annuli requires caution because any retardation in growth may leave a mark. The stress of spawning, movement from fresh- to salt water, parasitism, injury, pollution, and sharp and prolonged change in temperature may all leave marks on the scales similar to annuli. Scales grow in a direct relationship with body growth, making it possible to measure the distance between annuli and back calculate age at different body sizes. Other structures also show growth changes and can be used for aging, such as fin spines, otoliths, and various bones such as opercles and vertebrae (see Chap. 10, Age and Growth). For a detailed discussion of growth measurements and their utility, refer to the section on age and growth in a good fishery biology text such as Everhart and Youngs (1981).

Scale morphology can also be useful in identification of fragments such as scales found in archaeological kitchen middens or in stomach contents. An example of the latter is Lagler's (1947) key to scales of Great Lakes

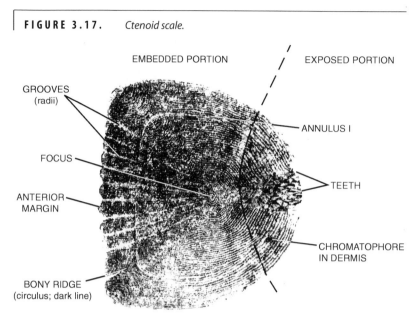

EMBEDDED PORTION

EXPOSED PORTION

GROOVES (radii)

ANNULUS I

FOCUS

TEETH

ANTERIOR MARGIN

CHROMATOPHORE IN DERMIS

BONY RIDGE (circulus; dark line)

From Lagler et al. 1977.

families. Scale morphology is also useful in classification, as shown by McCully's (1962) study of serranid fishes, Hughes's (1981) paper on flatheads, Johnson's (1984) review of percoids, Coburn and Gaglione's (1992) study of percids, and Roberts's (1993) analysis of spined scales in the Teleostei.

SUMMARY

1. Fishes have more skull bones than do higher vertebrates. The skull encloses and protects the brain and is composed of the neurocranium and the branchiocranium. The neurocranium is derived from the chondrocranium (the original cartilaginous braincase) and the dermatocranium (dermal bones derived from scales). Bony fish skull bones are divided into four regions: ethmoid, orbital, otic, and basicranial.

2. The branchiocranium consists of five series of endoskeletal arches (mandibular, palatine, hyoid, opercular, and branchial) derived from gill arch supports.

3. The notochord of primitive fishes is replaced by the vertebral column in Chondrichthyes and bony fishes. Vertebrae form around the notochord at intersections of myosepta with dorsal, ventral, and horizontal septa.

4. Posterior vertebrae support the caudal fin in most fishes. In teleosts, hypurals (enlarged haemal spines) support the branched principal caudal fin rays. Three basic types of caudal fins are 1) protocercal, the primitive undifferentiated caudal fin of adult lancelets,

agnathans, and larvae of more advanced fishes; 2) heterocercal, or unequal-lobed tail, in Chondrichthyes and primitive bony fishes; and 3) homocercal, or equal-lobed tail, found in most bony fishes.

5. Ribs (pleural ribs) attach to the vertebrae and protect the viscera. Intermuscular bones are segmental, serially homologous ossifications in the myosepta of teleosts.

6. Agnathans lack pectoral and pelvic girdles. Sharks have a coracoscapular (pectoral) cartilage with no attachment to the vertebral column. In bony fishes, the pectoral girdle lacks a vertebral attachment but is connected with the back of the skull by the posttemporal bone.

7. The dorsal, anal, and adipose fins form the median or unpaired fins. The median fins of agnathans are supported by cartilaginous rods, whereas chondrichthyan fins are supported by ceratotrichia (horny fin rays). In bony fishes, ceratotrichia are replaced during ontogeny by lepidotrichia, which are bony supporting elements derived from scales.

8. Primitive teleosts have a single dorsal fin composed of soft rays. Advanced teleosts usually have two dorsal fins. The anterior fin is composed of spines, and the posterior fin is composed of soft rays.

9. The skin and its derivatives, such as scales in fishes, provide external protection. The five basic types of scales are placoid, cosmoid, ganoid, cycloid, and ctenoid.

4

Soft Anatomy

The organs and organ systems between the skin and scales on the outside of a fish and the axial skeleton on the inside (see Chap. 3) are termed the soft anatomy. Soft anatomy includes the muscles, cardiovascular system, alimentary canal, gas bladder, kidneys, gonads, and nervous system. The sense organs, although part of the nervous system, are treated in Chapter 6 in their functional context as receivers and integrators of information. For a comprehensive treatment of soft anatomy, see Harder (1975).

MUSCLES

Fish muscle is structurally similar to that of other vertebrates, and fishes possess the same three kinds of muscles, but fishes differ in that a greater proportion (40% to 60%) of the mass of a fish's muscle is made up of locomotory muscle. Among the three types, **skeletal** muscle is striated and comprises most of a fish's mass, other than the skeleton. **Smooth** muscle is nonskeletal, involuntary, and mostly associated with the gut but is also important in many organs and in the circulatory system. **Cardiac**, or heart, muscle is nonskeletal but striated and is found only in the heart.

Jawless fishes have a simple arrangement of striated skeletal muscles. These primitive fishes have no paired appendages to interrupt the body musculature. Skeletal muscle behind the head is uniformly segmental and is composed of shallow W-shaped myomeres.

In jawed fishes, two major masses of skeletal muscle lie on each side of the fish, divided by the horizontal connective tissue septum. The **epaxial** muscles are the upper pair, and the **hypaxials** are the lower pair (see Fig. 8.1). A third, smaller, wedge-shaped mass of red muscle lies under the skin along the horizontal septum. This band of red muscle is poorly developed in most bony fishes but is much more extensive and used for sustained swimming in fishes such as the tunas (see Chap. 7, Endothermic Fishes).

Cheek Muscles

Seven principal muscles are involved in opening and closing the jaws and operculum during feeding and breathing. The major muscles are the **adductor mandibulae**, large muscles with several sections (Fig. 4.1, A1, A2, A3) that insert on the inner surface of the lower jaw and originate on the outer face of the suspensorium, the chain of bones that suspend the jaws from the neurocranium. The adductor mandibulae function to close the jaws (see also Fig. 8.3). The **levator arcus palatini** (LAP) occupies the postorbital portion of the cheek. The **dilator operculi** (DO), the **adductor operculi**, and the **levator operculi** (LO) insert on the opercle. The **adductor arcus palatini** (AAP) originates from the ventrolateral margin of the parasphenoid and underlies the orbit. The **adductor hyomandibulae originates** on the prootic and exoccipital and inserts on the hyomandibula. In addition, **pharyngeal** muscles, or **retractores arcuum branchialium**, run from the upper pharyngeal bones to the vertebral column and function in operating the pharyngeal jaws.

Fin Muscles

Muscles are arranged in pairs at the bases of the dorsal and anal fins: **protractors** erect the fins and **retractors** depress the fins. In addition, **lateral inclinators** function to bend the soft rays of the anal and second dorsal fins. For the paired fins, a single ventral **abductor** muscle pulls the fin ventrally and cranially. An opposing dorsal **adductor** muscle pulls the fin dorsally and caudally.

Eye Muscles

Extrinsic eye muscles move the eye within its orbit. Eye muscles are evolutionarily very conservative, in that most vertebrates have three pairs of these striated muscles: **inferior** or **ventral** and **superior** or **dorsal oblique**; **inferior** or **ventral** and **superior** or **dorsal rectus**; and **external** or **lateral** and **internal** or **medial rectus** (Fig. 4.2). Eye muscles are innervated by three cranial nerves: superior oblique by the trochlear (IV), external rectus by the abducens (VI), and the other four by the oculomotor (III). Posteriorly, the eye muscles insert into dome-shaped cavities called **myodomes** in actinopterygian fishes. A **suspensory ligament** above the lens and a **retractor lentis** muscle below form the focusing muscle of the eye.

Eye muscles have been converted into two remarkable structures in fishes: an electric organ in the electric stargazer (*Astroscopus*, Uranoscopidae) and heater organs in two groups of scombroid fishes. The upper edges of the four uppermost eye muscles form an electric organ in the electric stargazer. During development, the portion from the superior rectus loses its innervation from the trochlear

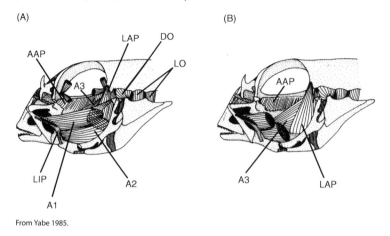

FIGURE 4.1.   *Cheek muscles of a sculpin,* Jordania zonope. *(A) Super-ficial musculature. (B) After removal of A1 and A2. A1, A2, and A3 = adductor mandibulae; AAP = adductor arcus palatini; DO = dilator operculi; LAP = levator arcus palatini; LO = levator operculi.*

From Yabe 1985.

nerve, and the portion from the external rectus loses its connection with the abducens, so that the electric organ in adult stargazers is innervated solely by the oculomotor nerve (Dahlgren 1927). Large stargazers can produce an electric discharge from these muscles strong enough to incapacitate a careless human handler. Their usual function is likely to stun prey or deter predators. In billfishes (Istiophoridae and Xiphiidae), the superior rectus has been converted into a heat-producing muscle that keeps the eye warm during incursions into deep, cold waters (see Box 7.1). In *Gasterochisma* (Scombridae), the external rectus is the muscle that becomes the heater organ (Block

1991), showing independent evolution of this character (see Chap. 7, Endothermic Fishes).

Smooth Muscle

Smooth muscle lines the walls of the digestive tract. It is arranged in bundles of longitudinal and circular muscles that work in opposition to one another to permit peristaltic transport of food. Smooth muscles are associated with the swim bladder and move products along the ducts of the reproductive and excretory tracts. The lens muscle of the eye, also a smooth muscle, moves the lens, dilating

FIGURE 4.2.   *The extrinsic eye muscles of a fish. The cranial nerves that supply the muscles are indicated by Roman numerals.*

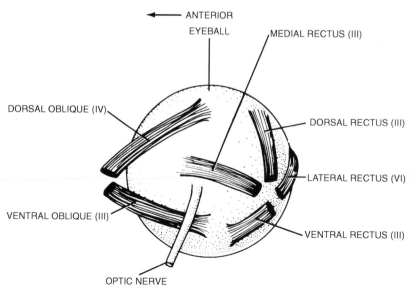

From Walker and Liem 1994.

or constricting it automatically in response to changing light.

Cardiac or heart muscle is dark red involuntary muscle. It is thickest in the walls of the ventricle.

Ligaments

Ligaments are nonelastic strands of fibrous connective tissue that serve to attach bones and/or cartilages to one another. Names of ligaments usually include their initial and terminal points. Some, however, are named after their shape or after persons. **Baudelot's ligament** is a strong white ligament that originates on the ventrolateral aspect of an anterior vertebra (usually the first) in lower teleosts or on the posterior part of the skull (usually the basioccipital) in advanced teleosts and inserts on the inner part of the cleithrum. Baudelot's ligament helps anchor the pectoral girdles to the sides of the fish.

White Muscle Versus Red Muscle

Faced with the conflicting demands of low-speed, economical cruising versus short bursts of maximum speed, fishes have solved the problem by dividing the locomotory system into two systems with different fiber types, white and red (Bone 1978; Webb 1993). White muscle makes up the majority of the postcranial body of most fishes. It is used anaerobically in short-duration, burst swimming but fatigues quickly. White muscle gets its color because its fibers lack myoglobin and because it has comparatively little vascularization and hence a limited oxygen supply. White muscle fibers are relatively large in diameter, up to 300 μm. White muscle fibers have relatively few, small mitochondria, with energy resulting from anaerobic glycolysis. Muscle glycogen is depleted rapidly during contraction, producing large amounts of lactates that may require up to 12 hours for full recovery after glycogen depletion (see Chap. 5, Respiration and Ventilation).

Red muscle usually forms a thin, lateral, superficial sheet under the skin between the epaxial and hypaxial muscle masses on each side of the fish. Red muscle is much better developed in muscles involved in sustained swimming, such as lateral red muscle in tunas and pectoral fin muscles in wrasses and parrotfishes. Red muscle is hard to fatigue because it is highly vascularized and is therefore provided with a rich oxygen supply. The red color is caused by abundant myoglobin. In contrast with white muscle, red muscle has small-diameter fibers (18 to 75 μm) and high blood volume (three times the number of capillaries of white muscle per unit weight). Mitochondria are large and abundant, and energy is supplied by the aerobic oxidation of fats. During exercise, little change occurs in muscle glycogen or in the buildup of lactates; recovery after exercise is rapid. The strong taste of the prominent lateral red muscle in tunas (*chiai* in Japanese) leads to its being picked out from cooked tuna prior to canning for human consumption. (It is canned for catfood, which is why tuna catfood smells, and tastes, so strong.)

Lamnid sharks and higher tunas (tribe Thunnini) have more and deeper portions of red muscle than other fishes.

A countercurrent heat exchanger system (see Chap. 7, Endothermic Fishes) between the arterioles and venules of the cutaneous artery and vein ensures that the heat produced by muscular contraction remains in those tissues and is not carried off by the circulatory system to be lost at the gills. In at least the bluefin tuna (*Thunnus thynnus*), this heat exchanger may function in actual regulation of body temperature (Carey and Lawson 1973). Cross sections of the body in representative scombrids show increasing development and internalization of the red muscles phylogenetically from mackerels to tunas (Sharp and Pirages 1979).

Some fishes, such as the scup (*Stenotomus chrysops*), also have another type of muscle. **Pink muscle** is intermediate between red and white muscle in myoglobin, giving it a pink color, and is also intermediate in the other descriptive and metabolic qualities detailed above (Webb 1993). Like red muscle, pink muscle is used for sustained swimming and is recruited after red muscle but before white muscle.

Another variation on muscle color and function occurs in the antarctic nototheniod family Chaenichthyidae (see Chap. 17, Polar Regions). Many chaenichthyids have blood but lack hemoglobin, leading to being called "bloodless." They even lack typical red muscle, instead having yellow muscle in the heart and in the adductor and abductor muscles of the pectoral fin. The protein composition of yellow muscle is similar to that of normal red muscle in fishes with hemoglobin (Hamoir and Geradin-Otthiers 1980).

Electric Organs

Fishes in six different evolutionary lines have developed the ability to amplify the usual electrical production associated with muscle contractions (see Chap. 6, Electroreception). The muscles involved in electrogeneration are modified skeletal muscles. Caudal skeletal muscles, and sometimes lateral body muscles as well, are modified for electrogeneration in the Rajidae, Mormyridae, Gymnotoiformes, and Malapteruridae. In torpedo rays (Torpedinidae, Narcinidae), hypobranchial muscles are involved, whereas an extrinsic eye muscle generates strong electrical discharges in the teleostean electric stargazer, *Astroscopus* (see above).

Only a few of the major muscles have been discussed here; see Winterbottom (1974) for a complete treatment.

CARDIOVASCULAR SYSTEM

The **cardiovascular system** serves all bodily functions but is most closely associated with respiration, excretion, osmoregulation, and digestion. The cardiovascular system is the system of arteries, veins, and capillaries that carry respiratory gases, wastes, excretory metabolites, minerals, and nutrients. The cardiovascular system of only a few fish species has been investigated extensively, most notably in hagfish, dogfish, skate, Port Jackson shark, trout, salmon, carp, cod, eel, and lungfishes (see Randall 1970; Satchell 1991; Farrell 1993).

Anatomy

The basic pattern of blood flow in fishes involves a single-pump, single-circuit system—heart to gills to body and back to heart (Fig. 4.3).

The heart is located posterior and ventral to the gills in all fishes, although it is located farther anterior in teleosts than in chondrichthyans. It lies in a membranous **pericardial cavity** that is lined with **parietal pericardium**. The basic fish heart consists of four chambers in series: Venous blood enters the 1) **sinus venosus** (a thin-walled sac) from the **ducts of Cuvier** and the **hepatic veins**; it next flows into the 2) **atrium**; then into the 3) **ventricle**, a thick-walled pump; and blood flows out of the heart into the 4) **conus** or <u>bulbus arteriosus</u> (Farrell and Jones 1992). The conus arteriosus is a barrel-shaped chamber invested with cardiac muscle, present in Chondrichthyes and lungfishes (Dipnoi). The muscular conus arteriosus is replaced by the bulbus arteriosus in actinopterygian fishes. The bulbus is a nonmuscular, onion-shaped elastic reservoir that is passively dilated with blood as it exits the ventricle. The bulbus dampens pressure oscillations, thereby providing continuous rather than pulsed blood supply to the body.

In lungfishes, the atrium and ventricle are partly divided by a partition, partially separating oxygenated and deoxygenated blood, a step toward development of the two-pump, four-chambered heart of higher vertebrates. This division is least complete in the Australian *Neoceratodus*, which is least dependent on atmospheric air, and is most complete in the South American *Lepidosiren*, which is most dependent on atmospheric air for respiration.

Heart valves prevent backflow of blood and maintain pressure in the circulatory system. Valves may be present between each of the sections of the heart. **Sino-auricular valves** (usually composed of both endocardial and myocardial muscle) separate the sinus venosus and atrium. **Auriculo-ventricular** or atrioventricular valves vary in number depending on the group: Chondrichthyes and most bony fishes have two rows of valves; the bowfin has

four rows, the paddlefish has five rows, whereas gars and bichirs (*Polypterus*) have six rows. These valves are absent in lungfishes, which do have valves in the conus arteriosus. The number of **ventriculo-bulbar** valves is related to the length of the conus. There is usually one or, rarely, two in bony fishes, two to seven in Chondrichthyes, and up to seventy-four in eight rows in gars. Valves outside the heart region can occur in various parts of the circulatory system of fishes, such as segmental arteries and veins in the caudal regions of the Port Jackson shark and teleost veins.

Blood is supplied to heart muscle from anterior hypobranchial arteries in Chondrichthyes, bony fishes, and *Neoceratodus*. In *Lepidosiren*, the coronary supply originates from the second afferent artery. Hagfishes have no special coronary circulation. Hagfishes differ in other regards. Nervous innervation of all fish hearts, except the hagfish, is from the vagus. Hagfishes also have several accessory hearts in parts of the venous system (Farrell 1993).

Heart size, as a proportion of body weight, is lower in fishes than in other vertebrates. Inactive fishes have very small hearts, making up less than 1 part per 1000 body weight. More active fishes have relatively large hearts. For example, in mackerels and tunas, the heart makes up 1.2 parts per 1000 body weight, and in flying fishes (Exocoetidae), the heart constitutes 2.1 parts per 1000 body weight.

Blood vessels of the gills and head.   The number of **afferent branchial arteries** bringing oxygen-deficient blood *to* the gills from the ventral aorta varies among different groups of fishes. Agnathans have seven to fourteen, the number depending on the number of gill pouches. Most Chondrichthyes have four, but sharks and rays with more gills have more arteries, such as *Hexanchus*, the six-gilled shark, which has five, and the seven-gilled shark, *Heptranchias*, which has six. Lungfishes have four to five afferent branchial arteries, whereas bony fishes have four (Fig. 4.4).

Efferent branchial arteries bring oxygenated blood *from* the gills to the rest of the body. These arteries merge to form the dorsal aorta, the largest and longest artery in

FIGURE 4.3.   *Block diagram showing the simplest type of fish circulatory system. Solid black vessels contain blood of lower oxygen content; white vessels contain blood with higher oxygen content. Arrows indicate direction of blood flow.*

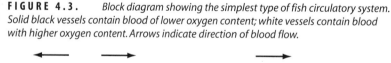
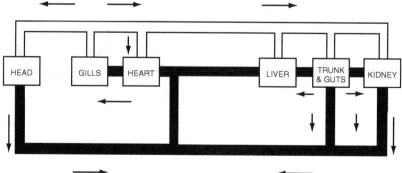

From Mott 1957.

FIGURE 4.4. *Gills and blood vessels of the head of a cod* (Gadus).

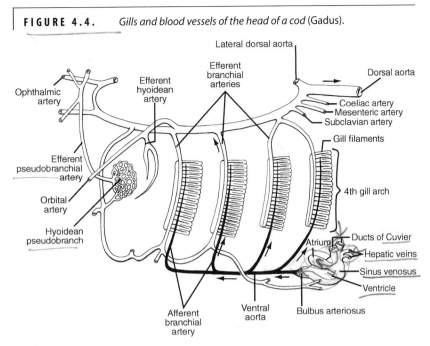

From Lagler et al. 1977.

a fish's body. Efferent branchial arteries number one per hemibranch in Chondrichthyes and one per holobranch in bony fishes. **Internal carotid arteries** run from the aorta to the brain. Major veins such as the facial, orbital, postorbital, and cerebral join into paired **anterior cardinal veins**, which empty into the **common cardinal** (also called the duct of Cuvier) and then into the heart. The jugular vein collects blood from the lower head and also empties into the common cardinal in Osteichthyes.

Blood vessels of the body. The **dorsal aorta** is the main route of transport of oxygenated blood from the gills to the rest of the body (Fig. 4.5). It lies directly ventral to the vertebral column in the trunk region and gives off major

vessels and segmental arteries. The **subclavian** artery goes to the pectoral girdle, the **coeliaco-mesenteric** artery supplies the viscera, and the **iliac** or **renal** artery supplies the kidneys. The dorsal aorta becomes known as the **caudal artery** upon entering the haemal canal of the caudal vertebrae.

The major return route of blood from most of the body is the **postcardinal vein**. It is best developed on the right side and empties into the common cardinal or ducts of Cuvier, then into the sinus venosus, and finally into the heart proper.

In the higher tunas (tribe Thunnini), an additional pair of large arteries, the **cutaneous arteries**, exit the dorsal aorta posterior to the coeliaco-mesenteric artery and run

FIGURE 4.5. *Main blood vessels of a bony fish.*

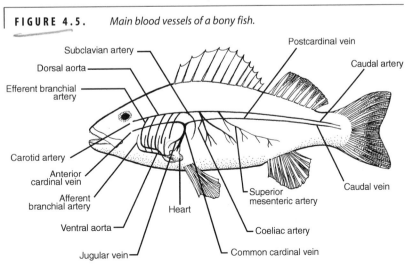

After Lagler et al. 1977.

BOX 4.1 *The Pseudobranch*

Many fishes have a **pseudobranch**, a small structure under the operculum composed of filaments like those in the gills. It was named pseudobranch, or false gill, because, unlike the true gills, the blood reaching it is oxygen-rich blood, not venous oxygen-deficient blood. The history of the pseudobranch and speculations on its function make an interesting story (Laurent and Dunel-Erb 1984). In the 1700s, Broussenot thought it had a respiratory function; in the 1800s, Hyrtl noted that it receives arterial blood, and Müller believed it was associated with vision. Three important morphologic features of the pseudobranch have fueled speculation about its function.

1. A pseudobranch covered with epithelium is rich in a respiratory substance, carbonic anhydrase. Is it, therefore, an endocrine organ?
2. The pseudobranch is associated with chloride cells (see Chap. 7, Control of Osmoregulation and Excretion). Does it have an osmoregulatory function?
3. The pseudobranch has rich nervous innervation. Could it have a sensory role?

The path of blood to and from the pseudobranch in Actinopterygii suggests that the pseudobranch is involved in providing oxygenated blood to the eye. Blood passes from the efferent hyoidean artery to the afferent pseudobranchial artery to the pseudobranchial capillaries to the ophthalmic artery to the choroid gland of the eye. The **choroid rete mirabile** is a large, discrete organ behind the retina of the eye. It is composed of several thousand capillaries arranged countercurrent to each other, a very effective mechanism for maximizing gas exchange. The pseudobranch, in combination with the countercurrent multiplier system of the choroid rete, modifies incoming oxygenated arterial blood by concentrating O_2 without building up CO_2.

Not all fishes possess pseudobranchs (e.g., adult eels, Anguilliformes, and catfishes, Siluriformes, lack them). However, these fishes are mostly nocturnal in habit and rely heavily on nonvisual senses. Hence it is not surprising that the complex circulatory apparatus that supplies highly oxygenated blood to the eye has been lost in these groups. Interestingly, larval eels do possess a pseudobranch, and it has been speculated that it serves a respiratory function in these larvae. It is generally accepted that the chondrichthyan spiracular gill is homologous to the actinopterygian pseudobranch.

---

laterally between the ribs. As these arteries approach the fish's skin, they divide into two vessels, each of which runs posteriorly, sending out arterioles to the underlying red muscle. After passing through an extensive network of capillaries—the countercurrent heat exchanger that retains metabolic heat in the red muscle—the cutaneous vein returns the unheated blood to the heart. Phylogenetically, the most advanced tunas show the greatest development of the subcutaneous circulatory system (Fig. 4.6).

Lymphatic system.   The lymphatic system is derived from the venous part of the blood vascular system and is similar to that of other vertebrates. Lymph is collected by paired and unpaired ducts and sinuses that empty into the main blood system. Agnatha have more connections to the venous system; they essentially have a hemolymph system. At least some species of Agnatha and bony fishes have contractile lymph "hearts." Chondrichthyes have lymph vessels but do not have sinuses or contractile lymph "hearts."

Blood.   Paralleling the trend in heart size discussed above, the volume of blood in teleosts is smaller than in Chondrichthyes, and both have lower blood volumes than tetrapods. Agnatha have the greatest volume among fishes. Blood itself is composed of plasma and blood cells. Plasma contains dissolved minerals, digestive products, waste products, enzymes, antibodies, and dissolved gases, but few detailed analyses of fish blood have been published.

Solutes in the blood function to lower its freezing point. Freezing point depression of blood plasma is $-0.5°C$ in freshwater bony fishes, $-1.0°C$ in freshwater Chondrichthyes, -0.6 to $-1.0°C$ in marine bony fishes, and $-2.2°C$ in marine Chondrichthyes versus $-2.1°C$ for seawater. Antarctic notothenioids (see Chap. 17) have additional blood antifreeze glycoproteins that reduce the freezing point of their blood to -0.9 to $-1.5°C$, with some notothenioids showing freezing points as low as $-3°C$.

Red blood cells (RBCs) account for nearly 99% of oxygen uptake. RBCs are nucleated, yellowish red oval cells in most fishes but are round in lampreys. Fishes have relatively fewer and larger RBCs than do mammals. Human RBCs measure 7.9 μm across, whereas fish cells range from 7 μm in some wrasses to relatively giant 36-μm cells in the African lungfish, *Protopterus*. Red blood cells are absent in notothenioids (see above) and in the leptocephalus larvae of eels.

FIGURE 4.6.   *Anterior arterial system in ventral view in the Scombridae, showing phylogenetic increase in development of the subcutaneous circulatory system (darkened vessels). Numbers indicate vertebrae; stippled areas show where pharyngeal muscles originate. (A) Wahoo (Acanthocybium). (B) Frigate tuna (Auxis). (C) Little tuna (Euthynnus). (D) Skipjack (Katsuwonus). (E) Longtail Tuna (Thunnus tonggol). (F) Albacore (Thunnus alalunga).*

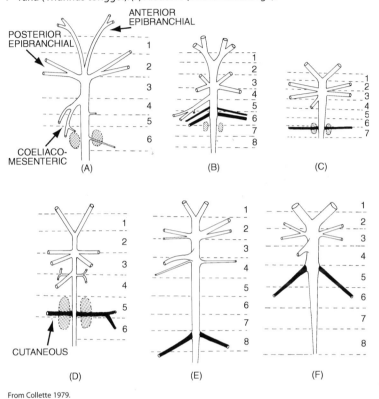

From Collette 1979.

ALIMENTARY CANAL

As in other vertebrates, the alimentary tract can be divided into anterior and posterior regions. The anterior part consists of the mouth, buccal cavity, and pharynx. The posterior part consists of the foregut (esophagus and stomach), midgut or intestine, and hindgut or rectum. Voluntary striated muscle extends from the buccal cavity into the esophagus, involuntary smooth muscle from the posterior portion of the esophagus through the large intestine. Barrington (1957), Kapoor et al. (1975), and Fange and Grove (1979) provide detailed accounts of the alimentary tracts of fishes.

In the Agnatha, the absence of true jaws is correlated with the absence of a stomach. Evolution of jaws permitted capture of larger prey, making a storage organ, the stomach, highly advantageous. Hagfishes and lampreys have a straight intestine, but the surface area of the intestine is increased in the lampreys by the **typhlosole**, a fold in the intestinal walls. Chondrichthyes increase the surface area of the intestine by means of a **spiral valve**, a sort of a spiral staircase inside the intestine (Fig. 4.7).

The anatomy of the digestive tract in bony fishes deserves additional description. The buccal cavity (mouth) and pharynx lack the salivary glands present in mammals. These areas are lined with stratified epithelium, mucous cells, and, frequently, taste buds. This area is concerned with seizure, control, and probably also selection of food.

The **esophagus** is a short, thick-walled tube lined with stratified ciliated epithelium, mucous-secreting goblet cells, and, often, taste buds. The anterior portion has striated muscles, the posterior part smooth muscles that produce peristaltic involuntary movement of food toward the stomach. The esophagus is very distensible, so choking is rare, but miscalculation of prey size or armament can lead to the death of the predator, as in the case of sticklebacks (Gasterosteidae) becoming stuck in the throats of pickerels (*Esox*).

In predaceous fishes, the **stomach** is lined with columnar epithelium with mucous-secreting cells and one type of glandular cell that produces pepsin and hydrochloric acid. Although usually a fairly simple structure, evolutionary modifications of the fish stomach have led to some unusual functions. For example, the stomach, not the gas bladder, is used for defense by blowfishes and porcupinefishes (Tetraodontidae and Diodontidae), by taking in water or air (see Box 19.1). The stomach is

FIGURE 4.7.   *Variation in intestinal length and other features among carnivorous and herbivorous fishes. (A) An herbivorous catfish (Loricariidae). (B) Spiral valve in cross section of intestine of a shark. (C) A carnivore, the northern pike (Esox lucius). (D) A carnivore, a perch (Perca).*

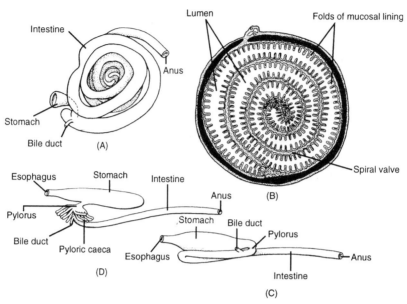

From Lagler et al. 1977.

modified into a grinding organ in sturgeons (Acipenseridae), gizzard shad (*Dorosoma*), and mullets (Mugilidae) and is used to extract oxygen in some South American armored catfishes such as *Plecostomus*.

Many gnathostome fishes lack true stomachs. This is a secondary condition with no simple ecological explanation (Kapoor et al. 1957). Fishes without true stomachs include chimaeras (Holocephali) and lungfishes (Dipnoi). This condition is best documented in the Teleostei, including minnows (Cyprinidae) such as the European *Rutilus*, killifishes (Cyprinodontidae), wrasses (Labridae; see Chao 1973), and parrotfishes (Scaridae). Characteristics of the stomachless condition are both cytologic and biochemical. Cytologically, no gastric epithelium or glands are present. The stratified epithelium of the esophagus grades into the columnar epithelium of the intestine. Biochemically, no pepsin or hydrochloric acid is produced, making it impossible to dissolve shells or bones.

The **intestine** of most fishes is lined with simple columnar epithelium and goblet cells. Usually no multicellular glands are present. The chief exception to this is the cods (Gadidae), which have small tubular glands in the intestinal wall.

Pyloric caeca are fingerlike pouches that connect to the intestine near the pylorus. Pyloric caeca may function in absorption or digestion. They vary in number from only three in a scorpionfish (*Setarches*) to thousands that form a caecal mass in tunas. The number of pyloric caeca is useful in the classification of some groups, like the Salmonidae.

The length of the intestine varies and is generally correlated with feeding habits (see Chap. 18, Herbivores). Carnivores such as pickerels (*Esox*) and perch (*Perca*) have very short intestines, one-third to three-quarters of body length. The intestine is much longer in herbivores and detritus feeders, 2 to 20 times the body length. In the herbivorous North American stoneroller minnow (*Campostoma*), the intestine is very long and even wrapped around the swim bladder. The important factor is not the actual length of the intestine but the internal surface area of the intestinal mucosa. In addition to sharks, as mentioned earlier, some primitive bony fishes such as the coelacanth (*Latimeria*), lungfishes, gars, and ladyfish (*Elops*) have a spiral valve intestine that increases the surface area internally (see Fig. 4.7).

The **hindgut** or **rectum** is not as well defined externally in fishes as it is in higher vertebrates. Generally, the muscle layer near the rectum is thicker than in anterior regions, and the number of goblet cells in the large intestine increases in the rectal region. In Chondrichthyes, the hindgut is lined with stratified epithelium, contrasting with a single cell layer in the midgut. An ilio-caecal valve between the small and large intestines is often found in teleosts, but this valve is absent in Chondrichthyes, Dipnoi, and *Polypterus*.

The liver and pancreas both participate in digestion. The **liver** develops as a ventral evagination of the intestine, as in other vertebrates. The anterior portion develops into the liver proper, the posterior portion into the gallbladder and bile duct. The liver also stores fat in some fishes. Before vitamins A and D were synthesized, cods and sharks were harvested for their liver oil, which is rich in these vitamins.

The **gallbladder** is a thin-walled temporary storage organ for the bile. It empties into the intestine near the pylorus by contraction of smooth muscles. **Bile** contains bile pigments (biliverdin and bilirubin) from the breakdown of blood cells and hemoglobin, and also contains fat-emulsifying bile salts, which may assist in converting the acidity of the stomach to the neutral conditions in the intestine.

The **pancreas** is both an endocrine organ and an exocrine organ that produces digestive enzymes. These enzymes include proteases such as trypsin, carbohydrases such as amylase and lipase, and, in some insect-feeding fishes, chitinase. The pancreas is a compact, often two-lobed structure in Chondrichthyes, is distinct in soft-rayed teleosts, but becomes incorporated into the liver as a hepato-pancreas in most spiny-rayed teleosts (except for parrotfishes). The anatomically and histologically diffuse nature of the pancreas makes it difficult to study pancreatic function in these advanced fishes.

The various parts of the alimentary tract work together in conjunction with the feeding habits of a fish. For example, de Groot (1971) presented an instructive study of the correlations between various organ systems and feeding in three families of flatfishes. The Bothidae, which are diurnal carnivores, possess a single loop of the intestine, heavily toothed gill rakers, small olfactory lobes of the brain, and large optic lobes. The Pleuronectidae are also diurnal but have complex loops of the intestine, less-toothed gill rakers, medium olfactory lobes, and large optic lobes. The Soleidae, which are nocturnal feeders, have more complex intestinal loops, few gill raker teeth, large olfactory lobes, and small optic lobes.

GAS BLADDER

The **gas bladder** (swim bladder) is a gas-filled sac located between the alimentary canal and the kidneys (Jones 1957; Marshall 1960; Steen 1970). It is filled with CO_2, O_2, and N_2, in different proportions than occur in air, making the term "air bladder" inappropriate. The original function of the gas bladder was probably as a lung, but in most fishes today it functions mainly as a hydrostatic organ that helps control buoyancy. It also plays a role in respiration, sound production, and sound reception in some fishes.

Embryologically, the gas bladder is a two-layered, specialized outgrowth of the roof of the foregut and possesses tissues similar to those of the foregut.

Tissue	Layer	Embryological Origin
a. peritoneal investiture	tunica externa	
b. collagenous layer		
		mesoderm
c. fibrous layer	tunica interna	
d. smooth muscle		
e. bladder epithelium		entoderm

The structures and mechanisms by which gases enter and are released from the gas bladder differ in the major groups of teleosts. The **pneumatic duct** is a connection between the gas bladder and the gut. **Physostomous** fishes retain the connection in adults, whereas **physoclistous** fishes lose the connection in adults, if it is present at all during development. In physostomous fishes, gas can be taken in and emitted through the pneumatic duct. The more primitive soft-rayed teleosts have the primitive physostomous condition, whereas the more advanced spiny-rayed fishes are physoclistous, lacking a pneumatic duct.

Another, more complex mechanism, which involves two distinct regions of the gas bladder, has evolved to allow gas exchange in these fishes (Fig. 4.8). The **anteroventral secretory region** contains the gas gland and the rete mirabile. The **gas gland** secretes lactic acid into the beginning of the capillary loop. This acidifies and reduces the solubility of all dissolved gases. A change of 1 pH unit releases 50% of the O_2 bound to hemoglobin. This raises the partial pressure of blood oxygen by the Root and Bohr effects (see Box 5.2).

The **rete mirabile**, or wonder net, is not really a net but a looping bundle of arterial and venous capillaries associated with the gas gland that functions as a countercurrent multiplier. The rete is better developed in deep-dwelling fishes that have longer retial capillaries, thus providing more surface area and allowing a greater multiplying factor. Rattails (Macrouridae) and ophidioids living at abyssal depths of 4000 m and more have retial capillaries 25 mm in length or more; shallow-water forms have retes only 1 mm long (Marshall 1971).

The posterodorsal resorptive region of the gas bladder is called the **oval**. It develops from the distal end of the degenerating pneumatic duct and consists of a thin, highly vascularized area. Circular muscles contract and close off the oval, preventing outflow of gases. Longitudinal muscles contract and expose the oval, permitting gas escape. The walls of the gas bladder are lined with a layer of cells containing crystals of guanine 3 μm thick, which decreases permeability by 40 times over an unlined membrane and thus limits gas escape except at the oval, when it is open.

The gas bladder of physostomous fishes receives blood from a branch of the coeliaco-mesenteric artery. Blood is returned to the heart through the hepatic portal system. The rete, oval, and gas bladder wall of physoclists are supplied by the coeliaco-mesenteric artery and returned by a vein from the hepatic portal system. The oval and bladder wall are also supplied by intercostal branches of the dorsal aorta and returned through the postcardinal system.

Nervous innervation of the gas bladder is sympathetic through a branch from the coeliaco-mesenteric ganglion and by branches of the left and right intestinal vagus (X) nerves. Cutting the vagus prevents gas secretion into the gas bladder. Gas secretion is also inhibited by atropine, a cholinesterase blocker. The gas gland has high cholinesterase activity, and the secretory fibers are probably cholinergic. Sensory nerve endings function as stretch receptors, responding to stretching or slackening of the gas bladder, thus providing information to the fish about the relative fullness of the gas bladder.

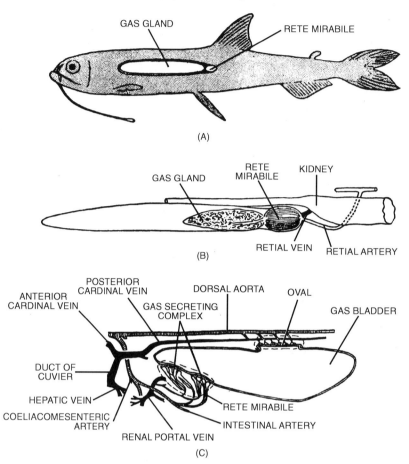

FIGURE 4.8. *The gas bladder. (A) Position of gas bladder in a deep-sea snaggle-tooth* (Astronesthes). *(B) Details of the gas bladder in* Astronesthes. *(C) Generalized blood supply of the gas bladder in physoclistous bony fishes.*

From Lagler et al. 1977.

KIDNEYS

The **kidneys** are paired longitudinal structures located retroperitoneally (outside of the peritoneal cavity), ventral to the vertebral column. Left and right kidneys are frequently joined together to form a soft black material under the vertebrae from the back of the skull to the end of the body cavity. The kidneys are one of the primary organs involved in excretion and osmoregulation (see Chap. 7, Excretion). Three kinds of kidneys are present in vertebrates: pronephros, mesonephros, and metanephros. The pronephros is present in larval fishes, the mesonephros is the functional kidney in Actinopterygii, and the metanephros is the kidney present in vertebrates more advanced than fishes. Kidney tubules are involved with moving sperm in some fishes, so the two systems are sometimes discussed as the urogenital system.

The **pronephros** has **nephrostomes**, anterior funnels that empty into the body cavity by way of pronephric tubules. Adult hagfishes have an anterior pronephros and

a posterior mesonephros, but it appears to be the mesonephros that is the functional kidney (Hickman and Trump 1969). Lampreys have a pronephros until they reach about 12 to 15 mm and then develop a mesonephros during metamorphosis. A pronephros is a transitional kidney that appears during ontogenetic development of actinopterygian larvae and then is replaced by a mesonephros as the fish grows.

The mesonephros is a more complex kidney that does not have funnels emptying into the body cavity. The **mesonephros** consists of a number of **renal corpuscles**, each composed of a glomerulus surrounded by a Bowman's capsule. The **glomerulus** receives blood from an afferent arteriole from the dorsal aorta. The glomerulus acts as an ultrafilter to remove water, salts, sugars, and nitrogenous wastes from the blood. The filtrate is collected in **Bowman's capsule** and then passes along a **mesonephric tubule** where water, sugars, and other solutes are selectively resorbed. Marine and freshwater fishes differ considerably in kidney structure, reflecting the very different problems faced by animals living in solutions of very different solute concentrations (see Chap. 7, Osmoregulation and Ion Balance). Freshwater fishes have larger kidneys with more and larger glomeruli, up to 10,000 per kidney measuring 48 to 104 μm across (mean of several freshwater species = 71 μm). The glomeruli of marine fishes are only 27 to 94 μm across (mean of several marine species = 48 μm).

Urine contains water plus creatine, creatinine, urea, ammonia, and other nitrogenous waste products. Only 3% to 50% of the nitrogenous wastes are excreted through the urine (see Table 7.2), and much of this is as ammonia; most of the rest is excreted as ammonia at the gills during the process of respiration. Some fishes have a storage organ for urine that has been called a "urinary bladder," but it is a posterior evagination of the mesonephric ducts, making it mesodermal in origin and not homologous with the entodermally derived urinary bladder of higher vertebrates.

Freshwater fishes produce copious amounts of highly dilute urine to avoid "waterlogging" by the large amount of water diffusing in through all semipermeable membranes (see Fig. 7.3). Marine fishes drink seawater to correct dehydration and excrete a low volume of highly concentrated urine. Most nitrogenous wastes are excreted extrarenally through the gills.

Some fishes are **aglomerular**, lacking glomeruli in their kidneys. At least 30 species of aglomerular teleosts are known from seven different families of mostly marine fishes, such as Batrachoididae, Ogcocephalidae, Lophiidae, Antennariidae, Gobiesocidae, Syngnathidae, and Cottidae (Hickman and Trump 1969; Bone et al. 1995). Aglomerular kidneys are unable to excrete sugars and are useful to physiologists studying the function of glomeruli. It would be particularly interesting to study kidney function of freshwater members of the above families to see how they meet the problem of bailing out excess water if they lack glomeruli in their kidneys.

GONADS

As in higher vertebrates, the sexes in fishes are usually separate (**dioecious**), with males having testes that produce sperm, and females having ovaries that produce eggs. "Fishes as a group exemplify almost every device known among sexually reproducing animals; indeed, they display some variations which may be unique in the animal kingdom" (Hoar 1969, 1). Only basic anatomy is treated here; other aspects of reproduction are discussed in Chapters 9, 10, and 20.

Testes

The **testes** are internal, longitudinal, and usually paired. They are suspended by lengthwise mesenteries known as **mesorchia**. The testes lie along the gas bladder when it is present. Kidney tubules and ducts serve variously among different groups of fishes to conduct sperm to the outside. Testes may constitute as much as 12% of body weight in some species at sexual maturity, although this proportion is usually smaller.

Agnathans have a single testis. Sperm is shed into the peritoneal cavity and then passes through paired genital pores into a urogenital sinus and out through a urogenital papilla.

Among Chondrichthyes, internal fertilization is universal, males using modified pelvic fins, termed **claspers**, to inseminate females. Sperm leave the testis through small coiled tubules, **vasa efferentia**, which are modified mesonephric (kidney) tubules. Sperm pass through **Leydig's gland**, which consists of small glandular tubules derived from the kidney. Secretions of Leydig's gland are involved in spermatophore production. The sperm then go through a **sperm duct**, which is a modified mesonephric duct, and into a **seminal vesicle**, a temporary storage organ that is also secretory.

Among bony fishes, the situation is similar, but no true seminal vesicles or sperm sacs are present. Marine catfishes (Ariidae), gobies (Gobiidae), and blennies (Blenniidae) have secondarily derived structures that have also been called seminal vesicles, but these are glandular developments from the sperm ducts and are not comparable to structures with the same names in higher vertebrates. These vesicles provide secretions that are important in sperm transfer or other breeding activities.

The lungfishes, sturgeons, and gars make varying use of kidney tubules and mesonephric (Wolffian) ducts (Fig. 4.9). In the bowfin (*Amia*), vasa efferentia bypass the kidney and go to a Wolffian duct. In *Polypterus* and the Teleostei, there is no connection between the kidney and gonads at maturity. The sperm duct is new and originates from testes. Thus the sperm duct of more primitive fishes such as the Chondrichthyes and the Chondrostei is not homologous with that in the Teleostei.

The tubular structure of the teleost testis has two basic types distinguished by the distribution of spermatogonia, the sperm-producing cells. In most teleosts, **spermatogonia** occur along the entire length of the tubules, but in atherinomorph fishes, the spermatogonia are confined to the distal end of the tubules (Grier 1981).

Ovaries

The **ovaries** are internal, usually longitudinal, and phylogenetically originally paired but are often variously fused and shortened. Sometimes only one ovary is present in adults. The number or relative lengths of the ovaries may be used as a taxonomic character in some fishes, such as the needlefishes. The ovaries are suspended by a pair of lengthwise mesenteries, the **mesovaria**. The ovaries are typically ventral to the gas bladder. Kidney tubules and ducts are not used to transport eggs. Ovary mass can be as high as 70% of body weight and tends to increase with body size of individual females.

Agnathan ovaries have the same basic structure as do the male testes. There is a single ovary, and the eggs are shed into the body cavity and then pass through paired genital pores and out through a urogenital papilla.

In the Chondrichthyes, eggs are shed into the body cavity, the **gymnovarian** condition. The ovarian capsule

FIGURE 4.9.   *(A–H) Representative types of urogenital systems in fishes. Upper series, males; lower series, females; black organs, mesonephric kidneys; stippled organs, testes; organs with circles, ovaries; stippled lines, vestigial structures; cl = cloaca; f = open funnel of oviduct; gp = genital papilla; l = Leydig's gland; md = mesonephric duct; n = nidamental gland; ov = oviduct; u = uterus; up = urinary pore; vd = vas deferens.*

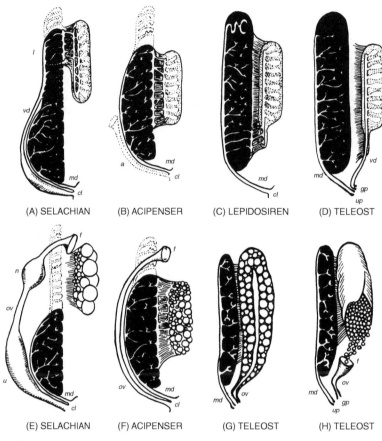

(A) SELACHIAN   (B) ACIPENSER   (C) LEPIDOSIREN   (D) TELEOST

(E) SELACHIAN   (F) ACIPENSER   (G) TELEOST   (H) TELEOST

From Hoar 1957.

is not continuous with the oviduct. The eggs enter the funnel of the oviduct, which is a **Müllerian duct**, not a modified mesonephric duct; it develops as a posterior continuation of the ovarian tunic. The anterior part of the oviduct is specialized to form a **nidamental** or shell gland where fertilization takes place. The nidamental gland secretes a membrane around the fertilized egg. In **oviparous** (egg-laying) taxa, the membrane is horny, composed of keratin. The nidamental gland may function as a seminal receptacle where sperm are nourished before fertilization. In **viviparous** (live-bearing) species, the posterior part of the oviduct is modified to form a **uterus**, which houses the developing embryo.

In bony fishes, the primitive gymnovarian condition is found in lungfishes, sturgeons, and bowfin. In gars and most teleosts, the lumen of the hollow ovary is continuous with the oviduct, termed the **cystovarian** condition. In trouts and salmons (Salmonoidei) and some other teleosts, the oviducts have been secondarily lost in whole or in part, so the eggs are shed into the peritoneal cavity and reach the outside through pores.

NERVOUS SYSTEM

The nervous system can be divided into the cerebrospinal and autonomic systems. The **cerebrospinal** system is composed of the central nervous system and the peripheral nervous system. The **central nervous system** is further subdivided into the brain and the spinal cord (Healey 1957; Bernstein 1970; Northcutt and Davis 1983). The **peripheral system** is composed of the cranial and spinal nerves and the associated sense organs (vision, smell, hearing, the lateralis system, touch, taste, and electrical and temperature detection; see Chap. 6). The autonomic nervous system is composed of sympathetic and parasympathetic ganglia and fibers.

Central Nervous System

Brain.   Fishes have small brains, on average being only 1/15 the size of the brain of a bird or mammal of equal body size. In the pickerel (*Esox*), the brain is only 1/1305 of body weight. The largest brains in fishes are in the elephant fishes (Mormyridae), constituting 1/52 to 1/82 of body weight. This large brain is associated with electroreception, as we shall see later. In the ocean sunfish (*Mola mola*), the spinal cord is even shorter than the brain. A 1.5-ton fish, 2.5 m long, has a spinal cord only 15 mm long.

The brain can be divided into five parts from anterior to posterior. The most anterior part is the **telencephalon**, (Fig. 4.10, tel), or forebrain, which becomes the cerebrum of higher vertebrates. Its function in fishes is primarily associated with reception and conduction of olfactory stimuli. The olfactory nerve (cranial nerve I) runs from the nostrils to the olfactory lobe of the brain. The olfactory lobe is large in Agnatha, huge in sharks such as the hammerheads (Sphyrnidae), and moderately large in smell-feeding teleosts such as catfishes (Fig. 4.10E).

The **diencephalon**, or 'tween brain (Fig. 4.10, dienc), lies between the forebrain and the midbrain and is also known as the saccus dorsalis. It functions as a correlation center for incoming and outgoing messages regarding homeostasis and the endocrine system. The **pineal body** is a hollow, invaginated, well-vascularized structure dorsal to the diencephalon and connected to it by a narrow hollow stalk. It frequently underlies a more or less unpig-

mented area of the cranial roof and is light-sensitive in some if not all fishes. Its functions are diverse, including light detection, circadian and seasonal clock dynamics, and color change, as well as some less well-understood properties. The pineal contains neurosensory cells that resemble cones in the retina. Photosensitivity of the pineal has been demonstrated by behavioral tests in rainbow trout. Light sensitivity of the pineal may allow it to play a navigation role in the cross-ocean migrations of large tunas such as the bluefin, *Thunnus thynnus* (Rivas 1954; Holmgren 1958; Murphy 1971). The pineal may regulate color change associated with background matching. It also produces an apocrine secretion containing glycogen. There is a possibility that the pineal may also play an endocrine role, in that it produces the hormone melatonin, implying a potential pineal-pituitary relationship.

The **mesencephalon**, or midbrain, is important in vision. The optic nerve (cranial nerve II) brings impulses from the eyes and enters the brain here. The midbrain is also a correlation center for messages coming from other sensory receptors. Fishes have two optic lobes, which are relatively large in sight-feeding species (Fig. 4.10C,D, tect).

The **metencephalon**, or hindbrain, functions in maintaining muscular tone and equilibrium in swimming. The

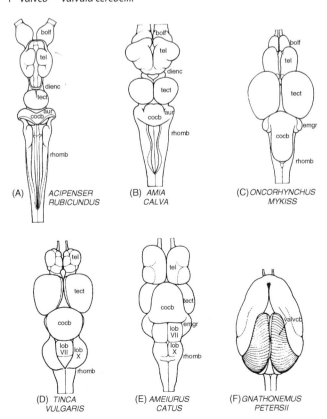

FIGURE 4.10. *Dorsal views of brains of representative fishes. (A) Sturgeon. (B) Bowfin. (C) Trout. (D) Minnow. (E) Catfish. (F) Elephant fish. Major brain parts from anterior to posterior: bolf = olfactory lobe; tel = telencephalon; dienc = diencephalon; tect = optic lobe; cocb = cerebellum; rhomb = myelencephalon; valvcb = valvula cerebelli.*

From Nieuwenhuys and Pouwels 1983.

cerebellum (Fig. 4.10, cocb), a large single lobe, is the largest component of the fish brain. Cranial nerve IV (trochlear) runs from the metencephalon to the eye muscles. The metencephalon is poorly developed in the Agnatha, being small in the lampreys (Petromyzontidae) and almost absent in the hagfishes (Myxinidae). In the elephant fishes (Mormyridae), the cerebellum is hypertrophied to form the valvula cerebelli (Fig. 4.10F, valvcb) which extend over the dorsal surface of the telencephalon. This large cerebellum is related to reception of electrical impulses.

The **myelencephalon** (Fig. 4.10, rhomb), brain stem, or medulla oblongata is the posterior portion of the brain and the enlarged anterior part of the spinal cord. Cranial nerves V through X arise here. The myelencephalon serves as the relay station for all the sensory systems except smell (cranial nerve I) and sight (cranial nerve II). It contains centers that control certain somatic and visceral functions. In bony fishes, it also contains respiratory and osmoregulatory centers.

A series of investigators has correlated brain morphology with ecology and behavior: H. M. Evans (1940) studied European freshwater species, H. E. Evans (1952) investigated four species of American minnows (Cyprinidae), and R. J. Miller and H. E. Evans (1965) studied brains of suckers (Catostomidae).

Peripheral Nervous System

Cranial nerves.   The 10 cranial nerves in fishes are similar to those in other vertebrates. Cranial nerve I, the **olfactory nerve**, is a sensory nerve that runs from the olfactory bulb to the olfactory lobes of the brain. The **optic nerve**, II, runs from the retina to the optic lobes. As in other vertebrates, cranial nerves III (**oculomotor**), IV (**trochlear**), and VI (**abducens**) are somatic motor nerves that innervate the six striated muscles of the eye: IV, the superior oblique; VI, the external rectus; and III, the other foureye muscles. Unlike other vertebrates, four cranial nerves (VII through X) innervate parts of the lateral line system. The **trigeminal**, V, is a mixed somatic sensory and motor nerve serving the anterior portion of the head. Cranial nerve VII, the **facial**, and VIII, the **acoustic**, usually join to form the **acousticofacialis** nerve, which then subdivides into four groups of mixed nerves serving the temporal and branchial regions of the head. Patterns of nerves, such as the ramus lateralis accessorius of the facial nerve (which innervates taste buds on the posterior head and body), have proved to be useful in assessing relationships of teleosts (Freihofer 1963). The **glossopharyngeal**, IX, is a mixed nerve that supplies the gill region. It often fuses with X, the anterior ramus of the vagus. The **vagus** is a mixed nerve connected to the body lateral line and viscera.

SUMMARY

1. Fishes have three kinds of muscles (skeletal, smooth, and cardiac, or heart, muscle) and have relatively more skeletal muscle than do other vertebrates.

2. In the locomotory system, white muscle forms most of the postcranial body and is used anaerobically for burst swimming but fatigues quickly. Red muscle usually forms thin, lateral, superficial sheets under the skin, is used in sustained swimming, and fatigues slowly.

3. The basic pattern of the cardiovascular system is a single-pump, single-circuit system that goes from heart to gills to body and back to the heart. Many fishes have a pseudobranch, a small structure under the operculum composed of gill-like filaments that may provide oxygenated blood to the visual system.

4. The anterior region of the alimentary tract consists of the buccal cavity (mouth) and the pharynx. The posterior region consists of the foregut (esophagus and stomach), midgut or intestine, and hindgut (rectum).

5. The gas or swim bladder is a gas-filled sac located between the alimentary canal and the kidneys. It develops from the roof of the foregut. A pneumatic duct connects the gas bladder and the gut in primitive teleosts (physostomous condition). Physostomous fishes can take gas in and emit it through the mouth and pneumatic duct. Advanced teleosts are physoclistous, losing the connection in adults. Physoclistous fishes have a secretory region containing a gas gland and a rete mirabile to produce gas, and an oval where gas is resorbed.

6. Kidneys, paired longitudinal structures ventral to the vertebral column, are one of the primary organs involved in excretion and osmoregulation. A pronephros is present in hagfishes and larval fishes, whereas a mesonephros is the functional kidney in Actinopterygii.

7. The sexes in fishes are usually separate, and the gonads are usually paired. Males have testes that produce sperm, and females have ovaries that produce eggs. In Chondrichthyes and primitive bony fishes, the eggs are shed into the body cavity, the gymnovarian condition. In gars and most teleosts, the lumen of the hollow ovary is continuous with the oviduct, the cystovarian condition.

8. The fish brain can be divided into five parts from anterior to posterior: 1) telencephalon, or forebrain, primarily associated with smell; 2) diencephalon, a correlation center for messages regarding homeostasis and the endocrine system; 3) mesencephalon, or midbrain, important in vision; 4) metencephalon, or hindbrain, which maintains muscle tone and equilibrium in swimming and has a large median lobe (cerebellum), which is the largest component of the fish brain; and 5) the myelencephalon, brain stem, or medulla oblongata, the posterior portion of the brain and enlarged anterior portion of the spinal cord that relays input for all sensory systems except smell and sight.

9. Fishes have small brains. The largest brains occur in the elephant fishes (Mormyridae), which have a large proportion of their brain devoted to electroreception.

5

Oxygen, Metabolism, and Energetics

One topic that fascinates fish researchers is the physiological mechanisms that fishes have evolved as adaptations to the aquatic environment. Regardless of specific habitat—be it swift mountain stream, slow bottomland river, lake or pond, deep ocean or coastal estuary—fishes face similar challenges. They need oxygen, nutrients, and energy; they create metabolic wastes that must be eliminated; they perceive and respond to stimuli; and they must maintain internal physiological stability to continue to function properly. In the following three chapters, we explore physiological mechanisms that fishes have evolved to meet these challenges. It should be kept in mind that we are addressing general principles for a diverse group of vertebrates and that many exceptions exist to the basic rules discussed.

RESPIRATION AND VENTILATION

The challenge to an active fish is to extract oxygen from the water and distribute it to the cells of the body fast enough to meet the demands of cellular metabolism and to prevent excess buildup of lactate. As in all eukaryotic life forms, fishes require oxygen to produce sufficient energy to support their metabolic needs. Oxygen helps maximize the amount of adenosine triphosphate (ATP) that can be generated from glucose, the primary metabolic fuel of cellular metabolism. Oxygen permits the aerobic completion of cellular respiration (glycolysis, the Krebs cycle, and oxidative phosphorylation). If oxygen is not present, oxidative phosphorylation and the Krebs cycle cannot proceed, and the only energy available from the metabolism of glucose is from the small amount of ATP released during the initial glycolysis reaction.

For glycolysis to continue producing some ATP, the pyruvate that is also produced is often converted to lactate and stored temporarily. However, if lactate levels get too high, glycolysis can be inhibited, and then no ATP will be produced and cellular metabolism will cease. When oxygen next becomes available, such as following bursts of activity, the stored lactate can be converted back to pyruvate and oxidative metabolism may proceed. However, lactate conversion bears a metabolic cost, because a period of elevated oxygen consumption is required to pay off the **oxygen debt** accumulated during the period when oxygen was lacking.

Water as a Respiratory Environment

Terrestrial organisms live in an oxygen-rich environment and therefore do not need to maximize their oxygen uptake efficiency. Water, however, contains considerably less oxygen than air—less than 1% by volume as opposed to nearly 21% for air. Fishes, therefore, must have an oxygen uptake mechanism that is very efficient. The relatively high density and viscosity of water (over 800 times more dense than air, over 50 times more viscous) mean that more energy is required simply to move water across the respiratory surfaces than is true of air. A fish may use as much as 10% or more of the oxygen that it gets from the water just to keep the breathing muscles going (Jones and Schwarzfeld 1974). For air-breathing animals, the relative cost is much lower, around 1% to 2%.

To further complicate matters, gas solubility in liquids diminishes with increasing temperature. Warm water, therefore, contains less oxygen than cool water, making the situation more difficult for warm-water fishes. The oxygen concentrations of saturated freshwater at 10°C, 20°C, and 30°C, for example, are 8.02, 6.57, and 5.57 mL O_2 per liter. Seawater at 10°C, 20°C, and 30°C contains 6.35, 5.31, and 4.46 mL O_2 per liter. The noticeable difference between fresh- and seawater at any given temperature is due to the diminished solubility of gases in water as the concentration of salts or other solutes increases. This **salting-out effect** is true for all water solutions, including natural aquatic environments, blood plasma, cytoplasm, or a glass of carbonated beverage (just add some table salt and see what happens). The combined effects of temperature and salinity make oxygen availability especially low in warm, marine environments.

Aquatic Breathing

The **gills** of most fishes are remarkably efficient at extracting oxygen from the water, in part due to the large surface area and thin epithelial membranes of the **secondary lamellae** (Fig. 5.1). Diffusion of gases across the gill mem-

warm water 2 O then cold

BOX 5.1   *Cheating on Oxygen Debt*

Animals need oxygen to maintain a high output of ATP for cellular metabolism. Although some of this ATP can come from the anaerobic process of glycolysis, most of it is produced by oxidative phosphorylation via the electron transport chain in the mitochondria of cells. This process requires the presence of oxygen, which acts as the final electron acceptor.

Some invertebrates can withstand long periods of oxygen deprivation because they have alternate biochemical pathways that permit the production of ATP without a dangerous drop in pH or increase in inhibitory metabolic by-products such as lactate. Among nearly all vertebrates, however, lactate formation appears to be the only metabolic pathway for dealing with anaerobic conditions—except among some fishes.

Goldfish (Cyprinidae) have an alternate mechanism that permits them to survive for months without oxygen (Hochachka and Mommsen 1983; Hochachka and Somero 1984). This can be quite useful in regions where goldfish are likely to be trapped under ice with little or no oxygen through a long winter. Rather than convert pyru-

vate to lactate, the pyruvate is converted to acetyl-CoA, which normally occurs under aerobic conditions so that acetyl-CoA is available for the Krebs cycle. Under anaerobic conditions, the acetyl-CoA is converted to acetate, which is then transformed to acetaldehyde by the enzyme acetaldehyde dehydrogenase. Acetaldehyde is converted to ethanol by ethanol dehydrogenase, and the ethanol can be excreted, thereby avoiding the problem of accumulating metabolic end-products (in essence, goldfish produce grain alcohol when trapped under ice during winter).

The fermentation of glucose to ethanol also consumes the hydrogen ions produced by the hydrolysis of ATP to adenosine diphosphate (ADP), thereby preventing a decrease in pH. The presence of these two important enzymes, acetaldehyde dehydrogenase and ethanol dehydrogenase, gives goldfish the ability to continue producing ATP by glycolysis without suffering the problems associated with decreasing pH and lactate buildup. And they do not have to pay back any subsequent oxygen debt.

---

brane is further enhanced by blood in the secondary lamellae flowing in the opposite direction to the water passing over the gills, thereby maximizing the diffusion gradient across the entire lamellar surface. This **counter-current flow** ensures that as the blood picks up oxygen from the water, it moves along the exchange surface to an area where the adjacent water has an even higher oxygen concentration. Therefore, oxygen diffuses from the water into the blood across the entire lamellar surface because the oxygen content of the blood inside the lamella is always less than that of the adjacent water.

Gills will only function efficiently if water is kept moving across them in the same direction, from anterior to posterior. This is accomplished in one of two ways. First, the great majority of fishes pump water across their gills by increasing and decreasing the volume of the **buccal** (mouth) **chamber** in front of the gills and the **opercular chamber** behind them. The expansion and contraction of these two chambers is timed in such a way that the pressure in the buccal chamber is almost always greater than the pressure in the opercular chamber, thereby ensuring that the water flows in the anterior to posterior direction throughout most of the breathing cycle (Fig. 5.2). Only for a very brief time does the pressure in the anterior (buccal) chamber drop below that in the posterior (opercular) chamber, causing a momentary flow reversal.

A second method of gill ventilation, called **ram ventilation**, consists simply of keeping the mouth slightly

open while swimming. The forward movement of the fish keeps water flowing over the gills. This is an efficient way to ventilate the gills because the work of ventilation is accomplished by the swimming muscles, but it can only be used by strong swimmers while they are moving at relatively high speeds. Some predatory pelagic fishes, such as tunas (Scombridae), rely exclusively on ram ventilation and must therefore swim constantly.

It had been thought that sharks also had to swim constantly in order to breathe. However, observations of so-called "sleeping" sharks on the ocean floor, including relatively sedentary species such as reef whitetip sharks and nurse sharks, indicate that they too use a gill pumping mechanism similar to the one described above for teleosts. Many larger fishes use ram ventilation while swimming at moderate to high speeds but rely on pumping of the buccal and opercular chambers while still or moving slowly. As speed increases, they can switch from gill pumping to ram ventilation (Roberts 1975a).

The total surface area of the gills is considerable. Not surprisingly, active fishes with higher metabolic demands generally have larger gill surface areas than less active fishes. For example, skipjack tuna are active pelagic predators and have about 13 cm^2 of gill area per gram of body weight (Roberts 1975b). Scup (Sparidae) are nearshore, active fish and have about 5 cm^2g^{-1}. Benthic, yet active, plaice (Pleuronectidae) have a little over 4 cm^2g^{21}, whereas the sluggish, benthic oyster toadfish (Ba-

FIGURE 5.1. *(A, B) The gill arches of a fish support the gill filaments (also called the primary lamellae) and form a curtain through which water passes as it moves from the buccal cavity to the opercular cavity. (C) As water flows across the filaments of a teleost, blood flows through the secondary lamellae in the opposite direction. (D) In elasmobranchs, even though septa create some structural differences in gill filaments, water flow across the secondary lamellae is still countercurrent to blood flow. (E) The countercurrent flow of water and blood at the exchange surface of the secondary lamellae ensures that the pO_2 in the water always exceeds that of the blood, thereby maximizing the efficiency of oxygen diffusion into the blood.*

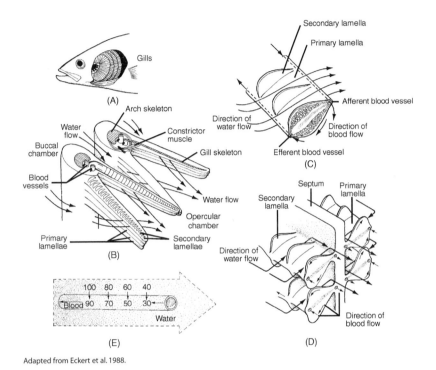

Adapted from Eckert et al. 1988.

trachoididae) has about 2 cm^{2}g^{-1}. Fishes with large gill areas control how much of the gills are receiving blood at any given time (see Jones and Randall 1978). By constricting or dilating blood vessels in the gill filaments, a fish can meet its oxygen needs without experiencing needlessly high osmotic stress. (Because the gill epithelium is so thin, water and ions also are exchanged with the surrounding environment; see Chap. 7, Osmoregulation and Ion Balance.)

Agnathans have a very different gill structure and rely on different means of ventilation. Hagfishes (Myxinidae) have a muscular, scroll-like flap known as a **velum** that moves water in through the single median nostril and over the gills (Fig. 5.3). When the hagfish's head is buried in food, water enters and leaves the gill area via the external opening behind the last gill pouch. Lampreys (Petromyzontidae) expand and contract the branchial area, causing water to flow in and out through the multiple gill openings. This method of ventilation is especially practical when the lamprey's buccal funnel is attached to the substrate or a host organism.

Although gills typically are identified as the respiratory organ of most fishes, any thin surface in contact with the respiratory medium is a potential site of gas exchange.

Cutaneous gas exchange (across the skin) can be important to some fishes, particularly in young fish whose gills have not yet developed fully. Newly hatched alevins of chinook salmon (Salmonidae) rely on cutaneous respiration for up to 84% of their oxygen (Rombough and Ure 1991). As the fish develop and their gills increase in size and efficiency, dependence on cutaneous respiration decreases to about 30% of total uptake in the fry and later stages. Adult eel (Anguillidae), plaice, reedfish (Polypteridae), and mudskipper (Gobiidae) gain about 30% or more of their oxygen through their skin (Feder and Burggren 1985; Rombough and Ure 1991).

Aerial Respiration

Fishes in over 40 genera, representing 13 orders, have some capacity to obtain oxygen from the air (Table 5.1). Some are facultative air breathers, only supplementing gill respiration when necessary. Others are obligate air breathers and must have access to the surface or they will drown. Most air breathers live in freshwater habitats in tropical areas of South America, Africa, or Asia (Kramer and McClure 1982). A few species are estuarine. Three genera (*Amia*, *Lepisosteus*, and *Umbra*) inhabit

FIGURE 5.2. *The timing of the expansion and contraction of the buccal (oral) and opercular cavities ensures that the pressure in the buccal chamber exceeds that of the opercular chamber throughout nearly all of the respiratory cycle. This creates a nearly steady flow of water from the buccal chamber to the opercular chamber, passing over the gill secondary lamellae, which have blood flowing through them in the opposite direction. Fish is viewed from below.*

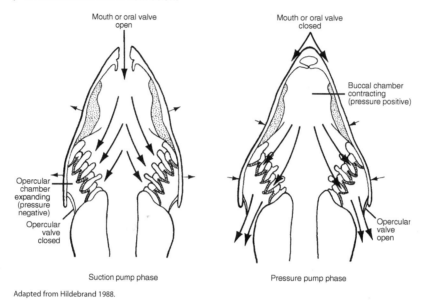

Adapted from Hildebrand 1988.

FIGURE 5.3. *(A, B) Hagfishes have one or more external gill openings on each side. Movement of the scroll-like velum draws water in through the nostril and pushes it through the pharynx and branchial pouches. Excurrent branchial ducts then direct the water to the gill openings. (C, D) Lampreys have multiple external gill openings on each side. Expansion and contraction of the branchial pouches provide ventilation through each external opening. This permits continued breathing while the mouth is attached to substrate or a host.*

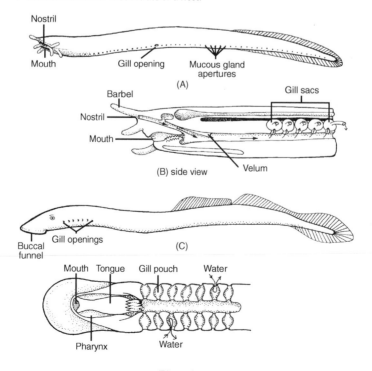

Table 5.1. Habitat and respiratory organs of various air-breathing fishes. Orders are in boldface.

	Habitat	Respiratory Organ
Ceratodiformes		
lungfish (*Neoceratodus*)	Aust; rivers	lung
Lepidosireniformes		
lungfish (*Lepidosiren*)	SAm; rivers	lung
lungfish (*Protopterus*)	Afr; rivers	lung
Lepisosteiformes		
gar (*Lepisosteus*)	NAm; freshwater	gas bladder
Amiiformes		
bowfin (*Amia*)	NAm; freshwater	gas bladder
Polypteriformes		
bichir (*Polypterus*)	Afr; freshwater	air sacs
reedfish (*Erpetoichthys*)	Afr; freshwater	air sacs
Osteoglossiformes		
arapaima (*Arapaima*)	SAm; freshwater	gas bladder
bony-tongue (*Heterotis*)	Afr; swamps	gas bladder
butterflyfish (*Pantodon*)	western Afr; freshwater	gas bladder
knifefish (*Notopterus*)	Asia; freshwater	gas bladder
gymnarchid (*Gymnarchus*)	Afr; swamps, rivers	gas bladder
Anguilliformes		
eel (*Anguilla*)	Eur, NAm, Asia, Afr; rivers, estuaries	skin
Elopiformes		
tarpon (*Megalops*)	NAm, SAm, Afr; coastal	gas bladder
Characiformes		
trahira (*Erythrinus*)	SAm; swamps	gas bladder
trahira (*Hoplerythrinus*)	SAm; swamps	gas bladder
lebiasinid (*Piabucina*)	SAm; swamps	gas bladder
Cypriniformes		
loach (*Misgurnus*)	Eur, Asia; rivers, pools	intestine
Siluriformes		
catfish (*Pangasius*)	Asia; freshwater	gas bladder
catfish (*Clarias*)	Afr, Asia; ponds, swamps	arborescent gill organ
catfish (*Saccobranchus*)	Asia; ponds, swamps	suprabranchial air sacs
armored catfishes		
(*Hoplosternum*)	SAm; rivers, pools	intestine
(*Plecostomus*)	SAm; swamps	stomach
(*Ancistrus*)	SAm; swamps	stomach
thorny catfish (*Doras*)	SAm; rivers, swamps	intestine
Gymnotiformes		
knifefish (*Hypopomus*)	SAm; swamps	opercular chamber
electric knifefish (*Electrophorus*)	SAm; rivers, swamps	mouth/pharynx
Esociformes		
mudminnow (*Umbra*)	Eur, NAm; stagnant water	gas bladder
Synbranchiformes		
swamp-eel (*Synbranchus*)	SAm; swamps	suprabranchial chamber
cuchia (*Monopterus*)	Asia; ponds, swamps	suprabranchial chamber
Perciformes		
labrisomid (*Mnierpes*)	SAm; rocky shores	skin

Table 5.1   (Continued)

	Habitat	Respiratory Organ
gobies		
(*Gillichthys*)	NAm; estuaries	mouth/opercular chamber
(*Periophthalmus*)	Asia; estuaries	skin, opercular chamber
gouramis		
(*Anabas*)	Asia, Afr; swamps	suprabranchial labyrinth
(*Macropodus*)	Asia; ponds	suprabranchial labyrinth
(*Betta*)	Asia; freshwater	suprabranchial labyrinth
(*Colisa*)	Asia; freshwater	suprabranchial labyrinth
(*Trichogaster*)	Asia; freshwater	suprabranchial labyrinth
(*Osphronemus*)	Asia; freshwater	suprabranchial labyrinth
snakehead (*Channa*)	Afr, Asia; ponds	suprabranchial chamber

Afr = Africa; Aust = Australia; Eur = Europe; NAm = North America; SAm = South America.
Adapted from Withers 1992; based on Dehadrai and Tripathi 1976, and Munshi 1976.

freshwaters in temperate North America. Air breathing is rather widespread among several tropical groups because high temperatures and high rates of decomposition dramatically decrease the amount of oxygen available. A thick forest canopy also shades many tropical aquatic habitats, thereby decreasing aquatic photosynthesis that adds some oxygen to the water.

Adaptations for aerial respiration are diverse, which is not surprising considering the wide taxonomic diversity of air-breathing fishes (see Table 5.1). This diversity in itself is not unexpected when we consider that fishes have had over 400 million years to address this problem. The fact that some solutions were found relatively early in the history of fishes made it possible for some vertebrates to move onto land by the end of the Devonian.

Gills are not well suited for aerial respiration because they collapse and stick together when not supported by the buoyancy of water. However, there are a few fishes that have modified gill structures that assist with aerial respiration, such as the modified treelike branches found above gill arches two and four of the walking catfish (Clariidae) or the complex platelike outgrowths of the gill arches of anabantoids such as the giant gourami (Osphronemidae) and several other Asian perciforms (Fig. 5.4). Most air breathers, however, rely on other organs for aerial respiration. Gills still are important for getting rid of metabolic wastes, such as carbon dioxide and ammonia, and for regulating ionic and acid-base balance (see Chap. 7), even among obligate air breathers with reduced gills.

Modified extrabranchial respiratory structures include highly vascularized surfaces such as the skin, mouth, and opercular cavity (see Table 5.1). Modifications of the gut, such as the stomach or intestine, also are used by some air breathers. And some species use modified gas bladders or lungs, both of which develop as diverticula off the gut (gas bladders evaginate dorsally whereas lungs evaginate ventrally). Gas bladders of most fishes are not well vascularized except in the regions designed for gas deposition or removal (discussed later in this chapter). However, several air breathers have highly vascularized and subdi-

vided gas bladders designed for gas exchange. These include *Arapaima* (Osteoglossidae), a South American osteoglossiform and the largest freshwater fish in the world, as well as the North American bowfin (Amiidae) and gars (Lepisosteidae). The lungfishes (Dipnoi) have true lungs. The Australian lungfish (Ceratodontidae) is a facultative air breather with a single lung, whereas the African and South American lungfishes (Protopteridae and Lepidosirenidae respectively) are both obligate air breathers with bilobed lungs (see Chap. 13, Subclass Dipnoi, The Lungfishes).

A strong reliance on air breathing in some fishes aids survival in oxygen-poor habitats, but it also helps some cope with drought. African lungfishes live in habitats that are subject to seasonal drought. When rivers and ponds dry up, these fish burrow into the sediment, dramatically slow their metabolism, and can remain in this torpid state for years. When the rains return and water levels rise, they leave their mud cocoons and become active (see Chap. 13). The walking catfish uses a different strategy when confronted with drought conditions. It "walks away" to find another pond, using a side-to-side lurching action supported by its stout pectoral spines.

GAS TRANSPORT

Oxygen enters the blood at the respiratory surfaces and is transported via the circulatory system to tissues and released (see Chap. 4, Cardiovascular System). Some oxygen simply is dissolved in the blood plasma. This is not enough, however, to support the level of metabolism of most large organisms, except in some antarctic icefishes (Channichthyidae). Most fishes need an oxygen-carrying protein to increase the overall capacity of the blood to transport oxygen. In fishes and other vertebrates, this protein is **hemoglobin**, and each of its four subunits can bind a single molecule of oxygen. Hemoglobin is carried within red blood cells, permitting intracellular biochemistry to operate in a manner that optimizes the function

FIGURE 5.4.   *(A) Lateral views of the gill arches of the walking catfish* (Clarias batrachus) *show the respiratory fans (fan), respiratory membranes of the suprabranchial chamber (resp mem), and treelike extensions (arborescent organs; arb org) that permit the fish to extract oxygen from air when it is out of water. (B) A cut-away view of the branchial region of the giant gourami* (Osphronemus goramy) *shows a labyrinth of platelike extensions to accomplish the same goal.*

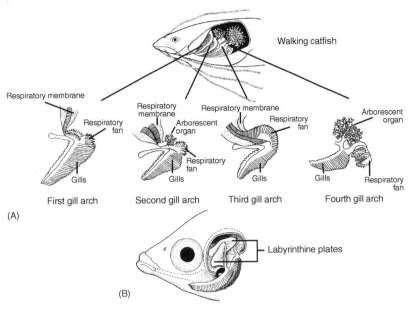

(A) Adapted from Munshi 1976. (B) Adapted from Peters 1976.

of hemoglobin but does not affect molecules carried in the plasma or in other cells in the bloodstream.

For hemoglobin to work well as an oxygen-transporting protein, not only must it bind oxygen at the respiratory surface, but it must also release it at the tissues elsewhere in the body. Hemoglobin must therefore alter its oxygen-binding ability under different conditions. Like many proteins, hemoglobin is sensitive to the physical and chemical conditions of its environment, such as temperature and pH. At the tissues, blood pH tends to be lowered by the presence of carbon dioxide, which combines with water to form carbonic acid (H_2CO_3). At the respiratory surfaces, however, carbon dioxide is released to the environment, thereby decreasing the level of carbonic acid in the blood and raising the pH. Hemoglobin's structure is altered somewhat by the changing pH conditions. Oxygen can bind more easily at higher pH but is released from its binding sites when pH decreases, a phenomenon known as the Bohr effect (Box 5.2).

Hemoglobin can be similarly affected by changes in temperature, with affinity for oxygen decreasing as temperature increases. This is one reason that typically coldwater fishes often cannot survive at higher temperatures, even if the oxygen content of the water is increased. At these higher temperatures, the structure of the fish's hemoglobin may be altered to the point where the fish simply cannot pick up enough oxygen at its gills and therefore could suffocate even though sufficient oxygen was present in the water.

Another interesting phenomenon seen in some fish

hemoglobins is known as the Root effect, which further enhances the release of oxygen by hemoglobin under low pH conditions (see Box 5.2). This phenomenon becomes very important in understanding the function of the teleost gas bladder, discussed later.

As fishes adapted to different habitats through evolution, their hemoglobins changed also. As a result, different fish hemoglobins show different temperature sensitivities. Antarctic fishes (Nototheniidae) possess hemoglobins that are effective at temperatures well below the effective temperature range of hemoglobins of temperate fishes (Hochachka and Somero 1973). Although most fish hemoglobins are sensitive to temperature changes, those of the warm-bodied tunas and lamnid sharks are not. This is adaptive, because blood temperatures in these fishes may increase as much as 10°C or more as blood travels from the gills to the warm swimming muscles (see Chap. 7). If the hemoglobin were not thermally stable, arterial blood might unload its oxygen as it warmed in the countercurrent heat exchanger, resulting in loss of oxygen to venous blood and depriving the highly active swimming muscles, which need the oxygen most (Hochachka and Somero 1984).

Some fishes, such as trouts (Salmonidae) and suckers (Catostomidae) have more than one type of hemoglobin. The different hemoglobins exhibit different degrees of sensitivity to decreased pH, therefore providing a "backup" system to ensure some oxygen transport even if blood pH drops considerably. If all of the hemoglobins exhibited the Bohr effect, a substantial drop in blood pH,

BOX 5.2   *Oxygen Dissociation Curves and the Bohr and Root Effects*

Oxygen dissociation curves are graphical depictions of the percent saturation of an oxygen-binding molecule, such as hemoglobin, at different oxygen concentrations. This concentration usually is expressed in terms of **partial pressure**, which is a means of expressing the amount of a gas in solution in a liquid. It is important to keep in mind that the partial pressure of oxygen (pO_2) only reflects the amount of oxygen in solution. Oxygen that is bound to a carrier molecule such as hemoglobin is not in solution and therefore does not contribute to the partial pressure.

The ability of hemoglobin to bind oxygen to its four binding sites is largely a function of the shape of the hemoglobin molecule, because this determines the accessibility of those binding sites to oxygen. One result of this is the sigmoid shape of the dissociation curve, which is due to the "cooperation" of the four hemoglobin subunits. It is relatively difficult for the first oxygen molecule to bind to its binding site, so oxygen loads slowly at first. After the first binding site is occupied, however, the structure of the hemoglobin molecule changes slightly, and the second binding site becomes more accessible. Therefore, the second oxygen molecule binds more quickly. This further alters the hemoglobin structure, making the third binding site even more accessible, and so on. Hence the four subunits of the hemoglobin molecule exhibit **subunit cooperativity** and help the molecule load up with oxygen quickly once the first molecule is bound. The curve levels off once all four binding sites are occupied with oxygen, because the hemoglobin is 100% saturated. Hemoglobin in hagfishes and lampreys consists of only a single subunit and therefore does not exhibit subunit cooperativity.

Another consequence of this dependence on molecular configuration is that environmental conditions such as temperature and pH can significantly affect hemoglobin's ability to bind oxygen. Like many other proteins, the configuration of hemoglobin may change under different temperature or pH conditions. If the structural change is great enough, oxygen cannot bind readily to its binding sites, and any oxygen that was bound might be released and go into solution. A decrease in pH thus results in the oxygen dissociation curve shifting toward the right, demonstrating that at a given partial pressure of oxygen in solution, hemoglobin will be less saturated at low pH than it would be at a higher pH (Fig. 5.5A). Consequently, it would take a higher pO_2 to completely saturate the hemoglobin at this lower pH.

In effect, hemoglobin has a lower affinity for oxygen under acidic conditions due to the configuration of the oxygen binding sites. This phenomenon is known as the **Bohr effect**. This sensitivity of hemoglobin to pH makes it an excellent oxygen-carrying molecule, because it binds oxygen at the respiratory surface where pH is a bit higher and unloads its oxygen at the tissues where pH is lower due to higher CO_2. Higher temperatures can also alter the configuration of hemoglobin and lead to the Bohr effect.

At low pH, some fish hemoglobins also exhibit an extreme change in configuration that prevents oxygen from binding to some hemoglobin binding sites regardless of how high the pO_2 gets. In this case, termed the **Root effect**, hemoglobin molecules will never become fully saturated, and the oxygen dissociation curve never reaches 100% saturation (Fig. 5.5B). The Root effect, therefore, results in a decreased capacity of hemoglobin to bind oxygen, whereas the Bohr effect results in a decreased affinity of hemoglobin for oxygen. Hemoglobin molecules that are affected by both will show a greatly diminished amount of bound oxygen under low pH conditions. Regardless, both Bohr and Root effects promote oxygen uptake at the gills and oxygen release in the tissues.

Independent of Bohr or Root effects, hemoglobins of different fish species may show different affinities for oxygen. For example, the oxygen dissociation curve of the toadfish is located to the left of the curve for the mackerel (Scombridae) (Fig. 5.5C; Hall and McCutcheon 1938). The higher affinity of toadfish hemoglobin makes it better adapted for low-oxygen environments. Mackerel, however, require more oxygen in their environment for their hemoglobin to become saturated enough to support their active lifestyle.

perhaps due to a burst of swimming activity, might inhibit oxygen loading at the gills (Brunori 1975; Hochachka and Somero 1984). Evolutionary changes in hemoglobin structure permit its use as an indicator of phylogenetic relationships among vertebrate groups (Box 5.3).

In addition to transporting oxygen, the blood must pick up the carbon dioxide that is produced in cellular metabolism and transport it back to the gills for release to the environment. If excess CO_2 is not removed, blood and tissue pH will drop low enough to interfere with normal metabolic processes. Let us consider in more de-

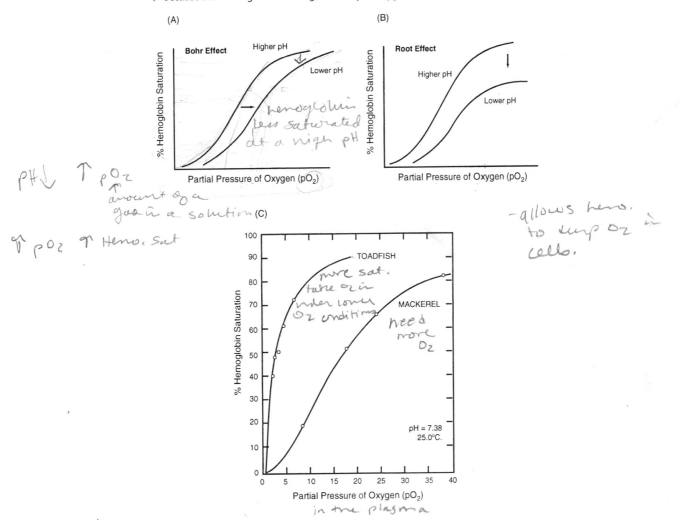

FIGURE 5.5. *Oxygen dissociation curves. Vertical axes indicate the percent of total oxygen-binding sites that are occupied by oxygen. The horizontal axes indicate the concentration of oxygen dissolved in the solution surrounding the hemoglobin, typically blood plasma. (A) A decrease in pH results in a shift of the curve to the right (the Bohr shift). (B) A decrease in pH may also prevent full saturation of hemoglobin with oxygen (the Root effect). (C) Toadfish can survive better than mackerel in low-oxygen conditions because their hemoglobin has a higher affinity for oxygen than mackerel hemoglobin.*

tail the specifics of CO_2 transport. This will, in turn, help us further understand how oxygen is delivered to the tissues.

Carbon dioxide can be carried in the blood in three different ways. A relatively small amount is simply dissolved CO_2 in the plasma. A greater amount is bound to proteins to form **carbamino compounds**. The greatest proportion is carried as **bicarbonate ion** (HCO_3^-) which results from the dissociation of carbonic acid. To understand this process, it is important to recognize that all reactions shown in Figure 5.6 are reversible.

It is also important to recognize the critical role of the enzyme **carbonic anhydrase**, which is present in red blood cells and catalyzes the reaction between CO_2 and water. At the tissues, CO_2 diffuses down its concentration gradient into the blood. In the plasma, some CO_2 com-

bines with water to form carbonic acid, which dissociates to bicarbonate and hydrogen ions. Most of the CO_2, however, is drawn into the red blood cells, where this same reaction is taking place at a faster rate due to the presence of carbonic anhydrase. The rapid production of H^+ from the dissociating carbonic acid inside the red blood cell causes the pH to drop. This, in turn, alters hemoglobin and causes the release of oxygen, which then diffuses out of the red blood cells and into the tissues. Some hemoglobin molecules now bind CO_2 to form carbaminohemoglobin. Some hemoglobin also binds some of the excess hydrogen ions, thereby preventing the blood pH from dropping too low.

The dissociating carbonic acid also causes the concentration of HCO_3^- inside the red blood cell to increase. Much of this HCO_3^- diffuses across the membrane of the

Hemoglobin Helps Reveal Vertebrate Bloodlines

Hemoglobin is a remarkable molecule for its ability to bind reversibly to oxygen, carry carbon dioxide, and help buffer the blood by binding hydrogen ions. Recently another use for this respiratory protein has been found—helping to understand phylogenetic relationships among vertebrates.

All jawed vertebrates have hemoglobin molecules made up of two alpha subunits and two beta subunits, whereas the hemoglobins of hagfishes and lampreys consist of only a single subunit. Over the course of vertebrate evolution, hemoglobin has changed from its ancestral form, so that the hemoglobin molecules of different species today differ somewhat in their amino acid sequences. Recent biochemical advances allow mapping of the entire amino acid sequences of hemoglobins, permitting their use in reconstructing parts of the vertebrate family tree. This technology has added to our understanding of the relationship of the coelacanth (Coelacanthidae) to other fishes and to the tetrapods.

The coelacanth is a sarcopterygian fish that was believed to have gone extinct 65 million years ago, until live coelacanths were discovered along the eastern coast of Africa beginning in 1938 (see Chap. 13, Box 13.2). Another group of sarcopterygians, the extinct Osteolepimorpha, are considered by many to be the ancestors of tetrapods (Chap. 11, Class Sarcopterygii). This would make the coelacanth the closest living relative to modern tetrapods.

Despite considerable fossil evidence, however, the position of the osteolepimorphs as tetrapod ancestors has been challenged (see Gorr and Kleinschmidt 1993). Some believe that lungfishes gave rise to the tetrapods, which would make the modern lungfishes the closest living tetrapod relatives. Still others hold that some groups of ray-finned fishes may be the closest living relatives to the tetrapods.

Comparisons of the hemoglobin of a coelacanth, one lungfish (Lepidosirenidae), one Port Jackson shark (Heterodontidae), several teleosts, and several amphibians support the position of the coelacanth as the closest living relative of the tetrapods (Gorr and Kleinschmidt 1993). In comparing the amino acid sequences in the beta chains of the hemoglobin molecules, Gorr and Kleinschmidt found about a 58% match between the coelacanth and the tadpole of a bullfrog (Ranidae). The other fishes had amino acid sequence matches of only about 45% to 48% when compared to the tadpole.

Comparison of the alpha chains yielded different results, but the authors argued that the relatively high and variable rate of evolutionary change in the alpha chain make it less suitable for inferring evolutionary relationships than the less variable beta chain. In support of the beta chain results, Gorr and Kleinschmidt also reported that amino acid sequences in two other very important parts of the hemoglobin molecule placed the bullfrog tadpole closer to the coelacanth than to any of the other fishes.

---

red blood cell and into the plasma, keeping intracellular HCO_3^- levels from getting so high that they would inhibit further CO_2 uptake. In response to this loss of negative ions from inside the cell, chloride (Cl^-) from the plasma diffuses into the red blood cell (the **chloride shift**), thereby balancing the distribution of charges (Cameron 1978).

The net result of all of these reactions is that the blood has taken up CO_2 and the pH has dropped, oxygen has been released from hemoglobin, the hemoglobin molecule itself has helped buffer against too much of a pH drop by taking up some CO_2 and H^+, and the bicarbonate level in the plasma has increased.

When the blood gets to the respiratory surface where CO_2 levels are low and oxygen levels are high, these reactions occur in the opposite direction, resulting in the release of CO_2, a slight increase in blood pH, and the binding of oxygen to hemoglobin within the red blood cells.

METABOLIC RATE

Metabolism is the sum total of all biochemical processes taking place within an organism. Since these reactions give off heat as a by-product, measuring the heat lost by an animal probably is the best way to measure its metabolism. This can be a difficult process, however. More often, some parameter related to metabolism serves as an indirect measure of metabolism. In fishes, the rate of oxygen consumption frequently is used as an indicator of metabolic rate, assuming no significant anaerobic metabolism takes place during the measurement period.

Metabolic rates can be influenced by a variety of factors, including age, sex, reproductive status, food in the gut, physiological stress, activity, season, and temperature. For this reason, it is useful to define metabolic terms. **Standard metabolic rate** is the metabolic rate of a fish while it is at rest and has no food in its gut. Experimen-

FIGURE 5.6.   *The uptake of carbon dioxide at the tissues is enhanced by the presence of carbonic anhydrase in the red blood cells. This enzyme catalyzes the conversion of CO_2 to H_2CO_3, which dissociates to form HCO_3^- and H^+. The increase in intracellular levels of H^+ causes a drop in pH, causing hemoglobin (Hb) to lose its oxygen (the Bohr effect). Hemoglobin can bind some CO_2 as well as some H^+ to help buffer against too great a drop in pH. Bicarbonate diffuses out of the red blood cell into the plasma, permitting further uptake of CO_2. To balance the loss of negative charges, Cl^- diffuses into the cell (the chloride shift). These reactions occur in reverse at the gills.*

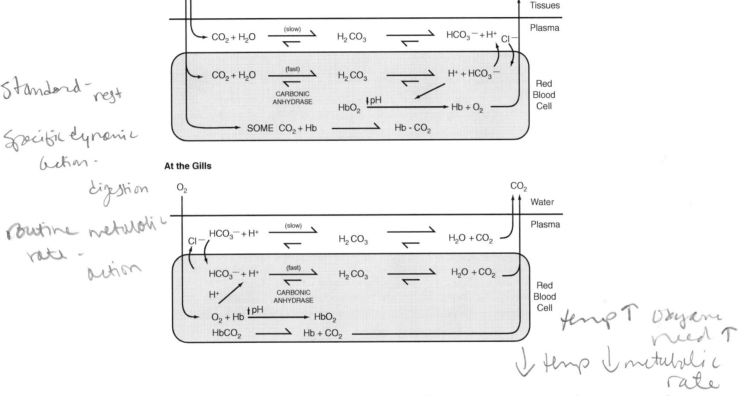

[handwritten margin notes:] Standard - rest / Specific dynamic action - digestion / routine metabolic rate - action / temp ↑ oxygen need ↑ / ↓ temp ↓ metabolic rate

tally, researchers should wait long enough after the fish's last meal to be sure that metabolism associated with digestion (called the **specific dynamic action**, or **heat increment**) is not a factor.

Fishes rarely remain still while metabolic rates are being measured, so the term **routine metabolic rate** is often used to indicate that the rate was measured during routine activity levels. The resulting estimates of metabolic rate are therefore higher than what might be expected for a resting fish. Sometimes researchers will measure metabolism at several levels of activity and extrapolate back to zero activity to estimate standard metabolic rate. Because metabolism is affected by temperature, the temperature should be recorded whenever measuring fish metabolism.

Metabolic rate increases with activity until a fish reaches the point at which it is using oxygen as rapidly as its gas uptake and delivery system can supply it. This is its **maximum** (or active) **metabolic rate**. The difference between the standard metabolic rate and the maximum metabolic rate at any given temperature is known as the **metabolic scope**. The concept of metabolic scope can be important in trying to understand a fish's metabolic lim-

its. Any factors that increase standard or routine metabolic rates, such as stress due to disease, handling, reproduction, or environmental conditions, narrows this scope and limits other activities.

In general, a fish tends to have lower metabolic rates at lower temperatures. Hence, as temperature increases, a fish's need for oxygen also increases. Because the availability of oxygen in water decreases with increasing temperature, warm conditions stress most fishes. This stress probably was an important selection factor favoring the evolution of air breathing in many tropical fishes.

Under laboratory conditions, fish acclimated to low temperatures consume less oxygen than fish of the same species acclimated to higher temperatures (e.g., see Beamish 1970; Brett 1971; Kruger and Brocksen 1978; DeSilva et al. 1986). The rates of many chemical reactions tend to increase with temperature. As the biochemical processes in the body increase so does the need for oxygen to provide the energy needed to support increased levels of cellular metabolism. Caution should be exercised, however, if one tries to use such studies to predict seasonal changes in metabolic rate. Under natural

[handwritten at bottom:] Standard metabolic rate — maximum metabolic rate = metabolic scope

environmental conditions, the gradual acclimatization of a fish to seasonal changes involves many physiological processes, each of which can have an impact on overall metabolism. Therefore, the results of temperature acclimation studies during a single season may not represent true seasonal changes in metabolic rates (Moore and Wohlschlag 1971; Burns 1975; Evans 1984).

Temperature–metabolic rate generalizations are based on studies of individual species acclimated to different temperatures. These generalizations should not be applied across species, especially those adapted to very different thermal environments. At low temperatures, for example, polar fishes have metabolic rates considerably higher than those of temperate species acclimated to the same low temperatures (Brett and Groves 1979). Metabolic rates of tropical fishes and those of temperate species acclimated to high temperatures differ only slightly.

Size can also have a considerable effect on metabolism. Not surprisingly, large fishes generally will have higher overall metabolic rates than small fishes, assuming other

factors such as activity are constant. However, the metabolic rate per unit of mass, often called the mass-specific metabolic rate or **metabolic intensity**, is higher for smaller fishes. This relationship seems to hold true for other animal groups as well, although the reason for this is not well understood.

Among the more metabolically costly things that a fish does is swim. Because water is 800 times denser than air, more energy is required to move through it. There is a trade-off, however, in that the density of water also provides buoyancy, so that fishes do not have to utilize as much energy fighting gravity as they would in a less dense medium.

Not surprisingly, oxygen consumption in fishes increases with swimming velocity. The increase is exponential, starting out quite slowly at first, but increasing dramatically at higher velocities (Fig. 5.7A). Such oxygen consumption curves probably underestimate the true metabolic cost of swimming at high speeds because of the increased use of anaerobic metabolism by swimming

FIGURE 5.7.   *The amount of oxygen used by stream fishes while holding position at different water velocities varies with fish morphology and lifestyle. (A) Water-column species, such as rainbow trout (Oncorhynchus mykiss), must increase swimming effort as water velocity increases. The resulting exponential increase in oxygen consumption rates with increasing velocity has been shown in numerous studies of swimming fishes. (B) Mottled sculpin (Cottus bairdi) are benthic fishes that lie on and cling to the substrate. Hence their oxygen consumption rates do not change with increasing water velocity. (C) Longnose dace (Rhinichthys cataractae) combine tactics. At low and moderate velocities, they remain on the substrate, and oxygen consumption rates do not change much. At higher velocities, however, they must swim, and oxygen consumption increases dramatically.*

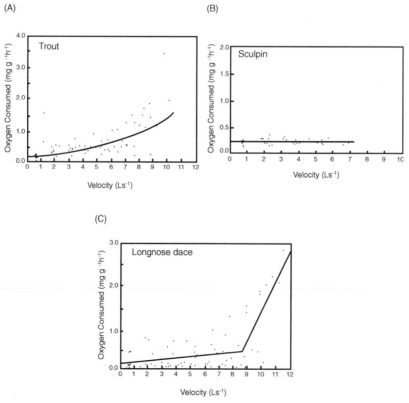

From Facey and Grossman 1990; used with permission.

muscles when not enough oxygen is supplied to support aerobic metabolism. Recall that if oxygen delivery is insufficient, these tissues will utilize anaerobic metabolism and build up an oxygen debt that must be repaid later.

The evolution of a torpedo-shaped, fusiform body undoubtedly is the result of its energetic advantages. Fin shape and placement are also important considerations, as well as body flexion during the act of swimming. The fastest, most active swimmers are streamlined, with high, thin caudal fins that oscillate rapidly while the rest of the body remains fairly rigid. This eliminates the drag that would be created by throwing most of the body into curves while swimming forward. The relationship among body shape, fin placement, and swimming style are addressed in more detail in Chapter 8 (see Locomotion: Movement and Shape).

Body shape and other morphological features are also important to the energetics of many benthic fishes. Bottom-dwelling stream fishes, for example, are able to hold their position in a high-flow environment without much energetic cost due to body shape and judicious use of their fins. Mottled sculpin (Cottidae) can use their pelvic fins to hold to the rocky substrate of swift mountain streams. They can even hold position in a plexiglass swimming tunnel, apparently by using their large pectoral fins to create downward force as the water flows over them (Facey and Grossman 1990). Their overall body shape of a large head and a narrow, tapering body may also help them remain on the bottom as water flows over them. These morphological adaptations give sculpins the ability to hold position in moderate currents without a significant energetic cost (Fig. 5.7B).

The bottom-foraging longnose dace (Cyprinidae) responds similarly at low to moderate velocities, showing no change in oxygen consumption. At higher velocities, however, it must resort to swimming to hold its position, and its oxygen consumption increases dramatically. This change in behavior breaks the oxygen consumption curve into two distinct segments (Fig. 5.7C).

ENERGETICS

Buoyancy Regulation

For fishes that are not benthic, maintaining a location up in the water column has the potential for being energetically expensive. However, this is usually not the case for most teleosts because of their ability to regulate buoyancy by regulating the size of the **gas bladder**, a flexible-walled, gas-filled chamber in the body cavity. This structure is often referred to as the "swim bladder," but it has more to do with saving energy by regulating buoyancy than with generating propulsive forces for the act of swimming (discussed in Chap. 8). The gas bladder also is important in hearing by some fishes (see Chap. 6, Hearing).

The need to regulate the volume of the gas bladder is a result of the effect of changing pressure as a fish changes depth. If a fish is neutrally buoyant at a given depth and descends in the water column, the increase in ambient pressure will decrease the volume of the gas bladder,

making the fish negatively buoyant (i.e., the fish sinks). If the fish continues to descend, the gas bladder shrinks even more and the fish would have to expend energy to prevent further sinking. Conversely, if a fish ascends in the water column, the gas bladder expands and the fish becomes positively buoyant. It would now have to either expend considerable energy to swim downward in the water column or continue to float toward the surface, with the gas bladder continuing to increase in size as the pressure decreases. An ability to regulate the volume of the gas bladder by the release or addition of gases would permit the fish to maintain neutral buoyancy at a variety of depths and save a great deal of energy.

The gas bladder is derived as an outpocket from the esophagus, and in some groups it retains its connection to the gut via the **pneumatic duct** (the **physostomous** condition). **Physoclistous** fishes, which include the higher teleosts (Paracanthopterygii and Acanthopterygii), have a closed gas bladder in which the pneumatic duct is sealed off. The structure and function of the teleost gas bladder is one of the more fascinating stories in animal physiology. We will consider this in two parts: gas release and gas addition.

Consider first the case of gas release. A fish swimming upward experiences increasing gas bladder volume. To remain neutrally buoyant, the fish must release some of the gas. In physostomes, gas can be released directly via the pneumatic duct. In some physostomes, however, such as eels (Anguillidae, Congridae), the pneumatic duct serves as a resorptive area for slow gas release via the blood but can release gas rapidly via the esophagus if necessary (Fig. 5.8A).

In physoclists, the gas must be released via the blood. Although most of the wall of the gas bladder is not permeable to gases because it is poorly vascularized and lined with sheets of guanine crystals, there is a modified area (called the oval in some species) where gas can diffuse into the blood when the gas bladder expands due to decreased pressure (Fig. 5.8B). The blood carries the excess gas to the gills where it is released to the surrounding water. Fishes regulate the loss of gas by controlling the flow of blood to the resorption area and by using muscles to regulate the amount of gas entering the resorptive region.

The addition of gas to the gas bladder is a bit more complex. As a fish descends, the volume of the gas bladder decreases due to increasing pressure, and the fish must add gas to maintain neutral buoyancy. A physostome could presumably swim to the surface, gulp air, and force it into the gas bladder via the pneumatic duct. This might work in shallow water, but in deep water the change in pressure makes this impractical, if not impossible. Hence a physostome is in the same predicament as a physoclist.

The addition of gas takes place by the diffusion of gases from the blood into the gas bladder at a special vascularized region of the bladder wall known as the **gas gland**. Since this process occurs by diffusion and not by active transport, a dramatic increase in the amount of gas in solution in the blood must occur. Where does this gas come from? How can the concentration get high enough to exceed the concentration of gas already in the gas

FIGURE 5.8.   *Schematic representation of the gas bladders of a physostome (A) and a physoclist (B). The pneumatic duct permits gas release via the esophagus in a physostome, whereas a physoclist must rely on a specialized area of the bladder wall for gas resorption. Both have gas glands with associated retia for gas addition. (C) Production of lactate and hydrogen ions by gas gland tissue triggers the hemoglobin's release of oxygen (the Bohr and Root effects) and a decrease in gas solubility (the salting-out effect). Countercurrent exchange of ions and dissolved gases in the rete creates very high gas pressures in the gas gland, thereby facilitating the diffusion of gases into the gas bladder.*

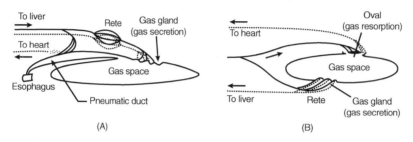

(A)   (B)

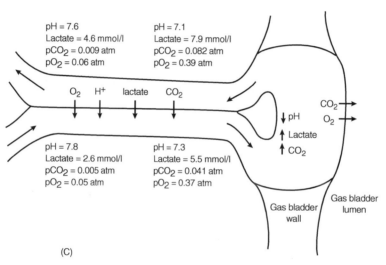

(C)

(A), (B) After Eckert et al. 1988. Data presented in (C) are for eels *(Anguilla)*, from Kobayashi et al. 1989, 1990.

bladder, thereby permitting the addition of gas by diffusion?

Three general physiological phenomena discussed earlier act together to bring this about (Fig. 5.8C). First is the effect of acidification on hemoglobin's ability to hold oxygen. The Bohr and Root effects cause unloading of oxygen from hemoglobin when pH decreases. This oxygen goes into solution in the blood, increasing the amount of dissolved oxygen. When this concentration of dissolved oxygen in the blood gets higher than the concentration of oxygen in the gas bladder, the gas diffuses down its concentration gradient, resulting in the addition of gas to the gas bladder.

The second phenomenon is the reduced solubility of gases in an aqueous solution as the concentration of lactate and hydrogen ions increases (the salting-out effect).

The third phenomenon is the efficiency of countercurrent exchange. The blood vessels leading to and from the gas gland are divided into a network of small capillaries that run countercurrent to one another. Such a bundle of capillaries is called a **rete mirabile** (wonderful net), or rete for short. The importance of the rete here is that high concentrations of gases and metabolic by-products are maintained in the area of the gas gland, because as these gases and by-products are carried away, they diffuse across the thin walls of the rete capillaries and into the incoming blood.

These three phenomena work together in the following manner. The tissues of the gas gland undergo anaerobic respiration, even in the presence of oxygen. This results in high levels of lactate and hydrogen ions, resulting in a decrease in pH. Gas gland cells also produce carbon dioxide, which combines with water to form carbonic acid, further reducing pH. The drop in pH triggers the Bohr and Root effects, causing hemoglobin to release oxygen and increasing the amount in solution in the blood. Elevated levels of plasma CO_2 also enhance the addition of this gas into the gas bladder (Pelster and Scheid 1992). As blood leaves the gas gland and travels through the rete, lactate, hydrogen ions, and dissolved gases diffuse down their concentration gradients into the blood coming toward the gas gland. Hence the counter-current arrangement of the rete capillaries helps build up the levels of diffusible gases in the gas gland. These get high enough to exceed the pressure of gases in the gas bladder, and gas addition occurs. Elevated lactate levels assist this process by decreasing the solubility of gases in the blood due to the increased ionic concentration (Alexander 1993).

But how can the blood give up oxygen in the gas gland and enter the rete with a higher partial pressure of oxygen than it had when it entered the gas gland, as shown in Figure 5.8C? Recall that partial pressure only indicates the amount of gas in solution and that oxygen bound to hemoglobin is not in solution and therefore is not accounted for in the partial pressure. For this reason, it is possible for blood to have a high total oxygen content but a low pO_2 if most of the oxygen is bound to hemoglobin. Conversely, blood can have a high pO_2 and at the same time have a low total oxygen content if all of the oxygen is in solution and little is bound to hemoglobin. So blood leaving the gas gland actually has less total oxygen than when it entered because some of the oxygen has diffused into the gas bladder. However, the partial pressure is higher because the oxygen that is present is in solution. Hemoglobin cannot bind much because the pH is low.

One other important factor is the timing of the release of oxygen by hemoglobin under acidic conditions (the Root-off shift) and the binding of oxygen by hemoglobin when pH increases (the Root-on shift). The Root-off shift occurs nearly instantaneously, whereas the Root-on shift takes several seconds. Therefore, hemoglobin in blood in the rete that is leaving the gas gland area does not increase its affinity or capacity for oxygen until it is already out of the rete. If the Root-on shift occurred more quickly, this hemoglobin probably would bind some of the oxygen and remove it from the rete and gas gland region, thereby decreasing the efficiency of this system.

Understanding how the rete mirabile functions to build up high gas pressures in the gas gland helps explain why fishes with long retes can build up higher gas pressures than those with short retes. Deep-sea fishes (see Chap. 17, The Deep Sea), which must deposit gas under high pressure conditions, tend to have longer retes than shallow-water fishes (Alexander 1993). The rete associated with the gas bladder of the American eel (Anguillidae) lengthens as the fish metamorphoses from its shallow-water, freshwater, or estuarine juvenile phase to its deep-water, oceanic reproductive phase (Kleckner and Kruger 1981).

Because the main purpose of the gas bladder is to maintain buoyancy at a given depth, several groups of teleosts find it more adaptive to have greatly reduced gas bladders, if they have one at all. Many benthic fishes, such as sculpins and flounders, either have gas bladders that are greatly reduced in size or lack gas bladders altogether. The absence of a gas float makes it that much easier to remain on the bottom. Fishes that constantly

swim and change depth rapidly and frequently, such as some tunas, also lack gas bladders. Herring (Clupeidae) are marine physostomes that lack a gas gland. However, since their high body lipid content also provides buoyancy, decreasing gas bladder volume with depth is less of a problem (Brawn 1962).

Gas bladders are found only in the bony fishes, so the elasmobranchs must utilize other means to reduce their density. A cartilaginous skeleton helps, because cartilage is much less dense than bone (the specific gravity of cartilage is 1.1, as opposed to 2.0 for bone). The constant swimming of pelagic sharks also helps prevent sinking by providing upward lift (see Chap. 8). Pelagic elasmobranchs also maintain high levels of low-density lipids in their greatly enlarged livers, which may make up 20% to 30% of their total body mass (Alexander 1993). Livers of other, more benthic sharks make up only about 5% of their body mass. The basking shark (Cetorhinidae) has a large liver that contains much squalene (specific gravity = 0.86), which is less dense than most other fish oils (specific gravities around 0.92). Wax esters are another type of low-density, oily compound (specific gravity = 0.86), that has been found in the livers of some benthopelagic sharks (Van Vleet et al. 1984).

Some teleosts also utilize lipids to reduce body density. The skin, muscles, and even the bones of the castor oil fish (Gempylidae) contain deposits of lipids, including wax esters (Bone 1972). Wax esters have also been found in the muscles and adipose tissues of the coelacanth (Nevenzel et al. 1966) and in some mesopelagic lantern-fishes (Myctophidae) that lose their gas bladders as adults (Capen 1967). Other tactics to reduce body density include reduced ossification of bone and increased water content of tissues. This is true in the lumpsucker (Cyclopteridae) (Davenport and Kjorsvik 1986), a coastal teleost, and in some bathypelagic species (Gonostomatidae and Alepocephalidae) (Denton and Marshall 1958; see Chap. 17, The Deep Sea).

Energy Intake

Fishes obtain the energy needed to meet metabolic demands through feeding. The diversity of feeding adaptations found among fishes is discussed in Chapter 8. The emphasis here is on postingestion processes.

Fishes have a fairly simple digestive system when compared to other vertebrates. Food is taken into the **mouth** and passed down the **esophagus** into the **stomach**. Secretion of mucus by the epithelial lining of the esophagus helps to lubricate the passage of food along the gut. Most fishes lack a mechanism for chewing food in the mouth, so food items are swallowed whole or in large chunks and much of the physical breakdown takes place in the stomach. Many fishes, such as minnows (Cyprinidae), suckers, croakers (Sciaenidae), cichlids (Cichlidae), wrasses (Labridae), and parrotfishes (Scaridae), have bony arches or toothed pads deep in the pharynx that are equipped with toothlike projections. These pharyngeal teeth grind up food before it reaches the stomach (see Chap. 8, Pharyngeal Jaws).

The stomach is often highly distensible and can store

food. Tough ridges along the internal wall of the stomach, along with contractions of the muscular wall, aid in the physical breakdown of foods. Acidic secretions of the stomach help to break down foods further; proteolytic enzymes also function more efficiently at lower pH. The combined physical and chemical activity of the stomach creates a soupy mixture that is released into the small intestine in small amounts.

Chemical digestion continues in the **intestine**, aided by **bile** from the liver, which helps emulsify lipids, and by secretions from the pancreas. **Pancreatic juice** contains bicarbonate to neutralize the acid from the stomach and a wide variety of enzymes to complete the process of chemical digestion.

Herbivorous fishes may depend in part on fermentation by symbiotic microorganisms in their guts to digest the plants they consume. Many of 27 primarily herbivorous tropical marine fishes from five families (Pomacanthidae, Kyphosidae, Scaridae, Siganidae, and Acanthuridae) showed elevated levels of short-chained fatty acids (SCFAs) in the posterior gut segments (Clements and Choat 1995). SCFAs are produced by microbial digestion of plant matter in the guts of terrestrial vertebrate herbivores. Most species examined also showed elevated SCFA levels in their blood, suggesting a direct contribution of metabolic fuel by microbial fermentation.

Fermentation digestion may not only benefit herbivorous fishes, however, as some planktivorous fishes studied by Clements and Choat (1995) also showed elevated SCFA levels. The relative contribution of gut microorganisms to digestion and nutrition in fishes deserves further study.

Some herbivorous fishes seem to rely on physical grinding or low stomach pH to break through plant cell walls (Lobel 1981). Of the 27 herbivores studied by Clements and Choat (1995), the 6 showing the lowest SCFA levels all possessed some mechanism for mechanically grinding ingested plant material.

The **small intestine** is also the primary site of absorption of the products of digestion, and mechanisms exist for maximizing this uptake. Elasmobranchs have a short, thick intestine with a large, spiraling fold of tissue (the **spiral valve**) to increase absorptive surface area. Teleosts generally have longer intestines, often with numerous side pouches (**pyloric cecae**) to increase absorptive area (Buddington and Diamond 1987). Herbivorous and microphagous teleosts have particularly long, often coiled, intestines to increase the opportunity to extract nutrients (see Fange and Grove 1979; Lobel 1981). Some of these fishes, such as the minnows, suckers, and topminnows (Cyprinodontidae), and several tropical marine fishes, including wrasses and parrotfishes, have reduced stomachs or lack them altogether (Fange and Grove 1979; Lobel 1981; Buddington and Diamond 1987). Although some nutrient absorption may continue in the large intestine, this last major portion of the gut functions primarily in water absorption.

Once basic metabolic demands are met, excess nutrients can be accumulated. **Carbohydrates** are stored as **glycogen** either in the liver or in muscle tissue. **Lipids** and **proteins** are also stored, resulting in an increase in mass that we refer to as growth. Lipids tend to accumulate either in the liver, in muscles, or as distinct bodies of fat in the visceral cavity. Protein often goes into tissue growth. All of these are potential energy sources when needed, although carbohydrates are metabolized first. In prolonged periods of starvation, such as during the migration of salmonids, body lipids and proteins will also be used. Stored lipids yield considerably more energy per gram than stored carbohydrates or proteins.

Energy Budgets

Energy budgets can aid in understanding the energy intake and utilization of an individual. The construction of an energy budget is a complex process because the energetic costs and benefits of all physiological activities must be accounted for if the budget is to provide a reasonably realistic view of how energy is being allocated. In addition, each individual organism is different. Consequently, energy budgets, like any physiological model, provide a broad conceptual framework rather than a precise prediction of what will happen in any particular organism. Energy budgets can, however, be quite useful in understanding how energy is allocated. In addition, energy budgets of individual fishes can be used to construct community energy budgets, thereby providing some understanding of energy flow through ecosystems.

Several methods can be used to determine the energetic content of food items, waste products, or components of fish growth such as muscle, fat, bone, or gametes (see Wootton 1990). The energetic costs of different activities can be estimated either by direct calorimetry (measuring the heat produced by an organism) or by some form of indirect calorimetry, such as measuring oxygen consumption (earlier in this chapter). We can then construct a conceptual model (Fig. 5.9) to represent how energy may be partitioned. The energy equation is often represented as

$$C = E + R + P$$

where C is the energy **consumed**, E is the energy **excreted**, R is the energy used in **respiration**, and P is the energy remaining for **production**.

Some of the potential energy in food will never be digested and is therefore lost in the feces. The proportion that is digested is sometimes represented by the **absorption efficiency** (or digestibility) and varies for different food types. Carnivorous fishes feeding on soft-bodied, highly digestible prey may have absorption efficiencies as high as 90% or more, whereas herbivores tend to have considerably lower absorption efficiencies (e.g., 40% to 65%; see Wootton 1990). In general, foods high in lipids and proteins have much higher absorption efficiencies than foods high in carbohydrates (Brett and Groves 1979).

Of the energy that is absorbed during digestion, some is subsequently lost through the excretion of nitrogenous wastes. An additional amount has, in a sense, already

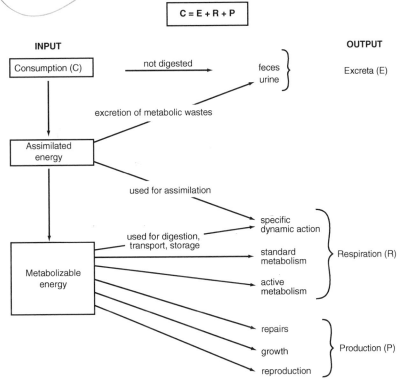

FIGURE 5.9. *Partitioning of energy consumed by a fish. Only energy not required to meet basic physiological needs (digestion, standard metabolism, repairs) or needed for activity is available for growth and gametes.*

$$C = E + R + P$$

Adapted from Videler 1993.

been used in providing the energy needed for digestion. This has been estimated to be about 10% to 20% of the energy value of ingested food, depending on the amount and type of food consumed (Jobling 1981). Larger meals and foods with higher protein content require more energy to digest and assimilate.

The remaining assimilated energy must be allocated among maintaining standard metabolism, swimming or other forms of activity, and production of gametes or new somatic tissue (growth). Only energy remaining after other physiological maintenance needs have been met is available for growth or reproduction. Therefore, any factors that increase other metabolic demands can ultimately decrease growth or reproduction.

Environmental factors affect the amount of energy needed to sustain standard metabolism. Increased temperatures will elevate metabolism and increase the need for energy. Other energetic costs include the maintenance of proper salt and water balance (osmoregulation) and the costs of health maintenance by the immune system. Energy requirements for basic maintenance may increase due to changes in salinity, or the diversion of energy to fighting infections, diseases, or parasites. In addition, exposure to contaminants that affect ion or water balance or that diminish the effectiveness of a fish's immune system can indirectly divert more energy away from growth and reproduction.

SUMMARY

1. Fishes need oxygen to provide energy for physiological function. In the presence of oxygen, far more energy can be derived from the metabolism of glucose than is possible in the absence of oxygen. Although anaerobic metabolism can provide some energy, it also results in the buildup of lactate, which can inhibit further metabolism.

2. Water's high density and viscosity, as compared to air, make it a difficult medium to move across respiratory surfaces. Water also contains considerably less oxygen than air, especially at elevated temperatures. Fish gills provide a large surface area for gas exchange, and the countercurrent flow of blood and water across the lamellae maximizes the efficiency of gas exchange by secondary diffusion. Some fishes have special adaptations to allow them to breathe air.

3. Blood transport of carbon dioxide and the release and binding of oxygen are closely linked because of hemoglobin's sensitivity to pH. At metabolically active tissues, high levels of carbon dioxide result in lower pH, which enhances the release of oxygen by hemoglobin. The loss of carbon dioxide to the surrounding water at the gills results in an increase in pH, enhancing hemoglobin's ability to bind oxygen.

↑ CO_2 ↓ pH — enhances release no oxygen

4. Metabolism is influenced by a wide variety of factors, including the presence of food in the gut, activity, age, sex, reproductive status, temperature, and season. Because of the impacts of these numerous factors, metabolic studies of fishes acclimated to controlled laboratory conditions may not accurately represent the metabolic rates of fishes in nature.

5. Many fishes that live in the water column use buoyancy control mechanisms, such as the addition or release of gases from the gas bladder, to save energy.

6. Energy in food is made available by digestion. Although some mechanical breaking down of food is accomplished in the mouth or pharynx of some fishes, most digestion takes place in the stomach and intestine. The intestines also function in nutrient uptake. Some fishes that feed on plants rely on symbiotic microorganisms in the gut to help break down their food.

7. Energy budgets indicate how the energy that is consumed is allocated. Some of the energy in food is not digestible and is subsequently excreted. Of the energy that is digested and absorbed, some must be used for basic metabolism and maintenance. Energy remaining after basic needs have been met can be used for growth and reproduction.

SUPPLEMENTAL READING

Evans 1993; Hoar and Randall 1976–1995.

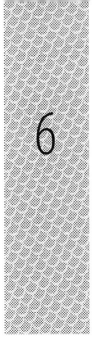

6

Sensory Systems

Sensory systems provide fishes with information from their external environment, thereby permitting appropriate responses to changing conditions. Sensory cells must act as signal transducers, receiving stimuli in one form and converting them to biological signals that can be transmitted via the nervous system. We will consider the following categories of fish sensory systems: **mechanoreception** (hearing, lateral line), **vision**, **chemoreception** (taste and smell), **electroreception**, and **magnetic reception**. Finally, we will consider **Mauthner cells**, a specialized neural network found in some fishes that is designed to translate sensory input into rapid action.

MECHANORECEPTION

In fishes, mechanoreception is largely involved in the detection of movement, particularly movement of the water in which a fish lives. Fishes have two major mechanosensory systems: the **inner ear** and the **lateral line system**. (Touch involves a variety of receptors that are not well defined.) **Sensory hair cells** of the inner ear are responsible for fish equilibrium and balance, as well as hearing (Fig. 6.1). The sensory hair cells of the lateral line system detect disturbances in the water, thereby helping a fish detect currents, capture prey, maintain position in a school, and avoid obstacles and predators (Popper and Platt 1993).

Equilibrium and Balance

Postural equilibrium and balance are maintained by the **pars superior**, a portion of a fish's inner ear that, in jawed fishes, consists of three **semicircular canals** and an additional chamber known as the **utricle** (Fig. 6.2). Lampreys (Petromyzontidae) have only two semicircular canals, and hagfishes (Myxinidae) have one. The semicircular canals are filled with a fluid (**endolymph**) and have sensory hair cells in their terminal ampullae. Changes in acceleration or orientation set the endolymph in motion and cause displacement of a gelatinous cupula that encloses the cilia of the hair cells. Lateral displacement of the cilia results in changes in the firing rate of the sensory neurons innervating the hair cells, thereby signaling the fish's brain of the changes in acceleration or orientation.

The utricle contains the lapillus, a solid deposit, or **otolith** ("ear stone"), which rests on a bed of sensory hair cells. The downward pull of gravity on the lapillus triggers impulses from the sensory cells and provides the fish with information regarding its vertical orientation in the water. The utricle works in coordination with the detection of light from above the fish by the retina of the eyes, and together they help keep the fish upright in the water (the **dorsal light reaction**). If a fish's lapilli are surgically removed and the fish is illuminated from the side, it will orient on its side with its dorsal surface toward the light source.

Hearing

Hearing in fishes is primarily the responsibility of the inner ear, including the utricle of the pars superior and the **saccule** and **lagena** of the **pars inferior** (see Fig. 6.2). Each chamber contains an otolith (the lapillus, sagitta, and astericus, respectively) and is lined with patches of tissue composed of sensory hair cells. An otolithic membrane provides a mechanical linkage between the otolith and the cilia of the sensory hair cells.

Most fish tissue is transparent to sound because its density is similar to that of the water. In other words, sound vibrations travel right through most of a fish's body. Structures that are significantly different in density, however, will vibrate differently from the rest of the fish's tissues and provide an opportunity for sensory detection of sound. As sound vibrations pass through a fish, the otoliths lag behind in their vibration due to their greater density. The relative difference in vibration between the fish's sensory hair cells and the otoliths excites the sensory hair cells and triggers action potentials in the sensory neurons of the auditory nerve.

Gas spaces in a fish, such as a **gas bladder**, can increase sensitivity to sound because sound waves cause slight increases and decreases in their volume. These movements of the gas bladder can then be transmitted to the inner ear for detection by the otolith organs. In cod (Gadidae), the gas bladder is close to the inner ear, and deflation of the bladder results in a decrease in hearing sensitivity (see Hawkins 1993). Herring (Clupeidae) have thin, hollow ducts extending anteriorly from the gas bladder that expand into gas-filled **bullae** in the inner ear, thereby increasing sensitivity to sound (see Enger 1967). Squirrelfish (Holocentridae) have an ante-

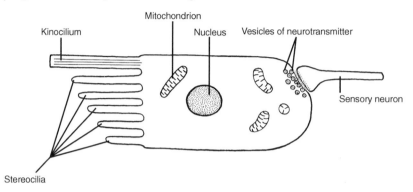

FIGURE 6.1.   *Mechanoreception involves sensory hair cells, which are found in the inner ears of fishes and other vertebrates. The apical surface of a sensory hair cell usually has numerous stereocilia and a single, much longer kinocilium. Deflection of the stereocilia toward or away from the kinocilium causes an increase or decrease in the firing rate of the sensory neuron innervating the hair cell at its basal surface.*

rior extension of the gas bladder that touches the wall of the sacculus. African mormyrids (Mormyridae), known for their sensitive hearing, have a gas bubble in the head that assists with sound detection (see Popper and Platt 1993).

The **otophysan** fishes (the "ostariophysans" in older literature) are the dominant fishes in freshwaters of the world (64% of all freshwater species) and have particularly acute hearing and pitch discrimination. This is due to a series of small bones, the **Weberian ossicles**. These modified vertebrae connect the anterior end of the gas bladder to the inner ear (Fig. 6.3), thereby conducting sound vibrations in much the same manner as the middle ear ossicles of mammals. The Weberian ossicles give the otophysans the highest sensitivity and greatest frequency range of hearing among fishes. Deflation of the gas bladder or interference with the Weberian ossicles causes decreased hearing sensitivity in dwarf catfish (Scoloplacidae) (Hawkins 1993).

Sharks are particularly sensitive to low-frequency sounds, including those below 10 Hz that are undetectable by humans. Sharks are most sensitive to pulsed sounds in this range, such as those emitted by the erratic swimming of an injured fish, and can localize such sounds at distances up to 250 m (Myrberg 1978; Myrberg and Nelson 1991).

Lateral Line System

Fishes can detect vibrations in the water that originate from or reflect off prey, predators, other fishes in a school, and environmental obstacles. Detection begins when sound waves displace **neuromasts** of the lateral line system (Bleckmann 1993; Popper and Platt 1993).

Neuromasts may be freestanding in the skin (superficial neuromasts) or located in channels beneath the scales of the trunk (the lateral line) and in dermal bones of the head (cephalic lateral line canals), which open to the surrounding water via small pores (canal neuromasts) (Fig. 6.4). The number and arrangement of lateral line pores, particularly those on the head, can be an important taxonomic feature in distinguishing similar species.

Neuromasts of the lateral line system function in the

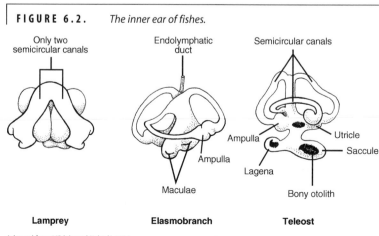

FIGURE 6.2.   *The inner ear of fishes.*

Adapted from Hildebrand (3d ed.) 1988.

FIGURE 6.3.   *A lateral view of the left side of the anterior portion of the vertebral region of an otophysan fish* (Opsariichthys, Cyprinidae). *The Weberian ossicles (tripus, intercalarium, scaphium, and claustrum) transmit sound vibrations from the gas bladder to the inner ear. The skull of the fish is to the left.*

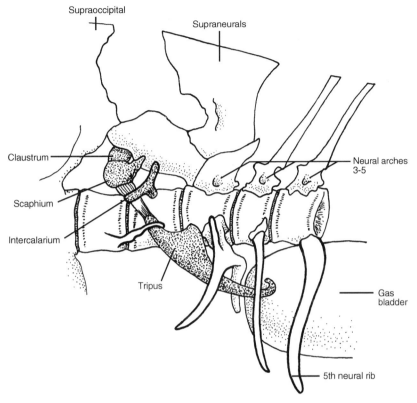

Adapted from Fink and Fink 1981; used with permission.

same basic manner as the sensory hair cells involved in hearing and equilibrium. Disturbances in the water around the fish or in the canals displace the gelatinous cupula, thereby bending the stereocilia and altering the firing rate of the sensory neurons.

The lateral line system is an old feature in the history of vertebrates, as indicated by its presence in fossil jawless fishes from the Silurian. The system is only useful in water and is therefore restricted to fishes and larval and permanently aquatic amphibians.

FIGURE 6.4.   *(A) Cross section of the lateral line on the trunk of a minnow (Cyprinidae) showing the distribution and innervation of neuromast receptors and the location of pores that connect the canal to the external environment. (B) Each neuromast is composed of several sensory hair cells, support cells, and innervating sensory neurons. The apical kinocilia and stereocilia project into a gelatinous cupula that overlays the entire neuromast.*

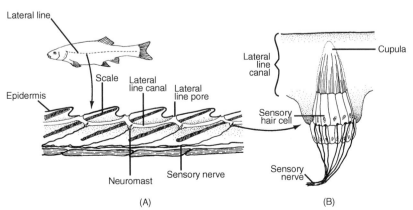

(A)   (B)

Adapted from Hildebrand (3d ed.) 1988.

VISION

The eyes of fishes are similar to those of all vertebrates, including those of humans (Fig. 6.5). Light must first pass through a transparent **cornea**. Because the optical properties of water are very similar to those of the liquid inside the eye, light does not have to bend as it travels into the eye of a fish in order to produce a clear image. Thus the cornea of fishes is much thinner than that of terrestrial vertebrates. (The cornea of terrestrial vertebrates assists in focusing by bending light as it travels from air into the eye.) The **iris** controls the amount of light entering the eye by increasing or decreasing the diameter of the **pupil**. Elasmobranchs can adjust the iris and change the shape of the pupil, but teleosts cannot (Guthrie and Muntz 1993).

The **lens** of fishes, which is more spherical in shape than the convex lenses of terrestrial vertebrates, focuses light on the **retina**. The lenses of elasmobranchs are somewhat flattened, whereas those of lampreys and bony fishes are spherical. Fishes focus on objects at different distances by moving the lens farther from or closer to the retina, whereas terrestrial vertebrates focus by changing the curvature of their lens, thereby altering its refractory power.

The outer layer of the eye (the **sclera**) is reinforced to protect the eye's delicate internal structures. The sclera of agnathans is fibrous and firm. Chondrichthyans have cartilaginous plates in their sclera, whereas teleosts frequently possess sclerotic bones. These are well developed in the mackerels and tunas (Scombridae) and particularly in the billfishes (Istiophoridae and Xiphiidae), which have a bony stalk extending part way back along the path of the optic nerve to the brain.

The **choroid** is a highly vascularized region between the protective sclera and the photoreceptive retina. It may contain a **tapetum lucidum**, a structure composed of reflective guanine crystals that enhances visual sensitivity under low light conditions. Tapeta lucida are found in most elasmobranchs, the Holocephali, the coelacanth, sturgeons, and some teleosts (Bone and Marshall 1982). The tapetum causes the reflective shine in the eyes of sharks and many nocturnal fishes.

Photoreception occurs in the retina, a complex structure composed of five layers. From the outermost layer to the innermost, these layers are as follows: 1) the pigment epithelium, which contains silvery argentine and melanin; 2) the photoreceptor layer, composed of rods and cones; 3) the bipolar layer, containing cells that form synapses with the photoreceptor cells; 4) the ganglion layer, composed of cells that form synapses with bipolar cells; and 5) the nerve fiber layer, containing cells that form synapses with the ganglion cells and send nerve fibers via the optic nerve to the brain. Light entering the eye of a fish, or any vertebrate, must pass through the pupil and lens, then the nerve, bipolar, and ganglion layers before being absorbed by the photoreceptors.

The retina has among the highest oxygen demand of any tissue in the body. In gars (Lepisosteidae), bowfin (Amiidae), and most teleosts, high oxygen levels seem to be maintained by a **choroid gland**. This U-shaped structure surrounds the optic nerve where it exits the eye. Blood flowing to and from the choroid gland travels through a rete mirabile (a countercurrent mechanism similar to that of the gas bladder; see Chap. 5, Buoyancy Regulation), maintaining high oxygen levels in the gland and assuring the retina of a plentiful oxygen supply. The

---

FIGURE 6.5.   *Cross-sectional view of the eye of a teleost.*

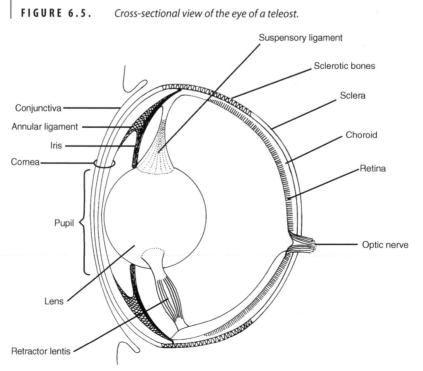

From Hildebrand (3d ed.) 1988.

choroid gland receives oxygenated blood from the pseudobranch, a gill-like structure on the inside surface of the operculum. Removal of the pseudobranch in trout (*Salmo*) results in decreased oxygen near the retina and a progressive loss of visual pigment (Ballintijn et al. 1977; see Box 4.1).

Like other vertebrates, fish retinas have two types of sensory cells, rods and cones. **Rods** are quite sensitive to low light levels. Crepuscular species (those that are active at dawn and dusk) have more rods than cones; deep-sea fishes and many nocturnal species have only rods. In bright light, the melanin of the epithelial layer protects the rods of teleosts. In dim light, **photomechanical movement** of the melanin changes the relative position of the epithelial and photoreceptive layers, exposing the rods to available light.

Some elasmobranchs and most teleosts have **cones**, photoreceptors responsible for vision in bright light. Cone-rod ratios are highest in diurnal fishes that rely on vision to find food. Four types of cones may be found in the eyes of fishes, each with its maximum absorption in a different part of the electromagnetic spectrum. Red cones respond best to light with a wavelength of about 600 nm (red light); green cones are most sensitive to wavelengths of about 530 nm (green light); and blue cones respond best to wavelengths around 460 nm (blue light). Ultraviolet cones exhibit peak activity at wavelengths of about 380 nm, in the ultraviolet range of the spectrum (Jacobs 1992; Hawryshyn 1992). Fishes with these cones can, therefore, see ultraviolet light.

Different species may have only two or three of these four types of cones, the cone type correlating well with light quality in the fish's habitat (see Guthrie and Muntz 1993). For example, shallow-water fishes usually have three cone pigments (red, blue, green) to cover the wide range of the spectrum that penetrates into the upper layers of the water column. Ultraviolet light attenuates rapidly with depth, so it is not surprising that fishes with cones sensitive to ultraviolet light frequently live near the surface. Marine fishes living at intermediate depths and freshwater fishes living near the bottom usually have two cone pigments that are most sensitive to the blue and green light that penetrates to these depths. Deep-sea fishes have rod pigments that are most sensitive to deep-penetrating short-wavelength light.

Not only can fishes detect a wide range of light wavelengths, including ultraviolet, but some fishes, such as anchovies (Engraulidae), cyprinids, salmonids (Salmonidae), and cichlids (Cichlidae), can detect polarized light. This probably enhances the contrast of objects viewed underwater, permitting a better view of predators, prey, and potential mates, as well as providing directional information for migrating fishes (Hawryshyn 1992).

Fishes in unusual or optically challenging habitats often demonstrate unusual eye morphologies that enhance their ability to see. For example, several families of deep-sea fishes, including the hatchetfishes (Sternoptychidae), giganturids (Giganturidae), pearleyes (Scopelarchidae), and barreleyes (Opisthoproctidae), have independently evolved elongate, tubular eyes. This shape, with a dense retina at the end opposite the large lens, probably enhances light gathering and binocular vision, and therefore depth perception. Not surprisingly, these eyes are oriented upward, either by their orientation on the fish or by the fish's orientation in the water column (see Munz 1971; Bone and Marshall 1982).

The eyes of mudskippers (Gobiidae) are well adapted for aerial vision. A strongly curved cornea and slightly flattened lens permit focusing out of water (Brett 1957). This structural adaptation, along with the location of the eyes on retractable stalks on the top of the head, allows these fishes to forage on tidal flats and exposed mangrove roots of the mangrove swamps in which they live. The eyes of the surface-dwelling South American "four-eyed fish," *Anableps* (Anablepidae), are adapted to permit simultaneous vision above and below the water (see Brett 1957). Each eye has two pupils (one above and one below the surface of the water), an oblong lens, and a retina that is divided into dorsal and ventral sections. Light entering from above the water's surface enters the upper pupil, travels through the short axis of the oblong lens, and focuses on the ventral retina. Conversely, light from below the surface enters the lower pupil, travels the long axis of the lens, and is focused on the dorsal retina.

CHEMORECEPTION

Fishes rely a great deal on their senses of smell (**olfaction**) and taste (**gustation**) to detect chemical cues in their environment. The distinction between olfaction and gustation is rather clear in terrestrial vertebrates. Volatile odorants travel in the atmosphere to the nose and trigger olfactory responses, whereas substances must be in solution and in contact with the taste buds in the mouth to initiate gustation. This distinction is less clear in fishes, however, because they live in water and all effective chemical stimuli are in solution.

Smell

A pair of blind **nasal sacs** contain the organs of smell in fishes. Jawed fishes have paired olfactory sacs with one or two **nares** leading into each sac. Hagfishes and lampreys have only a single naris and a median olfactory sac nostril (hence an alternate name for the lampreys, the Monorhina). In teleosts, the olfactory sacs are dead-end sacs that do not lead to the pharynx, except in a few cases such as stargazers (Uranoscopidae). Teleosts, therefore, do not breathe through their nares. This is also true of lampreys, in which the lone medial nostril leads to an olfactory chamber and rosette in a dead-end nasopharyngeal pouch. The nares of elasmobranchs are located ventrally on the snout and are not connected to the pharynx. In hagfishes, however, a nasohypophyseal duct connects with the pharynx so that hagfishes can smell water as it moves to the gills. Chimaeras (Holocephali) have paired nares that do connect to the oral cavity. This is also true of the lungfishes (Dipnoi).

Each olfactory sac is lined with a highly folded **olfactory epithelium**, often arranged in rosettes (Fig. 6.6).

FIGURE 6.6. *(A) External view of nares of a white sucker (Catostomidae). (B) The obvious flap of skin directs water across the sensory epithelium.*

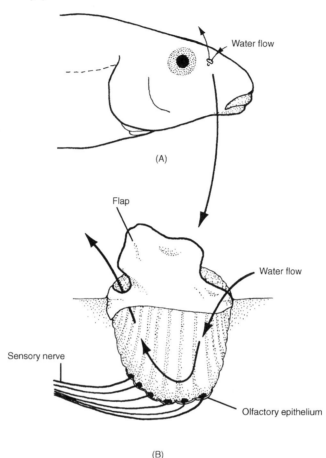

(A)

(B)

Adapted from Lagler et al. 1977.

Molecules of odorants bind to receptor proteins on membranes of the receptor cells in the sensory epithelium. The receptor cells then send nerve impulses to the brain (Hara 1993a).

The structure of the rosettes and olfactory sacs is related to the olfactory sensitivity of a fish. The more extensive the lamellar folding, the more surface area is available for sensory cells and the more sensitive the sense of smell. Freshwater eels (*Anguilla*) are known for their extremely keen sense of smell and have from 69 to 93 folds in each rosette. Perch (*Perca*), with less sensitive olfactory capabilities, have 13 to 18 folds in each rosette.

Fishes are extremely sensitive to certain types of chemicals. Amino acids, particularly those of fairly simple structure and with certain attached groups, are detectable by many fishes at concentrations of around 10^{210} moles per liter (see Hara 1993a). Other compounds that are detectable by some fishes at very low concentrations include bile acids (10^{-9} mol L^{-1}), salmon gonadotropin-releasing hormone (10^{-15} mol L^{-1}), and some sex steroids (10^{-12} mol L^{-1}). This ability to detect such small concentrations of certain chemicals makes olfaction valuable in homing (Box 6.1).

Olfactory cues can also be used to locate mates, and some fishes exhibit different olfactory sensitivities between sexes. In deep-sea anglerfishes (e.g., Ceratiidae), for example, males have enlarged olfactory organs, olfactory nerves, and olfactory lobes in the brain, whereas these structures are much less developed in females. In these species, females, which are much larger, are thought to release species-specific pheromones that the smaller, more mobile males use to locate them. Males then attach themselves to the females and spend the rest of their lives as parasitic sperm factories (see Chap. 17).

Taste

Taste buds are the receptor organs of the gustatory system in fishes (and in other vertebrates). Teleosts and elasmobranchs have taste buds in the mouth and pharynx. Teleosts also have taste buds on gill rakers and gill arches, and externally on the barbels and fins. Some catfishes (Ictaluridae) have taste buds distributed all over their bodies.

Taste buds are epithelial structures that tend to be rather bulbous in structure and protrude through the

BOX
6.1

BOX 6.1 *Following Their Noses*

Over 100 years ago, Buckland (1880) proposed that salmon use their sense of smell to guide them back to their natal (birth) stream. Not until the 1950s, however, did experimental work begin to provide data to support this hypothesis. In an early study supporting the **olfactory hypothesis** for salmon homing (Wisby and Hasler 1954), fish migrating up two tributaries of a larger stream were captured, marked, transported back downstream, and released. Eighty-nine percent of recaptured fish had moved back up the stream from which they had been captured. When the procedure was repeated with fish that had their nares plugged, only 60% of recaptured fish had returned to the stream from which they were captured. Olfaction was shown to be important to salmon homing in 16 of 20 similar experiments conducted by other researchers. Two of those studies also showed that fish that had been blinded behaved similarly to the control fish, indicating that vision was not essential for homing (Hasler and Scholz 1983).

Further studies have provided additional evidence supporting the olfactory hypothesis. In several studies in tributaries of Lake Michigan, coho salmon were exposed to low levels of synthetic chemicals while being raised in a hatchery prior to release. Two years later, as the adults migrated upstream to spawn, those same chemicals were dripped into tributaries. Fish recognized the scent and returned to the appropriately scented stream (Cooper and Hirsch 1982; Hasler 1983; Hasler and Scholz 1983).

It has become apparent through these and other studies that salmon imprint on the odor of their home stream during the smolt stage. The smolt stage is the phase of a salmon's life during which it migrates from its home stream toward the ocean or a large lake. Salmon rely so much on the scent of the home stream that they will "backtrack" downstream to pick up the scent if they happen to swim past, or are transported past, the proper tributary. This phenomenon should be kept in mind if hatchery-raised fish are to be released into the wild at the smolt stage in hopes of rebuilding naturally reproducing populations.

It had been proposed that returning adult fish are attracted by the smell of juveniles from the same stock (the pheromone hypothesis). This would promote stock fidelity by providing adults with a stock-specific odor to home in on as they searched for their natal stream. The experimental work mentioned earlier showed that salmon would imprint on synthetic chemicals, so the odor of close relatives was not necessary. However, under natural conditions salmon probably imprint on the combination of odors that characterize the natal stream. If stock-specific odors are a part of that smell, then they may be partly responsible for the return of adults to their natal streams.

---

epidermis to the epithelial surface. Each taste bud consists of basal cells, support cells, and from 5 to more than 60 gustatory receptor cells. Sensory neurons synapse with the gustatory cells at their basal surface. In gustation, as in olfaction, stimulus molecules bind to receptors on the sensory cells and trigger depolarization of the sensory neuron and the generation of action potentials that are carried to the gustatory centers of the brain (Hara 1993a).

Toxins, amino acids, and bile salts can stimulate gustatory receptors at sensitivity thresholds similar to those of olfactory receptors (see Hara 1993b).

ELECTRORECEPTION

A wide variety of fishes from different taxonomic groups can detect electricity in their environment. Elasmobranchs use their electroreceptive ability to home in on the weak electric fields that surround living prey. These fields are created by ionic differences between the living tissues of fishes or invertebrates and the surrounding water. Also, even the slightest muscular activity, such as the beating of a heart or the movement of respiratory muscles, emits a weak electrical current that is detectable. Dogfish (Scyliorhinidae) can detect a living flounder buried under 15 cm of sand and will attack a buried artificial prey item that emits a weak electrical current (Kalmijn 1971). Catfish also are able to locate live prey in this manner (Bleckmann 1993). Other electroreceptive fishes, such as the gymnotids of South America and the mormyrids of Africa, detect electrical fields created by their own electric organs or the fields created by other electrogenical fishes (Box 6.2).

Fish electroreceptor organs are structurally somewhat similar to the hair cells of the lateral line and balance organs of the inner ear, although they lack cilia. These sensory systems are probably phylogenetically related (Bleckmann 1993). Electrical fields in the fish's environment lead to changes in the flux of calcium ions across the membranes of the electroreceptive cells. This leads to a release of neurotransmitters from the receptor cell's base, which activate sensory neurons that carry impulses to the hindbrain.

There are two general types of fish electroreceptors.

Electric Organs

Electric organs are an example of how the electric properties of cell membranes can be put to a new use. Living cells maintain an electrical potential difference across their membranes, with a net negative charge on the membrane's inside surface. The current-generating cells of electric organs are referred to as **electrocytes**. In most electric fishes, electrocytes are disklike modified muscle cells, called **electroplaques**. When stimulated, ion flux across the cell membranes creates a small electrical current. The cells of the electric organ are arranged in a column and discharge simultaneously, producing an additive effect. A sizable stack of cells can produce a considerable current—like many small batteries connected in series (Feng 1991). Although electrocytes of most electric fishes are modified muscle cells, several species of South American black ghost knifefishes (Apteronotidae) utilize modified nerve fibers.

The generation and detection of weak electrical fields is particularly well developed in several groups of freshwater tropical fishes living in murky waters where vision is of limited use. These weakly electric fishes, such as the gymnotids of South America and the mormyrids of Africa, surround themselves with electrical fields created by emitting steady pulses or waves of electricity from their electric organs (see Fig. 6.8). The South American synodontid catfishes (Synodontidae) are also believed to produce and detect weak electrical fields for object detection or communication (Hagedorn et al. 1990). The electric organ is located dorsally on these catfishes and has apparently evolved from one of the muscles associated with sound production, which occurs by stridulation of the pectoral spines.

The production of weak electrical fields, as demonstrated in the gymnotids and mormyrids, requires considerable coordination by the central nervous system. In the gymnotids, pacemaker cells in the brain stem control the electric organs and spontaneously fire action potentials at regular intervals. The pacemaker cells of the mormyrids are located in the midbrain and can vary their rate of discharge (Withers 1992).

Although most electric fishes generate only mild electrical fields for communication and sensory purposes, others can generate currents strong enough to stun prey or ward off predators. The electric organs of the electric ray *Torpedo* (Torpedinidae) have about 45 columns of electrocytes (700 per column). The columns are oriented dorsoventrally, and the current is released dorsally because the dorsal surface of the organ and the overlying skin have lower resistance than the surrounding tissues. *Torpedo* can generate a discharge of 20 to 50 volts and stun prey 15 cm away (see Chap. 12, Feeding Habits). The electric eel *Electrophorus* (Electrophoridae), not a true eel but a close relative of the South American knifefishes, can generate pulses of over 500 volts with its several electric organs, the largest of which consists of about 1000 electrocytes. These organs are embedded in the fish's lateral musculature. The two electric organs of the electric catfishes (Malapteruridae) are located on either side of the body, and each contains several million electrocytes. These organs generate a current of about 300 volts. Other fishes that emit strong electrical currents include the stargazer (*Astroscopus*, Uranoscopidae), in which electroplaques are derived from ocular muscles.

Ampullary receptors, such as the ampullae of Lorenzini of the elasmobranchs, consist of receptor cells lying at the base of a canal that is filled with a conductive gel (Fig. 6.7). Ampullary receptors are the only electroreceptors found in sharks, which can detect the weak electricity emitted by prey but do not produce their own electrical fields. Ampullary receptors are sensitive to low-frequency AC or DC electrical stimuli. They are so sensitive that they respond to the small electrical fields generated by a fish's own movement through the earth's magnetic fields. This may permit a fish with ampullary receptors to use magnetic fields for navigation and orientation.

Fishes that generate and detect their own electrical fields have both ampullary and tuberous receptors, but there is a clear division of labor between the two receptor types. **Tuberous receptors** are not sensitive to low-frequency AC or DC currents but do respond to high-frequency AC currents. They are most sensitive to electric organ discharge (**EOD**) frequencies of the fish's own electric organs and therefore are of primary importance to the fish. Tuberous receptors are located in depressions of the epidermis and are covered with loosely packed epithelial cells (see Fig. 6.7). The fish's EOD frequency causes the tuberous receptor cells and their sensory neurons to generate a constant "background" rate of nerve impulses. A fish can detect objects moving into its electrical field (Fig. 6.8) when those objects cause a change in the field and alter the rate of impulses received by the brain, such as when the fish encounters an object with different conductance than the surrounding water.

FIGURE 6.7.   *Schematic diagram of the structure of tuberous (A) and ampullary (B) electroreceptive organs. Both organs are surrounded by layers of flattened cells that join tightly to one another. This helps prevent current from bypassing the organs. Tight junctions between the receptor cells and supporting cells help focus incoming electrical current through the base of the receptor cells, where they synapse with sensory neurons. Supporting cells in ampullary organs produce a highly conductive jelly that fills the canal linking the sensory cells to the surrounding water.*

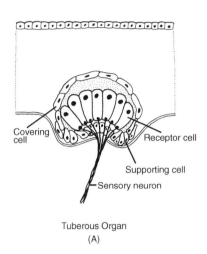
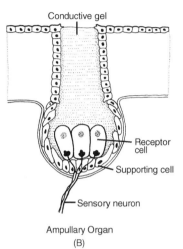

Tuberous Organ
(A)

Ampullary Organ
(B)

Adapted from Heiligenberg 1993, 144; drawing by H. A. Vischer.

Electric Communication

Most fishes that produce electricity use it for communication. Signals are species specific, and certain aspects of the EOD, such as amplitude, frequency, and pulse length, can be modified to exchange information about species, sex, size, maturation state, location, distance, and probably individual identification (Hagedorn 1986; Hopkins 1986).

Social interactions involving frequency shifts play an important role in dominance interactions in many electric fishes. South American gymnotiform knifefishes have individual characteristic waveforms to their EODs. In *Gymnotus carapo* (Gymnotidae), rapid increases and decreases in frequency indicate threat, whereas submissive individuals cease discharging. Within this species, individuals with higher EOD frequencies are consistently dominant. Male *Eigenmannia virescens* (Sternopygidae) will mouth fight until a dominance hierarchy is established, the ultimate

dominant male assuming the lowest discharge frequency. Females compete for spawning territories, and dominant females have the highest frequencies. In *Sternopygus macrurus* (Sternopygidae), mature males discharge at about 60 Hz, females at about 120 Hz, and immature animals emit at intermediate frequencies (Black-Cleworth 1970; Hopkins 1972, 1974; Hagedorn 1986).

The diverse mormyriform elephantfishes (Mormyridae) of Africa use EODs in species and sex recognition, to warn conspecifics of imminent social attack, to signal submission and courtship, and to maintain contact among shoal mates. Mormyrids receive information based both on the waveform of the EOD and on intervals between discharges. Variations in discharge interval of fractions of a millisecond are detectable by the fish. EODs are again both species and sex specific among different life history stages. Males typically have a pulse duration that is two to three times

FIGURE 6.8.   *Dorsal view of an* Eigenmannia *and its electrical field.*

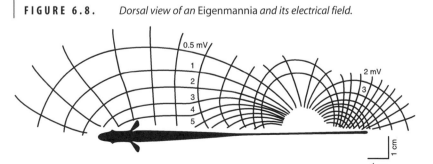

Adapted from Scheich and Bullock 1974

longer than that of females. Interactions include cessation and frequency modulation of EODs ("bursts," "buzzes," and "rasps"), echoing, and duetting. Males alternate their outputs with other males, whereas females synchronize their outputs with investigating males. A male can determine the sex of a conspecific by "listening" to the response to his electrical pulses. In direct analogy to gymnotiform behavior, the mormyriform *Gymnarchus niloticus* ceases discharging just prior to an attack on a conspecific, uses bursts of discharge pulses when aggressive, and modulates its frequency 1 to 30 Hz as a submissive gesture (Moller 1980; Hopkins 1986; Bleckman 1993). "Convergence between mormyrids and gymnotiforms . . . remains one of the most fascinating mysteries of electric communication" (Hopkins 1986, 560).

Social interactions include interference with a conspecific's electroreception. In *Gymnotus carapo*, dominant fish often shift their discharges to coincide with the short interval when a subordinate would be analyzing its own output, which could impair the subordinate's ability to electrolocate. Such interference is overcome in gymnotiforms such as *Eigenmannia* by a **jamming avoidance response** (JAR), in which neighboring fishes shift their EOD frequencies when they get near one another, thereby preventing the fish from interfering with one another's ability to electrolocate. Fish in a social group maintain a 10 to 15 Hz difference with their neighbors so that each individual has a "personal" discharge frequency (see Feng 1991). When several *Eigenmannia* were kept in separate tanks and all tanks were connected by electrical wires, the fish shifted their frequencies to an average separation of 7 Hz (Fig. 6.9). In several species, fish in a single social group have nonoverlapping frequencies (Bullock et al. 1972; Hagedorn 1986).

MAGNETIC RECEPTION

As mentioned earlier, some fishes, such as elasmobranchs, may have such a sensitive electroreceptive ability that they can detect the weak electrical fields they create as they move through the earth's magnetic field. This ability would provide these fishes with an indirect way of sensing the earth's magnetic field and give them directional information with respect to compass headings. Clearnosed rays (Rajidae) in the lab learned to orient in uni-

form DC fields weaker than the earth's magnetic field; the rays switched the location in which they searched for food when the electrical field around them was artificially reversed, suggesting that geomagnetic cues might be used in daily activities (Kalmijn 1978).

Some fishes, however, may be able to detect magnetic fields directly. Sockeye salmon and yellowfin tuna can detect earth-strength magnetic fields (Quinn and Brannon 1982; Walker 1984), and magnetite has been extracted from the heads of yellowfin tuna, chinook salmon, and chum salmon (Walker et al. 1984; Kirschvink 1985; Ogura et al. 1992). These fishes presumably sense magnetic fields directly, and the mechanism is related to the other mechanoreceptive sensory systems such as the inner ear and the lateral line system. The ability to discriminate among different field strengths and inclinations and to orient to the directional polarity of the earth's magnetic field would aid in magnetic compass orientation and navigation.

TURNING SENSORY RECEPTION INTO FAST ACTION

The sensory systems discussed thus far are designed to translate stimuli from the surrounding environment to neural signals that can be integrated and interpreted in a fish's brain. This can then result in a behavioral decision and, perhaps, an appropriate response. At times, however, the fish-eat-fish world demands an instantaneous reaction. A fish that spends an extra moment interpreting incoming stimuli and choosing an appropriate response may not survive until tomorrow. Animal nervous systems are structurally organized to deal with such instances through the use of **reflex arcs**, direct links between sensory neurons carrying incoming stimuli and motor neurons that result in action. A good example of a reflex in humans is the instant withdrawal of your hand from a hot or sharp object before your brain has had the opportunity to realize what has happened.

Mauthner Cells and the Startle Response

Some fishes have the ability to move extremely quickly when startled. This **startle response** consists of a rapid

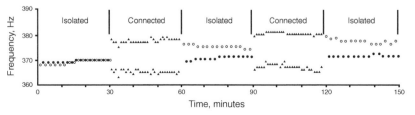

FIGURE 6.9.   *The JAR of two* Eigenmannia *kept in separate aquarium tanks. When electrically isolated, both fish converge on frequencies of about 370 Hz. When the tanks are connected electrically, the fish shift and maintain an approximately 10-Hz difference in the frequencies of their EODs.*

Adapted from Scheich and Bullock 1974, originally in Bullock et al. 1972.

FIGURE 6.10. *The Mauthner response, also called the C-type fast-start response. (A) Dorsal view photographs, taken 5 milliseconds apart, show the sequence of events during the startle response of a rainbow trout. The sequence occurs from left to right and from top to bottom, as in reading lines of text. The fish in frame 1 is already bent slightly to the right. The stimulus, a mallet blow to the side of the tank, occurs in frame 2. Frames 2 through 5 represent stage 1 of the response, in which the fish develops a C-bend to the right. Frames 6 through 20 are stage 2, during which the fish accelerates away from the stimulus. (B) Dorsal view silhouettes of various teleosts engaged in fast-start responses. The solid bars above each silhouette show the first two stages of the startle response.*

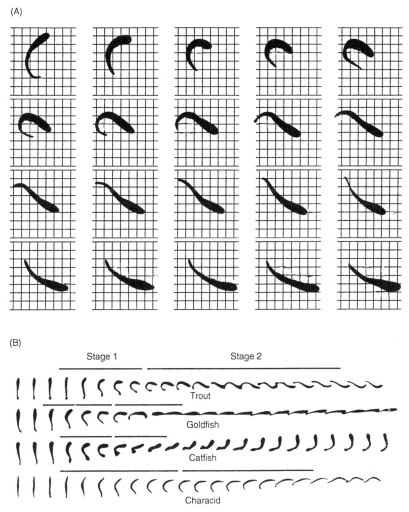

(A)

(B)

From Eaton and DiDomenico 1986; used with permission.

flexing of the body and accompanying flick of the tail, which together rapidly propel the fish away from the initiating stimulus. The entire response takes less that 100 milliseconds and occurs in two stages. In stage 1, which takes less than 20 milliseconds, the head and tail of the fish bend away from the stimulus; the fish's center of mass does not move. Stage 2, requiring about 80 milliseconds, involves a rapid tail beat as the posterior portion of the animal straightens out.

The speed of the startle response is made possible by the rapid conduction of nerve impulses along specialized nerve cells known as **Mauthner cells** (Eaton 1991). The combination of large diameter and presence of a myelin sheath gives the axons of these cells the fastest conduction velocity known among vertebrate neurons (50 to 100 m/sec).

The cell bodies of the two Mauthner cells lie within the brain stem, one on either side of the midline. They each receive sensory input directly from the visual, auditory, and lateral line systems from one side of the body and direct a response to the body musculature, thus bringing about the escape response. Bypassing "higher" brain areas saves time by eliminating additional synapses and information processing. Each Mauthner cell's large, myelinated axon crosses over to the opposite side of the brain stem before descending the entire length of the spinal cord. Therefore, the Mauthner cell receiving input from the fish's left side innervates the musculature along the fish's right side, and vice versa.

To prevent both sides of the fish from acting against one another, each Mauthner cell simultaneously receives inhibitory input from the sensory nerves and Mauthner cell on the opposite side of the body. Therefore, when the Mauthner cell on the right side of the brain stem is stimulated by sensory input from the fish's right side, it simultaneously stimulates contraction of the muscles on the left side of the fish and inhibits the Mauthner cell controlling contraction of the muscles along the fish's right side. The fish rapidly curls to its left, straightens its body and flicks its tail, and is propelled away from the perceived threat (Fig. 6.10).

SUMMARY

1. Sensory systems convert stimuli from a fish's environment into biological signals (nerve impulses) that can be integrated, interpreted, and acted upon.

2. Equilibrium, balance, and hearing are mechanoreceptive senses primarily located in various chambers of a fish's inner ear. The relative movement of fluid (endolymph) or solid deposits (otoliths) stimulates sensory hair cells that generate signals that are subsequently carried to the brain by sensory nerves. Hearing in some fishes is enhanced by the transmission of vibrations from the gas bladder to the inner ear via anterior extensions of the gas bladder or a chain of small bones known as the Weberian ossicles.

3. Disturbances in the water are sensed by neuromasts, clusters of sensory hair cells and supporting cells covered by a gelatinous cupula. Neuromasts may be free-standing in a fish's skin or located in canals beneath the scales along the trunk (the lateral line) or in the dermal bones of the head. Small pores allow vibrations from the surrounding water to enter these canals.

4. Fish eyes are structurally quite similar to those of terrestrial vertebrates, except that the lens is more spherical. The retinas of fishes contain rods, for vision in dim light, and cones, for color vision and vision in bright light. Different cones respond to light of different wavelengths of the spectrum visible to humans, and some fishes have cones capable of detecting ultraviolet light. Some fishes can also detect polarized light, thereby enhancing underwater vision and perhaps providing directional cues for migration.

5. Fishes rely a great deal on their chemical senses, smell and taste. The organs of smell are located in blind nasal sacs and open to the surrounding water via nares. Some fishes can detect very low concentrations of odorant molecules. Taste receptors are located in the mouth and pharynx, and some fishes have taste buds on their gill rakers and gill arches, and externally on barbels, fins, or elsewhere on the body.

6. Two general types of receptors are used by fishes to detect electricity in their environment. Ampullary receptors can detect the weak electricity generated by living prey organisms. Some teleosts possess specialized organs capable of generating an electrical field, which is subsequently received by tuberous receptors. These fishes utilize this sensory system to gather information about their environment and to communicate with conspecifics.

7. Some fishes can detect magnetic fields, thus providing valuable orientation information during migration. Sensitive electroreceptors may enable some fishes to detect the electricity generated by their own movement through the earth's magnetic field, hence providing an indirect means of magnetic reception. Other fishes, however, have demonstrated direct magnetic sensory abilities, and biologically produced magnetic particles have been found in regions of their skull believed to be the site of this sense.

SUPPLEMENTAL READING

Atema et al. 1988; Coombs et al. 1989; Douglas and Djamgoz 1990; Evans 1993; Farrell and Randall 1995; Hara 1992; Hoar and Randall 1969–1993; Tavolga et al. 1981.

7 Homeostasis

In this final chapter on fish physiology, we explore those processes that serve to maintain internal equilibrium, or homeostasis, thereby allowing other physiological systems to function properly. Specifically, we will investigate 1) the roles of the endocrine system and the autonomic nervous system in controlling various physiological responses; 2) the importance of body temperature and of thermal relationships between fishes and their environments; 3) the mechanisms involved in maintaining water, solute, and pH balance; 4) how fish immune systems defend the body against invasion; and 5) how various forms of physiological stress can compromise a fish's ability to maintain an internal steady state.

THE ENDOCRINE SYSTEM

Together with the nervous system, the endocrine system maintains communication among the various tissues in the body and regulates many physiological functions. Neural circuitry and the speed of action potentials make the nervous system rather direct and fast-acting. The endocrine system, however, is better suited for long-term regulation of physiological processes because its tissues release chemical signals (**hormones**) into the blood. These hormones travel throughout the body but only affect those cells with the proper molecular receptors.

Despite this apparent division of labor, there is considerable overlap between the nervous and endocrine systems—particularly in the control of various endocrine tissues by the brain. As endocrinological research on fishes and other animals advances, it has become increasingly difficult in some cases to distinguish separate roles for these two regulatory systems.

An Overview of Fish Endocrine Systems

Hormone function has been more thoroughly studied in mammals than in fishes. The following section, based on Wendelaar Bonga (1993), is an overview of the function of major fish hormones that have fairly well-defined functions.

The **pituitary** has often been viewed as the "master" gland of vertebrate endocrine systems. However, the **hypothalamus** of the brain controls the pituitary, and therefore ultimate control of much of endocrine function lies in the brain. The pituitaries of all jawed vertebrates have three functional regions. The **neurohypophysis** is continuous with the hypothalamus and is one area of overlap between the nervous and endocrine systems. This region of the pituitary consists primarily of the axons and terminals of neurons that originate in the hypothalamus. The **pars intermedia** and the **adenohypophysis** lie in contact with the neurohypophysis. In teleosts, extensions of the neurohypophysis penetrate into these two other regions, thereby establishing a physical link between the hypothalamus and both the pars intermedia and adenohypophysis (Wendelaar Bonga 1993). Chondrichthyans also show this intimate contact between the neurohypophysis and the adenohypophysis but only limited contact between the neurohypophysis and the pars intermedia. Agnathans apparently lack a direct link between the neurohypophysis and the rest of the pituitary; hagfishes lack a pars intermedia altogether.

The neurohypophysis is primarily the storage and release site of chemical messengers of the hypothalamus. **Neuroendocrine cells**, so named because they are neurons that function as endocrine cells, begin in the hypothalamus but extend into the neurohypophysis, where they release their hormones. These hormones are contained within vesicles, as is true of neurotransmitters of typical neurons, and the hormone is released by exocytosis when an action potential reaches the end of the axon.

Some hormones are released by these neuroendocrine cells into blood vessels and trigger effects elsewhere in the body. **Arginine vasotocin** and **isotocin**, for example, seem to be involved in protecting fishes from excess water loss in high-salt environments. Other chemicals released by the neurohypophysis regulate the function of cells of the adjacent adenohypophysis and pars intermedia. These are sometimes referred to as **releasing factors** or **releasing hormones**. Some of these chemicals diffuse to the intended target cells in immediately adjacent sections of the pituitary, whereas others travel the short distance to their target cells via blood vessels.

The adenohypophysis, largely under the control of the hypothalamus, manufactures and releases hormones that control many physiological functions elsewhere in the body, including many other endocrine tissues. **Prolactin** assists with osmoregulation in freshwater through its effects on the gills, skin, intestine, kidney, and bladder.

Growth hormone stimulates growth throughout the body and assists in the physiological adaptations of the gills necessary for salmonids to migrate from freshwater to the sea, including the promotion of chloride cell development and function (discussed later in this chapter). **Adrenocorticotrophic hormone** (ACTH) from the adenohypophysis regulates the synthesis and release of other hormones (corticosteroids) by the cells of the interrenal tissue. **Melanophore-stimulating hormone** (MSH), also from the adenohypophysis, is partly responsible for melanin production and color changes in some fishes.

Fishes are the only jawed vertebrates known to possess a **caudal neuroendocrine system**. Located at the caudal end of the spinal cord, this region of neuroendocrine cells, the **urophysis**, produces two hormones. **Urotensin I** stimulates the production of steroid hormones by the interrenal tissue. **Urotensin II** promotes the uptake of ions by the intestine when a fish is in seawater, and stimulates contraction of the heart, intestine, gonadal ducts, and urinary bladder.

The **thyroid** tissue of most fishes is scattered as small clusters of cells in the connective tissue of the throat region, as opposed to the rather discrete gland found in tetrapods. These cells produce **thyroxin**, which plays an important role in growth, development, and metabolism in many fishes. Thyroxin is quite important in helping to bring about the sometimes extreme morphological and physiological changes associated with metamorphosis (see Chap. 10, Transitions and Transitional Stages). For example, thyroxin promotes the transformation of flounder from "normal" fish larvae to flatfish with both eyes on one side of the head. It also initiates seaward migratory behavior and the accompanying osmoregulatory adaptations of juvenile salmonids, and the somewhat similar physiological changes seen in eels during their seaward spawning migration.

The **interrenal** tissues of fishes are homologous with the distinct adrenal glands of the tetrapods. The interrenal consists of two different types of cells, each of which produces different hormones. The **chromaffin cells** are located in the wall of the posterior cardinal vein in the pronephros of agnathans and the anterior, or head, kidney of dipnoans and teleosts. Chromaffin cells also are found in the heart of agnathans and the heart and intercostal artery of dipnoans. Elasmobranchs seem to have chromaffin cells limited to the axillary bodies of the autonomic nervous system.

Chromaffin cells produce and release the catecholamines **epinephrine** and **norepinephrine** under the influence of both the nervous system and hormones from other endocrine tissues. These hormones, which are associated with the mammalian "fight-or-flight" response, have similar strong effects on fish energy balance by increasing glucose production and release in the liver. The catecholamines enhance the delivery of oxygen to body tissues by increasing gill ventilation rates, and they increase the intracellular pH of red blood cells, thereby enhancing oxygen transport ability (see Chap. 5, The Bohr Effect). Catecholamines also increase blood perfusion of gill lamellae, thereby improving oxygen uptake. This increased blood flow to the gills also leads to increased ion exchange, which, along with the effect of cortisol described below, explains why stressed fishes may experience significant osmoregulatory imbalances (discussed later in this chapter).

The second group of interrenal cells are the **steroid-producing cells**, located primarily in the pronephric or head kidney region. These manufacture and release **corticosteroids**, including **cortisol**, which is important in energy metabolism because it stimulates the production and utilization of blood glucose for cellular energy. Cortisol is also important in osmoregulation and in responding to stress such as that brought on by handling or exposure to toxicants. The release of cortisol is probably associated with osmoregulatory imbalances and immune system depression brought on by stress.

Several other hormones are involved in osmoregulation in fishes. These include the **natriuretic peptides**, which are released by cells of the heart and aid in salt regulation, especially in saltwater fishes. **Angiotensin II**, triggered by the renin-angiotensin mechanism of the kidney (discussed later in this chapter), also is important. It initiates drinking in saltwater teleosts and helps fishes adapt to seawater by decreasing urine production, thereby limiting water loss. Angiotensin II also increases blood pressure throughout the body.

Unlike the terrestrial environment, most aquatic systems have abundant levels of calcium. To help prevent blood calcium levels from getting too high, calcium movement across fish gills is regulated by **stanniocalcin**, which is released by the **corpuscles of Stannius** in the kidney.

As in the tetrapods, glucose metabolism is strongly influenced by **insulin** from cells within the **pancreatic islets**. Insulin enhances the transport of glucose out of the blood and promotes glucose uptake by liver and muscle cells. It also stimulates the incorporation of amino acids into tissue proteins. **Glucagon** is produced by different cells within the pancreatic islets and seems to function in the breakdown of glycogen and lipids in the liver. It also promotes the transport of ions across the gills of marine teleosts.

Melatonin, produced by the **pineal** (near the top of the brain) and the retina of the eye, has a strong influence over many behaviors and physiological processes because it functions in the maintenance of circadian rhythmicity (see Chap. 22, Circadian Rhythms). This important hormone is secreted during the dark phase of daily light-dark cycles and helps regulate fish responses to daily and annual cycles of daylight. The pineal's endogenous circadian rhythmicity is adjusted by light input.

THE AUTONOMIC NERVOUS SYSTEM

Involuntary physiological functions, such as regulation of internal organ function, are at least in part controlled by the **autonomic nervous system** (ANS). Neural signals from the central nervous system (brain and spinal cord) travel to **ganglia** of the ANS, which are located either along the spinal cord or near or within the target organs (see Nilsson and Holmgren 1993). Signals then travel from these ganglia to the target tissues. Teleosts have well-developed chains of autonomic ganglia that run along either side of the spine in the anterior portion of the abdomen and into the head. In the posterior abdominal region, the two chains fuse to a single chain. The ganglia in the ANS are connected to each other, and they are connected to the central nervous system by cranial and spinal nerves. A particularly large ganglion (the coeliac ganglion) is located on the right side at the level of the third to fourth spinal nerves and is connected via nerves to the stomach, spleen, intestine, liver, and gas bladder.

The ANS is poorly developed in the agnathans, particularly in the hagfishes. Agnathans lack chains of autonomic ganglia, although ganglia have been identified elsewhere, such as in the dorsal aorta, which may be part of the ANS. Little attention has been given to the structure and function of the ANS in agnathans. Elasmobranchs have paravertebral ganglia that are connected in a somewhat irregular manner. The most anterior of these ganglia form the axillary bodies along with clusters of chromaffin cells in the pronephric region. An ANS ganglion also occurs in the head region of elasmobranchs (the ciliary ganglion) and sends nerves to the eyes.

In all major groups of fishes, heart rate is partly controlled by direct innervation from the ANS. In lampreys, elasmobranchs, and dipnoans, ANS stimuli help to decrease heart rate, whereas in most teleosts (except pleuronectid flatfishes) ANS influence can increase or decrease heart rate. The ANS can also indirectly affect heart rate through its stimulation of the release of catecholamines (epinephrine, norepinephrine) from the chromaffin cells. These endocrine cells are located in the posterior cardinal veins of the anterior kidney regions of hagfishes, lampreys, dipnoans, and teleosts, and in the axillary bodies of elasmobranchs. The ANS further affects fish cardiovascular systems through its control of vasoconstriction in both branchial (gill) and systemic vessels. This helps to regulate blood pressure, blood perfusion of the gills, and blood flow to the gut to promote digestion and absorption. Gas bladder volume, and therefore fish buoyancy, is regulated by ANS control of gas secretion and reabsorption by regulating blood flow to various parts of the gas bladder (see Chap. 5, Buoyancy Regulation, for a discussion of gas bladder function).

In addition to promoting digestion by increasing blood flow to the gut, the ANS also helps control the contraction of the smooth muscles of the gut wall, thereby helping food move through the gut. The dispersion and aggregation of pigment in melanophores are also partly controlled by the ANS, along with MSH from the anterior pituitary.

TEMPERATURE RELATIONSHIPS

Most people tend to think of fishes as "cold-blooded." This is an oversimplified view, however, and in some cases is simply not accurate. Most fishes are about the same temperature as the surrounding water. This temperature is not necessarily cold, as in the case of any fish in a warm, shallow pond or tropical lagoon. Similarly, a "warm-blooded" mammal such as an arctic ground squirrel is not warm when hibernating. For these reasons, the terms *warm-blooded* and *cold-blooded* are rather misleading when it comes to understanding the often complex thermal relationship between animals and their environments.

The terms *poikilotherm* (changing temperature) and *homeotherm* (stable temperature) also have been used to try to characterize animal thermal conditions. Here again, however, confusion can arise. The body temperatures of most fishes will change with their environment, but the ambient external temperature is often quite stable due to the thermal stability of water.

Perhaps a better way to approach this topic is to consider the source of an animal's body heat. Animals that rely primarily on external heat sources are referred to as **ectotherms**. These include most invertebrates, fishes, amphibians, and reptiles. **Endotherms**, however, generate and retain their own heat, as is the case of birds, mammals, and some interesting exceptions from the otherwise ectothermic groups, including certain fishes.

Ectotherms have considerably lower metabolic rates than endotherms, even at similar body temperatures. Therefore, they require less energy for basic maintenance and can survive on less food. Ectothermy may also leave more energy for other functions such as growth and reproduction because less energy is used for maintaining body temperature. At sufficiently low temperatures, metabolic rates of ectotherms become extremely low and growth is inhibited. Many ectotherms, therefore, cannot thrive in extremely cold climates, although there are some notable exceptions such as polar fishes and invertebrates (see Chap. 17, Polar Regions).

Many endotherms, such as birds and mammals, have high metabolic rates. The high energetic cost of maintaining basic metabolism requires a large food intake and leaves less energy for growth, reproduction, or other functions. The advantage to endothermy is that an elevated body temperature keeps thermally sensitive biochemical reactions operating efficiently, and the animal can utilize a wider range of thermal environments as long as sufficient food is available to maintain proper body temperature.

Endothermic Fishes

Some fishes maintain elevated body temperatures without the high energetic cost normally associated with high metabolic rates. They accomplish this by conserving heat generated by active swimming muscles. For example, some large pelagic predatory fishes such as tunas (Scombridae) and some mackerel sharks (Lamnidae) and thresher sharks (Alopiidae) maintain swimming muscle temperatures above ambient water temperature. Their internal temperatures remain fairly stable even as the fishes move from warm surface waters to colder deep water (Hazel 1993). Bluefin tuna keep muscle temperatures between 28 and 33°C while swimming through waters that range from 7 to 30°C (Carey and Lawson 1973). Yellowfin tuna maintain muscle temperatures at about 3°C above ambient water, whereas skipjack tuna keep muscles at about 4 to 7°C above ambient (Carey et al. 1971).

Such elevated muscle temperatures allow these fishes to maintain relatively high levels of swimming muscle activity while pursuing prey in colder waters, such as those found at greater depths and at nutrient-rich higher latitudes. The high level of muscle activity and greater efficiency at higher temperatures may result from higher power output of warmer muscles, an elevated metabolism of biochemical end-products such as lactate, and a temperature-induced Bohr effect (see Chap. 5, Box 5.2), which would assist in the unloading of oxygen from hemoglobin to the muscle cells (Hazel 1993).

These warm-bodied fishes conserve heat from muscular activity through adaptations of their circulatory systems. In a typical ectothermic fish, blood returns from the body to the heart and then travels to the gills for gas exchange (see Chap. 4). The large surface area and thin membranes of the gills permit heat to escape to the environment, so that when the blood leaves the gills it is the same temperature as the surrounding water. In a typical fish this blood would now travel down the core of the fish via the dorsal aorta, keeping the core body temperature about the same as the surrounding water (Fig. 7.1A). In the large tunas, however, the cool blood leaving the gills is mostly diverted to large peripheral vessels that run along the outside of the fish's body (Fig. 7.1B).

As the blood flows toward the large swimming muscles near the core of the body, it passes through a network of small blood vessels where it runs countercurrent to blood leaving these muscles, which has been warmed by the heat produced by muscle activity. This type of an arrange-

ment of blood vessels is referred to as a **rete mirabile** (wonderful net), as discussed in Chapter 5 for the gas bladder. The oxygenated blood is warmed as it passes through the rete and travels toward the swimming muscles. In this way, the heat generated by the activity of the large swimming muscles is kept within the muscles themselves and is not transported via the blood to the gills, where it would be lost to the surrounding water (Hazel 1993). Bigeye tuna can regulate their body temperature by utilizing the heat exchange mechanism only in colder water when it is needed (Holland et al. 1992).

In most fishes, the red muscle tissue responsible for sustained swimming is located laterally and just below the skin, where it readily loses heat to the water. In the tunas, however, red muscle is located more centrally, along the spinal column. This arrangement of the swimming muscles contributes to the unique and very efficient swimming style observed in the tunas (termed *thunniform*), in which the high, thin tail oscillates rapidly while the body remains rigid (see Chap. 8, Locomotory Types).

The evolution of thunniform swimming and the accompanying displacement of the red swimming muscles toward the body core put an insulating layer of less-vascularized white muscle between the heat-generating red muscle and the surrounding water (Block and Finnerty 1994). This muscle arrangement may have been a prerequisite for the development of the circulatory adaptations (retia) necessary to maintain elevated body temperatures (Block et al. 1993; Block and Finnerty 1994). Swordfish (Xiphiidae) also have their red swimming muscles more centrally located and also possess an associated heat exchanger (Carey 1990).

FIGURE 7.1. *(A) The circulatory system of a "typical" fish sends blood from the gills down the core of the fish, making it impossible to maintain an elevated core temperature in cold water. Arrows indicate direction of blood flow. (B) In the warm-bodied bluefin tuna (Thunnus thynnus), most blood from the gills is shunted toward cutaneous vessels near the body surface and is carried through a heat-exchanging rete en route to the warm swimming muscles.*

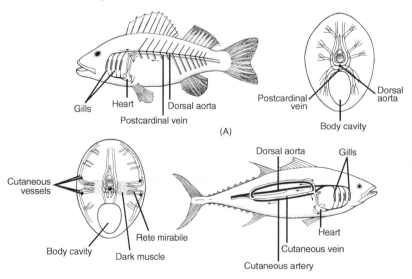

Figure from Carey 1973.

Smaller tunas also have retia for heat exchange, but they tend to be located more centrally, below the vertebral column (Stevens et al. 1974). Cool blood from the dorsal aorta is warmed as it passes through the rete and into the swimming muscles. It appears that this type of centrally located rete is found in smaller tunas that inhabit warmer oceans, whereas large tunas from colder regions have lateral retia, as shown in Figure 7.1B. Large mackerel sharks, such as the great white, mako, and porbeagle, maintain elevated visceral and body core temperatures with a heat-exchanging rete located anterior to the liver (Carey et al. 1981). Small retia have also been observed in the viscera and red muscle of two species of thresher sharks, suggesting that they too maintain elevated body temperatures by conserving heat generated by swimming muscles (Block and Finnerty 1994).

These countercurrent heat exchangers, combined with the heat produced by muscle activity and the large size of these fishes, make it possible for tunas and sharks to maintain relatively high muscle temperatures, which presumably enhances swimming performance. This presumption, though logical, has yet to be tested (Block and Finnerty 1994). The heat from the swimming muscles combined with another rete on the surface of the liver also maintains elevated gut temperatures (Hazel 1993), which may help to improve the efficiency of digestion. Some sharks and tunas, then, have found a way to take advantage of many of the benefits of endothermy by conserving and recirculating heat that would have been lost to the environment, thereby avoiding the additional metabolic costs of specialized thermogenic tissues.

Another use of endothermy in fishes is in warming parts of the central nervous system, especially the brain and eyes. This may be of even greater importance than maintaining elevated muscle temperatures, because heating the central nervous system permits utilization of deeper, colder habitats without suffering the negative effects that temperature fluctuation has on nervous system function (Block and Finnerty 1994). All endothermic fishes that have been studied warm some part of their central nervous system, suggesting that this may have been a strong factor in the evolution of fish endothermy.

Maintaining elevated temperatures only in certain parts of the body (**regional endothermy**) can be accomplished by the generation of heat by special thermogenic tissues and by circulatory adaptations that use blood warmed in other parts of the body. In "billfishes" (Xiphiidae, Istiophoridae), also pelagic predators, some eye muscles have lost the ability to contract and instead produce heat when stimulated by the nervous system (Box 7.1). In these fishes, the thermogenic organ seems to have developed for the particular purpose of generating heat for the brain and eyes. This is quite different from the maintenance of warm swimming muscles by tunas and sharks, which simply use circulatory adaptations to conserve the heat produced by muscle contraction. One member of the tuna family, the butterfly mackerel (*Gasterochisma melampus*), also utilizes thermogenic tissue to keep brain and eye temperatures elevated (Block et al. 1993). In this case, the thermogenic tissue is derived from a different eye muscle (the lateral rectus rather than the superior rectus), implying a distinctly separate evolutionary origin from that of

BOX 7.1 *Brain Heaters in "Billfishes"*

The largest, swiftest, widest-ranging teleosts are the marlins and sailfishes (Istiophoridae) and swordfish (Xiphiidae). These "billfishes" maintain elevated brain and eye temperatures, perhaps allowing them to hunt in cold water without experiencing a decrease in brain and visual function (Block et al. 1993). During development, one of the eye muscles (the superior rectus) develops the capability of generating heat without contracting. This is the result of a loss of the contractile filaments, which take up most of the volume of normal skeletal muscle cells, and a dramatic increase in the amount of mitochondria, which may take up as much as one-half to two-thirds of the cell volume of these specialized thermogenic cells. In addition, these modified cells have high levels of myoglobin, an oxygen-storing protein indicative of high metabolic activity. They also have an unusually large sarcoplasmic reticulum, the organelle responsible for calcium storage in skeletal muscles cells (see Chap. 4, Muscles).

It seems that the central nervous system stimulates these thermogenic cells in the same way that normal skeletal muscle cells become activated. The release of calcium from the sarcoplasmic reticulum does not, however, lead to contraction. Because there are no contractile proteins and associated calcium-binding proteins (see Chap. 4), this excess calcium is rapidly pumped back into the sarcoplasmic reticulum. Heat is released by the activity of these ion pumps (Fig. 7.2). In addition, the high levels of intracellular calcium may stimulate metabolic activity of mitochondria, resulting in additional heat production (Block 1991).

Interestingly, modified, noncontractile muscle cells also make up the electricity generating electroplaques of electric fishes (torpedo rays, knifefishes, electric eel, etc.) (see Chap. 6, Box 6.2). Hence two very different, specialized cell types—thermogenic and electrogenic—arise from alterations in developmental pathways associated with the basic muscle cell.

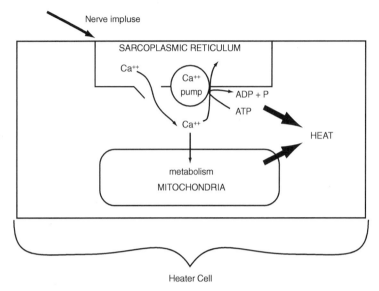

FIGURE 7.2. *Stimulation of the modified muscle cells of the billfish brain heater releases calcium from the sarcoplasmic reticulum (SR), which is then transported back across the SR membrane. This cycling of calcium at the membrane generates heat. It is speculated that the excess calcium may also stimulate mitochondrial metabolism, generating additional heat.*

Modified from Block 1991.

the "billfishes." Eye muscles of other tunas and the lamnid sharks do not appear to be modified as heater organs (Block and Finnerty 1994), but retia near the eyes apparently help maintain elevated eye and brain temperatures. The lamnid sharks also have a large vein that drains warm blood from the core swimming muscles to the spinal cord, thereby warming the central nervous system (Wolf et al. 1988).

Regional endothermy allows an animal to gain some of the advantages of endothermy, without the often higher energetic cost of maintaining an elevated overall metabolic rate.

Coping with Temperature Fluctuation

Two general terms are often used to describe physiological adjustments to environmental conditions. **Acclimatization** refers to adjustments made under natural environmental conditions, including seasonal changes in temperature, photoperiod, and associated hormones. However, fishes are frequently studied out of their natural environments and held under controlled laboratory conditions. Under these artificial conditions, the process of physiological adjustment, typically to a single parameter such as temperature, is referred to as **acclimation**. The absence of natural seasonal cues, such as changing photoperiod, may cause an artificially acclimated fish to respond somewhat differently than one that has been naturally acclimatized.

For example, laboratory-acclimated fishes typically have higher metabolic rates at higher temperatures (see Chap. 5), yet seasonal reproductive cycles cause naturally

acclimatized sunfish (Centrarchidae) to have higher metabolic rates in spring than in summer (Roberts 1964; Burns 1975). Other studies also have shown seasonal changes in metabolic rates that were independent of temperature in trout (Salmonidae) (Dickson and Kramer 1971), two minnows (Cyprinidae) (Facey and Grossman 1990), sunfish (Evans 1984), and sculpin (Cottidae) (Facey and Grossman 1990).

Fishes that experience changing environmental temperatures, such as those characteristic of diel or seasonal changes, have several cellular and subcellular mechanisms for adapting to the new set of conditions (see Hazel 1993). Many physiological adjustments are the result of switching on or off genes that are responsible for the manufacture of particular proteins. For example, acute heat stress initiates the synthesis of a group of proteins that have become known as **heat shock proteins**, or HSPs. These are believed to reconfigure proteins that become denatured at higher temperatures, thereby allowing them to function biochemically.

To compensate for the decreased rate of biochemical reactions at low temperatures, some fishes increase the concentration of intracellular enzymes. This can be accomplished by altering the rate of enzyme synthesis, degradation, or both. Increased cytochrome *c* concentration in green sunfish moved from 25°C to 5°C is due to a greater reduction in the degradation rate (60%) than in the rate of synthesis (40%) (Sidell 1977).

In some fishes, alternate enzymes (termed **isozymes**) may be produced to catalyze the same reaction more efficiently at different temperatures. Isozymes are regulated by switching on or off the different genes that control their production. Rainbow trout acclimated to 2°C versus 18°C exhibit different forms of acetylcholinesterase, an enzyme important to proper nerve function because it breaks down the neurotransmitter acetylcholine (Hochachka and Somero 1984). The ability of longjaw mudsuckers (Gobiidae) to tolerate rather wide ranges of temperatures is probably due to the fish's ability to regulate the ratio of isozymes of cytosolic malate dehydrogenase, an important enzyme in the Krebs cycle (Lin and Somero 1995).

Some fishes exhibit **allozymes**, alternate forms of the same enzyme that are controlled by different alleles of the same gene. Different populations of the species may exhibit higher or lower frequencies of the appropriate alleles depending on their geographic location. In one of the best-studied cases, livers of mummichog (*Fundulus heteroclitus*) along the east coast of the United States exhibited two allozymes of lactate dehydrogenase, an important enzyme in carbohydrate metabolism. In Maine populations, the frequency of the allele for the form more effective at colder temperatures is nearly 100%, whereas the frequency decreases progressively in populations farther to the south (Place and Powers 1979). In Florida, the alternate allele, which codes for the form more effective at higher temperatures, has a frequency approaching 100%.

Acclimation to cold temperatures includes modifications at the cellular and tissue levels also. Fishes, as well as other organisms, can alter the ratio of saturated and unsaturated fatty acids in their cell membranes in order to maintain some uniformity in membrane consistency (Hazel 1993). The proportion of unsaturated fatty acids, which are more fluid at colder temperatures (e.g., compare cold vegetable oil with butter), increases in those species that are active during winter. Other compensatory changes that accompany acclimation to lower temperatures include increases in the size of the heart and liver, and in the length of intestinal villi.

Decreased muscle performance at low temperatures can be compensated for at several levels of muscle function. Acclimation of striped bass (Moronidae) to low temperatures results in a substantial increase in the percent of red muscle cell volume occupied by mitochondria (Eggington and Sidell 1989) and an overall increase in the proportion of the trunk musculature occupied by red fibers (Jones and Sidell 1982). Muscle fibers of goldfish (Cyprinidae) show increased area of the sarcoplasmic reticulum at lower temperatures (Penney and Goldspink 1980). At colder temperatures, fishes may utilize more muscle fibers to swim at a particular speed than they use at warmer temperatures (Sidell and Moerland 1989). Because lower temperatures require the recruitment of more muscle fibers to sustain a given speed than is necessary at higher temperatures, maximum sustainable swimming velocities are lower at low temperatures (Rome 1990), suggesting another advantage of elevated body temperatures in the endothermic and predatory tunas and sharks.

Coping with Temperature Extremes

Extreme temperatures are dangerous to most living systems. Proteins, including the enzymes that catalyze critical biochemical reactions, are temperature sensitive. High temperatures may cause structural degradation (denaturation), resulting in partial or complete loss of function. Death can come quickly to a seriously overheated animal. Cold temperatures can slow critical biochemical reactions by reducing molecular movement and interaction.

Living in water generally protects fishes from extreme environmental temperatures. Nevertheless, even at moderately high temperatures, fishes encounter an additional problem associated with the aquatic environment—decreased oxygen availability due to limited gas solubility. When combined with elevated oxygen demand due to increased metabolic rate and a temperature-induced Bohr effect that interferes with hemoglobin function, high temperatures result in a physiologically stressful situation, as discussed in Chapter 5. Not surprisingly, few fishes tolerate high temperatures (see Chap. 17, Deserts and Other Seasonally Arid Habitats).

The physiological challenges of low temperature include compensating for the effects on cellular metabolism, nerve function, and cell membranes (Hazel 1993). Probably the greatest potential danger at very low temperatures is intracellular formation of ice crystals that can puncture cell membranes and organelles, leading to cell death. Intracellular ice formation also causes extreme osmotic stress, because as water freezes, solutes remain dissolved in a decreasing volume of cytoplasm, causing osmotic concentration to increase.

Freshwater fishes generally are protected from dangerously cold temperatures because freshwater freezes at 0°C but is densest at 4°C. Ice, therefore, forms on the surface of a lake or pond. Ions and other solutes depress the freezing point of intracellular fluid of most fishes to around −0.7°C, and freshwater fishes below the ice will not experience temperatures cold enough to freeze their body fluids. Hence freshwater fishes seldom need special physiological mechanisms to cope with potentially freezing conditions.

Marine fishes at high latitudes, however, are faced with different circumstances (see Chap. 17, Polar Regions). Seawater freezes at about −1.86°C, which is below the freezing point of the body fluids of most fishes. A marine fish could, therefore, find itself in a situation where the temperature of its environment is lower than the fish's freezing point—a potentially dangerous situation. Although some intertidal invertebrates and terrestrial vertebrates can tolerate internal ice formation, this is not a tactic utilized by fishes. Rather, fishes prevent ice formation through several different mechanisms.

One tactic involves a simple physical property of crystal formation. Crystals will not grow unless a "seed" crystal exists to which other molecules can adhere. Under controlled laboratory conditions, mummichogs were **supercooled** to about −3°C, well below their normal freezing point, without ice formation (Scholander et al. 1957). When touched with ice crystals, the mummichogs froze nearly instantaneously. This phenomenon of supercooling has been offered as an explanation of how some benthic fishes off the coast of Labrador survive in such a cold environment. The potential danger of contacting ice crystals is not a problem for the fishes off Labrador because they live on the bottom, where they are unlikely to encounter ice.

Polar fishes do come in direct contact with ice, however, and still do not freeze, indicating that they have developed physiological mechanisms to prevent internal ice formation (DeVries 1988). This protection generally involves the production of some type of biological antifreeze by the liver, a process controlled by genes that are activated by very low temperatures (Hazel 1993). Antifreeze compounds, usually glycoproteins, prevent ice crystals from growing once they begin to form and can bring the freezing point of some antarctic fishes, particularly the notothenioid ice fishes, down well below the freezing point of seawater (Raymond et al. 1989; Knight et al. 1991; Evans 1993).

The freezing point of body fluids can also be lowered by increasing the concentration of osmolytes (ions and other solutes). The higher the concentration, the lower the freezing point. Notothenioids do this and achieve a slight (tenths of a degree) lowering of the freezing point. Other fishes, such as rainbow smelt (Osmeridae) rely on increasing osmolytes and other mechanisms to lower their freezing points in seawater. Rainbow smelt utilize a combination of two ice prevention tactics. They have an antifreeze in their blood to help prevent ice crystal formation. At very low temperatures, however, this antifreeze apparently is not enough protection, so smelt produce glycerol to increase the osmotic concentration of the blood and intracellular fluids, thereby further decreasing the freezing point (Raymond 1992). At temperatures near the freezing point of seawater, the glycerol concentration is so high that the smelt are nearly isosmotic to the ocean. This increase in glycerol concentration is more apparent in the colder winter months and may account for the reported sweeter flavor of these fish during that time of year. Other fishes that live in areas that have markedly contrasting warm and cold seasons, such as tomcod (Gadidae), shorthorn sculpin (Cottidae), and winter flounder (Pleuronectidae), also exhibit increased levels of biological antifreezes during winter (DeVries 1974; Duman and DeVries 1974). Because glycerol and glycoprotein antifreezes are metabolically costly to produce, it makes sense to manufacture them only when needed.

Antarctic fishes of the suborder Notothenioidei must maintain year-round protection from freezing because their environment rarely gets above $-1.5°C$, even in summer. In most fishes, molecules as small as glycoprotein antifreezes would be lost in the urine. The fish would then need to produce more, at considerable energetic cost. The urine of notothenioids, however, does not contain these antifreezes because the kidneys of these fishes lack glomeruli, the small clusters of capillaries through which blood normally is filtered (kidney function, including aglomerular kidneys, is discussed later in this chapter).

Thermal Preference

The strong effect of temperature on biochemical and physiological processes drives fishes to select environmental temperatures at which they can function efficiently (Coutant 1987). Because different physiological processes may have different optimal temperatures, the temperature selected by a fish often represents a compromise, or "integrated optimum" (Kelsch and Neill 1990). Fishes probably select temperatures that maximize the amount of energy available for activity, or metabolic scope (the difference between standard and maximum metabolic rates, see Chap. 5; Fry 1971; Kelsch and Neill 1990). Of course, habitat selection in the wild involves a compromise between temperature requirements and other important factors, such as dissolved oxygen levels, food availability, and avoidance of predators and competitors (see Coutant 1987). Temperature is, however, a very strong determinant of habitat choice by some fishes.

Numerous laboratory investigations have shown most fishes select temperatures close to those to which they have become accustomed (see Kelsch and Neill 1990). There are a few exceptions, however. Chum salmon (Salmonidae) and blue tilapia (Cichlidae) show very narrow and constant temperature preferences regardless of acclimation temperature, and guppies (Poeciliidae) show a slight decline in preferred temperature with increased acclimation temperature (see Kelsch and Neill 1990).

The physiological ability to adapt to different temperatures to the point of shifting temperature preference may reflect the climate in which a species evolved (Kelsch and Neill 1990). Species that evolved in areas with substantial seasonal changes in temperature, such as the bluegill (Centrarchidae) of temperate North America, need the biochemical and physiological ability to shift temperature optima. More tropical species, such as guppies and tilapia, and cold-water fishes, such as salmonids, probably have not had to respond to selective pressures that would favor individuals that can make these kinds of adjustments.

Temperature preferences can change as fishes grow, so that different life stages of a given species utilize different thermal niches. For example, juvenile striped bass prefer temperatures around 25°C, whereas large adults will select cooler temperatures, around 20°C (Coutant 1985). This ontogenetic shift in temperature preference has important implications for the success of efforts to introduce this highly prized sport fish into various reservoirs and estuaries. A body of water that is ideal for the success and growth of young fish may be thermally unsuitable for large adults. These large "trophy" fish may congregate in small areas of slightly cooler water (often 18 to 20°C), such as near underground spring inputs or in the hypolimnetic waters of stratified lakes and reservoirs. Extreme crowding can lead to increased susceptibility to disease and overfishing. It can also lead to locally depleted food supplies and subsequent poor growth and reduced fecundity. The thermal preference may be so strong that starving fish will not leave cooler deep waters to feed on abundant prey in warmer surface waters (Coutant 1985).

Strong thermal preferences probably evolved to help fishes select habitats that offer them the best chances for growth and reproduction. However, this physiological constraint on habitat selection can become a liability, particularly in the face of human alterations of aquatic environments. In summer, the deep, cooler hypolimnion of warm reservoirs can be attractive to large striped bass. As summer progresses, however, these deep waters can become low in oxygen, leading to fish mortality. Coutant (1985) discusses evidence for and implications of this temperature-oxygen habitat squeeze on striped bass populations in various habitats, including freshwater and coastal systems.

Power plant cooling systems often discharge heated water into lakes and rivers, thereby altering their thermal structure. This can cause fishes to congregate in areas that may not be ultimately beneficial. For example, if the plant shuts down for a few days during the winter, fish that had become acclimated to the warmer water are suddenly left stranded in a cold environment and may die. Hydroelectric dams often release deeper, cooler water from an upstream reservoir. Fishes that congregate in these cooler hypolimnetic waters may be more susceptible, therefore, to being drawn through the turbines and injured or killed. The release of cooler water through a hydroelectric dam also can attract downstream fishes to the tailrace water during the warm summer months. The concentration of fish may create an attractive sport fishery, but it can also lead to overfishing and subsequent depletion of brood stock.

In some "pumpback" hydroelectric dams, large motors run turbines in reverse to push water back to the upstream side of the dam when power is not needed. When more electricity is needed, such as during periods of peak demand, this water is released again to generate electricity. The attraction of fishes to the foot of the dam during periods of power generation can set the stage for high fish mortality if those fishes are drawn through the turbines as water is pumped back to the upstream side of the dam.

The combination of cooler temperatures and high turbulence can cause water that is released to become supersaturated with gases, especially nitrogen and oxygen. The blood of fishes in this area may also become supersaturated because of gas diffusion across the highly permeable gill membrane. When these fishes move to warmer, less turbulent areas, the gases come out of solution and form bubbles in the blood. This **gas bubble disease** (similar to the bends in humans) can cause blocked and ruptured blood vessels, resulting in disorientation and death.

Thermal preferences also may cause fishes to congregate in areas with high levels of toxic pollutants, as has been reported for striped bass in the San Francisco Bay-Delta area. Uptake and bioaccumulation of some of these contaminants has been correlated with poor growth, high parasite loads, and decreased reproductive potential (Coutant 1985).

The impact of temperature preferences on fish habitat selection is a good example of the link between fish physiology and fish behavior and ecology. The effect of temperature preferences on the success of introduced striped bass also demonstrates the importance of basic physiological and behavioral information, as well as a thorough understanding of the habitat, when considering ecosystem manipulation or species introductions.

REGULATING IONS IN THE BLOOD AND BODY FLUIDS

Osmoregulation and Ion Balance

One of the most important homeostatic functions of living organisms is proper regulation of the internal osmotic environment. Deviation from the normal range can jeopardize proper physiological function through water loss or gain, the changing of internal ionic concentrations, and shifts in ionic and osmotic gradients.

Most fishes, like all other vertebrates, are **osmoregulators**. That is, they regulate their internal osmotic environment within a fairly narrow range that is suitable for proper cellular function, even if the external osmotic environment fluctuates. Fishes that can tolerate only small changes in the solute concentration of their external environment are referred to as **stenohaline**, whereas those with the ability to osmoregulate over a wide range of environmental salinities are **euryhaline**.

Freshwater teleosts are hyperosmotic to their environment (Table 7.1) and therefore tend to gain water and lose salts by diffusion across the thin membranes of the gills and pharynx (Fig. 7.3). Ions are also lost in the urine. If left unchecked, the fish's cells would swell and burst from the constant influx of water. To prevent this, freshwater fishes excrete a large volume of dilute urine and actively transport salts back into their blood. This is believed to take place at the **chloride cells** on the gill epithelium, although the mechanism for this transport is not clearly understood (Box 7.2).

Marine teleosts face the opposite problem (see Table 7.1). The high salt concentration of the ocean draws water out of the fish, and salts diffuse in across the permeable membranes. To counteract potential dehydration, marine teleosts drink seawater and actively excrete excess salts. Small, monovalent ions, such as sodium and chloride, are excreted by the chloride cells of the gills (see Box 7.2). Larger multivalent ions, especially magnesium and sulfate which are abundant in seawater, are excreted in the urine (see Fig. 7.3).

Nonteleost marine fishes have evolved other ways of coping with the marine environment. Hagfishes (Myxinidae) (see Chap. 13), the most primitive vertebrates, are **osmoconformers**, similar to many marine invertebrates. Their overall internal osmotic concentration is about the same as that of seawater (see Table 7.1). Because they live in fairly stable osmotic conditions near the bottom in relatively deep water, they do not have to contend with internal osmotic instability. Although the overall internal osmotic concentration of hagfishes is the same as the ocean, there are differences in the concentrations of some individual ions (see Table 7.1). There is no difference, however, in the concentration of the two major ions, sodium and chloride, giving hagfishes the highest concentrations of these physiologically important ions among the vertebrates.

Lampreys (Petromyzontidae), the other group of extant agnathans, are osmoregulators and appear to utilize osmoregulatory strategies very similar to those of teleosts

Table 7.1.   Plasma ionic concentrations (in milliosmoles per liter) of seawater, of freshwater, and of various fishes.

	Na	Cl	K	Mg	Ca	SO4	Urea	TMAO	Total
Seawater	439	513	9.3	50	9.6	26	0	0	1050
hagfish (*Myxine*)	486	508	8.2	12	5.1	3.0	—	—	1035
lamprey (*Petromyzon*)	156	159	32	7.0	3.5	—	—	—	333
shark*	255	241	6.0	3.0	5.0	0.5	441	72	1118
teleost (*Lophius*)	180	196	5.1	2.5	2.8	2.7	—	—	452
euryhaline teleost (*Pleuronectes*)	142	168	3.4	—	3.3	—	—	—	297
Freshwater (soft)	0.25	0.23	0.005	0.04	0.07	0.05	—	—	1
lamprey (*Lampetra*)	120	104	3.9	2.0	2.5	—	—	—	272
stingray (*Potamotrygon*)	150	150	—	—	—	—	1.3	—	308
teleost (*Cyprinus*)	130	125	2.9	1.2	2.1	—	—	—	274
euryhaline teleost (*Pleuronectes*)	124	132	2.9	—	2.7	—	—	—	240

*Na, Cl, urea, total from *Scyliorhinus canicula*; other data from *Squalus acanthias*.
From Evans 1993.

(Evans 1993), indicating convergence in osmoregulatory strategies between these distantly related fishes.

Marine elasmobranchs (see Chap. 12) have evolved another way of preventing osmotic stress in a salty environment. They maintain high levels of **urea** in their blood, as well as elevated levels of **trimethylamine oxide** (TMAO), which helps to stabilize proteins against the denaturing effect of urea. Urea and TMAO, along with somewhat elevated levels of sodium and chloride (as compared to teleosts), give elasmobranch blood an osmotic concentration even higher than that of seawater (see Table 7.1). Hence elasmobranchs actually gain water by diffusion across their gills. The few fresh- or brackish

water elasmobranchs maintain reduced levels of plasma urea, sodium, and chloride (Evans 1993).

Marine elasmobranchs rid themselves of excess sodium and chloride by active secretion via the rectal gland, which lies just anterior to the cloaca. Secretory tubules of the gland are lined with salt-secreting cells that are similar structurally and biochemically to the chloride cells of fish gills (see Box 7.2). The rectal gland produces a solution that has a higher NaCl concentration than the surrounding seawater, although it is isosmotic to the plasma. The NaCl solution drains into ducts leading to the lower intestine and is eliminated with other wastes.

FIGURE 7.3.   *Maintaining osmotic balance in fresh- versus seawater. (A) Freshwater teleosts must produce a large volume of dilute urine to offset the passive uptake of water across their gills. They must also actively transport ions into the blood at the gills to compensate for the loss of these ions to the dilute freshwater environment. (B) Marine teleosts passively lose water to their environment and gain salts by diffusion across their gills. They must, therefore, take in water through their food and by drinking seawater. Monovalent ions are actively transported out of the blood at the gills. Magnesium and sulfate ions, which are abundant in seawater, are excreted in the urine. Marine fishes conserve water by producing urine that is isosmotic to their blood.*

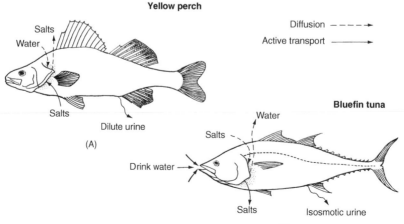

The elasmobranch strategy of offsetting high external ionic concentrations with elevated levels of organic osmolytes such as urea is also used by the ancient sarcopterygian, the coelacanth (Coelacanthidae), as well as the crab-eating frog (Ranidae), a marine amphibian (Withers 1992). The coelacanth also uses a rectal gland to secrete excess NaCl. The phylogenetic distance of these groups indicates another example of convergent evolution in the face of similar physiological challenges.

pH Balance

Like all animals, fishes must maintain blood and tissue pH within certain limits because many enzymes that control critical biochemical processes are pH sensitive. Low or high pH can alter the configuration of these molecules, inhibiting their function. Blood pH is largely affected by metabolic by-products such as CO_2, which forms carbonic acid (H_2CO_3) when in solution, and organic acids such as lactic acid from anaerobic metabolism. Terrestrial vertebrates primarily regulate pH through altering their respiration rate to regulate the amount of CO_2 in their blood. If CO_2 levels increase, blood pH decreases, and the animal responds by breathing faster, thereby eliminating more CO_2 and elevating pH. Conversely, if pH increases, a reduction in breathing rate will lead to more CO_2 being retained, thereby decreasing pH. However, this relatively simple mechanism does not work well for fishes.

Fish gills are quite permeable to CO_2, as discussed in Chapter 5. This high permeability creates a situation in which CO_2 levels in water are not much different than CO_2 levels in the fish's blood. In addition, CO_2 levels in water are often quite variable, depending on environmental conditions such as temperature. Fishes therefore cannot rely on altering their breathing rate either to increase or to decrease their blood CO_2 level to maintain fairly constant blood pH. Fishes instead rely on epithelial transport of ions that affect pH, such as H^+ and HCO_3^-. The primary responsibility for this seems to fall on the chloride cells of the gills, although transport across the skin and kidneys may play a minor role (Heisler 1993).

It was first suggested in the 1930s that ionoregulation and acid-base regulation took place simultaneously in the gills of fishes (Krogh 1939). It took nearly 30 years, however, for good evidence to begin to accumulate. Not only do the gills perform both functions, but it appears that in freshwater fishes the processes of pH balance and ion regulation are tightly linked, as ions that affect pH (H^+, HCO_3^-) are exchanged for important electrolytes (Na^+, Cl^-) by the chloride cells (see Stewart 1978, 1983; Cameron 1978, 1980; Wood 1991; Heisler 1993).

As an example, lowering the blood pH of arctic grayling (Salmonidae) increases Na^+ influx and decreases Cl^- influx at the gills (Cameron 1976). This is consistent with the fish's need to excrete more H^+, which is exchanged for Na^+, and a decreased need to eliminate HCO_3^-, which is exchanged for Cl^-. Conversely, raising blood pH results in increased Cl^- uptake, as HCO_3^- is eliminated to decrease pH. Chloride transport across isolated mummichog

opercular epithelium was enhanced by the addition of HCO_3^- to the solutions on both sides of the membrane (Zadunaisky 1984).

Studies of carp (Cyprinidae), however, suggested that Na^+ transport was not linked to the transepithelial movement of ions relevant to blood pH, although the movement of Cl^- across the gill epithelium was related (see Heisler 1993). It has also been suggested that some H^+ may be transported out of the cell and into the surrounding water without a direct link to other ions, with as much as 30% to 40% of acid or base excretion occurring independent of Na^+ or Cl^- transport (McDonald et al. 1991). This may be related to the relative permeability of the intercellular connections within the gill surface, which would allow the diffusion of H^+ from the gills in some species (McDonald et al. 1991).

As evidence of the apparent importance of chloride cells in pH regulation, chloride cells of rainbow trout showed increased surface area of their apical membranes and development of the tubular system from the basolateral membranes after as little as 2 days of exposure to conditions that induce a decrease of blood pH (acidosis) (Laurent and Perry 1991). Catfish (Ictaluridae) showed chloride cell proliferation during induced acidosis (Cameron and Iwama 1987).

There is still much to be learned about the role of chloride cells in regulating blood pH. For example, there is considerable controversy over whether H^+ or NH_4^+ is the ion primarily exchanged for Na^+. If NH_4^+ is significantly involved, then the issue of nitrogen excretion, addressed next, is added to the already complex issue of balancing pH and ionic regulation.

The efflux of ions that affect pH in exchange for an influx of Na^+ or Cl^- may not be the mechanism utilized by marine teleosts, as they must get rid of excess Na^+ and Cl^- that diffuses in from the environment. However, marine fishes certainly can eliminate excess H^+, as demonstrated in starry flounder (Pleuronectidae) and seawater-adapted coho salmon (Milligan et al. 1991). The mechanism of transport of ions that affect pH by marine fishes and its regulation have yet to be fully explained.

Excretion

Digestion and cellular metabolism create metabolic wastes that must be removed from the body. The breakdown of proteins presents a particular problem because the excess amine groups form **ammonia**, which is toxic at high concentrations. Fortunately for fishes, ammonia (NH_3) and its cation ammonium (NH_4^+) are soluble in water and can be lost by diffusion across the gills. Gills, therefore, are an important excretory organ for fishes. Some of the ammonia is converted to **urea** in the liver and either diffuses out at the gills or is excreted in the urine (Wood 1993).

In general, it appears that most teleosts rely on their gills to eliminate most nitrogenous wastes, predominantly in the form of ammonia (Table 7.2). Chondrichthyan fishes eliminate most of their nitrogenous wastes via the gills as urea. Diffusion of these wastes across the gills does, however, require immersion in water. Fishes

BOX 7.2 *Chloride Cells*

Over 50 years ago, large, ovoid, "chloride-secreting" cells were described on the gills of teleosts (Keys and Willmer 1932). These cells came to be known generally as chloride cells, but many years of study have demonstrated that they are responsible for more than chloride transport (see Zadunaisky 1984; Karnaky 1986; Wood 1991). Bone and Marshall (1982) suggested that "ionocyte" is more reflective of their role in regulating levels of several important ions. The structure and function of these remarkable cells have been studied in some marine and freshwater fishes, and a rather fascinating, though still incomplete, understanding of their physiology has emerged.

Chloride cells are generally columnar in shape and span the width of the epithelial membrane, with their apical surface in contact with the surrounding water. They have an extensive system of tubules that is continuous with the cell membrane of the basal and lateral surfaces and has many sites for the active transport of ions across the membrane. In effect, therefore, the cells have a very large surface area, with pockets of extracellular fluid surrounding and penetrating well into the cell itself. Chloride cells also contain many mitochondria, indicating a high capacity for metabolic activity (Fig. 7.4).

Chloride cells are located on the gills of teleosts, especially near the bases of the secondary lamellae and on the epithelial lining of the operculum. They also have been reported on the skin of the jaw of the longjaw mudsucker (Gobiidae) (see Zadunaisky 1984; Karnaky 1986). In marine teleosts, chloride cells occur together with **accessory cells** that interdigitate with the chloride cell. The chloride and accessory cells are joined by rather leaky connections that permit the diffusion of some ions across the membrane from the intercellular space to the surrounding water. The apex of the cells forms a mucus-filled pit in the epithelial surface (see Fig. 7.4).

The role of the chloride cells in osmoregulation of marine teleosts is understood largely through studies of the euryhaline mummichog (Fundulidae) (see Zadunaisky 1984; Karnaky 1986). Sodium is actively transported out of the cytoplasm into the extensive tubule system by sodium-potassium exchange pumps. This results in high Na^1 concentrations throughout the tubular and extracellular space and drives a secondary active transport system on the tubular membrane that transports Na^1 and Cl^2 from the extracellular/intratubular fluid into the cytoplasm.

The sodium is subsequently pumped out again by the sodium-potassium pump, but the chloride diffuses to the apical surface of the cell and through channels in the cell membrane into the apical pit (see Fig. 7.4). The high concentration of negatively charged chloride ions in the apical pit attracts positively charged sodium ions, which are drawn through the leaky connections between the chloride cells and their adjacent accessory cells. Chloride and sodium concentrations build up in the apical pit and subsequently diffuse into the surrounding seawater. Hence the marine fish has eliminated some of the excess sodium and chloride that it had gained by diffusion and ingestion.

The importance of chloride cells in the uptake of ions by freshwater fishes has only recently become more clear. Rainbow trout moved from freshwater to deionized water show proliferation of chloride cells on the gill filaments (Laurent and Dunel 1980), and brown trout residing in naturally ion-poor streams have more chloride cells than fish from streams with higher ionic concentrations (Laurent et al. 1985). Rainbow trout moved to NaCl-poor water show increased numbers of chloride cells and an increased surface area of chloride cell apical membranes (Perry and Laurent 1989).

---

that can survive out of water for extended periods convert ammonia to urea, which is less toxic and can be stored until the fish returns to the water. For example, African lungfishes (Protopteridae) produce ammonia when in the water but switch to urea production while estivating in a mud cocoon through long dry periods (see Chap. 13, Subclass Dipnoi: The Lungfishes; see Wood 1993).

The basic process of urine formation in most fishes is similar to that of other vertebrates. However, unlike most terrestrial vertebrates, fishes cannot produce urine that is more concentrated than their blood. In the kidneys, blood pressure forces water and small ions across the walls of small capillary beds, called **glomeruli**, and into the surrounding **Bowman's capsules**, which are the beginning of the kidney tubules (**nephrons**) (Fig. 7.5). As the filtrate travels along the nephron, water and important solutes are removed and added back to the blood. Waste products, excess ions, and other molecules that were not contained in the initial filtrate are added to the urine for elimination from the body. Urine drains from the nephrons into **collecting ducts**, and then to the **bladder**, where it may be held prior to being excreted. The urinary bladder may play an important role in salt and water balance by removing salts from the urine of freshwater

BOX 7.2 *(Continued)*

Although the precise mechanism of how chloride cells transport sodium and chloride into freshwater fishes has not been identified, changes in chloride cell ultrastructure suggest that transport sites are inserted into the apical membrane by the fusion of vesicles originating from the Golgi system (Laurent and Perry 1991). Although chloride cells of freshwater salmonids are similar to those of marine fishes, they generally occur singly, without the accompanying accessory cells seen in saltwater fishes.

These recent studies clearly indicate that chloride cells are important in regulating salt balance in freshwater salmonids. This is probably true of other freshwater fishes as well, but it should be noted that trout can adapt to either fresh- or salt water and may not be representative of stenohaline freshwater species.

Some interesting studies of euryhaline and diadromous fishes have shown how chloride cells change as fishes move from freshwater to the sea, or vice versa. Fishes moved from freshwater to seawater develop more chloride cell–accessory cell complexes on their gill and opercular epithelia, whereas movement from seawater to freshwater results in degeneration of these complexes (see Zadunaisky 1984; Laurent and Perry 1991). This physiological adaptation can take 24 to 48 hours and has been demonstrated in rainbow trout, guppies, and tilapia. In addition, development of Atlantic salmon and coho salmon from the freshwater parr stage to the seaward-migrating smolt stage is marked by increased numbers of large chloride cells with extensive tubular systems and accessory cells (Laurent and Perry 1991).

Although the fishes mentioned above can adapt to either fresh- or salt water, this adaptation must be a gradual process. Killifishes, however, are rather remarkable in their ability to adapt to rapid changes in salinity. Mummichog, for example,

maintain chloride cell–accessory cell complexes even in low salinities. Evidently they are able to survive rapid changes by regulating the ion permeability of the gill epithelium by altering the thickness of the connection between the chloride and accessory cells (Karnaky 1986). The connection is quite thick (less permeable) in low salinities, limiting ion loss by diffusion, and much thinner (more permeable) in higher salinities, permitting ion excretion. The ability to adapt to rather rapid changes in salinity is beneficial to estuarine fishes, such as the mummichog, because they live in habitats that undergo dramatic changes of salinity as ocean tides rise and fall.

As if the mechanisms of regulating salt balance are not complex enough, it has become clear that chloride cells also perform another important physiological role that entails ion transport—that of regulating blood pH. In fact, chloride cell transport of sodium and chloride appears to be tightly linked to transport of ions important to pH maintenance, such as H^l and HCO_3^-, in some fishes (see pH Balance, earlier in this chapter).

Neural and endocrine control of ion and pH regulation by chloride cells is not well understood, although it appears that the hormones prolactin and cortisol may be important in regulating chloride cell function (Zadunaisky 1984; Wood 1991; Wendelaar Bonga 1993; see Control of Osmoregulation and Excretion, later in this chapter). And although some work has been done, much has yet to be learned regarding the effects of environmental contaminants or pH disturbances on a fish's ability to regulate ion flux and internal pH. Any factors that affect gill morphology, mucus secretion, or the membrane-based ion pumps could interfere with ionic and acid-base regulation. The effects of various environmental factors on gill function is an active area of research (e.g., see *Physiological Zoology*, vol. 64 [1], January/February 1991).

---

fishes and removing water from and adding salts to the urine of saltwater fishes (see Evans 1993).

Different groups of fishes differ in the complexity of their kidney tubules (see Fig. 7.5). Hagfishes have glomeruli, but their blood pressure seems too low to force filtration into the nephron (see Evans 1993). Glomerular filtrate does form and drains almost directly into paired ducts. It is unclear whether the levels of some solutes are modified as the filtrate moves along the ducts and is excreted. Much remains to be learned about the function of hagfish kidneys. The lamprey kidney is more complex, with each tubule having several distinct segments. Most

water reabsorption takes place across the membrane of the final segment, the collecting duct. Estuarine and anadromous lampreys show decreased levels of renal filtrate formation and increased levels of water reabsorption when in brackish or salt water (Evans 1993).

Elasmobranchs have very complex kidney tubules, each being subdivided into several distinct functional regions. Elasmobranchs form a relatively large volume of renal filtrate because their high internal osmotic concentration allows them to absorb water even in salty environments. Active transport of salts into and out of the tubule in the different segments ensures that the fish excretes

Table 7.2.   Percent of nitrogenous wastes eliminated as ammonia nitrogen (amm) and urea nitrogen (urea) through gills and kidney of various fishes.

Fishes	Medium[a]	Gill		Kidney		Reference[d]
		Amm	Urea	Amm	Urea	
Agnatha						
lamprey (*Lampetra*)[b]	FW	95	0	4	1	(1)
Chondrichthyes						
dogfish (*Squalus*)[b]	SW	2	91	0	7	(3)
sawfish (*Pristis*)[b]	FW	18	55	2	25	(2)
Bony fishes						
carp (*Cyprinus*)[b]	FW	82	8	10	0	(4)
goldfish (*Carassius*)[c]	FW	79	13	7	1	(4)
catfish (*Heteropneustes*)[b]	FW	85	11	0	4	(7)
trout (*Oncorhynchus*)[b]	FW	86	11	1	2	(5)
cichlid (*Oreochromis*)[c]	FW	61	25	0	14	(6)
trout (*Oncorhynchus*)[a]	10% SW	56	32	10	2	(8)
mudskipper (*Periophthalmus*)[c]	25% SW	47	23	13	17	(9)
goby (*Boleophthalmus*)[c]	25% SW	61	14	11	14	(9)
poacher (*Agonus*)[c]	SW	41	9	43	7	(6)
sculpin (*Taurulus*)[c]	SW	63	4	20	13	(6)
wrasse (*Crenilabrus*)[c]	SW	67	2	28	3	(6)
blenny (*Blennius*)[c]	SW	35	18	39	8	(6)

[a]FW = freshwater; SW = seawater.
[b]Kidney excretion measured by urinary catheter. Therefore, any excretion via skin or gut would be included in the "gill" component.
[c]Kidney excretion measured by placing the fish in a chamber with a watertight curtain separating anterior (head and gills) and posterior sections. Therefore, any excretion via skin or gut is mostly included in the "kidney" (posterior) component.
[d]References: (1) Read 1968; (2) Smith and Smith 1931; (3) C. M. Wood and P. A. Wright, unpublished data; (4) Smith 1920; (5) C. M. Wood 1993; (6) Sayer and Davenport 1987; (7) Saha et al. 1988; (8) Wright et al. 1992; (9) Morii et al. 1978.
From Wood 1993.

excess water and waste products while retaining essential osmolytes, including the urea that it needs to maintain its high internal osmotic concentration.

Teleosts have kidney tubules that are less complex than those of elasmobranchs. In freshwater teleosts, the glomeruli and different tubule segments are well developed and produce a fairly large volume of dilute urine as a result of the constant influx of water from the environment. Sodium and chloride are reabsorbed in the distal segment and collecting duct, and little is lost in the urine. Further reabsorption of important salts can take place in the urinary bladder before urine is excreted (Evans 1993).

The kidney tubules of many marine teleosts are considerably reduced compared to freshwater teleosts. Some species, such as the antarctic notothenioids, lack glomeruli and the first segment of the proximal tubule altogether, whereas others simply have fewer glomeruli. This results in a low rate of filtrate formation and a decreased overall urine output, thereby helping prevent excess water loss. The aglomerular notothenioids form urine by actively transporting salts into the kidney tubule; water then follows by passive diffusion. This means of urine formation prevents the loss of important molecules, such as biological antifreezes (see earlier section on coping with extreme temperatures).

Control of Osmoregulation and Excretion

Several hormones help control kidney function in fishes. **Urotensins**, produced by the caudal neurosecretory system of the jawed fishes, are especially important in regulating salt and water balance in marine fishes (Wendelaar Bonga 1993). (The caudal neurosecretory system is a region of neurosecretory cells located near the last few vertebrae of the spinal column.) Urotensin II promotes uptake of ions in the intestine, whereas urotensin I stimulates the production of steroid hormones, including **cortisol**, by the interrenal tissue. Cortisol has an important osmoregulatory role in both freshwater and marine teleosts through its control of the number and activity of the chloride cells of the gills. This is especially important in fishes that migrate between fresh- and salt water. Cortisol production is also stimulated by stress due to handling, low pH, and some environmental toxins, which is why stressors often disturb the osmoregulatory balance of fishes.

Prolactin from the posterior pituitary appears to be important in maintaining the relatively low salt and water permeability of kidney tubules, bladder, gills, and gut epithelium. This low water and ion permeability helps fishes produce dilute urine and conserve essential ions when in freshwater (Wendelaar Bonga 1993).

FIGURE 7.4.   *The extensive tubular system and high density of mitochondria within chloride cells provide the surface area and energy for the rapid transport of ions. In marine teleosts, the combined effect of Na-K exchange pumps and NaCl uptake pumps along the tubular system increases intracellular Cl^- concentration. Cl^- then diffuses through channels in the apical membrane and into an apical pit. The high Cl^- concentration in the apical pit draws Na^+ through the relatively leaky connection between the chloride cell and its accessory cell. Na^+ and Cl^- next diffuse from the pit into the surrounding seawater.*

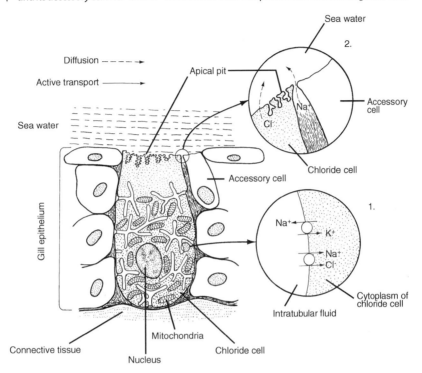

Catecholamines, such as epinephrine (adrenaline) and norepinephrine (noradrenaline), increase blood flow to the gill lamellae, thereby indirectly altering ion and water exchange across these highly permeable epithelial surfaces. Catecholamines are produced by the **chromaffin cells**, which are associated with the kidneys of jawed fishes, and have several effects associated with sudden increases in activity, including increased blood perfusion of the gills (Wendelaar Bonga 1993).

Natriuretic peptides, produced by cells of the heart, are another group of hormones controlling osmoregulation. These peptides appear to be more important in marine than in freshwater fishes, as they promote the secretion of salts by the chloride cells of the teleost gills and the rectal gland of elasmobranchs (Wendelaar Bonga 1993). Exposing euryhaline fishes to higher salinities increases levels of natriuretic peptides in the blood, whereas marine teleosts acclimated to lower salinities show reduced levels (Evans et al. 1989).

Osmoregulation and excretion are also regulated from within the kidney. Each individual nephron monitors its own filtrate and regulates the rate of filtrate production by controlling blood pressure within the glomerulus. This is accomplished by the **juxtaglomerular apparatus** (JGA). Some of these cells (the **macula densa**) monitor the sodium concentration in the distal segment of the nephron. Low sodium levels indicate that the filtration rate is low and filtrate is flowing slowly through the nephron. To compensate, other specialized cells (**juxtaglomerular cells**) lining the arteriole leading into the glomerulus release **renin**, an enzyme that triggers a series of biochemical reactions that result in an increase in plasma levels of **angiotensin II**. This hormone is a strong vasoconstrictor and brings about an increase in blood pressure, including the blood pressure in the glomerulus. The higher blood pressure results in an increase in the filtration rate and a subsequent increase in the rate at which the filtrate flows through the nephron. Angiotensin II also promotes drinking among marine teleosts (Hazon et al. 1989) and may have a role in elevating interrenal production of cortisol, another hormone with an important osmoregulatory role (see above). The renin-angiotensin mechanism apparently occurs in all jawed fishes (Wendelaar Bonga 1993).

THE IMMUNE SYSTEM

The immune system plays an important role in homeostasis by maintaining an individual's health in both non-

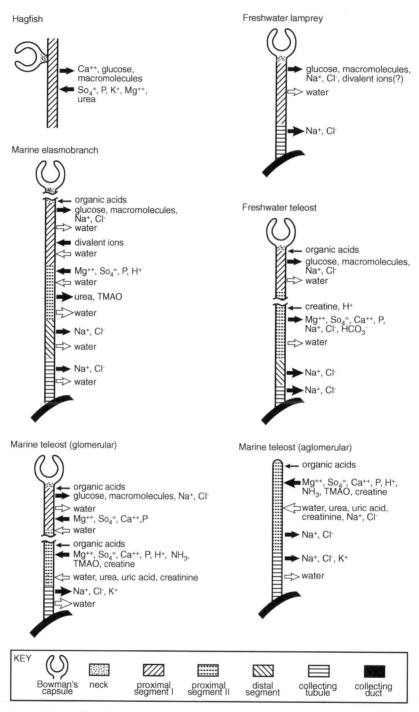

FIGURE 7.5.   *Schematic representation of the structure and function of nephrons of different fishes. Regions are named based on the structure of the teleost nephron; regions with similar function are named and shaded similarly in other fish groups. Open arrows indicate diffusion; closed arrows indicate active transport.*

Based on Hickman and Trump 1969.

specific and specific ways. **Nonspecific defenses** consist of immune factors that block invasion by potential pathogens. For example, the external layer of skin and scales is a physical barrier to infectious organisms. In addition, mucus secreted by fish epithelial cells is an effective non-specific defense. Its sticky, viscous consistency helps to trap microorganisms and inhibit their movement. Mucus also contains chemicals that inhibit the growth and effectiveness of prospective pathogens and that destroy bacteria (Ellis 1989; Anderson 1990). The volume of mucus secreted may increase in stressful situations, indicating a response on the fish's part to shield itself from potentially harmful chemicals, microorganisms, or other agents.

If pathogens breach this initial line of defense, a fish may develop a local inflammatory response that increases blood flow and delivers various types of disease-fighting white blood cells to the area (Roberts 1989). These include macrophages, which phagocytize invading cells, and natural cytotoxic cells, which destroy cells that have become infected with pathogens. Cytotoxic cells may also destroy abnormal cells that could become cancerous tumors with further growth.

The **specific response**, in contrast, involves the detection of an invader and the creation of specialized response mechanisms to identify and destroy it. The specific immune responses of fishes have not been well studied but appear to be similar to those of other vertebrates (Ellis 1989; Van Muiswinkel 1992). The organs primarily responsible for this response are the kidney, thymus, spleen, and intestine.

Invading compounds that initiate a specific response are called **antigens**. When an antigen invades the body, it is recognized as being foreign by two types of white blood cells (**B cells** and **T cells**) that have receptor proteins that bind foreign particles. If the antigen binds to a B cell, the B cell grows and divides, creating a clone of similar cells all capable of recognizing and binding to the activating antigen. In many cases, macrophages assist in the activation of B cells by ingesting pathogens, partially digesting them, and presenting antigenic components of the pathogen on the surface of the macrophage. Another type of white blood cell (**helper T cell**) binds to the presented antigen and also binds to and activates a B cell that has bound to another molecule of the same antigen.

Whether stimulated with or without the assistance of macrophages and helper T cells, the clone of activated B cells results in the production of plasma cells that produce **antibodies** that bind specifically to molecules structurally similar to the antigen that triggered B cell activation. These antibodies tag the antigenic particles for destruction by other components of the immune system—**macrophages**, which engulf and digest the tagged antigens, or complement proteins, which destroy tagged cells by puncturing their membranes.

The cloning process that yields antibody-producing plasma cells also produces **memory cells**, which remain in the bloodstream for extended periods. Memory cells help the animal's immune system react quickly if it encounters the same antigen in the future. Consequently, initial exposures to an antigen require longer for a response to develop and may lead to a greater degree of infection or disease, whereas subsequent exposures are dealt with quickly and the antigens are destroyed before they cause much trouble. Vaccinations, which have become important in fish culture, take advantage of memory cell development. By exposing fish to a less virulent form of a pathogen, the fish's immune system can defeat this initial infection and will retain memory cells to help it respond quickly and more effectively to subsequent exposures to a potentially more virulent form of the pathogen.

STRESS

The term *stress* has various meanings with regard to fishes (see Adams 1990). In a broad context, stress is any reaction to factors that drive physiological systems outside their normal range. Agents that cause stress include temperature, pollutants, pathogens, inadequate oxygen, and handling by humans.

Acute stress is often easy to detect because it rapidly exceeds the ability of organisms to cope and frequently results in death. Fish kills caused by oxygen depletion due to an algal bloom, or by poisoning due to chemical spills, result from acute stress. Sublethal or **chronic stress** is more common and more difficult to detect. Long-term exposure to low levels of stressors, either continuously or periodically, may not kill fish directly but can interfere with the long-term health, growth, and reproductive capabilities of individuals. This, in turn, can affect populations and assemblages.

Low pH, chlorinated hydrocarbons, and other chemical pollutants may impair reproduction by alteration of sexual maturation and spawning, diminished sperm motility, decreased egg survival and hatching, alterations in embryonic development, and decreased growth or increased mortality among larvae (Donaldson 1990). In some cases, the timing of exposure to the stressor is an important consideration. Exposure of salmonid and cyprinid eggs to pH 5 for only 30 minutes at the time of fertilization can be lethal (Gillet and Rombard 1986); such low pH levels are frequently encountered in acidified streams and lakes (see Chap. 25).

Chronic exposures may affect reproduction indirectly, such as reduced embryonic survival among eggs incubated at normal pH but taken from rainbow trout exposed to low pH (4.5 to 5.5) for 6 weeks prior to spawning (Weiner et al. 1986). Female Atlantic croaker (Sciaenidae) fed polychlorinated biphenyls (PCBs) have significantly smaller ovaries than control fish (Thomas 1988). White sucker (Catostomidae) from contaminated lakes have smaller eggs and reduced fecundity, and do not show an increase in fecundity with age when compared to suckers from a control lake (Munkittrick and Dixon 1988).

Environmental contaminants may alter and even reverse fish sexual differentiation by acting as endocrine disrupters (see Bortone and Davis 1994 for a review). For example, the combination of chemicals in pulp-mill effluent has been linked to alterations of hormone levels and maturity in white sucker and lake whitefish (Salmonidae), as well as precocious development of adult male traits in juvenile American eels (Anguillidae). In field and laboratory studies, female mosquitofish (Poeciliidae) exposed to pulp-mill effluent developed male characteristics, such as a gonopodium, the intromittent organ of males (Fig. 7.6). Other poeciliids, the least killifish (Cyprinodontidae) and the sailfin molly (Poeciliidae), also showed masculinized females in streams downstream from pulp-mill effluent discharge.

Several types of stress affect fish endocrine systems and result in suppression of immune systems, thereby decreasing a fish's ability to fight infections, diseases, and para-

FIGURE 7.6.   *(A) The anterior rays of the anal fin of a male mosquitofish (Gambusia affinis) are elongated to form the gonopodium, which is used to inseminate females. (B) The anal fin of a normal female mosquitofish does not have a gonopodium. (C) However, females can become masculinized by exposure to pulp-mill effluent and develop a gonopodium.*

Photographs by S. A. Bortone.

sites that normally might not present much of a physiological challenge (Anderson 1990; Dunier 1994). For example, exposure to noxious stimuli often results in increased levels of catecholamines (epinephrine, norepinephrine) and cortisol. Increased levels of these hormones help mobilize a fish's energy reserves in order to respond to the stressor. These endocrinological changes also can result in osmotic imbalances, decreases in white blood cell counts, and a diminished response to infectious agents (see Thomas 1990). If exposure to the noxious stimulus is acute, the effects often will persist for only a few days, and the fish may not experience long-term deleterious effects. Chronic exposure, however, can result in prolonged elevation of cortisol, which can increase susceptibility to disease by suppression of the immune system.

Mechanisms of immune suppression include interference with antibody production and inhibition of the proliferation of cells responsible for directly attacking pathogens. For example, macrophages from oyster toadfish (Batrachoididae), spot (Sciaenidae), and hogchoker (Soleidae) from an estuary contaminated with polycyclic aromatic hydrocarbons (PAHs) exhibited diminished immune function (Weeks et al. 1986; Seeley and Weeks-Perkins 1991). Juvenile chinook salmon injected with sublethal doses of PAHs or PCBs showed significantly diminished secondary immune responses (Arkoosh et al. 1994). Injection of PCB 126 at a dose of 1.0 mg/kg body weight suppressed the activity of nonspecific cytotoxic T

cells in channel catfish (Ictaluridae) (Rice and Schlenk 1995).

Several pesticides and metals also impede immune function in fishes (Dunier 1994). For example, atrazine has been linked to degeneration of macrophages in mullets (Mugilidae); mercury impairs the immune function of rainbow trout; and cadmium significantly alters macrophage-mediated immune functions in rainbow trout (Zelikoff 1994) and impairs antigen-binding function in bluegill (Centrarchidae) (Kusher and Crim 1991). Physical stress, such as that associated with handling by humans, can also hinder normal immune function, leaving fishes more susceptible to naturally occurring pathogens (Anderson 1990). Stress therefore can increase a fish's vulnerability to other harmful agents. And if fishes must expend more energy maintaining health, less energy will be available for growth and reproduction (see Chap. 5, Energy Budgets).

Indicators of Stress

Because chronic stress is not immediately lethal, it often goes undetected until its effects influence fish populations and community structure. In recent years, interest in early detection of stress in fishes has led to increased study of **biomarkers**, cellular and subcellular indicators of environmental stress. The principle behind the study of biomarkers is that stress can be detected at the subcellular and cellular level before it affects organismal or popula-

Macrophage Aggregates as Biomarkers

Macrophages are white blood cells that phagocytize and digest foreign cells as well as cellular debris from damaged cells. They are, therefore, important components of a fish's immune system. In some tissues, most notably the spleen, liver, and head kidney, macrophages cluster together in formations known as macrophage aggregates (MAs). Under standard histological preparation, with hematoxylin and eosin used as stains, MAs appear as yellowish masses, often with dark granules of the pigment melanin within the cells. For this reason, MAs are sometimes referred to as melanomacrophage centers (MMCs).

MAs are believed to be important as sites of storage of nondigested cellular debris, perhaps as a way to keep these materials isolated from other tissues. They also may act as "antigen traps" and be important in generating rapid and specific secondary immune responses (Ellis 1989).

The number and size of MAs seem to be affected by several factors, including fish size and age, disease, and nutrition (Blazer 1987; Blazer et al. 1994). Fishes exposed to environmental contaminants may show increased number and size of MAs. Winter flounder (Pleuronectidae) from a contaminated site had more and larger MAs in the liver and spleen than conspecifics from a clean reference site (Wolke et al. 1985). Similar patterns occurred in spleens of brown bullhead (Ictaluridae) (Blazer et al. 1994), whitemouth croaker (Sciaenidae) (Macchi et al. 1992), and rock bass (Centrarchidae) (Facey et al., unpublished data). Thermally stressed largemouth bass (Centrarchidae) also showed more numerous liver MAs than fish from a reference site (Blazer et al. 1987). It appears, therefore, that MAs are good biomarkers of various environmental stressors.

---

tion health (Adams 1990). Biomarkers, as well as biological indicators of stress at higher levels of biological organization, have been addressed in several recent books and symposia (e.g., see Adams 1990; McCarthy and Shugart 1990; Huggett et al. 1992; Peakall 1992; Couch and Fournie 1993).

Environmental stressors can result in the alteration of DNA and interfere with the molecular activity of some hormones (Thomas 1990). Exposure to many organic chemicals can result in increased levels of liver enzymes responsible for their detoxification and metabolism (see Jiminez and Stegeman 1990). The cytochrome P-450 enzymes, for example, seem to be induced by exposure to PAHs and PCBs. Exposure to high levels of heavy metals induces the production of metallothioneins, proteins responsible for binding these metals. Therefore, information about the levels and activities of these types of enzymes can provide information about the environment in which fishes live.

Chronic stress can result in a variety of changes in cellular and tissue morphology, particularly of the liver, gills, and spleen. The liver is the primary organ of detoxification, and benthic fishes in areas with contaminated sediments may show a greater incidence of liver tumors. Irregular nuclei, necrotic cells, and areas of cell regeneration also may be indicators of exposure to environmental contaminants. Liver abnormalities may subsequently affect other physiological functions due to the liver's importance in energy storage and metabolism, contaminant detoxification, and egg yolk production in females.

Gills are sensitive to various environmental stressors and are extremely important in gas exchange, excretion, osmoregulation, and acid-base balance. Acids, metals,

detergents, and other chemical contaminants may interfere with gill function in a number of ways. A fish's response to these factors may include the production of excess mucus and thickening of the gill epithelium, both of which impair gill function by inhibiting gas exchange and electrolyte and pH regulation (Hinton and Lauren 1990; Laurent and Perry 1991). Epithelial thickening can lead to fusion of secondary lamellae, thereby reducing gill surface area and further inhibiting respiration. Stress can also cause gill hemorrhage or separation of epithelial layers, which upsets osmoregulatory and pH balance by increasing ion permeability (see Laurent and Perry 1991). Even short-term stress from handling and confinement can alter gill permeability to water and sodium (McDonald et al. 1991). These effects on gill permeability require a fish to invest more energy in salt, water, and pH balance, thereby leaving less energy for other functions such as growth and reproduction (see Chap. 5, Energy Budgets).

The spleen also can indicate environmental stress because of its important role in fish immune systems, as indicated by the presence of **macrophage aggregates** (MAs) (Box 7.3). Although MAs are found in other tissues, such as the liver and head (anterior) kidney, they are especially numerous in the spleen. Splenic MAs tend to be larger and more numerous in fishes from contaminated areas than in similar-aged fishes from noncontaminated reference sites (see Box 7.3).

Through these and other mechanisms, it is becoming possible to detect stress from a variety of agents, thereby permitting early detection of potential impacts on fish physiology, health, growth, reproductive success, and community structure.

SUMMARY

1. Most long-term regulation of physiological processes in fishes is accomplished by the endocrine system. Many endocrine tissues are controlled by the pituitary, which is controlled by the hypothalamus of the brain. Physiological functions controlled by the endocrine system include osmoregulation, growth, metabolism, color changes, development and metamorphosis, and stress responses.

2. Involuntary physiological functions, such as heart rate, blood pressure, blood flow to the gills and gas bladder, and the contraction of the smooth muscles of the gut, are controlled by the autonomic nervous system.

3. Most fishes have body temperatures close to that of the water around them because of heat exchange at the gills. Some large pelagic predators, such as tunas and lamnid sharks, can maintain elevated body temperature by conserving the heat generated in active swimming muscles through countercurrent heat exchange. "Billfishes" use heat from special thermogenic tissue behind the eye to keep their eyes and brain warm while swimming in deep, cool water.

4. Seasonal changes in water temperature affect fish metabolism. Fishes can compensate for some change by altering the concentration or form of certain enzymes to maintain essential biochemical processes in cold conditions.

5. High water temperatures diminish the availability of oxygen in the water and can destroy physiologically important proteins such as hemoglobin and many enzymes. Hence few fishes can survive warm water temperatures. The temperature of seawater in polar regions drops below the freezing point of the blood of most fishes. To avoid freezing, many polar fishes rely on supercooling or biological antifreeze compounds.

6. The large surface area of the highly permeable gill membrane allows for considerable exchange of water and ions between a fish's blood and the surrounding water. To maintain a fairly stable internal osmotic condition, freshwater fishes produce dilute urine and take up ions through chloride cells in the gills. Saltwater fishes must drink seawater to replace water lost by diffusion. They also must eliminate excess ions through their kidneys and the chloride cells of the gill epithelium.

7. Most fishes eliminate nitrogenous wastes at their gills in the form of ammonia or ammonium. Fishes also produce urea, which is excreted in the urine. Kidney structure in fishes does not permit the concentration of urine to exceed the concentration of the blood plasma.

8. Osmoregulation in fishes is controlled by several hormones, including urotensins, cortisol, prolactin, and the catecholamines (epinephrine and norepinephrine). A renin-angiotensin mechanism, as described in mammals, also contributes to the regulation of osmoregulation by monitoring the concentration of urine as it is produced and influencing urine production by altering blood pressure.

9. A fish's immune system can also be seen as part of its attempt to maintain fairly stable internal conditions, because it acts to prevent the entry of pathogens or to destroy them if they do enter the body. The proper functioning of this system can be compromised by stress, such as that caused by handling or environmental factors, including certain contaminants. Stress from environmental factors can also result in a thicker mucus layer on a fish's gills, thereby inhibiting gas exchange. Some contaminants disrupt sexual differentiation because their structure mimics that of naturally occurring sex hormones.

SUPPLEMENTAL READING

Evans 1993; Hoar and Randall 1969–1995

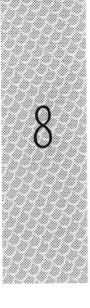

8

Functional Morphology of Locomotion and Feeding

"Structure without function is a corpse and function without structure is a ghost."

(Vogel and Wainwright 1969, 93)

Structure and function are inseparable. In the preceding five chapters, we have characterized the anatomy of fishes and described the function of various physiological systems. Such anatomic and physiological descriptions only make evolutionary sense when we understand their function, and function has not been ignored in the preceding introductory material. But structure-function relationships deserve more in-depth exploration. The study of how parts operate and how environmental selection pressures have influenced their construction and operation is variously referred to as functional morphology, physiological ecology, ecomorphology, and ecological physiology. These closely interrelated topics draw heavily on many disciplines besides anatomy and physiology, including physics, biomechanics, biochemistry, ultrastructure, structural engineering, developmental biology, population ecology, behavior, paleontology, and of course evolution.

Our goals in this chapter are to explore further the anatomic and physiological challenges that arise from living in water and to bring together and expand upon the subject matter introduced in the preceding chapters. We will focus on two general tasks in this chapter—locomotion and feeding—for examples of the intimacy and intricacy of structure and function; additional discussions that emphasize functional morphology can be found in several other chapters (e.g., see Chaps. 9 and 17 through 19). We can literally only skim the surface of this fascinating, interdisciplinary topic, and we strongly encourage interested readers to pursue the additional and more detailed information available in the cited references and suggested readings at the end of the chapter.

LOCOMOTION: MOVEMENT AND SHAPE

". . . [T]he gap between the swimming fish and the scientists is closing, but the fish is still well ahead."

(Lindsey 1978, 8)

Body shape and locomotory behavior in fishes are determined by the extreme density of water. Locomotory adaptations in terrestrial and flying animals strongly reflect a need to overcome gravity. In contrast, body and appendage shape in fishes reflects little influence of gravity, because gas bladders or lipid-containing structures make most fishes neutrally buoyant (see Chap. 5, Buoyancy Regulation). Fish locomotion is more constrained by the density of water and the drag exerted by it (Videler 1993).

Water is about 800 times more dense and 50 times more viscous than air. Locomotion through this dense, viscous medium is energetically expensive, a problem exacerbated by the 95% reduction in oxygen-carrying capacity of water as compared to air (see Chap. 5, Water as a Respiratory Environment). The chief cause of added energetic cost is drag, which has two components: **viscous** or **frictional drag**, involving friction between the fish's body and the surrounding water; and **inertial** or **pressure drag**, caused by pressure differences that result from displacement of water as the fish moves through it.

Viscous drag is not affected greatly by speed but more by the smoothness of a surface and by the amount of surface area, which is linked to body and fin shape; production of mucus reduces viscous drag. Inertial drag increases with speed and is therefore also intimately linked to body shape. Most fast-swimming fishes have a classic streamlined shape that minimizes both inertial and viscous drag. A streamlined body is round in cross section and has a maximum width equal to 25% of its length. The width-length ratio is 0.26 in some pelagic sharks, 0.24 in swordfish, and 0.28 in tunas. The thickest portion of a streamlined body occurs about two-fifths of the way back from the anterior end, another rule followed by large pelagic predators.

Interestingly, these same streamlined fishes are also slightly negatively buoyant and hence sink if they cease swimming. They often have winglike pectoral fins that are extended laterally at a positive attack angle, thus generating lift. They minimize drag by retracting paired and median fins into depressions or even grooves in the body surface; a sailfish houses its greatly expanded dorsal fin "sail" in a groove on its dorsal surface during fast swimming (Hertel 1966; Hildebrand 1982; Pough et al. 1984).

Most fishes swim by contracting a series of muscles on one side of the body and relaxing muscles on the other.

The muscle blocks, called myomeres, attach to collagenous septa, which in turn attach to the backbone and skin (Fig. 8.1). Depending on the swimming form involved (see below), contractions may progress from the head to the tail or occur on one side and then the other. The result of the contractions is that the fish's body segments push back on the water. Given Newton's third law of motion concerning equal and opposite forces, this pushing back produces an opposite reactive force that thrusts the fish forward. Forward thrust results from combined forces pushing forward and laterally; the lateral component is canceled by a rigid head, by median fins, and in some cases by a deep body that resists lateral displacement.

Locomotory Types

A general classification of swimming modes or types among fishes has been developed, building on the work of Breder (1926), Gray (1968), Lindsey (1978), and Webb (1984; Webb and Blake 1985). The chief characteristics of the different types are how much of and which parts of the body are involved in propulsion and whether the body or the fins undulate or oscillate. **Undulation** in-

volves sinusoidal waves passing down the body or a fin or fins; **oscillation** involves a structure that moves back and forth (Table 8.1). About 10 general types are recognized: anguilliform, subcarangiform, carangiform, modified carangiform (= thunniform), ostraciiform, balistiform, rajiform, amiiform, gymnotiform, and labriform; some of these are additionally subdivided. The names apply to the basic swimming mode of particular orders and families, although unrelated taxa may display the same mode, and many fish use different modes at different velocities.

The first four types involve sinusoidal undulations of the body. **Anguilliform** swimming—seen in most eels, dogfishes, other elongate sharks, and many larvae—occurs in fishes with very flexible bodies that are bent into at least one-half of a sine wave when photographed from the dorsal view (see Table 8.1). All but the head contributes to the propulsive force. As a wave proceeds posteriorly, it increases in amplitude. The speed (frequency) of the wave remains constant as it passes down the body and always exceeds the speed of forward movement of the fish because of drag and because of energy lost to reactive forces that are not directed forward (see above). To swim faster, faster waves must be produced. Anguilliform swim-

FIGURE 8.1.   *How fishes swim. (A) Lateral view of a spotted sea trout,* Cynoscion nebulosus, *with the skin dissected away to show the location of two myomeres on the left side. (B) The same myomeres as they appear relative to the backbone in a sea trout. The hatched region is the part of the myomere located closest to the skin; the dotted line shows the interior portion of the myomere where it attaches to the vertebral column. The anterior and posterior surface of each myomere is covered by a myoseptum made of collagen fiber in a gel matrix, shown as a slightly thickened line. (C) Cross section of a generalized teleost near the tail, showing the distribution of the various septa and their relationship to the backbone. Myosepta join to form median and horizontal septa. (D) How contractions produce swimming in a generalized fish. Progressive, tailward passage of a wave of contractions from the head to the tail* **push** *back on the water, generating* **forward thrust** *as one component of the* **reactive force**. *Sideways slippage (***lift***) is overcome by the inertia of the large surface area presented by fish's head and body.*

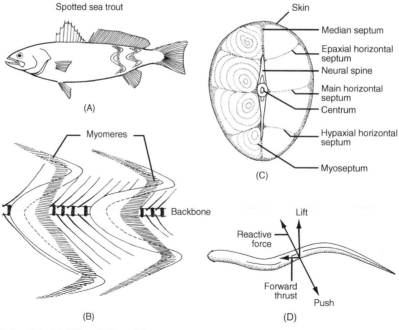

(C) From Wainwright 1983; used with permission.

Table 8.1.   Form, function, and locomotion in fishes. About 12 generalized types of swimming are recognized among fishes. The body part providing propulsion is indicated by cross-hatching; density of shading denotes relative contribution to propulsion. These locomotory patterns correlate strongly with body shape, habitat, feeding ecology, and social behavior. Convergence among unrelated fishes in terms of body morphology, swimming, and ecology demonstrates the evolutionary interplay of form and function. See Lindsey (1978), Beamish (1978), Webb and Blake (1985), Pough et al. (1984) for details.

	Swimming Type					
	Via Trunk and Tail				Via Fins	
		Via Tail				
	Anguilliform	Subcarangiform[a] Carangiform Thunniform	Ostraciiform	Tetraodontiform Balistiform Diodontiform	Rajiform[c] Amiiform Gymnotiform	Labriform[d]
Representative taxa	eels, some sharks, many larvae	salmon, jacks, mako shark, tuna	boxfish, momyrs, torpedo ray	triggerfish, ocean sunfish, porcupinefish	rays, bowfin, knifefishes	wrasses, surfperch
Propulsive force	most of body	posterior half of body	caudal region	median fin(s)	pectorals, median fins	pectoral fins
Propulsive form	undulation	undulation	oscillation	oscillation[b]	undulation	oscillation
Wavelength	0.5 to >1 wavelength	<1 (usually <0.5) wavelength			≫1 wavelength	
Maximum speed b.l./sec	slow to moderate 2	very fast to moderate 10–20	slow ?	slow ?	slow to moderate 0.5	slow 4
Body shape: lateral view cross section	elongate round	fusiform round	variable	variable often deep	elongate often flat	variable
Caudal fin aspect ratio	small medium to low	medium to large low to high	large low	small to medium low	variable low	large low
Habitat	benthic or suprabenthic	pelagic, w.c., schooling	variable	w.c.	suprabenthic	structure associated

b.l./sec - body lengths per second attainable; w.c. - up in water column.
[a]In subcarangiform types (salmons, cods), the posterior half of the body is used; carangiform swimmers (jacks, herrings) use the posterior third; and thunniform or modified carangiform swimmers (tunas, mako sharks) use mostly the caudal peduncle and tail (see text).
[b]Balistiform and diodontiform swimming is intermediate between oscillation and undulation; porcupinefishes use their pectoral fins.
[c]Rajiforms (skates, rays) swim with undulating pectoral fins; amiiforms (bowfin) undulate their dorsal fin; gymnotiform swimmers (South American knifefishes, featherfins) undulate their anal fin.
[d]Labriform swimmers use pectorals for slow swimming but use the subcarangiform or carangiform mode for fast swimming.
Line drawings from Lindsey 1978; used with permission.

mers are comparatively slow because of their relatively long bodies and the involvement of anterior regions in propulsion; the same segments that push back on the water waste energy by pushing laterally and also create drag, because water pushes on these bent sections as the fish moves forward.

Anguilliform swimming has its compensating advantages, including greater ability to move through dense vegetation and sediments and to swim backward. Anguilliform swimming in larval fishes, including species such as herrings that use carangiform swimming as adults, probably occurs because the skeleton of early larvae is unossified and the fish is exceedingly flexible and anatomically constrained from employing other modes (see Chap. 9, Larval Behavior and Physiology).

To get around the self-braking that occurs in anguilliform swimmers, faster-swimming fishes involve only posterior segments of the body in wave generation, using ligaments to transfer force from anterior body musculature to the caudal region. The progression of types from **subcarangiform** (trout, cod) through **carangiform** (jacks, herrings) to **modified carangiform** or **thunniform** (mackerel sharks, billfishes, tunas) entails increasing involvement of the tail and decreasing involvement of the anterior body in swimming.

One major advance in the carangiform and thunniform swimmers is the existence of a functional hinge that connects the tail to the caudal peduncle. This hinged coupling allows the fish to maintain its tail at an ideal attack angle of 10° to 20° through much of the power stroke. In anguilliform and subcarangiform swimmers, this angle changes constantly as the tail sweeps back and forth, producing less thrust at low angles and creating more drag at greater angles.

Thunniform swimmers also typically have a tail that originates from a narrow peduncle (= **narrow necking**) that is often dorsoventrally depressed and may even have lateral keels that streamline it during side-to-side motion. Narrow necking creates an overall more streamlined shape to the body and also reduces viscous drag and lateral resistance in that region of the body where they tend to be highest. The tail itself is stiff and sickle-shaped, being very narrow while quite tall. Such a large height-width ratio tail, referred to as a **high aspect ratio** tail, experiences minimal drag and is ideal for sustained swimming. The shape reduces viscous drag by reducing surface area and reduces inertial drag by having pointed tips that produce minimal vortices at their tips. The efficiency of the system is increased by tendons that run around joints in the peduncle region and insert on the tail, the joints serving as pulleys that increase the pulling power of the muscle-tendon network.

The thunniform mode of propulsion, involving a streamlined shape, narrow-necked and keeled peduncle, and high aspect ratio tail, has evolved convergently in several fast-swimming, pelagic predators, including mackerel sharks, tunas, and billfishes, as well as porpoises, dolphins, and the extinct reptilian ichthyosaurs. The fish and mammalian groups are also endothermic to some degree (Lighthill 1969; Lindsey 1978; Pough et al. 1989). Higher-speed, sustained swimming in the mackerel sharks and tunas is also made possible by the large masses of red muscle along the fish's sides (see Chap. 4, Red Muscle Versus White Muscle). Location of the red muscle close to the fish's spine allows the body to remain fairly rigid and also permits the retention of heat generated by muscle contraction. Hence thunniform swimming and endothermy are tightly linked.

Low aspect ratio, broad, flexible tails, such as those found in subcarangiform minnows, salmons, pikes, cods, and barracudas, are better suited for rapid acceleration from a dead start and can also aid during hovering by passing undulations down their posterior edge. Intrinsic muscles associated with the tail in low aspect ratio species help control its shape. Rainbow trout are able to increase the depth and hence produce a higher aspect ratio tail during high-speed swimming. **Fast-start** preda-

tors, such as gars, pikes, and barracudas, hover in the water column and then dart rapidly at prey. These unrelated fishes have converged on a body shape that concentrates the propulsive elements in the posterior portion of the body: Dorsal and anal fins are large and placed far to the posterior, the caudal peduncle is deep, and the tail has a relatively low aspect ratio. Maximum thrust from a high-amplitude wave concentrated in the tail region allows for rapid acceleration from a standing start (see Fig. 17.1).

Ostraciiform swimming, as seen in boxfishes and torpedo rays, is extreme in that only the tail is moved back and forth while the body is held rigid; the side-to-side movement of the tail is an oscillation more than an undulation. Boxfishes have a rigid dermal covering that extends back to the peduncle area. In the weakly electric elephantfishes, body muscles pull on tendons that run back around bones in the caudal peduncle region and insert on the tail, causing the fish to swim with jerky tail beats. Such an arrangement is thunniform in anatomy but more ostraciiform in function. Weakly electric fishes, such as the elephantfishes and South American knifefishes mentioned below, often have devices for keeping their bodies straight while swimming. This relative inflexibility probably minimizes distortion of the electrical field they create around themselves (see Chap. 6, Electroreception).

The last five swimming types employ median and paired fins rather than body-tail couplings. **Tetraodontiform** swimmers (triggerfishes, ocean sunfishes) flap their dorsal and anal fins synchronously; their narrow-based, long, pointed fins function like wings and generate lift (forward thrust) continuously, not just during half of each oscillation. **Rajiform** swimmers hover and move slowly via multiple undulations that pass backward or forward along the pectoral fins of skates and rays. In **amiiform** swimmers, undulations pass along the dorsal fin (bowfin, African osteoglossomorph *Gymnarchus*, seahorses), whereas in **gymnotiform** swimming, undulations pass along the anal fin (South American and African knifefishes or featherfins). Rajiform and related swimming modes are slow but allow for precise hovering, maneuvering, and backing. The frequency with which waves pass along a fin can be very high, reaching 70 Hz in the dorsal fin of seahorses.

Labriform swimmers (chimaeras, surfperches, wrasses, parrotfishes, surgeonfishes) row their pectoral fins, pushing back with the broad blade, then feathering it in the recovery phase. As some negative lift is generated during the recovery phase, these fish often give the impression of bouncing slightly as they move through the water. If rapid acceleration or sustained fast swimming is needed, labriform swimmers, as well as many other fin-based locomotors, shift to carangiform locomotion.

Three final aspects of locomotory types deserve mention. First, the distinctiveness of the different locomotory types suggests that they are specializations, and specialization for one function usually produces compromises in other functions. Fishes that specialize in efficient slow swimming or precise maneuvering usually employ undulating or oscillating median fins. The long fin bases necessary for such propulsion (e.g., bowfin, knifefishes, pipe-fishes, cutlassfishes) require a long body, which evolves at a cost in high-speed, steady swimming. Low-speed maneuverability can also be achieved with a highly compressed (laterally flattened), short body that facilitates pivoting, as found in many fishes that live in geometrically complex environments such as coral reefs or vegetation beds (e.g., freshwater sunfishes, angelfishes, butterflyfishes, cichlids, surfperches, rabbitfishes). These fishes typically have expanded median and paired fins that are distributed around the center of mass of the body and can act independently to achieve precise, transient thrusts, a useful ability when feeding on attached algae or on invertebrates that are hiding in cracks and crevices. But a short, compressed body means reduced muscle mass and poor streamlining, whereas large fins increase drag. Again, such fishes achieve maneuverability but sacrifice rapid starts and sustained cruising. Relatively poor fast-start performance may be compensated for by deep bodies and stiff spines, which make these fishes difficult to swallow (see Chap. 19, Discouraging Capture and Handling); they also typically live close to shelter. At the other extreme, thunniform swimmers have streamlined bodies, large anterior muscle masses, and stiff pectoral and caudal fins that are extremely hydrodynamic foils. They trade off exceptional cruising ability against an inability to maneuver at slow speeds.

Although specialists among body types can be identified, optimal design for one trait—sustained cruising, rapid acceleration, or maneuverability—tends to reduce ability in the other traits. Because most fishes must cruise to get from place to place and must accelerate and maneuver to eat and avoid being eaten, "the majority of fishes are locomotor generalists rather than locomotor specialists" (Webb 1984, 82).

Second, this generalist strategy means that few fishes use only one swimming mode. Many fishes switch between modes depending on whether fast or slow swimming or hovering is needed. In addition, most fishes have median fins that can be erected or depressed, adding a dynamic quality to their locomotion. A largemouth bass can erect its first dorsal and anal fins to gain thrust during a fast-start attack, then depress these fins to reduce drag while chasing a prey fish, then erect them to aid in rapid maneuvering. Most groups, with the exception of the thunniform swimmers, are capable of hovering in midwater by sculling with their pectoral fins or by passing waves vertically along the caudal fin. When hovering, some forward thrust is generated by water exhaled from the opercles; this force is countered by pectoral sculling. The fin movement involved in hovering may be difficult to detect, both by human observers and potential prey, because fishes that use these techniques often possess transparent pectoral fins.

Finally, note that not all fishes fit neatly into one of these categories (e.g., Box 8.1) and that many additional categories can and have been erected to accommodate variations among taxa (for more complete and alternative categorizations, see Lindsey 1978; Webb and Blake 1985; Videler 1993).

Specialized Locomotion

Certain highly derived forms of locomotion exist among fishes and do not fall into any of the general categories. A number of species walk along the bottom of the sea or leave the water and move about on land; these fishes have bodies that depart from a streamlined shape. Sea robins move lightly across sand bottoms using modified pectoral rays that extend out from the fin webs. They give the appearance of someone tiptoeing on many moving fingers. Antennarioid frogfishes and batfishes pull themselves along the bottom with movements of their modified pectoral and pelvic fins; their forward motion is aided by jet propulsion of water out their backward-facing, constricted opercles (Pietsch and Grobecker 1987).

Terrestrial locomotion is accomplished in a variety of ways. Climbing gouramies (Anabantidae) use paired fins and spiny gill covers to ratchet themselves along, whereas snakeheads (Channidae) row with their pectoral fins. So-called clariid "walking catfishes" move across land by lateral body flexion combined with pivoting on their stout, erect pectoral spines. Mudskippers (Gobiidae) swing their pectoral fins forward while supporting their body on the pelvic fins. They then push forward with the pectoral fins, like a person on crutches. Rapid leaps of 30 to 40 cm are accomplished by coordinated pushing of the tail and pectoral fins. Their unique pectoral fins are roughly convergent with the forelimbs of tetrapods, including an upper arm consisting of a rigid platelike region and a fanlike forearm and plantar surface (Gray 1968). Some species with anguilliform movement (moray and anguillid eels) are able to move across wet land employing their normal locomotion, which is analogous to the "serpentiform" terrestrial and aquatic movements of most snakes (Chave and Randall 1971; Lindsey 1978).

Aerial locomotion grades from occasional jumping to gliding to actual flapping flight. Many fishes jump to catch airborne prey (trout, largemouth bass); meter-long arowanas (Osteoglossidae) can leap more than a body length upward and pluck insects and larger prey from overhanging vegetation. Other fishes take advantage of the greater speeds achievable in air: Needlefishes, mackerels, and tunas leave the water in a flat trajectory when chasing prey, and salmon leap clear of the water when moving through rapids or up waterfalls. Hooked fish jump and simultaneously shake their heads from side to side in an attempt to throw the hook; such oscillation is less constrained by drag in air than in water and therefore allows more rapid and forceful to-and-fro movement. Prey such as minnows, halfbeaks, silversides, and mullets jump when being chased.

Fishes capable of flight include gliders such as the exocoetid flyingfishes and pantodontid butterflyfishes, as well as gasteropelicid hatchetfishes, which purportedly vibrate their pectoral wings to generate additional lift (Davenport 1994; see Chap. 19, Evading Pursuit). The anatomy of the marine flyingfishes is highly modified for flying. The body is almost rectangular in cross section, the flattened ventral side of the rectangle providing a planing surface that may aid during takeoff. The ventral lobe of the caudal fin is 10%

BOX
8.1

Swimming in Sharks: The Alternative Approach

Different fish lineages have evolved a variety of solutions to the challenges of locomotion in water. In the process, mutually exclusive specializations for cruising, rapid starts, or maneuverability have arisen (see above). The fossil record indicates that similar body morphologies and an apparent trend toward increasing concentration of activity in the tail region have appeared repeatedly during osteichthyan evolution (Webb 1982; see Chap. 11). These patterns and trends all capitalize on the substantial stresses that can be placed on a rigid, bony skeleton and the forces achievable by muscle masses attached directly or indirectly to bony structures. Elasmobranchs are phylogenetically constrained by a relatively flexible and comparatively soft cartilaginous skeleton. Evolution of locomotion in chondrichthyans has, not surprisingly, taken a different albeit parallel path.

Most elasmobranchs swim via undulations, either of the body (sharks) or of the pectoral fins (skates and rays). Most sharks swim using anguilliform locomotion, although the amplitude of each wave in the caudal region is greater in swimming sharks than in eels. This exaggerated sweep of the posterior region probably capitalizes on the increased thrust available from the large heterocercal tail of a shark. Exceptions to anguilliform swimming include the pelagic, predatory mackerel sharks, which have converged in body form and swimming type with tunas, dolphins, and ichthyosaurs (see above). Skates and rays also undulate, passing undulations posteriorly along their pectoral fins while the body is held relatively rigid. The exception in this group is the torpedo rays, which differ in that they have an expanded tail fin and swim via ostraciiform oscillations. In these strongly electrogenic rays, the pectoral region is unavailable for swimming because it is modified for generating electricity.

The mechanics of swimming in sharks are fascinating and somewhat controversial. Three topics have received the most attention, involving the functions of the median fins, skin, and tail during locomotion. Despite anguilliform locomotion, most sharks are active, cruising predators with relatively streamlined bodies. This would seem anomalous given the relatively low efficiency of the anguilliform mode and the apparent incompatibility of a fusiform body bent into long propulsive waves. However, sharks enhance the efficiency of their swimming mode in several ways.

Most sharks have two dorsal fins, the first usually larger than the second, separated by a considerable gap. The dorsal lobe of the pronounced heterocercal tail may be thought of as a third median fin in line with the dorsal fins, again separated from the second fin by a considerable gap. The distances between the three fins are apparently determined by the size of the fins, their shapes, and the waveform of swimming of the fish. Each fin tapers posteriorly, leaving behind it a wake as it moves through the water. This wake is displaced laterally by the sinusoidal waves passing down the fish, so the wake itself follows a sinusoidal path that moves posteriorly as the fish moves through the water. This wave is slightly out of phase with the fish's movements by a constant amount.

Calculations of the phase difference and wave nature of the wake suggest an ideal distance between fins that would maximize the thrust of the second dorsal and particularly of the tail. If timed correctly, the trailing fins can push against water coming toward them laterally from the leading fins. Such an interaction between flows would enhance the thrust produced by the trailing fin. Measurements of swimming motions and fin spacing in six species of sharks indicate just such an interaction (Webb and Keyes 1982; Webb 1984). Unlike bony fishes that use their median fins for acceleration and braking but fold them while cruising to reduce drag, sharks use their median fins as additional, interacting thrusters.

The energy provided with each propulsive wave of muscular contraction is additionally aided by an interaction between the skin and the body musculature of a shark. The skin includes an inner sheath of two layers of collagen fibers that are mechanically similar to tendons. The fibers in one layer lie at about a 60° angle to the fibers in the adjacent layer, thus creating a cylinder reinforced with a double helix of wound fibers, an exceptionally strong and incompressible—but readily bendable—structure (S. Wainwright 1988).

Inside the skin, hydrostatic pressure varies as a function of activity level. The faster the shark swims, the higher the internal hydrostatic pressure. Pressure during fast swimming is about 10 times what it is during slow swimming, ranging between 20 and 200 kiloPascals. Internal hydrostatic pres-

BOX
8.1   *(Continued)*

sure develops from unknown sources, probably due to changes in the surface area of contracting muscles relative to skin area and to changes in blood pressure in blood sinuses that are surrounded by muscle. The shark's body is therefore a pressurized cylinder with an elastic covering.

During swimming, the higher the internal pressure, the stiffer the skin becomes, which increases the energy stored in stretched skin. Body muscles attach via collagenous septa not just to the vertebral column but also to the inside of the skin (for this reason, it is exceptionally difficult to remove the skin from the muscle of a shark). As the muscles on the right side of the body contract, muscles and skin on the left side are stretched. The stretched skin is very elastic, but stretched muscle is less so. As muscles on the right side relax, the energy stored in the skin on the left side is released, aiding muscles on the left side at a point when they can provide relatively little tension. Therefore, the skin may act in initiating the pull of the tail across the midline and increase the power output at the beginning of the propulsive stroke.

The faster the shark swims, the greater the elastic recoil from the stretched skin. Muscles attach to the relatively narrow vertebral column of calcified cartilage but also attach to the much larger surface area of stiff, elastic skin that encompasses the shark from head to tail and in essence forms a large, cylindrical external tendon. Muscles pulling on the skin provide propulsive energy that probably exceeds the thrust derived from muscles attached to the vertebral column (Wainwright et al. 1978; Wainwright 1983).

Most of the power in shark swimming comes from the tail, but this tail is not symmetrical as it is in most bony fishes. The heterocercal tail, with its expanded upper lobe, would seem to provide a lifting force to the posterior end of the body. This lift should cause the body to rotate around its center of mass, plunging the anterior end in a perpetual dive (Fig. 8.2A).

One long-held explanation is that the flat underside of the head and particularly the broad stiff pectoral fins create lift at the anterior end to counteract the downward force. However, it seems inefficient for the tail and the pectoral fins to function against each other, the tail propelling and the pectoral fins continually braking the shark's progress. Given the 400-million-year success of elasmobranchs and the widespread occurrence of heterocercal tails in many previously speciose lineages of both bony and cartilaginous fishes, it is hard to imagine that heterocercal tails are inherently inefficient. This apparent dilemma has prompted an ongoing search for mechanisms that promote relatively straight forward propulsion.

The search has turned into something of a debate. The classic model, as described above, indicates that the tail pushes back and down on the water, creating an equal and opposite resulting force up and forward, which is counteracted by paired fins and head shape to produce forward movement. The classic model is supported by video analysis and dye-tracer studies (Ferry and Lauder 1996). These more recent findings refute earlier interpretations of photographs and selective amputation of fin parts of tails held in a test apparatus (Fig. 8.2B; Simons 1970; Thomson 1976, 1990), which suggested that forward thrust is generated by differential movements of the upper and lower lobes of the tail.

The answer to how sharks climb, dive, and turn—rather than pivoting around their center of balance—probably lies in their ability to continually adjust the relative angles of attack not only of their tail fin parts but also of their pectoral and other fins. Maneuverability in bony fishes usually involves deep, compressed bodies and use of median and pectoral fins; to accelerate, bony fishes increase the frequency of their tail beats. Sharks, with their streamlined bodies and relatively rigid fins, have taken a different evolutionary path to achieve maneuverability that may involve tail lobe dynamics and paired fin adjustments. Sharks change speed by altering tail beat frequency, but they also vary tail beat amplitude and the length of the propulsive wave passing down their body (Webb and Keyes 1982).

Sharks have taken the relatively inefficient anguilliform swimming mode imposed by their flexible bodies, and combined elastic skin, rigid but carefully spaced median fins, and a dynamic heterocercal tail to achieve an efficient compromise among cruising, acceleration, and maneuverability. The actual mechanics of swimming in sharks and bony fishes are still a matter of debate and research, but our growing understanding underscores the intricacies and importance of locomotory adaptations in fishes.

FIGURE 8.2.   *The two competing models that explain how forward locomotion is accomplished in sharks. The classical model interprets the shape of the heterocercal tail as generating a downward and backward thrust, which has a resultant force (FR) that is upward and forward. The resultant force would then cause the head of the shark to be pushed downward, but this motion is countered by the planing surfaces of the pectoral fins and head. In the alternative Thomson model, upper and lower lobes of the tail provide counteracting forces that drive the fish directly ahead.*

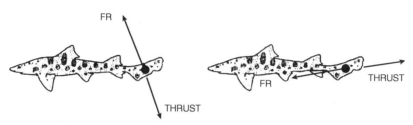

CLASSICAL MODEL   THOMSON MODEL

FR

THRUST   FR   THRUST

From Ferry and Lauder 1996; used with permission of the authors.

to 15% larger in surface area than the dorsal lobe and is the only part of the body in contact with the water during taxiing. The pectoral fins are supported by enlarged pectoral girdles and musculature. The pectoral fins differ from normal teleost fins in the shape of and connections between the lepidotrichia, and the pectoral fin rays are thickened and stiffened, giving the leading, trailing, dorsal, and ventral surfaces more of a winglike than a finlike construction. In some flyingfishes, pelvic fins also contribute lift and are appropriately modified.

Some other atheriniform fishes such as needlefishes and halfbeaks also propel themselves above the water's surface by rapidly vibrating their tail, the lower lobe of which is the only part still in the water. Some halfbeaks have relatively large pectoral fins and engage in gliding flight. Gradations of pectoral fin length and lower caudal lobe strengthening and lengthening among atheriniforms provide a good example of apparent steps in the evolution of a specialized trait, namely flying (Lindsey 1978; Davenport 1994).

FEEDING: BITING, SUCKING, CHEWING, AND SWALLOWING

Adaptations concerned with feeding clearly involve structures used in food acquisition and processing, such as jaw bones and muscles, teeth, gill rakers, and the digestive system. Less obvious, but also important, are morphologic adaptations in eye placement and function, body shape, locomotory patterns, pigmentation, and lures. The functional morphology of feeding deserves detailed exploration because of its intimate linkage to all aspects of fish evolution and biology.

For many fishes, a simple glance at jaw morphology, dentition type, and body shape allows accurate prediction of what a fish eats and how it catches its prey. Small fishes with fairly streamlined and compressed bodies, forked tails, limited dentition, and protrusible mouths that form a circle when open are in all likelihood zooplanktivores. This generalization holds for fishes as diverse as osteoglossomorph mooneyes, clupeomorph herrings, ostariophysine minnows, and representative acanthopterygian groupers (e.g., *Anthias*), snappers (*Caesio*), bonnetmouths (*Inermia*), damselfishes (*Chromis*), and wrasses (*Clepticus*).

Large, elongate fishes with long jaws studded with sharp teeth for holding prey, and with broad tails adjoined by large dorsal and anal fins set far back on a round body are piscivores that ambush their prey from midwater with a sudden lunge (see Chap. 18). An alternative piscivorous morphology includes a more robust, deeper body, fins distributed around the body's outline, and a large mouth with small teeth for short chases and engulfing prey; this is the "bass" morphology of many acanthopterygian predators such as kelp basses, striped bass, sea basses, black basses, and peacock bass, all in different families.

Generalized body shapes in predators do not exclude highly successful specialists that have arrived at very different solutions to catching mobile prey. Examples include lie-in-wait and luring predators (goosefishes, frogfishes, scorpionfishes, stonefishes, flatheads, death-feigning cichlids), cursorial predators that run down their prey (needlefishes, bluefish, jacks, mackerels, billfishes), electrogenic predators that shock prey into immobility (torpedo rays, electric eels), or fishes with either an elongate anterior or posterior region for slashing and incapacitating prey (thresher sharks, sawfishes, billfishes).

A strong correspondence between morphology and predictable foraging habits exists in most other trophic categories, including herbivores (browsers, grazers, phytoplanktivores), scavengers, mobile invertebrate feeders, sessile invertebrate feeders, and nocturnal planktivores, to name a few. Convergent solutions to similar selection pressures are a striking characteristic of the foraging biology of fishes (Keast and Webb 1966; Webb 1982).

Our emphasis here will be on the functional morphology of structures directly responsible for engulfing and processing food. Moderate detail is provided, but we can only superficially discuss the diversity in structure, action, and interconnection among the 30 moving bony elements and more than 50 muscles that make up the head region of most fishes.

Jaw Protrusion: The Great Leap Forward

Jaws evolved in fishes. The major difference between vertebrates and invertebrates is not so much the development of an ossified and constricted backbone; coelacanths, lungfishes, and sturgeons all lack distinct vertebral centra. The real advance that undoubtedly drove vertebrate evolution was the assembly of closable jaws used in feeding. The mechanics of jaw function and adaptive variation in jaw elements tell us a great deal about both how fishes feed and how fishes evolved.

As will be discussed in Chapter 11, one of the major advances made by, but not exclusive to, higher teleosts is the ability to protrude the upper jaw during feeding. Jaw protrusion makes possible the **pipette mouth** of the higher teleosts. Pipetting creates suction forces that can pull items from as far away as 25% to 50% of head length. Jaw protrusion also functions to overtake a prey item, extending the food-getting apparatus around the prey faster than the predator can move its entire body through the water. Attack velocity may thus be increased by up to 40%. As many as 15 different functions and advantages have been postulated for the protrusible jaw of teleosts. These advantages generally involve increased prey capture ability and efficiency but also suggest that antipredator surveillance and escape ability may be enhanced (Lauder and Liem 1981; Motta 1984).

The elements involved in jaw protrusion include the bones of the jaw (premaxilla, maxilla, mandible), ligamentous connections of these bones to the skull and to each other (premaxilla to maxilla, ethmoid, and rostrum; maxilla to mandible, palatine, and suspensorium; mandible to suspensorium), and several muscles, notably the epaxials, levator operculi, hypaxials, adductor mandibulae, and levator arcus palatini (Fig. 8.3).

During jaw protrusion, the entire jaw moves forward and slightly up or down. Protrusion in a generalized percomorph occurs as the cranium is lifted by the epaxial muscles and the lower jaw is depressed by muscles associated with the opercular and hyoid bone series. Movement of the mandible causes the maxillary to pivot forward, the suspensorium (the hinge joint that suspends the lower jaw from the cranium) contributing to maxillary rotation. The descending process of the premaxillary is connected to the lower edge of the maxillary, so the premaxillary is pushed forward, its ascending process sliding forward and down the rostrum.

The jaw is closed through the actions of the adductor mandibulae muscle on the mandible, the levator arcus palatini on the suspensorium, and the geniohyoideus on the hyoid apparatus. Many variations on this simplified description exist, differing among taxa in terms of twisting of jaw bones, points of attachment and pivot between structures, inclusion of other small bony elements, and actions of muscles and ligaments on particular elements (Motta 1984).

Jaw protrusion creates rapid water flow that carries edible particles, both small and large, into the fish's mouth. Suction velocity increases from 0 to as much as 12 m/sec in as little as 0.03 second (Osse and Muller 1980). Fishes that feed on such different prey as phytoplankton, zooplankton, macroinvertebrates, and other fishes utilize suction to capture prey; the larger the object, the more suction pressure must be produced to capture it. Suction feeding, also known as **inertial suction**, results from rapid expansion of the buccal (mouth) cavity, which creates negative pressure in the mouth relative to the pressure outside the mouth. Particles in the water mass ahead of the fish are carried into the mouth along with the water. The jaws then close, pushing the water out the gill covers but retaining the prey in the mouth. Gill rakers, jaw teeth, and teeth on various nonmarginal jaw bones (palate, vomer, tongue) act as mechanical barriers to escape out the opercular chamber.

Suction pressures vary during a feeding event in advanced percomorphs, increasing and decreasing four times. The four phases of suction feeding are preparation, expansion, compression, and recovery (Lauder 1983a, 1985). During **preparation**, as the fish approaches its prey, pressure in the buccal cavity increases as a result of inward squeezing of the suspensorium and lifting of the mouth floor.

The **expansion** phase is when maximal suction pressure develops; the mouth is opened to full gape via lower jaw depression, premaxillary protrusion, and expansion of suspensory, opercular, and mouth floor (hyoid) units. Expansion is the shortest phase during jaw activity, requiring only 5 milliseconds in some anglerfishes. The negative pressures generated during expansion can reach -800 cm H_2O (0.7 atmospheres) in bluegill sunfish, approaching the physical limits imposed by fluid mechanics. Such rapidly achieved low pressures cause cavitation, which involves water vapor suddenly coming out of solution and forming small vapor-filled cavities (the bubbles produced behind an accelerating boat propeller result from cavitation) (Lauder 1983a). The popping noise made during feeding by bluegill may result from the collapse of cavitation bubbles.

The **compression** phase occurs and pressure increases as the mouth is closed by reversing the movements of cranial bones, an activity that requires contraction of a different set of muscles (see Fig. 8.3B). The opercular and branchiostegal valves at the back of the head open up after the jaws close, which allows water but not prey to flow out of the buccal and opercular cavities. **Recovery** involves a return of bones, muscles, and water pressure to their prepreparatory positions.

Modifications of this basic plan underscore some rather spectacular derivations that allow specialized feeding activities. In cichlids, the suspensorium and maxilla are mechanically decoupled. Jaw protrusion occurs as a result of movement of the suspensorium, independent of the maxilla. The consequence of this decoupling of suspensorium and maxilla is that the jaw can be protruded via four different pathways: lifting the neurocranium, abducting the suspensorium, lowering the mandible, or swinging the maxilla. Cichlids make use of different combinations of jaw elements and protrusion pathways to feed on different prey types or in different habitats. High-speed motion picture analysis of jaw action indicates that

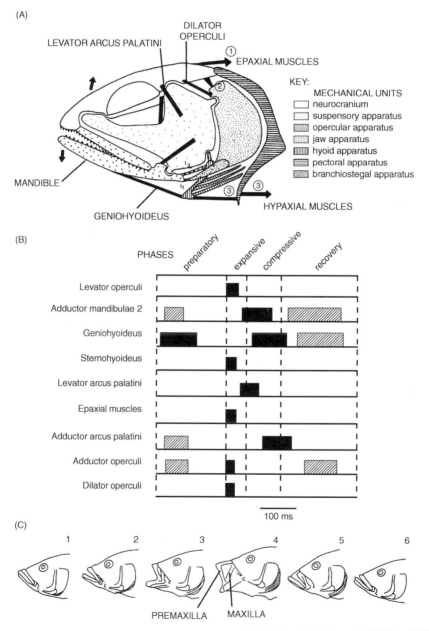

FIGURE 8.3.   *Opening, protrusion, and closing of the jaw in most percoids. (A) Jaw opening involves three major couplings of muscles, ligaments, and bones: 1) epaxial muscles that lift the cranium; 2) levator operculi muscles that move opercular bones up and out and help depress the mandible; and 3) hypaxial muscles that depress the mandible via actions of the hyoid apparatus. (B) Electrical activity of different muscle groups as measured during four phases of jaw opening and closing. Blackened bars represent major muscle activity; cross-hatched bars indicate occasional activity. Abductors move bones outward; adductors move bones inward. (C) The sequence of events during opening and closing of the jaw of a cichlid,* Serranochromis: *1 = preparatory; 2–4 = expansion; 5–6 = compression.*

(A), (B) slightly modified from Lauder 1985; (C) from Lauder 1985 after Liem 1978; used with permission.

some cichlids may use eight different feeding patterns in which they vary their gape, biting force, and amount of jaw protrusion depending on the prey type, location, and behavior. The cichlid jaw is the closest that fishes have come to a prehensile feeding tool. Cichlids show a diversity of foraging types unequaled in any other fish family (see Chap. 15). It is likely that the derived trait of a decoupled suspensorium and resulting trophic versatility have contributed greatly to their success (Liem 1978; Lauder 1981; Motta 1984; Liem and Wake 1985).

Fishes other than cichlids have reworked the basic elements of jaw protrusion and have evolved dramatic specializations that increase attack velocity or suction. As mentioned in Chapter 11, the pikehead, *Luciocephalus pulcher*, shoots its jaw out, increasing its attack speed from 1.3 m/sec to 1.8 m/sec. Little suction is generated during a strike. Extreme and rapid jaw protrusion in this species involves modified anterior vertebrae and massive epaxial muscles and tendons that run from the vertebrae to the posterior part of the cranium. Upward flexion of the head, made possible by a highly bendable neck, leads to extreme jaw protrusion. Other predators have converged on analogous neck-bending abilities to increase prey capture efficiency, including a characin and two cyprinids (Lauder and Liem 1981).

Suction pressure is produced via expansion of the buccal cavity. A generalized perciform such as the yellow perch increases its mouth cavity volume by a factor of six, creating a negative pressure capable of supporting a water column about 15 cm high. The apparent record for volume increase is held by a small (30 cm long), bizarre, elongate midwater fish, *Stylophorus chordatus*. *Stylophorus*, among its other oddities, has a tubular mouth and a membranous pouch that stretches dorsally from its mouth to its braincase. During feeding, the fish throws its head back and thrusts its tubular mouth forward. The mouth becomes separated from the braincase by a distance of about 1 cm, the intervening space being filled by the now expanded membranous pouch. Mouth volume increases almost 40-fold, creating pressures three times greater than in the generalized perch. The fish engulfs copepods as water rushes in at a calculated velocity of 3.2 m/sec, from as far away as 2 cm (Pietsch 1978).

Another extreme of jaw protrusion occurs in the tropical sling-jaw wrasse, *Epibulus insidiator* (Westneat and Wainwright 1989). Sling-jaws protrude their jaws up to 65% of their unextended head length, which is twice the

extension found in any other fish (Fig. 8.4). This extreme protrusion is accomplished via a major reworking of many jaw elements. Several bones in the sling-jaw's head have unique sizes and shapes, including the quadrate, interopercle, premaxilla, and mandible. Ligaments connecting these bones are unusually large, and a ligament found in no other fish links the vomer to the interopercle.

The modified bones undergo extreme and in some cases unique rotations during jaw protrusion; the lower jaw actually moves forward during protrusion, a departure from the depression movement seen in all other fishes. The sling-jaw shoots its mouth out at small fishes and crustaceans on coral reef surfaces, suctioning them into its mouth. It achieves a strike velocity of 2.3 m/sec, but all of this speed is contributed by the jaw, because the fish hovers almost still in the water while attacking prey. Extreme jaw protrusion in sling-jaws involves the evolution of unique bones and ligaments, but the muscles of the jaw and skull have shapes, functions, and sequences of activity that differ little from generalized perciforms. Novel jaw function is therefore accomplished by drastic modification of some structures and retention of primitive condition in others. The sling-jaw exemplifies a widely made observation about the evolutionary process: that every species represents a mosaic of ancestral and derived traits.

Suction feeding has evolved repeatedly during fish evolution and occurs in many nonteleosts as well as in primitive and specialized teleosts that are unable to protrude their jaws. Elasmobranchs, including skates, rays, and such sharks as nurse sharks, can generate suction forces as strong as 1 atmosphere (about 1 kg/cm^2) for feeding on buried mollusks or lobsters in reef crevices (Tanaka 1973). Lungfishes and bowfin among nonteleosts, and anguillid eels, salmons, pickerels, and triggerfishes among teleosts do not protrude their jaws but use inertial suction for feeding; sturgeons have independently evolved jaw protrusion and suction feeding.

FIGURE 8.4.   *Extreme jaw protrusion in the sling-jaw wrasse,* Epibulus insidiator. *The sling-jaw has novel bone shapes and extreme bone and ligament rotations, and has even invented a new ligament involved in jaw protrusion. (A) A 15-cm-long wrasse approaches its crustacean prey with its mouth in the retracted condition. Note that the posterior extension of the lower jaw, involving the articular and angular bones, extends as far back as the insertion of the pectoral fin. (B) During prey capture, the wrasse protrudes both its upper and lower jaws forward, extending them a distance equal to 65% of its head length. Jaw expansion creates suction forces that draw the prey into the mouth. Positions (A) and (B) are separated by about 0.03 second.*

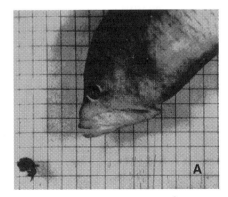 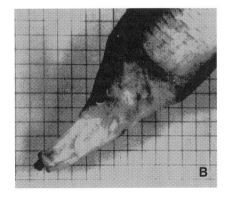

Suction in the nonprotruding species is often accomplished by rapid depression of the floor of the mouth. Triggerfishes and other tetraodontiform fishes such as boxfishes can reverse this flow and forcefully expel water from their mouths (Frazer et al. 1991). Alternate blowing and sucking are used to manipulate food items in the mouth during repositioning for biting. Blowing is also used for uncovering invertebrate prey buried in sand or for manipulating well-defended prey items. A Red Sea triggerfish, *Balistes fuscus*, feeds on long-spined sea urchins. The only spine-free region of the urchin is the oral disk around the mouth. Triggerfishes swim up to an urchin sitting on sand and blow a powerful jet of water at the urchin's base. The water stream lifts the urchin off the substrate and rolls it over, at which point the triggerfish bites through the now-exposed oral disk, killing the urchin (Fricke 1973). Triggerfishes also use blowing to uncover buried prey such as sand dollars. Blowing involves compression of the mouth via actions of muscles associated with the opercular, mandibular, and hyoid bones (Frazer et al. 1991; Turingan and Wainwright 1993).

Pharyngeal Jaws

Depression of the mouth floor also creates water flow toward the throat, thereby helping push food items posteriorly. Here the prey encounter a second set of jaws, the pharyngeal apparatus (see Chap. 11, Division Teleostei). Pharyngeal jaws evolved from modified gill arches and their associated muscles and ligaments. The lower pharyngeal jaws are derived from the paired fifth ceratobranchial bones, whereas the upper jaws consist of dermal plates attached to the posterior epibranchial and pharyngobranchial bones. Both jaws bear teeth that vary depending on the food type of the fish (see below). Dentition not only varies functionally among species that eat different food types but also may develop differently among individuals of a population as a function of the food types encountered by the growing fish. In the Cuatro Cienegas cichlid of Mexico, *Cichlasoma minckleyi*, fish that feed on plants develop small papilliform pharyngeal dentition, whereas those that feed on snails develop robust molariform dentition (Kornfield and Taylor 1983).

In their simplest action, pharyngeal jaws help rake prey into the esophagus. They may additionally reposition the prey, immobilize it, or actually crush and disarticulate it. These actions involve at least five different sets of bones and muscles working in concert, including 10 different muscle groups and bones of the skull, hyoid region, lower jaw, pharynx, operculum, and pectoral girdle. The main action is the synchronous occlusion (coming together) of the upper and lower pharyngeal jaws. In cichlids, the prey is crushed between the anterior teeth of both pharyngeal jaws, pushed posteriorly by posterior movement of both jaws, and then bitten by the teeth of the posterior region of the jaws (Lauder 1983a,b; 1985).

Pharyngeal pads and their function as jaws influence feeding in another important manner. Gape limitation, the constraint on prey size imposed by mouth size (see Box 18.2), is in part determined by oral jaw dimensions: A fish cannot eat anything it cannot get into its mouth.

But gape limitation is also influenced by pharyngeal gape. If a prey item is too large to pass through the pharyngeal jaws, it is also unavailable to the predator. Hence many predators can capture but not swallow a prey item because of pharyngeal gape limitation.

In small-mouthed species, such as the bluegill sunfish, oral and pharyngeal gape differ only by 20% to 30%. But in piscivores that use oral protrusion for prey capture, such as the largemouth bass, oral jaws may be twice the size of the pharyngeal jaws, which means that usable prey size is considerably smaller than what can be engulfed by the mouth. Posterior to the pharyngeal jaws is the throat, the width of which is determined by spacing between the cleithral bones of the pectoral girdles. Thus a predator can only eat prey that can pass through its oral jaws, pharyngeal jaws, and intercleithral space (Lawrence 1957; Wainwright and Richard in press).

A crucial function of the pharyngeal apparatus in many species is therefore to crush prey to a size small enough to pass through the throat. Here prey morphology comes into play, because prey that is just small enough to fit between the pads may be too hard to crush and is thus unavailable to the predator. This interplay of structure, function, and the constraints created by the pharyngeal apparatus is shown nicely in Caribbean wrasses that feed on hard-bodied prey (P. Wainwright 1987, 1988). Wrasses, along with other "pharyngognath" fishes such as parrotfishes and cichlids, have a highly modified pharyngeal apparatus that can crush hard-bodied prey. The size of the muscles that move the pharyngeal jaws differs among three species, the clown wrasse (*Halichoeres maculipinna*), slippery dick (*H. bivittatus*), and yellowhead wrasse (*H. garnoti*). In all three species, muscle mass and pharyngeal gape increase with increasing body size (Fig. 8.5).

At any size, clown wrasses have smaller pharyngeal musculature than do the other two species. Small slippery dicks and yellowhead wrasses can crush and eat snails that are unavailable to larger clown wrasses. Small clown wrasses cannot crush even small snails. These abilities are reflected in the natural feeding preferences of the species. Small clown wrasses feed preferentially on relatively soft-bodied crabs and other invertebrates; they shift to snails only after attaining a body length of 11 cm, when they eat hard-bodied prey that are smaller than those taken by equal-sized fishes of the other two species. Slippery dicks and yellowhead wrasses feed extensively on snails beginning at a relatively small fish body length of 7 cm. Pharyngeal crushing strength accounts for inter- and intraspecific differences in feeding habits in these fishes; competitive interactions and optimal prey characteristics other than shell strength have little if any influence.

Dentition

The prey a fish eats and how those prey are captured are often predictable from the type of teeth the fish possesses. Even within families, species differ considerably in their dentition types as a function of food type and foraging mode (e.g., butterflyfishes, Motta 1988; cichlids, Fryer and Iles 1972; surgeonfishes, Jones 1968). Here we focus

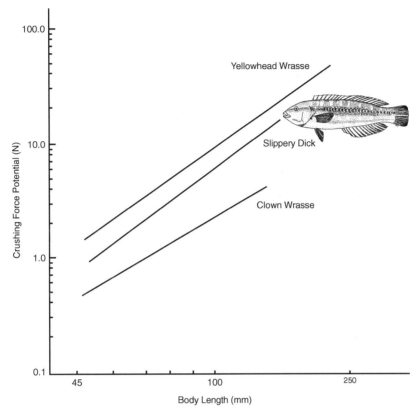

FIGURE 8.5.   *Crushing ability in three related wrasses as a function of body size. Larger wrasses can crush larger snails because of their stronger pharyngeal jaws, but differences among species also affect feeding preferences. Clown wrasses have relatively weak jaws and feed on relatively soft-bodied prey, particularly when the fish are younger. Slippery dicks and yellowhead wrasses have strong jaws and feed on shelled prey throughout their lives.*

From P. Wainwright 1988; used with permission of The Ecological Society of America. Slippery dick drawing from Gilligan 1989.

on general groups of foragers and how their dentition corresponds to food type.

Piscivores and feeders on other soft-bodied, mobile prey such as squid show five basic patterns of marginal (= oral or jaw) teeth.

1. Long, slender, sharp teeth usually function to hold fish (mako, sandtiger, and angel sharks, moray eels, deep-sea viperfishes, lancetfishes, anglerfishes, goosefishes). In some groups (e.g., goosefishes, anglerfishes; also esocid pikes), elongate dentition is repeated on the palatine or vomerine bones. These medial teeth point backward and may have ligamentous connections at their base, which allows them to be depressed as prey is moved toward the throat but prevents escape back through the anterior jaws.

2. Numerous small, needlelike **villiform** teeth occur in elongate, surface-dwelling predators such as gars and needlefishes, as well as more benthic predators such as lizardfishes and lionfishes.

3. Flat-bladed, pointed, **triangular** dentition is usually used for cutting off prey and is found in such fishes as requiem sharks, piranhas, barracudas, and large Spanish mackerels. Piranhas have teeth that are remarkably convergent in shape with those of many sharks (Fig. 8.6). In sharks, the lateral margins of bladelike teeth are often serrated, which enhances their cutting function when the head is shaken or the jaws are opened and closed repeatedly. Sharks and piranhas, as well as other characins, have also converged on replacement dentition. Tooth replacement occurs in a few other teleostean groups, including some salmons, wrasses, filefishes, and triggerfishes (Roberts 1967; Shellis and Berkowitz 1976; Tyler 1980).

4. Recurved, conical **caniniform** teeth with sharp points characterize such piscivores as bowfin, cod, snappers, and some sea basses. Sharp, conical dentition serves to grasp and hold. It reaches its extreme form in the almost triangular, fanglike, slightly flattened teeth of the African tigerfish, *Hydrocynus*.

5. Surprisingly, many highly predaceous piscivores have limited marginal **cardiform** dentition that has a rough sandpaper texture and consists of nu-

FIGURE 8.6.   *Convergence in dentition among predatory fishes. The triangular, razor-sharp teeth of a piranha,* Pygocentrus nattereri, *are remarkably similar in shape and action to those of many sharks. Note the small lateral cusps at the base of the teeth, a feature also shared with many sharks. Piranhas also replace their teeth as do sharks, but piranhas alternately replace all teeth in the left or right half of a jaw, rather than replacing individual teeth or rows of teeth. The teeth in the left side of the jaw (= right side of photo) have recently erupted.*

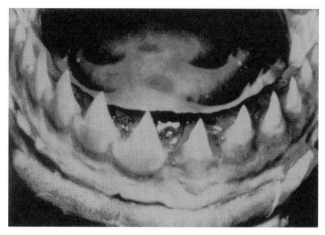

From Sazima and Machado 1990; used with permission.

merous, short, fine, pointed teeth (e.g., large sea basses, snook, largemouth bass, billfishes). The former species rely on large, protrusible mouths for engulfing prey fishes, whereas billfishes immobilize their prey by slashing or stabbing with the bill (see Box 18.1).

Often, a predator will have a mixture of dentition types, such as anterior canines followed by or intermixed with smaller, needlelike teeth (e.g., the pike-characin *Hepsetus*), or long canines intermixed with smaller conical teeth (some wrasses). Ultimately, and regardless of location in the mouth and whether teeth are of one or several types, primary dentition type reflects food characteristics. The primary biting teeth of ariid marine catfishes are palatine, not marginal, in location. Among 10 Australian species, piscivores have sharp, recurved palatine teeth; worm feeders have small, sharp recurved palatine teeth; and molluscivores have globular, truncated palatine teeth (Blaber et al. 1994).

Fishes that feed on hard-bodied prey, such as mollusks, crabs, and sea urchins, often have teeth and jaw characteristics that represent a separation of the activities of capturing versus processing prey. Many such fishes have strong conical dentition in the anterior part of their jaws for plucking mollusks from surfaces. The prey are then passed posteriorly to flattened, **molariform** teeth in marginal or pharyngeal jaws. Convergence is apparent when comparing mollusk-eating fishes from different taxa. Horn sharks (*Heterodontus*) have small conical teeth anteriorly that grade posteriorly into broad, rounded pads for crushing and grinding. Wolf-eels have strong, conical canines anteriorly and rows of rounded

molars posteriorly in each jaw. Similar anterior-posterior differences occur in freshwater drum, sheepshead, cichlids, and wrasses.

A suction versus chewing arrangement occurs in many fishes that feed on sand-dwelling mollusks. Suckers such as the river redhorse, *Moxostoma carinatum*, are ostariophysans in which the molarlike teeth occur on the pharyngeal arches. In ostariophysans, only the lower arch develops dentition, which usually occludes against horny pads in the roof of the mouth. In higher teleosts, the pharyngeal teeth are composed of both dorsal and ventral pharyngeal arch derivatives, such as in the redear or "shellcracker" sunfish, *Lepomis microlophus*. Analogously, stingrays suction mollusks off the bottom and then crush them in pavementlike dentition. Fishes that remove invertebrate prey such as sponges, ascidians, coelenterates, and chitons that are attached to surfaces tend to have powerful oral jaws either with incisor-like dentition (triggerfishes) or with teeth fused into a parrotlike beak (parrotfishes, pufferfishes). In parrotfishes, the beak bites off algae or pieces of coral that are then passed to the pharyngeal mill for grinding.

In addition to marginal, medial, and pharyngeal teeth, fishes have one other mouth region where hard structures aid in the capture or retention of prey. These are the gill rakers, which are bony or cartilaginous projections that point inward and forward from the inner face of each gill arch. As with the various teeth, gill raker morphology corresponds quite closely with dietary habits.

Piscivores and molluscivores, such as sea basses, black basses, and many sunfishes, tend to have short, widely spaced gill rakers that prevent escape of large prey out the gill openings. Fishes that eat zooplankton of large and intermediate size, such as bluegill sunfish and black crappie, have longer, thinner, and more numerous rakers. Feeders on small zooplankton, phytoplankton, and suspended matter have the longest, thinnest, and most numerous rakers; menhaden, *Brevoortia* spp., filter phytoplankton, detritus, and small zooplankters and have more than 150 rakers just on the lower limb of each gill arch.

Among related species, gill rakers differ according to diet. In North American whitefishes (Coregoninae), the inconnu (*Stenodus leucichthys*) feeds on small fishes and has 19 to 24 rakers; the shortnose cisco (*Coregonus reighardi*) feeds on mysid shrimp, amphipods, and small clams and has 30 to 40 rakers; whereas the cisco (*C. artedii*) eats small zooplankters, midge larvae, and water mites and has 40 to 60 rakers (Scott and Crossman 1973). In most filter-feeding fishes, particles are captured by mechanical sieving, whereby large particles cannot pass through narrow spaces between gill rakers. Electrostatic attraction, involving capture of charged particles on mucus-covered surfaces, is also suspected (Lauder 1985).

Mouth Position and Function

Mouth position, in terms of whether the mouth angles up, ahead, or down, also correlates with trophic ecology in many fishes (Fig. 8.7). The vast majority of fishes, regardless of trophic habits, have **terminal** mouths, which means that the body terminates in a mouth that

FIGURE 8.7.   *Correspondence among mouth position, feeding habits, and water-column orientation in teleosts. Fishes with supraterminal mouths frequently live near and feed at the surface, whereas fishes with subterminal mouths often scrape algae or feed on substrate-associated or buried prey. Fishes with terminal mouths often feed in the water column on other fishes or zooplankton but are also likely to feed at the water's surface, from structures, and on the bottom. See also Fig. 23.3*

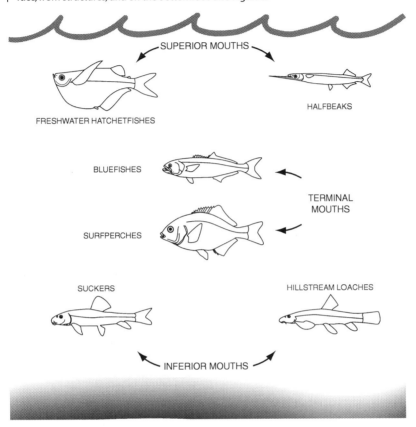

Fish drawings from Nelson 1994; used with permission.

opens forward. Deviations from terminal location usually indicate habitat and feeding habit. Fishes that swim near the water's surface and feed on items at the surface often have mouths that open upward, termed **superior** or **supraterminal** (e.g., African butterflyfishes, freshwater hatchet or flyingfishes, halfbeaks, topminnows). Some predators that lie on the bottom and feed on prey that swim overhead also have superior mouths (stonefishes, weeverfishes, stargazers).

Mouths that open downward, termed **subterminal** or **inferior**, characterize fishes that feed on algae or benthic organisms, including suckers, some North American minnows, suckermouth armored catfishes, Chinese algae eaters, some African minnows and cichlids, clingfishes, and loach gobies. Upside-down catfishes feed on the undersurfaces of leaves but do so while swimming upside down; not surprisingly, they have inferior mouths. Fishes that do not have to visually fix on their prey (e.g., algae-scraping clingfishes, catfishes, loaches, some cichlids), or that take somewhat random mouthfuls of sediments that are then sifted orally (suckers, mojarras), may gain an antipredator advantage by having an inferior mouth. A terminal mouth in such fishes would require that they angle head

down each time they scraped or sampled the benthos, which would make them less able to escape rapidly if surprised by a predator.

Specialized suctorial mouths characterize unrelated fishes that scrape algae from rocks, particularly if they also live in high-energy environments. This ecological grouping includes hillstream loaches, suckermouth armored catfishes such as the familiar *Plecostomus* of the aquarium trade, Southeast Asian algae eaters, and the loach gobies of Australia. The gyrinocheilid algae eaters live in swift streams where they rasp algae from rocks with their lips while remaining attached with their suctorial mouth. Gyrinocheilids have evolved an additional incurrent opening dorsal to the operculum that opens into the gill chamber. They breathe in through the dorsal opening and out through the operculum. Drawing water in through the mouth in the more normal manner would require the fish to detach from the substrate, at which moment it might risk being swept downstream. Mouths are not the only way for algae feeders to remain attached in wave-swept habitats. Gobiesocid clingfishes accomplish this via pelvic fins modified into a suction disk (Wheeler 1975; Nelson 1994).

SUMMARY

1. Functional morphology focuses on how structures work in the context of the daily tasks and interactions experienced by organisms. Locomotion and feeding offer many intriguing examples of the structure-function relationship. Locomotion in water presents very different physical challenges than are experienced by terrestrial animals. Density and drag are much greater in water, making locomotion energetically expensive and leading to the general hydrodynamic, streamlined shape of most fishes.

2. Swimming in fishes usually involves alternating contractions and relaxations of muscle blocks on either side of the body that result in the fish pushing back against the water and consequently moving forward. Many variations on this basic theme exist, and about 10 different modes of swimming have been identified that involve either undulatory waves or oscillatory back-and-forth movements of the body or fins. Body and fin shape correlate strongly with locomotory mode and habitat, the most extreme examples being the rapid-swimming, highly pelagic mackerel sharks, tunas, and billfishes with streamlined bodies and lunate, high aspect ratio tails.

3. Locomotory adaptations create trade-offs. Maneuverability is often achieved at a cost in fast starts and sustained speed and vice versa. Versatility is achieved by using different modes for different purposes (fin sculling for positioning, body contractions for fast starts and cruising), which causes most fishes to evolve generalist rather than specialized swimming traits. Highly specialized locomotion includes fishes that can "walk" across the bottom or on land, climb terrestrial vegetation, leap, glide, and even fly.

4. Sharks, being cartilaginous, cannot rely on muscles attached to a rigid bony skeleton for propulsion. They instead undulate via contractions of their body muscles, which are firmly attached to a relatively elastic skin; the skin functions as an external tendon and provides propulsive force by rebounding. Some propulsive force comes from changing hydrostatic pressure inside the cylinder of the shark's body. The spacing of the two dorsal fins aids the tail in propulsion, and both lobes of the tail work to provide forward thrust via mechanisms that are incompletely understood.

5. Food getting in fishes involves adaptations of the jaw bones and muscles, teeth, pharyngeal arches, gill rakers, and digestive system, as well as modifications in body shape, sensory structures, and coloration.

6. Food type can often be predicted from jaw and body shape and dentition type, regardless of taxonomic position. Zooplanktivorous fishes are usually streamlined, with compressed bodies, forked tails, and protrusible mouths that lack significant teeth. Lurking, fast-start piscivores are generally elongate, round in cross section, with broad tails, posteriorly placed median fins, and long, tooth-studded jaws that grab prey. Alternatively, many piscivores that pursue prey for short distances are more robust, with fins distributed around the body outline, and with large mouths for engulfing prey. Many specialists that depart from these norms can be found.

7. An important food-getting innovation among modern fishes, particularly in teleosts, was the development of protrusible jaws and the pipette mouth. Modifications to jaw bones, ligaments, and muscles allow a fish to shoot its upper jaw forward and increase the volume of the mouth cavity, both creating suction forces and increasing the speed with which a fish overtakes its prey.

8. In addition to anterior, marginal jaws and dentition on the roof of the mouth and tongue, teleosts have their gill arches modified into a second set of posterior, pharyngeal jaws. Pharyngeal jaws help move prey toward the throat and in many fishes serve to reposition prey for swallowing and for processing via crushing, piercing, and disarticulation. Pharyngeal teeth facilitate the eating of hard-bodied prey (mollusks, arthropods) and plant material.

9. Dentition type corresponds strongly with food type and is often repeated on the marginal jaws, vomer, palate, and pharyngeal pads. Piscivores and other predators on soft-bodied prey variously possess long, slender, sharp teeth; needlelike villiform teeth; flat-bladed triangular teeth; conical caniniform teeth; or rough cardiform teeth. Mollusk feeders have molariform teeth. Gill rakers also capture prey and may be numerous, long, and thin in plankton feeders, or widely spaced, stout, and covered with toothlike structures in predators on larger prey.

10. Mouth position also correlates with where a fish lives and feeds in the water column. Water-column feeders typically have terminal mouths that open forward, whereas surface feeders often have superior or supraterminal mouths that open upward. Fishes that feed on benthic food types have subterminal or inferior mouths that open downward and that may generate suction forces that allow a fish to attach to hard substrates while feeding.

SUPPLEMENTAL READING

Alexander 1983; Duncker and Fleischer 1986; Gerking 1994; Gray 1968; Hildebrand 1982; Hildebrand et al. 1985; Hoar and Randall 1978; Pough et al. 1989; Videler 1993; Vogel 1981; Wainwright et al. 1976; Webb and Weihs 1983.

9

Early Life History

In the next two chapters, we explore processes and attributes of fishes that are undergoing rapid growth and development. Our chief emphasis in this chapter is on the earliest stages of development: gametes, embryos, and larvae. In the next chapter, we focus on transitional stages between larvae and juveniles, and on various aspects of the growth process. Adult biology is explored primarily with respect to reproduction (determination, differentiation, maturation, longevity, and senescence). Related topics concerning the timing, effort, and behavioral interactions associated with reproduction are detailed in Chapters 20 and 23.

COMPLEX LIFE CYCLES AND INDETERMINATE GROWTH

Two general traits shared by most fishes set them apart from most other vertebrate taxa and also underlie many of their more interesting adaptations. These two traits are indeterminate growth and a larval stage. Many fishes emerge from an egg as a larva, which bears little anatomic, physiological, behavioral, or ecological resemblance to the juvenile or adult into which the fish will eventually transform. In fact, continual growth moves each individual through a progression of life history stages that differ in most traits, creating a spectrum of continually changing structures and characters upon which natural selection has operated.

Indeterminate growth describes the continual increase in length and volume that occurs in most fishes throughout their lives. Although this growth may slow considerably as a fish ages, the potential for continuing increase profoundly affects many, if not most, aspects of a fish's life. With regard to most traits, larger body size appears to confer an advantage, at least within a species. Reproduction is intimately tied to body size in terms of egg number and size, with larger females producing more and bigger eggs (see Chap. 23, Life Histories and Reproductive Ecology). Mate choice by both males and females often favors larger individuals, and larger fish are better able to defend a spawning territory (see Chap. 20, Sexual Selection, Dimorphism, and Mate Choice).

Swimming energetics and shoaling interact with body size: Fish tend to shoal with individuals of like size (see Chap. 19, Responses of Aggregated Prey), and larger fish can swim faster and migrate over larger distances (see Chap. 22, Annual and Supra-annual Patterns: Migrations). Predation rate is typically greater on smaller fish, and small fish may be constrained from feeding in profitable areas by predators or larger conspecifics. Indeterminate growth leads to size-structured populations in which different-size individuals essentially function as different species, the so-called ontogenetic niche (Werner and Gilliam 1984; see Chap. 23, Population Dynamics and Regulation).

Physiological limitations of small body size can be explained by allometric (proportional) growth of many structures, such as the increased visual acuity and sensitivity that occur as a fish grows. Foraging is affected by body size as well, not only because many fish are gape limited and hence only able to eat things they can swallow whole but also because many prey types are not available to young fishes until muscle attachment sites and muscle masses reach a size capable of overcoming prey defenses (see Chap. 8, Pharyngeal Jaws).

EARLY LIFE HISTORY: TERMINOLOGY

Given the diversity and complexity of stages, states, phases, or intervals in the early life history of fishes, it is not surprising that several classification systems have been developed to describe these stages, each differing slightly or greatly in terminology (Fig. 9.1). These schemes all attempt to subdivide about one dozen recognizable, general events during development into a coherent, descriptive progression. The simplest classification recognizes an egg (which after fertilization or activation contains a developing embryo), which hatches into a larva, which metamorphoses into a juvenile.

Subdivisions of this basic sequence generally involve end point events, some of which occur quickly, others gradually (see Fig. 9.1). Significant end point events include closure of the blastopore and lifting of the tailbud of the developing embryo; absorption of the yolk sac, independent feeding, and flexion of the notochord of the larva; development of fin rays, scales, and pigmentation; and changes in body proportions of the juvenile (Fig. 9.2). These general descriptions overlie a more complicated sequence of events involving changes in the anatomy, physiology, behavior, and ecology of a developing fish

FIGURE 9.1.   *Events during, and terminology describing, the early life history of teleost fishes. The three basic stages—egg, larva, juvenile—can be further subdivided depending on definable events that occur during development and growth. The top half of the diagram summarizes one commonly used set of terminology, particularly for pelagic marine larvae. Alternative systems for describing these events are given in the lower half of the diagram; the approach of Balon (second from bottom) may be more descriptive of many freshwater taxa.*

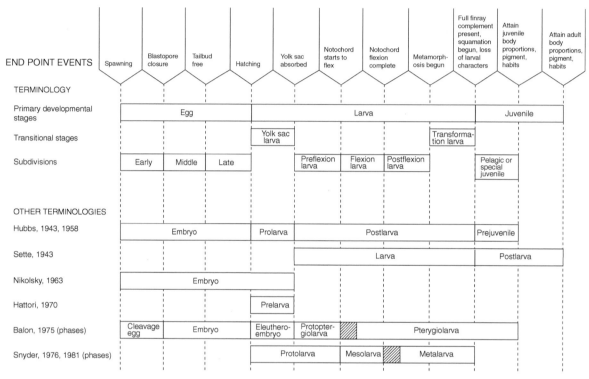

From Kendall et al. 1984; used with permission.

(Fig. 9.3). From a systematics standpoint, most fish species are readily distinguishable as such from the earliest stages. Early life history stages have consequently played an important role in fish systematics (Cohen 1984; Moser et al. 1984).

Part of the controversy over developmental terminology arises from the great diversity of embryonic and larval types, developmental rates, and transitional stages or events that exist among the more than 25,000 species of fishes. Attempts at generalization are frustrated by exception and nuance, and by whether research focuses on marine or freshwater species, pelagic or demersal young, live- or egg-bearers, and embryology or taxonomy. Some workers maintain that development is a continuous and gradual process and that designating exact stages is an arbitrary process. Others maintain that development is saltatory, that it occurs with periods of gradual change punctuated by significant events or thresholds that allow for rapid change, such as the shift from dependence on yolk or maternal secretions to independent, exogenous feeding. This disagreement will not be resolved in the short space available here, but the interested reader should consult references by Balon (1975, 1980, 1984), Richards (1976), and Kendall et al. (1984) for a review.

EGGS AND SPERM

Gametogenesis

Most fishes have paired gonads, although one member of the pair may be consistently larger than the other in some species or only one gonad may be functional. Hagfishes and lampreys are unique in that only one ovary develops, from fusion of two primordia in lampreys and from loss of one ovary in hagfishes (see Chap. 13). Unlike sharks and other vertebrates, testes and ovaries in jawless fishes and bony fishes develop from only the cortex of the peritoneal epithelium, not from both the cortex and medulla. Testes in immature males are typically reddish and take on a smooth texture and creamy white coloration as the fish matures and spawning time approaches. The testes generally account for less than 5% of body weight (see below). Follicles within the testes produce the developing spermatozoa (= **spermatogenesis**) through a series of meiotic and developmental transformations typical of vertebrates (spermatogonia, primary spermatocytes, secondary spermatocytes, spermatids, metamorphosis, spermatozoa) (Hoar 1969; Hempel 1979; Gorbman 1983; Adkins-Regan 1987; Jamieson 1991).

Fish sperm vary in size, shape, number of flagella (zero

FIGURE 9.2. *Stages during the early life history of the horse mackerel,* Trachurus symmetricus.

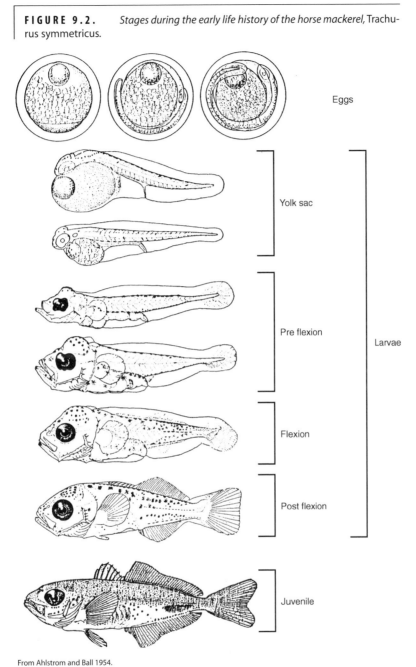

Eggs

Yolk sac

Pre flexion

Flexion

Post flexion

Juvenile

Larvae

From Ahlstrom and Ball 1954.

to two), and presence or absence of acrosomes and other structures. Fish sperm are highly diagnostic of higher taxa and of some species (Jamieson 1991; Fig. 9.4). Sperm heads range in length from about 2 μm (bowfin, burbot, medaka) to 70 μm (Australian lungfish). The caplike acrosome at the anterior end of most primitive fish sperm is lost in practically all neopterygian fishes (gars, bowfin, and teleosts). Two African families of osteoglossomorph fishes, the Mormyridae (elephantfishes or mormyrs) and Gymnarchidae, lack a flagellum. Their sperm may move by some form of amoeboid motion.

Typical ejaculates during spawning contain millions of sperm. Sperm is released in seminal fluid in species with external fertilization, or in packets called spermatophores in internal fertilizers. It is commonly stated (e.g., see Box 20.1) that males produce an excess of sperm and that consequently male reproductive success is limited more by access to females than by ability to produce gametes (the opposite is considered limiting in females). However, under circumstances where males mate daily over a prolonged breeding season, sperm depletion can occur, and mating may in fact be delayed until sperm stores are replenished (e.g., Nakatsaru and Kramer 1982; Jamieson 1991; see also Shapiro et al. 1994).

Oogenesis, the development of eggs, occurs within the ovary and also progresses through various stages involv-

FIGURE 9.3.   *Behavioral, physiological, and anatomic events during the postembryonic early life history of a representative teleost, the northern anchovy,* Engraulis mordax. *The horizontal axis at the bottom represents days after hatching; events noted are those occurring after the larva begins exogenous feeding, when it is no longer solely dependent on yolk for energy. Photopic refers to daytime vision; scotopic refers to nighttime vision. RBCs = red blood cells. Reynolds number refers to problems that small larvae have with water viscosity (see below, Larval Behavior and Physiology). Time to 50% starvation refers to how long larvae can live without feeding, based on half of the larvae in an experiment dying after a given number of days without food.*

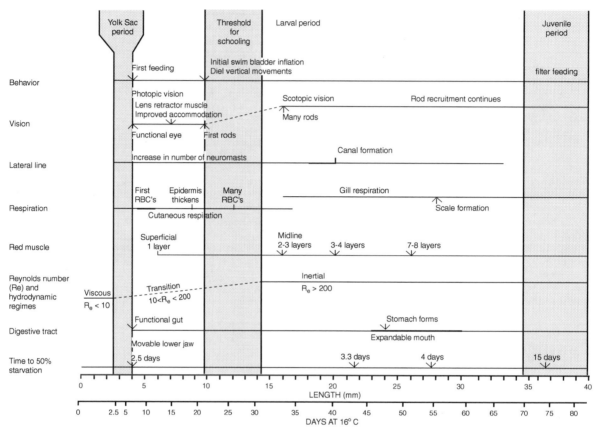

From Hunter and Coyne 1982; used with permission.

ing oogonia, primary and secondary oocytes, and finally ova or eggs (Wallace and Selman 1981; Fig. 9.5). Oogones develop from primordial sex cells in the germinal epithelium of the ovary wall. Proteinaceous yolk granules are deposited around primary oocytes during **vitellogenesis**, the precursors of yolk material being manufactured in the liver. Oil droplets are incorporated in the yolk. Ripe eggs pass from the ovary through the oviduct, which is a continuation of the ovarian tissue, to the outside via the cloaca. In elasmobranchs, no direct connection between ovary and oviduct exists, and hence eggs pass through the peritoneal cavity on their way to the oviduct. In several osteoglossomorph bonytongues, a loach (*Misgurnus*), anguillids, salmonids, and galaxiids, the oviduct is greatly reduced or absent, and eggs enter the body cavity prior to being shed (Blaxter 1969; Hempel 1979; Wootton 1990).

Females that have spawned are termed **spent**; their ovaries are bloody and contain residual eggs that are resorbed by the ovary. Egg resorption is a general process. Usually, most of the ripe eggs in an ovary are spawned, while a small proportion is resorbed and the proteins, fats, and minerals contained in them are reused by the female for maintenance, growth, or production of more eggs. The number resorbed varies greatly among and within species, depending on fish size, number of previous spawnings that season, and energy state of the female.

Fecundity, the number of eggs released by a female during a spawning bout or breeding cycle, varies from one to two in some sharks to tens of millions in tarpon, *Megalops atlanticus,* and European ling, *Molva molva,* to 300 million in the giant ocean sunfish, *Mola mola;* seasonal and lifetime fecundities can also be calculated. Most larger, temperate marine fishes produce tens of thousands to millions of eggs at a time. Fecundity generally decreases with increasing egg size and with increasing parental care but increases with body size in an individual. Mouth brooders such as sea catfishes produce only about 100 eggs at a time; livebearers such as the four-eyed fish, *Anableps,* contain only a dozen embryos. The relationship between egg number and body size is usually proportional to the mass of the female, reflecting the volume of a female's body that can carry the eggs. Hence egg number generally increases in relation to the third, fourth, or fifth power of the length of the female (see Fig. 23.1).

FIGURE 9.4. *Fish sperm and their utility in constructing phylogenies. (A) Schematic diagram of the sperm of the percichthyid* Macquaria ambigua; *entire structure is about 5 mm long. (B) Longitudinal section of the sperm of a coral trout,* Plectropomus leopardus *(Serranidae); a = axoneme; cc = cytoplasmic canal; dc = distal centriole; m = mitochondrion; n = nucleus; pc = proximal centriole. (C) Schematic diagrams of spermatozoa of nonneopterygian fishes and a cladogram based on sperm characteristics; the cladogram is basically similar to the general phylogeny presented in Chapter 13, showing the parallel evolution of sperm and other taxonomically useful characteristics.*

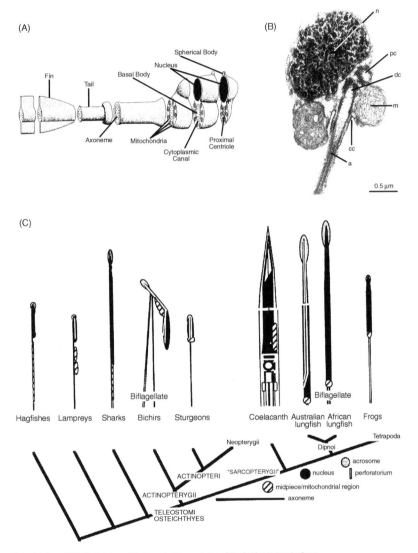

From Jamieson 1991 (C slightly modified); used with permission of Cambridge University Press.

Exceptions show the premium placed on assuring the survival of young rather than just on producing large numbers of eggs. In mouth-brooding cichlids, fecundity increases in relation to the square of the length of the female, because mouth size increases only linearly with increasing body length (Breder and Rosen 1966; Hempel 1979; Lowe-McConnell 1987).

Because of resorption, fecundity estimates based on counts of ripe eggs may not necessarily indicate true **fertility**, which is the number of viable offspring produced. Fecundity estimates for live-bearing fishes are fur-

ther complicated by the consumption of eggs by developing embryos, a form of maternal provisioning (see below).

Certain generalizations apply to fish eggs, with a strong correlation between habitat and the characteristics of fertilized eggs. "Most marine fishes, regardless of systematic affinities, demersal or pelagic habits, coastal or oceanic distribution, tropical or boreal ranges, spawn pelagic eggs that are fertilized externally and float individually near the surface of the sea" (Kendall et al. 1984, 11). Egg sizes for pelagic spawners range from about 0.5 mm (*Vinciguerria*, Photichthyidae) to 5.5 mm in diameter

FIGURE 9.5.   *Stages in the development of teleostean eggs, as shown by the multiband butterflyfish,* Chaetodon multicinctus. *(A) Primary oocyte growth and yolk development (vitellogenesis); N = nucleus; OD = oil droplets; PO = primary oocyte; YG = yolk granule; ZR = zona radiata of the vitelline membrane. (B) Yolk vesicle or cortical alveoli stage (left cell), early vitellogenesis (right cell); CA = cortical alveoli; F = follicle that holds oocyte. (C) Maturing egg, in which nucleus has migrated to cell periphery and yolk granules have coalesced; YGF = yolk granule fusion. (D) Postovulatory ovary, after eggs are shed; PF = postovulatory follicle, from which egg has been released.*

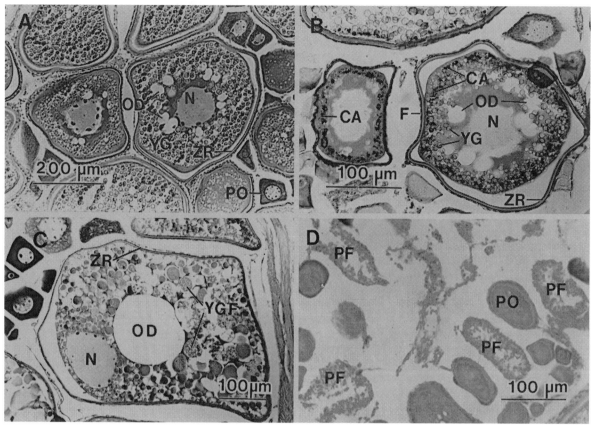

From Tricas and Hiramoto 1989; used with permission.

(moray eels), with a modal size of about 1 mm, generally spherical shape, and a single oil globule. This remarkable convergence among phylogenetically widespread taxa suggests a common, adaptive set of solutions to the selection pressures encountered by eggs that disperse passively in a near-surface environment and that contain an embryo dependent on yolk supplies for nutrition.

Most freshwater fishes and some coastal marine species diverge from this pattern and produce **demersal** eggs that are laid on the bottom. Many spawn in nests and engage in some form of parental care (see Chap. 20). Demersal eggs tend to be relatively large, with a diameter of 7 or 8 mm in salmon, anarhichadid wolffishes, and zoarcid eelpouts. The largest teleostean eggs are produced by mouth-brooding marine catfishes and range from 14 mm to 26 mm. Shark eggs are generally larger than osteichthyan eggs, whereas the largest bony fish eggs are produced by the coelacanth, *Latimeria chalumnae*, with a diameter of 9 cm. Demersal eggs often have thick chorions and special coatings or filaments that may provide mechanical protection and also help the eggs adhere to structures or to other eggs (ictalurid catfishes, North

American minnows, killifishes, freshwater perch, blennies) (Boehlert 1984; Matarese and Sandknop 1984).

Departures from spherical shape are found in the elongate eggs of some cusk-eels, anchovies, minnows, cichlids, parrotfishes, and gobies. Congrogadid eel blennies have strangely cross-shaped eggs, and eggs of some darters are deeply indented and appear almost heart shaped. Although usually smooth, the outer vitelline membrane of the egg, termed the **chorion**, may be sculptured (lizardfishes, deep-sea hatchetfishes, mullets, some flounders) or have filaments, stalks, or spines (myctophiform lanternfishes and many atherinomorphs such as killifishes, flyingfishes, topsmelts, sauries, and halfbeaks). Filaments often help eggs attach to other eggs or to structures such as seaweeds, as in flyingfishes.

Variation in other egg structures can help in species or family identification. The degree of segmentation and pigmentation of the yolk differs, with primitive teleosts such as many eels, herrings, and salmons having segmented yolks, whereas more advanced teleosts have homogeneous yolks. Pigmented yolks produce colorful eggs in gar, catfishes, and salmon. The occurrence, number, and location of oil globules in the yolk differ among species. Oil globules may serve as nutrition for embryos or as flotation mechanisms; when pigmented with melanin, they may help protect sensitive developing structures from harmful radiation (F. D. Martin, personal communication). Oil globules in some species go through highly predictable patterns of movement (Malloy and Martin 1982). Eggs also vary among species in terms of the space between the chorion and the yolk, termed the **perivitelline** (around the yolk) **space** (Page 1983; Boehlert 1984; Matarese and Sandknop 1984).

An interesting aspect of reproduction is the effort and energy that different species and individuals expend, which often correlates with the life history pattern a species has evolved (see Chap. 23, Life Histories and Reproductive Ecology). **Reproductive effort** includes food intake and its transfer to the gonads, as well as energy expenditure in somatic versus gonadal growth. In females, oocyte maturation involves mobilization of first lipids and then proteins from other parts of the body, such as fat deposits and body muscle, to the ovary. This maturation is also accompanied by as much as a 10-fold increase in oxygen consumption by the ovary until the final stage of oogenesis, when ovarian mass increases via water accumulation. True costs of reproduction must also account for energy expended during reproductive migrations, courtship, spawning, internal brooding, and other forms of parental care, among other factors. Energy expended in nongonadal growth can be substantial. For example, migratory female sturgeon and salmon use 80% to 90% of their body fat reserves during reproduction, much of which is expended during migration to the spawning grounds (Kamler 1991).

The difficulty of estimating reproductive effort accurately has led to the development of a variety of indices. Some indices depict characteristics of an individual at a given point in time (**instantaneous measures**), others over the reproductive life of an individual (**cumulative measures**). One popular and simple instantaneous measure is the **GSI**—variously called the gonadosomatic, or gonadal-somatic, or gonosomatic index—a calculation of the percentage of the body mass of an animal devoted to gonadal material. GSI can be calculated as the ratio of gonad weight to total body weight, or as the ratio of gonad weight to body weight minus gonad weight (and sometimes minus gut weight). Values vary greatly among species: cichlids, less than 5%; darters, 11% to 23%; pupfishes, 2% to 14%; American plaice, 5% to 20%; sticklebacks, 20%; salmonids 20% to 30%; and European eels, 47%.

The GSI of ripe females is a relatively accurate portrayal of effort for **total spawners** that spawn only once in a breeding season or lifetime, but it underestimates effort in **repeat**, **batch**, or **serial spawners**, because only a fraction of the eggs or oocytes a female will produce are present at any moment (Heins and Rabito 1986). GSIs for males are generally much smaller than for females, reflecting the lower effort directly expended in reproduction in males of many species. GSIs in male sticklebacks are 2% compared with 20% for females, and only 0.2% in male tilapia.

Female pike, *Esox lucius*, allocate 6 to 18 times more energy to ovaries than males do to testes.

Intraspecific variation in male GSI can reflect differences in reproductive tactics. In bluegill sunfish, *Lepomis macrochirus,* some males guard nests and attract females with which they spawn; other males lurk at the periphery of the nests and sneak into the territory to join a spawning nest guarder. GSIs for territorial males are about 1% but for sneakers are 4.5%. GSIs in male bluehead wrasses vary between 3% and 5%, the larger value characterizing males that previously engaged in group spawning, where sperm competition among males is likely (Gross 1982; Shapiro et al. 1994).

Measuring GSI during different parts of the year can indicate ovarian or testicular cycles and spawning periodicities. Multimodal values would indicate protracted spawning seasons, whereas a single maximum would indicate a more defined spawning season. Not surprisingly, GSI values generally reach their maxima just before spawning (Page 1983; Wootton 1990; Kamler 1991). GSI calculations have been modified and improved to account for differences in body size among females and for females that do not shed all eggs at once (DeMartini and Fountain 1981; Erickson et al. 1985).

Fertilization

In most fishes, fertilization occurs outside the body of the female. Regardless, fertilization occurs when a sperm penetrates or is permitted to enter the egg membrane via a funnel-shaped hole in the membrane called the **micropyle**. The micropyle sits above the animal pole of the egg and is too narrow to allow passage of more than one sperm. After sperm entry, the micropyle closes and the chorion tends to harden, which prevents **polyspermy**, or entry of more than one sperm. Sturgeon eggs have several micropyles and polyspermy occurs. Micropyle presence and size are diagnostic of different species. A zygote is formed when pronuclei of sperm and egg fuse (Hoar 1969; Hempel 1979; Matarese and Sandknop 1984; Jamieson 1991).

In species with external fertilization, gametes remain viable for less than a minute to as long as an hour, depending on temperature; longer viability generally occurs in colder water (Hubbs 1967; Petersen et al. 1992; Trippel and Morgan 1994). Little good information exists on the proportion of eggs fertilized during a natural spawning event. Petersen et al. (1992) ingeniously squirted fluorescent dye into the egg-milt cloud produced by spawning bluehead wrasse and then passed a fine-mesh net through the dye cloud. They estimated that on average 75% of the eggs spawned were fertilized and that this number declined if males and females were farther apart when they released gametes.

Females of some internally fertilized species are able to store sperm in the ovary. In the dwarf perch, *Micrometrus minimus* (Embiotocidae), newborn males are mature and inseminate but do not fertilize newborn females. The females store this sperm for 6 to 9 months until they mature and ovulate (Warner and Harlan 1982; Schultz 1993). Some species store sperm and use it to fertilize multiple batches of eggs. Females of the least killifish, *Heterandria formosa*, store sperm from one copulation for as long as 10 months and use it to fertilize as many as nine different developing broods of embryos, several of which may be developing simultaneously, a phenomenon known as **superfetation**.

In a few species, activation of cell division is not synonymous with fertilization. Some poeciliid livebearers are **gynogenetic**, in that females use sperm from males of other species to activate cell division, but no male genetic material is actually incorporated into the zygote (see Chap. 20, Gender Roles in Fishes). In some internally fertilized species, fertilization occurs, but development may be arrested after a few cell divisions and then resume when environmental conditions are more favorable for hatching. In some annual fishes, such as the South American and African rivulines, eggs are fertilized and then buried; they spend the dry season in a resting state known as diapause (Lowe-McConnell 1987; see Chap. 17, Deserts and Other Seasonally Arid Habitats).

Internal fertilization is universal among sharks but is limited to about a dozen families of bony fishes, most notably the coelacanth; a silurid catfish; brotulids; livebearers; goodeids; three genera of halfbeaks; the neostethids and phallostethids of Southeast Asia; scorpaenids in the genera *Sebastodes* and *Sebastes* (e.g., *Sebastes viviparus*); Baikal oilfishes; embiotocid surfperches; an eel pout, *Zoarces viviparus*; and clinids. An Asian cyprinid, *Puntius viviparus*, was originally described as a live-bearer, but subsequent examination of the type material indicated that predation on cichlid young, which were contained in the stomach, had been mistaken for developing young in the ovary (attesting to the value of depositing type material and voucher specimens in museums; see Chap. 2, Steps in Classification).

Internal fertilization requires that males possess an **intromittent organ** for injecting sperm. This structure has different names and is derived from different structures in different taxa. The pelvic fin of elasmobranchs is modified into **claspers**; the pelvic girdle, postcleithrum, and pectoral pterygial elements form the **priapium** of phallostethoids; the anal fin forms the **gonopodium** of cyprinodotoids such as goodeids, anablepid four-eyed fishes, jenynsiids, and poeciliid livebearers; brotulids and surfperches have an enlarged genital papilla. In some cardinalfishes, the female purportedly inserts an enlarged urogenital papilla into the male to receive sperm (Hoar 1969).

EMBRYOLOGY

After fertilization, the chorion of the egg stiffens, a process known as **water hardening**, which serves to protect the developing embryo. Embryogenesis in fishes proceeds as in most vertebrates. The embryo develops on top of the yolk; yolk is concentrated at the vegetative pole, and the fertilized egg is considered telolecithal. In hagfishes, elasmobranchs, and teleosts, cleavage is meroblastic, in that cell division occurs in the small cap of cytoplasm that will

develop into a fish, but not in the yolk. Lampreys have holoblastic cleavage, whereas bowfin, gar, and sturgeon exhibit an intermediate form, termed semiholoblastic. Cell division and differentiation continue in fairly predictable sequences, with many interspecific differences in the timing of appearance of different structures (Fig. 9.6; see Lagler et al. 1977 and Lindsey 1988 for more thorough discussions of fish embryology).

Organs and structures that are at least partially developed in many embryos prior to hatching include body somites, which are forerunners of muscle blocks; kidney ducts; the neural tube; optic and auditory vesicles; eye lens placodes; head and body melanophores; a beating heart and functioning circulatory system, much of which is linked to circulatory vessels in the yolk; pectoral fins and median fin folds, but not the median fins themselves; opercular covers, but not gill arches and filaments; otoliths; lateral line sense organs below which scales will later form; and the notochord. By the time of hatching, the head and the tail have lifted off the yolk, the mouth and jaws are barely formed, fin rays may be present in the caudal fin, but little if any skeletal ossification has occurred. The nonfunctional gut is a simple, straight tube, and the gas bladder is evident as a small evagination of the gut tube. Advanced embryos curl around on themselves, their bodies making more than a full circle. In fish

FIGURE 9.6.   *Embryonic development of the lake trout, Salvelinus namaycush. (A) At 17 days after fertilization (at 7.8°C), somites and brain ventricle are just beginning to form. (B) At 22 days, auditory vesicles and eye lens placodes have developed. (C) At 23 days, tail is lifting off and pericardial cavity is evident. (D) At 67 days, embryo is fully formed and about to hatch (at 4.4°C); note pigmentation, upturned notochord, gut tube, dorsal and ventral fin folds, pectoral fins, pigmented eye, well-developed yolk circulation, and large yolk sac.*

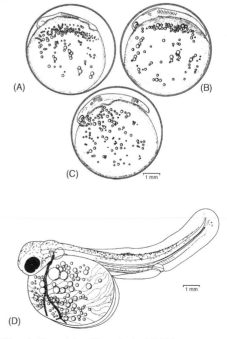

From Balon 1980; used with permission of Kluwer Academic Publishers.

that hatch at a relatively undeveloped stage, the eye is seldom pigmented, maintaining the transparency of the helpless, newly hatched free embryo with its very large and cumbersome yolk sac (Blaxter 1969; Hempel 1979; Matarese and Sandknop 1984; Lindsey 1988).

Just prior to emergence, hatching gland cells on the head of some fishes secrete proteolytic enzymes that aid in breaking down the chorionic membrane, or "egg shell." Thrashing movements of the tail and body aid in the hatching process. The time spent in the egg prior to hatching varies greatly among species, from a few days in striped bass, centrarchid sunfishes, some cichlids, and surgeonfishes, to a week or two in smelts, darters, mackerels, and flatfishes, to several weeks or months in salmons and sharks. In all fishes studied, hatching time decreases with increasing temperature (Blaxter 1969; Kendall et al. 1984). Hatching is a convenient landmark during early development, but it should be noted that the exact developmental stage at which the embryo frees itself from the egg varies greatly among species and even varies within species depending on the temperature and oxygen content of the water, among other factors (e.g., Balon 1975).

Determining the taxonomic identity of unfertilized or recently fertilized eggs can be a difficult task, often requiring electron microscopy. Embryos, especially those in advanced stages of development, are considerably easier to identify. Species-specific patterns of pigmentation often develop on the body and fin folds, particularly of marine species (e.g., codfishes, flatfishes). Other characters of use in identification include the general morphology of the head, gut, and tail; number of body myomeres; and existence of elongate or otherwise precocially developed fin rays (e.g., flyingfishes, ribbonfishes, sandfishes, weeverfishes) (Matarese and Sandknop 1984).

Species can be assigned to categories based on where embryonic development occurs and whether developing embryos are dependent on maternal versus yolk provisioning. If the mother lays eggs that then rely on yolk for nutrition, the species is **oviparous**, or egg-laying, as is the case in chimaeras, some sharks and skates, and most bony fishes. If the young develop inside the mother and the mother provides nutrition via a placental connection, secretions, or additional eggs and embryos eaten by developing young, the species is **viviparous**, or live-bearing. The young of viviparous fishes are generally born as juveniles, having bypassed a free larval stage. Viviparity characterizes about half of the 1000 species of Chondrichthyes and about 500 (2%) of the more than 20,000 species of bony fishes (many sharks, all rays, the coelacanth, eelpouts, brotulids, some halfbeaks, goodeids, four-eyed fishes, livebearers, rockfishes, Baikal oilfishes, surfperches, clinids, labrisomids). "Virtually every known form of vertebrate viviparity and possible maternal-fetal relationship may be found in fishes" (Wourms 1988, 4).

Some researchers recognize **ovoviviparity**, whereby young develop inside the mother but depend primarily on the yolk that was laid down during oogenesis (e.g., some sharks, scorpaenids, and, arguably, guppies). However, many intermediate conditions exist involving both yolk and maternal secretions, blurring the distinctions and making ovoviviparity difficult to define (Breder and Rosen 1966; Hoar 1969; Kendall et al. 1984; Wourms 1988 Nelson 1994). Although live birth usually follows internal fertilization, the two are not synonymous. Many elasmobranchs have internal fertilization but lay eggs that develop for many months outside the female. The priapium, a bizarre and complex organ of male phallostethids and neostethids, is used for holding the female during copulation and for fertilizing eggs just before they are laid (Breder and Rosen 1966; Nelson 1994).

Meristic Variation

Embryonic development generally progresses according to instructions laid down in the genetic blueprint, but the timing and even details of development are quite sensitive to environmental influences. Pollutants and chemical changes in the water often result in larval abnormalities and can be used to monitor environmental quality (for reviews, see von Westernhagen 1988; Weis and Weis 1989; Longwell et al. 1992). But even natural variation in temperature, oxygen content, salinity, light intensity, photoperiod, or CO_2 can affect development. Meristic traits such as numbers of fin rays, vertebrae, lateral scale rows, myotomes, and gill rakers are known to vary in relation to environmental conditions.

The pattern of meristic variation is not simple. The most commonly found relationship is for fin ray, vertebral, or scale numbers to increase with decreasing temperature (e.g., herrings, minnows, rainbow trout, grunions, killifishes, rivulines, darters). This inverse relationship exemplifies a general phenomenon, termed **Jordan's rule**, which applies to latitudinal effects on meristic numbers, although the actual determinant is water temperature (Lindsey 1988). However, an opposite pattern of increased meristic values with increased temperature has been observed for fin rays in guppies and plaice.

Another common pattern is the so-called V relationship, in which fewer meristic elements develop in fish raised at an intermediate temperature, but more elements are laid down at higher and lower temperatures (e.g., vertebrae or pectoral rays in brown trout, chinook salmon, rivulines, sticklebacks, paradise fish, snakeheads, plaice). An arched, L or A, pattern of higher numbers at intermediate temperatures has been observed for the fin rays of brown trout and chinook salmon. The actual quantitative difference between experimental groups raised at different temperatures is in the range 0.1% to 3% difference in number of elements per Celsius degree of temperature. For example, a 1% per degree difference in vertebral count over a 5°C temperature range for a fish with 100 vertebrae would translate into a 5-vertebrae difference between groups.

A critical or sensitive period often occurs during embryonic or larval development, when effects on meristic characters are strongest. Vertebral counts in brown trout are most sensitive to temperatures around the time of gastrulation and again as the last vertebrae are formed. Vertebral number is most sensitive before hatching in herrings and killifishes but later in paradise fish and plaice. Vertebrae form before fin rays and have an earlier sensitive period (Blaxter 1969). Meristic characters may also be sensitive to events prior to fertilization, such as the temperatures at which parents are held and the age of the parents (Lindsey 1988).

Causal mechanisms underlying these patterns are poorly understood. As an embryo develops, it differentiates via segment formation and grows via elongation. Environmental conditions may affect segment formation and elongation differently. Low temperatures may inhibit segment formation more than they inhibit elongation. Hence an embryo developing at low temperatures might be longer when differentiation occurred, causing more segments to be laid down and producing the pattern of more fin rays at lower temperatures (Barlow 1961; Blaxter 1969, 1984; Fahy 1982; Lindsey 1988; Houde 1989).

LARVAE

Biologically, the larval stage or its equivalent is probably the most thoroughly studied period of the early life history of fishes. This results from the identifiability of larvae, their relative abundance, and their importance in determining the distribution and later abundance of many species, particularly those of commercial value.

Larval life generally begins as the fish hatches from the egg and switches from internal, yolk reserves to external, planktonic food sources. The free-swimming young may still have a large yolk sac and be termed a **free embryo**, **eleutheroembryo**, or **yolk-sac larva** until the yolk is absorbed. **Fry** is a nonspecific term often used for advanced larvae or early juveniles; **swim-up stage** often refers to free embryos or larvae that were initially in a nest but have grown capable of swimming above the nest.

The larval stage continues until development of the axial skeleton, fins, and organ systems is complete (Fig. 9.7). Median fin rays develop, first as short, fleshy interspinous rays; true fin rays and spines develop later between the interspinous rays. Scales develop, first along the lateral line near the caudal peduncle, then in rows dorsal and ventral to the lateral line, and then spread anteriorly (Fig. 9.8). Once the full complement of scales is attained, the number remains fixed. The end of the notochord, termed the urostyle, flexes upward, and a triangular hypural plate develops just below it. Caudal rays grow posteriorly from the hypural plate elements. Characteristic larval pigmentation patterns develop, the eyes become pigmented, and the mouth and anus open and

become functional. Until gill filaments develop, the larva relies on cutaneous respiration, largely involving oxygen absorption across the thin walls of the primordial fin folds. The circulatory system is at first relatively open, and corpuscle-free blood is pumped through sinuses around the yolk and between the fin membranes (Russell 1979).

Pelagic larvae, which are particularly common among marine fishes, also characterize lake and river species in freshwater (e.g., whitefishes, temperate basses, centrarchid sunfishes, percid pike-perches, and perches). Larval periods in pelagic larvae vary widely in duration, from 1 to 2 weeks in sardines and scorpaenid rockfishes to about 1 month in many coral reef species to several months or even years in anguillid eels. A lengthy larval existence undoubtedly aids in the dispersal of these young to appropriate habitats (see Box 9.1).

Stream fishes, such as many minnows, darters, and sculpins, generally have demersal (bottom-living) larvae with short larval periods. Young fish remain on the bottom among rocks or vegetation until they develop reasonable swimming abilities, reflecting the turbulent conditions of streams that can injure larvae or carry them downstream to suboptimal habitats such as lakes or the ocean. Once muscular and skeletal features are formed, many are captured in stationary drift nets, indicating a juvenile dispersal phase, as opposed to larval dispersal in pelagic, lake, and river species. Larvae of some rocky intertidal fishes (e.g., sculpins, pricklebacks, gunnels, clingfishes) may not disperse offshore but instead actively spend their entire larval existence within 5 m of the shoreline, which could guarantee their return to a suitable habitat (Marliave 1986).

Larval periods are bypassed or very short in live-bearing fishes and in fishes that still possess considerable yolk reserves after hatching, such as salmonids. Arguably, no larval or juvenile phase exists in viviparous dwarf surfperch, *Micrometrus minimus*, males of which are sexually mature when born (Schultz 1993). Development that involves a larval phase with distinct metamorphosis into the juvenile stage is termed **indirect**. **Direct development** occurs if a larval stage is very brief or not definable, that is, if the fish hatches into a miniature but immature adult, as in many coastal marine forms and in many catfishes, salmoniforms, and cottids (Youson 1988).

Larval Feeding and Survival

A defining event in a larva's life is when yolk supplies are exhausted and the fish becomes dependent on exogenous

FIGURE 9.7.   *A recently-hatched, marine teleost larva, as represented by a 6 mm clingfish,* Gobiesox rhessodon. *Note development of mouth, eyes, median fin supports, melanophores, and upward flexion of the notochord at the base of the tail.*

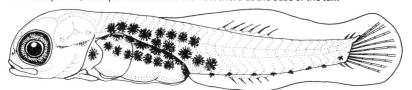

From Allen 1979; used with permission.

FIGURE 9.8.   *Development of scales in the black crappie,* Pomoxis nigro-maculatus, *showing the general pattern of scales developing initially near the tail and spreading anteriorly.*

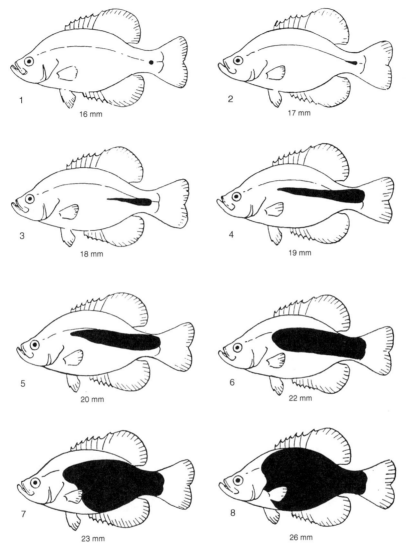

From Ward and Leonard 1954; used with permission.

food sources, usually in the form of small planktonic organisms such as diatoms, larval copepods and mollusks, and adult euphausiids, amphipods, ciliates, tintinnids, appendicularians, and larvaceans. Even fishes that are herbivorous as juveniles and adults are usually carnivorous as larvae, as in rabbitfishes (Bryan and Madraisau 1977). This generalization may reflect the difficulty with which energy and nutrients are extracted from plants.

The potential importance of food availability at the onset of exogenous feeding has guided our thinking about sources of larval mortality and the subsequent strength of year classes of fishes. Several influential hypotheses address the relationship among early larval biology, food availability, and adult population size in marine fishes. Hjort (1914) proposed the **critical period hypothesis**, which states that starvation at a critical period,

perhaps the onset of exogenous feeding, is a strong determinant of later year class strength. Blaxter and Hempel (1963) coined the phrase **point of no return** to describe when larvae, as a result of starvation, are too weakened to take advantage of food if it were available. Such irreversible starvation depends on larval condition and age: Well-fed, young anchovy may last only 1 to 2 days without food before irreversible starvation sets in, whereas healthy, older flatfish larvae may be able to go 2 to 3 weeks without food.

Cushing (1975) proposed the **match-mismatch hypothesis**, which suggests that the timing of reproduction in many marine fishes has evolved to place larvae in locales where food will be available, that is, that fish reproduction and oceanic production are synchronized. Since the cues of photoperiod and temperature that fish use to initiate spawning (see Chap. 22, Reproductive Seasonality) are not necessarily the same ones that determine plankton production, mismatches can occur and result in high larval mortality (May 1974; Russell 1976; Hunter 1981; Blaxter 1984; Houde 1987).

Because of the relationship between larval survival and later population size, the actual causes and patterns of larval mortality are of considerable theoretical and obvious practical interest. Literally billions of larvae are produced by most populations of marine fishes annually. In most species, more than 99% of these larvae die in their first year from the combined effects of starvation and predation; the average fish probably dies in less than a week (Miller 1988). Hence very minor shifts in mortality rates can have major implications for later year class strength and for recruitment into older, catchable size classes.

The temperatures at which larvae develop affect individual growth and development rates, metabolic rates, and energy requirements, all of which can influence mortality (Houde 1989). Across a 25°C temperature range characteristic of the difference between tropical and temperate conditions (5 to 30°C), mortality rates of marine larvae can vary fourfold, being highest at the higher temperatures. Growth rates at these higher temperatures at any given size are six times faster. Larval duration at the lower end of the temperature range typically exceeds 100 days, whereas at 25 to 30°C, metamorphosis occurs in 1 month or less.

One might expect that spending the least amount of time as a small, vulnerable larva would lessen the chances of both predation and mortality. However, the metabolic requirements of small ectothermic animals such as fish larvae increase in direct relationship to temperature. **Gross growth efficiency** (the ratio of weight increase to weight of food consumed) is constant despite temperature. A larva, because of its higher metabolic rate at higher temperatures, must consume more food to achieve the growth rate of a larva at lower temperatures. This is additionally compromised because gross growth efficiency declines with increased ingestion, and **assimilation efficiency** (how much of the food is actually useful to the larva) declines with increased temperature. To maintain the same average growth rate, a tropical larva has to eat three times more food than a temperate larva of the same species. Mismatching larvae with food availability therefore becomes more critical at higher temperatures.

Spawning patterns among species appear to represent adaptations to these temperature relationships. Tropical fishes typically spawn over an extended period, producing multiple batches of young, rather than releasing all their eggs in one large spawning session. This kind of "bet-hedging" strategy increases the probability that some larvae will encounter the kind of conditions necessary for successful growth, whereas a single spawning might lead to complete reproductive failure if food abundance were low, as it usually is in tropical pelagic areas. Temperate marine fishes that spawn in the summer, such as Atlantic mackerel and white hake, tend to spread their reproduction out over a longer period than do winter spawners such as herring and capelin at the same latitude (Houde 1989).

Pelagic larvae are particularly common among coral reef species. Whereas many nearshore temperate species have short larval periods or retain their larvae near the adult habitat, dispersal via a pelagic stage is almost universal among coral reef fishes. Of the 100 or so families that commonly inhabit coral reefs, 97 have pelagic larvae.

Tropical species with nonpelagic larvae provide instructive exceptions in that it is easy to postulate historical constraints or adaptive disadvantages to dispersal. Marine plotosid catfishes are a freshwater derivative family with highly venomous spines. The brightly colored young form dense, ball-shaped shoals and probably gain predator protection from this behavior, with a lack of dispersal helping to keep siblings together and facilitating the formation of monospecific shoals. Many of the viviparous brotulas live in fresh- or brackish water caves near coral reefs, a habitat that could be difficult to relocate by a settling larva. One species of damselfish, *Acanthochromis polyacanthus*, lacks a dispersed larval stage. It is also the only damselfish worldwide that continues to care for its young after they hatch. Other nondispersers include batrachoidid toadfishes, the monotypic convict blenny (Pholidichthyidae), and apparently reef species of the croaker family.

The adaptiveness of a floating larva for the other 97 families probably results in part from selection for avoidance of abundant predatory fishes and invertebrates in benthic habitats, larvae generally not settling until they have developed avoidance capabilities. Dispersal may also reflect 1) the possibility that successfully reproducing adults live in saturated habitats that offer few opportunities for settlement for their young and 2) the widespread and spotty distribution of coral reefs relative to immense oceanic expanses, necessitating the dispersal of offspring over as wide an area as possible (although evidence also exists to suggest that larvae are retained in nearshore gyres and currents and may actively return to parental regions; see Box 9.1). Pelagic dispersal is about the only thing these larvae have in common. Reproductive strategies include viviparity and oviparity, mouth-brooded eggs, nest builders, and demersal and floating eggs, some of the latter attached to seaweed. Larval periods range from 9 to more than 100 days; sizes at settlement, from 8 to 200 mm. Some settle prior to metamorphosis to a juvenile stage, some after, and one family (Schindleriidae) is even mature at the time of settlement (Leis 1991).

Larval Behavior and Physiology

Depicting larvae as largely passive greatly oversimplifies their behavioral capabilities, which diversify as they grow older (Noakes and Godin 1988). Making meaningful behavioral observations on very small, transparent larvae is understandably difficult, but successes point out the dynamics of larval development and the interdependence of morphology, physiology, and behavior.

Atlantic salmon larvae hatch from eggs buried in

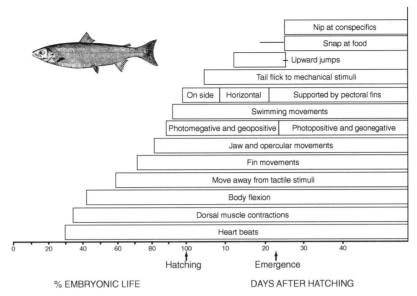

FIGURE 9.9.   *Anatomic and behavioral development of Atlantic salmon during embryonic life and the first 2 months after hatching. Note that even embryos are capable of avoiding aversive stimuli via fin and body movements. Emergence refers to yolk-sac larvae moving up out of the gravel of the nest.*

From Huntingford 1993 after Abu-Gideiri 1966 and Dill 1977; used with permission.

gravel, and the larvae remain in the gravel for almost another month. Even during this early period, obvious behavioral responses occur (Fig. 9.9). The larvae can make coordinated swimming movements involving both body and fins. A general geopositive and photonegative response, moving them down and away from light, keeps them buried. They also orient toward water currents. Emergence occurs as these responses reverse to photopositive and geonegative, causing them to swim up and out of the gravel but still into the current. They almost immediately react to food and interact with conspecifics, nipping at both, which helps obtain food and drive away potential competitors. They are also capable of swimming away if nipped.

Swimming capabilities in fish larvae in general are heavily dependent on the relationship of body size to water viscosity (its "syrupiness") and density (mass per unit volume). Small particles in a viscous medium such as water encounter tremendous friction forces with which larger particles do not have to contend. Such **Reynolds number** considerations (R_e = larval size $\times$ velocity $\times$ water density/water viscosity) mean that small larvae function at low Reynolds numbers (less than 1) and must swim sporadically and energetically, approaching 50 tail beats per second in anchovy and mackerel larvae, followed by long periods of rest. Larger larvae are able to swim and glide via inertial forces because viscosity is less of a factor (Purcell 1977; Hunter 1981; see Chap. 10, Body Size, Scaling, and Allometry).

Accuracy of feeding strikes has been shown to increase rapidly with development in cichlids as well as herrings, carp, and sea bass. Herring and anchovy larvae feed more

efficiently as they grow, improving their success rate from 3% to 10% shortly after hatching to 60% to 90% in the later stages of larval life (Blaxter 1969; Hunter 1981; Kamler 1991). In cichlids, many larvae and even juveniles are cared for by one or both parents. Many cichlid larvae form cohesive shoals almost immediately upon hatching, which obviously simplifies parental supervision. Some cichlid larvae feed on mucus produced by the parents and are attracted to the parents or to models with similar color patterns. By 25 to 40 days after hatching, shoals disband as individuals begin to charge, ram, and otherwise display at their siblings. Territory defense follows shortly thereafter (Huntingford 1993).

Feeding in mottled sculpin larvae, *Cottus bairdi*, is strongly dependent on lateral line development. Sculpin as early as 6 days after hatching are able to detect vibrations produced by brine shrimp larvae in the dark, using free neuromasts that are located superficially on the skin of the head. With development, many of these neuromasts become enclosed in subdermal canals, and sensitivity to minute disturbances, and ability to detect small prey, decrease (Jones and Janssen 1992).

More conventionally, ability to detect prey increases with age in most fishes, although some interesting specialized structures appear in larvae and then disappear during metamorphosis. In some weakly electric mormyrid elephantfishes of Africa, a special organ develops in larvae that produces electricity with a different discharge pattern than that shown in adults. This larval organ, which runs from the head to the caudal peduncle, degenerates after two months and is replaced by the adult caudal organ (Hopkins 1986).

Visual anatomy and ability also change with age. Many adult fishes have a duplex retina, one that contains both cones and rods (see Chap. 6, Vision). Cones function primarily by day, rods by night. The eye as it initially forms in embryos and larvae generally contains only cones, limiting feeding and other activities to daytime (e.g., herrings, many salmons and cichlids, soles). Herring do not develop rods until metamorphosis, which also coincides with the beginning of shoal formation, as does the development of the lateral line, which is also involved in shoaling behavior. Many aspects of visual capability— in terms of sensitivity to light, size of objects detected, resolving power, dark adaptation, and range of wavelengths detected—improve with growth in a number of species as a function of increased lens diameter, increased retina area, addition of rods or sometimes of cones, diversification and increased density of cones, and addition of visual and screening pigments.

Some fish larvae do have unique visual structures not possessed by later stages. Stalked, elliptical eyes exist in larvae of 14 families of marine teleosts, including anguilliforms, notacanthiforms, salmoniforms, and myctophiforms (see Fig. 9.10A). Stalked eyes can increase 10-fold the volume of water viewable from any given spot, which could save energy spent searching for food as well as help in detection of predators. Stalked eyes are lost during metamorphosis to the juvenile stage (Blaxter 1975; Weihs and Moser 1981; Hairston et al. 1982; Fernald 1984; Noakes and Godin 1988).

The larval phase is characterized by the onset of function of most of the organ systems an individual will use for the rest of its life, except for reproduction. The marked vulnerability of larvae to starvation and predation therefore decreases as these organ systems become functional. An obvious interrelationship exists between feeding ability and development of the jaws, digestive system, vision, and swimming musculature.

A less obvious pattern holds for development of predator avoidance. In white sea bass, *Atractoscion nobilis* (Sciaenidae), larvae between 3 and 7 mm long (4 to 23 days old) show little difference in predator escape behavior. Only about 25% of larvae react to approaching predators, usually by a startle response involving Mauthner cells, which are a pair of large, early developing interneurons that connect the hindbrain with motor neurons in the spinal column and cause the body to flex suddenly (see Fig. 6.10). Between 7 and 10 mm (23 to 30 days), the visual and acoustic systems of the larvae improve markedly. Visual acuity and accommodation improve, the optic tectum of the brain where visual information is processed develops, the gas bladder inflates, and the number of free neuromasts on the head and body increases. Both gas bladder and neuromasts are involved in detection of sound or of water displacement, perhaps allowing the larva to sense predatory motion as well as its own movements better. As a result, white sea bass larvae become much more adept at avoiding ambushing and hovering predators such as juvenile white sea bass and flatfishes. This behavior is augmented by a change to a demersal existence, which reduces the threat of predation by fast-swimming water-column plankti-

vores (anchovies, sardines, mackerel) to which they are still relatively vulnerable. Rapid development of neurosensory structures is therefore critical in the transition from a relatively passive target to a larva in which more than 75% actively avoid predators, over the course of less than 1 week (Margulies 1989). In herring (*Clupea harengus*) larvae, successful predator avoidance coincides with the appearance of lateral line neuromasts and the filling of the otic bulla with gas, both being structures associated with hearing (Blaxter and Fuiman 1990).

Larval Morphology and Taxonomy

Whereas eggs tend to be generally similar across many taxa, the larvae that emerge are strikingly distinct and often rather bizarre when compared with our expectations of fish morphology. The challenges to ichthyologists are identifying and linking larvae with their adult counterparts and understanding the adaptive significance of the various, seemingly incongruous structures that many larvae possess and then lose as they metamorphose into juveniles. One key is that larvae, although capable of locomotion, are at least initially relatively helpless and vulnerable. They are too slow to actively avoid most predators, other than those that also float with currents, such as various cnidarians with stinging cells.

Many predators on pelagic larvae are small, gape limited, and orient visually to their prey. To counteract these predators, larvae rely on structures that make them spiny and increase their body dimensions, or they mimic potentially noxious planktonic animals such as siphonophores (Fig. 9.10). Some structures such as extended fins, skin flaps, and gelatinous body coatings may also slow the sinking rate of larvae, keeping them in more nutrient-rich surface waters. Pigmentation patterns, which are often characteristic of larval stages and useful for identification, may screen harmful ultraviolet rays. This would apply particularly to the heavy melanistic pigmentation found on many species of **neustonic**, or surface-dwelling, marine larvae (Moser 1981).

Marine fish larvae have provided many mysteries to ichthyologists, and our knowledge of them has grown slowly and incrementally (Box 9.1). Some of this history is reflected in the names given larval stages, where many years of search were required to link larval and adult animals. Metamorphosis so radically alters many species that the two stages bear little obvious resemblance, attesting to different selective regimes and adaptations of different life history stages. The situation can be further complicated by stages intermediate between larvae and juveniles, sometimes called **prejuveniles**, which are also distinct.

For example, the amphioxides larva of branchiostomatid lancelets and the kasidoron larva of the deep-sea gibberfishes (Gibberichthyidae) once had familial status, the Amphioxididae and the Kasidoridae respectively. Generic status was initially given to a number of small fishes that are now identified as larvae or prejuveniles of well-known taxa, such as *Ammocoetes* (lampreys), *Leptocephalus* (anguilliform eels), *Tilurus* and *Tiluropsis* (notacanthoid halosaurs), *Querimana* (mullets), *Vexillifer* (pearlfishes),

FIGURE 9.10.   *Larval diversity in marine fishes. Fish larvae often bear little external resemblance to the adults into which they grow. Spines probably make larvae harder to swallow, whereas trailing appendages could mimic siphonophore tentacles and therefore be avoided by predators, or may aid in flotation by slowing the sinking rate of the larva. (A) A 26-mm lanternfish larva; note eyes on stalks and trailing gut. (B) A 17-mm lanternfish larva (Loweina); note elongate pectoral ray and dorsal and anal fin folds. (C) An 8-mm sea bass larva; note serrated dorsal and pelvic spines. (D) An 11-mm sea bass larva; note elongate dorsal spine. (E) A 64-mm exterilium larva of an unknown ophidiiform; note trailing gut. (F) An 8-mm squirrelfish larva; note spines on head and snout.*

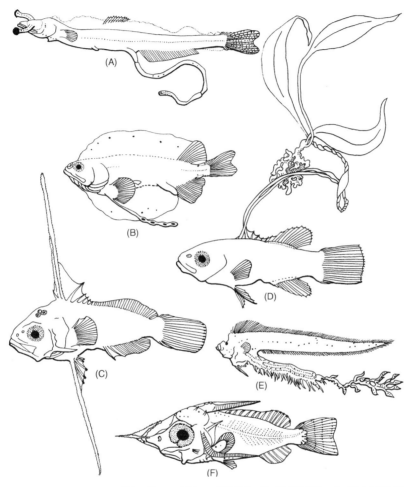

(A) from Moser and Ahlstrom 1974; (B) from Moser and Ahlstrom 1970; (C) from Kendall 1979; (D) from Kendall et al. 1984; (E) from Moser 1981; used with permission from Washington Sea Grant Program; (F) from McKenney 1959;

Rhynchichthys (squirrelfishes), *Dikellorhynchus* (tilefishes), *Acronurus* (surgeonfishes), and *Ptax* (snake mackerels). The scutatus larva of the big-eyed frogfish, *Antennarius radiosus*, was initially described as its own genus and species, *Kanazawaichthys scutatus*. These and other distinctive larval stages are still given separate descriptive names, such as the exterilium (external gut) stage of ophidioid cusk-eels (see Fig. 9.10E), the stylophthalmus (stalk-eyed) larva of idiacanthid black dragonfishes, and the flagelloserranus larval stage of sea basses with elongate, ballooning second and third dorsal spines (see Fig. 9.10D) (Richards 1976; Kendall et al. 1984; Pietsch and Grobecker 1987; Boschung and Shaw 1988; Eschmeyer 1990; Nelson 1994).

SUMMARY

1. Fishes experience indeterminate growth, growing throughout their lives, changing from larvae to juveniles to adults. They also change continually in terms of anatomic traits and ecological requirements and interactions. In most respects, large size confers physiological and ecological advantages on an individual.

2. Gametogenesis describes the development of sperm and eggs through a series of definable stages. Fish sperm vary in size, shape, and number of accessory structures (flagella, acrosomes). Fish eggs develop as oocytes that are invested with yolk. Females are able

---

| BOX 9.1 | *Getting from Here to There: Larval Transport Mechanisms* |

Many inshore marine fishes in temperate and tropical environments spawn offshore, but their larvae or juveniles use shallow habitats such as bays, mangroves, and other estuarine regions as nurseries. This characterizes many coral reef species and anguillid eels, croakers, porgies, bluefish, scorpionfishes, and flatfishes among temperate species. An important question therefore is how do such larvae, with their relatively limited swimming capabilities, move to shallow habitats?

For coral reef species, many of which spawn at reef edges, the overriding process appears to be one of luck: Larvae are carried by currents that flow past reef environments, coral reefs being basically small oases of shallow habitat embedded in the larger, deeper oceanic realm. Active orientation, directed movement, and habitat choice are implied by distributions of some species (see Leis 1991).

For species along continents, however, adults may spawn 100 km or farther offshore, and larvae must traverse the continental shelf to arrive at nursery grounds in 2 or 3 months. Given that the average fish larva can swim only about one to two body lengths per second, or about 1 km/day, active processes such as directed swimming will be too expensive energetically as well as too slow to transport larvae such extensive distances or to fight the strong, outflowing currents that characterize embayments, sounds, passes, and many other estuarine locales. Additionally, once larvae find their way to a nursery ground, they frequently remain in specific regions, again requiring that they fight strong tidal, river, and wind currents that should flush them back out to sea. Behavioral adaptations of the larvae themselves are therefore implicated in finding and remaining in appropriate habitats.

The larval transport phenomenon has three main components: movement toward shore, location of and movement into nursery areas, and retention in nursery areas (Fig. 9.11; Norcross and Shaw 1984; Boehlert and Mundy 1988; Miller

1988). Most interpretations of distribution patterns and behavior propose a combination of passive and active mechanisms during each phase. The degree of passivity decreases with age; young larvae rely largely on passive transport of the water mass in which they hatched, older larvae actively seek particular water masses with which they move, and older larvae or young juveniles swim actively.

The biggest mystery in larval transport has been the means by which larvae traverse hundreds of kilometers to get from offshore spawning grounds to inshore nursery areas (e.g., anguillid eels, bonefishes, menhaden, scorpionfishes, croakers, bothid and pleuronectid flatfishes). Spawning behavior of adults with respect to placement of eggs in favorable locales is obviously important. Spawning of many species on both west and east coasts of North America occurs in winter when wind-driven, onshore currents are most common. Most marine eggs are buoyant and drift toward the surface. This places them in surface layers that are pushed shoreward by winds either directly or as a result of Ekman transport, which involves wind-generated water movement that varies with depth. Some larvae may be carried by portions of major currents such as the Gulf Stream that spin off from the main current and move shoreward as "warm core rings."

Active behavioral mechanisms must influence distribution because larvae of different species that spawn in similar locales may wind up in different places. For example, mullet, bluefish, and dolphinfish spawn offshore, but larvae of the first two species use inshore nurseries, whereas dolphinfish larvae remain offshore. Vertical movements by larvae into upper surface waters could aid in retaining them in water masses that were moving shoreward (Norcross and Shaw 1984).

Shoreward movement is complicated by the larva's need to feed, so egg placement must also be in areas that are productive. Species that develop offshore often spawn in productive regions

---

to resorb unshed eggs and reuse the materials that went into egg production. Fecundity, the number of eggs a female produces, generally increases with increasing body size within a species and varies among species from one or two in some sharks to many millions in large teleosts.

3. Most marine fishes produce pelagic eggs that are fertilized externally and from which pelagic larvae

hatch. Most freshwater fishes deposit eggs on vegetation or on the bottom, often in nests.

4. Reproductive effort is a measure of the energy allocated to reproduction by an individual and can be characterized in part by the ratio of gonad weight to body weight, adjusted for the frequency with which an individual spawns and the phase of the spawning cycle.

BOX 9.1 *(Continued)*

that are relatively stable, such as gyres, upwellings, fronts, or other circulating patterns that retain larvae where food availability remains high (e.g., pollock, Dover sole). Some evidence suggests that coral reef fishes spawn at times and places that tend to retain them in local circulation patterns, which would promote their return to parental or nearby locales. For isolated island regions such as Hawaii, larval retention might be insurance against dispersal into vast and uninhabitable oceanic regions (Norcross and Shaw 1984; Lobel 1989).

After the journey toward shore, larvae frequently accumulate along shorelines and at mouths of bays and estuaries, only to be found later inside these regions. Such pulsed, directional movement against the general net flow of water out of an estuary may involve **selective tidal stream transport**, whereby small fish ride favorable currents and avoid unfavorable ones, usually by moving up into the water column on flood tides and down to the bottom during ebb tides (see Chap. 22, Tidal Patterns). This has been the suggested mechanism for movement into and retention within estuaries by young anguillid eels, herrings, shads, croakers, and plaice (Miller 1988).

Once in a nursery area, a fish must fight currents that would distribute it offshore or to less desirable inshore habitats. This becomes less of a problem as a juvenile fish grows larger and stronger and can actively choose locales or currents, but it remains a significant constraint for small larvae of species such as Atlantic herring (*Clupea harengus*). Herring spawn in the estuary itself, and larvae first move to upstream areas but later reside in downstream areas. This distribution is achieved by vertical movement with respect to different currents.

In most estuaries, surface waters are relatively fresh and move downstream, bottom waters are relatively saline and move upstream, and intermediate depths exhibit no net directional movement. In the St. Lawrence River estuary of Canada, young herring larvae remain near the bottom and are consequently carried upriver, whereas older larvae tend to move up and down twice daily and hence hold position in a relatively confined area of the estuary (Fortier and Leggett 1983). Species that spawn in estuaries (e.g., various herrings, cods, flatfishes, wolffishes, sculpins, gobies) tend to have large, demersal eggs and brief larval stages, all characteristics that would minimize export from the habitat (Hempel 1979; Norcross and Shaw 1984).

Larvae would have to respond to environmental cues that informed them when they were approaching appropriate or inappropriate habitats. Postulated cues that larvae might use to discriminate between water masses include odor, salinity, oxygen, turbidity, pH, geomagnetism, turbulence, light, food availability, temperature, and current speed and direction. Responses to these cues are likely influenced by tidal, circadian, or lunar rhythms (see Chap. 22). Much discussion has focused on whether larvae can in fact detect minor differences in these factors among water masses and orient appropriately, and this remains an area of active research (Boehlert and Mundy 1988; Miller 1988).

Many of our conclusions about larval transport remain conjectural because of a lack of confirming data. In addition, alternative explanations that view larvae as passive, drifting particles that regulate little more than their buoyancy readily explain certain aspects of their distribution. For example, rivers commonly enter estuaries and create ebb tides that are less saline than flood tides. Increased buoyancy in saltier water would tend to move drifting particles inshore with the flood, with no behavioral selection of water mass necessary (Miller 1988). Such purely passive drift cannot be totally dismissed in explanations of larval transport, although most evidence points to some type of behavioral regulation at most stages of development, at least once the egg stage is past.

5. Fertilization occurs external to the body of the female in most bony fishes and internally in most elasmobranchs and in about a dozen families of bony fishes. Females of some internally fertilized species can store sperm for several months. Males of internally fertilizing species possess an intromittent organ, modified from fins or cloacal tissue, for injecting sperm into the female.

6. Embryonic development in fishes proceeds as in most vertebrates. Many species are identifiable at or just after the time of hatching, based on pigmentation and fin development. Although most fishes hatch from eggs (oviparity), some undergo internal gestation and are born live (viviparity), with various intermediate conditions (generally referred to as ovoviviparity).

FIGURE 9.11.   *The general sequence of movement of marine larvae from offshore to inshore nursery grounds, as exemplified by events along the Oregon coast. Larvae are spawned offshore and carried onshore by shallow, wind-driven currents (e.g., Ekman transport). They then move alongshore by drifting with nearshore currents until encountering stimuli from estuaries, which they enter probably via selective tidal stream transport.*

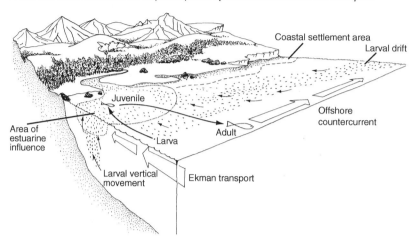

From Boehlert and Mundy 1988; used with permission.

7. Meristic (countable) traits may vary among developing fishes as a function of environmental conditions. Typically, fishes that develop in colder water lay down greater numbers of scales, fin rays, and vertebrae.

8. The larval period is most often the stage at which an individual disperses from the habitat of its parents. An important event in a larva's life comes when the yolk supply is exhausted and the young fish must forage on its own. Food availability during this critical period may determine future size of a population and may determine many aspects of spawning locale and timing. Most larvae die from starvation or predation during their first week of life. Physiological tolerance and sensitivity, ecological and behavioral competence, and survivorship all increase with increasing age.

9. Fish larvae often bear little external resemblance to the adults into which they grow. Larvae often possess large spines and trailing fins and appendages that may give them protection from predators. Many fish larvae and juveniles were originally given different specific, generic, and even familial names before they were linked with the adults into which they develop.

10. Many marine fishes spawn far offshore. The mechanisms by which their larvae move inshore to nursery habitats are a matter of debate. Onshore movement may involve passive transport via wind-driven surface currents. Movement into an estuary may involve selective tidal stream transport, where a small fish moves up into the water column on flood tides and then hugs the bottom during ebb tides.

SUPPLEMENTAL READING

Blaxter 1974; Breder and Rosen 1966; Hoar and Randall 1988; Jamieson 1991; Lasker 1981; Moser et al. 1984.

10 Juveniles, Adults, Age, and Growth

The numerical abundance of larvae of many species and the processes that reduce this abundance have been a major focus of ichthyological research. But growth and change continue throughout the life of a fish. Since most commercially important fishes are exploited as adults, juvenile growth and maturation to adulthood have been extensively studied. Additionally, many fishes undergo a postreproductive period when organ systems degenerate, providing a comparative model for studying the aging process and old age, topics of importance in human biology.

JUVENILES

Transitions and Transitional Stages

Hatching or birth and the onset of exogenous feeding represent two landmark events in the early life of a fish. Also of importance for many species is the change from larval to juvenile habitat, a transition that often involves settling from the water column and assumption of a near-benthic existence. Traditionally, the larval phase is considered to end and the juvenile phase begin as larval characters are lost and the axial skeleton, organ systems, pigmentation, squamation, and fins become fully developed and the fish looks essentially like a miniature adult. This transition can be brief and relatively simple, requiring minutes or hours in some damselfishes, or it can very long and complicated, taking several weeks in salmons, squirrelfishes, gobies, and flatfishes (see below, Complex Transitions) (Kendall et al. 1984).

Certain complex adaptations that essentially define major taxonomic groups do not appear until the juvenile phase. One example is the alarm reaction of the Ostariophysi (see Chap. 14, Superorder Ostariophysi; Chap. 19, Discouraging Capture and Handling). Minnows and other ostariophysan fishes have a characteristic escape response to alarm substance, a chemical released from the skin of injured conspecifics. The alarm reaction appears relatively late in development, after shoaling behavior develops and after fish can already produce alarm substance in their epidermal club cells. However, the alarm reaction is genetically hardwired. After 51 days posthatching, European minnows, *Phoxinus phoxinus*, react to alarm substance in the water the first time they encounter it, regardless of experience with predators (Magurran 1986).

Although eggs and larvae are by far the most vulnerable stages during the life history of an individual (see above and Chap. 23, Population Dynamics and Regula-

tion), attainment of the juvenile stage still requires successful food acquisition and predator avoidance abilities. The interplay between these factors is exemplified by juvenile brook trout, *Salvelinus fontinalis*. Brook trout do not undergo the smolt transformation characteristic of many other salmonids, as discussed below. Instead they hatch in spring and take up residence in small, shallow streams. Their chief task is acquiring sufficient energy stores during their first summer to get through the winter period of low food availability. Larger juveniles have a greater chance of surviving the first winter. To acquire energy and to grow, they must establish and defend a feeding territory. The best territories are in relatively shallow water, exposing the fish to both aquatic and aerial predators.

Predators can be avoided by remaining motionless, but motionless fish can't chase prey or repel territorial invaders. Feeding and fighting distract an individual from avoiding predators. Basically, smaller fish take more risks and tend to feed more extensively and openly, whereas larger fish are less willing to accept predation risks and are more willing to disrupt their feeding by taking evasive actions. The greater likelihood of winter starvation forces smaller juveniles to make the trade-off between predation risk and foraging differently from larger fish of the same age (Grant and Noakes 1987, 1988; see Chap. 19, Balancing Foraging Against Predatory Threat).

Transitional stages complicate the search for universally descriptive terminology about early life history. They also make it difficult to pinpoint when fish change from one developmental form to another. Transitional stages occur most commonly between larval and juvenile and between juvenile and adult periods. The transitional phase between larva and juvenile in reef fishes has been variously referred to as postlarval, late larval, new recruit, juvenile recruit, pelagic juvenile, and settler. The transitional phase may be variable in length, even within a species. This variability makes sense when it is realized that a young fish may not find habitat appropriate to its next stage simultaneously with its ability to make the transition into that stage. Hence if it were forced to settle from the plankton at day 35 of development, or at the moment that skeletal elements became ossified and fin rays fully developed, a larva that was still far out at sea might have no choice but to sink to the bottom several kilometers down and starve or freeze to death.

Variability in larval period is evident in larvae of the naked goby, *Gobiosoma bosci*, which settle from the plankton and take up a benthic, schooling existence for up to 20 days before transforming to solitary juveniles. Other gobies and a wrasse may have a 20- to 40-day window of opportunity during which they can search for appropriate habitat as larvae without transforming into the more sedentary juvenile form. Flatfishes can delay transformation to the juvenile form if they do not encounter appropriate juvenile habitat; they do this by alternating between settling on the bottom and swimming above it. Substrate preferences, which imply active search for appropriate habitat, have been observed in some larvae (e.g., surgeonfishes; Sale 1969). Direct observations of settling coral reef species indicate that such flexibility may be relatively widespread and that settlement and transition from larva to juvenile should not be viewed as a "parachute drop." Once settling competence is acquired, many larvae may have days or even weeks before they must settle and assume juvenile habits (Victor 1986; Breitburg 1989; Leis 1991; Kaufman et al. 1992, 114).

Complex Transitions: Smoltification in Salmon, Metamorphosis in Flatfish

Metamorphoses by definition imply major changes in the anatomy, physiology, and behavior of an animal. The transition from larva to juvenile in many fishes involves complex suites of change that frequently include major alterations in feeding habits and habitat. These alterations necessitate a breaking down and reworking of embryonic and larval structures and a rebuilding into adult structures that will function under very different environmental conditions.

A brief example involves sea lampreys (see also Chap. 13, Petromyzontiforms). Larval lampreys, termed ammocoetes, are sedentary, blind, freshwater animals that reside in burrows in silty bottoms and filter suspended matter from the water. At metamorphosis to the juvenile stage, this animal is transformed into a predator/parasite with a suctorial mouth, rasping tongue, salivary glands that secrete anticoagulants, functional eyes, tidal ventilation, and an ability to live in seawater (Youson 1988). Many other taxa could be mentioned, but details of two well-studied groups, salmons and flatfishes, will serve to exemplify the complexity of the reworkings that go into changing an animal adapted to larval existence into one adapted to meet the challenges of later stages in its life history.

Smoltification.   Widespread interest in salmonids has resulted in detailed knowledge and special terminology associated with different life history stages. Typically, salmon and trout spawn in gravel pit nests termed **redds**; the eggs hatch into **alevins** (yolk-sac larvae), which resorb the yolk and become **fry**. Fry develop species-typical patterns of vertical bars on their sides called **parr marks**, the fish now being called **parr**. After a few months or years, depending on species and population, the parr of anadromous species that spend their juvenile lives at sea (Pacific and Atlantic salmon, steelhead trout) head downstream as silvery **smolts**. The processes associated with the downstream migration of smolts are among the most intriguing and best studied biological aspects of the early life history of fishes.

Smoltification is a complex phenomenon involving reworkings of just about every characteristic of a young salmon. An interesting feature of the changes is that they are preparatory: They occur as the animal changes from a parr to a smolt in freshwater, anticipating the environmental conditions that the young fish will later encounter after it enters the ocean. Anatomically, smolts turn from countershaded and barred to silvery, which is a better form of camouflage in the open sea (see Chap. 19, Invisible Fishes). They also take on a slimmer, more streamlined shape that involves a reduction in condition as body lipids are reduced. The silvering results from an increase in the density of purine crystals, mostly guanine but also hypoxanthine, which are deposited beneath scales and deep within the dermis.

Despite loss of lipids, smolts are more buoyant than nonmigratory conspecifics. Increased buoyancy results from increased gas volume in the gas bladder, which may reduce the energetic costs of migration. The complexity of hemoglobins in the blood increases, affecting oxygen affinity and the Bohr shift, among other respiratory factors (see Box 5.2). These alterations prepare a migrating smolt for oceanic conditions that often include reduced availability of oxygen compared with the cold, turbulent waters of a stream or river. Many changes occur in gill function, including increased chloride cell number and changes in ion permeability and enzymatic activity. These changes anticipate the move from the hypoosmotic freshwater environment where ion loss is the major problem to the hyperosmotic marine environment where water retention is the major problem (see Chap. 7, Osmoregulation and Ion Balance) (Wedemeyer et al. 1980; McCormick and Saunders 1987; Hoar 1988).

Behaviorally, Atlantic salmon parr are first highly territorial in shallow water but then move into deeper water and form shoals, although a dominance hierarchy frequently exists in the shoals. Even this aggression decreases as fish start to move toward the sea. The movement is aided by a reversal in rheotaxis, the response to flowing currents that kept even embryos headed upstream (see above). Positive rheotaxis disappears as fish in large shoals drift downstream with the currents (Hoar 1988; Noakes and Godin 1988; Huntingford 1993). It is during smolting that young fish learn or imprint on the odor of their home stream, enabling them to identify it among hundreds of alternatives when they return from the sea during the spawning migration (see Chap. 22, Diadromy).

Many of the transformations that occur during smoltification can be linked to changes of circulating hormones (Fig. 10.1). Increases in corticosteroids, prolactin, and growth hormone affect lipid metabolism, osmoregulation, and mineral balance respectively. Cortisol and estradiol levels also increase. Thyroxin levels also increase naturally, and experimental injections of thyroid-stimulating hormone (TSH) can induce many of the physiological and behavioral events of smoltification, such as purine deposition, gill enzyme activity, increased swimming activity, body growth, and lipid consumption. These changes suggest that thyroid hormone, interacting with photoperiod and endogenous rhythms, plays an important role in the process (Hoar 1988; Huntingford 1993).

Smoltification is by no means fixed in terms of age in a species or even a population. Atlantic salmon may smolt at ages of 1 to 7 years, depending on temperature and latitude. Onset of smoltification in siblings may vary by as much as a year. Feeding opportunities appear to be the key determinant of the onset of smoltification. Well-fed individuals smoltify younger, although genetic differences in feeding activity may cause some fish to cease feeding and consequently delay smoltification. Some evidence indicates a size threshold, in that Atlantic salmon that do not attain a length of 10 cm by the autumn of the first growing season are less likely to smolt the next year. Rate of growth and age interact with this hypothesized

threshold length, in that faster-growing fish are more likely to smolt, and older fish may smolt at a smaller size.

Timing is also important, in that the smolt stage itself lasts a few weeks, and if a fish does not enter the sea, the process will reverse and the fish will return to the parr condition. Although Pacific salmon (*Oncorhynchus* spp.) and Atlantic salmon are considered anadromous and therefore undergo smoltification, these fishes can become landlocked and never migrate to sea, or some individuals within a population may bypass the smolt and migratory phases and remain behind in freshwater. When this happens in males, they may mature quickly at 1 year of age and spawn with females that return the next season (see Chap. 20, Alternative Mating Systems and Tactics). In Atlantic salmon, the proportion of such precocious males differs among populations, ranging from 5% to more than 50% of the males. The factors determining precocious maturation in male salmons are widely debated, with evidence suggesting that food availability or genetic factors are determinant (Thorpe 1987; Hoar 1988).

Asymmetric flatfish. Symmetry is an almost universal anatomic characteristic of animals. Most animals, regardless of phylum, exhibit bilateral symmetry in their morphology, having roughly mirror-imaged structures to the right and left of midline. Deviations from symmetry then become challenges to our ability to understand how and why biological systems function. Biologists seek to understand the causation and function of asymmetry at the

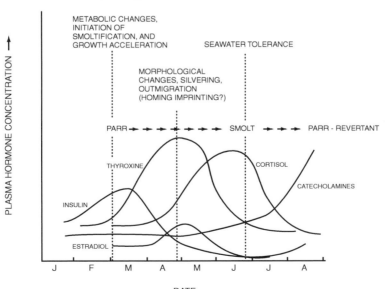

FIGURE 10.1.   *Sequence of events during smolting in coho salmon,* Oncorhynchus kisutch, *and their correspondence with changing levels of important hormones. Hormones implicated in the parr-smolt transformation tend to peak prior to, and may cause, the acquisition of the various anatomic, physiological, and behavioral traits that characterize the smolt stage. One group, the catecholamines, remains low through the smolting process but climbs if the fish does not migrate to sea and instead reverts to the parr stage.*

Drawing by W. W. Dickhoff in Hoar and Randall 1988; used with permission.

proximate level of genetic and environmental control of development and at the ultimate level of its possible adaptiveness.

Among the more startling examples of asymmetry is the "handedness" of flatfishes. The 11 families and about 570 species of pleuronectiforms (flounders, halibuts, soles, etc.) as a group are characterized by adults that lie on the bottom on one side of their body. Their flattened bodies are functionally analogous to many other benthic-living fishes such as angel sharks; skates; rays; banjo, suckermouth armored, and squarehead catfishes; ogco-cephalid batfishes; platycephalid flatheads; and some scorpionfishes. The major difference is that all the other groups are flattened in a dorsal-ventral plane (= **depressed**), whereas flatfishes are laterally flattened (= **compressed**) (people often depress cockroaches and compress mosquitoes).

Depressed fishes maintain their bilateral symmetry despite their extreme morphology. Most compressed fishes are deep-bodied, bilaterally symmetric species that swim in the water column and use their flattened bodies to increase maneuverability or to increase their body depth against predators (e.g., serrasalmine characins, centrarchid sunfishes, many pompanos, monodactylid fingerfishes, butterflyfishes, ephippid batfishes and spadefishes, surgeonfishes). Flatfishes are laterally compressed but lie on the bottom on either their right or left side and are therefore faced with the challenge of receiving sensory information from only half their sense organs, the other half being buried in the sand or mud. The most obvious accommodations to their unusual orientation can be seen in the structure and development of their visual apparatus.

Flatfishes begin life as normal, bilaterally symmetric, pelagic larvae. In the starry flounder, *Platichthys stellatus*, larvae emerge from the egg when about 3 mm long and begin exogenous feeding. For the next month or two, they lead normal pelagic lives, until they reach a length of 7 mm. Then metamorphosis to a compressed shape begins (size at metamorphosis varies between 4 and 120 mm in different flatfishes). Most bones are incompletely ossified at this time, which apparently makes the transformation easier. The anterior neurocranium, brain, and eye sockets (orbits) rotate (Fig. 10.2). This allows one eye to actually migrate across the top of the head. In some species of bothids and paralichthyids, the eye moves through a slit that appears between the skull and the base of the dorsal fin. The dorsal fin remains in the midline or, in some species, grows forward until the first spine sits anterior to the eyes. The entire process happens quickly, over about a 5-day period in starry flounders, in less than 1 day in some species.

Other asymmetries occur that reflect transformation to a benthic and compressed existence. The nasal organ on the blind side migrates to the dorsal midline. The blind side is usually unpigmented, may lack a lateral line, and has smaller pectoral and pelvic fins. Squamation frequently differs on the two sides. During metamorphosis, the semicircular canals undergo a 90° displacement, and the dorsal light reaction (see Chap. 6, Equilibrium and Balance) also changes appropriately for a fish lying on its side. At the time of metamorphosis or shortly thereafter, the fish takes up a benthic existence and loses its gas bladder.

As a rule, families are characterized by having both eyes on a particular side of the head. Hence lefteye flounders (Bothidae) lie on their right side and have both eyes on the left side, the right eye having migrated; this is termed the **sinistral condition**. Occasional freaks occur

FIGURE 10.2. *Progressive eye migration in a developing summer flounder,* Paralichthys dentatus. *When the flounder larva is about 10 mm long, the right eye begins to migrate over to the left side of the fish via a process that includes bone resorption and rotation of the fish's neurocranium. The entire process takes 3 to 4 weeks, during which time the larva grows 5 to 10 mm. Stippling in stages A through D indicates the position of the right eye on the right side of the body. Note other developmental changes, including development of eye structures, anterior migration of the dorsal fin, growth and elaboration of pectoral and pelvic fins, and mouth growth.*

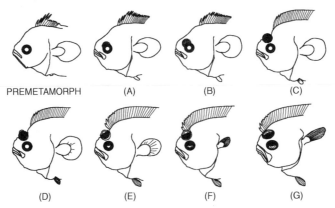

PREMETAMORPH   (A)   (B)   (C)

(D)   (E)   (F)   (G)

From Keefe and Able 1993; used with permission.

because of presumed developmental abnormalities, and individual members of right-eyed species may be left-eyed; in such individuals, viscera may also be twisted and color patterns abnormal. Regular variation in such "handedness" also occurs. Starry flounders, although members of the righteye (Pleuronectidae) family and usually right-eyed, or **dextral**, often include left-eyed individuals. In California, 50% of the individuals may be left-eyed, and in Japan 100% of these pleuronectids are left-eyed.

That these nonconformist individuals are in fact abnormal is evident in the development of their optic nerves. In all vertebrates, normal development results in a crossing of the optic nerves leading from the eye to the brain, such that the right side of the brain receives information from the left eye and vice versa. In left-eyed starry flounders, the optic nerve crosses twice, literally twisting around itself, a condition for which it is hard to assign any adaptive advantage. Experimental crosses of individuals from different populations have established that the determination of "handedness" in flatfishes is under complex genetic control, but no evidence exists to suggest that one side is adaptively better than the other (Policansky 1982a,b; Ahlstrom et al. 1984).

Adults

Determination, Differentiation, and Maturation

The development of a reproductively functional individual involves three very different processes—determination, differentiation, and maturation—that occur at different times during life history. **Sex determination** is the process by which the maleness or femaleness (**gender**) of an individual is decided, usually during early ontogeny. Determination can be either genetically or environmentally controlled. **Differentiation** involves the development of recognizable gonadal structures—ovaries or testes—in an individual, although maturing gametes are not necessarily present. **Maturation** implies the actual production of viable gametes, spermatozoa or ova. An individual's gender may be determined at fertilization, the fish may differentiate as a juvenile, but it is not technically an adult until it matures.

An interesting relationship exists among sex determination, sex change, and the existence of sex chromosomes in vertebrates. Birds and mammals have identifiable sex chromosomes. Male mammals are the **heterogametic** sex, possessing an X and a Y sex chromosome. The opposite holds for birds, in which the female is the heterogametic sex, known as Z-W heterogamety. In birds and mammals, gender is determined and fixed at fertilization. Little change occurs in the gender of individuals, nor are environmental conditions thought to affect sex determination. Some reptiles and amphibians have sex chromosomes; some do not. In turtles, males are generally produced at low temperatures; the opposite holds for crocodilians and lizards. In taxa in which such **environmental sex determination** (ESD) occurs, sex chromosomes are relatively rare (Gorman 1973; Francis 1992).

Sex chromosomes are rare among fishes, occurring in only 25 of the 700 or so species for which chromosome number and morphology have been described. Examples include several deep-sea families, including bathylagid smelts, sternoptychid hatchetfishes, neoscopelids, myctophid lanternfishes, and melamphaid ridgeheads. In shallow waters, heterogamety has been demonstrated in the following families: anguillid and conger eels; trout; lizardfish; bagrid, silurid, and loricariid catfishes; killifishes; livebearers; sticklebacks; sculpins; cichlids; gobies; and a tonguefish. Heterogamety among these families is fairly evenly divided between males and females (Gold 1979; Sola et al. 1981). As might be anticipated from a taxon in which less than 5% of all species possess sex chromosomes, sex determination in fishes appears to be exceedingly labile.

The exact stage at which gender is determined in fishes is controversial. Although genetic determination probably applies to most species, in many fishes, sex determination may not be fixed at fertilization or even during early ontogeny. Many fishes go through a prematurational sex change, differentiating but not maturing first as females, with some individuals later changing to males. This pattern is suspected or known from hagfishes, lampreys, minnows, salmonids, cichlids, butterflyfishes, wrasses, parrotfishes, gobies, and belontiid paradise fish. Such ambivalence is not altogether surprising when it is recalled that all gonads in agnathans and teleosts develop from a single structure, the epithelial cortex, which gives rise to ovaries in higher vertebrates. In sharks, ovaries develop from the cortex, and testes develop from the medulla; sharks consequently show no sexual lability.

The possibility for ESD exists in all the fishes listed above, with *environment* being understood to include climate, food availability, and social interactions. In the paradise fish, *Macropodus opercularis,* all individuals begin as females and some later differentiate as males, but all this occurs prior to maturation. Final determination is based on social status: Dominant individuals become male and subordinate individuals become female as a direct result of social interactions. ESD has also been documented in the ricefish or medaka (*Oryzias latipes*), poeciliid livebearers, rivulines, and Siamese fighting fish, all in response to temperature extremes (Francis 1992).

ESD is best understood in the Atlantic silverside, *Menidia menidia*. Northern populations have a limited spawning season and exhibit genetic determination, but southern populations have a longer season and are more sexually labile. Southern larvae spawned in the spring at low temperatures tend to become females, whereas those spawned in summer at higher temperatures become male. Spring-spawned individuals will have a longer growing period before the next spawning season than will late-spawned fish. Spring-spawned fish can therefore take advantage of the body size-egg number relationship and benefit more from larger body size as females than as males (Conover and Kynard 1981; Conover and Heins 1987).

Many species change sex after initial maturation, referred to as **postmaturational sex change** (see Chap. 20, Gender Roles in Fishes). Changing from functional female to functional male is most common in fishes and is termed **protogyny**; a change from male to female is **protandry**. Nonchangers are called **gonochores**. A few serranid hamlets are **simultaneous hermaphrodites**, producing viable sperm and eggs at the same time; only the rivuline *Rivulus marmoratus* fertilizes its own eggs (Soto et al. 1992). Some species of livebearers are **parthenogenetic**, having eliminated males from the reproductive picture. The environmental and social conditions promoting sex change, hermaphroditism, and parthenogenesis are detailed in Chapter 20.

Maturation and Longevity

Not surprisingly, age at first reproduction and longevity vary greatly among fishes (Finch 1990), making it difficult to identify patterns or draw conclusions. The adaptive significance of differences in age at first reproduction relates to trade-offs between committing energy to somatic growth versus reproduction; these trade-offs are discussed in Chapter 23 (Life Histories and Reproductive Ecology). Extremes in age at first reproduction include some embiotocid surfperches (the males of which are born with functional sperm) and sturgeons and some sharks (which may take 10 to 20 years to mature). Sturgeons may live 80 to 150 years. The slowest-maturing shark is the spiny dogfish, *Squalus acanthias*, a species well known to students of comparative anatomy. Spiny dogfish do not mature until 20 years of age and have the longest recorded lifespan of a shark, upward of 70 years. Many small stream fishes mature in 1 year, being reproductively active the spawning season after they hatch (e.g., most darters), although maturation may take longer in populations at higher latitudes. The record for naturally delayed reproduction among bony fishes is apparently held by American eels in Nova Scotia, which may not mature and undertake their spawning migration back to the Sargasso Sea until they are 40 years old (see Chap. 22, Catadromy).

Longevity patterns are only slightly more definable; with many exceptions, larger fishes generally live longer than smaller fishes. The oldest teleosts known are scorpaenid rockfishes of the northeastern Pacific. Radioisotopic and otolith analyses indicate that rougheye rockfish (*Sebastes aleutianus*) live for 140 years, silver-gray rockfish (*S. borealis*) for 120 years, and deep-water rockfish (*S. alutus*) for 90 years (Finch 1990; Leaman 1991). Among common sport species, European perch can live 25 years and largemouth bass can live 15 to 24 years (Radonski 1993). Numerous species live for a year or less, including the so-called annual fishes of South America and Africa, as well as North American minnows in the genus *Pimephales* (fathead, bullhead, and bluntnose minnows), several galaxiid fishes from Tasmania and New Zealand, retropinnid southern smelts, Japanese ayu, Sundaland noodlefishes (Sundasalangidae), a silverside, a stickleback, and a few gobies.

Death and Senescence

Death in fishes usually results from predation, accident, opportunistic pathogens, or accumulated somatic mutations that lead to a slow decline in health and an increased susceptibility to environmental factors. However, some fishes age via the programmed death process of **senescence** that is more typical of mammals such as ourselves. Senescence refers to age-related changes that have an adverse effect on an organism and that increase the likelihood of its death (Finch 1990). Senescence includes the metabolic and anatomic breakdown that occurs in older adult animals following maturation and reproduction.

Pacific salmon provide a dramatic example. Reproductively migrating fish in peak physical condition enter their natal river, mature, spawn, break down anatomically and physiologically, and die in a matter of weeks. Many of the anatomic and physiological changes that occur can be linked to the combined effects of overproduction of steroids and starvation. Interrenal cells, which are steroid-producing cells associated with the kidney and are homologous with the adrenal cortex of mammals, secrete corticosteroids, producing blood levels of these substances five or more times higher than normal levels. This **hyperadrenocorticism** results in rapid degenerative changes in the heart, liver, kidney, spleen, thymus, and coronary arteries; the latter degeneration is strikingly similar to coronary artery disease in humans. The digestive tract including intestinal villi degenerates, fat reserves are depleted, and feeding ceases. Fungal infections and reduced resistance to bacteria occur, indicating loss of immune function. A conflict between reproduction and survival is evident in the breakdown of the immune system: Elevated corticosteroids apparently serve to speed the mobilization of stored energy into reproductive activity but have the side effect of suppressing immune function. In naturally spawning Pacific salmons, these side effects are irreversible. Castrated males and females do not produce the elevated corticosteroids, do not spawn, but instead continue to grow to twice the length and live twice as long as intact fish. Precocious males, those that matured as parr and bypassed the smolt and marine phases, may survive spawning and breed again the next year (Finch 1990).

Equally spectacular senescence occurs in several other fish taxa. Reproduction in both parasitic and nonparasitic lampreys involves maturation accompanied by cessation of feeding and atrophy of most internal organs with the exception of the heart and gonads. Fats and muscle proteins are metabolized or transformed into gonadal products. Both males and females die shortly after spawning, probably from starvation. Anguillid eels live as juveniles for many years in rivers and lakes, then undergo a reproductive metamorphosis that includes enlargement of eyes, changes in body coloration and fin proportions, gut degeneration, and cessation of feeding. After a reproductive migration to the sea that may take them thousands of kilometers, all adults presumably die (see Chap. 22, Catadromy). Laboratory manipulation of hormone functions indicates that, as with salmons, rapid senescence results from elevated corticosteroids and starvation. During maturation, conger and snipe eels also experience gut atrophy and in addition lose their teeth. The ice goby (*Leucosparion petersi*), which enters freshwater to spawn and then dies, develops enlarged adrenals and splenic degeneration.

More gradual senescence has been observed in many

multiply spawning species, such as herrings, haddocks, guppies and other livebearers, annual killifishes, and medaka. Anatomic and physiological indicators of gradual senescence include reduced or even negative length and weight change, reduced egg output, corneal clouding, disordered scales, malignant growths such as melanomas, spinal deformities, and impaired regenerative capability. Such senescent changes are more common in small, short-lived species that mature at relatively early ages (Lindsey 1988; Finch 1990; Kamler 1992).

AGE AND GROWTH

Age

Many of the phenomena described above include fairly precise statements of the age of the fish involved. How are such ages determined? Although size is generally correlated with age, sufficient variation in size at any particular age exists in most species (see below), making it difficult to estimate one from the other with much precision, especially in long-lived or slow-growing fishes. Therefore, researchers interested in determining a fish's age look for structures that increase in size incrementally, in relation to some periodic environmental phenomenon. Many body parts meet this criterion, differing among fish species and among age groups (Fig. 10.3). The most commonly used techniques involve counting naturally occurring growth lines on scales, otoliths (statoliths in lampreys), vertebrae,

fin spines, eye lenses, teeth, or bones of the jaw, pectoral girdle, and opercular series; catch-and-release programs that include labeling growing structures with dyes or radioisotopes are used to validate the periodicity of growth. Representative growth patterns that can be used to age fishes include annual growth rings on scales and daily growth increments on otoliths (Brothers 1990).

Scales in most fishes begin to develop during the late larval stage or during metamorphosis to the juvenile stage. They arise as bony plates in the dermis. Bone-forming cells, termed osteoblasts, lay down layers of roughly concentric circles of bone, or **circuli**, along the midbody, starting in the region of the developing lateral line. Scales grow by accretion as more bone is added along their periphery, increasing in thickness but particularly in diameter. Diameter increase reflects body growth, in that circuli are closer together during periods of slow growth, such as winter at higher latitudes, and wider apart during rapid growth, such as during spring and summer, analogous to the growth rings of trees. This growth pattern creates alternating dark and light bands in the scale that correspond to periods of slow and fast growth respectively, particularly when viewed with transmitted light (e.g., backlit). In a habitat with distinct growing and nongrowing seasons, such as most temperate lakes, one thick band and one thin band constitute a year's growth. The band is referred to as an annular mark, or **annulus**. The number of annuli on a scale therefore gives a record of fish age in years.

In reality, many factors can confuse or interrupt annulus formation. Growth typically slows down when fish

FIGURE 10.3. *Methods for determining fish age. Growth lines are added periodically to hard body parts, but the best part to count differs among taxa. Scales, otoliths, fin rays, and vertebrae are the most commonly investigated structures. The body parts used for growth determination of oceanic pelagic fishes are shown in the figure.*

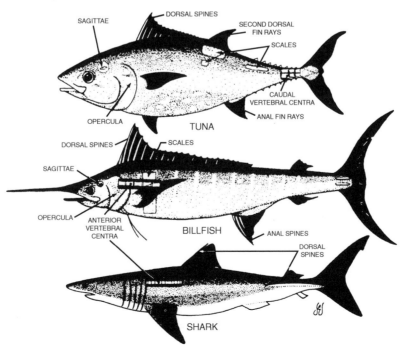

From Casselman 1983.

enter spawning condition, reflecting the allocation of energy away from growth and into gamete production and reproductive behavior. Decreased growth may also result from a decrease in feeding that occurs in many species that show parental care (see Chap. 20, Parental Care). Such **spawning checks** will appear as dense bands and can be mistaken for annuli, leading to overestimation of age. Bands resembling annuli, termed **false annuli**, can also result from multiple wet and dry seasons, as occur in many tropical locales, as well as from disease, recovery from injury, responses to pollutants, and forced periods of inactivity and nonfeeding. Feeding often slows down or ceases during summer periods of high water temperature and low oxygen.

Underestimation of age can result if scales do not begin to develop until the fish is a few years old, as in anguillid eels, or if older fish reach a growth asymptote and hence grow little if at all. Based on scale ages, Pacific sablefish (*Anoplopoma fimbria*) were generally thought to live 3 to 8 years and were managed as a fast-growing, short-lived, productive fishery. Subsequent studies, involving otolith sections and experimental injections of oxytetracycline into tagged fish, showed that the fish instead lived for 4 to 40 years, and as long as 70 years. However, older fish had essentially stopped growing, both in terms of increasing body and scale sizes, causing underestimation of age. These new analyses forced a major revision in the management strategies applied to the fishery and point to widespread realization that different parts of a fish's body can grow at different rates (e.g., Casselman 1990). Validation of the annular nature of growth rings on scales and other structures is now generally recognized as essential (e.g., Hales and Belk 1992). Validation involves injection of vital dyes or radioisotopes that are quickly incorporated into one circulus of a scale, followed by periodic examination, or some other method whereby the actual time interval represented by an incremental mark can be verified (Beamish and McFarlane 1983, 1987; Stevenson and Campana 1992).

The semicircular canals of the inner ear contain otoliths, which are calcareous structures of characteristic shapes and sizes depending on species. Otoliths form earlier than scales, often appearing in the otic capsules of embryos prior to hatching (Brothers 1984). Otoliths grow via accretion of layers of fibroprotein and calcium carbonate crystals. In many fishes, this deposition occurs on a daily basis, relatively independent of most environmental conditions. Hence a one-to-one correspondence of rings to days exists on the otoliths, allowing highly accurate estimates of fish age, particularly in larvae and juveniles (Pannella 1971; Brothers et al. 1976). The width of the daily increments can be a useful indicator of growth conditions and can offer valuable information on when significant events occur during the early life history of an individual, such as length of larval period (Brothers and McFarland 1981). Of the three otoliths, the saggita is usually largest and most useful for aging studies.

As fish age, otolith layers may grow too close together to resolve daily increments, but seasonal and annual records are still evident on these hard body parts. Changes in spacing or chemical composition of larger zones may indicate not only age but also when a fish moves from habitats that are more or less favorable for growth, as when eels migrate upriver or salmon smolts move from food-poor freshwater rivers to food-rich and saline estuaries (Fig. 10.4).

FIGURE 10.4.   *The correspondence between growth zones on an otolith and habitat use in an American eel,* Anguilla rostrata. *This sagittal otolith indicates that the eel was 16 years old when captured. It spent 3 years at sea or in the estuary of the St. Lawrence River (fast-growth nucleus zone), migrated upriver over a 2-year period (slow-growth transition zone), and finally took up residence in the upper St. Lawrence-Lake Ontario area (fast-growth edge zone). Habitat use was confirmed by measuring strontium-calcium ratios in the different zones of the otolith, using an electron microprobe associated with an electron microscope. Different ratios arise when an animal inhabits sea- versus freshwater.*

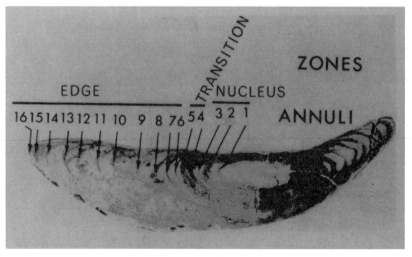

From Casselman 1983. American eel drawing from Bigelow and Schroeder 1953.

Growth

Among vertebrates, most growth in mammals and birds is **determinate**, ceasing after an individual matures. Lower vertebrates exhibit **indeterminate growth**, in that growth continues throughout the lifespan of an individual, although at a constantly decelerating rate. Hence older animals are generally larger, all other things being equal. The caveat here of equality among growth-controlling factors points out another crucial aspect of growth in fishes, namely, its plasticity. Size at age varies enormously in fishes, whether we are comparing species, populations, individuals within populations, or individuals within cohorts and clutches.

Just about any factor that might possibly influence growth has been shown to have an effect, including temperature, food availability, nutrient availability, light regime, oxygen, salinity, pollutants, current speed, predator density, intraspecific social interactions, and genetics (reviewed by Wootton 1990). These factors, often working in combination, create large variations in size of fishes of the same and different ages, leading to so-called size-structured populations, age differences in ecological roles and the ontogenetic niche, and cannibalism (Beverton and Holt 1959; Beverton 1987; see Chap. 23).

When plotted against age, growth curves for fishes appear asymptotic, meaning they tend to flatten out at older ages, although the degree of flatness varies greatly among and within species (Fig. 10.5). This variation forms the basis of an equation commonly used to describe individual growth in most fishes, known as the **von Bertalanffy growth equation**, which in its simplest form can be written

$$L_t = L_{max}(1 - e^{-gt})$$

where L is length, t is a point in time, L_{max} is the maximum or asymptotic length attained by the species, e is the base of natural logarithms, and g is the all-important constant that describes the rate at which growth slows. The von Bertalanffy equation is based on bioenergetic considerations, viewing growth as a result of anabolic and catabolic processes by which a fish takes in oxygen and energy to build tissues and uses up energy and tissue over its life. Many refinements of the equation have been made and alternatives proposed that take into account age- and weight-specific differences in growth, food consumption rates, temperature, and overall energy budgets (see Brett 1979; Gulland 1983; Weatherly and Gill 1987; Busacker et al. 1990; Wootton 1990).

The relationship between increasing mass and the length of the fish involves a power function. Mass increases as a function of the cube of the length of the fish because of the universal relationship between the volume and the surface area of a solid, with volume increasing

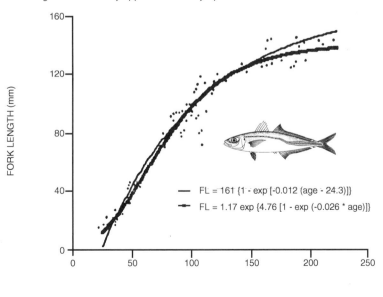

FIGURE 10.5. *Growth curves and their statistical description. The plotted lines indicate growth over time for the round scad,* Decapterus punctatus. *The thin line and the upper equation are calculated from the von Bertalanffy equation; the thicker line and lower equation are based on a related calculation, the Gompertz equation. The von Bertalanffy equation predicts asymptotic growth; the Gompertz equation predicts a sigmoidal curve where growth increases and then decreases. The two lines are statistically similar, showing how growth slows down with age and eventually approaches an asymptote.*

FL = 161 {1 - exp [-0.012 (age - 24.3)]}

FL = 1.17 exp {4.76 [1 - exp (-0.026 * age)]}

From Hales 1987; used with permission of the author. Round scad drawing from Gilligan 1989.

faster. Hence the equation for the relationship between mass and length is typically

$$M = aL^b$$

where M is mass, L is length, and a and b are constants. The length exponent, b, usually takes values of around 3.0, indicating that the fish is growing isometrically, that is, that its relative shape is remaining constant as it grows. Values greater or less than 3.0 indicate positive or negative allometric growth (see below) and can serve as an indication of the relative health or condition of the fish. The equation

$$K = W/L^3$$

where W is weight and L is length, can be used to calculate the **condition factor**, K, of a fish. Population or cohort measures of K can indicate whether populations or subgroups are growing or feeding at expected rates. Changes in an individual's condition factor could indicate periods of good versus poor feeding success, disease, or imminent spawning.

K is obviously a rough and simplified indicator of general condition and lags far behind any actual events causing changes in relative condition. More precise and accurate indices can be calculated, such as relative condition factor, relative weight, or covariance analysis of mass and length change; much debate exists over which is the best measure (Ricker 1975, 1979; Anderson and Gutreuter 1983; Wootton 1990; Cone 1992). Because condition factor tells about an individual's history rather than its recent experience, measures that minimize the time lag between cause and effect have been developed, including biochemical analysis of protein uptake rate, energy content or intermediary metabolism (RNA/DNA and ADP/ATP ratios), lipid content, and various chemical and biomarker indicators of stress (Busacker et al. 1990; Wedemeyer et al. 1990).

Body Size, Scaling, and Allometry

As emphasized repeatedly in this book, body size has an overriding influence on most aspects of fish biology. During ontogeny, fish can grow from a larva a few millimeters long to an adult several meters long. An individual must perform all life functions at all sizes in order to reach the next stage; hence size-related phenomena are constant selection pressures on growing fish. Central to discussions of size are the concepts of **scale** and **allometry**. Scaling refers to the structural and functional consequences of differences in size among organisms; allometry quantifies size differences among structures and organisms, and forms the basis of a quantitative science of size (Gould 1966; Calder 1984; Schmidt-Nielsen 1984).

Changes in scale, whether over ontogenetic or evolutionary time, involve alterations in the 1) dimensions, 2) materials, and 3) design of structures. A good example of scaling and its ramifications involves how an increase in body size affects the swimming speed and ability of large and small members of a species. The pelagic larvae of many marine fishes are small, elongate, and highly flexible, whereas adults take on a variety of shapes and swim-

ming modes (see Chap. 8, Locomotion: Movement and Shape). The larvae of many herrings are almost eel-like and swim slowly, but adults have much deeper bodies and swim faster via the carangiform mode, in which the tail is the primary propulsive region. An increase in overall body mass, a **dimensional change**, requires reworking of components. The internal skeleton changes from cartilage to bone, a **material change**. This corresponds to an increase in body musculature and a shift from anguilliform to carangiform swimming to take advantage of the stiffer nature of bone and the more efficient transfer of energy from contracting muscles to the propulsive tail. This shift also corresponds to a **design change** from elongate with a rounded tail to a deeper, streamlined body with a forked tail, which is a more efficient morphology for a carangiform swimmer.

Allometry as a concept underscores a basic fact of growth and scaling, namely, that the change in quantitative relationship between sizes and functions of growing body parts is seldom linear. Linear relationships take the form

$$y = ax$$

indicating that structure y changes as a constant function of structure x, with a being the proportionality constant. A doubling of the size of a fish will not necessarily lead to a doubling of its swimming speed. The relationship is more complex and depends on the measure of body size in question. For salmon, swimming speed increases approximately with the square root of the fish's length and with the one-fifth power of its mass (i.e., length$^{0.5}$, mass$^{0.2}$).

Allometric relationships are described by equations of the nature

$$y = ax^b$$

or

$$\log y = \log a + b \log x$$

The exponent b describes the slope of the line that results when the relationship between the structures is plotted on log-log paper. For simple, linear proportionalities, b equals 1, which is biologically rare. More often, b will take on positive or negative values for regression slopes greater or less than 1 respectively, indicating that a structure is increasing in size faster or slower than the increase in the trait to which it is being compared. The equations for swimming as a function of body size in sockeye salmon have exponents of 0.5 for body length and 0.17 for body mass (Schmidt-Nielsen 1983).

Numerous examples of allometric relationships in fishes can be given, emphasizing the far-reaching implications of size in fishes as well as convergence in selection pressures and solutions among disparate taxa. Focusing on locomotion and activity, the relative cost of swimming decreases with body size in most fishes, both within and among species. Such a relation indicates that it is more expensive for a small fish to move 1 g of body mass a given distance than it is for a larger fish (measured as oxygen consumed per gram of body mass per kilometer; $b = -0.3$). Heart size in fishes increases with body size in an almost linear fashion, taking on values of about 0.2% of body mass and having a slightly positive exponent (heart mass = $0.002 \times$ body mass$^{1.03}$).

Not surprisingly, surface area of the gills relates to activity level. Very active fishes such as tunas have comparatively more gill surface than sluggish species such as toadfishes. But within species and even among species, surface area of the gills (m^2) increases allometrically and positively with body size (kg), with an exponent of 0.8 to 0.9. Locomotion and respiration relate to feeding activity, which is eventually translated into growth. Gut length increases allometrically with body length in many species, with an exponent greater than 1. Growth rate also scales with size, being faster in larger species, with an exponent of 0.61 (measured as change in mass per day relative to adult body mass) (Schmidt-Nielsen 1983; Calder 1984; Wootton 1990).

Questions about size, scaling, and allometry are often linked to the idea of trade-offs, another recurrent theme in this book. What constraints are imposed on an animal by changing its size, both ontogenetically and evolutionarily? What are the advantages and disadvantages of being very small as opposed to being very large? Large size may confer many advantages, but an individual must be small before it is large. During growth, an individual must incur the costs of small size early in ontogeny as well as the energetic and efficiency costs of reworking its size and shape during growth. Juveniles of a large species are often inferior competitors to small adults of a small species. Rapid growth requires rapid feeding and high metabolic rate, which exposes a young fish to more predators and also often carries an increased risk of starvation. Size-related constraints also influence life history attributes such as whether a species will produce many small versus few large young, how extensive the parental care offered will be, and whether adults will mature quickly at small size or slowly at larger size.

One final topic with respect to size deserves mention. Fishes are supported by a dense medium, and their support structures do not reflect the constraints of gravity as much as the necessity to overcome drag. The shapes of fishes then become explainable in terms of drag reduction and which area of the body is used in propulsion, both intimately related to the mode of locomotion used. An important size-related attribute is the Reynolds number, a dimensionless calculation that accounts for the size of an object, its speed, and the viscosity and density of the fluid through which it moves (see Chap. 9, Larval Behavior and Physiology). Calculations of Reynolds numbers help explain swimming speed, body shape, and locomotory type. In very small fishes, including larvae, the effects of drag are so great and the Reynolds numbers so small that inertia is impossible to overcome. Larvae seldom glide, because their mass relative to water viscosity prevents them from developing inertia as they swim; they must continue to expend effort to gain any forward progress. However, their problems associated with overcoming inertia also mean that they are less likely to sink. Large fishes such as billfishes or pelagic sharks have high Reynolds numbers. They can use inertia to advantage and literally soar through the water, using their momentum to carry them forward.

THE ONTOGENY AND EVOLUTION OF GROWTH

Much of the emphasis in this and the preceding chapter has been on size relationships and the observation that indeterminate growth interacts intimately with many crucial aspects of fish biology. Growth processes—both general processes associated with length and mass increase and those related to changing body proportions—help explain many life history, behavioral, ecological, and physiological phenomena. We end this chapter by returning to the general question of how evolution has interacted with body growth processes to establish differences among life history stages and among species of fishes.

Ontogenetic Differences Within Species

Throughout the above discussions, we have emphasized anatomic, behavioral, and ecological differences among size classes of a species. Ontogenetic differences are detailed in several other chapters, such as the tendency for larger fish to occur deeper in a habitat or region and for populations to show age structure (see Chap. 23), for different size fish to interact with different sets of predators and prey (see Chaps. 18, 19), for shoals to be sorted by size (see Chap. 21), and for different size fish to have different foraging capabilities (see Chap. 8).

Additional examples of ontogenetic differences are probably not necessary. The major point here is that larvae have to be adapted to larval life, juveniles to juvenile life, and adults to adult life. These different stages often differ in habitat and ecology, and must function both during definable stages as well as during transitional periods. Adaptations appropriate to one stage may therefore create constraints for other stages. Young fish may be constrained by structures and proportions that are primarily adaptive in later life. For example, small juvenile largemouth bass are morphologically miniature adults. Instead of feeding on fishes, for which their morphology would be best suited, they eat relatively small zooplankton. This puts them in direct competition with juvenile and adult bluegill sunfishes, which are constructed to feed on zooplankton throughout their lives and hence have a competitive advantage over the juvenile largemouths (Werner and Gilliam 1984; see Chap. 23, Population Dynamics and Regulation). Conversely, later stages may retain characteristics of early ontogeny that may constrain them (see below). Regardless, the differing selection pressures on larval, juvenile, and adult fish within a species help clarify the general occurrence of differently appearing and behaving individuals.

An additional conflict exists during ontogeny, brought about by the need for each stage to be immediately functional. Although the life history of a fish appears as a continuum of events from birth through maturation to death, with each phase preparing the fish for the next, some evidence exists to suggest that adaptation to one phase actually inhibits progression into the next. For example, smolting and maturation in salmonids appear to be conflicting processes. Atlantic salmon that smolt rapidly at 1 year of age may mature much later than fish that bypass the smolt stage and remain behind in freshwater. Administration of male hormones to young male masu salmon, *Oncorhynchus masou*, inhibits smoltification but causes maturation; castration of older fish causes them to undergo many of the transformations of smolting. The complexity, timing, and changes in habitat that occur during an animal's life cycle may function not only to prepare an individual for later phases but also to overcome inhibitory or conflicting influences of previous phases (Thorpe 1987).

Evolution via Adjustments in Development: Heterochrony, Pedomorphosis, and Neoteny

Adjustments in developmental rates or timing may be a major way in which new species and even higher taxa evolve from old (Cohen 1984; Mabee 1993). Such a process may explain several phenomena, such as why some adult fish have apparent larval or juvenile traits, or why larval- or juvenile-appearing fishes are reproductively functional, or why closely related species may differ primarily in the duration of an early life history stage, in the time at which a particular structure changes, or in the rate at which different structures grow.

Such alterations in the time of appearance and the rate of development of characters during ontogeny are referred to as **heterochronic** events. Heterochrony results from modification of regulatory genes and processes. Juvenile traits in an adult animal are termed **pedomorphic** (child form); if juveniles become sexually mature, they are **neotenous**. Both pedomorphosis and neoteny are brought about as a result of heterochrony (Gould 1977; Youson 1988). Whether pedomorphosis or neoteny, or a related heterochronic phenomenon, produced a particular trait or condition is difficult to determine; regardless, distinguishing among possibilities is not critical to appreciating heterochrony as a major evolutionary process (Fig. 10.6).

Heterochronic changes in transitions between developmental stages, such as the timing of metamorphosis from embryo to larva or from larva to juvenile, is one way that new species evolve (Youson 1988). Variation in duration of larval life affects age or length at metamorphosis. Elopomorphs as a group are characterized by unique leptocephalus larvae that remain as larvae for long periods, up to 3 years in European eels. In cladistic terms, this synapomorphy defines the group. Long larval life may be related to the apparently unique ability of leptocephali to absorb dissolved organic matter from the water across a very thin epithelium (Pfeiler 1986). Within the elopomorphs, further variations in developmental rate characterize distinctive species and may suggest processes that led to their separate evolution. For example, tarpon (*Megalops atlanticus*) metamorphose at earlier ages and smaller lengths (2 to 3 months, 30 mm) than most other elopomorphs.

Intrageneric separation may have resulted from variation in larval period length in North Atlantic eels, *Anguilla rostrata* and *A. anguilla*. The different larval durations, less than 1 year in American eels and 2 to 3 years in European eels, have a critical effect on their respective distributions. Both species spawn at the same time and same Sargasso Sea locale (see Chap. 22, Catadromy). However, American leptocephali transform into juveniles and at the settle all along the Atlantic coast of North America. European larvae accompanying them in the Gulf Stream are unprepared to metamorphose and are therefore carried past North America and on to Europe. Hence a heterochronic shift in timing of larval metamorphosis could help keep the two species spatially separated.

Evolution of many lamprey species may have also occurred via heterochronic shifts. Many nonparasitic lamprey species can be easily paired with ancestral, parasitic forms. The major differences between ancestor and descendant species involve the length of larval versus adult life, with derived, nonparasitic forms typically having much longer larval periods, rapid metamorphosis, and a relatively short, nonfeeding adult reproductive phase (see Fig 13.4). A delay in the time of metamorphosis would result in just such a difference, essentially creating small adults that retained many larval characters but were reproductively mature (Youson 1988; Finch 1990).

Exceedingly small fish species may have evolved via heterochrony. Two of the smallest fish species known, a goby, *Trimmaton nanus*, and a cyprinid, *Danionella translucida*, reach sexual maturity when only 10 mm long. They retain such larval features as incomplete squamation, limited pigmentation, and partial ossification of the skeleton (Winterbottom and Emery 1981; Roberts 1986; Noakes and Godin 1988). One family, the subtropical and tropical Pacific schindleriids, has many neotenic characters, including a functioning pronephros (the early embryonic, segmented kidney of fishes that is drained by the archenephric duct rather than by the ureter), a transparent body, and large opercular gills. *Schindleria praematura* attains sexual maturity when 1 cm long, prior to settling from the plankton (Leis 1991; Johnson and Brothers 1993; see Chap. 15, Suborder Gobioidei). Characteristics of the deep-water, pelagic ceratioid anglerfishes indicate that they evolved from shallow-water, benthic species via neoteny that involved an extended pelagic larval or juvenile phase. Many ceratioids have a gelatinous balloonlike skin as adults, which is also a pelagic larval trait. Mature males are distinctly larval in appearance and are parasitic on females; larval-like males also occur in the deep-sea black dragonfishes and in the goby genus *Crystallogobius* (Moser 1981; see Chap. 17, The Deep Sea). Although large size confers an advantage in many species, the production of new species via heterochronic shifts that lead to retention of small body size shows that small size is also advantageous under certain conditions.

SUMMARY

1. Although we recognize specific phases during the ontogeny of an individual, the transitions that occur between phases often require long time periods and

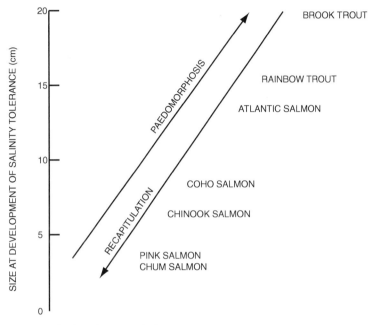

FIGURE 10.6. *Salmonids differ in the minimum size at which they develop the necessary salinity tolerance to undergo the parr-smolt transformation. Observed differences among species could be explained by heterochronic shifts in the development of this trait. Such shifts might have changed the timing of the onset of the various physiological processes involved in salinity tolerance. It is not known what the ancestral condition was, and so either acceleration or deceleration of timing could be responsible. Hence either pedomorphosis (increasing size at smoltification) or recapitulation (decreasing size at smoltification), or both, could have affected the evolution of this trait.*

From McCormick and Saunders 1987; used with permission.

can involve complex changes and reworkings of the anatomy and physiology of a fish. Examples of complex and protracted transitions include smoltification, when young salmon move from freshwater to the ocean, and metamorphosis, when flatfish change from symmetric, planktonic larvae to asymmetric, bottom-dwelling juveniles.

2. Reproductive development includes three very different processes—determination, differentiation, and maturation. Gender determination in most fishes is probably under genetic control and occurs at the time of fertilization. In some fishes, environmental conditions such as temperature can affect determination. Differentiation occurs when recognizable ovaries or testes appear in an individual. Maturation is synonymous with achieving adulthood and occurs when a fish produces viable sperm or eggs. Complicating this picture are many fish species that undergo postmaturational sex reversal, changing from functional females to functional males (protogyny) or from male to female (protandry). A few species are simultaneous hermaphrodites, functioning as males and females at the same time. Self-fertilization is exceedingly rare in fishes.

3. Age at first reproduction and longevity vary greatly among fishes. Some male surfperches are born mature, whereas some sharks, sturgeons, and eels may not mature until they are older than 20 years. Longevity also ranges from less than 1 year in annual fishes to more than 100 years in sturgeons and rockfishes. Death usually occurs as a result of accident or disease, but some fishes such as lampreys and salmons show programmed death (senescence) similar to that observed in mammals.

4. The age of a fish can be determined by counting growth rings on otoliths, vertebrae, fin spines, and other hard body parts. Such growth rings are usually added annually to a structure, but climatic and other environmental factors can lead to variation that can provide information about habitat shifts during ontogeny. Daily growth increments are often detectable on the otoliths of young fishes, allowing back calculation to actual spawning dates.

5. Size at a particular age varies greatly in fishes. Growth curves that describe size-age relationships can be calculated using a number of equations, the von Bertalanffy growth equation being a commonly used indicator. The condition of a fish, calculated by dividing body mass by length, is one indicator of the kind of growth conditions an individual has experienced.

6. As a fish grows, the dimensions of its body change, as do the materials used in construction; together these changes represent a change in the design of the individual. Changes in the relationship between body parts during growth or between body functions and body size can often be described by an allometric equation; in most instances, the relationship is nonlinear.

7. Fishes undergo indeterminate growth and have complex life cycles. At each stage during its life, a fish must function physiologically, ecologically, and behaviorally. Such function is compromised by transitional periods, by alterations that are preparatory for the next stage, and by traits retained from previous stages. Evolution may occur through adjustments in the timing or rate of ontogenetic development. Such heterochronic changes include rapid maturation of juveniles (neoteny) or retention of juvenile characteristics in adults (pedomorphosis) and may explain the evolution of some of the smallest fish species as well as speciation in lampreys, elopomorphs, salmons, and deep-sea anglerfishes.

SUPPLEMENTAL READING

Finch 1990; Gould 1977; Hoar and Randall 1988; Pitcher and Hart 1982; Ricker 1975; Weatherly and Gill 1987.

PART 3

Taxonomy, Phylogeny, and Evolution

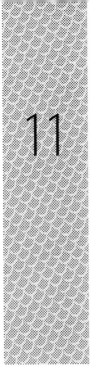

11

"A History of Fishes"

Fishes were the first vertebrates. Understanding the evolutionary history of fishes is therefore important not only for what it tells us about fish groups but also for what it tells us about evolution of the vertebrates and ultimately our own species. Innovations during fish evolution that were passed on to higher vertebrates include dermal and endochondral bone and their derivatives (vertebral centra, bony endoskeletons, braincases, teeth), jaws, brains, appendages, and the internal organ systems that characterize all vertebrate groups today. During 500 million years of evolution, fishes colonized and dominated the seas and freshwaters and eventually emerged, at least for short periods, onto land. Major clades prospered and vanished or were replaced by newer groups with presumably superior innovations.

Extant (living) fishes therefore represent the most recent manifestations of adaptations and lineages that have their roots in the early Paleozoic. The more than 25,000 species of extant fishes constitute only a fraction of the diversity of fishes that has existed historically. Many of the extinct forms are exotic in their appearance, whereas others are remarkably similar to living forms, at least in external morphology.

A major challenge to ichthyology involves unraveling the evolutionary pathways of both modern and past fish taxa in the process of determining relationships among groups. Which of the many fossil groups represent ancestral types, which were independent lineages that died out without representation in modern forms, and what are the links between and among groups of the past and present? What do fossilized traits tell us about ancient environments? Where do similarities represent inheritance, convergence, or coincidence among extinct and living groups? And how have past adaptations influenced and perhaps constrained present morphologies and behaviors? The focus of this chapter is on fishes that lived during the Paleozoic and Mesozoic eras, and on modifications that occurred during the evolution of different, major extinct groups, leading to the dominant bony and cartilaginous fishes of today. We deal first with jawless fishes, then with ancestors of modern bony fishes because these occur earlier in the fossil record, and finally with the last major group to evolve, the cartilaginous sharks, skates, rays, and chimaeras.

AGNATHANS

The very first fishes undoubtedly arose from invertebrate protochordates, perhaps a urochordate or cephalochordate. However, the first fishes left no fossil record, and their form and relationships are a mystery (but see Box 11.1). By

Phylum Chordata[a]
  Subphylum Vertebrata
    Superclass Agnatha
      Class Pteraspidomorphi (Diplorhina)
        Order [†]Arandaspidiformes
                [†]Pteraspidiformes (Heterostraci)
                [†]Thelodontiformes (Coelolepida; position uncertain)
      (?Class Myxini, Order Myxiniformes [modern hagfishes, problematic])[b]
      Class Cephalaspidomorphi (Monorhina)
        Order [†]Cephalaspidiformes (Osteostraci)
                [†]Anaspidiformes
                [†]Galeaspidiformes
                (Petromyzontiformes [modern lampreys])[b]
    Superclass Gnathostomata
      Class [†]Placodermi
        Order [†]Arthrodiriformes
                [†]Antiarchiformes
                [†]Ptyctodontiformes
      Grade Teleostomi
      Class [†]Acanthodii
      Class Sarcopterygii
        Subclass Coelacanthimorpha (Actinistia; coelacanths)
        Sub- (or infra-) class Dipnoi (lungfishes)[c]
        Subclass [†]Osteolepimorphi (rhipidistians)[d]
      Class Actinopterygii
        Subclass Chondrostei
          Order [†]Paleonisciformes
                  Polypteriformes (modern bichirs, reedfish; problematic)
                  Acipenseriformes (modern sturgeons, paddlefishes)
        Subclass Neopterygii
          Order Semionotiformes (modern gars)
          Order Amiiformes (modern bowfin)
          Division Teleostei (modern bony fishes)

[†]Extinct group.
[a]Classification based primarily on Nelson 1994.
[b]Ancestry of modern hagfishes and lampreys is debated (see below and Chap. 13).
[c]Nelson 1994 combines lungfishes with extinct porolepimorphs to form an unnamed subclass.
[d]Rhipidistians are polyphyletic, with unresolved relationships.

FIGURE 11.1.   *Periods of occurrence of major fish taxa based on the fossil record. Thickened portions of lines indicate periods of increased diversity within a group; differences in line thickness apply only to relative diversity within a taxon, not among taxa. Line thickness represents generic diversity, except for coelacanths, in which case line thickness shows relative numbers of species. No interrelatedness is implied by proximity of lines. Approximate millions of years before present (MYBP) since the beginning of each time period are shown at the bottom. Time periods are not drawn to scale (e.g., the Cretaceous lasted almost 50 million years longer than the Silurian, but both are given equal space). Fossils are lacking for myxiniforms and petromyzontiforms during the Mesozoic. Neopterygian lineage represents five highly successful, extinct pre-teleostean orders, most notably the semionotiforms and pycnodontiforms, that flourished during the Mesozoic and left no apparent descendants, except perhaps modern gars. A quick glance at the figure reveals why the Devonian is commonly referred to as the Age of Fishes: During the middle Devonian, most major groups discussed in this chapter were in existence.*

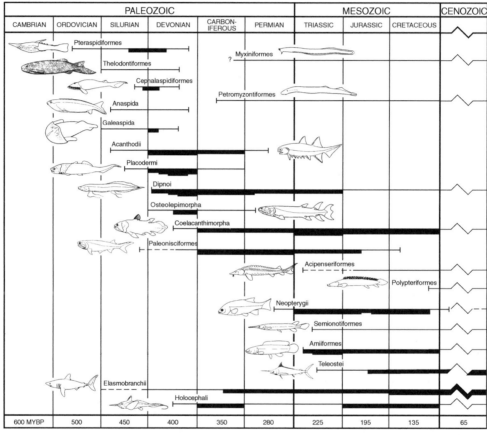

Data largely from Carroll 1988, Pough et al. 1989, and Benton 1993 and references therein, except for Coelacanthimorpha (Forey 1991) and Dipnoi (Campbell and Barwick 1987).

the time the first fossils appear in late Cambrian/early Ordovician deposits, roughly 500 million years ago (mya) (Fig. 11.1), complex tissue types have evolved. These early fragments are mostly disarticulated bony head shields and require some speculation to reconstruct. Clearly recognizable specimens, named *Sacabambaspis janvieri*, date to 470 mya from Bolivia (see Fig. 11.2A) (Gagnier et al. 1986; Gagnier 1989). This and related **agnathous** (jawless), finless forms inhabited shallow seas or estuarine habitats in tropical and subtropical regions of the Gondwanan and Laurasian supercontinents (see Chap. 16). Their innovations include true bone (probably evolved independently in several ancestral groups) and a muscular feeding pump. The former adaptation, which existed only as an external covering, would have provided protection from predators to which softer ancestors were more vulnerable, as well as

serving as a metabolic reserve for calcium and phosphate and an insulator of electrosensory organs (Northcutt and Gans 1983; Carroll 1988). A muscular feeding pump would have been more efficient for moving food-bearing water through a filtration mechanism than was the ciliary feeding mechanism of protochordates.

The first fishes were historically termed **Ostracoderms** (shell-skinned), in reference to the bony shield that covered the head and thorax. But Ostracoderm is now considered an artificial designation that includes two distinct classes, the **Pteraspidomorphi** and **Cephalaspidomorphi**. Both classes *may* be represented today by jawless fishes, namely, the hagfishes (pteraspidomorphs) and the lampreys (cephalaspidomorphs); however, only the lamprey-cephalaspidomorph relationship is well supported (see below and Chap. 13).

FIGURE 11.2. *The earliest known fishes were jawless pteraspidomorphs with armored head shields. Pteraspidomorphs included such small, primitive forms as (A) Arandaspis from Australia, as well as more advanced forms such as (B) Pteraspis from Devonian Europe. Thelodonts were another, arguably pteraspidomorph order of small fishes, such as (C) Phlebolepis, and (D) the most recently discovered fork-tailed thelodonts of northwestern Canada.*

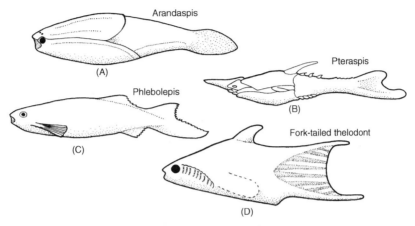

(A) After Rich and van Tets 1985; (B), (C) after Moy-Thomas and Miles 1971; (D) after Wilson and Caldwell 1993.

Pteraspidomorphi

Pteraspidomorphi (or Diplorhina = "two nares") derive their alternative name from impressions on the inside of the head plates indicating two separate olfactory bulbs in the brain. Pteraspidomorphs were jawless filter feeders in both marine and freshwater environments; they occurred from the late Cambrian until the end of the Devonian. Two orders of pteraspidomorphs are often recognized, the pteraspidiforms and the thelodontiforms.

Pteraspidiforms (or heterostracans = "those with a different shell") had dermal armor that extended from the head almost to the tail, necessitating swimming by lashing the tail back and forth, much like a tadpole. The tail in most forms was **hypocercal**, in that the notochord extended into the enlarged lower lobe of the tail. Their body form, armor, and tail morphology suggest that pteraspidiforms plowed the bottom, pumping sediments into the ventral mouth and filtering digestible material through the pharyngeal pouches. The armor is generally sutured and shows growth rings, indicating incremental growth. Early pteraspidiforms were small (ca. 15 cm), although some later species reached 1.5 m.

Six families and more than 50 genera of pteraspidiforms are generally recognized (Denison 1970; Carroll 1988). Primitive forms, such as the Ordovician *Arandaspis*, *Astraspis*, and *Sacabambaspis*, had symmetric tails, full body armor, and multiple branchial openings (Fig. 11.2A). Later groups, such as *Pteraspis* from the Lower Devonian (Fig. 11.2B), were quite different and had hypocercal tails, fused dorsal and ventral head plates, and single branchial openings. The Devonian also produced highly derived forms, such as the sawfishlike *Doryaspis* and the tube-snouted, blind *Eglonaspis*. Trends in development of pteraspidiform lineages include reduction of armor through fusion of plates, narrowing of the head shield, and development of lateral, presumably stabilizing, pro-

jections (**cornua**). These changes all suggest strong selection for increased mobility and maneuverability. While these anatomic changes were taking place, pteraspidiforms invaded freshwater habitats (Carroll 1988).

Arguably related to the pteraspidomorphs were the contemporaneous **thelodontiforms** (nipple teeth), also known as coelolepids (hollow scales) (Fig. 11.2C). These diminutive (10 to 20 cm), fusiform, jawless fishes were covered with denticles rather than bony plates. The small denticles resemble the placoid scales of modern sharks. Thelodonts share with pteraspidiforms a hypocercal tail and similar eye and gill positions, including a median pineal opening. The existence of dentine in the dermis of both groups may also be a shared trait, but the denticles of thelodonts are substantially different from the plates of the pteraspidiforms, thus confusing the picture. In addition, thelodonts had dorsal and anal "fins." Their mode of life was probably similar to pteraspidiforms, namely skimming and filtering small organisms from bottom sediments while swimming, although their fusiform body shape and terminal mouths suggest they may have been water-column swimmers. More recent discoveries in northwestern Canada indicate that some thelodonts were shaped like minnows or pupfishes and had compressed bodies, forked tails, and a stomach (Wilson and Caldwell 1993; Fig. 11.2D). Early thelodonts appear in marine deposits, but later groups invaded freshwater.

Cephalaspidomorphi

Cephalaspidomorphi (or Monorhina = "single nostril") first appeared in the Upper Silurian, approximately 100 million years after the appearance of the pteraspidiforms. Their alternative name refers to the single, median, slitlike opening—the nasohypophyseal foramen—in the anterior region of the head shield, associated with the pineal body. Three distinct groups and more than 30 genera of cephalaspidomorphs arose, all of which disappeared by the end of the Devonian. They were similar to the pteraspidomorphs in that they lacked jaws and paired true fins, and had two semicircular canals, numerous pairs of gill arches, and dermal bone. However, they differed from pteraspidomorphs in a number of important characteristics.

The best known cephalaspidomorphs are a predominantly freshwater group, the **Cephalaspidiformes**, comprising six families (Fig. 11.3A). These were abundant and diverse fishes; nearly 100 species just in the genus *Cephalaspis* have been described (see Jarvik 1980). Rather than acellular bone, cephalaspidiform armor was cellular. An-

other cephalaspidiform innovation, also evolved in jawed vertebrates, is ossification of the endoskeleton. In addition, cephalaspidiform head shields are sutureless and lack any apparent growth rings. In fact, all fossils of a species are the same size, suggesting a naked (nonfossilizing), growing larval form that metamorphosed into an armored adult of fixed size, paralleling the developmental history of modern lampreys (which, however, lack armor). The head shield included one medial and two lateral regions (sensory fields) of small plates sitting in depressions and connected to the inner ear by large canals, for which either an acousticolateralis, electrogenerative, or electroreceptive function has been suggested (Moy-Thomas and Miles 1971; Carroll 1988; Pough et al.

FIGURE 11.3.   *Cephalaspidomorphs were diverse jawless forms that appeared during the Silurian and lasted into the Devonian. The largest order was the cephalaspidiforms, including (A) Hemicyclaspis. Anaspids, such as (B) Pharyngolepis, were convergent in body form with the pteraspidomorph thelodonts. (C) Thin sections of head shields of cephalaspidiforms clearly show brain differentiation and cranial nerves (Roman numerals), organized similarly to modern lampreys.*

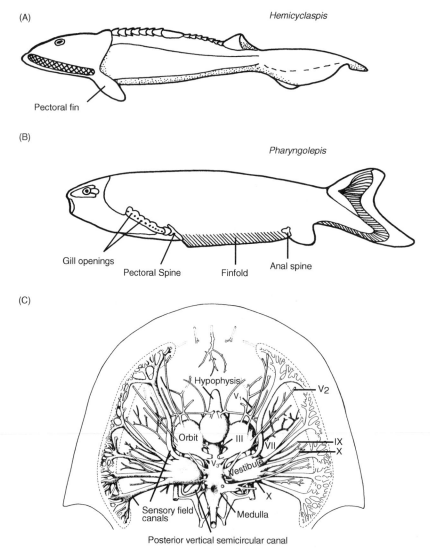

(A)   *Hemicyclaspis*

Pectoral fin

(B)   *Pharyngolepis*

Gill openings   Pectoral Spine   Finfold   Anal spine

(C)

Hypophysis   V2

V1

Orbit   III   VII   IX   X

V3   Vestibule

Sensory field   Medulla
canals

X

Posterior vertical semicircular canal

(A), (B) After Moy-Thomas and Miles 1971; (C) from Stensiö 1963; used with permission of Scandinavian University.

BOX 11.1 *Unraveling the Mystery of the Conodonts*

Between late Proterozoic and late Triassic times (600 to 200 mya), a group of animals known as **conodonts** (cone-shaped teeth) arose, proliferated, and died in seas worldwide. The fossil remains, referred to as conodont elements, consist of toothlike structures generally about 1 mm long and made from calcium phosphate (Fig. 11.4A). Their existence has been known since the mid-1800s, and their abundance and diversity, involving more than 200 recognized genera, have allowed them to serve as stratigraphic landmarks in determining the age of fossil beds. The unfossilized soft body parts of the animal surrounding the elements have been the subject of considerable conjecture and debate for most of the past century.

The conjecture has been fanciful at times. Conodont elements have been claimed as parts of cnidarians, as the copulatory structures of nematodes, as the radular teeth or other parts of snails and cephalopods, as jaws of annelid worms, as arthropod body parts, as teeth of chaetognaths and various fish groups, and even as calcareous algae. More conservative authors have generally placed the animals in a separate, extinct phylum, the †Conodonta, with uncertain relationships (Clark 1987).

The debate heated up in 1973 with the discovery of impressions of the body outline of an animal that contained conodonts in its midsection and that possessed a finlike structure at the posterior end and suggestions of a nerve cord near the anterior end (Melton and Scott 1973). A new phylum, the †Conodontochordata, was erected but later collapsed when it was concluded that whatever the animal was, it had eaten several different conodont-bearing animals. Ten years later, a more definitive body outline containing conodonts was found in Carboniferous beds in Scotland. Subsequent Scottish fossils plus fossils from Wisconsin and South Africa have given us a better idea of the real conodont animal (Briggs et al. 1983; Smith et al. 1987).

The real animal is apparently a true chordate, with V-shaped muscle blocks, a bilobate head, a compressed body, axial lines suggestive of a notochord, and unequal tail fins supported by raylike elements (Fig. 11.4B). Total body length was 4 cm to as large as 40 cm. The conodont elements are contained in the head region and apparently functioned as teeth. Eyeballs and extrinsic eye muscles, chevron-shaped muscle blocks, and apparent bone cells in a dermal skeleton of some species strongly suggest that not only were conodonts chordates but may even be classified as vertebrates (Janvier 1995; Purnell 1995; Gabbott et al. 1995).

What are the affinities then of this ancient, highly successful, tooth-bearing, primitive chordate/vertebrate? The most popular analysis at present is that the conodont-bearing animals were on the line that led to the myxine hagfishes. The body form of hagfishes and the reconstructed conodont animal (see Fig. 11.4B) are strikingly similar. The dentition of hagfishes consists of vertical rows of grabbing, horny teeth arranged on a V-shaped, flexible dental cartilage (see Chap. 13). The bite of a hagfish is three-dimensional: The cartilage can be protruded, flattening the teeth against the food item. When the cartilage is retracted, the rows of teeth come together, biting off pieces of food, much like closing a book (Dawson 1963; Jarvik 1980; Krejsa et al. 1990a,b). The conodont animal may have captured small prey with teeth that closed together in a manner analogous to that of hagfishes. Once in the mouth, prey could have been cut or ground up by different elements in the apparatus.

Most significantly, these findings push the fossil record for presumed fish groups back past the Cambrian and into the Proterozoic. Such an early beginning for fishes is logical. The first truly fishlike fossils that appear in the Ordovician occur across a wide geographic range and already possess layers of complex bony material (calcified cartilage, bone, dentine, and enamel) differentially distributed on the body. Intermediate steps in the development of such mineralized tissue must have taken considerable time to occur. The phosphatized tooth elements of the conodonts provide a transition from the toothless, soft-bodied filter feeding of protochordates to the mineralized bone of the early pteraspidiforms. Alignment of the conodont animal with the earliest fishes also strengthens the conclusion that evolution in feeding structures was critical to the success of the first and later vertebrates.

1989). Paired lateral appendages—"pectoral fins" without internal skeletal support—were common, and the tail was heterocercal, which may have made skimming along the bottom easier by counteracting the upward lift that the lateral appendages would have generated.

The internal anatomy of the cephalaspidiform head shield is remarkably well known. Swedish paleontologist Erik Stensiö and colleagues painstakingly sectioned rocks containing cephalaspidiforms and worked out the anatomic details of the braincase and cranial nerves (Fig. 11.3C). One result of this work is the discovery of similarities in the cranial anatomy of cephalaspidiforms and

FIGURE 11.4. *(A) Conodont apparatus. The various elements (A–G) are approximately 1 mm long and occur on the right (dextral) and left (sinistral) sides of the head region of the conodont animal, functioning as the feeding apparatus. (B) The 350-million-year-old, 40-mm-long conodont animal.*

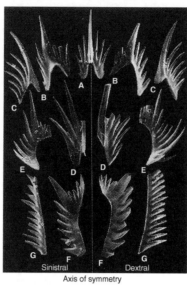

(A)

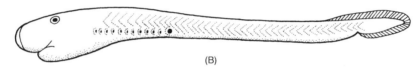

(B)

(A) From Clark 1987; used with permission. (B) As reconstructed by Aldridge and Briggs 1989.

extant lampreys, including parallels in the shape of the olfactory lobes, diencephalon, and myelencephalon; the relationship of the hypophyseal sac to the olfactory opening (a common nasohypophyseal opening); the relative sizes of right and left ganglia; the alternation of cranial nerves; the separation of dorsal and ventral nerve roots; the location of blood vessels; the existence of two vertical semicircular canals; and other details (Moy-Thomas and Miles 1971).

These similarities, despite a separation of more than 400 million years, suggest common ancestry if not direct ascendancy. However, recognizable cephalaspidiforms died out during the Devonian, and the first accepted petromyzontiform (lamprey) fossils do not appear until the mid-Carboniferous, nearly 50 million years later. These first lampreys, *Hardistiella* and *Mayomyzon* (see Fig. 13.5), were much more similar to living lampreys than to the cephalaspidiforms, although the more primitive *Hardistiella* possessed a hypocercal tail and an anal fin, characteristics of anaspid cephalaspidomorphs, but also of thelodonts, just to confuse the picture.

The other two cephalaspidomorph orders are the **anaspids**, with four families, and the more recently discovered, less well-known **galeaspids**, with two families. As in cephalaspidiforms, anaspids originated in nearshore marine habitats and gradually entered freshwater. Parallel trends in the evolution of cephalaspidiforms/anaspids

and pteraspidiforms/thelodonts are striking. As with the thelodonts, anaspids were small (15 cm), relatively fusiform fishes, with a hypocercal tail, terminal oval mouth, and dorsal and anal fin folds. The anaspid body was covered largely with overlapping, tuberculate scales (compare Fig. 11.3B with Fig. 11.2C, *Phlebolepis*). Anaspids existed from the mid-Silurian to late Devonian. One advance compared with thelodonts was the development of flexible, lateral, finlike projections that had muscles and an internal skeleton, thus giving these small fishes considerable maneuverability for their apparently suprabenthic existence.

Galeaspids had both cephalaspidiform and pteraspidiform affinities, but the median nasohypophyseal opening and the structure of their ossified endochondral braincase suggest they were an independently evolved cephalaspidomorph group, representing a unique evolutionary radiation of at least nine genera confined to what is today southern China (Halstead et al. 1979; Janvier 1984; Pan Jiang 1984).

Later Evolution of Primitive Agnathous Fishes

Although much has been written about possible descendants of the early agnathans, additional discussion of the interrelationships of these primitive groups is beyond the scope of this book, mainly because authorities disagree as to where the relationships lie. Different authors consider different characters as ancestral, derived, or convergent and consequently arrive at different conclusions about relationships between and among pteraspidomorphs, cephalaspidomorphs, myxinoids, petromyzontiforms, and gnathostomes. For reviews of this controversy, the reader is referred to Jarvik 1980; Carroll 1988; Pough et al. 1989; Forey and Janvier 1993; and Long 1995.

GNATHOSTOMES: EARLY JAWED FISHES

Although debate is common concerning relationships among jawless fishes, few authors have suggested that a known agnathous group was ancestral to the first jawed fishes that arose during the early Silurian. No intermediate fossils between jawed and jawless forms have been found—early fossils of jawed fishes had jaws, teeth, scales, and spines. The origins of jaws and the other structures that characterized the early gnathostomes are lost in the fossil record, belonging to some group about which we know nothing.

Instead, we have an abundance of fossil material that gives us a clear picture of the diversity of forms that the innovation of jaws must have permitted. The evolutionary importance of true jaws cannot be overemphasized: "[P]erhaps the greatest of all advances in vertebrate history was the development of jaws and the consequent revolution in the mode of life of early fishes" (Romer 1962, 216). This revolution included a diversification of the food types that early fishes could eat. Large animal prey could be captured and dismembered, and hard-bodied prey could be crushed. Agnathous fishes were probably limited to planktivory, detritivory, parasitism, and microcarnivory. With the advent of jaws, both carnivory and herbivory on a grand scale became possible, as reflected in the size of the fishes that soon evolved. Jaws also allowed for active defense against predators, leading to de-emphasis on armor, which in turn meant greater mobility and flexibility. This increase in agility was greatly enhanced by the development of paired, internally supported pectoral and pelvic appendages, "the most outstanding shared derived character of the gnathostomes besides the jaw" (Pough et al. 1989, 235).

Class Placodermi

Although a unique, separate group with no apparent living descendants, **placoderms** (plate-skinned) deserve discussion because of their tremendous success and diversity. Their name refers to the peculiar bony, often ornamented, plates that covered the anterior 30% to 50% of the body. Most placoderms had depressed, even flattened bodies, suggesting benthic existence. They may have preyed upon, and eventually replaced, pteraspidiform and cephalaspidiform fishes. As in ostracoderms, placoderms occurred first in marine deposits but later moved into freshwater. As in both ostracoderms and acanthodians (see next section), many placoderm groups show an evolutionary trend toward reduction in external armor,

leading to a mobile existence in the water column. Placoderms had ossified haemal and neural arches along the unconstricted notochord and three semicircular canals. Placoderms arose in the late Silurian, flourished worldwide in the Devonian, and disappeared by the early Carboniferous.

Eight or nine orders, 30 families, and about 50 genera of placoderms are recognized (Fig. 11.5). **Arthrodires** (jointed neck) appear first in the fossil record and were the largest order, containing 20 families. They possessed a unique hinge at the back top of the head between the braincase and the cervical vertebrae, termed the **craniovertebral joint**. This joint allowed opening of the mouth by both dropping the lower jaw and raising the skull roof, thus increasing gape size. As the group evolved, this joint became larger and more elaborate, and dentition diversified into slashing, stabbing, and crushing structures. Arthrodires were among the largest of the placoderms: *Dunkleosteus* (Fig. 11.5E) was more than 2 m long, with a head more than 1 m high. Their large size and impressive dentition implicate the arthrodires as major predators of Devonian seas.

None of the other placoderm orders attained the success of the arthrodires. The **antiarchs** (e.g., *Bothriolepis*, Fig. 11.5C) were predominantly freshwater, heavily armored, benthic fishes with a spiral valve intestine and jointed, arthropod-like pectoral appendages that had internal muscularization. Ptyctodontids greatly resembled modern holocephalans in body form (Fig. 11.5B) and are the first known fishes to possess apparent male intromittent organs in the form of claspers associated with the pelvic fins. Rhenanids were extremely dorsoventrally depressed and bore a striking resemblance to modern skates, rays, and angel sharks (e.g., *Gemuendina*, Fig. 11.5D), although these lateral fins were too heavily armored to be undulated or flapped in the manner characteristic of modern skates and rays.

Although the craniovertebral joint of many placoderms afforded increased jaw mobility compared to forms with a fixed upper jaw, both placoderms and acanthodians lacked replacement dentition. Placoderm "teeth" consisted of dermal bony plates attached to jaw cartilage. This bone was often differentiated into sharp edges and points but was subject to breakage and wear, with no apparent repair or replacement mechanism. Placoderm jaw morphology and hinging also prohibited them from developing suction forces when feeding. The innovations of serially replaced teeth and of jaws that could create suction characterize the fish taxa that evidently replaced the placoderms and acanthodians.

Placoderms have been postulated as ancestral to some modern fishes, but a placoderm ancestry for subsequent gnathostome groups is unlikely given the unique traits that characterized the placoderms. The diagnostic dermal plates of the skull and trunk have no counterparts in bony fishes. Jaw musculature in placoderms is inside the palatoquadrate, rather than external to the jaw bones as in all modern gnathostomes. Placoderm "tooth" structure is entirely different, and the attachment of the jaw to the braincase is unique. Placoderms must be viewed as a

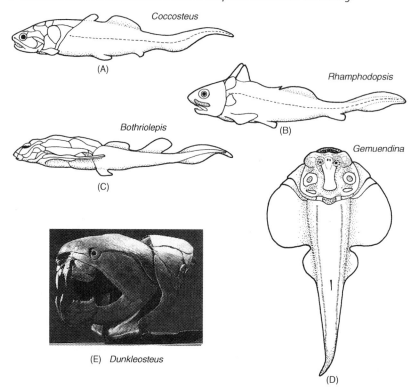

FIGURE 11.5. *Placoderms. (A) The coccosteomorph* Coccosteus, *(B) the ptycto-dontid* Rhamphodopsis, *(C) the antiarch* Bothriolepis, *(D) the rhenanid* Gemuendina, *and (E)* Dunkleosteus, *a giant arthrodire placoderm from the Devonian. The meter-high head of* Dunkleosteus *was followed by a proportionately large body, but actual lengths are unknown because fossilized remains of the posterior skeleton are lacking.*

(A)–(D) After Jarvik 1980 and Stensiö 1963; (E) photo by Chip Clark; used with permission.

highly successful, diverse group of fishes that have left only a legacy of intriguing fossils.

ADVANCED JAWED FISHES I: TELEOSTOMES (OSTEICHTHYES)

One compelling argument against an ancestor-descendant relationship between the most primitive jawed fishes and later forms is their simultaneous appearance in the fossil record. Placoderms, acanthodians, actinopterygians, and chondrichthyans all debut in mid- to late Silurian deposits (e.g., see Fig. 11.1). The primitive forms for which we have fossil information are already far advanced beyond the character states of any presumed common ancestor. While we tend to view the more advanced fishes as improvements over the primitive taxa, in part because the latter are represented today, placoderms and acanthodians existed literally side by side with the "more advanced" forms for more than 100 million years. At some point, for climatic or biological reasons that are unclear, the innovations of the more derived gnathostomes, or the evolutionary constraints placed upon the more primitive groups, led to a replacement of one group by the other. The result was an incredible series of explo-

sions of species belonging to four or five very different lineages, derivative forms of which are still alive today.

The first bony fishes, known by the unranked taxonomic term **Osteichthyes** (bony fishes) and represented by fragments in the late Silurian, shared cranial, scale, and fin similarities with the acanthodians and may have had a common ancestor. By mid-Devonian, two distinct lineages of osteichthyans had developed, the classes **Sarcopterygii** (fleshy or lobe fins) and **Actinopterygii** (ray fins). (The grade Teleostomi includes sarcopterygians, actinopterygians, and the more primitive acanthodians, because all three groups possessed three otoliths.)

Osteichthyans shared numerous characteristics, including the bone series in the opercular and pectoral girdles, the pattern of their lateral line canals, fins supported by dermal bony rays, a heterocercal tail with an epichordal (upper) lobe, replaceable dentition, and a gas bladder that developed as an outpocket of the esophagus. Sarcopterygians diversified into two groups, the monophyletic Dipnoi (Schultze and Campbell 1986) and a complex group often referred to as rhipidistians that includes the subclass Osteolepimorpha (Carroll 1988; Pough et al. 1989; Nelson 1994). Actinopterygians underwent tremendous multiple radiations, the two major groups being the Chondrostei (including modern sturgeons, paddlefishes, and perhaps bichirs) and the Neopterygii, which includes the modern bowfin, gars, and teleosts.

Class Acanthodii

The first group of relatively advanced, jawed fishes to arise was the **Acanthodii**, or "spiny sharks," in the early Silurian. Their Latin name refers to the stout median and paired spines evident in most fossils. Their similarity to sharks is largely superficial, and few recent authors feel they are related to modern chondrichthyans (but see Jarvik 1980). Acanthodians were generally small (20 cm to 2 m); occurred in both salt and freshwater, mostly in Laurasia (see Box 16.1); and had cartilaginous skeletons, a body covered with small, nonoverlapping scales, large heads, and large eyes. Their streamlined, round bodies, reduced armor when compared to ostracoderms, subterminal mouths often studded with teeth (including teeth inside the mouth and on the gill rakers), and fin placement all indicate they were water-column, not benthic, feeders. Given the success of ostracoderms in benthic habitats, it is not surprising that the next fish group to evolve would occupy the relatively unexploited water column.

Three orders and perhaps 60 genera have been described, many from isolated spines and teeth (Carroll 1988). All three orders show interesting parallels in evolution. Early acanthodians had multiple gill covers, broad unembedded spines anterior to all fins except the caudal, as well as additional spine pairs between the pectoral and pelvic fins (Fig. 11.6A). More advanced species had single gill covers and have lost the ancillary paired spines, the remaining spines being thinner and embedded in the body musculature (e.g., *Acanthodes*, Fig. 11.6B). Some specialized lineages were toothless and had long gill rakers, indicating a planktivorous habit. Acanthodians possessed a third (horizontal) semicircular canal and neural and haemal arches associated with the unconstricted vertebral column. Because of these and other shared, derived traits (otoliths, lateral line canals, ossified operculum, branchiostegals, cranial and jaw series, including the new interhyal bone), acanthodians are generally considered to be a sister group to the bony fishes (Osteichthyes; Osteichthyes + Acanthodii = Teleostomi), which means they share a common, immediate ancestor (Lauder and Liem 1983; Maisey 1986). Acanthodians survived until the early Permian, outlasting the major ostracoderm groups by 100 million years.

Class Sarcopterygii

Subclass Coelacanthimorpha.   Fossil **coelacanthimorphs** (or Actinistia) appeared in the middle Devonian and are not known after the late Cretaceous. They occurred worldwide in both marine and freshwater. The fossil record of the group is extensive: More than 90 species belonging to 47 genera and perhaps seven families are recognized, with a maximal diversity during the early Triassic (Fig. 11.7). Coelacanths (hollow spine) are in many respects more specialized than other sarcopterygians, possessing a unique spiny rather than a lobate first dorsal fin; a three-lobed caudal fin with a middle fleshy, fringed lobe; a rostral organ involving a rostral cavity with several openings on the snout associated with electroreception; and lacking internal choanae, cosmine in the scales, branchiostegals, and a maxilla.

Most evolution in the group occurred during the Devonian, and later species are surprisingly unchanged in body shape and jaw morphology from the early representatives, although trends of change (reduction in some bones, increases in others) have occurred (Cloutier 1991). Prior to the discovery of a living coelacanth in 1938 (see Box 13.2), coelacanthimorphs were of interest primarily

FIGURE 11.6.   *Acanthodians. (A)* Climatius, *a primitive acanthodian with multiple gill covers and multiple, unembedded spines. (B) The more advanced* Acanthodes, *with fewer, thinner, more deeply embedded spines, a single gill cover, and a more symmetric caudal fin.*

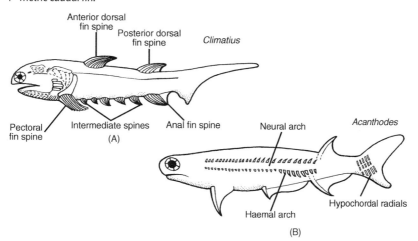

After Moy-Thomas and Miles 1971.

FIGURE 11.7.   *Phylogenetic relationships among the coelacanths, showing occurrence of major genera and apparent ancestor-descendant relationships. Coelacanths are among the best-studied ancient groups, stimulated in part by the discovery of a living species after a 65-million-year hiatus in the fossil record. Approximate time before the present since the beginning of each time period is given in the left hand column.*

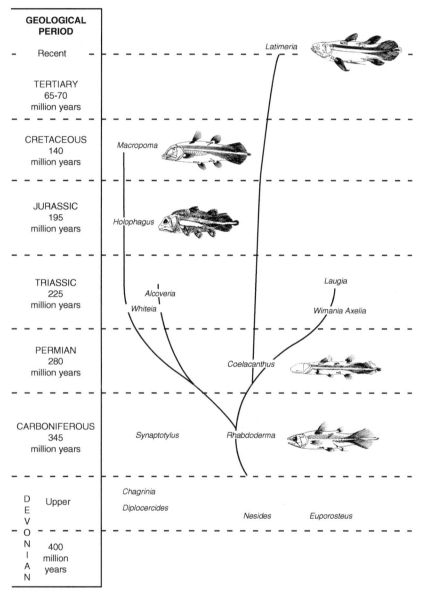

Redrawn from Balon et al. 1988, after Thys van den Audenaerde 1984; used with permission.

to paleontologists as a specialized, extinct group notable for its conservatism and its relationship to the osteolepiforms, the reputed ancestors of tetrapods.

Sub- (or infra-) class Dipnoi.   The **Dipnoi** (double-breathing), represented today by six species of modern lungfishes (see Chap. 13), evolved in marine environments during the early Devonian. Primitive lungfishes were characterized by two dorsal fins; fleshy, scale-covered, paired, leaflike **archipterygial** fins with a bony central axis and with fin rays coming off the central axis; a lack of teeth on the marginal jaw bones, but with tooth plates inside the

mouth, and with the premaxilla, maxilla, and dentary missing; a solid braincase; and a pore-filled, cosmine coating on the dermal bones that covered the skull and scales and that may have been associated with electroreception (Fig. 11.8). Later species moved into freshwater, and trends in lungfish evolution include loss of the first dorsal fin, fusion of median fins (second dorsal, caudal, and anal) to form a symmetric tail, elaboration of tooth plates and development of replaceable dentition, replacement of ossified centra with cartilage, fusion of skull bones, and concomitant loss of the cosmine covering. The five extant lungfishes of South America and Africa take these modifi-

FIGURE 11.8. *(A)* Scaumenacia, *an Upper Devonian lungfish from eastern Canada. (B) Tooth plates from a fossil lungfish,* Ceratodus, *from the Upper Triassic (below, approximately 5 cm wide) and from the extant Australian lungfish,* Neoceratodus. *The Australian lungfish is considered to be more similar to ancestral forms than are the living African and South American species.* Neoceratodus *toothplate is mounted on a piece of modeling clay.*

(A)

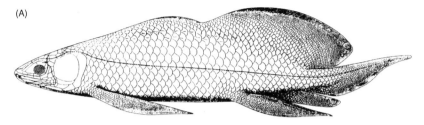

(B)

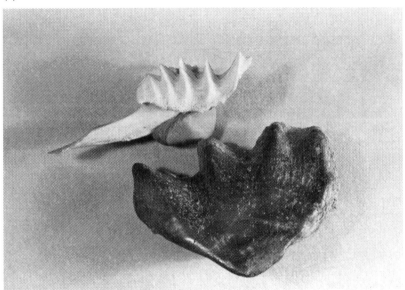

(A) From Jarvik 1980; used with permission. (B) Photo by G. Helfman.

cations to the extreme, having diphycercal tails and eel-like, largely cartilaginous bodies. The modern Australian lungfish is more similar to the heavier-bodied dipnoans of the Paleozoic and Mesozoic.

Lungfishes underwent extensive diversification during the Devonian, evolving more than 10 families, 50 genera, and 100 species, 80% of which occurred during the Upper Devonian (Marshall 1987). Numbers diminished substantially during the Carboniferous. Many lungfish species are known only from fossilized tooth plates, but some of these plates are found in characteristic lungfish burrows, indicating that air breathing and **estivation** (entering torpor during periods of drought) evolved as early as the Devonian (Moy-Thomas and Miles 1971). The modern families of lungfishes are well represented in the fossil record and have long evolutionary histories. Fossils of the Australian family, Ceratodontidae, appear in early Triassic deposits and can be found on all continents during the Mesozoic. Some ceratodontids were quite large; a North American Jurassic species, *Ceratodus robustus*, was 4 m long and may have weighed as much as 650 kg (Robbins

1991). The modern genus *Neoceratodus* occurs as early as the Upper Cretaceous in Australia. The lepidosirenid lungfishes of Africa and South America represent a family that goes back to the late Carboniferous, but members of the two extant genera do not appear until the Eocene and Miocene, on the same continents where they occur today (Carroll 1988).

Much controversy has swirled around the ancestry of the Dipnoi as well as a possible dipnoan ancestry for terrestrial vertebrates (see reviews in Carroll 1988; Pough et al. 1989). Some of this speculation originated with the early misidentification of lungfishes as amphibians (see Chap. 13). More recent arguments have focused on shared aspects of the lungs, limblike fins, and internal nostrils (e.g., Rosen et al. 1981). However, workers in this area generally agree that the ancestry of tetrapods is more closely linked to another group of sarcopterygians, the subclass Osteolepimorphi.

Subclass Osteolepimorphi: tetrapod ancestors. Appearing simultaneously in the early Devonian with the dipnoans, and

FIGURE 11.9.   Eusthenopteron foordi, *a well-known osteolepiform and member of the lineage considered close to the probable direct ancestor to tetrapods. (A) Is the full restoration, (B) Shows neurocranium, endoskeleton, and fin supports. Note the large mouth, large symmetric tail, and posteriorly placed median fins, all characteristics of active predators.*

(A)

(B)

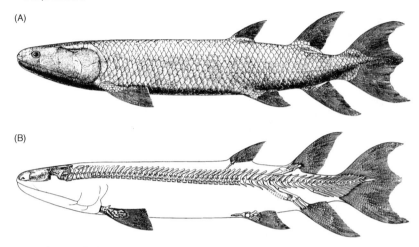

From Jarvik 1980; used with permission.

with an ancestry equally shrouded in mystery, are the **osteolipimorphans**, also known as **rhipidistians**. These large, predatory fishes were characterized by two dorsal fins, cosmine covering of the bones and scales, kinetic (jointed) skulls, lobed fins, and replacement teeth on the jaw margins. They appeared in the Devonian and remained common throughout the latter half of the Paleozoic; most forms disappeared by the end of the Permian. One order and seven families are generally recognized. They were large (up to 4 m long), cylindrically shaped predators that occurred primarily in shallow, freshwater habitats (Fig. 11.9). Evolutionary trends in the osteolepiforms include reduction in dermal bone thickness, a change from diamond- to round-shaped scales, and an increasingly symmetric tail. The latter trait is often considered indicative of a hydrostatic function for the gas bladder (Moy-Thomas and Miles 1971). Our knowledge of osteolepiforms in general, and *Eusthenopteron* in particular, results from exceedingly well-preserved material painstakingly prepared by E. Stensiö and associates. One specimen alone required 6 years of serial grinding and many more years of analysis to characterize the anatomy of the skull of this fish (Jarvik 1980).

Osteolepiforms are generally considered the most likely sister group of modern tetrapods, the chief candidate family being the Panderichthyidae. Focus has been placed on similarities between osteolepiforms and the early labyrinthodont amphibians in terms of many apparent derived homologies, including a dorsally sutured braincase, paired fins, dentition, internal nares, and vertebral accessories (Pough et al. 1989). Osteolepiforms had kinetic skulls that included a hinge joint between the anterior and posterior portions of the braincase. This hinging may have facilitated both wider opening and more rapid closing of the mouth, potentially increasing the size of prey and decreasing the likelihood of escape (Thomson 1967).

The paired fins of the osteolepiforms are very similar to those of the labyrinthodont amphibians of the Upper Devonian, such as *Ichthyostega* (Fig. 11.10). This fin type contains bones homologous to the proximal elements of tetrapod fore- and hindlimbs (humerus, radius, ulna; femur, tibia, fibula), unlike the axially arranged, leaflike archipterygial fins of the Dipnoi. The osteolepiform fin could provide improved body support for benthic locomotion, perhaps including movement across land. The teeth of both osteolepiforms and early amphibians were very similar, consisting of conical teeth with numerous infoldings of the dentine, termed labyrinthodont dentition. Both groups had **internal choanae**, in which the excurrent opening of nares is inside the mouth. They also had ossified neural spines that grew dorsally from ring-shaped, ossified, vertebral centra.

Class Actinopterygii

The primitive fish groups discussed so far are interesting for their antiquity and diversity, and for the effort required by paleontologists to slowly unearth and interpret features of their design. Yet these fishes bear little resemblance to most modern groups and are at most only distantly related to the familiar fishes of today. Speculation about the natural history, behavior, and ecology of extinct forms is based on very little information, much of it difficult to interpret. It is consequently difficult to imagine these animals as the living creatures that they once were. However, these difficulties do not apply to the ancestors of the Actinopterygii, the most successful of today's fishes. Although just as ancient as most of the other groups, primitive ray-finned fishes are similar in size and shape to many extant fishes, and many of their fossils are very well preserved. We can therefore equate fossil and extant actinopterygians in terms of descendancy, form, and possibly function.

FIGURE 11.10. *Comparative pelvic appendages of (A)* Eusthenopteron, *a Devonian osteolepiform fish; (B)* Ichthyostega, *a Devonian labyrinthodont amphibian; and (C)* Neoceratodus, *a modern lungfish. Notice the apparent homologous bone series of the osteolepiform and labyrinthodont limb, as compared with the less similar central axis and radials of the "archipterygial" lungfish fin.*

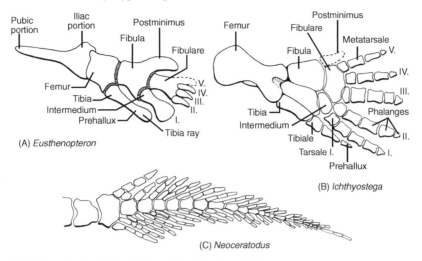

(A) *Eusthenopteron*

(B) *Ichthyostega*

(C) *Neoceratodus*

(A), (B) Redrawn from Jarvik 1980; used with permission. (C) From Semon 1898.

Subclass Chondrostei. *Paleonisciformes.* The origin of the Actinopterygii is once again obscure. Scale fragments appear in late Silurian marine deposits, which may mean that the group is older than the sarcopterygians and as old as placoderms and elasmobranchs. Only the acanthodians among the bony, jawed fishes are of greater antiquity, supporting speculation of an acanthodian ancestry for modern bony fishes. Complete fossil actinopterygians do not appear until the mid- to late Devonian, when the group had expanded into a variety of marine, estuarine, and freshwater habitats. These early fishes, most of which are placed in the order **Paleonisciformes**, suborder Paleoniscoidea, were relatively small (5 to 25 cm) and were distinguished from sarcopterygians by the presence of a single triangular dorsal fin, a forked heterocercal tail with no upper lobe above the unconstricted notochord, paired fins with narrow rather than fleshy bases, dermal bones lacking a cosmine layer, scales joined by a peg-and-socket arrangement and covered with ganoine (ganoid scales), relatively large eyes, and a blunt head (Fig. 11.11A,B). The term *ray fin* refers to the parallel endoskeletal fin rays that were derived from scales. These rays supported the median and paired fins, which were moved by adjacent body musculature. In contrast, the fins of the Sarcopterygii had a thick, bony central axis and muscles contained in the fin itself (see Fig. 11.10).

Paleoniscoids flourished throughout the late Paleozoic, evolving 20 families and 80 genera. Meanwhile, ostracoderms, acanthodians, and placoderms disappeared, and sarcopterygians diminished in abundance. This correlation suggests ecological interaction among groups and possible replacement of primitive jawed and jawless fishes with more advanced actinopterygian and chondrichthyan lineages. What innovations did the ray-finned fishes possess that gave them ecological superiority? The

available evidence strongly suggests that, once again, changes in jaw and fin structure leading to diversified feeding habits and increased mobility were critical to actinopterygian success and dominance.

Changes in the mechanics of jaw opening and closing during actinopterygian phylogeny have been the subject of intensive study (e.g., Lauder 1982; Lauder and Liem 1981; for reviews, see Carroll 1988; Pough et al. 1989). The highly ossified braincase of the early actinopterygians makes it possible to determine the origins, insertions, and approximate sizes of the different muscle masses involved in jaw function, from which we can estimate the forces in operation. During actinopterygian evolution, culminating in advanced teleosts, changes in the angles and connections between the skull case, dermal bones, muscles, and ligaments of the head and jaws have been most influential. In particular, the hyomandibula has been reoriented from oblique to vertical, the posterior end of the maxilla has been freed from the cheek bones, and the jaw musculature has increased in size and complexity. These changes increased the speed and strength of the bite. They also allowed for enlarging of the mouth both vertically and laterally. Hence when the mouth was opened, its volume increased and it assumed a more tubular shape. This changed the bite of a fish from a simple scissorslike action to a suction action. In the modified condition, when the mouth is opened, water and prey are sucked in; when the mouth is closed, instead of water being pushed back out through the jaws, flow continues posteriorly through the gill slits, thereby trapping prey inside rather than pushing it back out of the mouth. Transport of water over the gills during breathing may also have been facilitated by these modifications.

Apparent improvements in other skeletal components are no less important. Paleonisciform scales changed from

FIGURE 11.11.   *Actinopterygian fishes at different grades of development.
(A)* Moythomasia *and (B)* Mimia, *two primitive paleoniscoid fishes from the Upper Devonian, with thick rhomboidal scales extending onto the fins, broadly triangular dorsal
and anal fins, fulcral (ridge) scales along the back, a long mouth, and an asymmetric,
heterocercal tail. (C)* Parasemionotus, *a preteleostean neopterygian from the Triassic,
showing the thinner, rounded scales, more flexible fins, shorter mouth, and abbreviate
heterocercal tail. (D)* Eolates, *an advanced euteleost from the Lower Eocene, with characteristic teleostean diversified dorsal and anal fins, shortened vertebral column, premaxil-
lary-dominated upper jaw, and homocercal tail.*

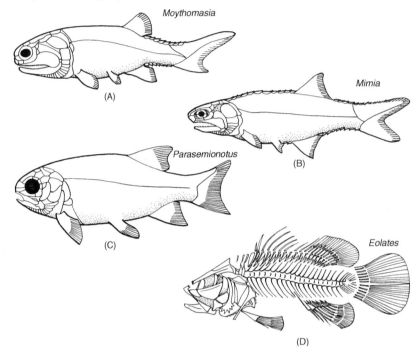

(A) After Jessen 1966; (B) After Gardiner 1984; (C) After Lehman 1966; (D) After Sorbini 1975.

heavy, interlocking, diamond-shaped units to thinner, lighter, circular, cycloid structures. This reduction was accomplished by elimination of the dentine, vascular, and ganoine layers. As paleoniscoid fins consisted of jointed scales, reduction in scale thickness meant increased flexibility in fins; fins became mobile structures composed of dermal fin rays that could be erected or lowered and also moved laterally. Associated with scale reduction was increased ossification of the vertebral column, leading to recognizable centra with dorsal and ventral accessory structures (neural and haemal arches). These accessory structures are closely linked to modifications in the caudal region, where a major trend has been toward an increasingly symmetric, homocercal tail (see Fig. 11.11). Caudal fin rays became supported by a series of ventral accessory, hypural bones.

All these changes during paleonisciform phylogeny imply increased reliance on locomotion, integrated in both escape and prey capture. Heavy ganoid scales offer passive protection against predators but do not function until a predator has already captured a prey individual, a risky event that most potential prey would undoubtedly rather avoid. Lighter scales mean a lighter, more flexible body, capable of more rapid swimming and quicker turns.

Greater reliance on a gas bladder for attaining neutral buoyancy has also been suggested, which also frees fins to provide propulsion and maneuverability. Weight reduction of fins allows them to better serve as propellants, or as brakes and flaps for swimming, stopping, and turning.

The correlation between dermal armor reduction and increased vertebral ossification may indicate a shift from reliance on an external, elastic-hydrostatic skeleton to an internal, muscular-tendonous system (see Chap. 8, Locomotion: Movement and Shape). Increased speed and mobility, combined with the already mentioned improvements in mouth structure, would mean that more advanced actinopterygians would be better not only at avoiding predators but also at capturing prey. These trends in paleoniscoid evolution reoccur later in more advanced actinopterygian lineages (Fig. 11.12).

Paleonisciforms are unrepresented among modern fish groups. Two extant groups, the polypteriform bichirs and the acipenseriform sturgeons and paddlefishes (see Chap. 13), are arguably related to the paleonisciforms, but these modern groups are too different and too poorly represented in the fossil record to permit direct comparison. Paleonisciforms, and 11 other extinct orders of primitive actinopterygians, are sometimes placed together in the

FIGURE 11.12. *Morphologic (and ecological) convergence in fish evolution. Paleoniscoids were ancestral to early neopterygians, which were ancestral to modern teleosts. Certain body designs or plans have apparently been repeatedly favored in actinopterygians, leading to convergent designs among unrelated lineages. These striking convergences in body shape and presumably function are depicted for representative paleoniscoids, early neopterygians, and teleosts. (A) Elongate piscivores with long, tooth-studded jaws and dorsal and anal fins placed posteriorly for rapid starts. (B) Compressed-bodied, predatory, shallow-water fishes with deeply forked tails and trailing fins. (C) Broad-finned bottom feeders with subterminal mouths. (D) Eel-like benthic forms. (E) Compressed, circular forms with large fins for maneuverability in shallow-water habitats with abundant structure (see also Chap. 8).*

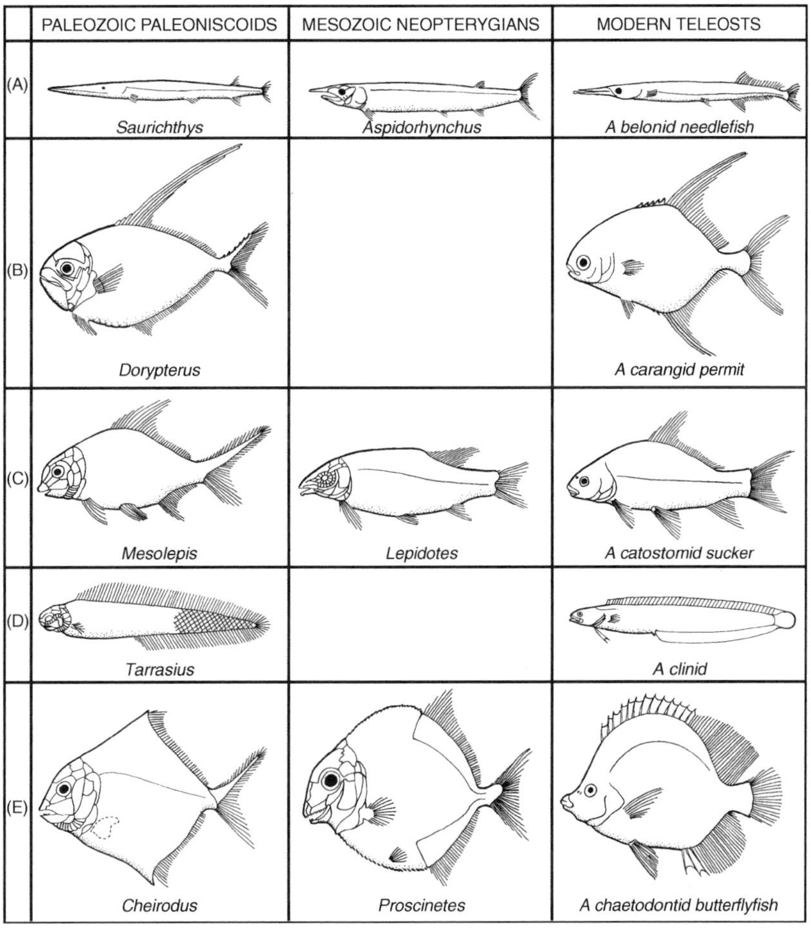

Adapted from Pough et al. 1989; not drawn to scale.

subclass **Chondrostei** (cartilaginous bony fishes) to recognize their anatomic similarities and to distinguish them from the more advanced neopterygians. Whether bichirs, sturgeons, and paddlefishes also belong to the same infraclass is a matter of considerable debate (see Box 13.3). The fossil record indicates that paleonisciform fishes were essentially extinct by the middle Mesozoic.

Subclass Neopterygii. "In their great numbers and degree of anatomical diversity, the modern ray-finned fishes may be considered the most successful of all vertebrates" (Carroll 1988, 136). Just as improvements in feeding and locomotion may have created competitively superior, primitive actinopterygians, continued evolution of these

same traits probably led to the replacement of early actinopterygians by more advanced forms. These descendants, termed **Neopterygii** (new fins), first appear in the fossil record during the Upper Permian and radiated first in the Triassic and Jurassic and then even more extensively in the late Cretaceous. Many of the orders of modern teleostean fishes, the dominant group of bony fishes alive today, are represented in this late Mesozoic radiation. In fact, of the 38 recognized living orders of teleosts, 18 have fossil records that date back to the Cretaceous; only 7 orders arose more recently than the Eocene (i.e., are younger than 50 million years old).

Preteleostean neopterygians included five orders, three of which are extinct. The two extant preteleostean groups, the **semionotiform** gars and **amiiform** bowfin (see Chap. 13), are intermediate between paleoniscoids and teleosts in a number of structures: Gars retain the ganoidlike scales of primitive neopterygians, bowfin have a primitive gular plate under the head, and both groups have identifiably heterocercal tail elements. In most other respects, they are quite specialized, as would be expected for fishes that have existed as recognizable taxa since the Mesozoic. They differ sufficiently in derived traits to justify their placement in separate orders, although some analyses indicate that similarities among gars, bowfin, and their fossil relatives justify placing them together in a separate group, sometimes referred to as the division Holostei (e.g., Olsen and McCune 1991). Neither gars nor bowfin are considered on a direct line to the teleosts. This distinction perhaps lies in the **Pachycormiformes**, a more primitive late Jurassic order of neopterygian fishes.

Division Teleostei. Teleosts (perfect bone) far outnumber all other living fish groups, accounting for more than 23,000 species, more species than in all other vertebrate classes combined. As Chapters 14 and 15 are devoted to characterizing different teleostean groups, they will only be briefly described here. For the present discussion, it is important to realize that teleostean evolution largely repeats and extends trends that originated with the ancestral paleoniscoids. Refinements in the structure and function of mouths and fins appear to explain much of the success of the group. Evidence of these trends is preserved both in the fossil record and in the ancestral traits retained by recognizably primitive teleostean superorders and orders. These trends are detailed in Figure 11.13 and summarized below.

Teleosts, despite their incredible diversity, form a definable group with a recognizable ancestry. Teleosts arose in the middle or late Triassic (200 mya), and teleostean evolution apparently involved four major radiations, three that each gave rise to distinct, primitive subdivisions, and a fourth that produced the major advanced groups alive today (between three and six other radiations died out during the Mesozoic). Multiple radiations imply that modern teleosts as a group could be polyphyletic, more a developmental grade than a single clade. Yet shared traits among the modern groups imply a monophyletic clade (Lauder and Liem 1983). Separate ancestors are postulated, but all may have belonged to the **pholidophorids** or **leptolepids**, early mainstream teleost groups, now extinct.

The first three radiations produced the **osteoglosso-morphs** (bonytongues), **elopomorphs** (tarpons and true eels), and **clupeomorphs** (herrings). These groups stand separately as subdivisions of the Teleostei, in contrast to the fourth radiation, known generally as the subdivision **Euteleostei**. Euteleosts are further divided into nine superorders (see Chaps. 14, 15; Fig. 14.1): Ostariophysi (minnows, catfishes, characins), Protacanthopterygii (a heterogeneous assemblage of salmons, smelts, and pikes), Stenopterygii (hatchetfishes and related deep-sea groups), Cyclosquamata (lizardfishes and related deep-sea groups), Scopelomorpha (lanternfishes), Lampridiomorpha (opahs and oarfishes), Polymixiomorpha (beardfishes), Paracan-

thopterygii (cods, anglerfishes), and Acanthopterygii, which includes the vast majority of spiny-rayed fishes and is further subdivided into the Atherinomorpha (topminnows, silversides, livebearers), Mugilomorpha (mullets), and Percomorpha (squirrelfishes, scorpionfishes, perches, basses, flatfishes, triggerfishes, etc.).

Although numerous derived traits characterize teleostean groups (see Fig. 11.13), trends in five areas can be readily linked to functional improvements that contributed to teleostean success. These trends include reduction in bony elements, repositioning and elaboration of the dorsal fin, change in placement and function of paired fins, structural modifications to and interaction between the caudal fin and gas bladder, and jaw improvements.

1. Reduction of bony elements (see Nelson 1994). Teleosts show a general reduction in bony elements compared to preteleostean groups. This reduction occurred through fusion or actual loss of bones. For example, higher teleosts have:
   a. Fewer, more ossified vertebrae (in general, 60 to 80 in many elopomorphs and clupeomorphs, 30 to 40 in ostariophysans, 30 to 70 in protacanthopterygians, 20 to 35 in paracanthopterygians, 20 to 30 in most percomorphs). A shorter, more ossified axial skeleton would allow for attachment of stronger trunk musculature, thus enhancing locomotion.
   b. Fewer vertebral accessories, such as intermuscular bones and ribs, and replacement of numerous small intermuscular bones with fewer, thicker zygopophyses (compare the boniness of fillets from a herring or trout with that from a tuna or flatfish).
   c. Fewer bones in the skull (e.g., orbitosphenoid missing in perciforms; 10 to 20 branchiostegals in osteoglossomorphs, elopomorphs, and clupeomorphs; 5 to 20 branchiostegals in ostariophysans and protacanthopterygians; 4 to 8 branchiostegals in paracanthopterygians and acanthopterygians).
   d. Reorganization and reduction in the number of bones of the tail, including fusion of the supporting bones (epurals, hypurals, centra) and reduction in the number of fin rays in the tail (principal fin ray number is 18 or 19 in lower teleosts, never more than 17 in perciforms; see also point 4 below).
   e. Reduction in the number of biting bones in the upper jaw from two to one. The maxilla becomes **excluded from the gape** in paracanthopterygians and acanthopterygians. In more primitive groups, it is a tooth-bearing bone, whereas in the two spiny superorders, it pivots with the elongate premaxilla to create a tubular mouth (see point 5a below).
   f. Reduction in the number of fin rays in paired fins (six or more soft pelvic rays in most lower teleosts, six or less in most paracanthopterygians, and one spine with five or fewer rays in most acanthopterygians).

FIGURE 11.13. *Phylogenetic relationships among actinopterygian fishes. The numbered characteristics defining the branching points (synapomorphies) are selected from a much larger list. Groups to the right of a branch point share the traits (although traits may be secondarily lost); groups to the left do not share the trait. Italicized numbers are unique derived traits (autapomorphies) particular to a group and not shared by other taxa. Pholidophoriforms are one of several possible groups ancestral to modern teleosts. Dagger (†) indicates extinct group. Additional details can be found in Lauder and Liem 1983; Pough et al. 1989; Nelson 1994; and papers cited in those publications. 1) Single dorsal fin; ganoin in scales, which have an anterior peglike process; pectoral fin with enlarged basal elements (propterygium); 2) fully ossified, sutureless adult braincase; 3) dentinous tooth cap, basal elements of pelvic fin fused, modifications to jaw and gill arch muscles; 4) dorsal fin spines uniquely flaglike, pectoral fin base platelike; 5) modifications to dermal elements of skull, pectoral girdle, and fins; spiracle penetrates postorbital process of skull; fins preceded by specialized scales (fulcra); 6) upper jaw bones fused; 7) number of endoskeletal elements supporting rays of median fins reduced to a 1:1 correspondence; caudal fin more symmetric, with reduction in upper lobe; dentition of upper pharyngeal consolidated into a tooth-bearing plate; clavicle reduced or lost; 8) vertebral centra convex anteriorly and concave posteriorly (opisthocoelous condition), elongate upper jaw largely constructed from infraorbital bones; 9) maxilla mobile, interopercle and median neural spines present; 10) jaw articulation involves quadrate and symplectic bones, gular plate present; 11) mobile premaxilla, posterior neural arches (uroneurals) elongate, ventral pharyngeal tooth plates unpaired; 12) particular combination of skull bones present (basihyal, four pharyngobranchials, three hypobranchials); 13) tooth plate on tongue bites against roof of mouth, intestine lies to the left of stomach; 14) two uroneural bones extend over the second tail centrum, epipleural intermuscular bones abundant in abdominal and caudal region; 15) ribbon-shaped (leptocephalus) larva; 16) neural arch of first tail vertebra reduced or missing, upper pharyngeal jaws fused to gill arch elements, jaw joint with unique articulation and ossification; 17) specialized ear-to-gas-bladder connection; 18) dorsal adipose fin and nuptial tubercles on head and body, first uroneural bones of tail have paired anterior membranous outgrowth. Additional characteristics of modern teleosts are given in Chapters 14 and 15.*

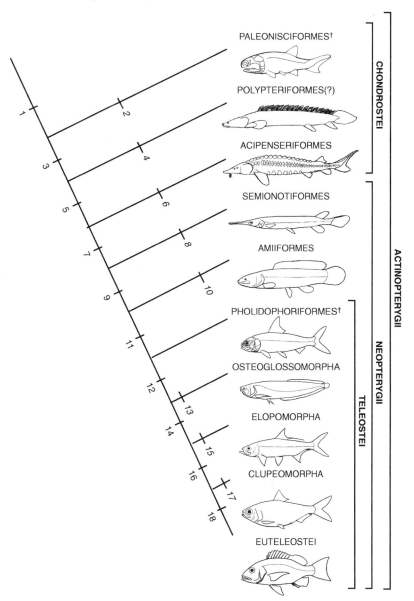

Line drawings are from Nelson 1994 used with permission.

g. Reduction in the amount of bone in the scales (compare the heavy cycloid scale of a tarpon, Megalopidae, or arapaima, Osteoglossidae, with the thin ctenoid scales of most paracanthopterygians and acanthopterygians). A trend toward reduction in armor is familiar by now, as it also occurred during the evolution of several groups (Table 11.1). One possible interpretation is that mechanical protection against predators was of paramount importance when several of these taxa arose, but a premium on mobility soon developed, because lighter, quicker fishes with improvements in both predator avoidance and food getting were favored.

2. Shifts in the position and use of the dorsal fin (Fig. 11.14). The dorsal fin in primitive teleosts is a simple, spineless, fixed, single, midbody keel that prevents rolling and serves as a pivot point for fishes that typically swim in open-water situations (e.g., mooneye, tarpon, bonefish, herrings, minnows, trouts). In higher teleosts, the trend is for the dorsal fin to become elongate and diversified. This is usually manifested as two fins, the anterior portion spinous and the posterior portion soft-rayed. Diversification of a fin into an anterior, hardened spinous portion and posterior, flexible portion maintains the protective function of the fin without sacrificing its role in maneuverability.

Stability is still provided when the fin is erect,

Table 11.1. Repeated trends in fish evolution. Although fishes represent diverse and heterogeneous assemblages assigned to at least five different classes, certain repeated trends have characterized the evolution of these groups or of major, successful taxa within them. The following list summarizes traits or characteristics common to the evolution of several groups.

1. Origin in oceans, radiation into freshwater: thelodonts, pteraspidiforms, cephalaspidiforms, anaspids, placoderms, dipnoans, actinopterygians, teleosts, elasmobranchs.

2. Feeding and locomotion improvements:
   a. Diversification of dentition: acanthodians, placoderms, dipnoans, paleoniscoids, teleosts, elasmobranchs.
   b. Increased caudal symmetry: dipnoans, osteolepimorphs, coelacanths, paleoniscoids, teleosts (reversed in pteraspidiforms and elasmobranchs).
   c. Decreased external armor: pteraspidiforms, acanthodians, placoderms, dipnoans, osteolepiforms, paleoniscoids, teleosts.

3. Bases of spines become embedded in body musculature: acanthodians, elasmobranchs.

4. Fusion of skull bones: pteraspidiforms, acanthodians, placoderms, dipnoans, teleosts.

5. Bone preceded cartilage as skeletal support: cephalaspidiforms (if ancestral to lampreys), dipnoans, acipenseriforms.

6. Electroreceptive ability: pteraspidiforms, cephalaspidiforms, acanthodians, placoderms, dipnoans, coelacanths, brachiopterygians, chondrosteans, teleosts, elasmobranchs (for extinct groups, based largely on morphology of pits and canals in head and body; reinvented in modern teleosts; see Chaps. 6, 13; and Pough et al. 1989).

but many other functions can be served. The erected spiny dorsal provides protection from predators by increasing the body dimensions of the fish; folding the spinous dorsal against the body enhances streamlining. Rapid raising and lowering of the dorsal serves as a social signal in many fishes (similar diversification and actions in the anal fin serve the same purposes). The soft dorsal, through its flexibility, can function as a rudder when slightly curved and as a brake when greatly curved. It can also provide mobility if sinusoidal waves are passed down its length (various knifefishes) or if it is flapped in conjunction with the anal fin (triggerfishes, ocean sunfish) (see Chap. 8, Locomotory Types).

Truly bizarre modifications of the dorsal fin are seen in many higher teleosts. In paracanthopterygian anglerfishes (see Chap. 14), the first spiny ray is modified into an elongate, ornamented lure to attract prey, whereas filaments and fleshy growths increase the resemblance that some fishes show to seaweed or other structures (e.g., sargassumfish, Antennariidae). Scorpionfishes (Acanthopterygii) use the spiny dorsal as a venom delivery system for protection against predators. Long, trailing filaments of probable social function (mate attraction, school maintenance) characterize many acanthopterygian fishes (e.g., carangids, angelfishes, cichlids). The familiar suction disk of the sharksucker, another acanthopterygian, is derived from the first dorsal fin.

3. Placement and function of paired fins. In basal teleosts, pectoral fins are oriented horizontally and located in the **thoracic** position, below the edge of the gill cover; pelvic fins occur at midbody in an **abdominal** location (Fig. 11.15). In this configuration, both fins act primarily as planes that help stabilize movement up and down (pitch) or from side to side (roll), as well as providing some braking force.

During teleostean phylogeny, pectoral fins move up onto the sides of the body, and their base assumes a vertical orientation; pelvic fins move into thoracic or even **jugular** (throat) position. These relocations have several apparent functions (see Webb 1982). Pectorals on the side can be sculled for fine movement and positioning, such as slow swimming, and hovering and backing in midwater. As these fins are often transparent, their use in locomotion might be less obvious to a potential prey animal than would be lateral undulations of the body. Placement of the pelvics forward helps in braking and reduces pitching; their location under the spinous dorsal, in combination with spinous armament, increases the effective body depth of a fish at the point at which it is most likely to be attacked by a predator (Webb and Skadsen 1980).

4. Caudal fin and gas bladder modifications. Actinopterygian evolution is characterized by a progressive increase in symmetry of the tail fin (see Fig. 11.11). Tail fins became externally and functionally homo-

FIGURE 11.14. *Diversification of the dorsal fin in modern teleosts. (A) Primitively, the dorsal fin is a single, spineless, subtriangular structure that serves as an antiroll device and pivot point during swimming, such as in the herrings (Clupeidae). However, this simple fin has been greatly modified in more advanced groups and can serve in locomotion, predator protection, and a variety of other functions. (B) In cods (Gadidae), three dorsal fins exist. (C) More commonly, a spiny anterior and soft-rayed posterior separation occurs, as in the squirrelfishes (Holocentridae). (D) In frogfishes (Antenariidae), modified dorsal spines serve as lures and as camouflage. (E) The sucking disk of the sharksucker (Echeneidae) is derived embryologically from the spiny dorsal fin.*

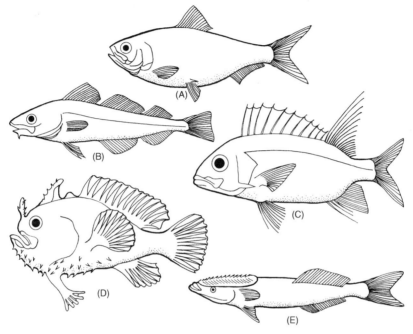

After Nelson 1994.

cercal fairly early in the group's history. Fossil impressions of the tails of late Paleozoic paleoniscoids show that upper and lower lobes were equal, in contrast to the heterocercal and abbreviate heterocercal tails of earlier paleoniscoids. Symmetry becomes more pronounced in the teleosts, reflecting the internal modifications that followed. These internal changes include notochord and body shortening and reworking of large bones and sets of bones that support the caudal fin rays. In particular, teleosts developed a series of hypural bones from several haemal arches. Some of these bones fused to form a ventral hypural plate, continuing a trend evident in paleonisciforms.

Fusion and reduction in the number of vertebrae, reduction in the number of intermuscular bones, and increased tail symmetry all correlate with a greater role of the caudal region in locomotion. The trend is toward an increased dependence on high-power caudal swimming, culminating in steadily swimming fishes with lunate tails, such as jacks (Carangidae), tunas (Scombridae), and billfishes (Istiophoridae) (Webb 1982). Primitive teleosts used sequential contraction of trunk musculature throughout the body, producing a wave

of contraction from head to tail (see Chap. 8). By focusing musculature contraction on the tail and its supporting structures, advanced teleosts could swim faster than more primitive teleosts that depended on sinusoidal movement of the body (Carroll 1988). Hydrodynamic attributes and implications of heterocercal and homocercal tails are discussed in Chapter 8 (Locomotion: Movement and Shape).

In apparent conjunction with tail and paired fin modifications, an additional teleostean trend is added control over gas bladder function. It can be debated whether gas bladders arose initially as breathing or buoyancy control structures (see Chap. 5, Buoyancy Regulation), but the latter function has taken precedence in teleosts. Living preteleosteans and primitive teleosteans have a physostomous gas bladder, in which a pneumatic duct connects the gas bladder with the gut and ultimately the mouth (see Chaps. 4, 5). The gas bladder is filled with gas by gulping air; gas is expelled largely via the same route. Fine adjustments are difficult in this system, and the fish is somewhat dependent on access to the surface.

More advanced teleosts (paracanthopterygians,

FIGURE 11.15.   *The phylogeny of paired fin locations in teleosts. The locations and functions of the pectoral and pelvic girdles have changed during evolution of the Teleostei. Pectoral fins move from a ventral to a lateral position, and the pectoral fin base changes its orientation from horizontal to vertical. Pelvic fins move from abdominal to thoracic and even jugular locations. Extant representatives of phases in this observed trend are represented by (A) an elopomorph (bonefish, Albulidae), (B) a primitive para-canthopterygian (troutperch, Percopsidae), and (C) a generalized acanthopterygian (cich-lid, Cichlidae). This trend is by no means absolute: Many specialized, relatively primitive teleosts have laterally placed pectorals (e.g., catfishes), and advanced teleosts may have pelvics in abdominal positions (e.g., atherinomorphs), but overall the trends describe a progressive change during teleostean phylogeny.*

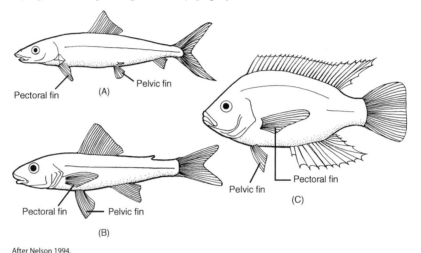

After Nelson 1994.

acanthopterygians) are physoclistous, having lost the pneumatic duct and the link to atmospheric air. They instead rely more on internally generated and absorbed gases to fill and empty the gas bladder and are capable of finer control of buoyancy (physostomous fishes have gas secretion capabili-ties, but these are usually not as refined as in physoclists; see Fig. 5.9).

Although gas bladders do not normally fossilize, the codevelopment of a gas-filled, internally con-trolled gas bladder, a homocercal tail, and paired, multifunctional, flexible fins is taken as a strong indication that they evolved as a suite of characters. A gas bladder makes an otherwise dense fish neu-trally buoyant, which means that small, precise ad-justments in body orientation and movement will not be counteracted by continual sinking. Thus a fish can remain at a fixed point in the water col-umn and turn on its body axis without moving forward.

This trend toward a combination of a function-ally homocercal tail and some sort of internal hydrostatic organ was previously evident in the three major sarcopterygian groups. Dipnoans, oste-olepimorphs, and coelacanths evolved symmetric, diphycercal tails. The first two groups had func-tional gas bladders (lungs), and early actinistians possessed a gas bladder (Lund and Lund 1985), al-though it is small and fat-filled in the living coela-canth. Interestingly, the feeding mode of the living

coelacanth involves hovering relatively still in open water by a combination of paired fin movements, middle caudal lobe movements, and buoyancy con-trol (Fricke et al. 1987, H. Fricke, personal commu-nication).

5. Feeding apparatus modifications. Two major changes have characterized the anatomy of foraging in teleosts.

a. Teleosts continue a trend seen in neopterygians with respect to increasing suction capabilities. Teleosts developed a protrusible **pipette mouth**, capable of generating powerful, directed, nega-tive pressures (see Figs. 8.3, 8.4). The pipette mouth results from enlargement of the muscles and modifications to the bones in the jaw appa-ratus, most notably the maxilla and hyo-mandibula, but also involves connections with the mandibular, opercular, and pectoral bone series. Early in teleostean evolution, the rear por-tion of the maxilla was freed from its connection with other cheek bones, allowing it to swing forward while also allowing the reoriented hyomandibula to move outward, thus increas-ing mouth volume. Skin folds developed along the lateral margins of the jaw bones, creating a hole-free tubular apparatus that prevents lateral escape by small prey.

In the more advanced groups (Paracanthop-terygii and particularly the Acanthopterygii), the premaxilla develops an **ascending process**, which is basically a vertical extension at its anterior tip that slides along the front of the skull, thus allowing the premaxilla to shoot forward at prey as the mouth is opened (see Chap. 8, Jaw Protrusion: The Great Leap Forward).

The end-product of action in this complex of bones, muscles, ligaments, and pivot points is very rapid expansion of the orobranchial chamber. In paracanthopterygian anglerfishes, mouth volume can increase 13-fold over the course of 7 milliseconds (Pietsch and Grobecker 1987). Maximal expansion of the gape is one direction these changes take, particularly in predators on other fishes. In feeders on zooplankton and other small prey, many advanced teleosts have if anything reduced gape width to increase suction power.

And some groups have maximized the speed of mouth extension without suction. For example, zooplanktivorous *Chromis* damselfish (Pomacentridae) can fully protrude their jaws in as little as 6 milliseconds to capture evasive copepods (Coughlin and Strickler 1990). During jaw protrusion in the pikehead, *Luciocephalus pulcher* (Luciocephalidae), head length is increased by one-third at a rate of 51 cm/sec, thus increasing the speed of attack by almost 40%, with no appreciable suction force generated; velocity increases of up to 89% have been recorded in largemouth bass (Nyberg 1971; Lauder and Liem 1981). Observations of mouth extension without suction have fueled a debate as to whether the pipette mouth developed primarily to generate suction power for inhaling prey or to extend rapidly the anterior portion of the body to overtake prey (Lauder and Liem 1981; Lauder 1982). Regardless, the primary selection pressure driving these modifications in the jaw was undoubtedly facilitation of prey capture.

b. Once prey are captured, they are passed back into the mouth to be manipulated. For soft-bodied prey, including most small fishes, manipulation primarily involves positioning the prey to facilitate headfirst swallowing, thus avoiding the improved teleostean fin spines that might cause choking or blockage if prey are swallowed tail first. Later digestion of fish prey requires little more than chemical breakdown in the gut.

However, many potential prey have extremely effective physical defenses that are relatively impervious to gut chemistry. The hard, calcareous shells of mollusks, the chitinous exoskeletons of crustaceans, and the cell walls of plants all require mechanical rupturing before digestive enzymes can have much effect. A protrusible mouth is effective for initial capture of prey, but because of the emphasis on fore-aft movement, protrusibility evolved at a sacrifice in up-and-down chewing motion. Therefore, mechanical rupturing must occur elsewhere. In teleosts, it has been the dentition and musculature of the **pharyngeal jaws** that have diversi-

fied to serve this chewing, crushing, and grinding function (Lauder 1982).

Pharyngeal pads lie posterior to the marginal jaws, just anterior to the esophagus (see Chap. 3, Fig. 3.9). They are derived from dermal tooth plates in the pharynx. During teleostean phylogeny, the function of these pads has elaborated from simple holding of prey prior to swallowing to manipulation and preparation that facilitates digestion. The pads have become armed with a variety of dentition types and have fused to dorsal and ventral elements of the gill arches. The branchial musculature has been reworked and a new muscular connection from the anterior vertebral column has been made, to bring upper and lower plates together in complex, powerful movements. Hence in acanthopterygian groups with this **pharyngognathous** condition, we find the mollusk-feeding croakers and drums (Sciaenidae) with molariform dentition, parrotfishes (Scaridae) with pharyngeal jaws capable of grinding up coral rock to expose the algae contained therein, and the highly successful cichlids with a variety of pharyngeal tooth and jaw arrangements that allow their food to be "crushed, triturated, macerated, compacted or in other ways prepared" (Liem and Greenwood 1981, 93; see Chaps. 8, 15).

Development and diversification of pharyngeal jaws and dentition have undoubtedly broadened the diet of teleosts to include hard-bodied prey and, more importantly, plant material; herbivory is essentially unknown in non-teleostean fishes. This diversification probably extended teleost foraging capabilities far beyond what was possible with the early actinopterygian dependence on more anterior jaw elements. It is more than coincidence that several of the most successful modern teleostean families (cyprinids, cichlids, labrids) have both highly protrusible front jaws and diversified pharyngeal jaws.

Although our emphasis above has been on identifying five general areas that changed during teleostean phylogeny, it is important to remember that these traits changed in concert, that anatomic trends during teleostean phylogeny represent a suite of adaptations. Modification of one trait probably enhanced the effectiveness of and was affected by the other traits. The greatest manifestation of the trends is evident in the acanthopterygian fishes, with their ctenoid scales, diversified yet spiny fins, fine maneuverability via pectoral fin sculling, physoclistous gas bladders, greatly expandable mouth volumes, and effective pharyngeal teeth. The end result "has been increased swimming speed combined with maneuverability . . . without significant loss of defensive structures" (Gosline 1971, 152). In other words, higher teleosts represent quick, spiny fish with a highly efficient feeding apparatus that can catch and eat small, hard prey items.

Also note that these trends generally describe different taxonomic groups but in no way preclude the possibility of a primitive group deriving specializations characteristic of a more advanced taxon. For example, true eels (a relatively primitive teleostean order) as well as other eel-like fishes, regardless of taxonomic position, have expanded dorsal and anal fins, greatly reduced or absent scales, and missing pelvic and even pectoral fins. Elaborate median fins are not found solely in advanced superorders. Many osteoglossomorphs are highly derived, specialized fishes that use their own electrical output to locate objects and locomote via an elongate dorsal (gymnarchids) or anal (gymnotids) fin (see Chap. 14).

Many adult deep-sea fishes (Stenopterygii, Cyclosquamata, Scopelomorpha; see Chap. 14), although belonging to relatively primitive superorders that should be characterized by physostomous gas bladders, are instead secondarily physoclistous, probably to prevent gas loss via the gut and because they never go to the surface to gulp air. Elaborate pharyngeal dentition, a hallmark of the Acanthopterygii, is used widely in the relatively primitive minnows and suckers (Ostariophysi; see Chap. 14). A protrusible mouth, brought on by an ascending, sliding premaxillary process, characterizes acanthopterygians and the closely related paracanthopterygians but was evolved independently and differently in a more primitive group, the ostariophysan cypriniforms, as well as in elasmobranchs and sturgeons. Environmental conditions determine the selection pressures operating on a lineage; groups that evolve more effective adaptations will be favored whether or not they are "breaking the rules" of teleostean phylogeny.

ADVANCED JAWED FISHES II: CHONDRICHTHYES

Class Chondrichthyes (cartilaginous fishes)
  Subclass Elasmobranchii (sharklike fishes)
    Superorder Cladoselachimorpha
      Order †Cladoselachiiformes (Cladoselachida)
    Superorder Xenacanthimorpha
      Order †Xenacanthiformes (Xenacanthida)
    Superorder Euselachii
      Order †Ctenacanthiformes
        †Hybodontiformes
  Subclass Holocephali (chimaeras)
    Superorder †Paraselachimorpha (five extinct orders)
    Superorder Holocephalimorpha (two extinct orders)
      Order Chimaeriformes (modern chimaeras)

†Extinct group.

The lineages of bony fishes can be traced with fair certainty back to the Silurian. Their success is evidenced by the diversity of forms found throughout the late Paleozoic and Mesozoic and because of the overwhelming dominance of teleosts today. However, another group of fishes also arose during the early Paleozoic that followed a very different course of development and that also radiated in the Mesozoic and is well represented today. These are the **Chondrichthyes** (cartilaginous fishes), a group that rap-

idly specialized as marine predators. Although traditionally thought of as primitive because of their cartilaginous skeleton, it turns out that many of the characters of modern Chondrichthyes are secondarily derived and represent specializations for a very different, parallel mode of life in water.

As with the sarcopterygian and actinopterygian divergence among the bony fishes, two major groups of chondrichthyans—the Elasmobranchii and the Holocephali—also evolved, although the connection between the two remains poorly understood. Our knowledge of chondrichthyan phylogeny is further hampered by a lack of fossil skeletal material. By its nature, cartilage does not fossilize readily, and hence our ideas concerning many basal groups rest on incomplete specimens. Accordingly, interrelationships among the Chondrichthyes are, once again, the subject of considerable debate.

Elasmobranchii

Definitive sharklike fossils first appear in the mid-Devonian, with fragmentary remains suggesting perhaps a late Silurian or even an Ordovician origin (Lund 1990). The **elasmobranchs** (plate or strap gills) have undergone several major radiations, with much controversy surrounding interrelationships. At least eight orders of elasmobranchs with origins in the Paleozoic arose and disappeared by the Triassic (Compagno 1990). Identification of the various lineages is based largely on tooth, scale, and spine morphology, and the fossil evidence indicates that, as with bony fishes, foraging and locomotor improvements characterize successive groups.

An early elasmobranch radiation (Fig. 11.16), characterized by **cladoselachid** sharks, had five gill slits and a terminal mouth. Their dentition, referred to as **cladodont**, consisted of multicuspid teeth in which the central cusp was usually larger. The teeth were made of enamel-covered dentine and were homologous with scales. These elasmobranchs were often large (2 m), pelagic, marine predators with an unconstricted notochord protected by calcified cartilaginous neural arches, and with small precaudal, lateral keels analogous to those found in modern pelagic sharks (Moy-Thomas and Miles 1971). The dorsal fins were often preceded by a spine that may have been supportive or protective in function. Caudal morphology was functionally symmetric, although the notochord extended into the dorsal lobe of the fin (Fig. 11.16A).

Cladoselachids were recognizably sharklike in appearance, but other groups resembled modern tarpon or even mammalian dolphins. Although predominantly large and marine (orodontids reached 4 m in length), some of these early sharks were as small as 15 cm (e.g., *Falcatus falcatus*). One group, the **xenacanthids**, invaded freshwater and assumed an eel-like morphology (Fig. 11.16C). Some xenacanthid sharks had pectoral fins reminiscent of the archipterygium of the dipnoans and probably were bottom dwellers. Other groups, such as the squatinactids and petalodontids, were skatelike. By the Carboniferous, sharks made up as much as 60% of the species of fishes in some shallow tropical habitats (Lund 1990). All told,

FIGURE 11.16. *Diversity in body form of Paleozoic sharks from the first major elasmobranch radiation. (A)* Cladoselache, *a cladoselachid. (B)* Denaea, *a symmoriid. (C)* Xenacanthus, *a xenacanthid.*

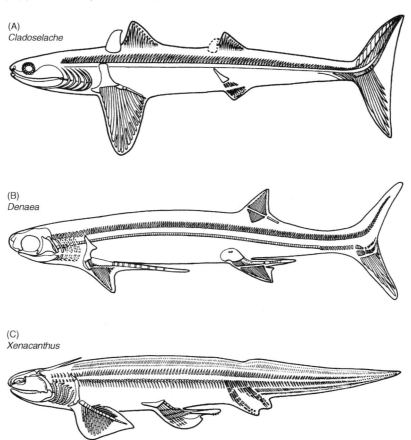

(A)
Cladoselache

(B)
Denaea

(C)
Xenacanthus

(A) From Shaeffer 1967; (B), (C) From Shaeffer and Williams 1977; used with permission.

more than 60 genera of sharks have been identified from this late Paleozoic radiation.

A second major radiation of primitive sharks occurred during the Mesozoic and involved sharks in the orders Ctenacanthiformes and particularly **Hybodontiformes** (Fig. 11.17). These orders are placed in the same superorder as modern sharks, the **Euselachii**. Unlike modern euselachians, hybodonts had terminal rather than ventral mouths. Hybodont teeth represented an innovation over more primitive sharks in that hybodonts had multicuspid teeth that were often differentiated into anterior grasping and posterior crushing types, functionally analogous to the marginal and pharyngeal teeth of modern teleosts. Hybodontoid paired fins were flexible and mobile, probably giving them a maneuverability that was not possible with the stiffer appendages of the earlier sharks. Caudal fins became increasingly heterocercal, a reverse of the trend seen in bony fishes. Paralleling a trend seen during acanthodian phylogeny, the spines that precede the dorsal fins became more deeply embedded in the body musculature.

While hybodonts were flourishing, a third major elasmobranch radiation during the mid-Mesozoic gave rise to modern sharks, skates, and rays (see Chap. 12). Fossils indicate an origin during the late Permian or early Triassic period; different authorities argue in favor of a hybodontid, a ctenacanthid, or an unknown ancestor. By the early Jurassic, recognizably modern sharks are found. One major distinction between modern and earlier sharks is the characteristic overhanging snout of modern euselachians, producing a ventral rather than terminal mouth. The overhanging snout results from an enlarged rostral area that encases a larger olfactory system. Modifications in jaw suspension, jaw-pectoral girdle linkage, and jaw opening muscles create a protrusible upper jaw and the generation of suction forces, paralleling the trend seen in teleosts. Calcified vertebral centra largely replaced the unconstricted notochord of earlier groups, and fin supports changed from multiple basal cartilages with cartilage radiating out to the fin margins to smaller, fused basal supports (usually three) and flexible, horny rays, termed **ceratotrichia**, supporting the web of the fin. This combination of vertebral and fin modifications should have provided for faster swimming and greater maneuverability.

Regardless of the radiation in question, several elasmobranch innovations probably gave them a selective advantage over the other early gnathostomes present at

FIGURE 11.17. *Sharks of the second and third major elasmobranch radiations. (A)* Ctenacanthus, *a ctenacanthid. (b)* Hybodus, *a hybodont. (C)* Squalus, *a modern squaliform shark.*

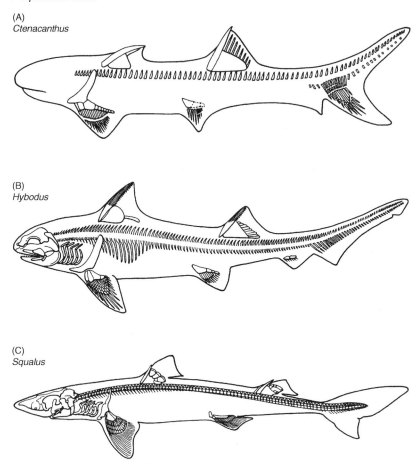

(A)
Ctenacanthus

(B)
Hybodus

(C)
Squalus

From Shaeffer and Williams 1977; used with permission.

the time. In contrast to the placoderms and most acanthodians, sharks quickly evolved a tooth replacement mechanism. Teeth grew in whorls or spiral bands (Fig. 11.18A), with the functional, exposed tooth backed up by several replacement teeth embedded in the jaw cartilage. As embedded teeth grew, they moved along the whorl until they erupted at the jaw periphery, only to be replaced later by younger teeth. Dentition replacement patterns differ among different lineages of modern sharks (see Chap. 12), but in all likelihood teeth were regularly shed and replaced spontaneously or following injury in primitive groups, as happens in modern elasmobranchs. This arrangement took on some relatively bizarre forms, as in the Permian edestoid shark, *Helicoprion* (Fig. 11.18B).

Other characteristics of modern sharks undoubtedly had their origins in late Paleozoic groups. Sexually dimorphic males with pelvic fins modified as intromittent organs for sperm deposition indicate that internal fertilization evolved early in chondrichthyans (see Fig. 11.16B,C). A strong dependence on electroreception, highly acute olfactory capabilities (associated with the longer snout that houses the nasal capsules), increased

buoyancy through oil accumulation in the liver (paralleling gas bladder evolution in bony fishes), and large brain and body size have all contributed to the position of modern elasmobranchs among the top predators in marine habitats (see Chap. 12). Prior to the radiation of teleosts in the late Mesozoic, elasmobranchs were the dominant forms among carnivorous fishes. Modern elasmobranchs apparently represent a reduced subset of the ecological and morphologic patterns that arose during the age of sharks.

Holocephali. A calcified cartilaginous skeleton and internal fertilization, among other traits, link the **Holocephali** (whole heads) with the elasmobranchs (see Chap. 12). Holocephalans arguably date back to the late Devonian, with definitive fossils not appearing until the Jurassic (Fig. 11.19). Origins of the group are a matter of debate, with some proponents linking them to ptyctodontid placoderms, whereas others favor an ancestry among early elasmobranchs. Regardless, holocephalans differ in many respects from elasmobranchs. Most notable is the position and structure of the gill chamber, which is located farther forward than in sharks and has a single opercular opening.

FIGURE 11.18. *Tooth replacement in sharks. (A) Cross section through the jaw of a modern shark, showing a functional tooth backed by rows of developing replacement teeth. Variations on this mechanism are found in many fossil groups. (B) Symphysial (middle) portion of the lower jaw of the late Paleozoic edestoid* Helicoprion, *with its spiral replacement tooth whorl.*

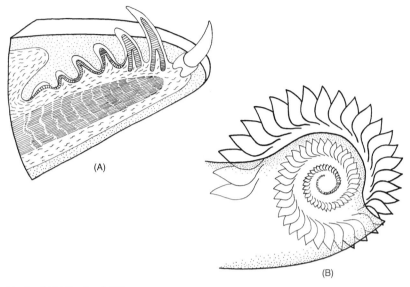

(A)

(B)

After Carroll 1988; Plough et al. 1989.

Holocephalans have nonprotrusible jaws, because the palatoquadrate (upper jaw) is fused to the braincase (**autostylic suspension**); in modern elasmobranchs the upper jaw gains mobility via a posterior hyomandibula and an anterior ligamentous connection to the chondrocranium (**hyostylic suspension**). Most fossil and all modern chimaeras and ratfishes have tooth plates on the jaw margins that continue to grow during ontogeny; iniopterygians, a group arguably included in the Holocephali, had replacement dentition. Tail form in holocephalans is variable but often of a diphycercal nature, hence the ratfish designation. Early holocephalans showed a diversity of form, and many were quite diminutive in size, not exceeding 10 cm in length. A chimaera in Greek mythology was an imaginary monster constructed of incongruous parts. The Holocephali represent both a modern and ancient enigma, because their parts indicate affinities with a variety of groups, or with no other groups at all. Again paralleling the situation in the bony fishes, holocephalans are to the Chondrichthyes what brachiopterygians are to the bony fishes. With some lucky fossil discoveries, we should develop a more meaningful synthesis of where the interrelationships of the Holocephali lie.

A HISTORY OF FISHES: SUMMARY AND OVERVIEW

As should be obvious, the gaps in our knowledge about fossil fishes and their relationships to one another and to modern groups are large and plentiful. Such gaps are the initiation points for future research. For starters, three

topics arise from these unanswered questions and deserve some exploration.

The Diversity of Fossil Fishes

We speak of the success of different ancient groups and compare among them and between modern teleosts and extinct forms. All of the preceding discussion is totally dependent on the fossil record. But how accurately does the fossil record represent the diversity of fossil fishes? How many fishes would we estimate are alive today if we were forced to rely on fossils? As of 1988, approximately 333, or about 8%, of modern teleostean genera are represented by Recent fossils (Carroll 1988; Nelson 1994). Significantly, the number of fossils available for study decreases with time because geologic processes tend to destroy fossilized material. Therefore, the fossil record of Recent fishes is at best an optimistic underestimator of the accuracy with which earlier groups are represented.

Fossilization is a chance procedure, compounded by the relatively small surface of the earth available for paleontologic discovery. Inadequacy of sampling is obvious when we realize that most fossils are recovered from only the top few meters of rock, and much of the surface land of the Mesozoic and early Cenozoic has been subducted by tectonic processes (see Chap. 16). Our limited sample size is aggravated by the inaccessibility of major areas of the earth's surface; recall that 70% of our planet is under water, where very little paleontologic exploration occurs.

Significantly, about 2400 species, or 10%, of living fishes occur in water deeper than 200 m (Cohen 1970), yet few of the recognized preteleostean groups are postulated as having occupied the deep sea. Deep-sea fishes,

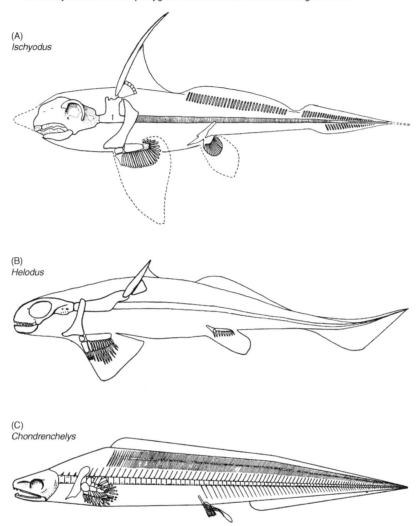

FIGURE 11.19.   *Extinct holocephalans. (A)* Ischyodus, *Jurassic. (B)* Helodus *and (C)* Chondrenchelys, *both Carboniferous. Note convergence in body form between* Chondrenchelys *and the actinopterygians* Tarrasius *and the clinid in Figure 11.12.*

(A)
Ischyodus

(B)
Helodus

(C)
Chondrenchelys

From Patterson 1965; used with permission.

regardless of taxon, are highly convergent in body form and structure (see Chap. 17); such adaptations should be obvious in fossils, and such fishes should be assignable to the deep-sea habitat. However, the fossil record for living deep-sea groups is understandably limited. For example, stomiiforms are among the most abundant of the deep-sea orders, with more than 50 recognized genera, but only 5 of these, or about 10%, have a fossil record (Nelson 1984; Carroll 1988). The deep sea is one of the most stable aquatic habitats on earth, and it seems unlikely that living in the deep sea is a teleostean innovation. Preteleostean diversity in deep-sea habitats is obviously underestimated.

These problems are further compounded with the realization that many fossil species are described based on a single, often fragmentary, specimen. How many of these fragments remain undiscovered, and more importantly, how many rare species never fossilized? In our search for

antecedents of modern groups, how does this selective preservation of forms affect our interpretations of lines of descent, particularly if there exists at best a 1-in-10 chance that an ancestor will fossilize? Our optimistic hope is that the fossil record is somehow a proportional and representative subsample of reality, that we accept that we have grossly underestimated the diversity of primitive fishes, and that many more future researchers will take up the challenges of paleontology.

The Tangled Web of Early Vertebrate Relationships: Primitive Does Not Necessarily Denote Ancestral

It is intellectually frustrating to have major living taxa (e.g., modern agnathans, jawed fishes, and gnathostomes in general) for which we can find no clear ancestral lineages. Such phylogenetic problems beg for solution. As a result, considerable effort has been expended attempting to link modern agnathan groups with Paleozoic forebears, and for that matter, modern gnathostomes with ancestors among the diversity of fishes that proliferated during the Devonian.

Plausible alternative explanations exist. First, the ancestors of modern groups may have died out without leaving fossil remains, at least none that we have found so far. Second, Paleozoic lineages may have been extinguished, period. Hence similarities between ancient and extant groups may result from convergence and perhaps some retention of primitive characteristics derived from a common, distant (unfossilized) ancestor. The latter scenario is perfectly reasonable given the rather advanced condition of bones in the agnathous fishes and of jaws and other supporting bony elements in the early gnathostomes when they first appear in the fossil record. Groups ancestral to these early lineages must have existed for millions of years but lacked the necessary mineralized structures to fossilize.

Extinction is a universal characteristic of species; it has been estimated that the average "lifespan" of a species is around 10 million years (Raup 1988). The mass extinctions that have occurred during the history of life (e.g., the Burgess Shale fauna in the Precambrian, and the Permo-Triassic and Cretaceous-Tertiary extinction events) have been particularly disastrous for shallow marine faunas, wiping out 50% to 100% of the species in existence at the time (Raup 1988). It seems reasonable to assume that fish lineages were as susceptible to mass extinctions as were contemporaneous invertebrate groups; declines in diversity of actinistians, amiiforms, hybodonts, holocephalans, and perhaps neopterygians at the end of the Cretaceous may attest to the vulnerability of fish groups to mass extinction.

Continuity in Fish Evolution

This chapter focuses on the antecedents of modern fishes. Implied in the organizational approach we have taken here is that fossil fishes should be dealt with separately from living forms. However, this separation is arbitrary and superficial. It is more a stylistic convenience for organizing a textbook than a statement of philosophy. Students of fish evolution should quickly recognize that modern fishes are extensions of fossil groups. As was pointed out earlier, the majority of modern fish families already existed in the Mesozoic, if not earlier (see Fig. 11.1). Although some primitive groups that are unrepresented today (e.g., ostracoderms, placoderms, acanthodians, osteolepimorphs, and paleoniscoids) probably deserve separate treatment from modern forms, it makes just as much sense to treat truly ancestral forms, such as primitive dipnoans, coelacanths, neopterygians, and chondrichthyans, together with their modern derivatives.

To paraphrase paleontologist A. R. McCune, Why should mode of preservation—in rocks or in alcohol—be the primary determinant of how we deal with a taxonomic group? If a modern, "primitive" species (e.g., the living coelacanth) were to become extinct through human neglect, would it immediately have to be placed only

in a discussion of extinct fishes? It is our hope that students of ichthyology will recognize the continuity that exists between primitive and advanced groups and view them not as separate entities but rather as a continuum of organic change within lineages.

The following chapters on chondrichthyans and living representatives of primitive taxa focus on species that have strong, direct ties to the extinct (we think) groups discussed above. Where one lineage grades into another is in reality an undefined segment in a line drawn in geologic time.

SUMMARY

1. Fishes have an ancestry that goes back at least 500 million years. Some fossil groups can be linked with extant taxa; some extant taxa lack obvious fossil antecedents; and numerous groups arose, prospered, and were extinguished.

2. The first fishes to fossilize occurred during the late Cambrian and lived into the Devonian. They lacked jaws but possessed bony armor and had a muscular feeding pump. Two classes are recognized—the pteraspidomorphs and cephalaspidomorphs. Both lineages lived in both marine and freshwater. Cephalaspidomorphs may have been ancestral to modern lampreys, based largely on striking similarities in the braincase and cranial nerves. Fossils recognizable as modern lampreys do not appear until the Carboniferous.

3. Hagfishes may have ancestry in primitive chordates known as conodonts, which are well known from toothlike structures that fossilized abundantly during Precambrian and later times but that could not be linked to any particular body form. Four-cm-long body outlines containing the conodont tooth apparatus were finally discovered in Scotland and Wisconsin in the 1980s.

4. The development of jaws was a critical step in the advancement of fishes. However, the ancestry of jawed fishes is unclear because no intermediate fossils between jawed and jawless forms have been found. Placoderms were early jawed fishes that arose in the Silurian, disappeared by the early Carboniferous, and left no apparent modern descendants. They had a bony, ornamented, platelike skin. Many were predators and achieved monstrous size. Many placoderms had a hinge at the back top of the head that allowed for greater opening of the mouth. Placoderm teeth consisted of dermal bony plates attached to jaw cartilage and could not be repaired or replaced.

5. The first advanced jawed fishes were the acanthodians, or spiny sharks, which are unrelated to modern sharks. Acanthodians were water-column swimmers. Many of their traits suggest they share a common ancestry with modern bony fishes, and they are often

placed with sarcopterygians and actinopterygians in the grade Teleostomi.

6. Two classes, the Sarcopterygii and the Actinopterygii, arose during the Silurian and Devonian and gave rise to modern bony fishes. Sarcopterygians diversified into coelacanthimorphs, dipnoans (lungfishes), and osteolepimorphs, whereas actinopterygians diversified into chondrosteans (modern sturgeons, paddlefishes, and maybe bichirs) and neopterygians (bowfin, gars, and teleosts). Lungfishes may be ancestral to tetrapods, but that distinction more probably belongs to the osteolepimorphs, in that they share skull characteristics, dentition, and fin patterns with primitive amphibians.

7. Actinopterygians arose during the Silurian. An early, successful group, the paleonisciforms, had a triangular dorsal fin, heterocercal tail, paired ray-supported fins with narrow bases, and ganoid scales. Important structural changes occurred in the jaw apparatus that strengthened the bite, increased the gape, and created suction forces. Mobility also improved with lightened scales, vertebral ossification, and an increasingly symmetric tail. Paleonisciforms may be ancestral to modern chondrosteans.

8. Neopterygian, or modern ray-finned, fishes are the most successful of all vertebrates. They first appeared in late Permian times and radiated extensively during the Mesozoic. Two extant preteleostean groups are the semionotiforms (gars) and amiiforms (bowfin). Teleostean evolution largely repeats and extends trends that originated with the ancestral paleoniscoids, particularly with respect to advances in jaw and fin structure and function.

9. The earliest teleosts were the pholidophoroids and leptolepids. Four distinct lineages arose from these ancestors: the bony-tongue osteoglossomorphs; the tarpon and true eel elopomorphs; the herringlike clupeomorphs; and the euteleosts, which contain most modern bony fishes. Five major trends characterize teleostean evolution: reduction of bony elements, shifts in position and function of the dorsal fin, placement and function of paired fins, caudal fin and gas bladder modifications, and improvements to the feeding apparatus.

10. Chondrichthyans (cartilaginous fishes) include two subclasses, the elasmobranchs (sharklike fishes) and the holocephalans (chimaeras); the affinities of the two groups are poorly understood. Sharklike elasmobranchs first appeared in the Silurian, underwent tremendous diversification, and are represented today by a comparatively depauperate group of specialized predators. Earlier successful radiations include the cladoselachids, xenacanthids (a freshwater group), and hybodonts. Modern, euselachian sharks arose during the Mesozoic, showing improvements in jaws, dentition, vertebrae, and fins that paralleled locomotory and feeding changes in bony fishes.

11. Holocephalans may date back to the Devonian. They differ from elasmobranchs by having nonprotrusible jaws in which the upper jaw is fused to the braincase, and by having a single opercular opening. Their ancestry is a mystery.

SUPPLEMENTAL READING

Carroll 1988; Gosline 1971; Jarvik 1980; Lauder and Leim 1983; Liem and Lauder 1982; Long 1995; Moy-Thomas and Miles 1971; Norman and Greenwood 1975; Pough et al. 1989.

12

Chondrichthyes: Sharks, Skates, Rays, and Chimaeras

The existence of three distinct classes of fishes alive today is testament to the past and current success of the original vertebrate lineages. Although cartilaginous fishes meet our basic definition criteria of gills, scales, fins, and aquatic existence that qualify them as fishes, the 800 species of living chondrichthyans possess many unique traits that underscore their divergence from other fish groups. And their large size and predatory habits make them intrinsically interesting to humans because of their potential not only as our prey and perhaps competitors, but even as predators upon us.

Class Chondrichthyes (cartilaginous fishes)*
  Subclass Elasmobranchii (sharklike fishes)
    Superorder Euselachii ([†]ctenacanth, [†]hybodont, and modern sharks
        and rays)
      Order Heterodontiformes (8 species, marine): Heterodontidae
          (bullhead, horn sharks)
      Order Orectolobiformes (31 species, marine): Parascylliidae
          (collared carpet sharks), Brachaeluridae (blind sharks),
          Orectolobidae (wobbegongs), Hemiscylliidae (bamboo sharks),
          Ginglymostomatidae (nurse sharks), Stegostomatidae
          (zebra shark), Rhincodontidae (whale shark)
      Order Carcharhiniformes (210 species, mostly marine):
          Scyliorhinidae (catsharks), Proscylliidae (finback catsharks),
          Pseudotriakidae (false catshark), Leptochariidae
          (barbeled houndshark), Triakidae (houndsharks), Hemigaleidae
          (weasel sharks), Carcharhinidae (requiem sharks), Sphyrnidae
          (hammerhead sharks)
      Order Lamniformes (16 species, marine): Odontaspididae
          (sand tiger sharks), Mitsukurinidae (goblin shark),
          Pseudocarchariidae (crocodile shark), Megachasmidae
          (megamouth shark), Alopiidae (thresher sharks), Cetorhinidae
          (basking shark), Lamnidae (mackerel sharks)
      Order Hexanchiformes (5 species, marine): Chlamydoselachidae
          (frill shark), Hexanchidae (cow sharks)
      Order Squaliformes (74 species, marine): Echinorhinidae
          (bramble sharks), Dalatiidae (sleeper sharks), Centrophoridae,
          Squalidae (dogfish sharks)
      Order Squatiniformes (12 species, marine): Squatinidae
          (angel sharks)
      Order Pristiophoriformes (5 species, marine): Pristiophoridae
          (sawsharks)
      Order Rajiformes (456 species, marine and freshwater): Pristidae
          (sawfishes), Torpedinidae (electric rays), Narcinidae, Rhinidae,
          Rhinobatidae (guitarfishes), Rajidae (skates), Plesiobatidae
          (deepwater stingray), Hexatrygonidae (South African stingray),
          Dasyatidae (stingrays), Urolophidae (round stingrays),
          Gymnuridae (butterfly rays), Myliobatidae (eagle rays),
          Mobulidae (manta rays)

*Classification after Compagno 1984; Robins et al. 1991; and primarily Nelson 1994.
[†]Extinct group.

SUBCLASS ELASMOBRANCHII

Although often portrayed as "primitive" fishes, modern sharks, skates, and rays are highly derived, specialized fishes that differ dramatically from the abundant, diverse elasmobranchs that dominated marine habitats through much of the Mesozoic (see Chap. 11). Many traits that characterize elasmobranchs—such as a cartilaginous skeleton, placoid scales, internal fertilization, replacement dentition, and multiple gill slits—appeared early in the 400-million-year history of the group. However, modern sharks, skates, and rays exhibit tremendous variation in these and other characteristics, and have developed additional anatomic, life history, and behavioral adaptations that set them apart from bony fishes and make them surprisingly vulnerable to human exploitation. Only in recent years has the uniqueness and vulnerability of elasmobranchs received recognition and adequate attention.

Definition of the Group

Modern elasmobranchs are generally large (> 1 m) predatory fishes with a calcified but seldom ossified skeleton, including distinctive calcified vertebral centra. They differ from bony fishes in that the skull lacks sutures and their teeth are not fused to the jaws but are instead embedded in the connective tissue of the jaws. Teeth, which have the same embryonic origin as and may be derived from placoid scales (see Chap. 3, Modifications of Scales), are replaced serially; such replacement is rare in bony fishes. The biting edge of the upper jaw is formed by the palatoquadrate cartilage rather than by the maxillary or premaxillary bones. The palatoquadrate is free from the braincase, creating a protrusible upper jaw during feeding. The mouth is subterminal or ventral. Nasal openings are ventral and incompletely divided by a flap into incurrent and excurrent portions; bony fishes generally have completely separated, dorsally positioned nasal openings.

Fin rays in elasmobranchs are soft, horny, unsegmented **ceratotrichia**. Typical sharks, skates, and rays usually have five, and sometimes six or seven, external gill slits on each side. The first gill slit of elasmobranchs is often modified as a spiracle, supported by the hyoid arch and first functional gill arch. Elasmobranchs lack lungs and gas bladders but possess large, buoyant livers and spiral valve intestines. Internal fertilization is universal to the group; males possess pelvic-fin-derived intromittent organs (**myxopterygia**, or **claspers**), and females either lay eggs or nourish embryos internally for several months before giving birth. Metabolic waste products in the form of urea and trimethylamine oxide (TMAO, an ammonia derivative) are concentrated in the blood and serve in osmotic regulation (TMAO is actively excreted by the kidneys of teleosts). A single cloaca serves as an anal and urogenital opening.

Historical Patterns

Most orders of living chondrichthyans appeared by the Upper Jurassic, and all orders appeared by the end of the Cretaceous. Some extant genera have been found in Upper Cretaceous deposits, with little change in some species since the Miocene (Compagno 1990). Most extant groups have evolutionary histories much younger than actinopterygian and neopterygian fishes (see Chap. 11, Advanced Jawed Fishes II: Chondrichthyes). All noneuselachian elasmobranchs are extinct: Cladoselachians died out by the Permo-Triassic transition, and xenacanths died out during the Triassic. Other primitive groups such as edestoids, ctenacanths, and hybodonts disappeared during the Mesozoic.

Although some living sharks have morphologies similar to ancestral Paleozoic and Mesozoic species, these similarities reflect convergent design. The modern groups are very different with respect to features of the cranium, vertebral column, fin skeletons, tooth structure, and squamation. The greatest departure from a generalized body form exists in the highly successful rajiforms, the most advanced of which are the large-brained, filter-feeding manta and devil rays (Mobulidae). Among the sharklike elasmobranchs, the most derived species include the sleeper sharks (Squalidae), angel sharks (Squatinidae), and saw sharks (Pristiophoridae) (Compagno 1990).

Modern Euselachian Diversity

Nearly 800 species of euselachians exist today, including 350 sharklike species and 450 skates and rays (Fig. 12.1). Among the sharks, the **requiem** or ground sharks (Carcharhiniformes) make up more than half the species and are particularly diverse in tropical and subtropical, nearshore habitats. Offshore, pelagic sharks include **lamniform** species such as mako, white, thresher, and basking sharks, whereas the **squaliform** dogfishes are particularly successful in the North Atlantic, North Pacific, and deep-sea regions. The **rajiforms**, distinguished from sharks primarily by having their pectoral fins fused to the sides of the head and having ventral rather than lateral gill slits, are concentrated in two suborders. Among rajiforms,

skates (Rajoidei) are most abundant in deep water and at high latitudes, whereas the stingrays (Myliobatoidei) are most diverse in tropical, inshore waters.

Amid this diversity, certain general patterns emerge that emphasize the unique traits and fascinating adaptations of elasmobranchs. These trends include 1) large size, 2) a marine habitat, 3) mobility, 4) slow metabolism and slow growth, 5) predatory feeding habits, 6) reliance on nonvisual senses, 7) low fecundity and precocial (independent) young, and 8) vulnerability to exploitation (see Compagno 1990; Gruber 1991).

Body size.   When compared with bony fishes, sharks as a group have always been relatively large. Modern sharks range from the 15-g, 16- to 20-cm dwarf dogshark (*Etmopterus perryi*, Squalidae) to the 12,000-kg, 12-m-long (or even larger) whale shark, *Rhincodon typus* (Rhincodontidae), the largest fish in the world. At least 90% of living sharks exceed 30 cm in body length, 50% reach an average length of about 1 m, and 20% exceed 2 m (Springer and Gold 1989). Maximum size of sharks, particularly the maximum size reached by the superpredatory white sharks (Lamnidae), is a subject plagued by misinformation and exaggeration (Box 12.1).

Large size is intimately linked with the feeding and reproductive ecology of sharks. As predators on other fishes, including other elasmobranchs, large size confers an advantage in terms of greater swimming speed during pursuit or long-distance cruising and allows for larger mouth size and larger jaw muscle attachment. Such traits make sharks effective predators on smaller fishes and also decrease their own vulnerability to predators, either via rapid escape or active defense. It is suggested that sharks larger than 1 m are relatively immune to shark predation, and it is not surprising that birth sizes of many sharks are close to the 1-m critical length (e.g., sand tiger, Odontaspididae; white and longfin mako, Lamnidae; dusky, Carcharhinidae). Sharks that give birth to smaller young often have relatively large litters or short intervals between reproduction (e.g., Atlantic sharpnose, Carcharhinidae; scalloped hammerhead and bonnethead, Sphyrnidae).

Predation also affects nursery ground location and interacts with growth rate. Sharks that drop their pups in offshore or beachfront areas that are frequented by large sharks tend to have relatively rapid growth rates of 30 to 60 cm during the first year (e.g., thresher, Alopiidae; shortfin mako; blue, tiger, spinner, sharpnose, Carcharhinidae; bonnethead). Sharks that release their young in relatively predator-free inshore nursery areas such as bays, sounds, estuaries, or shallow reef flats tend to grow only 15 cm in the first year (e.g., bull, sandbar, and lemon, Carcharhinidae; scalloped hammerhead) (Branstetter 1990, 1991).

Habitats.   Most elasmobranchs are marine organisms of relatively shallow temperate and particularly tropical waters, although all oceans except the Antarctic have one or more species. Most inhabit continental and insular shelves and slopes: 50% of all species occur in less than 200 m of water, and 85% occur in less than 2000 m. Only about 5% dwell in the open ocean, including a few raji-

FIGURE 12.1.   *Taxonomic distribution and representative orders of the approximately 800 species of modern sharks, skates, and rays. (A) Sharklike fishes in eight orders constitute 43% of modern euselachian species, with the carcharhiniform (ground or requiem) sharks outnumbering all other orders combined. (B) Raylike rajiforms or batoids make up 57% of the Euselachii, dominated by skates and stingrays; five representative families are shown.*

(A) Sharks:

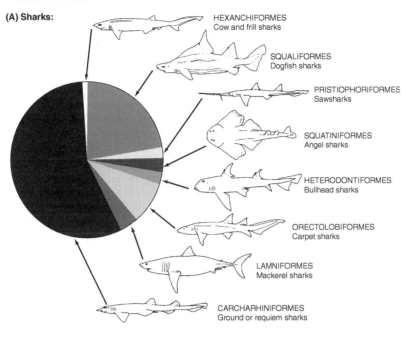

HEXANCHIFORMES
Cow and frill sharks

SQUALIFORMES
Dogfish sharks

PRISTIOPHORIFORMES
Sawsharks

SQUATINIFORMES
Angel sharks

HETERODONTIFORMES
Bullhead sharks

ORECTOLOBIFORMES
Carpet sharks

LAMNIFORMES
Mackerel sharks

CARCHARHINIFORMES
Ground or requiem sharks

(B) Rajiforms or Batoids

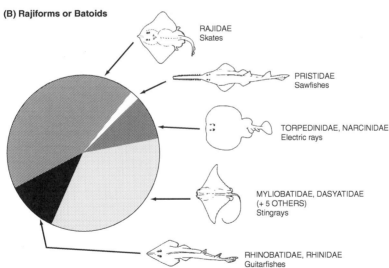

RAJIDAE
Skates

PRISTIDAE
Sawfishes

TORPEDINIDAE, NARCINIDAE
Electric rays

MYLIOBATIDAE, DASYATIDAE
(+ 5 OTHERS)
Stingrays

RHINOBATIDAE, RHINIDAE
Guitarfishes

Modified from Compagno 1990; used with kind permission from Kluwer Academic Publishers.

forms such as the manta ray; another 5% occupy freshwater. Truly cold-water, high-latitude sharks (not counting those that live in the perpetually cold, deep sea) are limited to the relatively large hexanchids (sixgill sharks), basking sharks and sleeper sharks (Squalidae), smaller squaliform sharks, and some catsharks (Scyliorhinidae).

Although bony fishes inhabit ocean depths to 8000 m (see Chap. 17, The Deep Sea), sharks do not occur as deep. The Portuguese shark, *Centroscymnus coelolepis*, is the record holder, occurring down to 3690 m (Clark and Kristof

1991). Deep-sea sharks, those frequently captured in aphotic (lightless) waters below 1000 m, include Portuguese, kitefin, lantern, sleeper, and roughskin dogfish sharks (Squalidae), cat sharks and false cat sharks (Scyliorhinidae, Pseudotriakidae), and the sixgill. Sharks have not occupied habitats characterized by other environmental extremes of temperature, oxygen, turbulence, drought, and salinity, as have bony fishes. A few sharks, such as the *Pentanchus* catsharks, inhabit deep-water basins that are characterized by relatively low oxygen levels

How Big Are White Sharks?

Maximum sizes of shark species are a matter of much speculation and imagination. Researchers tend to be conservative and therefore accept only documented measurements with an accurate measuring tape and weighing scale, preferably accompanied by the preserved specimen, or at least by a photograph with a ruler for scale. However, very large animals are difficult to preserve and harder to store, and photographs can be doctored or just misleading because of problems with parallax. Hence verified maximum sizes and reported maxima ("bigger than the boat") vary considerably.

For example, the longest recorded whale shark is 12 m, but the species is known to grow much larger, perhaps as large as 18 m. Basking sharks (Cetorhinidae) have been reliably measured at 9.76 m, but lengths of 12 to 15 m have been reported. Other accepted (vs. reputed) lengths for large predatory shark species include shortfin mako (3.3 m vs. 4.0 m), great hammerhead (5.5 m vs. 6.1 m), thresher shark (5.7 m vs. 7.6 m), greenland shark (6.4 m vs. 7 m), and tiger shark (5.9 m vs. 7.4 and 9.1 m) (Springer and Gold 1989; Herdendorf and Berra 1995).

Nowhere is the potential for sensationalism greater than in the case of the great white shark, *Carcharodon carcharias*. "Verified" lengths reported for this shark often include an Australian record of 36.5 ft. (11.1 m). Some authors have taken the liberty of rounding off that measurement to 40 ft. (12.3 m). Reexamination of the teeth and jaws

from the reputed 36.5-ft. specimen suggest that it was in fact only 16.5 ft. (5 m) long and that the reported length resulted from a typographical error. The largest reliably measured white shark was a 19.5-ft. (6-m) female caught off Western Australia in 1984. This length stands in contrast to a photograph published in *The Guiness Book of Animal Facts and Feats* (Wood 1982) of a purported 29.5-ft. (9.1-m) Azores shark, but the photograph suggests a much smaller animal, and no verification of the measurements has been possible.

Extrapolations from jaw dimensions of known sharks indicate that bite marks on dead whales could come from sharks larger than 20 ft. (> 6 m), and several specimens in the 7-m range (all female) have been reported, but no such giants have been authenticated (Randall 1973, 1987; Ellis and McCosker 1991). The heaviest white shark reliably weighed had a mass of 3324 kg (7312 lbs.) (Springer and Gold 1989). White sharks are born at a length of around 100 cm (41 in.) and a mass of 13 kg (28 lbs.) (Ellis and McCosker 1991).

If the extant white shark can attain a length of 6 m and weigh in excess of 3000 kg, then how large was the biggest member of the genus, the megatooth shark, *Carcharodon megalodon*, that lived during the Miocene, between 15 to 4.5 million years ago? Teeth from this giant are common at many fossil-bearing locales; enamel heights (the vertical distance from the base of the enamel por-

and relatively high temperatures and salinities. These species have elongate gill regions and expanded gill filaments (Compagno 1984).

At least two families of rays contain truly freshwater fishes, seldom if ever entering marine conditions (Compagno 1980). The river stingrays (Potamotrygoninae) include about 14 species restricted to freshwaters of Atlantic drainages of South America. Members of two genera (*Dasyatis* and *Himantura*) of the large stingray family Dasyatidae inhabit rivers in Africa, Southeast Asia, and New Guinea. A few carcharhinid sharks enter freshwater periodically, the most notable example being the bull shark, *Carcharhinus leucas*. Bull sharks have been captured as far as 4200 km up the Amazon River and more than 1200 km up the Mississippi River at Alton, Illinois (Thorson 1972; Moss 1984). Bull sharks regularly traverse the 175-km-long Rio San Juan between the Caribbean Sea and Lake Nicaragua in Central America and occur in other rivers and lakes of Mexico and Central and South America. Bull sharks are the most likely perpetrators of attacks on humans in rivers worldwide (e.g., Coad and Papahn

1988). The largetooth sawfish, *Pristis perotteti* (Pristidae), also moves regularly between lakes and the ocean in Central and South America and has established genetically distinct, reproducing populations in some lakes (Montoya and Thorson 1982; Thorson 1982).

Sharks and sawfishes that move freely between marine and freshwater are able to adjust the osmotic concentration of their blood appropriately. Whereas in salt water the major problem is salt accumulation and water loss, in freshwater the problems reverse (see Chap. 7, Osmoregulation and Ion Balance). A bull shark in freshwater reduces the salt concentration of its blood by about 20% and urea concentration by about 50%. This cuts osmotic pressure of the blood in half, to about 650 milliosmoles. To accommodate water still flowing in from the environment due to osmotic pressure, urine production increases 20-fold. Salt concentration of the urine is reduced by the same factor, thus conserving salts. The rectal gland, which functions in salt water to secrete excess NaCl back into the environment, shuts down. All of these changes are quickly reversed upon return to the ocean.

BOX 12.1   *(Continued)*

tion to its tip; see Fig. 12.2) in excess of 100 mm are not unusual (the largest white shark teeth are about 60 mm high); the largest *C. megalodon* tooth found had an enamel height of 168 mm (Compagno et al. 1993).

Paleontologists, and others, have assembled these teeth into reconstructed jaws of this shark and then extrapolated to total body length based on jaw dimensions. These reconstructions have been notoriously inaccurate. The most famous was produced by the American Museum of Natural History in 1909 (Fig. 12.2A). The jaws of this reconstruction were oversized for a couple of reasons. First, the preparators created a wider-than-accurate jaw by using all symphyseal (midline front) teeth of equal size across the jaws; most sharks, including *C. carcharias*, have smaller teeth at the sides. Second, the cartilaginous jaw of a shark is generally no broader than the enamel height of the biggest tooth. In the American Museum reconstruction, cartilage breadth was four times enamel height, creating a larger jaw. The two errors produced a jaw about 30% larger than it should have been, which created a larger shark.

Length estimates extrapolated from that jaw, influenced by tooth size-body length ratios of the mismeasured 36.5-ft. Australian specimen, have ranged between 60 and 100 ft. (18.5 to 31 m), which has been rounded to 120 ft. (37 m) in some popular books. It was a megatooth shark that terrorized the New England town of Amity in Peter Benchley's (1974) novel *Jaws*. Given that snout length is 6% of total length in white sharks, and assuming the ill-fated swimmer on the cover of the paperback version of the novel is 5.5 ft. (1.7 m) tall, the Amity shark was a conservative 68 ft. (21 m) long. Bruce, the mechanical shark used in the movie version of *Jaws*, depicted a white shark about 24 ft. (7.3 m) long (Stephens 1987).

Recent reconstructions of *C. megalodon* (Fig. 12.2B) have used more quantitative methods in estimating size, such as the statistical relationships of tooth enamel height and jaw dimensions to body length and mass from known white shark specimens (Fig. 12.2C). Extrapolating from *C. carcharias* to *C. megalodon*, a megatooth shark with a tooth enamel height of 168 mm would be about 16 m (52 ft.) long and weigh approximately 48,000 kg (105,000 lbs.) (Compagno et al. 1993). If body proportions were similar to those of extant white sharks, the jaws would be more than 1 m (3.25 ft.) wide, the dorsal fin would be 1.4 m (4 ft.) high, and the tail would be 1.75 m (6 ft.) high. The megatooth shark, although certainly the largest shark to ever live, occurred during the Eocene to Miocene with other relatively gigantic predators, including the speartooth mako, *Isurus hastalis* (estimated at 6 m long), a hemigaleid, *Hemipristis serra*? (5 m), as well as the extant great white shark (Compagno 1990).

The truly freshwater potamotrygonine stingrays are restricted to low salinity conditions. Unlike all other chondrichthyans, they have lost the ability to concentrate urea (although the enzymes for urea production still occur in their livers), and they lack functioning rectal glands. These stingrays die in water with more than 40% the salt concentration of seawater (about 15 ppt salt). More typical marine elasmobranchs, including those that spawn in estuaries, can survive salinities as low as 50% of seawater if acclimated slowly. Unlike bull sharks, these typical elasmobranchs achieve an osmotic balance by reducing only the urea concentration of their blood, without reduction in salt concentration (Thorson et al. 1967, 1973; Moss 1984; Thorson 1991).

Movement and home ranges.   Water as a medium for locomotion exacts a high energetic price on any organism. The long evolutionary history of the elasmobranchs is characterized by the development of anatomic and physiological traits that appear to favor movement at the lowest possible energetic cost. Many of the features of the integument, fins, buoyancy devices, and swimming behavior of sharks, as well as short- and long-term movement patterns, reflect possible adaptations to these energetic constraints.

Most elasmobranchs have heterocercal tails (see Box 8.1), with asymmetry in both the internal support and external appearance. The typical heterocercal tail is associated with an active lifestyle above the bottom, as in most requiem sharks and hammerheads. However, diversity in tail-fin shape is considerable (Bone 1988). Symmetric tails preceded by lateral keels characterize high-speed, pelagic sharks such as the mako, white, and porbeagle (Lamnidae); large body size and symmetric tails characterized the presumably pelagic, predatory edestoid sharks of the Carboniferous (see Chap. 11). Lateral keels are also found convergently on such pelagic predators as tunas (Scombridae) and billfishes (Istiophoridae).

Extreme heterocercality is usually found in relatively inactive, benthic sharks, such as wobbegongs and nurse sharks (Orectolobidae) and catsharks that essentially lack a lower tail lobe. A specialized, extreme heterocercal tail occurs in active swimmers such as the thresher sharks, in

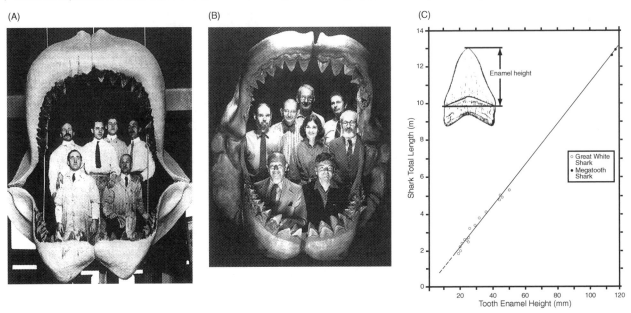

FIGURE 12.2.   *Reconstructing the jaws and estimating the size of the extinct megatooth shark,* Carcharodon megalodon. *(A) Jaw reconstruction as inaccurately prepared in 1909. Jaws are about one-third too large. (B) More recent reconstruction by the Smithsonian Institution, suggesting a body length of about 13 m. (C) Calculating the lengths of great white and megatooth sharks. Total body length is directly related to maximum tooth size (enamel height) in great white sharks; hence body length can be estimated for sharks from which only teeth are available (open circles). Assuming a similar relationship for the extinct megatooth shark, placement of two large teeth along the same regression line (closed circles) suggests a body length of about 13 m; the largest tooth found would have come from a 16-m-long shark. The equation for calculating total length from tooth height is total length (m) ≅ 0.096 × (enamel height, mm).*

Data from Randall 1973, Compagno et al. 1993; (A) From American Association for the Advancement of Science, Copyright 1971; used with permission. (B) Photo by Chip Clark, National Museum of Natural History, Smithsonian Institution; used with permission.

which the dorsal tail lobe, which may constitute 50% of the body length, is purportedly used for herding and stunning prey. Many rays lack a tail fin (e.g., stingrays; eagle rays, Myliobatididae; manta rays) and swim by flapping or undulating their scaleless pectoral fins. Angel sharks show reversed symmetry, with the lower tail lobe enlarged.

The placoid scales of sharks have a morphology that apparently serves a streamlining function (Fig. 12.3). Unlike the relatively flat scales of many bony fishes, each placoid denticle of a shark has a pedestal and an expanded top, which often has ridges running parallel to the body of the shark. It has been postulated that this particular shape, which is mimicked in the winged keels of high-performance sailboats, helps minimize drag by reducing turbulence along the shark's body. Additionally, the shape and arrangement of adjoining scales in active sharks may also aid in prey capture. Hydrophones detect much less noise from swimming sharks than from swimming bony fishes, suggesting that turbulence reduction

FIGURE 12.3.   *The role of scales in drag reduction in sharks. (A) The characteristic fluted pedestal or mushroom shape of the scales of many pelagic sharks, such as this mako shark, apparently serves to absorb turbulence, which reduces pressure drag. (B) Cross-sectional representation of placoid scales, showing reduction of turbulence along the body. Strength of water flow corresponds to thickness of black arrows.*

(A) From Bigelow and Schroeder 1948; used with permission. (B) From Moss 1984; used with permission.

also enhances stealth in sharks. Not surprisingly, the scales of benthic and slow-swimming sharks, as well as many rays, lack apparent streamlining features and instead are enlarged for protection or are absent (e.g., bramble shark, Echinorhinidae; thorny skate, Rajidae) (Moss 1984). The sting or barb on the dorsal surface of the tail of a stingray is a modified, elongate placoid scale with serrate edges and a venom gland at its base.

Maintenance of a constant depth is potentially expensive if a fish must constantly swim to overcome gravity. Bony fishes control their buoyancy by filling or emptying a gas bladder (see Chap. 5, Buoyancy Regulation). Sharks have arrived at a completely different solution to the challenge of buoyancy control; they both reduce their body weight and fill their body with low-density substances. Weight reduction comes largely from a skeleton made of cartilage, which has 55% the specific gravity of bone (1.1 vs. 2.0). Buoyancy can also be enhanced by the oils contained in the large liver, which may constitute up to 30% of the weight of the fish. Deep-sea squaloid sharks and some pelagic species such as whale, basking, and white sharks have large livers that are as much as 90% oil. These sharks are almost neutrally buoyant (e.g., a basking shark that weighs 1000 kg in air weighs only 3.3 kg in water, or about 0.3% of its air weight). Typical bony fishes with their gas bladders deflated have in-water weights about 5% of air weights (i.e., a 100-kg fish weighs 5 kg in water). The oil in a shark's liver is primarily squalene, which has a specific gravity of 0.86 (the specific gravity of seawater is 1.026) (Baldridge 1970, 1972; Moss 1984; Bone 1988).

An oil-based buoyancy system may have advantages over a gas-filled system. To maintain neutral buoyancy, bony fishes use energy to fill or empty the gas bladder or they will sink rapidly or float uncontrollably. The physics of gas secretion and absorption limit the rate at which bony fishes can change depth. Oil, unlike gas, is incompressible and provides constant buoyancy regardless of depth and pressure. Sharks can therefore move up and down repeatedly over tens or thousands of meters on a daily basis without having to make adjustments in their buoyancy control mechanism (e.g., white sharks, Carey et al. 1982; blue sharks, Carey and Scharold 1990; sixgill sharks, Carey and Clark 1995). Such rapid and continual vertical migration is uncharacteristic of most bony fishes. An additional energetic advantage of an oil-filled liver is that it serves as an energy reserve; many sharks metabolize their liver oils when starved.

Activity spaces, or home ranges, of most sharks are poorly known. Due to recent advances in electronic telemetry, sharks can be fitted with ultrasonic transmitters that track them through several days and even weeks of activity (Nelson 1990). Juvenile lemon sharks less than 1 m long in Bimini initially patrol shallow beach areas of about 70 hectares (0.7 km²). This area increases with age: A 1.8-m-long shark may have a range of 18 km², and a 2.3-m-long shark's range encompasses more than 90 km². Habitats change with age also, as the shark moves from the shallow nursery area to open sand flats and finally to the reef and beyond (Morissey 1991). Daily homing movements, involving daytime aggregation at seamounts and nighttime foraging in open water several kilometers away, have been demonstrated in scalloped hammerheads, *Sphyrna lewini* (Klimley and Nelson 1981).

On a larger scale, many sharks make extensive movements and migrations on the order of hundreds and even thousands of kilometers. Based on tag-and-recapture data, sharks can be classified as local, coastal, or oceanic. Local sharks, including bull, nurse, and bonnethead sharks, spend the majority of their adult lives in a relatively confined, nearshore area of a few hundred square kilometers. Coastal species, such as sandbar, blacktip, and dusky sharks, stay near continental shelves but move 1600 km or more. Sandbar sharks move between the northern Atlantic region of the United States and the Yucatan Peninsula of Mexico, a distance of about 5600 km. Oceanic species may cross entire ocean basins. Maximum distances traveled by oceanic species in the North Atlantic include 6000 km (blue shark), 4000 km (mako) and 2500 km (oceanic whitetip, Carcharhinidae). These are straight-line measurements and obviously underestimate actual distances covered. Additionally, some sharks undertake round-trips. Blue sharks make return trips between North America and Europe, a distance that exceeds 16,000 km (Casey and Kohler 1991).

Metabolism and growth rate: Life in the slow lane. Many aspects of the biology of sharks point to a strong emphasis on efficient energy use when compared with bony fishes. In addition to the anatomic features such as fin and scale morphology mentioned above, physiological attributes of sharks indicate a premium placed on energy conservation. Resting metabolic rates of a 2-kg spiny dogfish average 32 mg O_2 per kilogram of body weight per hour, about one-third of the average resting rate for comparably sized teleosts. In dogfish, active metabolic rate is only triple that of resting rate, whereas teleost active rates often go up 10-fold (Brett and Blackburn 1978). Extrapolated prey intake rates indicate that a 2-kg dogfish would need only 8 g of fish prey per day for maintenance, whereas a similar-sized salmon would require four times that amount.

Comparable data for larger sharks are unavailable, for obvious logistic reasons. Estimated oxygen consumption suggests that sharks consume about one-half the oxygen of equivalent-sized bony fishes (Moss 1982). For example, calculations of energy consumption have been made for a 4.5-m great white shark that was harpooned with a temperature-recording ultrasonic transmitter (Carey et al. 1982). Based on the rate at which the temperature of the shark's muscle mass changed as it passed through a thermocline (region of rapid temperature change), its oxygen consumption rate was calculated to be about 60 mg O_2, or 0.2 kilocalorie per kilogram of body weight per hour (a 60-kg human consumes about 1.6 kilocalories per kilogram of body weight per hour). Low metabolic requirements may translate into reduced energetic needs compared to bony fishes. White sharks feed commonly on dead whales. Based on the caloric value of 30 kg of whale blubber found in a 940-kg white shark's stomach and on the above calculations of metabolic rate, it was estimated that white sharks can maintain themselves by feeding only once every 6 weeks (Carey et al. 1982).

Food consumption rates vary greatly among shark species but are apparently most closely related to degree of activity (Stillwell 1991). Relatively sedentary sharks, such as nurse sharks, consume 0.2% to 0.3% of their body weight per day and digest an average meal over at least 6 days. Moderately active sharks, such as sandbar and blue sharks, consume 0.2% to 0.6% of body weight per day, digesting a meal in only 3 to 4 days. Very active sharks such as the mako, which are "warm-bodied" (see Chap. 7, Temperature Relationships), eat 3% of their body weight per day, digesting their meals in 1.5 to 2.0 days. Translated into annual consumption rates, the mako eats about 2 kg/day, or about 10 to 15 times its body weight per year. Although such figures appear large, they are about one-half the annual consumption of an individual teleost, showing the relative energy efficiency of feeding in sharks.

To the list of possible energy-saving mechanisms of sharks can be added an intriguing but as yet puzzling characteristic—namely, heat conservation. Lamnid sharks, and to a lesser extent thresher sharks, are able to conserve some of the heat generated during muscle contraction and thereby maintain their muscle and stomach temperatures at about 7 to 10°C above ambient water conditions (Carey et al. 1982; McCosker 1987). Heat generated during muscle contraction in most sharks is dissipated, because cold, oxygenated blood coming from the gills moves into the deeper parts of the body. In lamnid sharks, a countercurrent exchange arrangement helps to warm arterial blood flowing from the gills (see Chap. 7). This is not true thermal regulation as happens in birds and mammals or even tunas; body temperature is not constant but varies with external temperature.

The potential adaptive significance of this elaborate structure and process is a matter of conjecture. Higher body temperatures may permit maintenance of a higher metabolic level and generation of more muscle power, thus facilitating capture of fast-swimming prey (including endothermic marine mammals), and may increase the rate at which food is digested (Carey et al. 1985; Bone 1988), all of which might extend the ability of these large predators to invade cool waters at high latitudes (Block and Finnerty 1994).

Low metabolic demands may be linked to relatively slow growth and long lifespans. After an initial rapid growth phase of 15 to 60 cm increase per year (see above), growth in most sharks slows considerably. Growth rates of juveniles and adults of 12 species of medium- to large-sized sharks averaged only 5 cm/yr (Thorson and Lacy 1982; Branstetter and McEachran 1986). Some oft-cited values indicate that many sharks are long-lived: 70 to 100 years for spiny dogfish, 53 years for Australian school sharks (Carcharhinidae), 60 years for whale sharks, 50 years for lemon sharks, and 30 years for sandbar sharks (Smith and Gold 1989). Age estimation in sharks is frustrated by a lack of retained, calcified structures; growth rings in vertebrae are the most commonly used indicator of age when more direct measurements are unavailable (Cailliet 1990).

Feeding habits.   Sharks are apex predators throughout the world, stationed at the top of the food webs in which they occur. All elasmobranchs are carnivorous, taking live prey or scavenging on recently dead animals. No evidence of herbivory or detritivory exists, and the only departures from feeding on relatively large prey are the huge, filter-feeding, zooplanktivorous basking, megamouth, and whale sharks and manta rays. For most shark species, bony fishes constitute 70% to 80% of the diet. Feeders on other prey types include carcharhinid tiger, bull, and Galapagos sharks (other sharks); hammerheads (stingrays); white, tiger, and cookie cutter sharks (marine mammals); tiger sharks (seabirds); and leopard, nurse, and green dogfish sharks and most skates and rays (invertebrates).

The characteristic ventral mouth of most sharks (exceptions include the megamouth, frill, whale, and angel sharks and the manta rays) is apparently linked to dependence on bite strength rather than suction during shark feeding. Nurse sharks, as well as some heterodontiform sharks and many rajiforms, utilize suction extensively while feeding, but negative pressures are produced by expansion of the enlarged orobranchial cavity rather than enlargement of the mouth via jaw protrusion. Underslung jaws in sharks provide larger regions for muscle attachment than in fishes whose jaw bones extend all the way to the tips of their snouts. The shark configuration allows for both protrusion of the palatoquadrate (upper jaw) and the generation of the powerful biting forces required to cut through the skin, bone, and muscle of their prey. Hence sharks are not limited to prey that can be swallowed whole, whereas the vast majority of bony fishes are gape limited and risk suffocation if they attack prey larger than they can swallow whole.

Separation of the palatoquadrate from the cranium serves as more than a diagnostic character for identifying elasmobranchs. The freed palatoquadrate is responsible for protrusion during feeding (Fig. 12.4). Upper jaw teeth are retracted during nonfeeding periods, which aids in streamlining of the head profile. Protrusion during the bite means that, initially, only the upper jaw and not the much heavier snout and cranium have to be moved, which accelerates the bite without requiring a larger muscle mass. Protrusion may also optimize the angle at which teeth enter the prey. Hence prey can be efficiently cut by the closing action of the jaws and serrate teeth, rather than merely grasped. In many carcharhiniform and lamniform sharks, lower jaw teeth are spikelike, whereas upper jaw teeth are flat-bladed with serrated edges. In a typical jaw-closing sequence, the lower jaw is lifted first, impaling the prey, and then the upper jaw is brought down, once or repeatedly. Once prey are grasped, many sharks also shake their head and upper body, or even rotate the entire body about its long axis. During this movement, the upper jaw teeth slice through the prey, eventually removing a chunk of flesh (Moss 1977, 1984; Tricas and McCosker 1984). Jaw protrusion is extreme in some rajiforms due to the loss of ligamental connections between the palatoquadrate and cranium. These fishes dexterously pick up food objects from the bottom with

FIGURE 12.4.   *Head and jaw movements associated with feeding in the white shark,* Carcharodon carcharias. *(A) Normal resting position. (B) Snout is lifted and lower jaw depressed, achieving maximal gape. (C) Lower jaw is lifted forward and upward and palatoquadrate (teeth-bearing upper jaw) is rotated forward and downward, thereby closing the jaws (the bite). (D) Snout is dropped back down and palatoquadrate is retracted to resting position. The bite (component C) occurs quickly, requiring on average 0.08 second.*

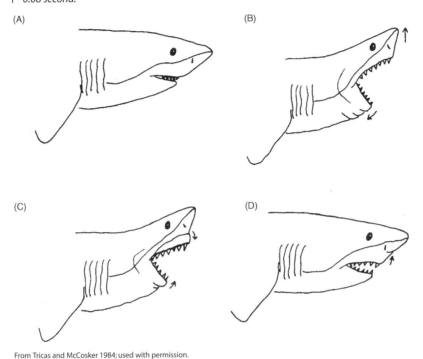

From Tricas and McCosker 1984; used with permission.

their jaws by a combination of suction and grasping (Moss 1977, 1984).

One departure from this norm occurs in some squaloid sharks, in which the lower teeth are more bladelike than the uppers and are arranged in a flat band across the lower jaw, perpendicular to the axis of the body. To this group belongs the highly specialized, 40-cm-long, bioluminescent cookie cutter shark, *Isistius brasiliensis* (Squalidae), the only ectoparasitic elasmobranch (Fig. 12.5). Craterlike wounds about 4 cm across are frequently found on tunas, dolphins, whales, and even megamouth sharks, all animals that spend part of their time at mesopelagic depths (200 to 1000 m). These wounds match precisely the mouth size of the cookie cutter shark, and it has been deduced that the small sharks lurk among the small mesopelagic organisms and prey on the larger predators that move into this region to feed (Jones 1971). The symmetric craterlike wounds may be formed when the shark attaches to the side of its victim and then spins about its long axis, removing a flat cone of tissue (E. Clark, personal communication).

Dentition morphology, an important taxonomic characteristic for identifying species, correlates strongly with food type in most sharks. Predators on fishes and squids, such as mako, sand tiger, and angel sharks, have long, thin, piercing teeth for grasping prey, which they often swallow whole. Most requiem sharks, which are also pi-

scivores, have such piercing teeth in the lower jaw and bladed teeth with finely serrate edges for cutting prey in the upper jaw (see above). Sharks and rays that feed on hard-shelled prey such as mollusks and large crustaceans have specialized, broad dentition for crushing (Fig. 12.6). Whereas most sharks engage only the anterior, peripheral row or rows of teeth while feeding, in mollusk-feeders, most of the posterior rows participate in the crushing action. Filter-feeding sharks typically have greatly reduced teeth that may function minimally during feeding; they instead use their gill rakers to trap small prey. In bony fishes, teeth are attached directly to the bone of the jaws, whereas in sharks the teeth are embedded in the connective tissue. Sharks have teeth only on their jaw margins, not attached to other bones and structures in the mouth, as in bony fishes. Shark teeth are basically enlarged scales, derived embryologically from the same tissues as the dermal denticles.

Dentition replacement is characteristic of all sharks, although detailed information on patterns of replacement is limited to a few species. Some sharks, such as white sharks and hammerheads, replace teeth individually as they are worn out or lost. In contrast, spiny dogfish and cookie cutter sharks apparently replace entire rows. Replacement occurs regardless of use; nonfunctional teeth in interior rows continue to grow and move forward, eventually displacing or replacing functional teeth. Teeth

FIGURE 12.5. *Cookie cutter sharks. (A)* Isistius brasiliensis, *the cookie cutter shark, is a small (about 40 cm) tropical species that lives at mid-ocean depths and parasitizes tunas, other fishes, and marine mammals, gouging circular plugs of flesh out of their sides with its specialized dentition. (B) The congeneric largetooth cookie cutter shark,* I. plutodus, *has the largest teeth for its body size of any known shark. Its teeth are twice as large relative to body size as a great white shark's teeth.*

(A)

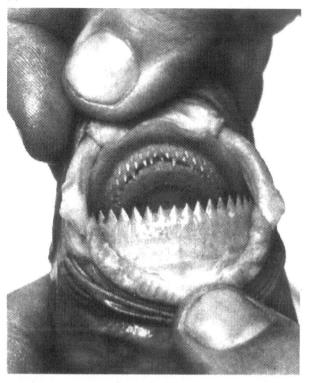

(B)

(A) Photo by C. S. Johnson, from Springer and Gold (1989); used with permission of the Smithsonian Institution Press. (B) From Compagno (1981); used with kind permission from Kluwer Academic Publishers.

also grow as the shark grows; hence teeth in the internal, nonfunctional rows are larger than the functional teeth about to be replaced. Turnover rates have been calculated for a few species in captivity. Lemon sharks replace a row of functional teeth about every 8 days, and sand tigers replace a tooth every 2 days. Given these numbers, it has

FIGURE 12.6. *Flattened, pavement, or molariform, crushing teeth characterize sharks and rays that feed on hard-bodied prey. (A) Lower jaw of a nurse shark,* Ginglymostoma cirratum *(Ginglymostomatidae), which feeds on lobsters. (B) Lower jaw of a cownose ray,* Rhinoptera bonasus, *a predator on clams.*

(A)

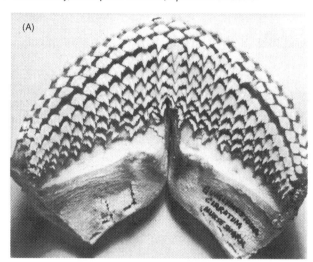

(B)

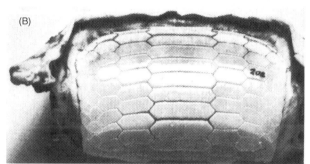

(A) From Moss (1972); used with permission. (B) From Case (1973); used with permission.

been estimated that a shark may produce on the order of 30,000 teeth during its lifetime (Moss 1967; Springer and Gold 1989; Overstrom 1991).

Some sharks use structures other than jaw teeth to capture prey. Thresher sharks possess a long, scythelike upper caudal lobe. Threshers herd fish into tight schools by circling and splashing with the tail, and then stun prey with quick whips of the tail. Tail use in prey capture is also implicated by the many thresher sharks that are caught by the tail on baited hooks (Compagno 1984). Two unrelated families of elasmobranchs have evolved armed rostral regions used in prey acquisition. Pristid sawfishes (Rajiformes) and pristiophorid sawsharks (Pristiophoriformes) both possess bladelike snouts armed with lateral teeth (modified denticles), which they slash laterally to stun and disable prey. The lateral projections of the rostral cartilage of hammerhead sharks have long intrigued anatomists. Recent observations by divers indicate that hammerheads, which tend to specialize on stingrays, may use the hammer to pin their stingray prey to the bottom while taking bites from the margins of the stingray's disk (Strong et al. 1990; see Box 18.1).

A spectacular nonoral prey capture device is evident

in the strong electrical discharges of torpedo rays (Torpedinidae). These rajiforms possess modified hyoid and branchial musculature capable of emitting electrical discharges of up to 50 volts and 50 amperes, producing an output approaching 1 kilowatt. Although electrical discharges can occur when the torpedo is disturbed and hence may serve as a predator deterrent, the more interesting function appears to be in prey capture. Torpedos will lie on the bottom and ambush prey during the daytime, but at night they swim or drift slowly in the water column. Upon encountering a potential prey fish, they envelop it with their pectoral fins and emit pulsed electrical outputs that are modified in terms of rate and duration in response to prey reaction. The stunned, immobilized prey fish is then pushed toward the torpedo's mouth via water currents produced by undulations of the pectoral disk (Hixon and Bray 1978; Fox et al. 1985; Lowe et al. 1994). Torpedos differ from most other rajiforms in that they possess a well-developed caudal fin that is used during locomotion. Most rajiforms use their pectoral fins for locomotion, but this anatomic region has been usurped for electric organ function in torpedinids.

Filter feeding occurs in four different groups of sharks and rays, and in each it has taken a slightly different evolutionary route. The megamouth shark, *Megachasma pelagios* (Megachasmidae), is the most spectacular shark species discovered in the twentieth century (Fig. 12.7). Megamouth sharks feed on euphausiid krill and jellyfish. Megamouth cartilage is poorly calcified and body musculature is relatively soft, suggesting a sluggish filter feeder with a terminal mouth. Prey are ingested via suction and then captured on dense gill raker papillae (Compagno 1990).

Manta rays, which can be over 6 m wide, guide small crustaceans and fishes into the pharynx with their cephalic horns; the prey are then caught on pharyngeal filter ridges. The cephalic horns are anterior subdivisions of the pectoral fins, making manta rays the only extant vertebrates with three functioning pairs of limbs. Whale sharks use a suction and draining mechanism to feed on microscopic plankton or small fish, which are filtered through cartilaginous rods (Compagno 1984, 1990). Whale sharks have been found with whole tuna in their stomachs, pur-

portedly captured by orienting vertically in the water column with the head above the water surface and then dropping down, thereby causing a vacuum that sucks water and prey into the mouth (Springer and Gold 1989).

Basking sharks use passive filtration in the form of gill raker denticles and mucus to capture very small zooplankters. Interestingly, basking sharks shed their gill rakers in the winter and regrow them next spring, presumably as an energy-saving device during the time of the year when food is largely unavailable at the high latitudes they inhabit. It has been suggested that basking sharks enter a winter torpid state and lie inactively on the bottom, although this phenomenon remains undocumented (Compagno 1984).

Although sharks swim actively throughout the diel cycle and will capitalize on opportunities to feed at any time, available data indicate that the primary foraging times of most species are during twilight and nighttime. Such crepuscular/nocturnal foraging has been confirmed by catch statistics and more recently by ultrasonic telemetry and direct observation. Catches of sharks on baited hooks are greater by night than by day. Activity cycle data from telemetered lemon sharks indicate a doubling of swimming speed during twilight as compared to daytime or nighttime speeds (Morissey 1991). Increased nocturnal swimming has been found in gray reef, blue, scalloped hammerhead, horn, angel, and swell sharks, as well as torpedo rays (Myrberg and Nelson 1991). In contrast, the great white shark is primarily a diurnal feeder, at least where its major prey is diurnally active marine mammals (Klimley et al. 1992; Klimley 1993).

Sensory physiology. As most sharks are primarily nocturnal foragers, it is not surprising that nonvisual senses are particularly well developed. Olfactory sensitivity has long been recognized as extreme in sharks. Fish extracts can be detected by lemon sharks at levels as low as 1 part per 25 million parts seawater, and in blacktip and gray reef sharks at 1 part per 10 billion, equivalent to about one drop dispersed in an Olympic-sized swimming pool (50 m long × 25 m wide × 2 m deep). This sensitivity, which is greatest to proteins, amino acids, and amines, is achieved at several levels of operation. Water flows into the shark's

FIGURE 12.7. *The megamouth shark,* Megachasma pelagios. *First captured in 1976 at 160 m depth in open water, this large (4–5 m) zooplanktivorous shark is known from only 7 specimens from tropical and subtropical Pacific, Indian, and Atlantic Ocean locales.*

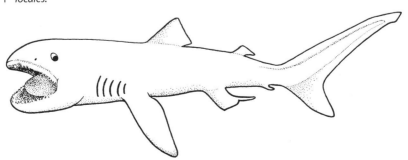

After Compagno (1984); used with permission of the Food and Agriculture Organization of the United Nations.

nasal sacs located under the front of the snout, perfusing the underlying, large olfactory organs. Receptor cells in the organs receive stimuli, which are then transmitted via the olfactory nerve to the olfactory bulb and lobes of the forebrain for integration. Olfactory lobes and bulbs in the shark brain are greatly enlarged.

Sharks are exceptionally adept at following odor trails in the water, even without currents to guide them upstream to an odor source (Montgomery 1988). Odor localization may be accomplished by comparing the intensity of stimulation in either chamber. This scenario suggests one possible function of the expanded rostral cartilage of hammerhead sharks. Lateral displacement of the nostrils on the margins of the hammer might provide improved stereo-olfaction, making odor localization easier. Olfaction also plays a role in breeding, as evidence exists that sexual behavior is mediated by chemical sexual attractants (pheromones) produced by females (Hueter and Gilbert 1991).

Contrary to conventional wisdom, sharks have good vision. As a group, sharks tend to be slightly myopic (farsighted). The shark retina is dominated by rods, as would be expected of fishes that are active chiefly at night and twilight. Nocturnal sensitivity is enhanced by an additional layer of reflective guanine platelets behind the retina, called the **tapetum lucidum**. The tapetum reflects light back through the eye (hence the "eye shine" of sharks and other nocturnal animals), which allows light entering the eye to strike the retinal sensory cells twice, increasing the likelihood of detection. Platelets are angled such that reflected light passes through the same receptors as incoming light, thus preserving acuity. Sharks also have low densities of cone cells in their retinas, which typically increase daytime acuity and are associated with hue discrimination (color vision). Color vision has been demonstrated in lemon sharks. Sharks differ from most other fishes, and are more similar to mammals, in that they possess a pupillary response to changing light levels, which means they can regulate the amount of light entering their eyes. In many batoids, an additional eyelidlike structure called the **operculum pupillare** expands under bright light conditions to shade the eye and perhaps increase responsiveness to movement.

Acoustic sensitivity and sound localization capabilities are also greatly developed in sharks. Sharks hear sounds below 600 Hz (the fundamental of E below high C on a piano), including infrasonic sound below 10 Hz. Hearing is centered in three chambers within the inner ear that contain the **maculae**. The maculae contain specialized nerve cells that are linked to small granules of calcium carbonate that vibrate in response to sound stimuli. The chambers are connected to the environment via two small pores, the endolymphatic ducts, on the top of the shark's head. A fourth macula, the **macula neglecta**, is located below an opening in the skull. The three major maculae are apparently involved in detecting low-frequency pulsed sounds, whereas the macula neglecta may provide information on the direction of the sound (Hueter and Gilbert 1991; Myrberg and Nelson 1991).

In addition to sensing vibratory cues in the surrounding medium (sound), sharks are also sensitive to minor variations in water displacement. This distant-touch sensitivity is accomplished via mechanoreceptors distributed both along canals such as the lateral line and as independent pit organs scattered about the body. Distant-touch sensitivity also has a directional component, suggesting that as a shark moves through the water, it can detect the presence, location, direction of movement, and relative speed of moving objects that displace water or of stationary objects that reflect water moving off the body of the shark.

Sharks have an additional channel for sensory input that is anatomically and developmentally related to distant touch and hearing—namely, **electroreception** (see Chap. 6). Input for electroreception begins at numerous small pores spaced precisely on the shark's head, snout, and mouth. The pores lead to conductive, jelly-filled canals that terminate in ampullary receptor cells termed **ampullae of Lorenzini** (not Alloplurus of Lixiniri, as was once suggested on an ichthyology exam). The receptor cells, which are anatomically similar to hair cells of the lateral line, fire in response to weak electrical fields, sending afferent fibers via the lateral line nerve to regions in both the mesencephalon and telencephalon of the brain. The function of the ampullae, which are an obvious external feature on most sharks, remained a substantial mystery prior to the discovery of electroreceptive capabilities in sharks.

Sensitivity to weak electrical fields is an ancestral trait in bony fish lineages that was lost and then re-evolved in neopterygians (see Chap. 6, Electroreception). Elasmobranchs as a group have retained and fine-tuned their electrical sensitivity to a level apparently unequaled by any other animal group. Many biological activities have as an integral component the generation of weak electricity. Most notable are muscular contraction, such as heart function and breathing, nerve conduction, and the voltage created by ionic differences between protoplasm and water. A resting flatfish (Pleuronectiformes) creates a low-frequency direct current bioelectrical field with a strength of more than 0.01 μV/cm (one-hundredth of a microvolt) measured 25 cm away.

Experiments have shown that predatory sharks use weak electrical cues, and ignore strong olfactory cues, to home in on prey. The electrical sensitivity of large sharks is truly amazing. Human sensitivity is on the order of 0.1 volt. Sharks have demonstrated detection thresholds of 1×10^{-9} V/cm, or one-billionth of a volt, approximately 10 times more sensitive than the 0.01 μV output from prey and sensitive enough to detect the electrical output of a standard D-cell flashlight battery several kilometers away (assuming no background geomagnetic interference), or the bioelectrical output of a human 1 to 2 m away.

Sharks are sufficiently electrosensitive that, theoretically, they can detect the earth's magnetic field and currents induced by their swimming through that field. Sharks could therefore determine their compass headings during transoceanic migrations. This idea has yet to be confirmed, although stingrays in the lab can learn to orient in uniform direct current fields weaker than the earth's field. In learning trials, stingrays also reverse the location where they search for food when the electrical field around them is artificially reversed, suggesting that geomagnetic cues can be used in normal daily activities (Kalmijn 1978). Magnetite in the inner ear has been implicated as the center of geomagnetic orientation (Vilches-Troya et al. 1984).

It has been noted that the ampullary electroreceptors are geometrically centered around the mouth of many elasmobranchs. This positioning could allow a shark or ray to home precisely on a potential food source solely by electroreception, effectively aligning the food in the "sights" of its receptor field and then engulfing the prey. In this way, sharks can detect prey buried in the sand or sit motionless in the dark and snap up prey that swim nearby (Kalmijn 1982; Tricas 1982).

Extreme sensitivity to environmental stimuli is of no use to an animal unless the information can be collected, processed, and acted upon. Such integration is the role of the central nervous system, particularly the brain. Not surprisingly, the ratio of brain to body weight in sharks is greater than for most bony fishes, except perhaps the electrogenic elephantfishes (Mormyridae) (see Chap. 14). The ratio of brain to body weight in sharks is actually more comparable to that of "higher animals" such as some birds and marsupials (Fig. 12.8). Requiem and mackerel sharks have large forebrains and complex cerebellums; eagle and stingrays have the most complex brains (Northcutt 1977).

Reproduction and development.   A few generalizations can be made about shark reproduction. Ages at maturation vary widely among shark species but are typically older than for most teleosts. Most sharks mature in 6 to 18 years, although much greater ages are not uncommon. For example, lemon sharks (*Negaprion brevirostris*) in south Flor-

ida mature at an average age of 24 years, and spiny dogfish (*Squalus acanthias*) in British Columbia mature at an average age of 35 years (Saunders and McFarlane 1993). Sharks produce relatively few, large young with a long gestation period. Newborns are small replicas of the adults; no larval stage exists. No parental care is given after egg laying or birth, with the possible exception of a bullhead shark (Heterodontidae), in which the female picks up her recently laid eggs in her mouth and wedges them in crevices in rocks or plants. Fertilization in sharks is internal. Male sharks possess intromittent organs in the form of modified pelvic fins termed myxopterygia or claspers (the term *claspers* apparently arose from Aristotle's misconception that the structures were used to hold the female, rather than to inseminate her). Maternal-embryonic relationships in advanced sharks rival that in mammals, contrasting sharply with the simpler reproductive features of most bony fishes (see Chap. 20, Parental Care).

Three basic modes of postfertilization development and embryonic nutrition can be identified that characterize families or orders of sharks (Wourms et al. 1988; Compagno 1990; Pratt and Castro 1991; Wourms and Demski 1993). The primitive condition is considered to be egg laying (**oviparity**), with the embryo deriving all nutrition from its large yolk reserves. About 40% of all living elasmobranchs are oviparous, including bullhead sharks (Heterodontiformes), many nurse sharks (Orectolobiformes), as well as all skates (Rajiformes).

Unlike bony fish eggs, shark and skate egg cases, termed **mermaid purses**, are large (2 to 4 cm) and are protected by a tough, keratinoid (horny) shell secreted by a maternal nidamental or shell gland. The egg case, which contains a single embryo, is attached to seaweeds or other structures. The embryo develops for a relatively long period (several weeks to 15 months) and emerges at a relatively large size. The protective nature of mermaid purses

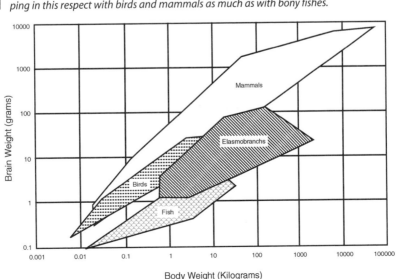

FIGURE 12.8.   *Sharks have relatively large brains for their body size, overlapping in this respect with birds and mammals as much as with bony fishes.*

From Springer and Gold (1989), based on Northcutt (1977) and Moss (1984); used with permission of the publisher.

is attested to by instances of their being transported in air for several days and still hatching out healthy young (Lagler et al. 1977). Mermaid purses are not free from predation. They have been found in the stomachs of other sharks, flatfishes, and elephant seals, and are actively preyed upon by boring gastropod mollusks such as whelks (Cox and Koob 1993). A small number of sharks retain the encased eggs in the oviduct before laying them (termed **retained oviparity**). Clutch size in oviparous sharks is difficult to assess because females may lay two eggs at a time repeatedly over several months (Linea-weaver and Backus 1970).

Other reproductive modes are apparently derived from oviparity. About 70% of living sharks and all ray species bear live young. In about half of the live-bearing species, the developing embryo is retained in the uterus and nourished with yolk provided by a yolk sac that is attached directly to the digestive system of the embryo (termed **ovoviviparity**, **yolk-sac viviparity**, or **aplacental viviparity**). This form of development characterizes frill and cow sharks (Hexanchidae), dogfish sharks, angel sharks, some carcharhinids and orectolobids (e.g., tiger and nurse sharks), the whale shark, as well as guitarfishes (Rhinobatidae), sawfishes, and torpedo rays among the batoids (rhinobatids, pristids, torpedinids). Whale sharks, once thought to be oviparous, have been found with as many as 300 embryos inside, making that species the most fecund shark known. A gravid female may contain smaller embryos still in their egg cases, whereas larger individuals are free-living. Whale shark pups are born at around 60 cm long (Joung et al. 1996).

Some ovoviviparous sharks have evolved an additional form of nourishment. After about 3 months, yolk reserves are exhausted, and the young then feed directly on eggs ovulated by the female (**oophagy**). Oophagous sharks include threshers, whites, makos, and sand tigers. Sand tigers have carried this one step further. The first embryo to consume its yolk then turns on its siblings and eats them (**embryophagy**) before assuming an oophagous existence (Gilmore et al. 1983; Gilmore 1991). At birth, litter size in sand tigers is two large (1 m) young, one in each uterus.

The most complex developmental pattern, **placental viviparity**, characterizes advanced members of the most diverse modern family of sharks, the Carcharhinidae, as well as the closely related hammerheads (Sphyrnidae) (Fig. 12.9). Yolk is consumed, and then the spent yolk sac attaches to the uterine wall to form a **yolk-sac placenta**. In a construction strongly analogous to the mammalian condition (although involving different embryonic tissues), the stalk of the yolk sac, which is attached to the embryo between the pectoral fins, forms an umbilical cord that transports nutrients and oxygen to the embryo and carries metabolic wastes to the mother. In some sharks, such as the sharpnose and hammerheads, the umbilical cord diversifies further and develops **appendicula**, or outgrowths, which serve as additional sites for exchange of materials, including uptake of nutrients in histotroph (uterine milk) secreted by special cells in the wall of the uterus. Uterine milk may be absorbed through the skin and mouth and also via modified gill filaments that exit from the spiracle and gill slits of the developing embryo. In the myliobatoid stingrays and manta rays, nourishment is solely via ingestion of uterine milk without a placental connection (**uterine viviparity**). Clutch sizes in viviparous sharks range from 2 (sand tigers, threshers, longfin mako) to as many as 70 to 135 (tiger and blue sharks), with an average around 8 to 10 (Branstetter 1990).

FIGURE 12.9.   *Placental viviparity in advanced sharks. A newborn Atlantic sharpnose shark (Carcharhinidae) showing the umbilical cord with outgrowths (appendicula) for nutrient uptake. The cord terminates in a placenta that attaches to the uterine wall of the mother.*

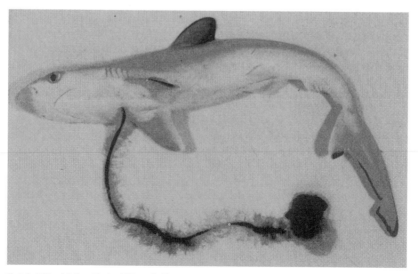

Photo by W. Hamlett. From Hamlett 1991; used with permission.

BOX 12.2 *Shark Conservation*

Many of the life history features of sharks make them exceedingly vulnerable to overexploitation by humans. As apex predators, they seldom occur in the large numbers that characterize more resilient fish species (e.g., herrings and mackerels feed much closer to the base of the marine food chain). Importantly, sharks replace themselves slowly. Slow growth and maturation rates, long gestation periods, long intervals between reproductive bouts, and relatively small clutch sizes suggest that shark species have evolved in circumstances where juvenile mortality is typically low compared to the much more prolific bony fishes on which sharks feed. Under natural circumstances, sharks have few predators aside from other sharks, and most of these can be avoided by attaining large size. The unfortunate outcome of this particular life history strategy is that shark populations are not capable of overcoming the high mortality rates imposed by commercial exploitation.

The history of commercial shark fishing is a history of collapsed fisheries (Pike 1991). Examples include fisheries for thresher sharks in California, school sharks in Australia, spiny dogfish in the North Atlantic, porbeagles off Newfoundland, basking sharks off Ireland, bull sharks and sawfish in Lake Nicaragua, and soupfin sharks off the U.S. Pacific coast.

Two examples typify the boom-and-bust pattern that characterizes commercial exploitation of shark populations. The porbeagle (*Lamna nasus*) fishery of the western North Atlantic has been well documented. The fishery was wiped out in only 7 years. In 1961, uncontrolled exploitation began, and about 3.5 million pounds were caught. Catch peaked only 3 years later at 16 million pounds, then crashed. By 1968, only a few hundred sharks could be found. Twenty years later, populations had not returned to pre-exploitation levels. Before 1937, about 6000 individual soupfin sharks, *Galeorhinus zyopterus* (or 600,000 pounds), were landed annually in California. With the development of a market for shark liver oil, this fish-

ery expanded rapidly to 9 million pounds (90,000 fish) in 1939, fell to 5 million pounds in 1941, and by 1944 was back to 600,000 pounds, despite continued intensive effort. Importantly, catch rates fell from 60 sharks per set in 1939 to 1 shark per set in 1944, indicating a significant decline in the population. Thirty years later, populations had still not returned to pre-1939 levels (Moss 1984; Anderson 1990; Gruber and Manire 1991).

Most shark populations cannot withstand a fishing mortality even as low as 5% removal of the existing population each year (Hoff and Castro 1991). The U.S. maximum sustainable yield (MSY) of sharks has been estimated at 16,250 metric tons, while total mortality exceeds the MSY by nearly 6000 metric tons annually. Between 1986 and 1990, commercial shark landings doubled every year in the United States. All indicators suggest that North American shark populations are in decline and that management plans, including a moratorium on the capture of some species, are desperately needed.

The repeated scenario of large initial catches, rapid decline, and slow, if any, recovery highlights the need for careful management of all exploited shark stocks. Interestingly, sharks exhibit a remarkably close relationship between stock size and recruitment. Fisheries managers can predict future recruitment into populations based on existing reproductive stocks. This degree of predictability characterizes few other commercial species. The unfortunate fact is that shark fishing remains one of the least regulated commercial fishing activities. Management plans have been proposed but not implemented in many countries. But even with implementation, shark conservation is complicated by shark biology. Local management is not sufficient to protect shark populations because so many species undergo long-distance movements through international waters. International efforts at conservation, which are historically difficult to negotiate, are crucial. Write your congressperson.

Gestation periods in sharks average 9 to 12 months but may be as short as 3 or 4 months (bonnethead) or as long as 2 years (spiny dogfish) to perhaps 3.5 years (basking shark). Production of young apparently exacts a large energy cost on females, which appear emaciated at the end of the gestation period. Females of many species reproduce only in alternate years, suggesting that it

takes at least a year for the female to recover from her last clutch. Such long reproductive intervals slow the potential rate at which shark populations can grow (Box 12.2).

Certain accommodations must be made in live-bearing sharks to facilitate passage of relatively large young through the birth canal. The expanded lateral lobes of the

FIGURE 12.10.   *Sexual dimorphism in the skin thickness of sharks. On the left is a cross section through a male blue shark, on the right a female. Female sharks often have thicker skin than males, probably because during mating males typically bite and hold females.*

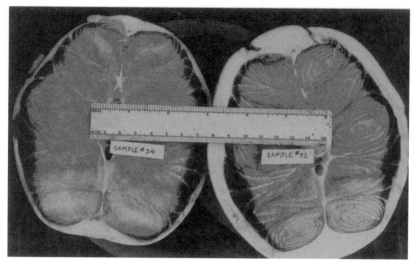

Photo by H. W. Pratt. From Pratt and Castro 1991; used with permission.

cranium of hammerhead sharks are soft and pliable at birth and then stiffen shortly afterward. Spines on embryonic spiny dogfish are covered with pads of tissue until after the young are born.

Courtship and copulation, in the few species for which observations exist, are complex and commonly require the male to hold the female in his teeth while the claspers are inserted. Bite scars are common on adult females but appear to do little actual damage because the skin of females is typically much thicker than that of males. In blue sharks, the female's skin is three times as thick as the male's and, more importantly, thicker than the male's teeth are long (Fig. 12.10). Females of some species can store sperm in the shell gland for years.

Subclass Holocephali

Superorder Holocephalimorpha (4 extinct suborders and modern chimaeras)
Order Chimaeriformes (modern chimaeras) (31 species, marine): Callorhynchidae (plownose chimaeras), Chimaeridae (shortnose chimaeras or ratfishes), Rhinochimaeridae (longnose chimaeras)

Chimaeras

Most of the characters that define the elasmobranchs also describe the Holocephali, indicating a common albeit unknown ancestor, perhaps among the placoderms. Some analyses have concluded that holocephalans are more primitive than elasmobranchs. Chimaeras, also known as ratfishes or rabbitfishes, share with sharks a cartilaginous skeleton and male intromittent organs, a sutureless skull, ceratotrichial fins, and spiral valve intestine, and similarly lack lungs and gas bladders, also

using an oil-filled liver for buoyancy. All chimaeras are oviparous, laying a 10-cm-long egg with horny shell similar to that of sharks. Development is again direct, without a larval stage (Bigelow and Schroeder 1953; Compagno 1990).

In contrast to sharks, chimaeras have upper jaws that are immovably attached to the braincase. The name Holocephali (whole head) refers to this fusion of palatoquadrate and neurocranium in modern chimaeras. Teeth differ from those of sharks, being continually growing, crushing or cutting plates instead of replaceable dentition. As in bony fishes, a single gill flap covers four internal gill openings, rather than the five or more external gill slits of sharks. Chimaeras lack distinct vertebral centra and a spiracular gill opening (embryonic chimaeras have a spiracle). Instead of a cloaca, they possess separate anal and urogenital openings. Males have sharklike pelvic claspers, and some species have an additional clasper, termed a **tentaculum**, on the head (Fig. 12.11). Holocephalans lack scales except for small dermal denticles along the midline of the back and on the claspers of the males. The first dorsal fin, with its poison-laden spine, is erectable, not fixed. The body generally tapers posteriorly to a pointed tail; chimaeras locomote via a combination of body undulations and pectoral fin flapping.

Eleven genera and 58 species of chimaeras are recognized. Adult size ranges from 60 to 200 cm, with females often larger than males. Chimaeras are cool-water, marine fishes. Although geographically widespread, low-latitude species occur in deeper water. As a group they are mostly found below 80 m to as much as 2600 m and are usually captured close to the bottom. They feed mostly on hard-bodied benthic invertebrates, which they crush with their tooth plates.

Surprisingly little is known about the general biology and natural history of modern holocephalans. Extant spe-

FIGURE 12.11. *Modern holocephalans. (A)* Chimaera cubana, *a 50-cm-long Caribbean chimaerid; note pelvic claspers and also frontal clasper, or tentaculum, on the forehead of the male. (B) Head of a callorhinchid chimaera,* Callorhinchus milii *(the elephantfish), showing the unique hoe-shaped proboscis.*

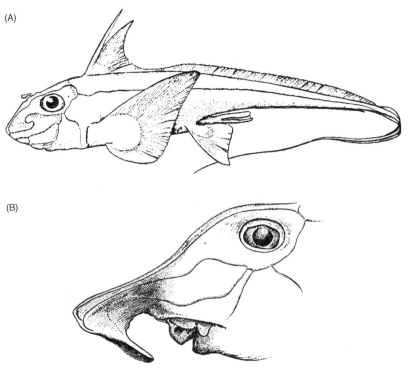

(A)

(B)

From Bigelow and Schroeder 1953; used with permission.

cies represent a small fraction of a previously successful and diverse group (see Chap. 11 and Lund 1991). Although only 3 families in a single order are alive today, eight fossil orders, represented by perhaps 20 families, lived during Paleozoic and Mesozoic times. The modern families have fossil records dating back to the Jurassic and Cretaceous (Carroll 1988).

SUMMARY

1. The Chondrichthyes contain two subclasses, the elasmobranchs (sharks, skates, and rays) and the holocephalans (chimaeras). Elasmobranchs are represented by one living superorder, the Euselachii, which contains eight orders and 350 species of sharklike fishes and one order and 450 species of skates and rays. Carcharhiniform requiem sharks and squaliform dogfish sharks are the most diverse shark orders, and rajid skates and dasyatid stingrays are the most diverse rajiform families.

2. Elasmobranch vertebrae consist of calcified cartilage. Teeth are replaced throughout life and are not fused to the jaws. The upper jaw is not fused to the braincase, and the mouth is usually subterminal. Fin rays consist of horny ceratotrichia. Most elasmobranchs have five external gill slits. Fertilization is internal. In general, elasmobranchs are marine, mobile predators

that grow slowly and have slow metabolism, rely on nonvisual senses, produce small numbers of young, and are extremely vulnerable to commercial exploitation.

3. Sharks are relatively large, many exceeding 1 m in length. The largest sharks are the whale shark (> 12 m), basking shark (9 m), and hammerhead, thresher, and tiger sharks (5 to 6 m). The largest verified great white shark was 6 m long and 3300 kg. The extinct megatooth shark was 16 m long and weighed approximately 48,000 kg. Some deep-water dogsharks are smaller than 20 cm.

4. Although most sharks live in shallow, marine habitats, some dogfish sharks inhabit the deep sea, to a depth of 4000 m. Two families of rays inhabit freshwater, and sawfishes and bull sharks frequently move into rivers, the latter occurring as far as 4000 km from the ocean. Elasmobranchs that inhabit freshwater osmoregulate to counteract influx of water and loss of salts.

5. Sharks are active predators with relatively large home ranges. Many coastal and oceanic species undertake migrations of 1000 to 16,000 km. Locomotion is very efficient. Placoid scales are shaped to minimize drag, and dorsal fins work in conjunction with the heterocercal tail and pectoral fins to maximize propulsion. The large, lipid-filled liver provides buoyancy. Sharks

have slower metabolism and require less food for a given body weight than do bony fishes. White sharks may only need to feed once every 6 weeks. Slow metabolism leads to slow growth and old age in many species.

6. Except for the largest sharks and manta rays, which are plankton feeders, most sharks use their protrusible jaws and sharp, often serrated teeth to dismember prey. Teeth are replaced every few days; a shark may produce 30,000 teeth during its lifetime. Feeding specializations include suction feeding and molariform teeth for crushing mollusks (many rays), elongate tails or snouts for striking and incapacitating prey (thresher sharks, sawfishes, sawsharks), and muscles modified for electricity production to stun prey (torpedo rays).

7. Sharks have good vision, particularly at night. They are exceptionally sensitive to chemical stimuli, can localize sound, and can detect weak electrical or geomagnetic cues, which they use to localize prey and perhaps to navigate in the open ocean. Sharks have relatively large brains compared to bony fishes.

8. Sharks mature at a relatively old age, between 6 and 18 years. Fertilization is internal; some species lay eggs, whereas others gestate young internally. Gestation is long, and young are small replicas of adults. No parental care is given after birth, but female investment during gestation is very high, particularly in those species with complex placental structures. Relatively few (2 to 300) young are produced at a time.

9. Because of slow growth, slow maturation, and low fecundity, sharks are very susceptible to overfishing. Shark fisheries typically boom then quickly bust and do not recover, as exemplified by the porbeagle fishery of the western North Atlantic and the soupfin shark fishery in California. North American shark populations are in decline, and a general moratorium on capture of many species is needed.

10. Holocephalans (chimaeras, ratfishes, rabbitfishes) include 3 genera and 31 species of cartilaginous fishes. They differ from sharks by having the upper jaw fused to the braincase, a single gill cover, separate anal and urogenital openings, and an erectable dorsal spine. They are entirely marine, inhabiting shallow to moderate depths. As in the sharks, holocephalans were much more diverse during the Paleozoic and Mesozoic than they are today.

SUPPLEMENTAL READING

Budker 1971; Castro 1983; Compagno 1984; Ellis 1976; Ellis and McCosker 1991; Gruber 1991; Hodgson and Mathewson 1978; Lineaweaver and Backus 1970; Moss 1984; Pratt et al. 1990; Springer and Gold 1989; Stevens 1987; Shuttleworth 1988; Wourms and Demski 1993.

13

Living Representatives of Primitive Fishes

A small number of anatomically primitive and unusual fish species occur on all the earth's major continents, often in tropical or subtropical, swampy habitats. These fishes represent the last remaining representatives of groups that dominated aquatic environments during the Paleozoic and Mesozoic periods. The ancestry of several species can easily be traced to otherwise extinct groups, whereas other species are obviously highly derived, specialized fishes whose close affinities can only be surmised from anatomic similarities. These "living fossils" include some of the most spectacular and controversial ichthyological discoveries of the past two centuries and typify the dictum that every species has a mixture of ancestral and derived characteristics.

JAWLESS FISHES: LANCELETS, HAGFISHES, AND LAMPREYS

Phylum Chordata
Subphylum Cephalochordata
Order Amphioxiformes (lancelets) (22 species, marine, tropical and temperate): Branchiostomatidae, Epigonichthyidae
Subphylum Vertebrata
Superclass Agnatha
Class Myxini (living hagfishes) (about 43 species, temperate marine)
Order Myxiniformes: Myxinidae (hagfishes)
Class Cephalaspidomorphi
Order Petromyzontiformes (living lampreys) (41 species, temperate freshwater and anadromous): Petromyzontidae (lampreys)

Amphioxiforms

It can be argued that lancelets are not fishes because they lack many diagnostic characters. However, cephalochordates—which are sometimes referred to as invertebrate chordates along with urochordates and hemichordates—seldom receive treatment in invertebrate textbooks. Their evolutionary and anatomic affinities are much closer to the vertebrates (see Northcutt and Gans 1983), and lancelets are studied primarily by ichthyologists, providing justification for their inclusion in an ichthyology textbook.

Lancelets are small (up to 3 cm), slender organisms that as adults occupy sandy, usually shallow bottoms in all major oceans (Fig. 13.1). They commonly bury in the sediments with just the anterior portion of the body protruding from the bottom. Lancelets filter diatoms and other small food items from the water via cilia that transport water through the mucus-laden mouth and pharynx and out the atriopore. Food-trapping mucus is produced by the **endostyle**, a pharyngeal organ that also functions

in iodine uptake and may therefore be homologous to the thyroid of higher vertebrates (Nelson 1994). Spawning in most species occurs in early summer; larvae metamorphose after 2 to 5 months. Larvae are free-swimming, planktonic animals that have the mouth and anus on the left side of the body. The mouth eventually moves to the middle, but the anus remains on the left side. Larvae can be very abundant at times, reaching densities of 1000/m^3 in regions of upwelling. The larvae settle on sandy or sandy-shell bottoms, mature in 2 to 3 years, and live as adults for 1 to 4 years, depending on the species. Both larvae and some immature adults undergo diel vertical migrations, moving to surface waters at night (Bigelow and Perez Farfante 1948; Boschung and Shaw 1988).

Lancelets are intriguing in their lack of many typical chordate structures. They differ from conventional fishes by lacking most parts of a head (e.g., no cranium, brain, eyes, external nostrils, or ears); hence they are sometimes called acraniates. Lancelets are also without vertebrae, scales, genital ducts, a heart, red blood cells, hemoglobin, and specialized respiratory structures (gills), and have only one cell layer in the epidermis. Lancelets have up to 25 pairs of gonads versus 1 in lampreys and hagfishes and 2 in most other fishes. The number of internal gill clefts increases throughout life in lancelets, whereas in fishes the number is fixed at birth. The notochord, a definitive chordate structure, extends beyond the anterior end of the dorsal nerve tube (i.e., beyond the "brain"), to the anterior end of the body.

For many years, an additional family of lancelets, the Amphioxididae, was recognized. However, the only specimens found were pelagic and immature, with the mouth on the left side. It was proposed that this was a neotenous species that retained larval traits in the adult animal or that it was a species that remained in a larval state for an extended period and then matured quickly, mated, and

197

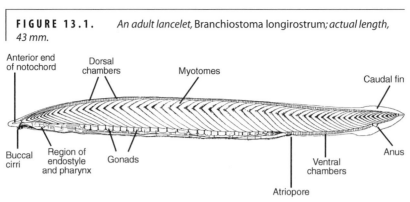

FIGURE 13.1.   *An adult lancelet,* Branchiostoma longirostrum; *actual length, 43 mm.*

Modified from Boschung 1983; used with permission.

died. Amphioxidids are in fact the larvae of a species of *Branchiostoma* that produces two types of larvae: an amphioxus type with a short larval period that takes up a benthic existence as an adult, and a larger amphioxides type that has a prolonged larval period and also lives in the plankton as an adult (Boschung and Shaw 1988).

Lancelets are economically valuable in two respects. *Branchiostoma lanceolatum,* a European and Mediterranean species, is the "amphioxus" commonly used in biology laboratories as a study animal. Besides its utility as an example of primitive chordate features, it is morphologically convergent with the ammocoete larva of the lampreys (see below), thus providing comparative material and popular questions for laboratory practicals. Lancelets are also targeted by a seasonal fishery in southern China, where they are dredged from the bottom with scoops at a rate of about 30,000 kg, or 1 billion lancelets, annually.

Lancelet taxonomy is based largely on numbers of myotomes (muscle blocks) and dorsal chambers. The limited fossil record for cephalochordates is uninformative on the subject of early evolution or affinities of the group. The only known cephalochordate fossil is from an early Permian deposit in South Africa (Oelofsen and Loock 1981), long after cephalochordates could have given rise to more advanced chordates.

Hagfishes and Lampreys: Evolutionary Relationships

Ancestor-descendant relationships of jawless fishes are fraught with controversy. Just about every conceivable permutation on relationship among hagfishes, lampreys, and jawed fishes has been proposed at some time, including hagfishes, lampreys, or gnathostomes as ancestral to the other groups (see Hardisty 1982). The possible ancestors of jawless fishes are well represented in the Silurian and Devonian. Although direct linkages are impossible to make, many authors have speculated that the jawless cephalaspidiforms are ancestral to modern lampreys and, less confidently, that jawless pteraspidiforms are ancestral to modern hagfishes (see Chap. 11). Regardless, the extinct groups are very different from one another, as are extant jawless fishes from either ancestral group and from each other. Although traditionally treated as related orders in the subclass Cyclostomata (round mouths), similarities in the body morphology of modern hagfishes and lampreys are likely the result of convergent evolution. It is probably wisest to deal with them individually and independently, and appreciate them for the unique yet primitive organisms that they are.

Lampreys and hagfishes share a host of anatomic, physiological, and biochemical traits that suggest common ancestry, but what remains uncertain is how far back the two groups separated. Recent authors have stressed an even greater number of differences between lampreys and hagfishes. Although both groups are scaleless, lampreys lack the mucus-producing capability of hagfishes. Lampreys have one or two dorsal fins supported by radial muscles and cartilage, whereas hagfishes have a single continuous caudal fin. Lampreys have a more terminal mouth, the mouth in hagfishes being more a subterminal. Lampreys have a larval stage; hagfishes have direct development.

In adult lampreys the external opening of the nasohypophysis is dorsal, and the tract ends internally in a blind sac above the branchial region; in hagfishes, the external opening is terminal, and the internal opening is into the pharynx. Lampreys have two semicircular canals, hagfishes only one. Lampreys have a pineal organ and functional eyes; hagfishes possess neither. Lampreys possess lateral line neuromasts that are touch sensitive; these are lacking in hagfishes. All lampreys have 7 gill openings; hagfishes vary between 5 and 14. Although the tongue possesses keratinous (horny), replaceable teeth in both groups, it is anatomically and functionally different. Myxinoids use the tongue for biting and tearing, whereas lampreys use it for rasping and suction (see Box 13.1). These and other differences in embryologic, skeletal, neuromuscular, respiratory, cardiovascular, endocrinologic, osmoregulatory, chromosomal, and reproductive features all point out the disparate nature of the two groups (Hardisty 1982).

Myxiniforms

Hagfishes, otherwise known as slime eels or slime hags, derive their alternative names from the copious mucus they produce via 70 to 200 ventrolateral pairs of slime

FIGURE 13.2. *Hagfishes. (A) Adult Atlantic hagfish,* Myxine glutinosa, *38 cm long. Portholelike structures along the side are mucous glands. (B) Ventral view of the head region of an Atlantic hagfish. Upper orifice is the nasal opening, lower orifice is the mouth. (C) The lingual (tongue) teeth of a hagfish. (D) Hagfish egg, approximately 40 mm long. (E) A hagfish pressing a knot against the side of its prey to gain leverage when tearing off flesh.*

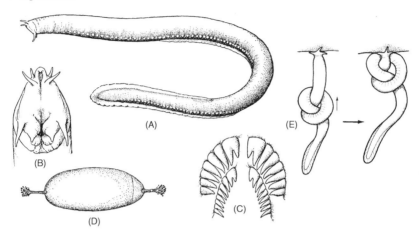

(A)–(D) From Bigelow and Schroeder 1948, used with permission. (E) After Jensen 1966.

glands (Fig. 13.2A). The slime glands contain both mucous cells and thread cells, the latter being a unique trait in hagfishes that may strengthen the slime. Each slime gland is surrounded by connective tissue and striated muscle fibers that help exude the slime upon stimulation. The mucus itself consists of a protein plus a carbohydrate that binds to water and expands to form a loose jelly. A 50-cm hagfish is capable of filling an 8-liter bucket with slime in a matter of minutes. The actual functions of slime production are poorly understood. Some authors have suggested 1) that hagfishes produce slime when attacking dying fish, perhaps hastening suffocation of the prey by clogging its gills; 2) that mucus could protect the hagfish from digestive enzymes when feeding inside the body of a prey animal; 3) that mucus is repulsive to other scavengers such as sharks and thus serves to overcome competition; and 4) that slime may help stabilize burrow walls in the muddy bottoms on which hagfishes live (Bigelow and Schroeder 1948; Brodal and Fange 1963; Hardisty 1979; Smith 1989).

Hagfishes typically produce slime in response to being disturbed or handled. Mucus may therefore also function as an antipredator device, perhaps by making the fish too slippery to handle or by clogging the gills of a potential predator and threatening it with suffocation. Slime has its drawbacks, however. A hagfish covered in its own slime will suffocate after a few minutes. Hagfishes rid themselves of slime by tying an overhand knot in their tail and then sliding the knot forward along the body, pushing the mucus ahead until the knot and mucus reach the anterior end and the fish can back away from the slime mass. A hagfish is also capable of backflushing its gills and nostril with water to rid them of slime (Conniff 1991).

Hagfishes are highly specialized animals, belying their typification as "primitive" fishes. They possess four rudimentary hearts: a primary, three-chambered branchial or systemic heart posterior to the gills, and three auxiliary, single-chambered hearts located just behind the mouth (the paired cardinal heart), at midbody (the portal heart), and at the end of the tail (the paired caudal heart). These multiple pumping stations beat at different rates; the branchial and portal hearts contract via intrinsic muscles, whereas the cardinal and caudal hearts are squeezed by surrounding, extrinsic skeletal muscle. The auxiliary hearts are necessary to reestablish blood flow in venous vessels after blood leaves several sinuses where blood flow slows. The sinuses take the place of capillary beds, giving the hagfish a partially open circulatory system, more an invertebrate than a vertebrate trait. Contraction of body wall musculature during activity also aids in pushing blood from sinuses into adjoining vessels. Hagfish blood is isosmotic with seawater, making it about three times saltier than the blood of bony fishes and lampreys. Hagfish kidneys are much simpler than those of other fishes, including lampreys, which may explain why hagfishes are restricted to a narrow range of salinities (see Chap. 7, Excretion) (Jensen 1966; Hardisty 1979, 1982).

Other hagfish peculiarities characterize the respiratory, digestive, immune, and sensory systems. Oxygen uptake in hagfishes occurs both at the gills and at capillary beds in the skin. Unlike most fishes, hagfishes inspire water through their nostril and then pump it via the mouth to the gill sacs. Cutaneous skin respiration comes into play when a hagfish has its nostril and gills buried deep in the carcass of a prey fish. Cutaneous respiration is undoubtedly facilitated by the oxygen-rich nature of the cold waters that hagfishes normally frequent. Hagfishes lack a true stomach, having instead an intestine that begins at the pharynx and ends at the anus, with an anterior muscular subdivision that prevents water inflow. Hagfishes do not bleed when their skin is cut, nor do such wounds become infected. Hagfishes have an immune system that produces antibodies, but they lack a defined thymus, spleen, or bone marrow, which are the usual sites of antibody production in vertebrates.

Hagfishes also lack complete eyes but have photosensitive receptors in their head (which contain retinal structures but no lens and are probably incapable of image formation) and cloacal region. Argument over whether the eyeless condition represents a degenerate character or whether the lineage ever possessed true eyes has been solved recently. Fossil material from the late Carboniferous indicates that Paleozoic hagfishes possessed more-developed eyes than do Recent forms (Bardack 1991). Apparently, visual sensory input has been lost over time in the deep, dark habitats that hagfishes occupy. Food is found largely through olfaction and touch, with the six barbels around the mouth serving both functions.

Hagfishes are nocturnal predators on a wide variety of small, benthic invertebrates but are better known for their scavenging behavior (Shelton 1978; Smith 1989). Hagfishes have an endearing habit of entering a dead or dying fish or other animal via some orifice or by digging through the skin and then consuming their prey from the inside, leaving only the skin and bones and making burial at sea a less-than-appealing proposition. The knot-tying action that hagfishes use to deslime their bodies is also employed during feeding. A hagfish that grasps a prey item in its mouth will pass a knot forward along its body and then press the knot against the prey as a means of levering off a piece of flesh (Fig. 13.2E). Such knot feeding is also seen in moray eels (see Box 18.2). Food is removed from large objects via a pinching motion that brings the horny teeth on the cartilaginous tongue together (Fig. 13.2C; see Box 11.1).

Reproduction in hagfishes remains something of an enigma. Both sexes contain only a single gonad, rather than the paired gonads found in most gnathostomous fishes. In immature animals, this gonad is differentiated anteriorly as ovarian tissue, posteriorly as testicular tissue. Upon maturation, one cell type prevails, and no evidence of functional hermaphroditism has been found. Fertilization is thought to be external, since males possess no intromittent organ and females never contain fertilized eggs. Females produce eggs in batches, depositing about twenty to thirty 1.5- to 2.5-cm-long, heavily yolked, sausage-shaped eggs covered by a horny shell. These comparatively large eggs attach to each other and to the ocean floor. Incubation takes about 2 months, development is direct with no larval stage, and the young emerge as 45-mm-long replicas of the adults. Most hagfishes show no obvious seasonality in spawning. However, actual spawning times, frequencies, places and behaviors, embryological details, ages at maturity, and reproductive lifespan are unknown for most species. A cash prize for information on the reproductive habits of *Myxine glutinosa*, established in 1854 by the Copenhagen Academy of Sciences, remains unclaimed.

Hagfish species occur almost worldwide in temperate and cold temperate ocean waters above 30° latitude in both hemispheres (Hardisty 1979). Few hagfish species occur shallower than 30 m, being limited by both the low salinities and high temperatures found at shallower depths; 34 parts per thousand appears to be the minimum

salinity tolerated, and 20°C the maximum temperature tolerated (Krejsa et al. 1990). The few tropical species occur in deep water, hagfishes having been captured as deep as 2700 m. Until recently, they had no commercial value and were largely viewed as nuisance species that scavenged on more valuable fishes (Box 13.1).

Hagfish taxonomy is based on arrangement of efferent gill ducts (one vs. more excurrent openings), number of slime pores, and tentacle and dentition patterns. Some authors recognize two families: the Myxinidae, with a single external gill aperture, and the Eptatretidae, with multiple external gill openings. Maximum lengths range between 25 and 80 cm. Recent analyses indicate several undescribed species in areas where only a single species was thought to occur (e.g., Wisner and McMillan 1990).

The only fossil hagfish known is a small, 7-cm-long specimen, found in Pennsylvanian (300 mybp) deposits in Illinois (Bardack 1991). This species, notable for its functional eyes, anteriorly placed gill pouches, and apparent lack of slime pores is otherwise very similar to extant forms. Its discovery underscores the conservative nature of the hagfish lineage, a clade that may trace its ancestry into pre-Cambrian times via the conodonts (Krejsa et al. 1990a,b) (see Box 11.1).

Petromyzontiforms

Whereas hagfishes have a bad reputation because of their scavenging habits, many lampreys are parasitic on other vertebrates. Lampreys superficially resemble hagfishes in general body form (Fig. 13.3). As with hagfishes, lampreys lack constricted vertebrae, the body being supported by a notochord. They also lack paired fins, jaws, a sympathetic nervous system, and a spleen. They too are scaleless, have a single nostril, and have horny teeth on the tongue. However, adult lampreys possess functional eyes, dorsal fins, an additional semicircular canal, a cerebellum, separate dorsal and ventral roots of the spinal nerves (an innovation among vertebrates), and a spiral-like rather than a straight intestine (Hardisty 1982).

Among the most striking differences between the two jawless fish groups is mode of reproduction. Whereas hagfishes presumably spawn repeatedly during their lives and produce a few large eggs each time, lampreys produce many small eggs and the adults die after spawning. Fecundity varies from about 1000 eggs in **nonparasitic** species to a few hundred thousand in the larger **parasitic** species. Hatching in lampreys occurs after 12 to 14 days, and young emerge as a 6-mm larval **ammocoete**.

In all lampreys, the free-living, blind, toothless ammocoete typically burrows into the bed of a silty stream or river (Fig. 13.3D). Ammocoetes, which were not definitively linked to adult lampreys until the mid-1800s, sit with their heads protruding from the bottom, filtering microscopic organisms from the water column by capturing them on mucus produced in the pharynx. This mode of feeding, possession of an endostyle that later develops into a thyroid gland, and the structure of the pharynx are strongly reminiscent of adult lancelets, supporting hypotheses of a relationship between the groups (although lancelets move water through the pharynx via ciliary ac-

BOX 13.1   *"Eelskin" Boots*

Marketing forces being what they are, it is surprising (or unsurprising) that few consumers know that the source of the popular "eelskin" wallets, purses, and briefcases is in fact the hagfish. Leather workers in South Korea developed a method for tanning hagfish skin in the late 1970s (Conniff 1991). The product, often marketed as "conger eel," is a soft, supple yet strong, thin leather of considerable economic value: An attaché case retails for U.S.$300, a golfbag for U.S.$1000. A substantial fishery, valued at U.S.$100 million annually, has developed off Korea, Japan, and surrounding waters. More than 1000 boats and dozens of leather-processing plants were involved in the mid-1980s. Hagfishes are caught primarily at night with baited bamboo or plastic traps at depths of 30 to 500 m, where the principal species captured are *Paramyxine atami*, *Myxine garmani*, and *Eptatretus burgeri* (Gorbman et al. 1990).

Effort peaked in 1986, when daily catches averaged 5000 kg per boat. However, this success soon fell to less than 1000 kg per boat-day, as hagfish populations experienced overfishing pressure from the working fleet. Boats also routinely lost 200 traps per month, creating a tremendous competing "ghost fleet" that was still catching and killing hagfish. Given the low fecundity and apparent infrequent reproduction of hagfishes, and a total lack of knowledge of population size or replacement rate, an unregulated effort was bound to lead to a collapsed fishery.

In the late 1980s, Korean leather companies sought other sources of hagfish leather to feed a growing worldwide demand. Fisheries opened up along the west coast of North America and the Atlantic coast of Canada, where commercial catches were previously nonexistent. In California alone, 1989 landings of Pacific hagfish, *Eptatretus stouti*, exceeded 2 million kg and involved boats from 19 different ports (Nakamura 1991). Biological information has lagged behind the economic efforts, and regulatory legislation has been slow in development. Eastern Pacific hagfishes, however, produce a thinner skin of lesser quality and durability than the western species, which has affected the desirability of eelskin products (Gorbman et al. 1990).

What are the possible ecological consequences of overexploitation of hagfish populations? Hagfishes are not viewed as particularly charismatic by most people, and it is unlikely that the environmental movement in any country will adopt the hagfish as a symbol of the need for preservation efforts and of loss of biodiversity. However, hagfishes can be exceedingly abundant in some areas; densities for the Pacific hagfish have been estimated at 443,000/km^2 ($\approx$ 0.5 hagfish/m^2) (Nakamura 1991). Hagfishes may therefore be a critical component of soft-bottom benthic regions, the most abundant habitat type in the world's oceans. In addition, hagfish and sea lampreys are an important component of the diet of several pinnipeds (seals and sea lions), animals of definite concern to the informed public. Substantial reductions in hagfish populations would have unpredictable ecological consequences, an unhappy situation with a familiar ring.

tion, whereas ammocoetes pump water via pharyngeal musculature). Ammocoetes can achieve high densities, on the order of 30/m^2, under favorable conditions (Beamish and Youson 1987). Ammocoetes may live and feed for up to 7 years, achieving a maximum size of about 10 to 15 cm.

Transformation to the adult stage takes place in summer and autumn in most species. In nonparasitic species, called **brook** or **dwarf lampreys**, the adult is no larger than the ammocoete. After an adult lifespan of about 6 months, during which no feeding occurs, the adult spawns and dies. In the parasitic species, free-living ammocoetes turn into parasitically feeding adults that may live for 1 to 3 years before spawning and dying.

Parasitic species are relatively large, up to 90 cm in length. Parasitic adults attach to the sides of fishes using the toothed **oral disk**, rasp a hole in the skin, and live off the body fluids of their host (some species are more predatory, taking pieces of flesh, fins, bones, and internal or-

gans) (Fig. 13.3B,C). Blood loss by the host can be substantial, amounting to 30% of the weight of the lamprey per day, and frequently leads to the death of the host. Parasitic lampreys can contribute significantly to the mortality of host species. The river lamprey, *Lampetra ayresi*, may kill 18 million kg of herring and 10% of the salmon off coastal British Columbia annually (Beamish and Youson 1987).

These natural levels of mortality may have relatively little effect on host population success under normal conditions, but where lampreys have been accidentally introduced, the effects can be catastrophic. The marine lamprey, *Petromyzon marinus*, invaded the upper Laurentian Great Lakes of North America via man-made canals that connected the lakes with the Atlantic Ocean. Marine lampreys have contributed to the decline or extirpation of several fish species, such as lake trout and white fishes. Extensive lamprey control strategies, involving chemical larvicides and various methods for trapping adults in

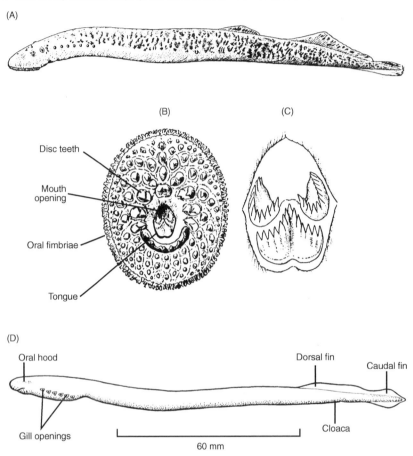

FIGURE 13.3.   *Lampreys. (A) Adult parasitic marine lamprey,* Petromyzon marinus, *about 45 cm long. (B) Oral disk of the marine lamprey showing the disk teeth used in holding on to prey. (C) Central mouth of* P. marinus, *showing the lingual (tongue) teeth used to rasp a hole in the prey. (D) Ammocoete larva of* Lampetra fluviatilis.

(A)–(C) From Bigelow and Schroeder 1948; (D) After Hardisty 1979; used with permission.

spawning tributaries, have apparently helped reduce lamprey populations (Smith 1971; Hanson and Swink 1989).

Lamprey evolution provides a rare glimpse of ongoing speciation processes: Ancestral and derived species exist contemporaneously. Twenty nonparasitic species, or about half of all lampreys, can be matched to ancestral parasitic forms. In some instances, a parasitic river lamprey ancestor has evidently given rise to two, three, or even four nonparasitic brook lamprey species. In each pair, the ammocoetes are almost indistinguishable. Although adults of the nonparasitic species are smaller, both species often have the same number of myomeres. Dentition in the adults of the parasitic species is relatively constant in number and shape and is functionally hooked and sharp. In the nonparasitic, nonfeeding adult species of a pair, dentition is variable and blunt. Derivation of nonparasitic forms apparently occurred via extension of the larval period and a shortcutting of the metamorphosis process (Fig. 13.4). Parasitic species may spend 4 years as larvae and then take 2 years to feed and mature. A corresponding nonparasitic form has a larval period of 6 years, followed by a relatively short (6 months or less) maturational period.

Hybridization between members of a pair is exceedingly rare; this reproductive isolation is maintained largely by size differences. Nonparasitic males, being smaller, cannot coil around and squeeze parasitic females while their vents are in proximity to one another, actions that are necessary for extrusion of eggs and proper fertilization (see below). Adoption of a shortened, nonparasitic mode of adult existence may expand the range where a species can occur: Small brooks may have abundant food resources for larvae but an insufficient supply of potential host fishes for a feeding adult. The phenomenon of repeated, parallel evolution of such **paired**, or **satellite**, **species** is unique among vertebrates (Hardisty and Potter 1971; Vladykov and Kott 1979; Beamish and Neville 1992).

Anatomic differences between lampreys and hagfishes are strongly reflected in their different foraging tactics. Unlike hagfishes, lampreys have circumferential teeth on the oral disk that aid in grasping live prey (see Fig. 13.2). The olfactory and respiratory pathways are necessarily separated because lampreys feed and breathe while attached to the exterior of their prey. The nasohypophyseal opening carries water to a blind olfactory organ dorsal to the gill pouches. Attachment involves sealing the contact

FIGURE 13.4.   *Comparative life histories of a species pair of European lampreys. The upper panel is the parasitic ancestor, the river lamprey* Lampetra fluviatilis; *the lower panel is the nonparasitic derived species, the brook lamprey* L. planeri. *Evolution of nonparasitic from parasitic forms involves a lengthening of the larval phase and a shortening of the maturational period. Onset of metamorphosis is denoted by M; unshaded areas represent nonfeeding periods.*

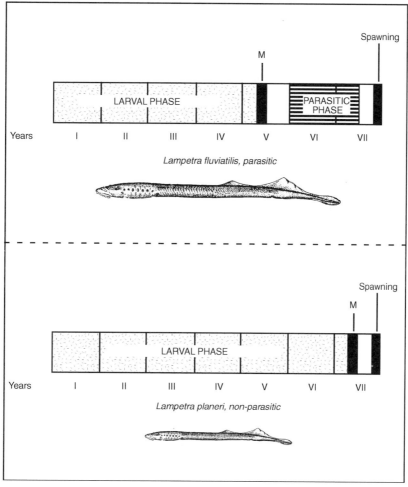

Modified from Hardisty 1979; used with permission.

region between prey and oral disk via secretion of mucus and the reduction of pressure in the buccal cavity. A vacuum is created as muscles in the mouth and pharyngeal region expel water out the gill openings. A velar flap then seals the branchial chamber off from the buccal and pharyngeal regions, thus maintaining low pressures in the buccal cavity, allowing water to be pumped in and out of the gills for breathing purposes while also keeping food out of the branchial chamber. Negative pressure in the mouth helps maintain the hold on the prey and also promotes flow of body fluids from the prey to the lamprey once the rasping tongue has gone to work. Uptake of blood and fluids is further aided by anticoagulants in the lamprey's saliva (Hardisty 1979).

Blood circulation in the adult lamprey differs from that of the hagfish in several respects. Although lamprey circulation is also "open" in that it is characterized by si-

nuses connecting arterial and venous systems, these sinuses are not as prevalent as in the hagfish. The primary lamprey sinuses are located in the branchial region and are associated with blood-gas exchange. Lampreys lack the multiple hearts of hagfishes, having instead a large, single, vagally innervated heart in a pericardial cavity, located in the typical fish position posterior to and upstream of the gills. Blood pressure is five times higher in lampreys than in hagfishes.

The reproductive biology of lampreys is well known compared to that of hagfishes. Lampreys also undergo a period of sexual intermediacy when both testicular and ovarian tissues can be found in the developing single gonad, but this period is confined to the ammocoete. During sexual maturation, lampreys undergo radical behavioral, anatomic, and physiological changes that parallel those found in salmon and anguillid eels, other families that die after reproduction (see Chap. 10, Death and Senescence; Chap. 20, Lifetime Reproductive Opportunities). These changes cause or at least foretell an inability to live beyond the spawning period. Feeding ceases, the gut atrophies, osmoregulatory function shifts, dentition deteriorates, the body shrinks, the eyes and liver degenerate, hematopoesis (blood production) decreases, and lipid and glycogen stores are reduced. Secondary sex characters, such as thickened fins and genital papillae, are formed.

Upon maturation, adults undertake a spawning migration again reminiscent of the migrations of salmon and eels. Distances moved may range from a few kilometers in nonparasitic or landlocked species to more than 1000 km in anadromous species that move from the ocean to freshwater. Spawning locales are typically the upper regions of streams where bottom types are dominated by gravel and cobbles. These locales are often used in successive years by new generations of lampreys, although little evidence exists to suggest that lampreys return, salmonlike, to spawn in the stream of their birth. Males construct a nest pit by attaching to a rock with the sucker and then carrying the rock downstream or by holding onto large cobbles and thrashing the downstream area. The pit thus created is ringed by large and small cobbles. A female will also take part in nest construction, but site selection appears to be initiated by the male. Nonparasitic species may engage in group spawning in a single nest, whereas larger, parasitic species may engage in pair spawning and male defense of the nest site.

When the nest pit is finished, the female attaches to one of the large upstream cobbles, the male attaches to the anterior portion of the female and coils his body around hers, and the two thrash while the male squeezes the eggs out of the female. Spawning occurs repeatedly over 2 to 9 days and both sexes die within a few days after spawning. Eggs remain in the nest pit for about 2 weeks, and larvae remain in the nest for an additional week. The ammocoetes then drift downstream to areas of slow current and silty or muddy bottoms, where they burrow and begin feeding (Hardisty and Potter 1971; Hardisty 1979).

Lampreys, like hagfishes, are cool-water species that seldom occur at latitudes below 30° or in water temperatures above 20°C in either hemisphere. Only two low-latitude lamprey species are known, and both occur at high elevations in Mexico, forming an interesting mirror image to the submergence of tropical species among the entirely marine hagfishes. Nonparasitic forms are entirely confined to freshwater, whereas parasitic forms may occupy freshwater or may be anadromous. Anadromous species hatch in freshwater where they live as larvae, move into coastal marine habitats as metamorphosed adults, and then return to freshwater to spawn (Hardisty 1979, 1982; Nelson 1994).

Lamprey taxonomy is based largely on mouth and dentition characteristics. Lampreys are taxonomically unique in that they have the largest diploid chromosome number of any vertebrate, between 140 and 170 in many Northern Hemisphere species. Some authors recognize three separate families of living lampreys: the Petromyzontidae of North America and Europe, and the Geotriidae and Mordaciidae of South America, Australia, and New Zealand. Two species of an extinct family, the Mayomyzontidae, have been found in Carboniferous deposits of North America. The more primitive *Hardistiella montanensis* had a hypocercal tail and lacked an oral sucker, whereas *Mayomyzon pieckoensis* was relatively small and lacked teeth everywhere except the tongue. *Mayomyzon* is notably similar to modern petromyzontids despite its 280-million-year antiquity (Fig. 13.5) (Nelson 1994).

FIGURE 13.5. *Reconstruction of the Carboniferous lamprey,* Mayomyzon piecko-ensis, *from Illinois. The fossil, seen in lateral view, bears a striking resemblance to modern petromyzontid lampreys. Several recognizable anatomic features are outlined in black: A.c = annular cartilage; D.t = digestive tract; E = eye; G.p = gill pouch; L = liver; L.w = lateral wall of braincase; O.c = otic capsule; Ol.c = olfactory capsule; P.c = piston cartilage.*

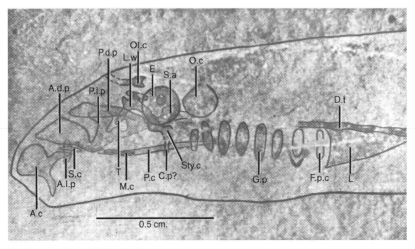

PRIMITIVE BONY FISHES

Grade Teleostomi
    Class Sarcopterygii
        Subclass Coelacanthimorpha
            Order Coelacanthiformes (Actinistia): Latimeriidae (coelacanth,
                1 species, marine)
        Sub- (or infra-) class Dipnoi[a]
            Order Ceratodontiformes: Ceratodontidae (Australian lungfish,
                1 species, freshwater)
            Order Lepidosireniformes: Lepidosirenidae (South American
                lungfish, 1 species, freshwater), Protopteridae (African lungfishes,
                4 species, freshwater)
    Class Actinopterygii
        Subclass Chondrostei
            Order Acipenseriformes: Acipenseridae (sturgeons, 24 species,
                coastal and freshwater), Polyodontidae (paddlefishes, 2 species,
               freshwater)
            Order (or Subclass) Polypteriformes (Cladistia)[b]: Polypteridae
                (bichirs and reedfish, 11 species, freshwater)
        Subclass Neopterygii
            Order Semionotiformes (Lepisosteiformes): Lepisosteidae (gars,
               7 species, fresh- and brackish water)
            Order Amiiformes: Amiidae (bowfin, 1 species, freshwater)

[a]Dipnoi + extinct Porolepimorpha = an unnamed subclass; see Chapter 11.
[b]Taxonomic position of the Polypteriformes is problematic (see Box 13.3).

Subclass Coelacanthimorpha: The Living Coelacanth

If you open an ichthyology text published prior to World War II (e.g., Günther 1880; Jordan 1905; Norman 1931), you will find passing mention of a relatively obscure group of extinct fishes that represented a side branch of the lineage that presumably gave rise to the tetrapods. These were the coelacanths, conservative sarcopterygian fishes that had gone unchanged in many respects since the Devonian (Jarvik 1980). Their fossil record stretched nearly 350 million years, from the Middle Devonian, 400 million years ago (mya), to the end of the Cretaceous, 66 mya, when they and the dinosaurs disappeared.

Imagine the world's surprise when, just before Christmas 1938, a living coelacanth was trawled from a depth of 70 m off the east coast of South Africa (Smith 1939, 1956). **Latimeria chalumnae,** described by J. L. B. Smith (Box 13.2), retains many of the characteristics that had defined the coelacanths since the establishment of their lineage: a thin bony layer encasing the vertebral spines and fin rays (hence the name *coel-acanth,* meaning "hollow spines"); an unconstricted and unossified notochord, modified as a strong-walled elastic tube; fleshy, lobed pectoral, pelvic, anal, and second dorsal "archipterygial" fins (Sarcopterygii); a symmetric, three-lobed, diphycercal tail with an epicaudal lobe; relatively large, thick, bony scales; a double gular plate under the lower jaw; a dorsal intracranial articulation (joint in the braincase that functions to increase gape size); and numerous other osteological features (Fig. 13.6).

The first and subsequent specimens also confirmed speculation about other aspects of coelacanth biology, including reproductive mode. One paleontologic finding of a Jurassic species showed skeletal impressions of small coelacanths inside a larger one, suggesting that coelacanths were viviparous (Watson 1927). A later fossil indicated eggs inside a coelacanth, suggesting oviparity and implicating cannibalism in the case of Watson's specimen. More recently, dissections of gravid female *Latimeria* indicate 5 to 26 well-developed young with yolk sacs or yolk-sac scars (Smith et al. 1975; Bruton et al. 1992). It is now felt that *Latimeria* is a lecithotrophic livebearer, in which young developing in the oviducts gain all their nutrition from their large, attached yolk sac (Fricke and Frahm 1992). Watson's original interpretation was correct.

Latimeria, and coelacanths by extension, are not ancestral to the tetrapods. Osteolepimorphs are the most likely ancestral group; *Latimeria* is in a separate subclass, the Coelacanthimorpha, sometimes referred to as the Actinistia (see Chap. 11). Osteolepimorphs apparently had well-developed lungs, as befits a tetrapod ancestor. In contrast, *Latimeria* has a fat-filled gas bladder that is no more than a vestigial outpocket of the gut. It is obviously used for hydrostatic control and is not a functional lung, not surprising for a fish that lives at depths between 100 and 250 m and seldom if ever ventures near the surface.

The blood vessel that drains the gas bladder returns blood to the sinus venosus at the back of the heart, as in other fishes. In tetrapods, this vein carries oxygenated blood to the left side of the heart and then to the rest of the body. The heart itself is characteristically fishlike in that it has no divisions into left and right sides. The gut has a spiral valve, also typical of primitive fishes and not found in tetrapods; *Latimeria*'s spiral valve has parallel spiral cones rather than a scroll valve as found in ancestral gnathostomes. *Latimeria* lacks internal choanae (nostrils with an excurrent opening into the roof of the mouth); osteolepimorphs and tetrapods possess internal choanae.

Recent behavioral findings have further clarified our understanding of *Latimeria*'s ecology. J. L. B. Smith (1956) called the coelacanth "Old Four Legs," in reference to the leglike appearance of the paired fins. This led to speculation that *Latimeria* literally walked along the bottom on its pectoral and pelvic fins. Motion pictures taken from small submarines indicate that *Latimeria* almost never touches the bottom (Fricke et al. 1987, 1991). Instead it drifts in the water column with the currents, sculling with its paired fins in an alternating diagonal pattern: When the left pectoral and right pelvic fins are moved anteriorly, the right pectoral and left pelvic fins move posteriorly. This is the pattern of locomotion shown by tetrapods and, interestingly, also by the lungfish *Protopterus* when moving across the bottom with its paired fins (Greenwood 1987).

Like most primitive fishes, *Latimeria* is highly electrosensitive, detecting weak electrical currents via a unique series of pits and tubes in the snout called the **rostral organ**. This structure bears similarities to enlarged ampullae of Lorenzini of sharks (Bemis and Heatherington 1982; Balon et al. 1988). During underwater observations, weak electrical currents were induced in a rod placed near drifting *Latimeria,* and the fish typically oriented to the rod in a vertical, head-down manner. As is characteristic of many nocturnally active fishes, the living coelacanth

---

BOX
13.2
The Living Coelacanth, at Least for Now

When Marjorie Courtenay-Latimer went down to the docks of East London, South Africa, to wish the crew of the trawler *Nerine* a happy Christmas, she could not have had a notion of how this friendly gesture would completely change her life and the course of twentieth-century natural science. Captain Goosen had saved several fishes from his recent catch that he thought she might want for the East London Museum's collections. Included in the pile was a curious, 5-foot-long fish that was "pale mauvy blue with iridescent silver markings.... Was it a lungfish gone balmy?" (Courtenay-Latimer 1979, 7).

Ms. Courtenay-Latimer sent Dr. J. L. B. Smith, a South African chemist turned ichthyologist, a rough drawing and description of the fish (Fig. 13.7). The Christmas mail and summer rains delayed communication between Courtenay-Latimer and Smith, and it was almost 2 weeks before a telegram arrived from Smith desperately urging Courtenay-Latimer to preserve as much of the fish as possible. Smith suspected the fish was a coelacanth, but it seemed so implausible. Unfortunately, the size of the fish, the summer heat, and bad luck conspired against them, and only the skin was preserved and mounted by a taxidermist. On February 16, 1939, Smith finally managed to drive to East London and view the mount and confirm that the fish was without doubt, "scale by scale, bone by bone, fin by fin . . . a true Coelacanth" (Smith 1956, 41). Smith named the fish *Latimeria chalumnae* in honor of Courtenay-Latimer and the Chalumna River off which the fish was captured. The hunt for a second, more complete specimen began immediately.

Despite a sizable promised reward, and despite an intensive collecting effort along much of the eastern coastline of Africa and deep-sea trawling around the world, a second specimen was not obtained for 14 years. The second coelacanth was slightly different in that it lacked a first dorsal fin

and a caudal fringe, probably having lost them to a shark. Smith erected a new genus, *Malania*, in honor of D. F. Malan, then prime minister of South Africa, who loaned Smith a plane to fly to the capture locale and snatch the fish away from French authorities. As Malan was also the architect of the racial separation doctrine of apartheid in South Africa, Smith's patronymic was viewed as a distasteful political expediency by many outsiders. Later analysis and additional specimens confirmed that only one coelacanth species existed; *Malania* was abolished in favor of *Latimeria*.

The second specimen and all but two subsequent specimens have been caught off the coast of the Comoro Islands (now the Federal Islamic Republic of the Comoros), a small island group in the Indian Ocean that lies between the island of Madagascar and Mozambique in East Africa. The fish are captured by hook-and-line fishermen off the western coasts of two islands, Grand Comoro and Anjouan. The fish is usually captured as bycatch of the fishery for oilfish (*Ruvettus pretiosus*, Gempylidae). The coelacanth, which has the native name *gombessa*, is not a desirable food fish (the often-cited fact that the scales are used for roughening bicycle tire tubes is erroneous; Stobbs 1988).

The fish are limited to areas of relatively recent, steep lava flows that are perforated with small caves. By day the fish rest in caves at depths between 180 and 250 m (Fricke et al. 1991). In the evening, they move into deeper water (200 to 500 m) to feed on small fishes, which they capture via a suction-inhalation mechanism, much like a giant sea bass (Fricke and Hissmann 1994). The relatively restricted depth range may relate to temperature preferences of 18 to 23°C, which may in turn be dependent on the oxygen saturation properties of coelacanth blood (Hughes and Itazawa 1972).

Specimens range in size from 42 to 183 cm in

---

forms daytime resting aggregations, with as many as 17 fish occupying a single small cave at a time. The fish have large, overlapping home ranges and return to the same caves repeatedly (Fricke et al. 1991). These observations suggest that the electrical sense of *Latimeria* could serve not only for prey detection but also for nocturnal navigation while moving through the complex lava slopes that these fish inhabit (Bemis and Hetherington 1982).

Coelacanths have an extensive, well-studied fossil record, dating back to the mid-Devonian (see Fig. 11.7). As

many as 121 different species have been described, of which 90 are probably valid. Actinistians reached their highest diversity in the early Triassic and disappeared from the fossil record in the late Cretaceous (Cloutier and Forey 1991).

Sub- (or Infra-) Class Dipnoi: The Lungfishes

Lungfishes are well represented in the fossil record on all major continents, including Antarctica. They arose early in

BOX
13.2
(Continued)

length and weigh from 1 to 95 kg, the largest individuals being female (Bruton and Coutouvidis 1991). Age estimates indicate that coelacanths live from 20 to as much as 40 to 50 years (Bruton and Armstrong 1991). Given the average large size (150 cm) and age of females, reproduction probably does not begin until relatively late in life. Based on dissections of gravid females, clutch size is small (5 to 26 young per clutch), and gestation is long, perhaps 13 months. Combined with presumed late maturation, reproduction rates are undoubtedly slow.

Given restricted habitat preferences, limited geographic range, and low catch numbers (average 5.5 per year), it is quite probable that the Comoro Island coelacanths represent the only population of coelacanths in the world, aside from occasional strays such as the individuals trawled up off South Africa in 1938, Mozambique in 1991, and southern Madagascar in 1995. Population size is understandably difficult to determine, but estimates of 200 to 500 or fewer individuals have been made from the available information (Fricke et al. 1991; Bruton and Armstrong 1991; Fricke and Hissmann 1994).

Between 1952 and 1992, at least 173 individuals were captured, most as research and display material for museums (Bruton and Coutouvidis 1991). Unfortunately, a black market for coelacanths has also developed because of the animal's freak appeal and because of the purported medicinal value of the fluid contained in the highly modified notochord. The twisted logic that drives this market is that since coelacanths as a group have survived for so long, coelacanth tissue can prolong the life of a human into whom it is injected (Stobbs 1988; Bruton and Stobbs 1991). Celebrity has transformed a by-catch fishery into a directed fishery; a single coelacanth is worth U.S.$150, or about 3 to 5 years' income to a local fisherman. The fish are eventually sold for

U.S.$500 to U.S.$2000 on the open market. The directed fishery was eliminated when the Comoran government outlawed the capture of coelacanths, but incidental captures still occur at the rate of 5 to 10 fish per year, which could represent as much as 5% of the adult population captured annually (H. Fricke, personal communication).

All these circumstances—slow growth and maturation, small clutch size, limited habitat and geographic range, limited recruitment, small and perhaps decreasing population size, intense exploitation—indicate that coelacanths are particularly vulnerable and threatened by extinction. International conservation efforts have recently been initiated: The coelacanth is listed as an endangered species by the International Union for the Conservation of Nature, thereby outlawing commercial trade by signatory nations. A Coelacanth Conservation Council has been formed to coordinate and promote research on and conservation of coelacanths; J. L. B. Smith Institute of Ichthyology, Private Bag 1015, Grahamstown 6140, South Africa; see also WWW.dinofish.com on the Worldwide Web).

Recommendations have been made to minimize the incidental catch and eliminate any directed fishery for coelacanths, including the capture of specimens destined for aquaria and museums. The Coelacanth Conservation Council has proposed that the coelacanth be adopted as the international symbol of aquatic conservation, equivalent to the panda's status for terrestrial conservation.

Coelacanths occupy a unique place in the consciousness of man: they represent a level of tenacity and immortality which man will never achieve during his short stay on earth. For the sake of future generations, can man afford to cause the coelacanth, which represents a group that has survived for over 400 million years, to become extinct within 50 years of discovering it? (Balon et al. 1988, 274).

---

the Devonian and were fairly common until the late Triassic. Today, they are represented by three genera that date back to the Cretaceous, with populations in South America, Africa, and Australia. Lungfishes possess a mosaic of ancestral and derived traits that initially clouded their taxonomic position (Conant 1987). The South American species, *Lepidosiren paradoxa*, reveals in its specific name some of the confusion that its mixture of traits must have caused. It was first described in 1836 and was thought to be a reptile because of the structure of its lung and the place-

ment of the nostrils near the lip. An African species, *Protopterus annectens*, was discovered the next year and was proclaimed to be an amphibian based on its heart structure. Both species were very different from fossil lungfishes, and relationships between extant and extinct forms were not obvious. After about 30 years of debate, the probable piscine nature of the lungfishes was generally accepted, with recognition that the fishes had "singularly embarrassed taxonomists" (Duvernoy 1846 in Conant 1987).

A general and distinctive characteristic of lungfishes is

FIGURE 13.6.   *The living coelacanth,* Latimeria chalumnae, *an extant member of a group thought to be extinct for 65 million years. Discovered by science in 1938, and restricted to volcanic slopes of the Comoro Islands, the entire species is believed to number less than 500 individuals and is recognized internationally as endangered. Some traits that distinguish coelacanths from other living fishes are noted.*

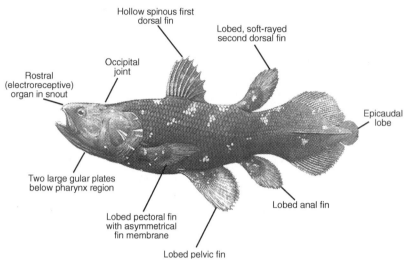

Drawing by S. Landry, from Musick et al. 1991; used with permission.

FIGURE 13.7.   *Marjorie Courtenay-Latimer's drawing and description of the first coelacanth, as sent to J. L. B. Smith. Key features pointed out by Courtenay-Latimer included bony plates on the head and the extra median lobe in the caudal fin.*

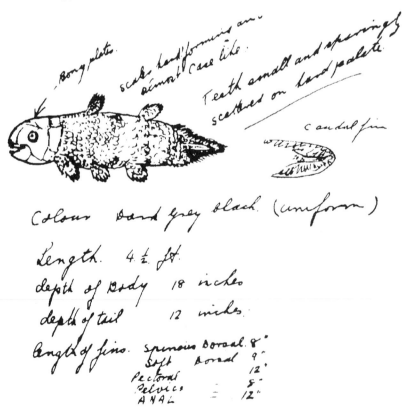

From Smith 1956; used with permission.

the existence and location of massive tooth plates (see Fig 11.8). Teeth are not attached to the jaw margins as in most other living fishes but instead occur only on interior bones, such as the vomer and pterygoid bones of the upper jaw and the prearticular bones of the lower jaw (Bemis 1987). These tooth plates are often quite massive and apparently function in crushing aquatic insects, crustaceans, and particularly mollusks; the tooth plates are better developed in the Australian than in the South American and African species. Because it is the tooth plates that most commonly fossilize, they form the basis of much of our understanding of evolution in the group (see Fig. 11.8).

The living African and South American lungfishes are placed in the families Lepidosirenidae and Protopteridae (Fig. 13.8A,B). The four **African species** of the genus *Protopterus* are widely distributed through central and southern Africa, occurring in both lentic (still) and lotic (flowing) habitats of major river systems, including a variety of swamp habitats (Greenwood 1987). Maximum sizes range from 44 cm (*Protopterus amphibius*) to 180 cm (*P. aethiopicus*). Young fishes are active by night; adults are active by day. Food includes a variety of hard-bodied invertebrate taxa, with mollusks predominating (Bemis 1987). Protopterids are obligate air breathers throughout their postjuvenile life, obtaining 90% of their oxygen uptake via the pulmonary route.

African lungfishes are best known for their ability to survive desiccation of their habitats during the African dry season. Such **estivation** behavior, as described for *P. annectens*, involves construction of a subterranean mud cocoon (Greenwood 1987). As water levels fall, the lungfish constructs a vertical burrow by biting mouthfuls of mud from the bottom, digging as deep as 25 cm into the mud. As the swamp dries up, the lungfish ceases taking breaths from the water surface, coils up in the burrow with its head pointing upward, and fills the chamber with secreted mucus. This mucus dries, forming a closely fitting cocoon, and the fish becomes dormant (Fig. 13.9). This dormant period normally lasts 7 or 8 months but can be extended experimentally for as much as 4 years in *P. aethiopicus*. During estivation, lungfish rely entirely on air breathing, lower their heart rate, retain high concentrations of urea and other metabolites in the body tissues, metabolize body proteins, and lose weight. With the return of rains, the lungfish emerges from the burrow and resumes activity, which includes cannibalizing smaller lungfish that have also just emerged from their burrows.

African lungfishes build burrow-shaped nests, often tunneling into the swamp bottom or bank. Eggs and young are guarded by one parent, presumably the male. The male has no specialized structures to aid in oxygenating the water in the nest as reported for *Lepidosiren* (see below), although male *P. annectens* have been observed "tail lashing" near the nest, which may serve the same purpose. Young African lungfishes have external gills (Fig. 13.10), one of the traits that caused many nineteenth-century biologists to consider them amphibians.

The **South American species**, *Lepidosiren paradoxa* (see Fig. 13.8B), is considered to be the most recently derived member of the family (Greenwood 1987). Surprisingly lit-

FIGURE 13.8.   *Modern lungfishes. (A) An African lungfish,* Protopterus annectens, *one of four species in the genus. (B) The South American lungfish,* Lepidosiren paradoxa, *showing the vascularized pelvic fins that develop on males during the breeding season. (C) The Australian lungfish,* Neoceratodus forsteri.

(A)
Protopterus

(B)
Lepidosiren

(C)
Neoceratodus

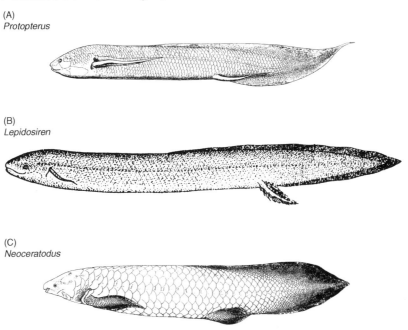

(A), (C) From Jarvik 1980; (B) from Norman 1931; used with permission.

FIGURE 13.9. *An African lungfish estivating in its mud-and-mucous cocoon, viewed from the ventral surface of the fish.*

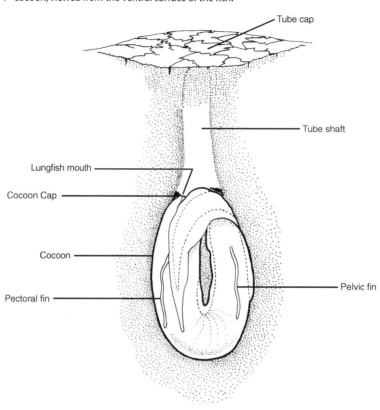

Redrawn from Greenwood 1987.

tle is known of its natural history compared to the African and Australian species. It occurs in swampy regions of the Amazon and Parana river basins (Thomson 1969) and grows to about 1 m in length. As with *Protopterus*, adults have reduced gills and are obligate air breathers. Estivation in burrows occurs but is poorly documented. *Lepidosiren* is best known for its reproductive behavior, although conjecture exceeds information. Eggs are deposited in a burrow nest as in *Protopterus* and guarded by the male. During the breeding season, after egg deposition, the male's pelvic fins develop vascularized filaments, in apparent response to increased testosterone levels in the male's bloodstream (Cunningham and Reid 1932; Urist 1973). These sexually dimorphic structures are purportedly used to supplement the respiratory needs of the young by helping to oxygenate the water in the burrow, although actual behaviors and measurements during breeding have yet to be described. Young lepidosirenid lungfishes are not obligate air breathers, a trait that may reduce their exposure to a variety of predators while they are small and exceedingly vulnerable.

The **Australian** species, *Neoceratodus forsteri* (see Fig. 13.8C), was the last lungfish to be described scientifically, in 1870. It has a very limited native distribution, restricted primarily to the Brisbane, Burnett, Fitzroy, and Mary river systems and several small reservoirs of northeastern Australia (Kemp 1987). Among living lungfishes, *Neoceratodus* is closest to the ancestral forms in many anatomic respects, including a large (up to 112 cm long),

relatively stout body, large cycloid scales covering the entire body, flipperlike "archipterygial" fins, pectoral fin inserted low on the body, a broad diphycercal tail, and an undivided lung. The fish feeds in the late afternoon and maintains activity during the night, capturing benthic crustaceans, mollusks, and small fishes that it crushes with its distinctive tooth plates.

Unlike the South American and African species, *Neoceratodus* is a facultative air breather that relies on gill respiration under normal circumstances. Its lung may in fact serve more as a hydrostatic than a respiratory organ (Thomson 1969). Uptake of oxygen through the skin occurs, at least in juveniles. No special adaptations to avoid desiccation have been observed, and the fish must be kept moist and covered by wet vegetation or mud to survive out of water.

Male and female Australian lungfish show only slight dimorphic coloration during the breeding season and are otherwise indistinguishable. In the native riverine habitat, spawning is rather unspecialized, involving deposition of eggs on aquatic plants in clean, flowing water at any time of the day or night. Fish spawn in pairs, when females deposit 50 to 100 eggs per spawning; no parental guarding occurs. Development of the young is direct and gradual, with no obvious larval stages or distinct metamorphosis. Young are born without external gills. As a result of habitat destruction, pollution, and perhaps interactions with introduced species, *Neoceratodus* populations

FIGURE 13.10. *A young African lungfish. Arrow indicates the external gills that misleadingly caused lungfishes to be classified as amphibians.*

From Herald 1961; used with permission from Chanticleer Press, Inc, New York.

have apparently declined dramatically. The fish is protected by the Queensland government, and its capture and sale are restricted by the Convention on International Trade in Endangered Species (CITES). Fish have been transplanted into several Queensland rivers and reservoirs to aid the species' recovery.

The seventh lungfish. No discussion of extant lungfishes would be complete without at least brief mention of *Ompax spatuloides*, Australia's other lungfish (Fig. 13.11). *Ompax* was described based on a 45-cm-long specimen served to the director of the Brisbane Museum during a trip to northern Queensland in 1872, 2 years after the scientific discovery of the first Australian lungfish. The fish was reported to occur syntopically with *Neoceratodus* in a single water hole in the Burnett River. It had a body covered with large ganoid scales, small pectorals, and an elongate, depressed snout, "very much the form of the beak of the *Platypus*" (Castelnau 1879, 164). The director had a sketch made of the fish but ate it nonetheless. The sketch and notes were sent to a prominent regional ichthyologist, Count F. de Castelnau, who described the species and speculated that it was most closely related to the gars of North America. *Ompax* appeared in Australian faunal lists as a ceratodontid for 50 years, even though a second specimen was never found. Finally, in 1930, an anonymous report appeared in a Sydney newspaper recounting how the "fish" had been fabricated from the nose of a platypus, the head of a lungfish, the body of a mullet, and the tail of an eel (Herald 1961). Ichthyology's Piltdown man had been unmasked.

Class Actinopterygii, Subclass Chondrostei: Sturgeons and Paddlefishes

The most primitive living actinopterygian fishes are highly derived, relict species that bear little resemblance to extinct forms. These are the sturgeons and paddlefishes, two families that probably diverged from each other during the Jurassic but still share a number of characteristics such as a cartilaginous skeleton, heterocercal tail, reduced squamation, more fin rays than supporting skeletal elements, unique jaw suspension, and a spiral valve intestine. Although largely cartilaginous, their skeletons are secondarily so: Ancestral, early Mesozoic "chondrosteans" (more correctly paleoniscoids) were bony.

FIGURE 13.11. *The seventh living "lungfish," *Ompax spatuloides*. This is the illustration that appeared in the original species description by Castelnau (1879). It shows 1) lateral view, 2) dorsal view of the head, and 3) presumably a cross section of the bill, but unlabeled in the original illustration.*

Ompax Spatuloïdes

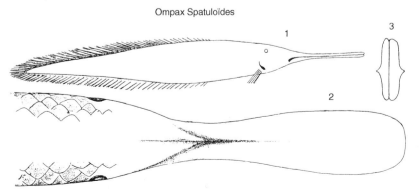

From Castelnau 1879.

Acipenseridae. All 24 species of **sturgeons** are in the family Acipenseridae and are restricted to the Northern Hemisphere (Binkowski and Doroshov 1985; Williams and Clemmer 1991; Nelson 1994). Four genera are recognized: *Acipenser*, *Huso*, *Scaphirhynchus*, and *Pseudoscaphirhynchus*. All species spawn in freshwater, although some species move seasonally between marine and freshwater. Species restricted to freshwater include the North American lake sturgeon (*Acipenser fulvescens*) and three river sturgeons (*Scaphirhynchus* spp.), the latter occurring only in larger rivers such as the Mississippi and Missouri. Anadromous species, those spending part of their lives at sea but returning to freshwater to spawn, include the Atlantic sturgeon, *Acipenser oxyrhynchus*, the white sturgeon, *A. transmontanus* (the largest North American freshwater fish, attaining a length of 3.8 m and a weight of 630 kg), and the beluga of eastern Europe and Asia, *Huso huso* (the largest freshwater fish in the world, attaining a length of 8.6 m and a weight of 1300 kg, and not to be confused with the toothed whale of the same common name). As with other anadromous species (see Chap. 22, Diadromy), landlocked populations of sturgeons can develop.

Anatomically, sturgeons can be identified by the four barbels in front of the ventrally located mouth, five rows of bony scutes (large bony shields) on a body otherwise covered with minute ossifications, a heterocercal tail, an elongate snout, a single dorsal fin situated near the tail, no branchiostegal rays, and a largely cartilaginous endoskeleton, including an unconstricted notochord (Fig. 13.12). Although generally slow-swimming feeders on benthic invertebrates, sturgeons have a protrusible mouth that can be extended very rapidly, allowing larger individuals to feed on fishes (Miller 1987). Vision plays at best a minimal role in prey detection, with touch, chemoreception, and probably electrolocation via rostral ampullary organs being more important (Buddington and Christofferson 1985).

The life history traits of sturgeons make them unique and susceptible to overexploitation by humans. They are exceptionally long-lived: Beluga have been aged at 118 years, and white sturgeon at 70 to 80 years (Nikol'skii 1961; Scott and Crossman 1973; Casteel 1976). As is often the case with long-lived vertebrates, sexual maturity is attained slowly. In Atlantic sturgeon, both sexes mature after 5 to 30 years, the older ages characterizing individuals at higher latitudes. After maturation, females may only spawn every 3 to 5 years (Smith 1985), and even longer

intervals may characterize other sturgeon species. Fecundity is relatively high: Ovaries may account for 25% of the body mass of a female. Hence a 200-kg female's ovaries can weigh 50 kg, containing 1.6 million eggs. Sturgeon eggs are the most valuable kind of caviar, making an individual female worth several thousand dollars. Sturgeons are also commercially valuable as a smoked product. The gas bladder can be processed into isinglass and used for gelatin and clarifying agents, and as a commercial art glue. Natural predators beyond the juvenile stage are rare; parasitic lampreys are one of the few organisms capable of attacking an adult sturgeon (Scott and Crossman 1973).

One trait common to sturgeons worldwide is a decline in numbers due to dam building, habitat destruction, and overexploitation. Large Atlantic sturgeon were at one time sufficiently abundant in North American coastal rivers so that navigation by canoes and small boats was sometimes hazardous, particularly given the fish's habit of leaping 1 to 2 m out of the water. Commercial landings exceeded 3 million kg annually in 1890, but 100 years later, landings were reduced by 99% and many stocks faced local extinction (Smith 1985). The shortnose sturgeon, *A. brevirostrum*, of Atlantic coastal rivers is listed as an endangered species, and lake sturgeon have been extirpated from a large part of their native range (ironically, lake sturgeon disappeared from the Sturgeon Falls area of the Menominee River, Wisconsin, around 1969; Thuemler 1985). It is thought that three species endemic to the Aral Sea may be extinct due to extensive drying of that huge body of water (Birstein and Bemis 1993; see Chap. 25, Competition for Water).

Population crashes in sturgeon have resulted from a combination of habitat degradation and the unique life history characteristics of these large, slow-maturing fishes. Spawning is hampered by siltation and contamination of clean, gravel and rock areas, and by dam construction that limits access to spawning sites. The spawning period in several species may be very short, on the order of 3 to 5 days, and if environmental conditions are inappropriate, spawning may be abandoned for that year (Buckley and Kynard 1985). Recruitment of new fish into the population is further prevented by overharvest of mature individuals as well as sexually immature fish, sometimes as a result of incidental by-catch of juveniles in gill nets set for anadromous shad or salmon. Given late maturation and the infrequency of spawning, stocks driven to low numbers have a difficult time recovering.

FIGURE 13.12. *An Atlantic sturgeon,* Acipenser oxyrhynchus. *Note the rows of bony scutes on the body, distinct heterocercal tail, and elongate snout with barbels preceding the ventral mouth.*

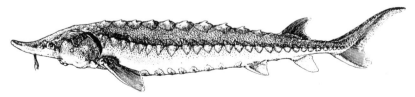

From Vladykov and Greeley 1963; used with permission.

By 1990, commercial fishing for Atlantic sturgeon was severely limited or outlawed throughout its range, and active programs for culturing several species were implemented (Binkowski and Doroshov 1985).

Acipenseroid fishes are generally regarded as highly modified descendants of paleoniscoids that lived during the Permian and Triassic. Early, recognizable sturgeon fossils date to the Upper Cretaceous (Wilimovsky 1956). A related, extinct family, the Chondrosteidae, is known from fossils from the Lower Jurassic to Lower Cretaceous periods.

Polyodontidae. **Paddlefishes** also date back at least to the Upper Cretaceous, but only two species remain: the paddlefish of North America, *Polyodon spathula*, and the Chinese paddlefish, *Psephurus gladius*. They have larvae similar to those of sturgeons and retain the heterocercal tail, unconstricted notochord, largely cartilaginous endoskeleton (with ossified head bones), spiracle, spiral valve intestine, and two small barbels. They differ from the acipenserids in most other respects. The bony scutes are missing, and the body is essentially naked except for patches of minute scales. Paddlefishes are not benthic swimmers but instead move through the open waters of large, free-flowing rivers, feeding on zooplankton or fishes.

The **North American paddlefish**, or spoonbill cat (Fig. 13.13A), prefers rivers with abundant zooplankton. Adult paddlefish typically swim through the water both day and night with the nonprotrusible mouth open, straining zooplankton and aquatic insect larvae indiscriminately through the numerous, fine gill rakers. Food size is limited by gill raker spacing, as small zooplankters escape the mechanical sieve of the paddlefish's mouth (Rosen and Hales 1981). This picture of the paddlefish as a passive filterer is confused by the occasional benthic and water-column fishes, such as darters and shad, found in its stomach (Carlander 1969). Small juveniles, in which neither gill rakers nor the paddle are well developed, pick individual zooplankters out of the water column. The function of the rostral paddle, which accounts for one-third of the body length in adults, remains poorly understood. It has been erroneously suggested that paddlefish use their paddle to dig in the bottom for food. The paddle may function in directing food-bearing water into the mouth. More recent interpretations suggest that the abundant ampullary receptors on the surface of the paddle and operculum serve to detect biologically generated electricity (Fig. 13.13B,C). Such receptors could be used for detecting concentrations of food organisms in the water column, although this function remains undocumented (Russell 1986).

North American paddlefish may live for 30 years and attain a length of 2.2 m and a mass of 83 kg, although fish of this size are now exceedingly rare. Diminishing populations are evidenced by changes in the species' range. Although currently restricted to the Mississippi River drainage system, populations of paddlefish historically occurred in the Laurentian Great Lakes and have been extirpated from at least four states (Gengerke 1986). Causes of population decline are similar to those affecting

FIGURE 13.13. *Paddlefishes. (A) A North American paddlefish,* Polyodon spathula. *(B) The rostral paddle of the North American paddlefish in dorsal view; arrows indicate position of the eyes. (C) Area at lower left of (B) enlarged, showing the stellate bones (sb) that support the paddle, and the ampullary organs, which are the dark circular holes in the paddle that reportedly serve as electroreceptors. (D) A 1-m-long Chinese paddlefish,* Psephurus gladius, *a rare and poorly known chondrostean restricted to the Yangtze river system of China.*

(A)

(B)

(C)

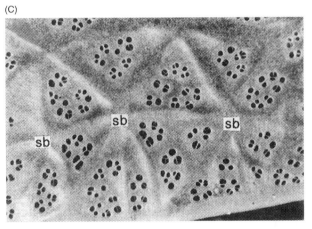

(D)

(A) Photo by W. Roston; In Warren and Burr 1994; (B)–(D) From Grande and Bemis 1991; used with permission.

sturgeon. Paddlefish are long-lived (to 30 years) but do not mature until they are 7 to 9 (males) or 10 to 12 (females) years old; they spawn only at 2- to 5-year intervals. Loss of spawning habitat—fast-flowing, clean gravel bottoms—is a major problem. Appropriate spawning areas are degraded by damming, which decreases water flow and leads to siltation. Man-made reservoirs are productive feeding habitats for adults but do not provide appropriate spawning areas. Paddlefish are sought commercially and recreationally for their flesh and eggs; overfishing has frequently been implicated in population declines (Russell 1986).

The **Chinese paddlefish**, *Psephurus gladius* (Fig. 13.13D), is the more primitive of the two species and differs primarily in head and jaw morphology and body size. The paddle is narrow and more pointed, not broad and rounded. *Psephurus* also has fewer but thicker gill rakers that resemble those of sturgeons, has a protrusible mouth, and grows larger (over 3 m and 700 kg, erroneously reported to 7 m). It inhabits the Yangtze river system of central China and feeds primarily on small, water-column and benthic fishes (Nichols 1943; Nikol'skii 1969; Liu and Zeng 1988). *Psephurus* is highly prized for its caviar but is now rare. It is considered the most endangered fish in China because of overfishing, habitat destruction, and dam construction, which blocks adults from reaching spawning grounds. Relatively little is known about its biology (Liu and Zeng 1988; Grande and Bemis 1991; Birstein and Bemis 1995).

The fossil record for polyodontids is limited to three known species and some fragments, all from the Upper Cretaceous, Paleocene, and Eocene of North America. An extinct paddlefish, *Crossopholis*, is more like *Psephurus* than *Polyodon*, suggesting that the filter-feeding lifestyle of the North American paddlefish is the more derived condition (Grande and Bemis 1991).

Subclass/Order Brachiopterygii/ Polypteriformes/Cladistia: Bichirs and Ropefish

Taxonomic relationships among and within most relict groups, both in terms of affinities with other living fishes and identification of ancestral lineages, are reasonably well understood. Lungfishes, coelacanths, chondrosteans, gars, and bowfin all have well-defined, relatively extensive fossil records with which modern species can be associated. In addition, derived traits are either unique to a group or shared with other groups in ways that confirm evolutionary hypotheses of relationship (see Table 13.1). Although healthy debate on the details of relationship among these fishes exists, most researchers agree on the general patterns of interrelatedness.

No such consensus applies to the brachiopterygians. Different workers have placed them in three major bony fish subclasses—the Dipnoi, Sarcopterygii, or Actinopterygii—based on similarities with those groups (Patterson 1982). Other taxonomists have emphasized unique characteristics and placed them in their own subclass, the **Brachiopterygii**. The fossil record provides few clues to their ancestry, because the few available fossils are restricted to Africa, indicate relatively little anatomic change, and do not appear until the late Cretaceous, long after the evolutionary histories of the other groups are well established.

Modern brachiopterygians are represented by two genera confined to west and central tropical Africa, including the Congo and Nile river basins (Fig. 13.14). Ten species and numerous subspecies, referred to as **bichirs**, belong to the genus *Polypterus*; the remaining species is the **ropefish** or **reedfish**, *Erpetoichthyes* (formerly *Calamoichthyes*) *calabaricus*. Bichirs grow to 120 cm in length, ropefish to 90 cm. All species are predatory and inhabit shallow vegetated and swampy portions of lakes and rivers.

In poorly oxygenated water, bichirs are obligate air breathers and will drown if denied access to the surface. Bichirs are unique in that they use their dorsally placed spiracles to exhale (not inhale) spent air from the highly vascularized and invaginated lungs; the spiracles serve no apparent aquatic respiratory function (Abdel Magid 1966, 1967). Polypterids are additionally unique in that they inhale through their mouths by **recoil aspiration** (Brainerd et al. 1989). They use the elastic energy stored in their integumentary scale jacket during exhalation to power inhalation of atmospheric air. The existence of similar bony scale rows in some Paleozoic amphibians suggests that the evolution of air breathing and perhaps eventual terrestriality may be linked to the recoil aspiration that originated in fishes.

Controversy over taxonomic position arises because brachiopterygians exhibit superficial anatomic traits that have been used to justify their inclusion in almost every one of the major taxa discussed in this chapter (see Table 13.1). Brachiopterygians possess lobelike fins (a sarcopterygian trait), ganoid scales (a paleoniscoid or lepisosteiform trait), two gular plates (as does the coelacanth), spiracles (in common with sturgeons), feathery external gills when young and double ventral lungs (in common with lepidosirenid lungfishes), a modified heterocercal tail (as in gars and bowfin), and a spiral valve intestine (shared by all major groups).

However, the internal construction of many of these seemingly shared primitive characteristics is very different from those of other taxa. The external gills are only analogous to, not homologous with, the gills of young lungfishes. The tail is heterocercal in structure but symmetric in external appearance; the medial and lower portions are created by rays coming off the ventral surface of the notochord, unlike any other fishes. In fact, brachiopterygians possess fins that set them apart from all other major taxa. Bichirs are also referred to as **flagfins**, because each of the 5 to 18 dorsal finlets consists of a vertical spine to which are attached horizontal rays, giving them a flag-and-pole appearance. In all other ray-finned fishes, the dorsal fin rays emerge as vertical bony elements from the body of the fish. The pectoral fin is lobe-shaped but constructed differently from the lobe fins of lungfishes and crossopterygians or, for that matter, any other fish, living or extinct. The supporting structures of the pectoral fin are shaped like a wishbone with a flat plate (Fig. 13.14B). A. S. Romer, a leader of modern vertebrate paleontology, referred to the polypterid pectoral fin as a "peculiar and overelaborated development" (Romer 1962, 198). And recoil aspiration breathing is performed by no other known extant group.

The actual taxonomic position of the Brachiopterygii is a matter of active debate. Some of this debate hinges on similarities between brachiopterygians and the extinct paleoniscoid fishes, members of which were probably an-

FIGURE 13.14. *Brachiopterygians. (A) A 29-cm-long bichir,* Polypterus palmas polli, *from the Ivory Coast. Note the lobelike pectoral fin base and the horizontal flaglike fin rays that extend from the distal portion of each dorsal fin spine. (B) The "peculiar and overelaborated" pectoral fin of a bichir, showing the wishbonelike basal structure (propterygium and metapterygium) that supports the radials and fin rays.*

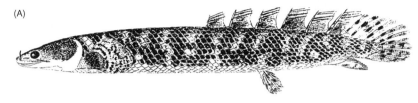

(A)

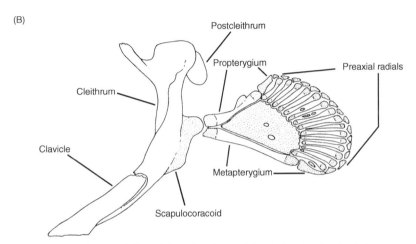

(B)

(A) From Hanssens et al. 1995; used with permission. (B) From Rosen et al. 1981; Courtesy of the Department of Library Services, American Museum of Natural History.

cestral to modern actinopterygians (see Chap. 11, Class Actinopterygii). *Polypterus* is considered by some workers to represent a surviving paleoniscid, in essence a living fossil, like *Latimeria.* The remainder of the debate involves anatomic similarities and differences among extant, primitive fishes. Where in the Actinopterygii bichirs and ropefish fit, if at all, may be determined by the next generation of ichthyologists (Box 13.3).

Infraclass Neopterygii, Order Semionotiformes (or Lepisosteiformes): The Gars

The gars and bowfin are descendants of the paleoniscoids, which dominated fresh- and marine waters for 200 million years from the mid-Devonian into the Mesozoic era. Traditionally, gars and bowfin were considered members of the order Holostei, which also included a variety of extinct fishes. However, most recent analyses conclude that holosteans are paraphyletic, more a grade of development than a true clade. The two modern groups differ in many important respects, and their relationships to the paleoniscoids, and position in the lineage leading to modern teleosts, are a matter of discussion. Both groups are considered neopterygian because of shared jaw, tail, and dermal armor characteristics. Some workers consider gars to be more primitive and place them in their own division

(Ginglymodii); some analyses see *Amia* as the more primitive group; and other analyses place gars and bowfin together as the sister group to the teleosts (Normark et al. 1991; Olsen and McCune 1991).

All seven species of living gars are in the family Lepisosteidae, genus *Lepisosteus* (Fig. 13.16). These elongate, predatory fishes are restricted to North and Central America and Cuba; five species occur east of the Rocky Mountains in North America, and the remaining species occur in Central America. Gars typically inhabit backwater areas of lakes and rivers, such as oxbows and bayous. Oxygen tension in such habitats is often low; gars can and in fact must breath atmospheric air at these times, using their compartmentalized, highly vascularized gas bladder as a lung (Smatresk and Cameron 1982; Smith and Kramer 1986).

Gars have entirely ossified skeletons. Their primitiveness is evident in their hinged, diamond-shaped, interlocking ganoidlike scales and abbreviate heterocercal caudal fin. Ganoin is an enamel-like material on the upper surface of the scale; it characterized the squamation of the Paleozoic and early Mesozoic paleoniscoids, which are thought to be ancestral to (or a sister group of) modern teleostean groups. The gars have retained this primitive trait. The same logic applies to the caudal skeleton. Abbreviate heterocercal tails characterized the later

**BOX
13.3**   *Just What Is a Bichir?*

Three possible conclusions concerning the taxonomic position of the Brachiopterygii are possible, depending on whether one focuses on similarities or differences. The decision reached in large part reflects the emphasis of the individual researcher and points out the dynamic and often contentious process by which systematists arrive at an understanding of evolutionary relationships among major taxa. If differences are emphasized, one can simply admit confusion and declare the group *incertae sedis*, of unknown taxonomic position. However, this conclusion ignores the wealth of anatomic information that has been obtained on the fishes and their obvious close relationship to actinopterygians.

Another approach is to focus on the specialized, unique traits of the brachiopterygians, such as their unusual external gills, peculiar fins, and chromosomal arrangement (see Table 13.1). These traits indicate a long, independent evolution of the group from a possible paleoniscoid ancestor. This approach suggests that brachiopterygians be placed on separate but equal footing with other major groups, such as in their own subclass or infraclass (Denton and Howell 1973; Jarvik 1980).

A third alternative is to place them within the Actinopterygii, with whom they share many similarities, but as a separate order, a view held by many authorities (e.g., Patterson 1982). In cladistic terms, they can be considered a sister group of Recent actinopterygians, having branched off from the paleonisciforms independently of the modern ray-finned fishes. Hence the Actinopterygii consist of two modern groups: the **Cladistia** (brachiopterygians) and **Actinopteri** (chondrosteans and neopterygians) (Fig. 13.15).

Part of the difficulty in this analysis stems from an attempt to compare among living fishes, each well adapted to environmental conditions of the recent past. Many anatomic traits, in fact those most critical to systematic analyses, are retained from ancestors, whereas other traits represent recent derivations that have evolved in response to conditions greatly changed from the ancestral selection pressures. Hence we have the existence of a mosaic of primitive and derived traits in every living species, homology and analogy intertwined, with difficulty in knowing the proportions of the two.

Each attempt at linking a trait in brachiopterygians with a counterpart trait in another group then becomes a possible apples-and-oranges comparison. External gills in young bichirs and lungfishes (and amphibians) may have evolved independently and in parallel because ecological and physiological conditions compromise the effectiveness of lungs in very small vertebrates. Or such gills could represent the retention of the ancestral condition in one group and *de novo* evolution in the other. Or it could represent common ancestry. Cladistic analysis offers a means of separating parallelisms from derivations in many cases. But without a fossil record of a specific trait, our conjecture often remains just that. The debate, as with many in systematics, will continue unresolved until sufficient paleontologic evidence is discovered to make a more informed decision.

"holosteans" but have given way to the homocercal tail of the teleosts. A constricted and ossified notochord may be a derived innovation in lepisosteids rather than an indication of ancestral status to teleosts, since lepisosteid vertebral centra are essentially unique among living fishes. Gar centra are **opisthocoelous**, being concave on their posterior surface and convex on the anterior surface, allowing for a ball-and-socket articulation. Most fishes have **amphicoelous** vertebrae, in which both surfaces are concave; only one blenny species, some tailed amphibians, and a few birds have opisthocoelous vertebrae (Suttkus 1963; Wiley 1976). The name Ginglymodi refers to the hinged articulation between gar vertebrae.

The alligator gar, *Lepisosteus spatula*, is the largest member of the family and one of the largest freshwater fishes in North America. It attains a length of 3 m and a weight of 140 kg (Suttkus 1963). Although most gars are considered water-column predators on other fishes and often hover just below the surface, alligator gar also feed extensively on bottom-dwelling fishes and invertebrates, and scavenge on benthic food (Seidensticker 1987). Alligator gar, as well as other species, frequently enter estuarine regions (Suttkus 1963).

Comparatively little is known about the life history and general biology of gars. This is unfortunate, because they are ecologically important in many fish assemblages, often becoming quite abundant in rivers and backwaters. Gars are additionally interesting in that they are the only freshwater fishes in North America with toxic eggs. The eggs are distinctly green in color and can cause sickness and even death when eaten by chickens and mice. However, the possible ecological function of this toxicity, and whether it actually affects fish or invertebrates that might feed on the eggs, remain undetermined (Netsch and Witt 1962).

Nine fossil species of gars are recognized, dating back to the Lower Cretaceous of North America, Europe, Africa, and India, and indicating a widespread Pangean distribution (Wiley 1976).

FIGURE 13.15.   *A cladistic description of phylogenetic relationships among the higher bony fishes. Cladistians (brachiopterygians) form a distinctive, primitive sister group to the Actinopteri (- chondrosteans and neopterygians). Cladistians and actinopterans share 14 primitive anatomic traits, whereas chondrosteans and neopterygians share 9 additional derived traits not possessed by cladistians.*

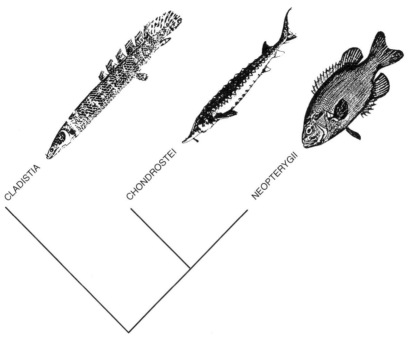

CLADISTIA   CHONDROSTEI   NEOPTERYGII

From Patterson 1982; used with permission.

Order Amiiformes: The Bowfin

The bowfin, *Amia calva,* is generally considered more derived than the gars (Fig. 13.17). It retains the abbreviate heterocercal tail and rudimentary spiral valve intestine but has teleost-like amphicoelous vertebrae as well as cycloid scales, a scale type in which the ganoid and dentine layers have been lost, leaving only a reduced bony layer. The bowfin's cycloid scales are much like those found in teleosts (see Jarvik 1980). The bowfin's head is exceptionally bony, invested in massive dermal bones that are greatly reduced in teleosts. The bowfin is distinct among all living fishes in possessing a **single, median gular plate** on the underside of the head. It is the only nonteleostean fish to swim via undulations of its long dorsal fin, which allows it to move slowly both forward and backward with stealth. Rapid swimming is accomplished by more conventional body and tail movements (Scott and Crossman 1973; Becker 1983).

The bowfin is widely distributed throughout much of the eastern half of North America from southern Quebec and Ontario to eastern Texas. It is most common in vegetated lakes and backwater areas of large rivers, occupying deeper waters by day and moving into shallows at night to feed. It has numerous, caninelike and peg-shaped teeth, strong jaw musculature, large size (to 1 m and 9 kg), and opportunistic, predatory habits. Bowfin feed on invertebrates, fishes, frogs, turtles, snakes, and small mammals, which they engulf via suction, whereas gar impale food on their small, sharp teeth (Lauder 1980).

FIGURE 13.16.   *A Florida gar,* Lepisosteus platyrhincus, *showing the distinctive elongate, tooth-studded snout and posteriorly placed dorsal and anal fins characteristic of this family of North and Central American predators.*

From Suttkus 1963; used with permission.

FIGURE 13.17.   *(A) The bowfin,* Amia calva, *member of a monotypic order endemic to North America. Note the elongate dorsal fin that is used in slow forward and backward locomotion. (B) Neurocranium and axial skeleton of a bowfin. Note upward flexion of the vertebral column in the caudal region, producing the abbreviate heterocercal tail.*

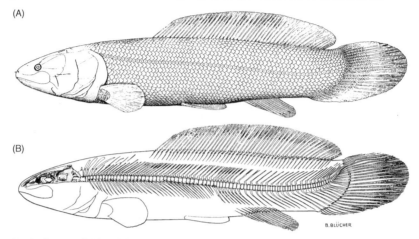

(A)

(B)

From Jarvik 1980; used with permission.

Bowfin males build nests in shallow water by clearing a circular depression on the bottom about 0.5 m across. The males also engage in parental care, guarding the young vigorously until they are relatively large (10 cm). The male has a distinct black spot at the base of its caudal fin; such nonseasonal sexual dimorphism does not occur in other living primitive bony fishes, although males and females differ in many teleosts.

Amia is incapable of surviving in warm, deoxygenated water without access to atmospheric oxygen. As in gars,

Table 13.1.   Characteristics of extant relict fishes. Presence (1) or absence (2) of a trait, or its condition, is indicated in the body of the table. Shared characteristics among unrelated forms are strong evidence of convergent evolution, since these groups have long histories of demonstrated, separate evolution.

	Lungfishes			Chondrosteans				
Trait	Australian	South American/African	Coelacanth	Sturgeons	Paddlefishes	Gars	Bowfin	Polypterids
Scales	cycloid	cycloid	cycloid[a]	scutes[b]	–[c]	ganoid	cycloid	ganoid
Gular plates	–	–	–	–	–	–	+	–
Spiracle	–	–	–	+	+?	–	–	+
Larva ext gills	–	+	–	–	–	–	–	+
Lungs[d]	sing vent	dbl vent	fat-filled gb	dorsal gb	dorsal gb	vasc gb	vasc gb	dbl vent
Spiral valve	+	+	+	+	+	+ (remnant)	1 (remnant)	+
Centra	–	–	–	–	–	+[e]	+	+
Tail	diphy	diphy	diphy	hetero	hetero	abb hetero	abb hetero	hetero[f]
Lobed fins	+	–	+[g]	–	–	–	–	+
Electroreceptors	+	+	+	+	+	–	–	+
Chromosomes								
2n	32–38	38/34	"60"[h]	112	120	68	46	36
micro	0	0	—	48	72	26	0	0
longest	30	30	—	4	5	3	3	12

abb = abbreviate; dbl = double; diphy = diphycercal; ext = external; hetero = heterocercal; micro = microchromosomes; gb = gas bladder; vasc = vascularized; vent = ventral.
[a]Coelacanths are often erroneously reported as having cosmoid scales; no extant fishes have scales containing cosmine (Jarvik 1980).
[b]Sturgeon have five longitudinal rows of bony scutes, plus "dermal ossifications" scattered around the body (Vladykov and Greeley 1963, 25).
[c]Paddlefish are mostly naked, with four types of scales (fulcral, rhomboid, round-based, and denticular) scattered on the head, trunk, and tail; the histology of these scales is unclear. Trunk scales are more abundant on *Psephurus* than on *Polyodon* (Grande and Bemis 1991).
[d]Outpocketings of the esophagus are gas bladders but are often called lungs when their primary function is breathing atmospheric air.
[e]Gar centra are opisthocoelous (concave on the rear face, convex on the front).
[f]The lower lobe of the brachiopterygian tail is created by rays coming off the the ventral surface of the notochord.
[g]Coelacanth fin bases are lobed except for the first dorsal.
[h]Chromosomal characteristics of coelacanths remain undetermined; the chromosome number has been estimated based on DNA/cell content and presumed phylogenetic trends (Dingerkus 1979).

bowfin gulp air, which is passed to a highly vascularized gas bladder. Some controversy has developed over whether bowfin are capable of lungfishlike estivation in drying conditions. Anecdotal evidence suggests that bowfin can bury in mud and survive for periods of weeks (e.g., Green 1966), whereas experimental laboratory findings suggest that bowfin are physiologically incapable of surviving more than 3 to 5 days of air exposure (McKenzie and Randall 1990). Definitive field manipulations have yet to be performed.

Amiiform fishes have been distinct since the early Triassic, and the genus *Amia* dates back at least to the late Cretaceous. During the Mesozoic, amiids were quite diverse, comprising 11 genera, occurring in North and South America, Europe, Asia, and Africa. One Eocene giant, *Maliamia*, attained a length of 3.5 m (Patterson and Longbottom 1989).

CONCLUSIONS

Trends in the characteristics of the living members of ancient groups, and comparisons with the recently successful teleosts, raise a number of intriguing questions. As anatomically and taxonomically diverse as these relict fishes are, certain convergent similarities in morphology, behavior, and ecology suggest interesting evolutionary patterns that may have characterized the evolution of major fish groups (Table 13.1). The success of extant lungfishes, gars, the bowfin, and the enigmatic bichirs in swampy, seasonally evaporating, tropical or semitropical environments underscores the question of evolutionary succession among major fish lineages. Are we dealing here with relegation of these remnant, competitively inferior groups to marginal habitats, or are we faced instead with the continued superiority of ancient groups in the habitats where they originally evolved and in which they had an evolutionary head start? What explains retention or independent evolution of the spiral valve in most of these primitive groups and the Chondrichthyes, but its replacement in higher bony fishes with a linear intestine? The same can be asked about electroreception. Why is it retained in primitive groups (except *Amia* and the gars) but lost in most modern higher taxa, except for a few that have independently re-evolved and elaborated the electrical sense (see Chap. 6)?

Are the trends that characterize fish evolution in general (see Table 11.1)—such as reduction in bony armor, development of the pipette mouth and pharyngeal dentition, elaboration of the dorsal fin, relocation of pelvic and pectoral girdles, an increasingly symmetric caudal fin—necessarily improvements on the primitive design? If so, how have the relict species managed to hold on in the face of what should be superior competition and predation by the "more" successful, improved teleosts? And finally, why have these few species, among the thousands of ancestral species and their derivatives, survived so long while their relatives succumbed to the ultimate fate of all organisms?

SUMMARY

1. Teleosts are the most successful fishes alive today, but a few highly derived species of several primitive groups represent the successful fishes of the past. These are the lancelets, hagfishes, lampreys, coelacanth, lungfishes, sturgeons, paddlefishes, bichirs, gars, and bowfin.

2. Cephalochordate lancelets are arguably fishes that lack most chordate structures. They are filter-feeding bottom dwellers. Lampreys and hagfishes are jawless fishes that are probably convergently similar. Differences in mouth position, tooth and tongue morphology, embryology, pineal complex, and gill structure suggest separate ancestries. Hagfishes are entirely marine, high-latitude predators and scavengers that lack larvae; they produce copious slime and can tie themselves into knots. Commercial "eelskin" comes from hagfishes. Lampreys are primarily freshwater, temperate, parasitic fishes with complex life cycles. Numerous nonparasitic species have evolved from parasitic ancestors.

3. Coelacanths were thought to have gone extinct 65 mya, until a live one was captured in 1938 off South Africa. Today, only a small, endangered population of 200 to 500 fish exists in the Comoro Islands. The living coelacanth is very much like its Paleozoic ancestors, with lobe fins, diphycercal tail, hollow spines, a specialized notochord, jointed skull, its young born alive, and tetrapod-like locomotion.

4. Living lungfishes are a small subset of a widely distributed, diverse Paleozoic and Mesozoic subclass. The Australian lungfish is most like earlier species; the South American and African species are highly derived in many respects. Lungfishes lack jaw teeth but have unusual tooth plates on the mouth roof and floor. African lungfishes can estivate in dried mud for up to 4 years.

5. The most primitive actinopterygian fishes are the highly derived, relict, chondrostean sturgeons and paddlefishes. They share many traits (cartilaginous skeleton, heterocercal tail, few scales, numerous fin supports, unique jaw suspension) but differ in most respects. Sturgeons are large, freshwater and anadromous, long-lived fishes of North America, Europe, and Asia that are highly prized for their eggs (caviar) and have been heavily overfished. Two species of paddlefishes occur in large rivers of North America and China. Paddlefishes have a long snout that may be used to detect weak electrical fields.

6. The bichirs and ropefish of Africa have been variously placed with the lungfishes, lobefins, and rayfins because they have larvae with gills, lobelike fins, ganoid scales, and a modified heterocercal tail. But they have uniquely constructed median, caudal, and paired fins and an unusual chromosomal arrangement, causing some taxonomists to place them in their own subclass, the Cladistia or Brachiopterygia.

7. Two living orders represent close ancestors of teleosts. The lepisosteiform gars are predaceous fishes that occur in North and Central America, where they occupy backwaters and swamps. They breath atmospheric oxygen via a highly vascularized gas bladder. Unusual traits include interlocking ganoid scales, opisthocoelous vertebral centra (convex anteriorly, concave posteriorly), an abbreviate heterocercal tail, and poisonous eggs. Closer to the teleosts is the monotypic bowfin (Amiiformes). Bowfin are restricted to eastern North America. They can also breath atmospheric air and are predaceous. Bowfin have cycloid scales, biconcave vertebra, a large gular plate, and an elongate dorsal fin; males guard the young for an extended period.

SUPPLEMENTAL READING

Balon et al. 1988; Bemis et al. 1987; Bigelow and Perez—Farfante 1948; Binkowski and Doroshov 1985; Birstein and Bemis 1995; Brodal and Fange 1963; Dillard et al. 1986; Foreman et al. 1985; Hardisty 1979; Hardisty and Potter 1971–1982; Jarvik 1980; McCosker and Lagios 1979; Musick et al. 1991; Smith et al. 1980; Smith 1956; Thomson 1969b, 1991.

14

Teleosts at Last
I: Bonytongues Through
Anglerfishes

By far the dominant living fishes are members of the division Teleostei. The name teleost means roughly "perfect bone," referring to their evolutionary position as the most advanced of the living bony fishes. Bone mass in teleosts is reduced from the preteleostean condition, but internal cross struts in the bone give it exceptional strength without great mass. Teleosts account for 96% of all living fishes, constituting the world's major fisheries. They inhabit the widest range of habitat types and show the greatest variation in body plans and foraging and reproductive habits of any fishes. By comparison, all the more primitive extant groups introduced in Chapters 12 and 13 are carnivorous and occur in a limited number of habitats. Elasmobranchs are 99% marine, whereas lungfishes, gars, bowfin, and sturgeons are largely big river or swamp dwellers. Teleosts in contrast occur in every imaginable fresh- and marine water habitat, from ocean trenches to high mountain lakes and streams, from polar oceans at −2°C to alkaline hot springs at 41°C, from torrential rivers and wave-tossed coastlines to stagnant pools.

There are flying, walking, and immobile teleosts, and annual teleosts that emerge from resting eggs when it rains and then breed and die. There are teleosts that brood their eggs and young in their mouths, others that lay eggs inside clams, and some that jump out of the water to lay eggs on the undersides of terrestrial plant leaves and periodically splash them to keep them moist. Trophically, teleosts feed on other fishes, carrion, invertebrates, mammals (including man), scales, eyes, eggs, and zooplankton. Importantly, teleosts are the only group of fishes that utilize plant material in all its forms, including phytoplankton, cyanobacteria, algae, detritus, and vascular plants and their seeds. The only truly endoparasitic vertebrates are teleosts. Some teleosts produce either light or electricity. Teleosts are the most diverse and diversified taxon of all the vertebrates, having radiated into more niches and adaptive zones than all the other vertebrate groups combined.

It is obvious that detailed information cannot be given on even a subset of the 23,600 living teleostean species. Our objectives in this chapter are to provide a feeling for 1) what characterizes a teleost and separates it from the more primitive fishes discussed earlier, 2) which characteristics separate different taxa within the teleosts and represent evolutionary advances within the division, 3) which groups have been successful in what regions and habitat types, and 4) what are some of the more interesting species and adaptations in this exceptionally successful group.

Our focus is on living fishes, but it should be recalled that teleosts have existed since the Mesozoic and that the taxonomy of many of these groups is strongly influenced by characteristics of relatives known only from fossils.

TELEOSTEAN PHYLOGENY

Teleosts first arose in the early Mesozoic (probably mid-Triassic, ca. 200 million years ago) from a neopterygian ancestor, possibly a pachycormiform (see Chap. 11, Subclass Neopterygii). The earliest teleosts were probably pholidophorids or leptolepids, groups that consisted of several families and that may have been ancestral to more than one of the main lineages of teleosts, including the elopomorphs and osteoglossomorphs. The important point to remember, reiterating the phylogenetic account given in Chapter 11, is that modern teleosts arose during four major radiations that produced the subdivisions **Osteoglossomorpha**, **Elopomorpha**, **Clupeomorpha**, and **Euteleostei**, the latter being by far the largest.

A listing of teleostean families is unavoidable, in part to appreciate their tremendous diversity but also because most fishes encountered anywhere in the world will belong to one of the 38 orders and 426 families (and 4064 genera) of teleosts. Despite their amazing diversity, teleosts share a number of characters that indicate common ancestry, particularly in the more advanced subdivi-

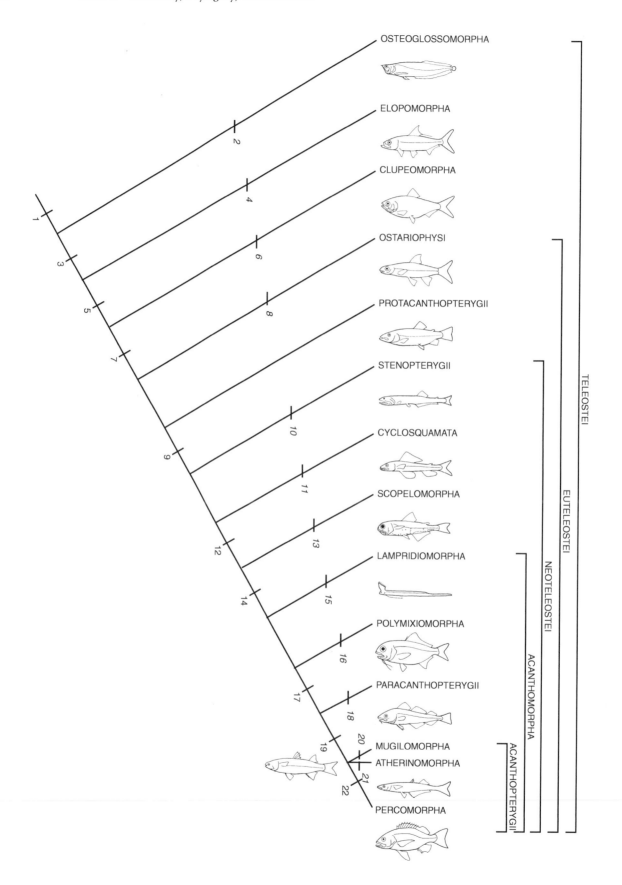

sion of the euteleosts, or "true teleosts" (see below). The primary shared derived (synapomorphic) characters that unite the teleosts involve numerous bones of the tail and skull (Fig. 14.1). Importantly, the **ural neural arches** of the tail are elongated into **uroneural bones**. This means that in the tail base region, the neural arches that sit dorsal to the vertebral column fuse into elongate bones termed uroneurals. These new bones serve as basal supports for the rays that form the upper lobe of the tail fin and thus help stiffen it; their number and shape change during teleostean phylogeny. In the skull, among other characters, teleosts have a mobile premaxillary bone rather than having the premaxilla fused to the braincase. A mobile premaxilla is essential for upper jaw protrusion and allows a fish to shoot its mouth forward during prey capture, creating suction pressures and also overtaking prey.

In sum, major changes that define the teleosts contributed to the advances in locomotion and feeding that apparently led to their success, as detailed in Chapter 11. Most of the characteristics described here are discussed in more detail in Wheeler (1975), Berra (1981), Carroll (1988), and Nelson (1994). The overall classification followed, many of the characteristics described, and the numbers of species provided for different orders are based on Nelson (1994).

A SURVEY OF LIVING TELEOSTEAN FISHES

Subdivision Osteoglossomorpha

Osteoglossiformes (217 species): Osteoglossidae (bonytongues), Pantodontidae (butterflyfish or African flyingfish), Hiodontidae (mooneyes), Notopteridae (featherfin knifefishes), Mormyridae (elephantfishes), Gymnarchidae

The **osteoglossiform** bonytongues and their relatives are generally considered the most primitive living teleosts (Fig. 14.2). They occur in freshwater on all major continents except Europe, although only Africa has more than a few species (see Chap. 16, Archaic Distributions). Although chiefly a tropical group, two species (the mooneye, *Hiodon tergisus*, and goldeye, *H. alosoides*) occur in major river systems of northern North America. The arapaima or pirarucu of South America (*Arapaima gigas*) is one of the world's largest freshwater fishes, reaching a length of 2.5 m. Anatomically, the order gets its name from well-developed teeth on the tongue, which occlude (bite) against similarly toothed bones (parasphenoid, mesopterygoid, ectopterygoid) in the roof of the mouth.

The South American arowana (*Osteoglossum bicirrhosum*), the African flyingfish (*Pantodon buchholzi*), the notopterid featherfins or Old World knifefishes, and the mormyrid elephantfishes are popular aquarium species. Mormyrids, the most speciose family in the subdivision with 200 species, and the related *Gymnarchus niloticus* possess a highly evolved electric sense that involves both the production and detection of weak electrical fields, an appropriate sense for fishes that are nocturnally active and typically occur in turbid waters. The electric sense is used to localize objects and is also important during social interactions (see Chap. 6, Electroreception; Chap. 21, Electric Communication). Mormyrids have the largest cerebellum of any fish and a brain size-body weight ratio comparable to that of humans; the mormyrocerebellum is the neural center for coordinating electrical input. Mormyrids have a large learning capacity and are purported to engage in play behavior, a rarity among fishes.

◀ **FIGURE 14.1.**   *Phylogenetic relationships among living teleosts. The numbered characteristics defining the branching points (synapomorphies) are selected from a much larger list; groups to the right of a branch point share the traits (although traits may be secondarily lost); groups to the left do not share the trait. Italicized numbers are unique derived traits (autapomorphies) particular to a group and not shared by other taxa. Characteristics 1 through 7 repeat those found in the cladogram of Figure 11.13. Monophyly and definition of some groups are a matter of debate, and no synapomorphies are given; for example, the Protacanthopterygii (esociforms pikes, osmeriform smelts, salmoniform salmons) remain a problematic group that lacks well-defined, unifying characteristics. Additional details can be found in Lauder and Liem 1983; Pough et al. 1989; Nelson 1994; and papers cited in those publications. 1) Mobile premaxilla, posterior neural arches (uroneurals) elongate, ventral pharyngeal tooth plates unpaired; particular combination of skull bones present (basihyal, four pharyngobranchials, three hypobranchials); 2) tooth plate on tongue bites against roof of mouth, intestine lies to the left of stomach; 3) two uroneural bones extend over the second tail centrum, epipleural intermuscular bones abundant in abdominal and caudal region; 4) ribbon-shaped (leptocephalus) larva; 5) neural arch of first tail vertebra reduced or missing, upper pharyngeal jaws fused to gill arch elements, jaw joint with unique articulation and ossification; 6) specialized ear-to-gas-bladder connection; 7) dorsal adipose fin and nuptial tubercles on head and body; first uroneural bones of tail have paired anterior membranous outgrowth; 8) anterior vertebrae and ribs modified to connect gas bladder to inner ear (Weberian apparatus), epidermal cells produce alarm substance; 9) first vertebra articulates with three bones of the skull (basioccipital and the two exoccipitals), retractor dorsalis muscle connects vertebral column with upper pharyngeal jaws, hinged jaw teeth capable of depression posteriorly; 10) unique photophore histology and tooth attachment; 11) unique gill arch structure involving second and third pharyngobranchials; 12) fifth upper pharyngeal tooth plate and associated internal levator muscle missing; 13) upper pharyngeal jaw dominated by third pharyngobranchial; 14) configuration of rostral cartilage and its ligamentous connection to premaxilla; lateral ethmoids joined to vomer; 15) uniquely protrusible upper jaw; 16) ligament connecting palatine and premaxilla in a unique position; 17) expanded premaxillary processes; 18) dorsal (neural) spine attached to second preural vertebra; 19) branchial retractor muscle (retractor dorsalis) inserts only on third pharyngobranchial, well-developed ascending process of premaxilla allows increased jaw mobility, ligament supporting pectoral girdle (Baudelot's ligament) originates on basioccipital of skull rather than on first vertebra; 20) no direct connection between pelvic girdle and cleithrum of pectoral girdle; 21) jaw protrusion occurs without ball-and-socket joint between palatine and maxilla, fourth pharyngobranchial lost; 22) pelvic girdle attached to pectoral girdle, anterior pelvic process displaced ventrally, pelvic fins have one spine and five soft rays. Additional characteristics are given in this and the following chapter.*

FIGURE 14.2.   *Old World osteoglossomorphs. (A) A mormyrid elephantfish,* Gnathonemus petersi, *from Africa. (B) A notopterid featherfin or knifefish,* Chitala chitala, *from Asia.*

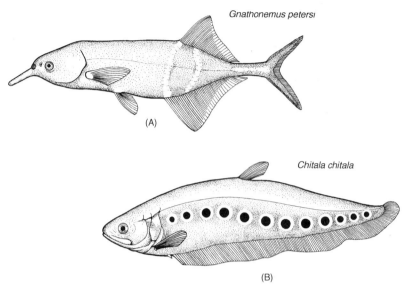

Gnathonemus petersi

(A)

Chitala chitala

(B)

After Paxton and Eschmeyer 1995.

Subdivision Elopomorpha

Elopiformes (8 species): Elopidae (tenpounders, ladyfishes), Megalopidae (tarpons)

Albuliformes (29 species): Albulidae (bonefishes), Halosauridae (halosaurs), Notacanthidae (spiny eels)

Anguilliformes (738 species): Anguillidae (freshwater eels), Heterenchylidae, Moringuidae (spaghetti eels), Chlopsidae (false morays), Myrocongridae, Muraenidae (moray eels), Synaphobranchidae (cutthroat eels), Ophichthidae (snake eels, worm eels), Colocongridae, Derichthyidae (longneck eels), Muraenesocidae (pike congers), Nemichthyidae (snipe eels), Congridae (conger eels), Nettastomatidae (duckbill eels), Serrivomeridae (sawtooth eels)

Saccopharyngiformes (26 species): Cyematidae (bobtail snipe eels), Saccopharyngidae (swallowers), Eurypharyngidae (gulpers, pelican eels), Monognathidae

A distinct pelagic larval form, termed a **leptocephalus**, unites this speciose marine group (Fig. 14.3). Leptocephali are typically in the shape of a willow leaf or ribbon, and many of them shrink during metamorphosis to the juvenile form. For many years, the correspondence between larval and adult species was not made, and hence the two life history stages were placed in very different taxa (see Chap. 9, Larval Morphology and Taxonomy). The leptocephali of elopiform tarpons and bonefishes have a forked tail, whereas eel larvae have a pointed tail. Leptocephali are exceedingly long-lived, remaining as larvae for as long as 2 to 3 years in some anguillid species (see Chap. 22, Catadromy). During this time, they disperse over large oceanic expanses, feeding perhaps on dissolved organic matter that they absorb through their skin. They are thin and fragile in appearance, this effect being heightened by a lack of red blood

cells, which makes them translucent. Elopomorphs are also distinguished by a reduction in the number of uroneural bones in the tail as compared to osteoglossomorphs and by the development of thin, riblike epipleural intermuscular bones that extend from the vertebral column into the surrounding trunk musculature. These are the small bones in the meat of primitive teleosts (conger eels, herrings, carps, trouts) that make them difficult to fillet and eat. Long epipleural and epineural ribs become less common in higher teleosts such as paracanthopterygians and acanthopterygians, which rely more on stouter, more firmly attached zygapophyses. These differences make both cleaning and eating easier.

The **elopiform** ladyfishes and tarpons retain a primitive characteristic, namely, a gular bone or splint on the underside of the throat; this structure is well developed in the more primitive bowfin, coelacanth, and bichirs but is lost in all other teleosts (except perhaps for an anabantoid, the pikehead). Other justifications for considering elopiforms as primitive teleosts include 1) a large number of branchiostegal rays in the throat (10 to 35 vs. 5 to 7 in many higher teleosts); 2) inclusion of the maxilla in the gape, giving them two biting bones in the upper jaw rather than one; and 3) heavy, bony scales that contain ganoin, a bone layer otherwise only found in gars and bichirs. The Atlantic tarpon, *Megalops atlanticus*, is a legendary game fish that reaches a length of 2.5 m and a mass of 150 kg. A large (65 kg) female may contain more than 12 million eggs, making tarpon one of the most fecund fishes. **Albuliform** bonefishes are also popular game fishes that occupy sandy flats in shallow tropical waters. The notacanthoids (halosaurs and spiny eels) are an offshoot suborder between the

FIGURE 14.3.   *Elopomorphs. (A) A tarpon,* Megalops atlanticus. *(B) A 7-cm-long leptocephalus larva of a ladyfish,* Elops saurus.

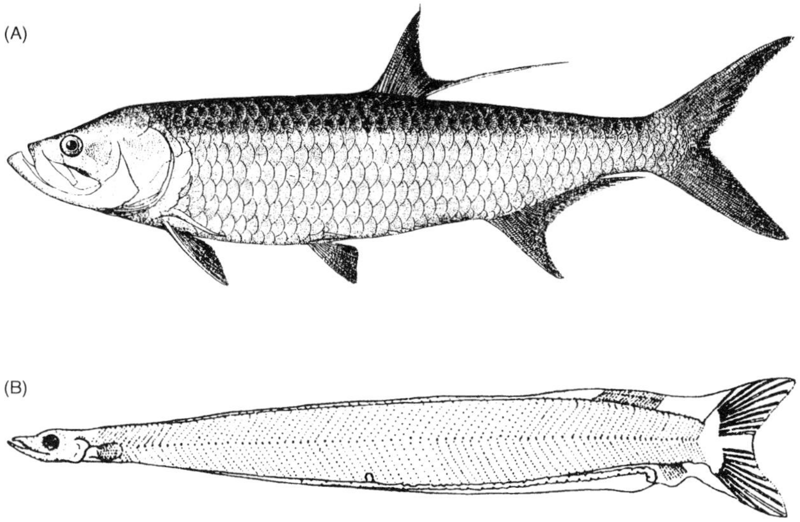

(A)

(B)

From Hildebrand 1963.

albuliforms and the anguilliforms; they develop from leptocephalus larvae but otherwise stand in marked anatomic and ecological contrast to other members of the albuliforms. These deep-sea, benthic eels occur as deep as 5000 m, making them among the deepest-living fishes known.

The 15 families of **anguilliforms** are "true" eels (i.e., those with a leptocephalus larva), as distinguished from the approximately 45 other families of eel-like fishes that have converged on an elongate body and other anatomic and behavioral traits (see Box 18.2; Chap. 23, Habitat Use and Choice). An eel-like body facilitates forward and backward movement into and out of tight places and sediments. Some anguilliforms are open-water, pelagic forms, despite the relatively slow locomotion imposed by an anguilliform swimming mode (see Chap. 8, Locomotory Types). Anguilliforms are distinguished by loss of the pelvic girdle and by a modified upper jaw that is formed by fusion of the premaxilla, vomer, and ethmoid bones.

The 15 species of anguillid eels are catadromous, spawning at sea but spending most of their lives feeding and growing in freshwater. Muraenid moray eels and their relatives (218 species) are largely marine, tropical and warm-temperate species best known from coral reefs. Their sinister appearance results in part because their sedentary habits require them to hold their fang-studded jaws open while actively pumping water over their gills and out a constricted opercular opening. Although capable of inflicting serious wounds, morays are more dangerous as agents of ciguatera food poisoning, a toxin that originates in a dinoflagellate alga and is magnified in piscivores that eat prey contaminated with the toxin.

The congroid eels (491 species) include 1) fossorial (burrowing) forms such as garden eels, worm eels, and snake eels (the latter burrow into sediments backward with a hardened, pointed tail); 2) deep-sea mesopelagic and bathypelagic families, including longneck eels, snipe eels, and sawtooth eels; and 3) benthic conger eels, which are very similar ecologically to morays. One family of congroids, the synaphobranchid cutthroat eels, contains a facultative parasitic species, the snubnose parasitic eel, *Simenchelys parasiticus*. Although often a scavenger, *Simenchelys* sometimes burrows into the flesh of bottom-living fishes such as halibut. Two 20-cm-long individuals were found lodged in the heart of a longline-captured, 500-kg shortfin mako shark, where they had been feeding on blood. Histologic features of the inhabited heart suggested that these eels had possibly been living in the shark's heart prior to its capture, pointing to a truly parasitic relationship (Caira et al., unpublished manuscript).

The last order of elopomorphs are the truly bizarre **saccopharyngiform** deep-sea gulper and swallower eels and their relatives (Fig. 14.4) (see Chap. 17, The Deep Sea). These species are distinguished not only by elaborate, extreme specializations of the head and tail, including an extremely long jaw, but also for a lack of features normally found in teleosts. Among the structures missing from different species are the symplectic and opercular bones, branchiostegals, maxilla and premaxilla, vomer and parasphenoid, scales, pelvic or pectoral fins, ribs, pyloric caeca, and gas bladder. Some early authors argued that saccopharyngoids were not really bony fishes. Nelson (1994, 115) considered the saccopharyngoids "perhaps the most anatomically modified of all vertebrate species." Another saccopharyngoid family, the monognathids, contains species with rostral fangs and apparent venom glands, a unique feature among fishes (Bertelsen and Nielsen 1987).

FIGURE 14.4. *A gulper or pelican eel,* Eurypharynx pelecanoides. *Ironically, these highly specialized, 40-cm-long bathypelagic fish feed on surprisingly small prey, which they capture by opening their huge, dark mouths although it probably generates little suction pressure. The related swallower eels feed on prey larger than themselves. Gulper eels are unique among teleosts because they have five gill arches and six visceral clefts (Nelson 1994).*

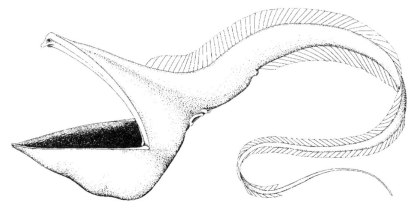

From Briggs 1974; used with permission of McGraw-Hill, Inc.

Subdivision Clupeomorpha

Clupeiformes (357 species): Denticipitidae (denticle herring), Engraulidae (anchovies), Pristigasteridae, Chirocentridae (wolf herring), Clupeidae (herrings)

Among the most abundant and commercially important of the world's fishes are the herringlike **clupeiforms**. Almost all are open-water, pelagic, schooling forms, four-fifths of which are marine. Clupeomorphs are distinguished by a gas bladder that extends anteriorly up into the braincase and contacts the utriculus of the inner ear and in some extends posteriorly to the anus; the air bladder also has extensions to the lateral line canals. This **otophysic** (ear-to-gas-bladder) condition apparently increases the hearing ability of these fishes by increasing their sensitivity to low-frequency (1 to 1000 Hz) sounds as compared to other fishes. Low-frequency (3 to 20 Hz) sounds that they detect are typically those produced by tail beats of other fishes, such as neighbors in a school and attacking predators (Blaxter and Hunter 1982). Clupeomorphs also typically possess a series of sharp, bony scutes along their ventral edge, and some also have scutes anterior to the dorsal fin. These scutes may make these fishes harder for predators to capture and swallow, although direct proof is lacking.

Of phylogenetic significance, the subdivision Clupeomorpha (modern clupeiforms and extinct, related orders) possess evolutionary advances over elopomorphs in terms of a modified joint at the posterior angle of the jaw (angular fused to articular rather than to retroarticular) and caudal skeleton reduction (reduced first ural centrum and reduction to six in the number of hypural bones). These derived traits foreshadow the continued changes in jaw and tail structures that occur during the evolution of higher teleostean groups.

The anchovies are relatively elongate zooplanktivorous clupeoids with large mouths made possible by an elongate maxillary bone that extends considerably behind the eye. Anchovies range in size from a minute Brazilian species (*Amazonsprattus*, 2 cm) to a piscivorous riverine New Guinea anchovy (*Thryssa scratchleyi*, 37 cm). The largest clupeids are the chirocentrid wolf herrings, *Chirocentris dorab* and *C. nudus*, with an Indo-Pacific to South Africa distribution. Wolf herrings are herrings gone mad. They reach a length of 1 m (the next largest clupeoid is a 60-cm Indian clupeid) and have fanglike jaw teeth plus smaller teeth on the tongue and palate, which they use to capture other fishes. The largest family in the subdivision is the Clupeidae, which includes 181 species of herrings, round herrings, shads, alewives, sprats, sardines, pilchards, and menhadens. Clupeids can be marine, freshwater, or anadromous, with landlocked forms common (e.g., anadromous shads, alewives, and herrings have become established in lakes and reservoirs and in rivers trapped between dams). Probably the best-known fish fossil in the world is *Knightia*, a freshwater Eocene herring from the Green River shale formations of Wyoming (Fig. 14.5).

SUBDIVISION EUTELEOSTEI

All the teleosts above the level of the clupeomorphs are often placed together in the subdivision **Euteleostei**, the "true" teleosts. This designation underscores the conclusion that teleostean phylogeny involved four major radiations, the first three producing relatively primitive and separate groups (osteoglossomorphs, elopomorphs, clupeomorphs) and the fourth containing a vast (391 families, 3795 genera, 22,262 species) assemblage of more advanced, euteleostean fishes.

The monophyly of the euteleosts remains an area of debate, because unique shared derived characters common to all or most members are lacking. One feature common

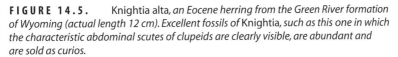

FIGURE 14.5. Knightia alta, *an Eocene herring from the Green River formation of Wyoming (actual length 12 cm). Excellent fossils of* Knightia, *such as this one in which the characteristic abdominal scutes of clupeids are clearly visible, are abundant and are sold as curios.*

to many euteleosts, itself an enigma, is the presence of an **adipose fin**. This small, rayless fin sits posterior to the first dorsal, often just anterior to the caudal peduncle. It can be a small vertical flap or bump, or it can be a long, substantial structure confluent with the caudal fin, as in many catfishes. It occurs commonly among the characiforms (characins and relatives); catfishes; smelts; the deep-sea stomiiforms, aulopiforms, and scopelomorphs; the salmoniforms (salmons, trouts, whitefishes); and troutperches. It is lacking from higher fishes, particularly those with a spiny first dorsal. Its functions are a mystery. Another characteristic shared by many but not most euteleosts is the presence of epidermal **breeding turbercles** and dermal **contact organs**, small bumps on the fins and bodies that develop during the breeding season in 25 families (Collette 1977; see Chap. 20, Sexual Selection, Dimorphism, and Mate Choice). Most primitive euteleosts (and some elopomorphs) also possess paired stegural bones, which are membranous outgrowths associated with the neural arches of the vertebral centra of the tail base.

Superorder Ostariophysi

Series Anotophysi
Gonorhynchiformes (35 species): Chanidae (milkfish), Gonorhynchidae (beaked sandfishes), Kneriidae, Phractolaemidae (snake mudhead)
Series Otophysi
Cypriniformes (2660 species): Cyprinidae (minnows, barbs, carps), Gyrinocheilidae (algae eaters), Catostomidae (suckers), Cobitidae (loaches), Balitoridae (river loaches)
Characiformes (1340 species): Citharinidae, Hemiodontidae, Curimatidae, Anostomidae, Erythrinidae (trahiras), Lebiasinidae, Ctenoluciidae (pike characids), Hepsetidae, Gasteropelecidae (freshwater hatchetfishes), Characidae (characins)

Siluriformes (2405 species): Diplomystidae, Ictaluridae (North American freshwater catfishes), Bagridae, Olyridae, Cranoglanididae (armorhead catfishes), Siluridae (sheatfishes), Schilbeidae, Pangasiidae, Amphiliidae (loach catfishes), Sisoridae, Amblycipitidae (torrent catfishes), Akysidae (stream catfishes), Parakysidae, Chacidae (squarehead, angler, or frogmouth catfishes), Clariidae (air-breathing catfishes), Heteropneustidae (airsac catfishes), Malapteruridae (electric catfishes), Ariidae (sea catfishes), Plotosidae (eeltail catfishes), Mochokidae (squeakers, upside-down catfishes), Doradidae (thorny catfishes), Ageneiosidae (bottlenose or barbelless catfishes), Auchenipteridae (driftwood catfishes), Pimelodidae (long-whiskered catfishes), Cetopsidae (whalelike catfishes), Helogeneidae, Hypophthalmidae (lookdown or loweye catfishes) Aspredinidae (banjo catfishes), Nematogenyidae, Trichomycteridae (pencil or parasitic catfishes), Callichthyidae (callichthyid armored catfishes), Scoloplacidae (spiny dwarf catfishes), Loricariidae (suckermouth armored catfishes), Astroblepidae (climbing catfishes)
Gymnotiformes (62 species): Sternopygidae (glass knifefishes), Rhamphichthyidae (sand knifefishes), Hypopomidae, Apteronotidae (ghost knifefishes), Gymnotidae (naked-back knifefishes), Electrophoridae (electric eel or electric knifefishes)

Freshwater habitats worldwide are dominated in terms of numbers of both species and individuals by ostariophysans, which account for about 64% of all freshwater species. Ostariophysans include such disparate taxa as milkfish, minnows, carps, barbs, suckers, loaches, piranhas, tetras, catfishes, and electric eels. However, two unique traits characterize most members of this massive taxon.

1. With the exception of the gonorhynchiforms, ostariophysans possess a unique series of bones that connect the gas bladder with the inner ear (gonorhynchiforms are sometimes considered ano-

tophysans, the remainder of the superorder being referred to as otophysans).

2. All ostariophysans produce and respond to chemical alarm substances that signal the injury of a conspecific.

The former characteristic, termed the **Weberian apparatus**, involves a set of bones derived from the four or five anterior (cervical) vertebrae and their neural arches, ribs, ligaments, and muscles (see Fig. 6.3). The group gets its name from this complex structure (*ostar* = "small bone," *physa* = "bladder"). When sound waves contact the fish, the gas bladder vibrates, and this vibration is passed anteriorly to the inner ear, being amplified by the intervening Weberian ossicles (see Chap. 6, Hearing). Presumably, improved hearing also occurs in unrelated taxa that have convergently evolved connections between the gas bladder and the inner ear, either by an otophysic extension of the gas bladder anteriorly (elephantfishes, clupeoids, cods, roosterfish, porgies, some cichlids) or by a bony connection involving the pectoral girdle or skull (squirrelfishes, triggerfishes).

The second shared derived trait that helps define the Ostariophysi is the **alarm response**, which involves 1) production of an alarm substance (*Schreckstoff*) and 2) a behavioral reaction to the presence of the substance in the water (*Schreckreaktion*) (see Chap. 19, Discouraging Capture and Handling). The alarm substance is given off when specialized dermal club cells are ruptured, as when a predator bites down on a prey fish. Nearby individuals, most likely schoolmates, sense the chemical in the water and take a variety of coordinated escape actions, depending on the species. Possession of the alarm response was a factor contributing to the inclusion of the gonorhynchiforms within the Ostariophysi.

Some ostariophysans lack one or both parts of the response for apparently adaptive reasons. Piranhas lack the alarm reaction, which makes sense given that many of their prey are also ostariophysans and it would be counterproductive for a predator to flee each time it bit into prey. Some nocturnal, nonschooling, or heavily armored ostariophysans lack both parts of the alarm response, including blind cave characins, electric knifefishes, and banjo and suckermouth armored catfishes. An interesting seasonal loss of the production end of the response occurs in several North American minnows. Nest building and courtship in these fishes often involves rubbing by males against the bottom and between males and females, during which time the skin and its breeding tubercles may be broken. It would be less than helpful to the male if he produced a substance that frightened females away during these activities. Males resume the production of alarm substance in the fall, after the breeding season. As with the Weberian apparatus, convergent evolution of alarm substances and responses has occurred in other teleostean groups, including sculpins, darters, and gobies (Smith 1992).

The five major orders within the Ostariophysi are too diverse to allow much detail here, and many aspects of their biology are treated in other chapters of this book. The most primitive order within the superorder is the

Gonorhynchiformes, which includes the milkfish, *Chanos chanos*, and three or four relatively small tropical families. Milkfish, an important food fish in the Indo-Pacific region, are often cultured in brackish fishponds, where juveniles are raised to edible size on an algae diet. Gonorhynchiforms have been problematic systematically because of a mixture of ostariophysan and nonostariophysan traits (i.e., they lack complete Weberian ossicles but do show modifications of the first three vertebrae suggestive of the weberian system, but all this may be the result of secondary loss of the gas bladder; they do have the alarm response and other ostariophysan osteological features).

The **Cypriniformes** is the largest order and probably contains the most familiar species of the superorder. The Cyprinidae, the largest family of freshwater fishes and the second largest family (after the gobies) of all fishes, contains 2000 of the 2600 cypriniform species. Among the better-known cyprinids are the minnows, shiners, carps, barbs, barbels, gudgeons, chubs, dace, squawfishes, tench, rudd, bitterling, and bream, and such popular aquarium fishes as the Southeast Asian "sharks" (redtail black shark, bala shark), goldfish, koi (domesticated common carp), zebra danios, and rasboras. Zebra danios, *Brachydanio rerio*, have become the standard research animal for studying developmental genetics (Eisen 1991). Zoogeographically, cyprinids are most diverse in Southeast Asia, followed by Africa, North America (where there are 270 species), and Europe. Interestingly, cyprinids are replaced ecologically in South America by characins (see Chap. 16, Similarities Between South America and Africa).

It is in the cyprinids that we see the first real development of pharyngeal dentition, a second set of jaws in the throat region that are derived from modified, tooth-bearing pharyngeal arches (see Chap. 8, Pharyngeal Jaws). Cyprinids are also the first to develop a highly protrusible upper jaw and to eliminate the maxillary bones from the biting bones and gape of the mouth, both trends that are increasingly developed in more advanced teleostean taxa (see Chap. 11, Division Teleostei). Exclusion of the maxilla from the gape is a characteristic of all fishes higher than the salmoniforms, although the bite of salmoniforms and their relatives involves the maxilla. The exclusion versus inclusion of the maxilla in cyprinids versus salmoniforms has led to some controversy over which group is more advanced. Current thinking is that the sum total of evidence favors salmoniforms as the more advanced clade; maxilla inclusion in salmoniforms may be a secondarily evolved trait.

Some cyprinids have chromosomes in the polyploid condition, an unusual occurrence among fishes. The normal diploid (2n) condition of most cyprinids is 48 or 50, although tetraploid (2n = 100), hexaploid, and octaploid species occur, as is the case for the goldfish (Buth et al. 1991). Polyploidy is linked with large size in minnows, the world's largest species (the Southeast Asian *Catlocarpio siamensis* and the Indian mahseer, *Tor putitora*) reaching 2.5 to 3 m in length. The largest minnow in North America is the piscivorous Colorado squawfish, *Ptychocheilus lucius*. Exceptional size also appears to be associated with predatory habits in other cyprinids, as implied by the scientific names of such large (2 m) species as *Elopichthys bambusa* and *Barbus esocinus*. However, most cyprinids are quite small (< 5 cm), and the smallest freshwater fish in the world is a Burmese cyprinid, *Danionella translucida*, which matures at 10 mm.

Gyrinocheilid algae eaters are interesting fishes because of modifications to the mouth and gill apparatus that allow them to scrape algae off rocks in flowing waters. Water is inhaled dorsally and exhaled ventrally through small apertures in the gill opening, the mouth having been modified into a sucking organ that helps them cling to rocks while scraping off algae. Suckers (Catostomidae) include about 70 species of relatively large (50 to 100 cm), chiefly North American fishes; one species, *Myxocyprinus asiaticus*, occurs in eastern China. Species include the buffaloes, quillback, carpsuckers, blue sucker, redhorses, jumprocks, and the extinct harelip sucker. Most suckers are benthic feeders in flowing water with inferior mouths and plicate or papillate lips. An exception is the lake suckers (*Chasmistes*), which are midwater planktivores with terminal mouths. Suckers can be quite confusing taxonomically because they frequently hybridize. Potential confusion is not helped by some of the scientific and common name combinations in this family, such as the quillback, *Carpiodes cyprinus*, and the river carpsucker, *C. carpio*, which can be mixed up with the common carp, *Cyprinus carpio*, which is a cyprinid.

Loaches (Cobitidae) are 110 species of predominantly Eurasian fishes that have their highest diversity in Southeast Asia. Included are popular aquarium fishes such as the kuhli, clown, and skunk loaches; the weatherfishes; and the golden dojo. Weatherfishes (*Misgurnus*) obtained their name because they become restless when atmospheric pressure drops preceding a storm. Their sensitivity to barometric fluctuations may somehow relate to their air-breathing abilities, which involve gulping atmospheric air and passing a bubble to the intestine, where gaseous exchange occurs. Balitorid river or hillstream loaches include many species specialized for living in fast-flowing mountain streams of India and Southeast Asia. Their paired fins tend to be enlarged and oriented horizontally, with adhesive pads on their ventral surfaces. In addition, their bodies are depressed dorsoventrally and their mouths are ventral, all anatomic adaptations to life in swift or turbulent water.

The **characiforms** are another large group of primarily tropical ostariophysans characterized (usually) by an adipose fin, well-armed mouths and replacement dentition (e.g., piranhas; see Chap. 8, Dentition), and ctenoid scales, as opposed to the cycloid scales found in lower groups. This is a remarkably diverse order anatomically and ecologically, including predators, zooplanktivores, scale eaters, detritivores, and herbivores; the latter category includes fishes that feed on seeds, leaves, and fruits. Characiforms may be surface, water-column, or benthic dwellers, although most species are found in midwater, many in shoals. Body sizes range from very small (13-mm adult tetras) to quite large (e.g., 1.5-m-long tigerfishes); body shapes range from long, slender almost darterlike benthic fishes (e.g., darter characins, *Characidium*) to deep-bodied, compressed piranhas and hatchetfishes.

Numerous popular, colorful aquarium fishes belong to this order, including *Distichodus*, *Prochilodus*, headstanders, spraying characins, freshwater hatchetfishes, blind characins, pencilfishes, tetras (*Cheirodon*, *Hemigrammus*, *Hyphessobrycon*, *Micralestes*, *Paracheirodon*), and silver dollars, as do important food fishes (*Prochilodus*, *Colossoma*, *Brycon*). Because of its large size and tropical nature, the order has undergone considerable taxonomic revision, much of which is still in progress. Currently, 10 families are recognized, although past classifications have recognized as few as 1 and as many as 16 families. The great majority of species (ca. 1150) are South American; about 200 are African; a small number live in Central America; and one species, the Mexican tetra, *Astyanax mexicanus*, extends naturally into southwestern Texas. Another 10 species, including piranhas, have been introduced into the United States.

The most primitive characiforms are the 100 species of African citharinids (*Distichodus*, *Citharinus*), attesting to an African origin for the order and the connection between Africa and South America prior to the breakup of Gondwana in the Mesozoic. African characiforms also include the advanced alestiine characids (about 100 species). Two of the largest characiforms are predaceous African species. The pike characin, *Hepsetus odoe* (Hepsetidae), is an impressive predator that reaches 65 cm in length and has fanglike teeth. It is remarkably convergent with the alestiine tigerfishes, *Hydrocynus* spp. (Characidae), which reach almost 2 m in length and over 50 kg in mass.

Aside from various popular aquarium species, certainly the best-known, or at least most notorious, characids are the piranhas. This subfamily, the Serrasalminae, contains about 60 species, some of which are predatory (*Serrasalmus*), others that are scale-eating opportunists and specialists (*Pygocentrus*, *Catoprion*), and some that are largely herbivorous, such as the pacus and silver dollars (*Colossoma*, *Metynnis*) (Sazima 1983; Nico and Taphorn 1988). Despite their reputation and potential for doing damage, many purported attacks on humans by piranhas actually result from postmortem scavenging on drowning victims (Sazima and Guimaraes 1987). The large herbivorous species are important food fishes in the Amazon basin. They are also important dispersers of seeds during the wet season, particularly because they use their massive dentition to husk seeds, which may aid germination (Goulding 1980).

The diversity of catfishes (**Siluriformes**) comes as a surprise to most everyone (Fig. 14.6; see Burgess 1989). Approximately 34 families of catfishes are recognized, and it is not surprising that catfish systematics is unsettled. Commonalities among the families include fusion, reduction, or loss of a number of skull bones found in lower teleosts, including the maxilla; teeth on the roof bones of the mouth (vomer, pterygoid, palatine); an adipose fin, sometimes with rays or a spine; an unsegmented, spinelike ray at the front of both the dorsal and pectoral fins that in some families produces a toxin and is also often accompanied by a shorter spine that helps lock the larger spine in the erect position; a lack of scales, often combined with the presence of bony plates or tubercles; small eyes (and nocturnal, benthic foraging); and one to four pairs of barbels associated with both the upper and

FIGURE 14.6.   *Selected catfishes, showing some of the array of body types and shared characteristics among the 34 families.*

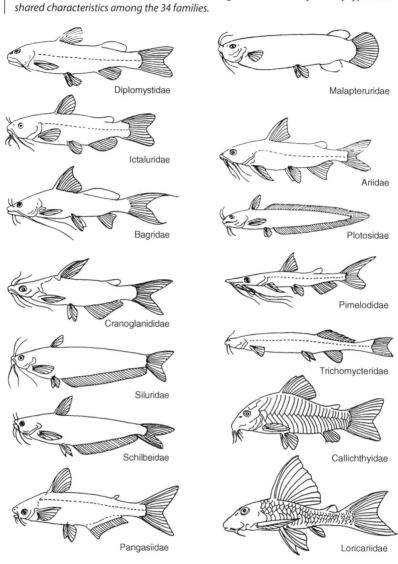

Diplomystidae

Malapteruridae

Ictaluridae

Ariidae

Bagridae

Plotosidae

Cranoglanididae

Pimelodidae

Siluridae

Trichomycteridae

Schilbeidae

Callichthyidae

Pangasiidae

Loricariidae

Drawings by John Quinn in Burgess 1989; used with permission of TFH Publications.

lower jaws that serve both chemosensory and tactile functions.

Catfishes are known from all continents, including Antarctica during the Oligocene. They reach their greatest diversity in South America, where the two largest families occur (loricariids with 550 species and pimelodids with 300 species). The most primitive catfishes are the South American diplomystids, which have a well-developed, toothed maxillary bone, a trait they share with the extinct Eocene hypsidorids. As with the vast majority of ostariophysans, most catfishes are confined to freshwater. However, two families of catfishes are primarily marine, the widespread sea catfishes (Ariidae) and the highly venomous Indo-Pacific eeltail catfishes (Plotosidae). The plotosids have diversified in several respects. Juvenile lined catfishes, *Plotosus lineatus*, are tightly schooling, diurnally active fishes, in contrast to the solitary, nocturnal behav-

ior of most other families. Both families contain members that occur in freshwater, such as the plotosid tandan catfishes of Australia (*Tandanus*), which probably represents secondary evolution of the use of freshwater habitats (i.e., the freshwater ancestor gave rise to a marine species that then reinvaded freshwater). Marine plotosids have highly venomous spines and relatively bold coloration, suggesting an aposematic, warning function (see Chap. 19, Evading Pursuit), which does not appear to deter their chief predators, the equally venomous sea snakes (Voris 1978).

Most catfishes are naked, lacking true scales. But in some families, different parts of the body are covered with individual or overlapping bony plates (armorhead catfishes, loach catfishes, sea catfishes, thorny catfishes, callichthyid armored catfishes, suckermouth armored catfishes), or thornlike projections, tubercles, or "odontodes" (sisorids, thorny catfishes, aspredinid banjo catfishes, spiny dwarf catfishes).

Some of the world's largest freshwater fishes are catfishes, including the predatory European wels (*Siluris glanis*, Siluridae) at 5 m and 330 kg, the herbivorous Asian Mekong catfish (*Pangasius gigas*, Pangasiidae) at 3 m and 300 kg, and a 3-m long-whiskered pimelodid (*Brachyplatystoma filamentosum*) of South America. The largest catfishes in North America are the flathead and blue catfishes, *Pylodictis olivaris* and *Ictalurus furcatus,* which reach about 1.5 m and 50 to 68 kg. Very small catfishes such as spiny dwarf catfishes (Scoloplacidae) and whalelike catfishes (Cetopsidae) are only 20 to 30 mm long as adults. Some small catfishes are notably unpleasant. The pencil or parasitic catfishes (Trichomycteridae) of South America include species that eat mucus and scales from other fishes or that pierce the skin or gill cavities of other fishes and feed on blood. At least one species, a candiru, *Vandellia cirrhosa*, is reputed to swim up the urethra of bathers and lodge itself there with its opercular spines, necessitating surgical removal (Burgess 1989). This behavior results in part because trichomycterids are rheophilic, which means they tend to swim "upstream" into gill cavities, or occasionally into a urethra.

Most catfishes are benthic, but some silurid sheatfishes, schilbeids, ageneiosid bottlenose catfishes, and hypopthalmid lookdown catfishes normally swim above the bottom (the relatively bizarre lookdowns are probably filter feeders). Occupation of the water column has produced some striking convergences in species that hover in open space. For example, the glass catfish, *Kryptopterus bicirrhus*, of Southeast Asia is a 10-cm-long transparent silurid with one pair of long barbels that protrude outward from its head, a long anal fin, no adipose fin, a forked tail, and only a single small ray in its dorsal fin (hence the name *kryptopterus*, or "hidden fin"). It tends to hover tail down in the water column, often in shoals. *Physailia pellucida*, the African glass catfish, is a 10-cm-long transparent schilbeid with four pairs of long barbels that protrude outward from its head, a long anal fin, a small adipose fin, a forked tail, and no dorsal fin. It also tends to hover tail down in the water column, often in shoals.

A number of unique modifications occur in the different families, far too many to detail here (see Burgess 1989). Among these traits include accessory air-breathing structures and terrestrial locomotion in air-breathing catfishes, airsac catfishes, and callichthyid armored catfishes (see Chap. 5, Aerial Respiration; Chap. 17, Deserts and Other Seasonally Arid Habitats); generation of strong electrical impulses in the African electric catfishes (see Chap. 6, Electroreception); climbing ability in climbing catfishes and jet propulsion in banjo catfishes (see Chap. 8, Locomotion: Movement and Shape); use of lures in angler catfishes (see Chap. 18, Pursuit); and mouth brooding of large eggs in sea catfishes (see Chap. 20, Parental Care).

The most advanced ostariophysans are the **gymnotiforms**, which show internal anatomic similarities to the siluriforms and probably share a common ancestor. However, gymnotiforms are distinct from catfishes and all other ostariophysans, as well as from all other teleosts except mormyrids and gymnarchids, in that they produce

and receive weak electrical impulses. Gymnotiforms are known collectively as South American knifefishes because of their strong resemblance to the African knifefishes (Notopteridae); the latter are osteoglossiforms and have electrogenic relatives among the mormyrids and gymnarchids but do not produce electricity themselves.

Gymnotiforms are restricted to Central and South America and consist of 62 species in six families. Anatomically, they are characterized by an elongate, compressed body; an extremely long anal fin that reaches from the pectoral fin to the end of the body; no dorsal or caudal fin; small eyes (and nocturnal foraging); and electrogenic tissue combined with modified lateral line organs for detecting weak electrical fields. The electrogenic tissue is derived from modified muscle cells in five families and, curiously, from nerve cells in the apteronotid ghost knifefishes, which also depart from the rest of the order by having a distinct tail fin. Electrical output in gymnotiforms is continual at high frequencies, as compared to the pulsed, low-frequency output of mormyrids. The electrical output is very weak, on the order of fractions of a volt, except in the monotypic 2-m-long electric eel (*Electrophorus*), which puts out a weak field for electrolocation purposes and strong pulses for stunning prey or deterring predators.

Superorder Protacanthopterygii

Esociformes (10 species): Esocidae (pikes, pickerels), Umbridae (mudminnows)
Osmeriformes (236 species): Argentinidae (argentines or herring smelts), Microstomatidae, Bathylagidae (deep-sea smelts), Opisthoproctidae (barreleyes), Leptochilichthyidae, Alepocephalidae (slickheads), Platytroctidae, Osmeridae (smelts), Salangidae (icefishes or noodlefishes), Sundasalangidae (Sundaland noodlefishes), Retropinnidae (New Zealand smelts, southern graylings), Lepidogalaxiidae (salamanderfishes), Galaxiidae
Salmoniformes (66 species): Salmonidae (whitefishes, graylings, chars, trouts, salmons)

Protacanthopterygians as a group have undergone repeated revision, taxa being removed and added to the superorder as new information is obtained and old data are reinterpreted. The name is retained here because it provides an organizational category for a series of apparently related fishes (Fig. 14.7).

The **esociforms** consist of two temperate families of freshwater fishes in which the maxillary bone is included in the gape but is toothless, and in which the median fins are located relatively far back on the body. Esocids include the northern and Amur pikes, the muskellunge, and the grass, redfin, and chain pickerels. The northern pike, *Esox lucius*, has the most widespread natural distribution of any completely freshwater fish, occurring across the northern portions of North America, Europe, and Asia (circumpolar distribution). The family goes back into the Cretaceous, and one Paleocene species is very similar to the northern pike, despite a 62-million-year separation. The mudminnows and blackfishes (Umbridae) are also scattered across northern North America, Europe, and Asia. The central mudminnow, *Umbra limi*, has a remarkable ability to

FIGURE 14.7. *Protacanthopterygians. (A) An esocid, the muskellunge,* Esox masquinongy. *(B) An osmerid, the capelin,* Mallotus villosus *(sexually mature female). (C) A salmonid, the chinook salmon,* Oncorhynchus tshawytscha *(mature female above, mature male below).*

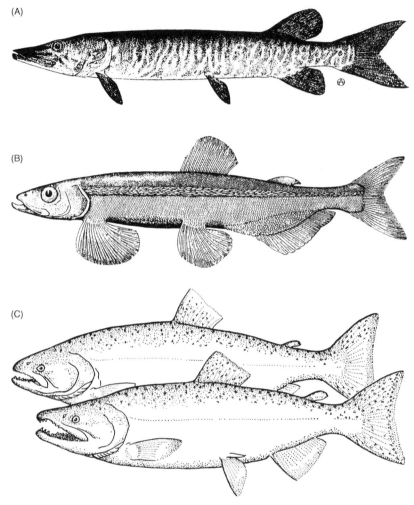

(A), (C) From Scott and Crossman 1973; (B) from Bigelow (1963); used with permission.

survive in high-latitude lakes that become ice-covered and oxygen poor through much of the winter (see Chap. 22, Seasonal Cycles).

Osmeriforms represent a diverse group of marine, freshwater, and diadromous (migrating between freshwater and the sea) species that inhabit waters ranging from very shallow to the ocean's depths. Osmeriforms are generally small, silvery, elongate fishes that swim in the water column. They have a single, soft-rayed dorsal fin and sometimes an adipose fin. The maxilla is included in the gape and is toothless, with a few exceptions. Although their pelvic fins are abdominal, as is the case for most of the preceding teleostean groups, several osmeriforms have their pectorals located higher on the body than is common in the more primitive groups. More than half of the osmeriforms are in the seven families of the suborder Argentinoidei, many of which are deep-sea inhabitants (see Chap. 17). The barreleyes are bizarre-looking deep-sea

fishes with elongate tubular eyes that point upward. The platytroctids are also deep-sea fishes that exude a blue-green luminous fluid from a papilla located at the anterior end of the lateral line, perhaps analogous to squid ink. The second suborder, Osmeroidei, contains two superfamilies, one for the osmerid smelts and some relatives, the second for some interesting Southern Hemisphere families. Osmerids include commercially important species such as capelins (*Mallotus*), eulachons (*Thaleichthys*), Asian ayu (*Plecoglossus*), and rainbow smelt (*Osmerus*), many of which are superficially similar to the more advanced silversides (Atherinidae).

The superfamily Galaxioidea includes species that dominate cold freshwater environments of the Southern Hemisphere. Retropinnids are small (10 to 35 cm), marine and anadromous (spawn in freshwater, grow in the sea) fishes of New Zealand and Australia known as southern smelts and southern graylings; they sometimes establish landlocked populations in lakes. Lepidogalaxiid salamanderfish (Fig. 14.8) are benthic-living, elongate fish that inhabit seasonally dry ponds of southwestern Australia, where they bury in the mud and live in a torpid state after ponds dry up, reemerging with the next rains (see Chap. 17, Deserts and Other Seasonally Arid Habitats). Salamanderfish lack eye muscles but instead have a flexible neck joint that allows them to bend their neck at a right angle to the side, a very unusual ability in fishes. Neck bending is made possible by large gaps between the back of the skull and the first cervical vertebrae and between the first and second cervical vertebrae. Salamanderfish have also apparently reinvented the physostomous gas bladder, in that they lack a normal gas bladder but instead have a gas-containing structure made up of simpler mesentery-like tissue and collagen fibers (Berra and Allen 1989; Berra et al. 1989). The remaining family of galaxiids includes the various whitebaits that constitute important commercial fisheries in New Zealand.

Galaxiids have complex life cycles, exhibiting all major types of diadromy, including anadromy, catadromy, and amphidromy (see Chap. 22, Diadromy). Some are also semelparous, spawning only one time before dying (see Chap. 23, Life Histories and Reproductive Ecology). Galaxioids as a group have suffered numerous extirpations and extinctions as a result of the stocking of nonnative trouts (see Chap. 25, Species Introductions).

Salmoniforms are an important group aesthetically, commercially, and ecologically, and a fascinating group evolutionarily. They are the focus worldwide of ichthyology classes and of popular and technical books. Different taxonomic treatments recognize as many as three differ-

FIGURE 14.8.   *Neck flexibility in the Australian salamanderfish,* Lepidogalaxias salamandroides. *This unusual benthic fish is able to bend its neck sideways and downward due to a unique arrangement of spaces between the skull and the cervical vertebrae. A lack of ribs throughout the vertebral column probably aids in neck bending and also allows the fish to make sinuous movements. (A) A 35-mm-long salamanderfish in the bent-neck position. (B) Cleared-and-stained 49-mm salamanderfish showing intervertebral gaps and lack of ribs. (C) Comparison specimen of the related* Galaxiella munda *(Galaxiidae, 46 mm), showing the tightly coupled vertebrae and more elongate ribs. Note also the well-developed pelvic girdle of* Lepidogalaxias, *which is used as a prop during resting.*

(A)

(B)

(C)

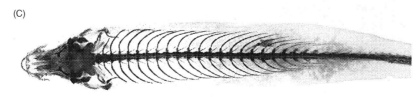

From Berra and Allen 1989; used with permission.

ent families (coregonid whitefishes, thymallid graylings, salmonid salmons and relatives), but recent work suggests that all salmoniforms should be united in one family, the Salmonidae. Their general anatomic similarities include an adipose fin, no spiny fin rays, a triangular flap at the base of the pelvic fin (**pelvic axillary process**), gill membranes free from the ventral side of the head, maxilla included in the gape, a physostomous gas bladder, and vertical barring (parr marks) on the sides of most young. Internally, the last three vertebrae angle up toward the tail, and a **myodome**, or area of the skull where the extrinsic eye muscles insert, is present.

The whitefishes and ciscoes consist of approximately 32 species of relatively large-scaled salmonids that lack teeth on the maxillary bone. They are zooplanktivorous fishes in high-latitude lakes of North America and Eurasia that show a great deal of within-species variation, termed *complexes*, from lake to lake. Several North American species have been decimated due to introduced predators, competitors, and parasitic lampreys (see Chap. 25). The graylings, *Thymallus*, are a smaller group of about 5 spe-

cies of riverine fishes that are easily identified by an elongate, flowing dorsal fin.

The subfamily Salmoninae contains seven Eurasian and North American genera (*Brachymystax, Acantholingua, Salmothymus, Hucho, Salvelinus, Salmo,* and *Oncorhynchus*) that differ from other salmonids by having small dorsal fins, small scales, and teeth on the maxillary bone. Most species of economic importance are in the latter three genera, although the Siberian taimen, *Hucho taimen*, is the world's largest salmonid at 2 m and 70 kg (a commercially caught individual weighed 114 kg). North American Salmoninae are currently divided into three genera and approximately 20 species, the names and relationships of which have been the subject of considerable debate (Fig. 14.9, Box. 14.1). The chars (or charrs) include the lake, brook, and bull trout, arctic char, and dolly varden, all in the genus *Salvelinus*. The northernmost-living freshwater fish in the world is the arctic char, *Salvelinus alpinus*, of Lake Hazen, on Ellesmere Island in Canada (80° N); many char are anadromous, moving into the sea to feed and grow and then back into freshwater to spawn.

FIGURE 14.9. *Phylogeny of the salmonids. A cladogram of most living salmonids based on life history traits shows the evolution of the various species. The same cladogram is constructed if anatomic and biochemical traits are used. The first four lineages represent separate genera (Thymallus, Brachymystax, Hucho, Salvelinus); Atlantic salmon and brown trout are in the genus Salmo; and the remaining eight species (those above trait 4) are all in the genus Oncorhynchus. The more primitive coregonine whitefishes would come off to the far left of the cladogram and are not shown. The following life history and reproductive characteristics are the shared derived characters that were used to construct the cladogram. The listed characters correspond to the numbered branch points in the cladogram. Groups to the right and above the number possess the trait; those to the left do not. 1) Egg diameter greater than 4.5 mm; females dig redds (nests); large males have hooked jaws (kype) during breeding season; 2) fall spawners; 3) commonly undergo long oceanic migrations; 4) most spawners undergo irreversible hormonal changes; 5) spring spawners; 6) anadromous forms die after spawning; 7) nonmigratory individuals tend not to reproduce; 8) most smolt in first year, some go to sea as even younger fry; 9) juveniles are strong schoolers, parr are slender; 10) freshwater phase reduced; young migrate soon after emerging from gravel.*

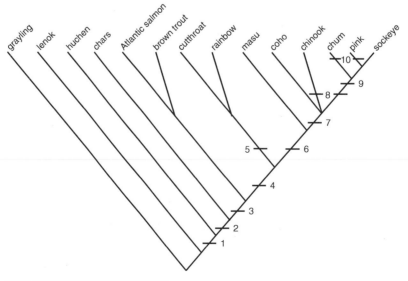

BOX 14.1	*What's in a Name: The Saga of the Rainbow Trout*

The rainbow trout, for better or worse, has probably been more actively stocked and cultured than any other fish species. It is consequently one of the best-known fishes worldwide. The species has been known for most of this century as *Salmo gairdneri*. Although salmonid taxonomists have argued over the correct name of the rainbow trout since 1792, the general public and most of the fisheries and ichthyological communities were unaware of such controversy. Hence considerable surprise and consternation arose when the American Fisheries Society accepted a decision by its Committee on Names of Fishes in 1988 to change the scientific name of the rainbow trout to *Oncorhynchus mykiss*. Such a change probably ranks second to the *Brontosaurus* versus *Apatosaurus* debate in terms of number of people affected (Gould 1990). The history and justification of the rainbow trout name change are good examples of the dynamics of scientific nomenclature and advance (Kendall 1988; Stearley and Smith 1993) and were clearly explained to the fishery biology community by Smith and Stearley (1989).

The change has resulted from an interaction among 1) older taxonomic debates concerning application and interpretation of the rules of nomenclature, 2) some recent discoveries involving biochemical and anatomic investigations on salmonids, and 3) how cladistic analysis has modified our view of evolutionary relationships. First, the rainbow trout turns out to be the same fish as an Asian species, the Kamchatka trout, *Salmo mykiss*, which was named by Walbaum in 1792. During the early scientific explorations of the North Pacific, wide-ranging salmonids were "discovered" repeatedly, one species being assigned many different names. Between 1792 and 1862, Walbaum, Richardson, and Suckley all independently described what we today recognize as the rainbow trout. This duplication resulted in part because salmonids are notoriously variable in anatomy, color, and behavior. The confusion was not alleviated by rainbow trout having two very different life history patterns, a sea-run form, known as the steelhead, and a landlocked form, known as the rainbow trout.

Additional confusion over the geographic ranges of rainbow and cutthroat trout (*S. clarki*) only clouded the issue more, and it wasn't until Okazaki (1984) that vertebral and scale counts, karyotypes, and electrophoretic data were brought together to demonstrate convincingly that *S. mykiss* and *S. gairdneri* from both sides of the Pacific were the same fish. Since the Kamchatka trout was described first by Walbaum, its specific name, *mykiss* (from an indigenous Kamchatkan name) has historical priority. Still unsettled was the question of which was the appropriate generic name.

The dilemma of generic names has been clarified by recent osteological (particularly skull and jaw bones) and biochemical studies of a variety of salmonids. These investigations indicate that the Pacific species of trout (rainbow, cutthroat, golden, Mexican golden, Gila, and Apache trouts) are more closely related to the Pacific salmons (*Oncorhynchus* spp.) than they are to the Atlantic salmonids, namely the Atlantic salmon (*Salmo salar*) and the brown trout (*S. trutta*). Since any accurate classification should reflect the true relations among the species, the most logical choice is to put all the western and Pacific species, be they "trout" or "salmon," in one genus, and the Atlantic species in another.

This conclusion is really nothing new; Regan in 1914 proposed just such a realignment. But Regan's ideas were unfortunately rejected by later workers who focused on characters that varied between sexes or who felt that primitive characters retained among taxa (symplesiomorphies) should form the basis for classification. However, the emergence of cladistics, with its emphasis on shared derived (synapomorphic) traits as stronger indicators of evolutionary relationships, has helped swing the debate. The argument boils down to an emphasis on similarities due to lack of change versus similarities due to the evolution of shared, specialized characters (see Chap. 2, Approaches to Classification). Regan's original analysis, based on skull characters, is borne out by biochemistry and other osteological features, as well as by a number of life history, reproductive, coloration, and behavioral characters (see Fig. 14.9). The result is that the ichthyological community has accepted *Salmo* as the generic name for the Atlantic species, *Oncorhynchus* for Pacific trouts and salmons, and *Oncorhynchus mykiss* as the scientific name for the rainbow trout.

The remaining salmonines are the Atlantic basin salmon and trout (*Salmo salar*, the Atlantic salmon; *S. trutta*, the European brown trout; and three other European species), and the 11 species of Pacific basin trouts and salmons in the genus *Oncorhynchus* that spawn in the fall; two of these are endemic to Japan. Pacific trouts and salmons include narrowly distributed, landlocked forms, such as golden and gila trouts, and species that are spectacularly anadromous, such as the coho, chinook, chum, and pink salmons, some of which undergo oceanic migrations of thousands of kilometers before returning to their birth river to spawn and die (see Chap. 22, Representative Life Histories of Migratory Fishes). The actual number of species of Pacific salmons is a matter of considerable and important debate because of the wholesale destruction of stocks in various rivers of the Pacific Northwest region of the United States. Many of these stocks are reproductively isolated and genetically distinct and therefore are viewed as unique evolutionary units for conservation purposes.

NEOTELEOSTS

All of the fishes above the salmoniforms are considered to be **neoteleosts**, a category without formal rank that lies somewhere between a subdivision and a superorder (Fig. 14.10). Seven neoteleostean superorders are recognized, including five relatively small, specialized deep-sea and pelagic superorders (discussed at length in Chap. 17, The Deep Sea) and two very diverse, advanced superorders, the paracanthopterygians and acanthopterygians. Neoteleosts as a taxon are considered to be monophyletic on the basis of four skull and jaw characters possessed by the different members that are lacking in most primitive fishes.

1. The manner in which the vertebral column connects to the back of the skull changes. In the neoteleosts, the first vertebra articulates with three bones of the skull (basioccipital and the two exoccipitals), whereas in more primitive teleosts it articulates with only the unpaired basioccipital.
2. A muscle, the retractor dorsalis, connects the vertebral column with dorsal elements of the upper pharyngeal jaws and pulls those jaws posteriorly.
3. A shift occurs in the insertion position of another muscle, one of the internal levators that originates on the base of the skull and lifts the pharyngeal jaws.
4. A unique hinged manner of attaching teeth to the jaws develops, allowing the tooth to be depressed toward the back of the mouth.

In addition, a trend toward more anteriorly located pelvics and more laterally located pectorals is evident during neoteleostean phylogeny, and acellular bone (skeletal material lacking bone cells) occurs in most euteleosts, whereas more primitive groups have bone cells (Smith 1988).

Superorder Stenopterygii

> Stomiiformes (321 species): Gonostomatidae (bristlemouths), Sternoptychidae (marine hatchetfishes), Photichthyidae (lightfishes), Stomiidae (barbeled dragonfishes)
> Ateleopodiformes (12 species): Ateleopodiidae (jellynose fishes)

Stomiiforms are all deep-sea fishes of the mesopelagic and bathypelagic regions (open water, with depths between 200 m and about 4000 m). They are often characterized by long teeth, large mouths, histologically unique photophores (light organs) that include a duct, and a

FIGURE 14.10. *Phylogenetic relationships among formally recognized, advanced teleostean superorders above the level of the Protacanthopterygii. Along the diagonal line of the cladogram are commonly used designations that do not have formal rank but that are generally distinguishable by shared derived traits as discussed in the following accounts.*

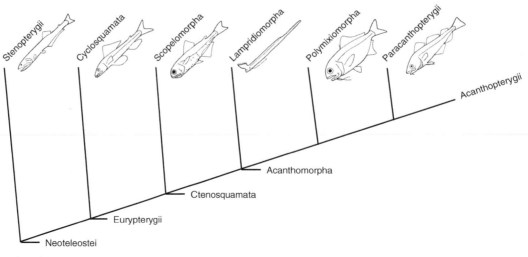

Based on Nelson 1994.

peculiar ventral adipose fin ahead of the anal fin in some. Gonostomatid bristlemouths in the genera *Cyclothone* and *Vinciguerria* may be the most abundant and widely distributed vertebrates on earth. The sternoptychid marine hatchetfishes possess several structural specializations that emphasize the vertical plane and body compression. The mouth opens vertically, the photophores point down, a preopercular spine points upward, the pelvic bones are oriented vertically, the lateral compression of the body is heightened by an abdominal keel-like structure, and a bladelike structure preceding the dorsal fin is made up of dorsal pterygiophores that project through the back of the fish. Pterygiophores normally serve as the basal support for median fins, not as elements of a fin itself. Idiacanthine black dragonfishes have a larval form with eyes at the ends of elongated stalks. Stomiiforms show many traits common to other, unrelated deep-sea fishes that are generally viewed as convergent adaptations to the light- and food-limited conditions of the deep sea (see Chap. 17, The Deep Sea).

The **ateleopodiform** jellynose fishes are an unusual group of bulbous-headed, elongate species that swim just above the bottom in deep water. Their skeleton is largely cartilaginous. Their large and pointed head, exaggerated anal fin, and relatively pointed tail are all traits they share with other deep, benthopelagic fishes such as chimaeras, spiny eels, halosaurs, eucla cods, rattails, and grenadiers (see Chap. 17, The Deep Sea).

Superorder Cyclosquamata

Aulopiformes (219 species): Giganturidae (telescopefishes), Aulopodidae (aulopus), Chlorophthalmidae (greeneyes), Ipnopidae (spiderfishes and tripodfishes), Scopelarchidae (pearleyes), Notosudidae (waryfishes), Synodontidae (lizardfishes), Pseudotrichonotidae, Paralepididae (barracudinas), Anotopteridae (daggertooth), Evermannellidae (sabertooth fishes), Omosudidae, Alepisauridae (lancetfishes)

Fishes more advanced than stenopterygians are sometimes referred to as **eurypterygians**. The most primitive superorder in this group is the Cyclosquamata (cycloid scales), which contains one order of almost entirely deep-sea fishes. **Aulopiforms** have a surprisingly extensive fossil record (approximately eight families and 20 genera) for a group that today inhabits primarily open water and deep sea. The deep-sea groups include the truly bizarre giganturid telescopefishes (Fig. 14.11A), which undergo a spectacular metamorphosis during which the premaxillary, palatine, orbitosphenoid, parietal, symplectic, posttemporal, supratemporal, hyoid, and cleithral bones in the head region are lost, as are several gill arches, the gill rakers, and the gas bladder. The larvae possess an adipose fin, pelvic fin, and branchiostegal rays, which are also lost during metamorphosis. The adults possess large tubular eyes, a huge mouth, flexible teeth, an expandable stomach, pectoral fins located exceptionally high on the body above the eyes, loose skin, and a peculiar tail with the ventral lobe extending far beyond the dorsal lobe.

Chlorophthalmid greeneyes are the first fishes encountered in this survey that are hermaphroditic. All of the more primitive groups so far have been distinctly gonochoristic, which means that an individual is only one sex throughout its life. Hermaphroditism of some form is surprisingly common in higher teleosts (see Chap. 10, Adults; Chap. 20, Gender Roles in Fishes). This dichotomy between primitive and advanced teleosts suggests that sexual lability—either in terms of initial sexual determination or of an ability to change sex later in life—represents a derived trait that became differentially retained or independently evolved after the neoteleostean or eurypterygian level of development was reached.

Included among the ipnopids are the often-illustrated spiderfishes or tripodfishes, which have greatly elongated pectoral, pelvic, and caudal rays that they use for resting on soft sediments of the deep ocean floor (Fig. 14.11B). The best-known shallow representatives of this order are the synodontid lizardfishes, which are common benthic inhabitants of coral reefs worldwide (and whose name should not be confused with the synodid upside-down catfishes). Lizardfishes are closely related to the secondarily pelagic Bombay ducks, which form an important fishery in the Indian Ocean. The most advanced family in the order is the alepisaurid lancetfishes, which are large (to 2 m) mesopelagic predators on other fishes. Lancetfishes are distinguished by their large, sail-like dorsal fin that extends from the head almost to the caudal peduncle (function unknown). These fishes, which look very much like scombroid snake mackerels, have proven a great boon to deep-sea taxonomists, as several species of mesopelagic fishes have been described from the stomach contents of alepisaurids (see Chap. 17, The Deep Sea).

Superorder Scopelomorpha

Myctophiformes (240 species): Neoscopelidae, Myctophidae (lanternfishes)

All fishes above the level of the Cyclosquamata have lost the fifth pharyngeal tooth plate and the muscle that lifts it (Johnson 1992). These advanced groups are often termed the **Ctenosquamata** in reference to the predominance of ctenoid scales among them. One order of deep-sea fishes, the **myctophiforms**, makes up the superorder Scopelomorpha (Fig. 14.12). Myctophiforms have lost the fifth pharyngeal tooth plate but most still have cycloid scales, justifying their primitive status among ctenosquamates. Their relative primitiveness is also shown in their retention of an adipose fin, but they are advanced in that the maxilla is excluded from the gape. The myctophid lanternfishes, with about 235 species, are an important group of mesopelagic deep-sea fishes in terms of diversity, distribution, and numbers of individuals. They occur in all seas, from the Arctic to the Antarctic, and are the prey of numerous other fishes as well as of marine mammals. They make up a large fraction of the deep-scattering layer (DSL), a diverse assemblage of fishes and invertebrates that lives at mesopelagic depths (below 200 m) during the day and migrates toward the surface at dusk. Myctophid taxonomy is often based on otolith

FIGURE 14.11.   *Telescopefish and spiderfish, two aulopiform fishes from the deep sea. (A) Giganturid telescopefishes are mesopelagic, water-column dwellers. (B) Ipnopid spiderfishes are deep-sea benthic dwellers.*

(A)

(B)

(A) From Walters 1964; (B) After Heezen and Hollister 1971.

structure and species-specific patterns of photophores, characters that are even preserved in some fossils.

ACANTHOMORPHA: THE SPINY TELEOSTS

The appearance of true fin spines, rather than hardened segmented rays, marks a major evolutionary step in the evolution of bony fishes. True spines occur in the dorsal, anal, and pelvic fins of higher teleosts. Spines develop when the two halves of the primitively paired and jointed dermal fin rays fuse into a single, unsegmented structure. Several other characteristics, mostly associated with improved locomotion and feeding, also mark the ascendancy of teleosts above the ctenosquamate level of development (see Chap. 11, Division Teleostei). Locomotion was im-

FIGURE 14.12.   *A myctophid lanternfish,* Diaphus mollis, *about 4 cm long. Round structures along ventral half of body are light-emitting photophores.*

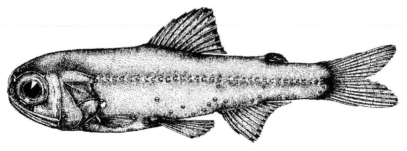

From Nafpaktitis et al. 1977; used with permission.

proved by strengthening of vertebral accessories (zygapophyses), providing body stiffening and better attachment for muscles. These changes allowed a shift from slow, sinusoidal motion of the entire body to rapid oscillation of the tail region, driven by tendons attached to the tail base (see Chap. 8, Locomotion: Movement and Shape). The tail itself also underwent considerable modification. Pharyngeal teeth diversified, and the maxilla shifted from a tooth-bearing bone to a structure that helps pivot the premaxilla, making mouth protrusion and suction more effective. These advances probably made possible the explosive radiation of spiny-rayed teleosts, known collectively as **acanthomorphs**, during the early Cenozoic.

Superorder Lampridiomorpha

Lampridiformes (19 species): Veliferidae, Lampridae (opahs), Stylephoridae (tube-eye or thread-tail), Lophotidae (crestfishes), Radiicephalidae, Trachipteridae (ribbonfishes), Regalecidae (oarfishes)

As is often the case, the most primitive members of a major taxon retain certain ancestral traits but possess others indicative of advanced status. **Lampridiforms** lack true spines, but the maxilla helps move the premaxilla and bears no teeth. The connection between the two upper jaw bones and the manner in which they slide to protrude the mouth are unique, so much so that the group used to be referred to as the Allotriognathi (strange jaws). The seven families of lampridiforms are almost all open-water oceanic fishes with unusual body and fin proportions. Opahs are relatively large (to 1.8 m, 70 kg), oval-shaped, colorful pelagic predators on squids and other fishes. The 30-cm-long tube-eye (*Stylephorus*) is capable of an almost 40-fold enlargement of the volume of its mouth during feeding,

which is probably a record among vertebrates (see Chap. 8, Jaw Protrusion: The Great Leap Forward).

The remaining four families contain relatively large and rare, elongate, pelagic fishes with long dorsal fins. Crestfishes and radiicephalids have ink sacs that they discharge through their cloaca. The oarfish, *Regalecus*, reaches a confirmed length of 8 m and perhaps as much as 12 m, making it the longest extant teleost; uncertainty arises because intact specimens are seldom found (Fig. 14.13). Despite their size, they are apparently planktivores. Their name comes from bladelike expansions at the end of the pelvic fins. Of the few specimens that have been encountered, many were obtained after storms, when the bodies of these strange-looking fishes are tossed up on beaches. The oarfish, with its bluish silvery body, scarlet dorsal crest of elongate fin rays, and deep red fins, is likely responsible for many sea serpent sightings, particularly those referring to monsters "having the head of a horse with a flaming red mane" (Norman and Fraser 1949, 113).

Superorder Polymixiomorpha

Polymixiiformes (5 species): Polymixiidae (beardfishes)

The taxonomic status of this enigmatic family has been the subject of considerable debate (see Johnson and Patterson 1993; Nelson 1994). Beardfishes possess advanced characters such as four to six true spines in the dorsal fin and four spines in the anal fin, and their pelvic fins are located fairly forward on the body. Yet they retain two sets of intermuscular bones, the epineurals and epipleurals. At different times, they have been classified with either the primitive paracanthopterygian percopsiforms (troutperches and relatives) or primitive acanthopterygian

FIGURE 14.13. *The oarfish,* Regalecus glesne, *the world's longest teleost, reaching lengths of 8 m to as much as 12 m. Whole specimens are rare.*

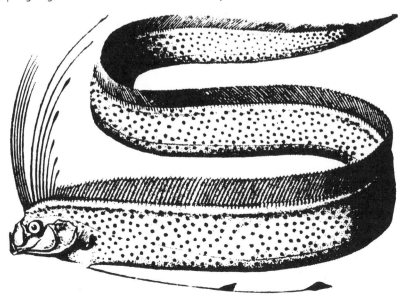

From McCulloch 1927; used with permission.

beryciforms (squirrelfishes and relatives), but their combination of primitive, advanced, and several unique characteristics complicates determination of their relationships. They are currently placed near the base of the two major radiations of spiny-rayed fishes. Beardfishes have large eyes and chin barbels, are about 30 cm long, and live at moderate depths (200 to 600 m).

Superorder Paracanthopterygii

Percopsiformes (9 species): Percopsidae (troutperches), Aphredoderidae (pirate perch), Amblyopsidae (cavefishes)
Ophidiiformes (355 species): Carapidae (pearlfishes), Ophidiidae (cusk-eels), Bythitidae (viviparous brotulas), Aphyonidae, Parabrotulidae (false brotulas)
Gadiformes (482 species): Ranicipitidae (tadpole cod), Euclichthyidae (eucla cod), Macrouridae (grenadiers or rattails), Steindachneriidae (luminous hake), Moridae (morid cods), Melanonidae (pelagic cods), Macruronidae (southern hakes), Bregmacerotidae (codlets), Muraenolepididae (eel cods), Phycidae (phycid hakes), Merlucciidae (merluccid hakes), Gadidae (cods)
Batrachoidiformes (69 species): Batrachoididae (toadfishes, midshipmen)
Lophiiformes (297 species): Lophiidae (goosefishes), Antennariidae (frogfishes, sargassumfish), Brachionichthyidae (handfishes), Chaunacidae (coffinfishes, sea toads), Ogcocephalidae (batfishes), Caulophrynidae, Neoceratiidae, Melanocetidae, Himantolophidae (footballfishes), Diceratiidae, Oneirodidae, Thaumatichthyidae, Centrophrynidae (deep-sea anglerfish), Ceratiidae (sea devils), Gigantactinidae, Linophrynidae

Paracanthopterygians represent a major side branch during the evolution of advanced acanthomorphs. They are weakly defined by a number of characters, chiefly involving the caudal skeleton and holes in the skull through which cranial nerves pass (Nelson 1994). Their ecological success is largely as benthic, marine fishes that are nocturnally active or live in permanently darkened waters, such as the bathypelagic region of the deep sea or in caves. Only about 20 relatively primitive paracanthopterygians, out of a total 1200 species, live in freshwater. Many paracanthopterygians have sonic muscles on their gas bladders and produce sounds.

Percopsiforms are small (< 20 cm), freshwater fishes, eight of nine of which live in eastern North America. The troutperch, *Percopsis omiscomaycus*, and its Columbia River congener, *P. transmontana*, are the most advanced fishes we will encounter with an adipose fin. This seemingly primitive condition is interesting given that modern percopsiforms possess several traits suggesting a reversal of the evolutionary trends of advanced teleosts, including those found in fossil percopsiforms. The modern species have fewer fin spines, more vertebrae, and a more posteriorly located pelvic girdle than occurred in fossil forms. The aphrododerid pirate perch, *Aphrododerus sayanus*, is a swamp dweller with the distinction of having its anus move from just anterior of the anal fin in juveniles to the throat region of adults, for functional reasons that remain mysterious. The amblyopsid cavefishes are 10 species of highly modified, often blind and scaleless forms that show numerous adaptations for cave life (see Chap. 17, Caves).

In **ophidiiforms**, the pectoral fins are high up on the body and have a vertical orientation. The pelvic fins, when present, are located anteriorly under the head in what is termed the mental or jugular position. Pelvic fin loss in this group probably relates to their eel-like bodies; many eel-like fishes, regardless of taxonomic position, have reduced or absent pelvic fins and girdles (see Chap. 23, Habitat Use and Choice). Ophidiiforms inhabit what have to be viewed as marginal or at least exceptional habitats for fishes. Carapid pearlfishes are **inquilines** (tenants), living inside the body cavities of starfishes, sea cucumbers, clams, and sea squirts; some are actually parasitic, feeding on the internal organs of their hosts. Pearlfishes are apparently unique among fishes in that they have two distinct larval stages, a "vexillifer" pelagic stage, followed by a "tenuis" demersal stage during which they search for a host. Ophidiid and bythitid cusk-eels and brotulas include blind species in freshwater caves of the Caribbean basin and Galapagos Islands, and infaunal coral reef species that hide deep within crevices (Fig. 14.14A). The depth record for a fish is held by a neobythitine cusk-eel, *Abyssobrotula galatheae*, taken 8370 m down in the Puerto Rico Trench (see Chap. 17, The Deep Sea). Bythitid brotulas and parabrotulid false brotulas are livebearers, a rare derivation among paracanthopterygians.

The **gadiforms** include some of the most important commercial fishes in the world, such as the cods, haddocks, hakes, pollocks, and whitings (Fig. 14.14B). Gadiforms lack true spines but have experimented with fin rays. The long dorsal fin is relatively diversified compared with most primitive groups; it is often divided into two or three parts, an anterior ray that is sometimes spinous (grenadiers) or elongate and even filamentous (morid cods, codlets, eel cods). The true cods (Gadinae) have three dorsal fins and two anal fins. Pelvic fins are thoracic or jugular in position and are sometimes modified into filaments with a possible sensory function (e.g., eucla cod, codlets, physid hakes). Many species have chin barbels (grenadiers, morid cods, eel cods, physid hakes, cods), a convergent trait in benthic or near-benthic (surpabenthic, benthopelagic) fishes. Gadiforms are marine fishes with the solitary exception of the burbot, *Lota lota*, which is a lake gadid of Holarctic (high latitude, Northern Hemisphere) distribution. The commercially important Atlantic cod, *Gadus morhua*, is the largest species in the order, reaching lengths of 1.8 m and weighing over 90 kg, although fish over 10 kg are rare due to extensive overfishing. The largest food fishery in the world is for North Pacific walleye pollock, *Theragra chalcogramma*, with a harvest in excess of 6 million tons in 1989.

Batrachoidiforms are well-camouflaged, benthic marine fishes with eyes placed high on the head, flattened heads and large mouths, relatively elongate dorsal and anal fins, multiple lateral lines, and only three pairs of gills (rather than the usual five pairs). Their dorsal fins have two or three stout spines. The midshipmen (*Porichthys*) have 600 to 800 lateral photophores, an unusual trait for a shallow-water fish. Fishes in this order are often quite vocal, producing a variety of sounds with their gas bladders. Male midshipmen have been the focus of complaints by houseboat dwellers in San Francisco Bay during the (midshipmen's) breeding season, when the males produce a sustained, low-frequency hum (Ibara et al. 1983).

FIGURE 14.14. *Paracanthopterygians. (A) An ophidiiform, the bearded brotula,* Brotula barbata. *(B) A gadiform, the Atlantic cod,* Gadus morhua.

(A)

(B)

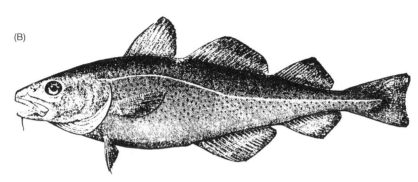

From Jordan 1905.

In the venomous toadfishes of the advanced subfamily Thalassophryninae, dorsal and opercular spines are part of a complex system that injects a powerful venom. Toadfishes are unusual zoogeographically because they are a shallow, warm-water family that is most diverse in the Americas, whereas most tropical marine families have their greatest diversity in the Indo-Australian region (Collette and Russo 1981; see Chap. 16, Major Marine Regions). Three South American species are restricted to freshwater. In morphology, ecology, and venom production, toadfishes are convergent with the scorpaeniform stonefishes and perciform weeverfishes.

The most advanced group within the Paracanthopterygii is the **lophiiforms**, a diverse and often bizarre-looking order of marine fishes that are primitively benthic, shallow-water dwellers but have evolved many highly modified, open-water, deep-sea forms. Many if not most of them use a modified first dorsal spine as a lure for catching smaller fish. The basal group is the lophiid goosefishes, known commercially as monkfish or poor man's lobster. Goosefishes occur on both sides of the Atlantic and also in the Pacific and Indian Oceans. The western North Atlantic goosefish, *Lophius americanus*, can exceed 1 m in length and 40 kg in mass and has a huge mouth with long, recurved teeth that point back into the mouth. Goosefishes prey on other fishes and on diving seabirds.

The antennariid frogfishes also rest on the bottom and are well-camouflaged, globose fishes that can walk across the bottom on their pectoral and pelvic fins (an old name for the lophiiforms, Pediculati, refers to the elbowlike bend in the pectoral and the footlike appearance of the

pelvic fins). The esca, or lure, of frogfishes can be quite ornate, mimicking a small fish, shrimp, or worm (Fig. 14.15). When not waved in front of potential prey, the esca sits in a protective depression between the second and third dorsal spines. If the esca is bitten off, it apparently can regenerate back to its species-specific form (Pietsch and Grobecker 1987).

The ogcocephalid batfishes (Fig. 14.14B) are among the least fishlike fishes around (other candidates include sea horses, shrimpfishes, boxfishes, and ocean sunfishes; see Chap. 15). The flattened, rounded head accounts for more than half the length of the body; it tapers quickly behind the expanded pectoral fins, giving the fish the appearance of a rounded axe with a short handle. Batfishes alternate walking on their pectorals with swimming via jet propulsion of water expelled from their round, backward-facing opercular openings. As modified as the batfishes are, they are rivaled in strange appearance by the 11 families of deep-sea anglerfishes, superfamily Ceratioidea. The ceratioids are the most successful fishes of the vast bathypelagic region, comprising 150 species. Among their other derived traits, many species have very small males that fuse to and become parasitic on the larger females (see Chap. 17, The Deep Sea).

SUMMARY

1. The 24,000 teleost species occupy almost all aquatic habitats. Teleosts arose in the Mesozoic and radiated as modern osteoglossomorphs, elopomorphs, clupeo-

FIGURE 14.15. *Paracanthopterygians. (A) A 6-cm-long bloody frogfish,* Antennarius sanguineus, *from the Galapagos Islands. Note the elaborate esca, or lure, at the end of the modified first dorsal spine, or illicium. (B) Dorsal view of a batfish,* Dibranchus spinosa. *Opercular openings are the teardrop-shaped holes at about midbody.*

(A)

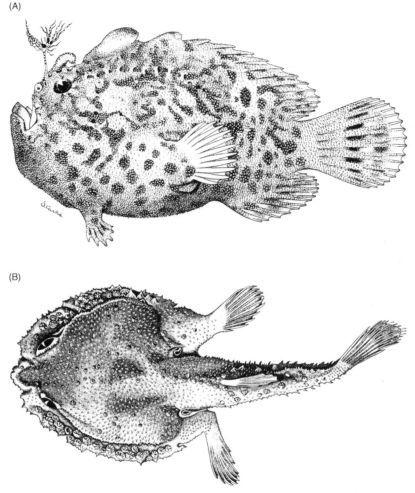

(B)

(A) Drawing by C. L. Starks in Heller and Snodgrass 1903; (B) from Briggs 1974; used with permission of McGraw-Hill, Inc.

morphs, and euteleosts. Today, 38 orders and 426 families are recognized, defined largely by skull and tail modifications that improved feeding and locomotion.

2. Osteoglossomorphs are chiefly tropical, freshwater fishes in which tongue teeth bite against the mouth roof. The African elephantfishes produce and detect weak electrical fields.

3. Elopomorphs are characterized by a ribbon-shaped leptocephalus larva and include tenpounders, tarpons, bonefishes, spiny eels, and true eels. Elopomorphs are predominantly marine fishes that occur from very shallow to very great depths. The anguilliform true eels include 45 families of elongate fishes, some of which are catadromous, spawning at sea but growing in freshwater.

4. Clupeomorphs are generally small, schooling fishes

of pelagic marine and occasionally freshwater habitats. They are characterized by an ear-to-gas-bladder connection and by bony scutes on the belly. The herrings and anchovies are exceedingly important fisheries species.

5. The fourth subdivision, the Euteleostei, contains 92% of the families and 95% of the species of teleosts. Many euteleosts possess an adipose fin and breeding tubercles or contact organs. The species-rich Ostariophysi contains predominantly freshwater fishes such as minnows, suckers, characins, loaches, catfishes, and South American knifefishes. Ostariophysans have modified anterior vertebrae—the Weberian apparatus—that aid in hearing. They produce and respond to alarm substances. Cypriniforms possess pharyngeal jaws and dentition used in manipulating and crushing prey and vegetation. Characins are highly successful South American and African fishes

such as piranhas and tetras. Siluriform catfishes include more than 30 families of primarily benthic, nocturnal fishes with barbels, spines, and adipose fins. Gymnotiform knifefishes have converged with the osteoglossomorph elephantfishes in the production and detection of weak electrical fields.

6. Protacanthopterygians are loosely related marine, freshwater, and diadromous fishes that include the pikes, smelts, and salmons. Esociform pikes and pickerels are Northern Hemisphere predators. Osmeriforms are generally small, silvery, elongate, water-column-dwelling fishes such as freshwater smelts, deep-sea barreleyes and slickheads, and Southern Hemisphere salamanderfishes and galaxiids. Whitefishes, graylings, salmons, and trouts constitute the salmoniforms, characterized by an adipose fin, a triangular flap at the base of the pelvic fin, and the configuration of the last three vertebrae. Many salmonids undergo tremendous migrations between fresh- and salt water during their lives.

7. All fishes more advanced than salmoniforms are called neoteleosts. The seven superorders share similarities in the articulation between the skull and first cervical vertebra, two muscles that move the pharyngeal jaws, and jaw teeth that can be depressed posteriorly. The first three superorders of neoteleosts are primarily deep-sea or pelagic fishes. Stenopterygians, which include bristlemouths and marine hatchetfishes, are deep-sea fishes with long teeth, large

mouths, and peculiar photophores (light organs). Cyclosquamates are primarily deep-sea forms, such as the bizarre giganturid telescopefishes and tripodfishes, but also include the shallow-water lizardfishes. Scopelomorphs are primarily lanternfishes, which have species-specific photophore patterns.

8. Neoteleosts above scopelomorphs possess true fin spines and are termed acanthomorphs. Other acanthomorph advances include strengthened vertebral accessories and tail structures that improve swimming, pharyngeal tooth diversification, and improved jaw protrusion. Lampridiomorphs (tube-eyes, oarfish) and polymixiomorphs (beardfishes) are primitive acanthomorphs; oarfish may exceed 8 m in length and are the world's longest teleost.

9. The superorder Paracanthopterygii consists of mostly marine, benthic, nocturnal fishes, including very deep-sea ophidiiforms, commercially important cods, acoustically active toadfishes, and the diverse bathypelagic lophiiform anglerfishes. Some anglerfish males are much smaller than and parasitic upon the larger females.

SUPPLEMENTAL READING

Carroll 1988; Nelson 1994; Paxton and Eschmeyer 1995; Wheeler 1975; Berra 1981.

Teleosts at Last
II: Spiny-Rayed Fishes

Most modern bony fishes belong to a single advanced superorder, the Acanthopterygii. The group is so diverse and its members so important from all standpoints that a full chapter is needed to discuss them, although no one chapter or one book can do them justice. Several genera and families, such as the sticklebacks, livebearers, darters, black basses, perches, butterflyfishes, cichlids, damselfishes, tunas, and billfishes are the subjects of one or several books themselves, and so the treatment below is understandably cursory. Again, the phylogeny and taxonomy presented here as well as many of the aspects of biology of different groups are taken largely from Nelson (1994), a reference that should be consulted for additional details.

SUPERORDER ACANTHOPTERYGII

Given the remarkable diversity of the higher spiny-rayed fishes—approximately 13,500 species in 251 families—it is a tribute to their successful suite of adaptations that they are generally recognized as a coherent group (Fig. 15.1). Although controversy about relationships and taxonomic position among the various orders and families abounds, certain generalities can be made about the group as a whole and the characteristics that define it. Two primary innovations are shared by most lineages of acanthopterygians.

1. Upper jaw mobility and protrusibility are maximal in this group. This is achieved by the development of a dorsal extension of the anterior tip of the premaxilla, termed the **ascending process**. This process slides along the rostral cartilage on the snout of the fish, shooting the upper jaw forward and downward. Protrusion is aided by a camlike connection between the maxilla and premaxilla, the maxilla rotating and helping push the premaxilla forward (Lauder and Liem 1983; see Chap. 11, Division Teleostei).
2. Pharyngeal dentition and action reach their highest level of development. Improved function is aided by a redistribution of the attachments of muscles and bones in the pharyngeal apparatus. The retractor dorsalis muscle (see Chap. 14) now inserts on the third pharyngobranchial arch, and the upper pharyngeal jaws are supported principally by the second and third epibranchial bones.

Acanthopterygians also typically have ctenoid scales (with numerous exceptions); a physoclistous gas bladder; maxilla excluded from the gape; two distinct dorsal fins, the first of which is spiny and the second of which is soft-rayed; pelvic and anal fins with spines; pelvic fins located anteriorly, containing one leading spine and five or fewer soft rays, and pectoral fins placed laterally on the body; and an externally symmetric tail fin supported by fused basal elements. A number of other trends in feeding, locomotion, and predator protection characterize the higher spiny-rayed fishes and show progressive change during acanthopterygian phylogeny. Most of these were discussed in Chapter 11 and will only be summarized here as particularly good examples or striking exceptions are encountered among the taxa. An important point to be remembered is that these are the most advanced and diverse of today's fishes, dominating the shallow, productive habitats of the marine and many lake environments.

Series Mugilomorpha

Mugiliformes (66 species): Mugilidae (mullets, gray mullets)

The mullets are a family of nearshore, catadromous fishes of considerable economic importance and of some taxonomic controversy. Their distinctly separated spiny and soft dorsal fins and spines in the pelvic and anal fins in part justify their inclusion with the other acanthopterygians (Fig. 15.2). They are considered primitive in that some have cycloid scales or scales intermediate between cycloid and ctenoid, and the pelvic girdle lacks any direct ligamentous or bony connection to the cleithral region of the pectoral girdle. In most higher groups, the two girdles are connected. Many mullets are detritivorous, feeding on the organic silt that covers the bottom and digesting the minute plants and animals in such ooze with a gizzardlike stomach. Mullets frequently leap from the water for inexplicable reasons; one study showed that the frequency of

FIGURE 15.1.   *Phylogeny of acanthopterygian or higher spiny-rayed fishes. Most recognized clades are given ordinal or higher status and do not have accepted common names. The common names given are of better-known representatives. See Figure 14.1 for characters that define branching points in the phylogeny of the Acanthopterygii.*

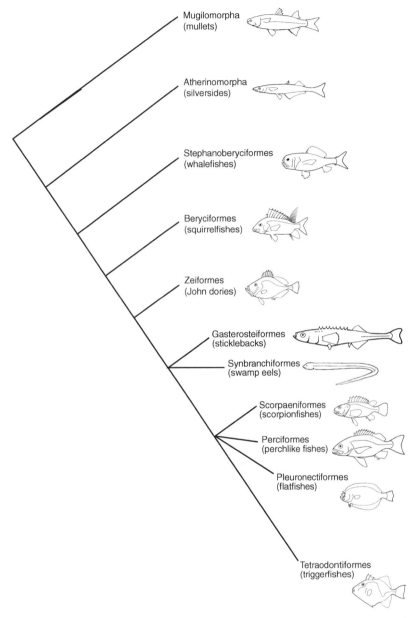

After Nelson 1994.

such jumping increases when dissolved oxygen levels are low (Hoese 1985).

Series Atherinomorpha

Atheriniformes (285 species): Bedotiidae, Melanotaeniidae (rainbow-fishes), Pseudomugilidae (blueeyes), Atherinidae (silversides), Notocheiridae, Telmatherinidae (sailfin silversides, Celebes rainbowfishes), Dentatherinidae, Phallostethidae
Beloniformes (191 species): Adrianichthyidae, Belonidae (needlefishes),

Scomberesocidae (sauries), Exocoetidae (flyingfishes), Hemiramphidae (halfbeaks)
Cyprinodontiformes (807 species): Aplocheilidae (rivulines), Profundulidae (Middle American killifishes), Fundulidae (topminnows, killifishes), Valenciidae, Anablepidae (foureye fishes), Poeciliidae (livebearers), Goodeidae (splitfins), Cyprinodontidae (pupfishes)

The most successful fishes at the surface layer of the ocean and of many freshwater habitats are in the three orders of the Atherinomorpha. Such well-known surface dwellers

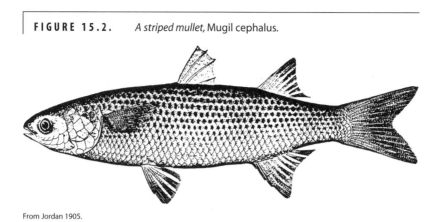

FIGURE 15.2.   *A striped mullet,* Mugil cephalus.

From Jordan 1905.

as silversides, needlefishes, sauries, flyingfishes, half-beaks, killifishes, topminnows, and livebearers all belong to this group. Anatomically, the atherinomorphs are set aside from the rest of the acanthopterygians in part because they have a unique way of protruding the jaw. The premaxilla does not articulate directly with the maxilla. Protrusion instead occurs by an intervening linkage between premaxilla and maxilla via the rostral cartilage. Atherinomorphs typically have terminal or superior mouths, as would be expected of surface-feeding fishes. Internal fertilization and live bearing of young have evolved repeatedly within the group; many of the egg-laying families have chorionic filaments that protrude from the egg and help it attach to plants and other structures.

Within the order **Atheriniformes** are eight families of generally small, silvery fishes with two distinct dorsal fins. The melanotaeniid rainbowfishes of Australia and New Guinea are strongly sexually dimorphic freshwater fishes. Males have brighter colors and longer fins than females, traits that make them popular aquarium species. Such pronounced sexual dimorphism is rare in more primitive groups outside of the breeding season, except in some deep-sea fishes. The atherinid silversides are widespread freshwater and marine fishes that normally occur in schools in shallow water. Atherinids include the grunions (*Leuresthes* spp.) of southern and Baja California, which ride waves up beaches to spawn in wet sand every 2 weeks during summer (see Chap. 22, Semilunar and Lunar Patterns). Sexual determination is under environmental control in some atherinids (see Chaps. 10, 20). A radiation of 18 atherinid species, including apparent piscivorous species, has developed in lakes of the Mexican plateau (see Box 15.1). The small (< 4 cm) phallostethids are peculiar Southeast Asian atheriniforms in which the pelvic girdle and other structures of the males are modified into a complex clasping and intromittent organ for holding onto females and fertilizing their eggs internally. Females lack a pelvic girdle; they are also unusual in that they lay fertilized eggs rather than having young develop internally, which is the more normal course for fishes with internal fertilization.

Beloniforms are predominantly silvery, marine fishes active at and sometimes above the surface of the water (Fig. 15.3). The adrianichthyids include the medakas or ricefishes, *Oryzias*, which are used extensively in genetic,

embryologic, and physiological investigations. The suborder Belonoidei contains species with a number of anatomic features that show precursors and intermediate conditions during the evolution of rather specialized traits. The lower lobe of the caudal fin in primitive beloniforms has more principal rays than the upper lobe. A rounded or square tail in primitive groups has changed into a forked tail fin with a slightly elongate lower lobe in the belonoid needlefishes, species which periodically leave the water in short, arcing leaps. The lower lobe is very pronounced in the exocoetid flyingfishes, which use it as a sculling organ to accelerate during takeoffs and to extend their gliding flights, which can last hundreds of meters (Davenport 1994; see Chap. 19, Evading Pursuit).

A tendency for elongation of the lower or both jaws occurs in all beloniform groups, expressed as a garlike prey capture structure in piscivorous needlefishes and as unequal jaw lengths in sauries and particularly in half-beaks. During development, different families show different developmental rates for the two jaws before the adult condition is reached, suggesting that evolution within the group has involved alterations in the developmental rate of the jaws (= heterochrony) (Boughton et al. 1991; see Chap. 10, Evolution via Adjustments in Development: Heterochrony, Pedomorphosis, and Neoteny). For example, the lower jaw of some juvenile needlefishes is at first longer than the upper jaw, which later catches up. In some flyingfishes, the lower jaw is at first elongate, but later in life both jaws are essentially equal in length and neither projects forward. Halfbeaks, despite their predatory appearance, use their elongate lower jaw to feed on floating pieces of sea grasses; some freshwater species take insects at the water surface.

The **cyprinodontiforms** are a major group of freshwater fishes, many of which show a high tolerance for saline and even hypersaline conditions (see Chap. 17, Deserts and Other Seasonally Arid Environments). They are largely surface swimmers, preying on insects that fall into the water, which they detect using lateral line pores on the upper surface of the head. Life history traits in different cyprinodontiform families take on extreme conditions. The aplocheilid rivulines of South America and Africa are annual fishes that live in temporary habitats, spawn during the rainy season, and then die, their genes being preserved in eggs that lie in a resting state in bottom muds until the

FIGURE 15.3. *Atheriniforms. (A) A belonid needlefish,* Tylosurus crocodilus. *(B) A cyprinodontid striped killfish,* Fundulus majalis, *male above, female below. (C) The four-eyed fish,* Anableps.

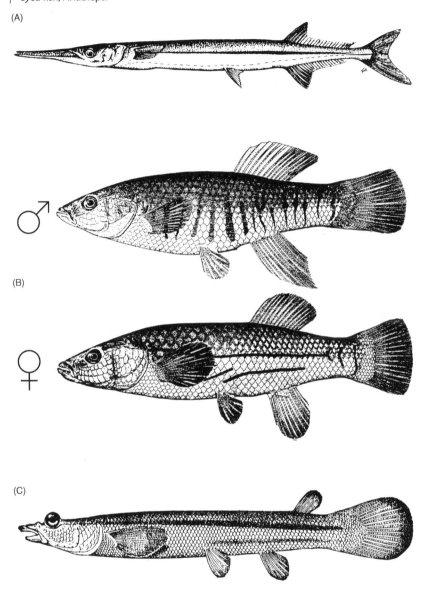

(A)

(B)

(C)

(A) From Collette 1995; used with permission; (B) from Bigelow and Schroeder 1953; (C) from Jordan 1905.

next rains (see Chap. 17). *Rivulus marmoratus* of south Florida and the West Indies is the only fish species known to be self-fertilizing. Sexual dimorphism reaches extremes in the elongate, brightly colored median fins of male rivulines (e.g., lyretails, panchax), poeciliids (sailfin mollies, guppies, swordtails), and pupfishes. Some of the live-bearing poeciliids are "species" that originated through hybridization and today do not include functional males; females instead use males of other species to activate embryogenesis, male genetic material being excluded from future generations (Meffe and Snelson 1989; see Chap. 20, Gender Roles in Fishes). Some goodeids have a placenta-like connection between the mother and the internally developing young (see Chap. 20, Parental Care, Fig. 20.5).

The cyprinodontid pupfishes are environmentally tol-

erant fishes that can live in water of highly variable salinity and temperature, characteristics that have allowed them to invade fluctuating environments such as salt-marsh and desert ponds (see Chap. 17). Many species of *Orestias* have evolved in Andean lakes (see Box 15.1), including Lake Titicaca, which at 4570 m above sea level is the highest natural body of water populated by fishes.

Anablepid four-eyed fishes, a phylogenetically intermediate family within the cyprinodontiforms, are unique anatomically in eye structure. Four-eyed fishes are surface dwellers that swim with their protruding eyes half out of water. The pupil of the eye itself is physically divided into dorsal and ventral halves, the upper half capable of forming focused images of objects in air and the lower half simultaneously forming images of objects underwater.

SERIES PERCOMORPHA

Stephanoberyciformes (86 species): Melamphaidae (bigscale fishes, ridgeheads), Gibberichthyidae (gibberfishes), Stephanoberycidae (pricklefishes), Hispidoberycidae, Rondeletiidae (redmouth whalefishes), Barbourisiidae (red whalefish), Cetomimidae (flabby whalefishes), Mirapinnidae, Megalomycteridae (largenose fishes)

Beryciformes (123 species): Anoplogastridae (fangtooths), Diretmidae (spinyfins), Anomalopidae (flashlight fishes, lanterneye fishes), Monocentridae (pinecone fishes), Trachichthyidae (roughies, slimeheads), Berycidae (alfonsinos), Holocentridae (squirrelfishes)

Zeiformes (39 species): Parazenidae, Macrurocyttidae, Zeidae (dories), Oreosomatidae (oreos), Grammicolepididae, Caproidae (boarfishes)

Gasterosteiformes (257 species): Hypoptychidae (sand eel), Aulorhynchidae (tubesnouts), Gasterosteidae (sticklebacks), Pegasidae (seamoths), Solenostomidae (ghost pipefishes), Syngnathidae (pipefishes, seahorses), Indostomidae, Aulostomidae (trumpetfishes), Fistulariidae (cornetfishes), Macrorhamphosidae (snipefishes), Centriscidae (shrimpfishes)

Synbranchiformes (87 species): Synbranchidae (swamp eels), Chadhuriidae, Mastacembelidae (spiny or tiretrack eels)

Scorpaeniformes (1271 species): Dactylopteridae (flying gurnards), Scorpaenidae (scorpionfishes, rockfishes), Caracanthidae (orbicular velvetfishes), Aploactinidae (velvetfishes), Pataecidae (Australian prowfishes), Gnathanacanthidae (red velvetfishes), Congiopodidae (racehorses, pigfishes), Triglidae (sea robins), Bembridae (deep-water flatheads), Platycephalidae (flatheads), Hoplichthyidae (ghost flatheads), Anoplopomatidae (sablefishes), Hexagrammidae (greenlings), Normanichthyidae, Rhamphocottidae (grunt sculpins), Ereuniidae, Cottidae (sculpins), Comephoridae (Baikal oilfishes), Abyssocottidae, Hemitripteridae, Agonidae (poachers), Psychrolutidae (fathead sculpins), Bathylutichthyidae, Cyclopteridae (lumpfishes), Liparidae (snailfishes)

Perciformes (9300 species)

  Suborder Percoidei (2860 species): Centropomidae (snooks), Chandidae (Asiatic glassfishes), Moronidae (temperate basses), Percichthyidae (temperate perches), Acropomatidae (temperate ocean basses), Serranidae (sea basses), Ostracoberycidae, Callanthiidae, Pseudochromidae (dottybacks), Grammatidae (basslets), Plesiopidae, Notograptidae, Opistognathidae (jawfishes), Dinopercidae, Banjosidae, Centrarchidae (sunfishes, black basses), Percidae (darters, perches), Priacanthidae (bigeyes), Apogonidae (cardinalfishes), Epigonidae (deep-water cardinalfishes), Sillaginidae (sillagos), Malacanthidae (tilefishes), Lactariidae (false trevallies), Dinolestidae (long-finned pike), Pomatomidae (bluefishes), Nematistiidae (roosterfish), Echeneidae (remoras), Rachycentridae (cobia), Coryphaenidae (dolphinfishes), Carangidae (jacks, pompanos), Menidae (moonfish), Leiognathidae (ponyfishes, slimeys, slipmouths), Bramidae (pomfrets), Caristiidae (manefishes), Emmelichthyidae (rovers), Lutjanidae (snappers, fusiliers), Lobotidae (tripletails), Gerreidae (mojarras), Haemulidae (grunts), Inermiidae (bonnetmouths), Sparidae (porgies), Centracanthidae, Lethrinidae (emperors), Nemipteridae (threadfin breams), Polynemidae (threadfins), Sciaenidae (croakers, drums), Mullidae (goatfishes), Pempheridae (sweepers), Glaucosomatidae (pearl perches), Leptobramidae (beachsalmon), Bathyclupeidae, Monodactylidae (moonfishes, fingerfishes), Toxotidae (archerfishes), Coracinidae (galjoen fishes), Drepanidae, Chaetodontidae (butterflyfishes), Pomacanthidae (angelfishes), Enoplosidae (oldwife), Pentacerotidae (armorheads), Nandidae (leaffishes), Kyphosidae (sea chubs), Arripidae (Australasian salmon), Teraponidae (grunters, tigerperches), Kuhliidae (flagtails, aholeholes), Oplegnathidae (knifejaws), Cirrhitidae (hawkfishes), Chironemidae (kelpfishes), Aplodactylidae (marblefishes), Cheilodactylidae (morwongs), Latridae (trumpeters), Cepolidae (bandfishes)

  Suborder Elassomatoidei (6 species): Elassomatidae (pygmy sunfishes)

  Suborder Labroidei (2200 species): Cichlidae (cichlids), Embiotocidae (surfperches), Pomacentridae (damselfishes), Labridae (wrasses), Odacidae, Scaridae (parrotfishes)

  Suborder Zoarcoidei (318 species): Bathymasteridae (ronquils), Zoarcidae (eelpouts), Stichaeidae (pricklebacks), Cryptacanthodidae (wrymouths), Pholidae (gunnels), Anarhichadidae (wolffishes), Ptilichthyidae (quillfish), Zaproridae (prowfish), Scytalinidae (graveldiver)

  Suborder Notothenioidei (122 species): Bovichthyidae, Nototheniidae (cod icefishes), Harpagiferidae (plunderfishes), Bathydraconidae (antarctic dragonfishes), Channichthyidae (crocodile icefishes)

  Suborder Trachinoidei (212 species): Chiasmodontidae, Champsodontidae, Pholidichthyidae (convict blenny), Trichodontidae (sandfishes), Pinguipedidae (sand perches), Cheimarrhichthyidae, Trichonotidae (sand divers), Creediidae (sand burrowers), Percophidae (duckbills), Leptoscopidae (southern sandfishes), Ammodytidae (sand lances), Trachinidae (weeverfishes), Uranoscopidae (stargazers)

  Suborder Blennioidei (732 species): Tripterygiidae (triplefin blennies), Dactyloscopidae (sand stargazers), Labrisomidae, Clinidae, Chaenopsidae (pikeblennies, tubeblennies, flagblennies), Blenniidae (combtooth blennies)

  Suborder Icosteoidei (1 species): Icosteidae (ragfish)

  Suborder Gobiesocoidei (120 species): Gobiesocidae (clingfishes)

  Suborder Callionymoidei (137 species): Callionymidae (dragonets), Draconettidae

  Suborder Gobioidei (2120 species): Rhyacichthyidae (loach gobies), Odontobutidae, Eleotridae (sleepers), Gobiidae (gobies), Kraemeriidae (sand gobies), Xenisthmidae, Microdesmidae (wormfishes, dartfishes), Schindleriidae

  Suborder Kurtoidei (2 species): Kurtidae (nurseryfishes)

  Suborder Acanthuroidei (125 species): Ephippidae (spadefishes), Scatophagidae (scats), Siganidae (rabbitfishes), Luvaridae (luvar), Zanclidae (moorish idol), Acanthuridae (surgeonfishes)

  Suborder Scombrolabracoidei (1 species): Scombrolabracidae

  Suborder Scombroidei (136 species): Sphyraenidae (barracudas), Gempylidae (snake mackerels), Trichiuridae (cutlassfishes), Scombridae (mackerels, Spanish mackerels, tunas), Xiphiidae (swordfish), Istiophoridae (billfishes)

  Suborder Stromateoidei (65 species): Amarsipidae, Centrolophidae (medusafishes), Nomeidae (driftfishes), Ariommatidae, Tetragonuridae (squaretails), Stromateidae (butterfishes)

  Suborder Anabantoidei (81 species): Luciocephalidae (pikehead), Anabantidae (climbing gouramis), Helostomatidae (kissing gourami), Belontiidae (gouramis), Osphronemidae (giant gouramis)

  Suborder Channoidei (21 species): Channidae (snakeheads)

Pleuronectiformes (570 species): Psettodidae, Citharidae, Bothidae (lefteye flounders), Achiropsettidae (southern flounders), Scophthalmidae, Paralichthyidae, Pleuronectidae (righteye flounders), Samaridae, Achiridae (American soles), Soleidae (soles), Cynoglossidae (tonguefishes)

Tetraodontiformes (339 species): Triacanthodidae (spikefishes), Triacanthidae (triplespines), Balistidae (triggerfishes), Monacanthidae (filefishes), Ostraciidae (boxfishes, trunkfishes, cowfishes), Triodontidae (threetoothed puffer), Tetraodontidae (puffers), Diodontidae (porcupinefishes, burrfishes), Molidae (ocean sunfishes)

The most advanced euteleostean clade is the **Percomorpha**, a diverse and varied taxon that contains more than 12,000 species of largely marine families, although several successful freshwater groups also belong in this lineage. Percomorphs have in common an anteriorly placed pelvic girdle that is connected to the pectoral girdle directly or by a ligament; the pelvic fin also typically has an anterior spine and five soft rays, with larger numbers of rays occurring in primitive percomorph taxa.

At the base of the percomorphs are two orders of

FIGURE 15.4.   *The orange roughy,* Hoplostethus atlanticus, *a trachichthyid beryciform.*

Photo by G. Helfman.

either deep-sea or nocturnal fishes, the stephanoberyci-forms and the beryciforms. These large-headed, round fishes have many percomorph characteristics, except that the tail fin has a primitively large number of rays (18 or 19) as compared to the 17 caudal rays that typify most advanced percomorphs. The primitive **stephanoberyci-forms** (gibberfishes, pricklefishes, cetomimoid whale-fishes) are largely deep-sea forms characterized by luminescent organs, weak or absent fin spines, and re-duced squamation. **Beryciforms** often have the large eyes typical of nocturnal fishes and possess strong spines on the head or gill covers. Included among the beryci-forms are such relatively shallow-water luminescent

forms as pinecone fishes and flashlight fishes, reef forms such as squirrelfishes, and the commercially important orange roughy, *Hoplostethus atlanticus* (Fig. 15.4). Beryci-forms are well represented in the fossil record, dating back to the late Cretaceous. **Zeiforms** are a confusing assortment of primitive marine percomorphs that have highly protrusible mouths and a unique caudal skeleton. Included in this order are such commercial species as the European John dory, *Zeus faber*.

Acanthopterygians as a group are not as well repre-sented in the deep sea as are more primitive taxa. After the stephanoberyciforms and primitive beryciforms, few deep-sea fishes occur. The most advanced beryciforms, such as the squirrelfishes, are also the shallowest dwellers. It is possible that as more advanced clades arose within the Euteleostei, their specializations made them competitively superior to primitive groups, and the younger taxa displaced the older out of productive shallow-water habitats and into the less productive deep-sea region. Alternatively, overriding trends within the Acanthopterygii include the development of stout spines and other hard structures (bony skull crests, spined scales, dermal ossifications), which may have been difficult to reverse. Since a common convergence among deep-sea forms is the loss or reduction of hard body parts, acanthopterygians may have been phylogenetically constrained from developing the energy-saving traits necessary for existence in the deep sea.

Gasterosteiforms are generally small marine and freshwater fishes with dermal armor plates, small mouths, and unorthodox propulsion (Fig. 15.5). Sticklebacks are among the world's most intensively studied fishes behaviorally, physiologically, ecologically, and evolutionarily

(Wootton 1984; Bell and Foster 1994). Although only seven stickleback species are recognized, separate populations often diverge in anatomic traits and may constitute distinct genomes. The extent of predator avoidance, spines, and dermal plates often vary in relation to the threat from predators experienced by a population. The suborder Syngnathoidei includes several unusually shaped fishes encased in bony rings, including pegasid seamoths and syngnathid pipefishes, sea dragons, and seahorses. In the primitive solenostomid ghost pipefishes, the female carries developing eggs in a brood pouch formed by pelvic fins fused to the ventral body surface. Syngnathids are the only vertebrates in which the male literally becomes pregnant. An evolutionary gradient of degrees of male parental care exists within the family and corresponds to the recognized phylogeny of the group. Pipefish taxonomy is based in part on whether eggs are embedded in or attached to the male's ventrum, whether the pouch is sealed or open, and whether plates

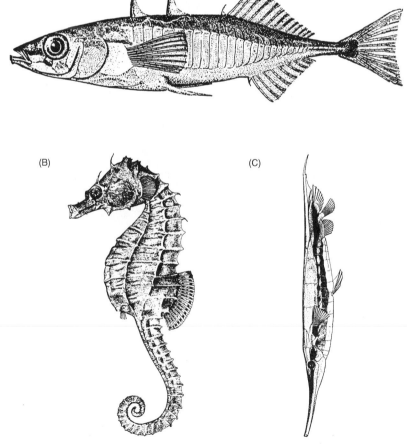

FIGURE 15.5. *Gasterosteiforms. (A) A three spine stickleback,* Gasterosteus aculeatus. *(B) A seahorse,* Hippocampus hudsonius. *(C) A 13-cm centriscid shrimpfish,* Aeoliscus strigatus. *Centriscids are extraordinary gasterosteiforms that often hover head-down among sea urchin spines, where they are particularly well camouflaged. The first dorsal spine forms the posterior end of the body, while the second dorsal, caudal, and anal fins are directed downward.*

(A)

(B)   (C)

From Jordan 1905.

or membranes protect the eggs. In primitive species, eggs are attached externally to the male's ventral surface, where they develop and hatch. In more advanced species, the eggs are deposited within a pouch and are fertilized; the embryos develop within the pouch, where they obtain protection, oxygenation, osmoregulation, and nutrition from the male. Such "role reversal" in reproductive behavior includes females that actively court and compete for males (e.g., Rosenqvist 1990). Locomotion is accomplished by rapid undulation of the small dorsal fin.

Exceptional fishes within the gasterosteiforms are the aulostomoid trumpetfishes and cornetfishes, which are elongate, large (to 1 m), lurking and stalking piscivores with very expandable mouths. The intriguing centriscid shrimpfishes of the Indo-Pacific are small, extremely compressed fishes with the shape and proportions of an edible pea pod encased in thin bone (see Fig. 15.5). Due to an almost right-angle flexure in the vertebral column, their second dorsal, caudal, and anal fins all point ventrally, and they tend to swim with their dorsal edge leading while oriented head down. The fish typically hover head

down among the spines of long-spined sea urchins, where they are protected and difficult to see due to their thinness and a long black lateral stripe.

The **synbranchiforms** are a small order of primarily freshwater, eel-like fishes. The synbranchid swamp eels are air-breathing fishes in Africa, Asia, and Central and South America. They have many unusual derivations, including loss of pectoral, pelvic, dorsal, anal, and, in some, caudal fins. Synbranchids also have a unique upper jaw arrangement in which the palatoquadrate attaches at two points to the skull, termed **amphystilic suspension** and not known in any other teleosts.

The **scorpaeniforms** are a large order of predominantly marine fishes (Fig. 15.6). Most have spines projecting from different bones on the head, including a posteriorly directed spine derived from a bone below the eye, giving them the name "mail-cheeked fishes." Many scorpaeniforms lack scales, but this may be more part of a general suite of adaptations to benthic living than a phylogenetic trait. The dactylopterid flying gurnards have huge pectoral fins that they expand as they walk along

FIGURE 15.6.   *Scorpeaniforms. (A) A treefish,* Sebastes serriceps, *one of the numerous sebastine rockfishes of the North Pacific. (B) The lumpfish,* Cyclopterus lumpus, *of the North Atlantic.*

(A)

(B)

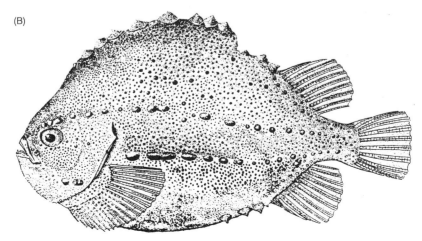

(A) From Jordan 1905; (B) from Bigelow and Schroeder 1953.

the bottom on their elongate pelvic fins; it is unlikely that adult "flying" gurnards ever leap out of the water or for that matter ever swim far above the bottom.

The scorpaenid scorpionfishes and rockfishes are a diverse group of benthic marine fishes with large mouths and venomous spines in their dorsal, anal, and pelvic fins. The sebastine rockfishes are an important commercial group of often live-bearing and long-lived species of the temperate North Pacific (e.g., Boehlert and Yamada 1991). Other subfamilies within this group include the colorful and venomous lion or turkeyfishes (e.g., *Pterois*), as well as the camouflaged and highly venomous stonefishes (*Synanceia*). The hexagrammid greenlings are littoral zone and kelp-associated fishes endemic to the North Pacific. The family includes the highly edible and predatory lingcod, *Ophiodon elongatus*.

The only freshwater scorpaeniforms are in the suborder Cottoidei, which includes the cottid sculpins of North American headwater streams and tide pools, as well as a species flock of comephorid oilfishes and other cottoid species in Lake Baikal in Asia (see Box 15.1). Many cottoids lack scales but have prickly skin. The cabezon, *Scorpaenichthys marmoratus*, of the Pacific coast of North America is unique among nontetraodontiform teleosts in having toxic eggs. The most advanced scorpaeniforms are the cyclopteroid lumpfishes and snailfishes. The globose lumpfishes have bony tubercles arranged in rows around their body and a sucking disk made from modified pelvic fins, an unusual trait for fishes that do not frequent high-energy zones. The lumpfish of the North Atlantic, *Cyclopterus lumpus*, is highly prized for its caviar and has been seriously depleted in parts of its range. Liparid snailfishes, which also have pelvic suction disks, occur broadly geographically and ecologically. They are found in most oceans from the Arctic to the Antarctic and can inhabit tide pools or benthic regions deeper than 7000 m.

ORDER PERCIFORMES

The largest order in the Percomorpha, and for that matter of vertebrates, is the **Perciformes**, containing 148 families and nearly 9300 species, more than one-third of all fishes. Our discussion will focus on selected families within the 18 perciform suborders (family names appear in boldface type in the following accounts). As might be expected in such a diverse taxon, the classification of the perciforms is a subject of much debate (see Johnson 1993). The success of perciforms is greatest in, but by no means limited to, coral reef habitats, where 6 of the 8 largest families abound (gobies, wrasses, sea basses, blennies, damselfishes, cardinalfishes). Two other large families, the cichlids and the croakers, reach their maximum diversity in tropical lakes and nearshore temperate marine habitats respectively. The fossil record for perciforms dates back to the early Cenozoic, and recognizable members of most suborders had evolved by the Eocene, indicating very rapid evolution and diversification over a period of about 20 million years (Carroll 1988).

Suborder Percoidei

The largest perciform suborder is the **Percoidei**, containing 71 families and 2860 species (Fig. 15.7). Percoids, in contrast with lower teleosts such as ostariophysans and protacanthopterygians (and continuing trends in acanthopterygians and percomorphs), are characterized by 1) the presence of spines in the dorsal, anal, and pelvic fins; 2) two dorsal fins (never an adipose fin); 3) ctenoid scales; 4) pelvic fins in the abdominal position; 5) laterally placed and vertically oriented pectoral fins; 6) maxilla excluded from the gape; 7) physoclistous gas bladder; 8) absence of orbitosphenoid, mesocoracoid, epipleural, and epicentral bones; 9) acellular bone; and 10) never more than 17 principal caudal fin rays.

The basal families of percoids are what would generally be considered basslike fishes. **Centropomids** are primarily large, piscivorous fishes of lakes, estuaries, and nearshore regions, including the snooks (*Centropomus*) of tropical America, and the barramundi of Australia and Nile perch of Africa (*Lates* spp.). Nile perch and their relatives have been widely introduced in African lakes and have caused decimation of native endemic cichlids (see Chap. 25, Species Introductions). The **moronid** temperate basses include lake-dwelling and anadromous predators in North America, such as the white bass and striped bass (*Morone* spp.). The closely related **acropomatid** (sometimes referred to as polyprionid) temperate ocean basses include such commercially important species as the Atlantic wreckfish (*Polyprion americanus*) and the giant sea bass of California (*Stereolepis gigas*), the latter reaching lengths of 2 m and weighing up to 250 kg.

The sea bass family **Serranidae** contains 450 species and is one of the largest fish families. It contains a tremendous diversity of sizes and shapes of fishes that have three spines on the opercle but may differ in many other characters. Serranids vary in size from 3-cm-long planktivorous anthiines to the 3-m-long, 400-kg jewfish, *Epinephelus itajara*, which eats lobsters and small turtles as well as fishes. Three subfamilies are currently recognized, the first two (Serraninae and Anthiinae) consisting of mostly small forms such as hamlets, sand perches, and the colorful *Anthias*. The subfamily Epinephelinae is defined by long, stout or filamentous dorsal and/or pelvic fin spines in the larvae (see Chap. 9, Larval Morphology and Taxonomy). It contains most commercially important species such as groupers, hinds, coneys, gag, and scamp but also includes diminutive and striking basslets and the chemically protected soapfishes (e.g., *Rypticus*, *Grammistes*), which exude a soaplike toxin from their skin when disturbed. Many serranids are hermaphroditic, usually starting as female and then later becoming male (protogyny), although some hamlets function simultaneously as either sex. To confuse the issue, most species known as basslets belong to the related family **Grammatidae**, which also includes the neon-colored royal gramma that lives under ledges in the Caribbean and is popular among divers and aquarists.

Two important percomorph families in North American freshwaters are the centrarchids and the percids. The family **Centrarchidae** contains about 29 species, including the numerous sunfishes, crappies, and rockbasses (*Le-*

FIGURE 15.7. *Representative percoid fishes. (A) A centropomid snook,* Centropomus undecimalis. *(B) A serranid black sea bass,* Centropristis striata. *(C) A centrarchid smallmouth black bass,* Micropterus dolomieui. *(D) A percid darter, the log perch,* Percina caprodes. *(E) An echeneid sharksucker,* Echeneis naucrates, *with a top view of the first dorsal fin that form a suction disk. (F) A carangid rough scad,* Trachurus lathami. *(G) A chaetodontid foureye butterflyfish,* Chaetodon capistratus.

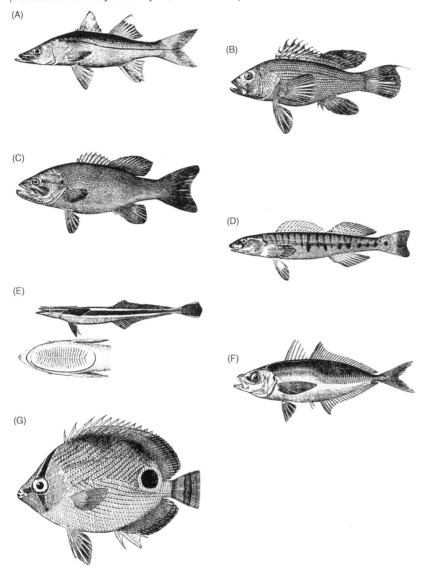

(A), (C), (D), (G) From Jordan 1905; (B), (E), (F) from Bigelow and Schroeder 1953.

pomis, Pomoxis, Ambloplites) as well as the seven black basses in the genus *Micropterus* (e.g., largemouth, smallmouth, etc., plus two undescribed species in the southeastern United States). Centrarchids are the dominant carnivores in most lakes in the United States and southern Canada and are also known for their nesting behavior, the males digging and defending circular nests on the bottom through much of the summer. Centrarchids are native to the region east of the Rocky Mountains with the exception of the Sacramento perch, *Archoplites interruptus*. They have been widely introduced elsewhere.

The family **Percidae** is one of the most successful nonostariophysan freshwater families in the world. At least 162 species of percids exist, 150 of which occur in North America. The dominant lake forms are larger species such as the yellow perch (*Perca flavescens*) and walleye (*Stizostedion vitreum*) (Colby 1977; Craig 1987). Yellow perch have their counterpart in the Eurasian perch, *P. fluviatilis*; three species of *Stizostedion* pikeperches also occur in Europe (Collette and Banarescu 1977). North American streams contain at least 145 species of darters, mostly in the genera *Percina* and *Etheostoma*. The greatest diversity is in the southeastern United States; Tennessee alone contains 90 darter species. These small, benthic fishes feed primarily on aquatic insect larvae and other invertebrates in fast-flowing, clean water where males defend nesting rocks and court females. During the breeding season, the males take on color patterns that rival the brightest poster colors of tropical fishes (e.g., Page 1983; Etnier and Starnes 1993). Siltation and other forms of pollution and habitat modification have placed numerous darters in danger of extinction.

The **apogonid** cardinalfishes are a speciose (ca. 320 species) family of small (< 10 cm), nocturnal coral reef fishes. Their large eyes, large mouths, distinctly separated dorsal fins (the second with a single spine), deep bodies, and relatively pointed heads distinguish them from most other reef fishes that hover motionlessly just above or in structure. Cardinalfishes mouth brood their eggs, the male or female being responsible in different species. Many cardinalfishes live in close association with invertebrates and use them as refuges (see Chap. 21, Interspecific Relations: Symbioses and Parasitism). A few Indo-Pacific species enter estuaries, and several New Guinea species are restricted to freshwater.

The **malacanthid** tilefishes are a marine group that inhabits burrows. Sand tilefishes (e.g., *Malacanthus*) are tropical species that live over shallow, sandy areas and dig complex burrows, which they reinforce with shell and coral fragments, pieces of which are piled in a mound at the burrow's entrance. Their engineering activities create hard-bottom patches that are used by small fishes and invertebrates that would otherwise not colonize sandy regions (see Chap. 24, Fishes as Producers and Transporters of Sand, Coral, and Rocks). The larger temperate latiline tilefishes (e.g., *Caulolatilus*) are commercially sought species that inhabit large burrows in deeper, soft-bottom regions, although it is unknown whether they construct the holes themselves.

The next 10 or so percomorph families (pomatomids through caristiids) are generally active, marine, water-column dwellers with relatively compressed, often silvery bodies. The larger species are piscivores, and the smaller ones are zooplanktivores. The cosmopolitan **pomatomid** bluefish, *Pomatomus saltatrix*, which occurs in most major ocean basins except the eastern Pacific, has a well-deserved reputation for voraciousness (see Hersey 1988). This aggregating predator will enter a school of prey fish and slash and dismember far more individuals than are actually eaten; attacks on humans unfortunate or foolish enough to be in the water during such feeding frenzies are well documented.

The family **Nematistiidae** is sometimes combined with the next four families (remoras, cobia, dolphinfishes, and jacks) to form a clade known as carangoids. The colorful roosterfish, *Nematistius pectoralis*, is a monotypic piscivore of warm, eastern Pacific, inshore areas. It looks like an amberjack with a cockscomb of seven elongate dorsal spines. It is actively sought as a game fish and attains lengths of 1.5 m and can weigh 50 kg. The **echeneid** remoras or sharksuckers are a highly specialized group of percomorphs in which the first dorsal fin has been modified into a sucking organ for attachment to sharks, billfishes, whales, turtles, and an occasional diver. Some remoras, such as the large (1 m) *Echeneis naucrates*, are frequently seen free-swimming, whereas smaller species are almost always attached to hosts and may clean their gills as well as feed on scraps from the predatory host's meals. The monotypic cobia, *Rachycentron canadum* (**Rachycentridae**), looks and swims like a large (to 1.5 m) remora without a suction disk. Sport fishers frequently locate cobia by fishing near manta rays, but the nature of the association is unexplored.

The **coryphaenid** dolphinfishes include two species of open-water, surface-oriented predators that are often found in association with floating structure or seen chasing flyingfishes. Their golden coloration has earned them the Spanish name *dorado*, and their color-changing habits are legendary. Male dolphinfish (bulls) have a square head profile involving expansion of the bony portion of the supraoccipital region (forehead), whereas in females the forehead slopes more gradually; such obvious skeletal sexual dimorphism is rare in acanthopterygians. The closely related **carangid** jacks and pompanos are a large (140 species) family of tropical nearshore and pelagic predators and zooplanktivores that range in size from the small scads (*Decapterus, Selar, Trachurus*) to the large amberjacks and pompanos (*Seriola, Caranx*). Carangids tend to be slightly to very compressed, the extreme occurring in the lookdown, *Selene vomer*, which literally disappears when it faces an observer head-on. Carangids of all sizes are often found in shoals, and evidence of cooperative hunting exists in a few large species (e.g., *Caranx melampygus*; see Chap. 18, Attack and Capture). Carangids engage in a form of highly efficient and powerful locomotion, termed *carangiform swimming*, involving side-to-side movement of primarily the tail; thrust is transferred from the body musculature to the tail via tendons that cross the caudal peduncle region (see Chap. 8, Locomotion: Movement and Shape).

The next group of families (lutjanids through nemipterids) are generally heavy-bodied, tropical fishes that swim near the bottom and feed on large invertebrates and fishes (with some notable exceptions). The **lutjanids** are a large family (125 species) of ecologically diverse but generally carnivorous marine fishes inhabiting shallow to moderate depths in tropical and warm temperate seas and estuaries (e.g., Polovina and Ralston 1987). The typical snapper is a fairly large (to 1 m), heavy-bodied, suprabenthic, nocturnal or crepuscular predator with large canine teeth, such as the gray snapper, red snapper, mangrove snapper, or mutton snapper (*Lutjanus* spp., *Pristopomoides* spp.). However, many snappers live in the water column and are more streamlined, including the vermilion and yellowtail snappers (*Rhomboplites, Ocyurus*). One subfamily, the caesionine fusiliers (often placed in their own family, the **Caesionidae**), are small, streamlined, and brilliantly colored planktivores with forked tails and protrusible mouths that school near drop-offs and reef edges on coral reefs of the Indo-West Pacific.

Lobotid tripletails get their name from the unusual arrangement of soft dorsal, soft anal, and caudal fins of nearly equal size and rounded shape, the dorsal and anal placed posteriorly on the body overlapping the tail fin. Tripletails are heavy-bodied, basslike fishes of estuarine and freshwaters worldwide in temperate and tropical waters. Adults can reach 1 m in length and are uncommon; juveniles are more frequently encountered floating leaflike on their sides in mangrove regions.

The silvery mojarras (**Gerreidae**) are common inhabitants of sandy or silty regions near coral reefs and other shallow, warm-water habitats worldwide; some species enter freshwater. Their body shape and feeding habits are somewhat incongruous. They have a forked tail and a mouth that protrudes slightly downward, which might suggest zooplanktivory, but they are typically observed foraging head down with their extremely protractile mouths extended into the bottom sediments. When they emerge, they typically expel clouds of sediment out their gill openings, having retained benthic invertebrates with their gill rakers.

Grunts (**Haemulidae**) are moderate-sized coral reef fishes that are unusual in that they are more diverse in the New rather than the Old World tropics. Most grunts form shoals as juveniles; some, such as the porkfish, *Anisotremus*, which is common in the Florida Keys, continue to shoal as adults. Grunts are typically nocturnal feeders on benthic or grassbed-associated invertebrates, undertaking distinctive migrations between daytime resting and nighttime feeding regions (see Box 21.2). The **inermiid** bonnetmouths are haemulid derivatives adapted for zooplanktivorous feeding. They have the slender bodies, forked tails, and protractile mouths typical of many zooplanktivorous fishes.

Porgies (**Sparidae**) are gruntlike in appearance but are more diversified in their feeding than the haemulids. The western Atlantic sheepshead, *Archosargus probatocephalus*, has massive pharyngeal dentition used for crushing hard-bodied prey such as mollusks. Several sparids, such as the pinfish, *Lagodon rhomboides*, feed extensively on plants, making this one of the few percoid families to include strongly herbivorous species (herbivory becomes more common in more advanced groups). Together with the **centracanthids**, **lethrinids**, and **nemipterids**, the porgies form a group of related families known as sparoids; the sparids are the only family in the group to inhabit the western Atlantic region.

The next three families contain fishes that are frequently seen swimming just above and probing into the bottom with modified appendages. **Polynemid** threadfins are tropical marine fishes with highly specialized pectoral fins that are divided into two parts. The upper webbed portion is located laterally and shaped like a normal pectoral fin, whereas the ventral portion consists of three to seven long, unconnected rays that extend down from the throat region and are used to feel for prey on the bottom. The mouth is subterminal, as befits a bottom feeder.

Some **sciaenid** croakers also have a subterminal mouth. Many species have one or several small chin barbels. Sciaenids are a widespread tropical and temperate family and are particularly diverse in the southeastern United States. This large family (270 species) includes such important commercial and sport fishes as the red drum (spot-tail bass), black drum, croakers, weakfish, sea trouts, kingfishes, white sea bass, corbinas, and the endangered Mexican totoaba. The common name for the family comes from its sound production habits, which involve vibration of muscles attached to the gas bladder. As is frequently the case with sound-producing fishes, sciaenid otoliths are exceptionally large. The role of acoustic stimuli in the biology of sciaenids is also reflected in their very extensive lateral line, which extends posteriorly onto the tail and anteriorly as numerous pits and canals on the head. Although predominantly a marine family, freshwater species are common in South America, and one species, the freshwater drum, *Aplodinotus grunniens*, is common throughout the Mississippi River and adjacent drainages of North America and into Central America.

The third bottom-oriented family is the **mullid** goatfishes. This tropical family of medium-sized, nearshore marine predators has two highly prehensile chin barbels, which the fishes use to probe bottom sediments for prey. Their foraging activities frequently flush invertebrates from the sand, and it is not unusual to see wrasses and carangids following goatfishes and capturing escapees.

Monodactylid fingerfishes and **toxotid** archerfishes are brackish-water families of chiefly Indo-Pacific distribution. The monodactylids, or monos, are popular aquarium fishes. Their silvery white, laterally compressed bodies are exaggerated by extremely tall dorsal and anal fins, making some species twice as deep as they are long. Adult *Monodactylus* lack a pelvic fin, although juveniles possess one. They are convergent in shape and ontogenetic pelvic loss with the very compressed carangid Germanfish, *Parastromateus niger*. The archerfishes are a well-known and unique group of small, surface-dwelling estuarine and freshwater fishes that feed actively on terrestrial prey. Insects are shot out of overhanging vegetation with bullets of water produced by compressing the gill covers and shooting water drops along a groove created by the tongue and palate. This behavior is all the more fascinating because the fish corrects for the curving trajectory of its propelled droplets as well as correcting for light refraction at the water's surface, its eyes being submerged during hunting (Dill 1977).

The monodactylids and toxotids form, with the next six families, an unranked group known as the Squammipinnes, a name that refers to the rows of scales that cover the base of the dorsal and anal fins. The best-known families in this group are the butterflyfishes and angelfishes. The **chaetodontid** butterflyfishes include 114 tropical shallow-water species. Their center of diversity is in the Indo-Pacific region, where about 100 of the species occur. The tropical Atlantic contains 13 species, and the eastern Pacific has only 4 species. In many people's minds, butterflyfishes are synonymous with coral reefs (Burgess 1978; Motta 1989). Butterflyfishes are colorful and swim conspicuously about the reef during the daytime, often in pairs or small shoals, residing for long periods on the same reefs and with the same partners. Trophically, they fall into several categories of microconsumers, feeding either on coral polyps, small invertebrates hidden in crevices in the reef, tube worms, or zooplankton. Anatomically, they are deep-bodied, highly compressed forms, their body shape being exaggerated by stout dorsal, pelvic, and anal spines and a slightly to greatly elongated snout region.

Closely related to and often mistaken for butterflyfishes are the similarly or larger-sized angelfishes (**Pomacanthidae**). A major distinguishing feature between the two is the existence of a stout, posteriorly projecting spine at the angle of the preopercular bone and the absence of a pelvic axillary process in angelfishes. Many angelfishes undergo dramatic ontogenetic color changes, several species having confusingly similar but striking patterns as juveniles that change to species-specific and still-striking adult patterns. Larger species, such as the Caribbean French and gray angelfishes, frequently form pairs. Trophically, angelfishes differ from butterflyfishes in consuming sessile, benthic invertebrates such as sponges, tunicates, and anthozoans. Some species are known to follow sea turtles and feed on their feces, which may explain their disconcerting habit of hovering near designated latrines associated with undersea habitats such as Tektite and Hydrolab, before such submarine structures had internal plumbing. Again, about three-quarters of angelfish species occur in the Indo-Pacific.

Most of the remaining families in the suborder Percoidei are relatively small. Among the more speciose groups are the **nandid** leaffishes of South American, African, and southern Asian freshwaters, which are best known for their striking morphologic resemblance to floating leaves. They enhance this deception behaviorally by drifting slowly through the water toward unsuspecting prey, which they engulf with a remarkably expandable mouth. **Kyphosid** sea chubs (also called rudderfishes) are a herbivorous family of 42 reef species that swim actively in shoals relatively high above the reef compared to most other herbivores. Kyphosids are unique among fishes in that at least two western Australian species contain symbiotic bacteria in their guts that break down algae via fermentation (Rimmer and Wiebe 1987). Although a predominantly tropical family, two temperate derivative species, the opaleye, *Girella nigricans*, and the halfmoon, *Medialuna californica*, extend into California waters. Some authors feel that the kyphosids and the next four families (**arripids, teraponids, kuhliids,** and **oplegnathids**) form a monophyletic grouping of closely related fishes. Terapon id grunters are a marine and freshwater family containing 45 species that have a unique means of producing sounds. Paired muscles run from the back of the skull to the dorsal surface of the gas bladder; in other sound-producing fishes that utilize muscles to vibrate the gas bladder, such as gadoids, triglids, and sciaenids, the muscles are derived from trunk musculature and originate in the body wall.

The next five families are placed in the superfamily Cirrhitoidea, united by having five to eight elongate and unbranched lower rays in their pectoral fins. The colorful **cirrhitid** hawkfishes are small- to medium-sized reef predators that are best known for sitting absolutely still on tops of corals in seemingly conspicuous locales, waiting for potential prey fishes either to not notice them or to habituate to their presence (a Hawaiian hawkfish is the only predator known to have eaten a cleaner wrasse). Hawkfishes look like a cross between a small sea bass and a scorpionfish, but they are readily identified by the presence of filamentous tufts or cirri at the top of each spine in the first dorsal fin and by the elongate pectoral rays characteristic of the superfamily. One fairly deep-water reef species, the longsnout hawkfish, *Oxycirrhites typus*, occurs almost exclusively in black coral trees. Its prefer-

ence for deep-water habitats may more closely reflect depletion of its preferred perch, which has been removed from accessible shallow locations due to the jewelry trade. The last family of percoids are the **cepolid** bandfishes, which vary in their morphology from elongate, eel-like forms to fairy basslet-like, deep-water forms.

Suborder Elassomatoidei

The **elassomatid** pygmy sunfishes are interesting because of their strong convergence with the true sunfishes (Centrarchidae) and their miniaturization (maximum length is 45 mm, but several species are smaller than 20 mm). The six elassomatid species are primarily dwellers of swampy habitats in the southeastern United States (e.g., Everglades pygmy sunfish, Okefenokee pygmy sunfish). Many males take on iridescent blue coloration during the breeding season.

Suborder Labroidei

Although some of the 16 remaining perciform suborders contain very speciose families (e.g., blennies and gobies), by far the most successful suborder numerically is the Labroidei. Labroids are predominantly tropical, marine fishes (e.g., damselfishes, wrasses, parrotfishes), with a few species in the first two families inhabiting warm temperate waters. Two additional families, the surfperches and odacids, are temperate and marine. The most successful family in the suborder is the tropical freshwater cichlids, although no single biological generality applies to all members of this fantastically speciose and varied family of fishes. The six families are united primarily on the basis of pharyngeal jaw morphology, involving features of both upper and lower jaws. Pharyngeal adaptations for handling a diversity of prey types have contributed substantially to the success of several labroid families.

Cichlids are viewed as the basal group of the suborder (Fig. 15.8). Among the more than 1300 cichlid species are many aquarium fishes that have achieved popularity because of their small size, colorfulness, and willingness to behave and breed within the confines of an aquarium. Familiar South American species include freshwater angelfishes (*Pterophyllum*), discus (*Symphysodon*), oscars (*Astronotus*), convict cichlids (*Cichlasoma*), peacock bass (*Cichla*), and gravel-eaters (*Geophagus*). One species, the Rio Grande cichlid, *Cichlasoma cyanoguttatum*, gets as far north as Texas. Numerous cichlids, particularly African tilapias, have been deliberately or accidentally introduced in Florida, California, and Hawaii, chiefly for aquaculture purposes. Although diverse in the New World (ca. 300 species), the great majority of cichlids occur in Africa, where they have radiated explosively into numerous species flocks (Box 15.1). Old World cichlids include fishes in the genera *Haplochromis*, *Lamprologus*, *Oreochromis*, *Pseudotropheus*, *Sarotherodon*, and *Tilapia*. Most cichlids build nests; many African forms brood the eggs and young in the mouth of either the male or, more usually, the female. A few species occur in India and Sri Lanka, where they commonly inhabit estuaries (e.g., *Etroplus*). Cichlids are convergent in behavior, morphology, and ecology with centrarchid sunfishes. The two families can be distinguished by sunfishes having two nostrils on each side of the head and a continuous lateral line, whereas cichlids have only one nostril on each side and a broken lateral line (Fryer and Iles 1972; Keenleyside 1991; among many other books).

The embiotocid surfperches look very much like some of the larger, deep-bodied cichlids. However, they are an entirely (with one exception) marine family of small- to medium-sized, inshore fishes that occur most commonly around kelp beds, rocky reefs, surf zones, and tide pools. Twenty-one of the 24 species occur along the Pacific coast of North America; the other three live in Korea and Japan. They are the only labroids that are livebearers, the female giving birth to fully developed, large young. In a few species, males are even born reproductively mature (see Chap. 9, Fertilization). Trophically, some species are specialized as zooplanktivores, whereas most pick invertebrates from the bottom or off plants.

The **pomacentrid** damselfishes are generally smaller, more colorful, tropical marine equivalents of the surfperches (Fig. 15.10). Many damselfishes are herbivorous (e.g., *Stegastes* spp.), whereas others are zooplanktivorous and show the usual adaptations associated with life in the water column above structure, namely a fusiform body, forked tail, and highly protrusible mouth (e.g., *Chromis* spp.). Herbivorous damselfishes are typically territorial, guarding a small patch of reef substrate in which they feed, hide, and, in the case of males, court females and guard developing eggs. No posthatching care of young is shown except in one Indo-Pacific species, *Acanthochromis polyacanthus*. Some species are intimately associated with invertebrates, such as the anemonefishes (*Amphiprion, Premnas*) (see Chap. 21, Interspecific Relations: Symbioses and Parasitism). This is a tropical family containing 315 species that reaches its highest diversity in the Indo-Pacific region, with a few species in each ocean basin occurring in warm temperate waters (e.g., the garibaldi, *Hypsipops rubicunda*, of California) (Emery and Thresher 1980; Allen 1991).

The next three families are closely related and are considered members of a single family by some workers. The largest family is the wrasses, **Labridae**, a remarkably diverse and widespread marine taxon of at least 500 species that occurs in all tropical seas (see Fig. 15.10). Many temperate and even cool temperate members occur in both the Pacific and Atlantic Oceans, such as the eastern Pacific California sheephead and senorita (*Semicossyphus, Oxyjulis*), the western Atlantic tautog and cunner (*Tautoga, Tautogolabrus*), and several eastern Atlantic wrasses (*Labrus* spp.). Wrasses range in size from 5 cm (many species) to the giant Maori wrasse of the Indo-Pacific, *Cheilinus undulatus*, which can be 2.3 m long and has the unlikely diet of cowries and crown-of-thorns starfish. Pharyngeal jaws are especially diversified among labrids, with several species able to handle well-protected prey such as crabs, mollusks, and echinoderms. Labrids typically bounce along in the water column using labriform locomotion, a paddling motion of their pectoral fins (see Chap. 8, Locomotory Types). To feed, they stop momentarily above the bottom to capture prey with protrac-

tile jaws and stout teeth, pick zooplankters out of the water column, or remove external parasites from other fishes. The razorfishes (*Hemipteronotus, Xyrichtys*) are very compressed and escape disturbances by diving rapidly into bottom sediments. Wrasses as a group are strongly diurnal and enter sandy bottoms or reef crevices at night to sleep. Many wrasses change sex, most starting off life as females and later changing to very differently colored and shaped males (see Chap. 20, Gender Roles in Fishes).

The **odacids** are a small family of 12 species limited to the temperate waters of New Zealand and southern Australia. They are intermediate in appearance between wrasses and parrotfishes, having the elongate wrasse body form but the nonprotractile jaws and fused teeth of parrotfishes. The **scarid** parrotfishes include 83 species of tropical marine fishes best known for their fused teeth that form a parrotlike beak. The beak is used for biting off algal fronds or pieces of dead coral, which are then passed to massive pharyngeal mills for grinding and extracting algal cells from the coral matrix. As with wrasses, parrotfishes generally change color and sex, from initial phase females to terminal phase males. Parrotfishes are generally larger than wrasses, with some species, such as the blue and rainbow parrotfishes of the Atlantic (*Scarus* spp.) and the bumphead parrotfish of the Pacific (*Bulbometapon*), attaining 1 m in length.

Suborder Zoarcoidei

The zoarcoids as a group are generally elongate fishes of the North Pacific that occupy benthic habitats ranging from tide pools to abyssal depths. The **zoarcid** eelpouts are eel-like fishes with round heads, long dorsal and anal fins, and pointed tails. Some eelpouts give birth to live young after eggs develop internally, a condition referred to as ovoviviparity. Zoarcids inhabit soft bottoms at moderate to great depths (20 to 3000 m). Most species occur in the North Pacific, some occur in the North Atlantic and deep tropical regions of both oceans, and about 10% of the 220 species occur in the southern oceans near Antarctica. In contrast, the **stichaeid** pricklebacks and **pholid** gunnels are most common in intertidal and shallow, nearshore habitats, primarily in North Pacific waters. Members of both families are primarily microcarnivores, although two pricklebacks, *Cebidicthys violaceus* and *Xiphister mucosus*, are herbivorous year-round, an uncommon trait at high latitudes for any fishes (Horn 1989).

The **anarhicadid** wolffishes or wolf-eels are the anatomic and ecological equivalents of moray eels at high latitudes (Fig. 15.11). Attaining lengths of 2.5 m and weighing 45 kg, these large benthic predators of the North Pacific and Atlantic often live under rocks. They have large anterior conical canines and massive lateral and palatine molars for catching and crushing crustaceans, clams, sea urchins, and fishes.

Suborder Notothenioidei

The notothenioids are commonly referred to as the icefishes. The suborder is restricted primarily to high latitudes of the Southern Hemisphere, with greatest diversity in

BOX
15.1

Explosive Speciation and Species Flocks

When a volcano erupts, an earthquake causes uplifting, a landslide blocks a river or divides a lake, or drought and flood cycles accompany longer-term climatic changes, freshwater habitats can become isolated, separating small numbers of fishes from their conspecifics, predators, and competitors. The reproductive future of such an isolated individual or individuals is usually rather bleak, as potential mates and resources may also be in short supply in the newly created habitat. But such events occasionally lead to an evolutionary bonanza, as evidenced by several so-called species flocks or species swarms.

A **species flock** is a group of closely related species that share a common ancestral species and are endemic to an isolated region such as a single lake or island (Greenwood 1984). Flocks evolve when a newly created habitat with essentially open niches is colonized and the colonizing species experiences a relaxation of the many selection pressures that previously kept its population in check. Descendants of the founders disperse, differentiate, and fill the open niches. The end-product is rapid speciation—a literal explosion of speciation events—and the production of several descendant species in the area.

Some of the most spectacular assemblages of fishes and other animals worldwide represent the end-products of **explosive speciation**. Nonfish examples include such showcases of evolution as the fruit flies, land snails, and honeycreepers of the Hawaiian Islands, Darwin's finches in the Galapagos Islands, and the amphipods of Lake Baikal in Asia. Fish examples are numerous and instructive, and involve many different teleostean as well as some non-teleostean taxa. In addition, the fossil record shows us that the process of explosive speciation has operated dramatically in the distant as well as recent past.

The best-known fish flocks occur among the cichlid fishes of the three Great Lakes of Africa: Malawi, Tanganyika, and Victoria. The first two lakes are long, deep, rift valley lakes, situated along a split in the earth's crust that is nearly 3000 km long. Lake Victoria, the largest lake in Africa, is more round and shallow, having been created as rivers were blocked by slow uplifting of the basin. These lakes vary in size between 28,000 km² and 69,000 km², putting them within the size range of the Laurentian Great Lakes of North America (19,000 km² to 82,000 km²). Although of similar sizes, the species diversities of the lakes are totally dissimilar. The five Laurentian Great Lakes together contain about 235 fish species, whereas their African counterparts contain almost five times that amount, or approximately 1000 species. The African Great Lakes each contain more

fish species than any other lakes in the world (Fryer and Iles 1972; Axelrod and Burgess 1976; Greenwood 1981, 1991; Ribbink 1991).

The exact number of species in each African lake is difficult to determine because the region is remote from academic institutions that specialize in fish taxonomy, which means that many species remain to be collected or described. In addition, anatomic differences among some species are subtle, and there is good evidence that environmental degradation and introduced species have recently wiped out many species (see Chap. 25, Introduced Predators). Approximations for the three lakes indicate the following distributions:

	Lake Malawi	Lake Tanganyika	Lake Victoria
Total species	>495	240	>290
Cichlids	>450	165	>250
Endemic cichlids	445	163	247

Conservatively, there are approximately 60 genera and 1000 species of cichlids in the three lakes (estimates vary widely), as well as smaller flocks in smaller lakes and in rivers of central Africa (e.g., Greenwood 1991; Stiassny et al. 1992). The important points here are that most of the fishes in each lake are cichlids, most or all of the cichlids are endemic, and in each lake it is likely that most or all of the endemic cichlids share a common ancestor (Meyer et al. 1990; Meyer 1993). The morphologic and ecological divergence from such an ancestor, is also astounding given that Tanganyika and Malawi are only 2 to 10 million years old. The largest and smallest cichlids in Lake Tanganyika are *Boulengerochromis microlepis*, a predator, which attains a length of 80 cm, whereas small planktivores of the genus *Lamprologus* may be only 4 cm long as adults. The difference in mass between adults of the two genera is about 8000-fold. Morphologic diversity includes African cichlids that artificially resemble many different families of teleosts and occupy habitats and niches that parallel those of the other teleosts (Fig. 15.8). More spectacular still, sedimentation, radiocarbon dating, and mitochondrial DNA data all indicate that Lake Victoria may have been completely dry as recently as 12,500 years ago, which means that 300 endemic species there evolved in a very short period (Johnson et al. 1996).

Trophically, African cichlids do it all. Trophic groups include species that specialize in eating phytoplankton, sponges, sediments, periphyton, leaves, mollusks, benthic arthropods, zooplankton, fish scales and fins, fish eyes, eggs and

BOX 15.1 *(Continued)*

embryos, and other fishes. Major anatomic adaptations associated with different trophic habits are found in the lips, marginal dentition, gill rakers, and particularly the pharyngeal jaws of the different trophic groups. Two flocks within flocks occur in Lake Malawi, where each subflock has differentiated into a particular feeding type. Approximately 27 (but perhaps as many as 200) species of closely related *mbuna* cichlids live over rocky areas and feed on the algae and associated microfauna of the algae, a food type known as *aufwuchs*. An additional flock of approximately 17 *utaka* cichlids live together and feed on zooplankton (Fryer and Iles 1972; Ribbink 1991).

Whereas the African cichlids form the largest species flocks among fishes, other examples are often as dramatic. The following is a partial list of well-known flocks and some of their interesting characteristics.

1. The oldest extant species flock of fishes occurs in Lake Baikal, Russia, the oldest and deepest lake in the world. Here, sculpinlike cottoid fishes have differentiated into perhaps three families and approximately 50 species. Highly derived members of this group include 2 species of live-bearing, pelagic comephorid Baikal oilfishes, a marked difference from the ancestral benthic, egg-laying sculpins. Lake Baikal has also produced a flock of amphipods and has an endemic, freshwater species of monk seal.

2. As many as three genera and 18 species of cyprinids form a flock in Lake Lanao of the Philippines, which sits above an uplifted waterfall 18 m high. This dramatic and controversial flock includes fishes with "supralimital" jaw specializations, indicating that derived species have jaw characteristics outside the normal variation found within the rest of the family. The flock is also unique in that the presumed ancestor, *Puntius binotatus*, still occurs in lowland streams below the waterfall. The validity of the flock and its traits is obscured by destruction of holotypes during World War II and subsequent, multiple introductions of game and forage species that have displaced the native fishes; only three of the original cyprinids still occur in the lake (Kornfield and Carpenter 1984).

3. Eighteen species of atherinid silversides occur in a few lakes of the Mesa Central of Mexico. Diversification in this group includes a wide range in adult sizes and feeding types, from relatively typical, small (6 cm) zooplanktivores to piscivorous giants 30 cm long with specific names like *lucius* and *sphyraena* ("pike" and "barracuda") (Barbour 1973; Echelle and Echelle 1984).

4. A complex of flocks occurs among killifishes in Lake Titicaca and surrounding lakes of the Peruvian and Bolivian Andes. Most of these species belong to one widespread genus, *Orestias*. Because several lineages are involved, the killifish assemblage is actually made up of several flocks rather than a single flock, the largest being about 15 species. The species have diversified into deep-water, midwater planktivorous, piscivorous, miniaturized, and broad-headed forms, a departure from the surface-dwelling, insect-feeding killifish norm (Parenti 1984).

5. Eight species of coregonid ciscoes evolved from a common ancestor in the Laurentian Great Lakes of North America. It is also likely that smaller flocks of coregonids have arisen in lakes of the western United States, Canada, and northern Europe (Smith and Todd 1984).

6. Approximately 13 species of cyprinid fishes in the genus *Barbus* co-occur in Lake Tana of Ethiopia. They are unusual in part because 8 of the 13 species are piscivores, which is an unorthodox feeding pattern for minnows (Nagelkerke et al. 1994).

7. Historical continuity is evident in flocks of semionotid fishes that occupied the rift valley lakes of what is now the northeastern coastline of North America between North Carolina and Nova Scotia (McCune et al. 1984; McCune 1990). Semionotids were very successful Jurassic neopterygians on a direct line to modern gars (Lepisosteidae) (see Chap. 11, Subclass Neopterygii). Semionotid flocks of up to 17 species formed and were extinguished repeatedly as lakes filled and evaporated on a 21,000-year cycle over a period of 33 million years. Within the genus *Semionotus*, body shape varied substantially, including elongate pikelike forms, rounded sunfishlike forms, and intermediate shapes (Fig. 15.9). The setting and speciation patterns directly parallel flock formation in the modern African rift valley lakes. Other fossil flocks include a radiation of 8 sculpin species during the Pliocene in Lake Idaho in the western United States (Smith and Todd 1984).

The diversity of species that exist, their ecological relationships and innovations, and the repeatability of the process are remarkable examples of speciation and adaptation, even occurring in taxa that we do not normally think of as highly variable, speciose, or particularly rapidly evolving.

(continued)

BOX 15.1   *(Continued)*

How can such speciation occur, especially in lakes that may be only a few thousand years old? A small lake near the edge of Lake Victoria may provide a clue. Lake Nabugabo sits 3 km away and 15 m above Lake Victoria, draining into the larger lake through a swamp. Six species of *Haplochromis* cichlids occur in Nabugabo, five of which are endemic and have close relatives in Victoria.

Charcoal dates from former strandlines indicate that Nabugabo is only 4000 years old and that Victoria has repeatedly risen and overflowed into Nabugabo and other surrounding lakes, providing colonists for the smaller lakes. As waters receded, these colonists were isolated from competitors and predators and would have been able to oc-cupy the niches of the newly created small lakes, speciating in as little as 4000 years. As the larger lake rose again, the new species could now swim into the larger, ancestral lake, increasing its diversity if they were unable to interbreed with their former conspecifics. As this scenario was repeated in numerous small lakes around Victoria, the generation of many species that would eventually occupy the larger lake is imaginable. It is through such **allopatric** processes of isolation and differentiation that species flocks in most lakes are likely to have developed. Other processes, involving within-lake (**sympatric**) development of species with minimal dispersal, are also likely but more poorly understood (Greenwood 1991).

FIGURE 15.8.   *Diversity in body shape among African cichlids. These fishes belong to several different genera of cichlids, yet are roughly similar in body form to several other teleostean families. Cichlid genera and suggested convergences are (A) Tilapia versus a centrarchid sunfish; (B) Xenotilapia versus a malacanthid tilefish; (C) Serranochromis versus a serranid sea bass; (D) Xenotilapia versus a gobiid goby; (E) Boulengerochromis versus a lutjanid red snapper; (F) Telmatochromis versus a batrachoidid toadfish; (G) Rhamphochromis versus a centropomid snook; (H) Telmatochromis versus an opistognathid jawfish; (I) Julidochromis versus a labrid wrasse; and (J) Spathodus versus a scarid parrotfish.*

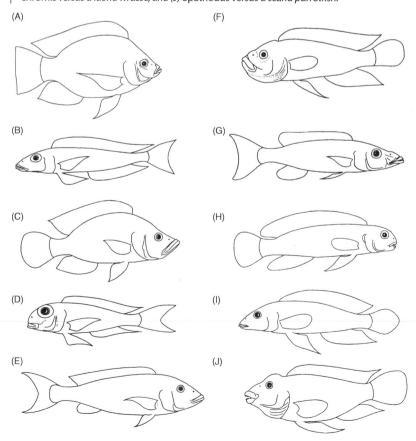

(A)

(B)

(C)

(D)

(E)

(F)

(G)

(H)

(I)

(J)

From Fryer and Iles 1972; used with permission.

FIGURE 15.9.   *Variation in morphology of an ancient species flock. (A) A completely reconstructed individual of the Jurassic neopterygian* Semionotus. *(B) Variation in body form of species related to* Semionotus *that probably co-occurred in a single lake.*

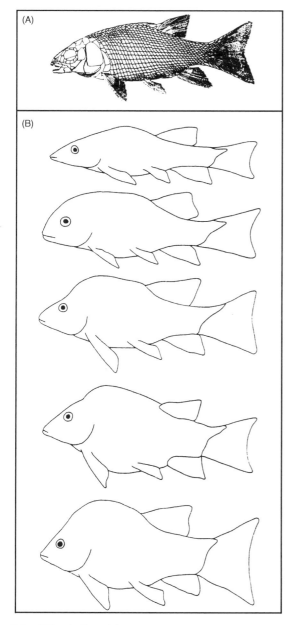

(A)

(B)

From McCune 1990; used with permission.

benthic habitats of Antarctica. These cold-water fishes show numerous physiological and behavioral adaptations to prevent their tissues from freezing, including the production of a variety of glycoprotein antifreezes (see Chap. 7, Coping with Temperature Extremes). Many of the fascinating characteristics of the biology of this group are detailed in Chapter 17 (Antarctic Fishes) and will not be repeated here. The **bovichthyids** of Australia, New Zealand, and southern South America are considered the stem group for the rest of the suborder. **Nototheniid** cod icefishes are predominantly benthic, with some secondarily

pelagic species that achieve neutral buoyancy by depositing lipids in their muscles and by reduction of skeletal material, two adaptations to water-column existence that occur convergently in other families derived from benthic ancestors (e.g., cottoid Baikal oilfishes, many deep-sea forms). The **channichthyid** crocodile icefishes are well studied because they lack red blood cells, hemoglobin, and myoglobin, making their blood and flesh colorless. These traits probably reflect the high amount of dissolved oxygen in cold antarctic waters.

Suborder Trachinoidei

Trachinoids are mostly benthic, questionably related marine fishes, several of which sit buried in the sand throughout the day or seek refuge in the sand when not feeding. **Chiasmodontid** swallowers depart from the suborder norm in being one of the few acanthopterygians to occupy mesopelagic and bathypelagic depths. They show convergent traits with other deep-sea fishes, including a large mouth, long teeth, slender jaw bone elements, distensible mouth and stomach, black coloration, and photophores. The family **Cheimarrichthyidae** consists of a single New Zealand species, *Cheimarrichthys fosteri*. It is known as the torrent fish, reflecting its daytime habitat in turbulent streams. Its body form, inferior mouth, large horizontally placed pelvics, and broad, flattened head converge with other swift-water fishes such as longnose dace, balitorine hillstream loaches, African kneriids, amphiliid loach catfishes, clingfishes, and rhyacichthyid loach gobies.

Trichonotid sand divers share a peculiarity with some elasmobranch rays by having protuberant eyes and a dorsal eyelid of sorts made up of an iris flap with strands that extend over the lens. Both groups rest on or bury in the sand in shallow water with only their eyes visible. **Ammodytid** sand lances are small, elongate, shoaling fishes that feed on zooplankton in the water column by day and spend nighttime buried in the sand. The **trachinid** weeverfishes are well-known eastern Atlantic and Mediterranean benthic fishes with highly venomous opercular and dorsal spines.

The **uranoscopid** stargazers are another venomous family, with two grooved spines and an accompanying gland sitting just behind the gill cover and above the pectoral fins. Stargazers also lie on the bottom or bury in the sand, with their dorsally located eyes exposed. Their incurrent nostrils are directly connected to the mouth, which may allow them to breathe while buried. A fleshy filament extends upward from the floor of the mouth and is used to lure prey. Stargazers include some of the only marine teleosts that are electrogenic, with strong pulses of electricity (up to 50 volts) being produced by highly modified extrinsic eye muscles. Stargazers discharge when captured and may also use electricity to stun prey. Stargazers are convergent in body form and habits with paracanthopterygian toadfishes and blennioid dactyloscopid sand stargazers.

Suborder Blennioidei

Blennioids are small, benthic, marine fishes of tropical and subtropical regions (Fig. 15.12). They generally pos-

FIGURE 15.10. *Representative labroids. (A) A pomacentrid damselfish, the sergeant major,* Abudefduf saxatilis. *(B) A labrid, the California sheephead,* Semicossyphus pulcher.

(A)

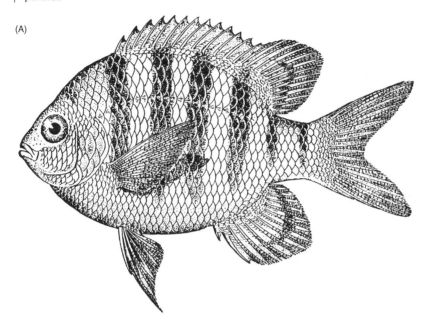

(B)

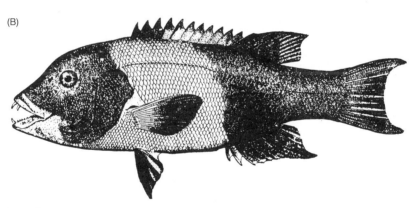

From Jordan 1905.

sess long dorsal and anal fins and fleshy flaps termed cirri on some part of the head. Triplefin blennies (**Tripterygiidae**) derive their name from having a soft dorsal fin plus a spiny dorsal divided into two parts (hakes and cods are the only other fishes with three distinctive dorsal fins). **Dactyloscopid** sand stargazers look like miniaturized (to 15 cm) uranoscopid stargazers with their oblique mouths and stalked eyes. They similarly have a specialized breathing mechanism probably related to their burying habits. In most fishes, water is brought into the mouth and out the gills by combined actions of buccal and opercular pumps (see Chap. 5, Respiration and Ventilation). Sand stargazers move water via a branchiostegal rather than an opercular pump. Fingerlike projections inside the mouth may keep sand out of the gills. Some dactyloscopid males care for eggs by carrying them under the axil of each pectoral fin. **Clinids**, also known as kelpfishes and fringe-

heads (including the intriguingly named sarcastic fringe-head), vary in size from 5 cm to the predatory giant kelpfish, *Heterostichus rostratus*, which reaches 60 cm. Clinids are shallow-water, benthic forms associated closely with structure in both temperate Southern and Northern Hemispheres. Some clinids give birth to live young.

Chaenopsid pikeblennies and tubeblennies are small tropical, New World fishes that are most often found living in or with corals. Many tubeblennies are essentially infaunal. Shortly after settling from the plankton, a tubeblenny will take up residence in an available polychaete worm tube and is likely to remain there for the rest of its life. The combtooth blennies of the family **Blenniidae** are very diverse, accounting for 345 species of mostly small, benthic fishes in tropical and subtropical waters worldwide. The comblike teeth are used to

FIGURE 15.11. *(A) A zoarcoid spotted wolffish,* Anarhichas minor. *(B) Skull of the related Pacific wolf-eel,* Anarrhichthys ocellatus, *showing the massive, diversified dentition of these predators.*

(A)

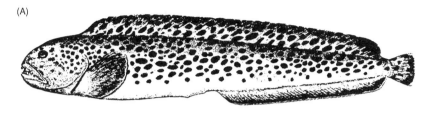

(B)

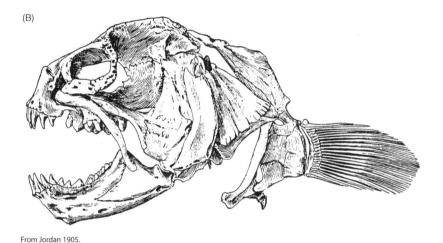

From Jordan 1905.

crop algae in many species. Many of the advanced nemophine sabre-toothed blenniids (*Aspidontus, Meiacanthus*) swim freely in the water column and are also involved in mimetic relationships with other fishes. The best-known example is the cleanerfish mimic, *Aspidontus taeniatus*. This sabre-toothed blenny strongly mimics the blue and black coloration and bobbing solicitation dance of labrid cleanerfishes, particularly of the cleaner wrasse, *Labroides dimidiatus*. When allowed to approach a posing host fish, rather than cleaning the host, *Aspidontus* bites off a piece of fin. Other sabre-toothed blennies attack passing fish and remove scales or pieces of fin (they attack prey as large as skindivers and generally attack once the diver has passed overhead, which is always a surprising, painful, and distinctly unsettling experience). Some of the sabre-toothed species are referred to as poison-fanged blennies because of their hollow lower canines that can inject a toxin. Poison-fanged blennies (*Meiacanthus*) may be mimicked by similarly colored blennies (e.g., *Ecsenius, Plagiotremus, Runula*), which thus

FIGURE 15.12. *A bleniid blenny,* Blennius yatabei, *from Japan.*

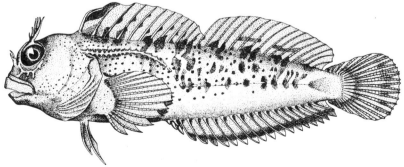

From Briggs 1974; used with permission of McGraw-Hill, Inc.

gain protection from predators (Losey 1972; Springer and Smith-Vaniz 1972).

Suborders Icosteoidei, Gobiesocoidei, and Callionymoidei

The monotypic family **Icosteidae** contains the very peculiar North Pacific ragfish, *Icosteus aenigmaticus*. Elliptical in shape and highly compressed, spineless, scaleless, without pelvic fins as an adult, and with a largely uncalcified cartilaginous skeleton, these 2-m-long pelagic predators look like a free-swimming flatfish with a limp body. They are a reported favorite prey of sperm whales. The **gobiesocid** clingfishes are a shallow-water to amphibious family of small marine fishes often found in high-energy wave zones (Fig. 15.13). The pelvic fins are modified into a sucking disk, the body is depressed, the head is rounded and flattened, and the skin is smooth and scaleless. They have unique pectoral girdle and vertebral rib arrangements. A relative giant in the family, the 30-cm Chilean *Sicyases sanguineus*, feeds on snails, barnacles, chitons, and other high intertidal prey as well as many kinds of algae, often preferring the wave-splashed supratidal region to more regularly inundated depths (or heights) (Paine and Palmer 1978). It is reputed to have aphrodisiac qualities. Dragonets (**Callionymidae**) are a diverse (130 species), chiefly marine family of small, shallow-water fishes in the Indo-West Pacific. Some species are pale white and live over sand, whereas others associated with hard bottoms are quite colorful. The family includes a popular aquarium species, the green and orange mandarinfish or splendid dragonet, *Synchiropus splendidus*.

Suborder Gobioidei

Gobioids are usually small, benthic or sand-burrowing fishes, mostly marine but with many freshwater species. Gobioids, as is common among benthic fishes, lack a gas bladder. Different families show differing degrees of fusion of the pelvic fin. Two families, the sleepers and the gobies, account for 95% of the more than 2100 species in the suborder, the latter family being by far the largest. In the stream-dwelling **rhyacichthyid** loach gobies of Indo-Australia, the flattened anterior third of the body in combination with the pelvic fins forms a sucking disk for holding position in fast-flowing water. The **eleotrid** sleepers are small- to medium-sized (to 60 cm), widely distributed estuarine and stream fishes of tropical and subtropical regions. They are often the major predators of

stream systems on oceanic islands such as Hawaii and New Zealand and have a complex life history that includes a marine planktonic larva, reflecting a probable marine ancestry for the family. Pelvic fins are usually separated.

The **gobiid** gobies (Fig. 15.14) are the largest family of marine fishes in the world, comprising more than 1875 species (the distinction of largest fish family overall is contested among gobies, cyprinids, and cichlids, all with around 2000 species). Gobies usually have their pelvic fins united, some species using them as a suction disk for clinging to hard substrates. Many species live on mud or sand or in association with invertebrates such as sponges, sea urchins, hard and soft corals, and shrimps (see Chap. 21, Interspecific Relations: Symbioses and Parasitism). The neon goby, *Gobiosoma oceanops*, is an important cleanerfish in the Caribbean and is convergent in coloration with labrid cleanerfishes of the tropical Pacific. Other derivative species include the essentially amphibious mudskippers (*Periophthalmus*, *Boleophthalmus*). Gobies are generally small (< 10 cm) and lay claim to the title of the world's smallest fishes. Diminutive species include an Indian Ocean species, *Trimmatom nanus*, that matures at 8 to 10 mm and several species in the genera *Eviota*, *Mistichthys*, and *Pandaka* that mature at about 10 mm. The largest goby is a western Atlantic/Caribbean form, the violet goby, *Gobioides broussenetti*, a purplish eel-like fish about 50 cm long with elongate dorsal and anal fins. Goby systematists include among their ranks the Emperor of Japan (e.g., Akihito 1986).

The remaining gobioids are mostly small, eel-like, tropical marine fishes that live on or in sand. The elongate **kraemeriid** sand gobies often rest in sand with just the head exposed, frequently in wave-tossed areas. The **microdesmid** wormfishes are mostly similar to kraemeriids, except for some spectacularly colored members of the subfamily Ptereleotrinae (e.g., *Ptereleotris*). These fishes are also known as hover gobies, dartfishes, or firefishes, names that describe their coloration and their habit of hovering above the bottom and diving rapidly into a burrow when disturbed.

The **schindleriids** are an enigmatic family of two species of small (2 cm) pelagic fishes that are neotenic, which means that they are essentially adults that retain larval traits (or larvae that have developed functional gonads). Retained larval characteristics in schindleriids include a larval-type kidney (pronephros), lack of pigmentation, and unossified skeleton. The extremely small size of many

FIGURE 15.13.   *A gobiesocid clingfish,* Gobiesox lucayanus, *from the West Indies.*

From Briggs 1974; used with permission of McGraw-Hill, Inc.

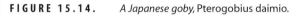

FIGURE 15.14.   *A Japanese goby,* Pterogobius daimio.

From Jordan 1905.

gobioids may represent convergent neotenic pathways (Johnson and Brothers 1993). Many fish families contain species that have evolved through heterochronic alterations in developmental sequences (see beloniforms above and Chap. 10); an example analogous to schindleriids involves nettastomatid duckbill eels, in which the leptocephalus larvae possess developed ovaries (Castle 1978).

Suborder Kurtoidei

The **kurtids** of the Indo-Malay and northern Australia regions are interesting because of the peculiar manner in which males care for the eggs. The males have a hooklike growth of the supraoccipital crest on the top of their heads to which the eggs are attached and where they are carried until hatching. The means of attachment and point at which fertilization occurs are apparently unknown.

Suborder Acanthuroidei

This suborder contains, with one exception, a group of medium-sized, compressed fishes with small mouths that usually form shoals over coral reefs or in nearby habitats. The **ephippid** spadefishes (*Chaetodipterus, Platax*) are conspicuous inhabitants of drop-offs and passes around reefs, although young spadefishes also frequent sandy beaches along the Atlantic coast of the United States. Juveniles of both genera look and act remarkably like floating leaves. The **scatophagid** scats bear a slight resemblance to serrasalmine piranhas, but as their name implies, their feeding habits tend more toward feces and detritus than to live prey. They are inhabitants of estuaries and the lower portions of rivers in the Indo-Pacific, where they are reputed to hang out near sewage outfalls.

Rabbitfishes (**Siganidae**) are reef, grassbed, and estuarine herbivores with a unique pelvic formula of I, 3, I, reflecting the hard spine at either end of the fin. Rabbitfishes are in fact very spiny fishes, their first dorsal spine projecting forward rather than upward, and many of the spines possessing a painful toxin (the forward-projecting spine frequently impales the uninitiated fisher). Although most rabbitfishes are countershaded species that shoal in sea grass and mangrove areas, reef-dwelling species such as the fox-face, *Siganus vulpinus,* converge in coloration and habitat with butterflyfishes and even form apparently monogamous pairs, as happens in several butterflyfishes.

The exceptional monotypic louvar, *Luvarus imperialis* (**Luvaridae**), is a pelagic derivative of the suborder. It is a large (to 1.8 m, 140 kg), nonshoaling fish with extremely high fecundity; a large female may contain nearly 50 million eggs. Louvars converge with the pelagic scombroids (see below) in having a lunate tail and a lateral keel on the caudal peduncle, and the posteriorly set dorsal and anal fins resemble the finlets of the tunas and mackerels. The head shape looks more like a dolphinfish, another pelagic species. Louvars feed on jellyfishes, salps, and ctenophores.

The moorish idol, *Zanclus canescens* (**Zanclidae**) is another monotypic species related to the surgeonfishes. It is a strikingly shaped and colored Indo-Pacific reef fish that is remarkably convergent with butterflyfishes in body form, coloration, and behavior, including elongate dorsal spines and projectile horns above the eyes, as in the butterflyfish genus *Heniochus*. The 72 species of **acanthurid** surgeonfishes, unicornfishes, and tangs are most easily distinguished by the knife blade present on the caudal peduncle. This blade is a modified scale and can exist as fixed, laterally projecting plates in *Prionurus* and the unicornfish genus *Naso*, or as single, forward-projecting knives that are exposed as the fish flexes its body. The blade is often covered with a toxic slime, with the strength of the toxin apparently directly related to the length of the blade. The peduncular blade makes surgeonfishes among the few fishes that should not be grasped by the tail. Unicornfishes derive their name from a long bony protuberance on the head of some species that serves an unknown function. Surgeonfishes are often beautifully colored fishes, the color changing with age. As a group they are herbivorous (except for planktivorous unicornfishes); species differ in dentition, jaw mechanics, and the body angles at which they remove algae from the reef.

Suborder Scombrolabracoidei

The monotypic *Scombrolabrax heterolepis* (**Scombrolabracidae**) is a peculiar, 30-cm-long, deep-water oceanic fish with a protractile jaw and a unique gas bladder arrangement that includes numerous bubblelike projections that fit into depressions of expanded vertebral accessories. Its relationships are puzzling, but it is probably closest to the gempylid snake mackerels.

FIGURE 15.15.   *Representative scombroid fishes. (A) A sphyraenid, the great barracuda,* Sphyraena barracuda. *(B) A scombrid, the albacore,* Thunnus alalunga.

(A)

(B)

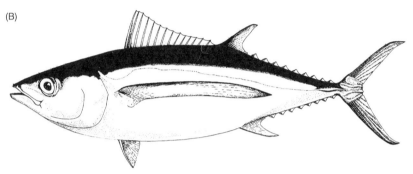

(A) Photo by G. Helfman; (B) from Briggs 1974; used with permission of McGraw-Hill, Inc.

Suborder Scombroidei

Scombroids include the unquestionably fastest and some of the largest predators in the sea, namely the tunas and billfishes. The suborder is characterized by a nonprotractile mouth, a secondary modification given the general trend in teleosts toward increasing protrusibility (Fig. 15.15). Several families have independently evolved some form of endothermy and heat conservation (see Chap. 7, Endothermic Fishes; Chap. 17, Pelagic Fishes). The barracudas (**Sphyraenidae**) are considered primitive members of this highly successful pelagic suborder. Twenty species of barracuda inhabit tropical and subtropical regions of the Atlantic, Pacific, and Indian Oceans. Most barracudas are schooling predators, an important exception being the usually solitary great barracuda, *Sphyraena barracuda*, the biology of which is surprisingly poorly studied. Great barracuda approach 2 m in length (a topic of considerable controversy); have fanglike, flattened teeth capable of slicing cleanly through most prey; and have the unnerving habit of following divers around the reef, motivated by either curiosity or territoriality.

The **gempylid** snake mackerels include 23 species of pelagic and deep-water predators characterized by an elongate body; large mouth with long teeth; a long, spiny dorsal fin; and a series of dorsal and ventral finlets just ahead of the tail. The family includes the cosmopolitan oilfish, *Ruvettus pretiosus*, a large (1.8 m, 45 kg) predator of moderate depths. It is sometimes referred to as the castor-oil fish because of the purgative quality of its meat. An active fishery for oilfish in the Comoro Islands off eastern Africa captures endangered coelacanths as bycatch (see Box 13.2). The **trichiurid** cutlassfishes look like

very compressed, silvery snake mackerels that have lost their pelvic fin, most of their anal fin, finlets, and most of the tail. The fanglike teeth belie a diet of large zooplankton, at least in smaller individuals.

The **scombrid** mackerels and tunas are highly adapted for a mobile, open-sea existence in terms of anatomy, physiology, and behavior (see Chap. 17, Pelagic Fishes). Primitive members of the group, such as the mackerels and Spanish mackerels and relatives (*Scomber, Scomberomorus*), tend to live closer to shore, whereas advanced members are highly pelagic and nomadic. Although chiefly a tropical and subtropical family, several species move into cold waters for feeding. Sizes range from relatively small, 50-cm mackerels (*Scomber, Auxis*) to the giant bluefin tuna, *Thunnus thynnus*, at 4 m and 500 kg. Most are schooling fishes of tremendous commercial importance (e.g., Sharp and Dizon 1978).

Chief predators on even relatively large tunas are the temperate and warm temperate **xiphiid** swordfish, *Xiphias gladius*, and the more tropical **istiophorid** sailfishes, spearfishes, and marlins (*Istiophorus, Tetrapturus,* and *Makaira*). The bill in both groups consists of an expanded premaxillary bone that is depressed and smooth in the swordfish and more rounded and prickly in marlins and their relatives. Other differences include no pelvic fins, a single caudal keel, and relatively stiff, sharklike pectoral, dorsal, and anal fins in the swordfish. The istiophorids in contrast have long pelvic filaments, flexible pectorals, double keels, and a long, depressible spiny dorsal that reaches its extreme expression in the sail of the sailfish, a structure of debated function. Controversy has also raged over how and whether billfishes utilize their bill for feeding, but recent observations indicate it can serve as a spear or as a cutlass or billy (see Box 18.1). Swordfish attain sizes of 530 kg, whereas both blue and black marlin grow to 900 kg.

Suborder Stromateoidei

Stromateoids are generally tropical and warm temperate fishes of the open sea that often associate as juveniles with floating or swimming objects, particularly with siphonophores and jellyfishes. A characteristic of the suborder is a thick-walled sac in the pharyngeal region that contains "teeth" made from hardened papillae. **Centrolophid** medusafishes bear a superficial resemblance to the icosteid ragfish. **Nomeid** driftfishes, as the name implies, hover around and under floating logs, siphonophores, jellyfishes, and seaweed as juveniles and occur in deeper water as adults. Juveniles of the man-of-war fish, *Nomeus gronovii*, live with impunity among and even feed on the stinging tentacles of the Portuguese man-of-war. **Ariommatids** are superficially similar to the carangid scads (i.e., *Decapterus*).

The **tetragonurid** squaretails are round fishes encircled by ridged scales and with a long caudal peduncle that has a single keel on either side formed from scale ridges. They feed on pelagic cnidarians and ctenophores, which they bite with specialized knifelike teeth. **Stromateid** butterfishes and harvestfishes are round or elliptical in profile with a forked tail and are similar in shape to some carangids; they are sometimes referred to as pompanos, a name more correctly applied to several carangids. As with many open-sea groups, butterfishes lack pelvic fins as adults (for unknown reasons).

Suborder Anabantoidei

Anabantoids are also called labyrinth fishes because of a complexly folded, auxiliary breathing structure derived from the epibranchial of the first gill arch located above the gills in the gill chamber (see Fig. 5.4). Functionally, the suprabranchial organ is the primary breathing structure for many species, as fish in well-aerated aquaria will die if not allowed to gulp air at the surface. In most anabantoids, the male exhales a nest of mucus-covered bubbles, among which eggs are laid and which he guards. The **luciocephalid** pikehead, *Luciocephalus pulcher*, is an elongate stalking predator on small fishes with a body form characteristic of other such piscivores (elongate jaws, slender body, dorsal and anal fins set far back on the body, rounded tail) (see Chap. 8, Locomotion: Movement and Shape; Chap. 18, Attack and Capture). As befits an advanced percomorph, pikeheads have the most protrusible mouth of any teleost. When feeding, the mouth is shot forward rapidly, surrounding the prey. Pikeheads have an interesting bone in the gular region of their throats that is analogous to the gular plate(s) of the primitive coelacanth, bowfin, bichirs, and some elopomorphs; whether this reinvented gular bone functions in oral incubation of eggs or in mouth protrusion is unclear (Liem 1967).

The next four families of gourami-like fishes are closely related. **Anabantid** climbing gouramis or climbing perches are African and Asian freshwater fishes that derive their name from their ability to move across wet ground (and supposedly even up wet tree trunks), jerking along by thrusts from the tail while the pectoral fins and gill covers act as props. The kissing gourami, *Helostoma temmincki*, is the sole member of the family **Helostomatidae**. The peculiar kissing behavior of this species is derived from its feeding habits, which involve scraping algae from surfaces using horny teeth on distinctive lips. The function of kissing, in which two individuals repeatedly press their open mouths against each other, is poorly understood.

The **belontiid** gouramis, fighting fishes, and paradise fishes have elongate pelvic fin rays that gouramis extend forward during exploratory behavior and when two individuals approach each other. Belontiids in the genus *Colisa* shoot water droplets at terrestrial insects, in a manner analogous to that of the toxotid archerfishes (Dill 1977). Bettas (Siamese fighting fish) are used extensively in behavioral and genetic studies. Males are exceedingly pugnacious toward each other. They are bred and fought like fighting cocks, making them one of the few fishes cultured for reasons other than food, appearance, or research. Fights to the death in the confines of an aquarium do not reflect real-life situations where a subordinate fish can flee from a dominant. The **osphronemid** giant gourami, *Osphronemus goramy*, reaches 80 cm in length and is a popular food fish that is cultured throughout Southeast Asia. Its air-breathing abilities make keeping it alive in fish markets very easy.

Suborder Channoidei

Snakeheads (**Channidae**) are highly predatory freshwater fishes of tropical Asia and Africa. They have a suprabranchial breathing organ reminiscent of that of the anabantoids and are primarily swamp dwellers. The robust, elongate bodies, long dorsal and anal fins, and ringed eyespot on the caudal peduncle give some channids a superficial resemblance to a bowfin. Some snakeheads reach more than 1.2 m in length and 20 kg in mass and are prized food fishes.

ORDER PLEURONECTIFORMES

Flatfishes are distinctive, compressed acanthopterygians that all share certain features, most noticeably a marked asymmetry that includes having both eyes on the same side of the head in juveniles and adults (Fig. 15.16). Flatfishes begin life as bilaterally symmetric, pelagic fishes, but during the larval period or shortly thereafter, one eye migrates to the other side and the fish settles to the bottom, lying on its blind side (see Chap. 10, Complex Transitions: Smoltification in Salmon, Metamorphosis in Flatfish). Eye movement is made more complicated by the position of the anterior portion of the dorsal fin, which often originates above or ahead of the eyes. Teeth, scales, paired fins, and pigmentation also typically differ between sides. Families are generally either right-eyed or left-eyed, depending on which eye stays put.

FIGURE 15.16. *A pleuronectiform, the bothid fourspot flounder,* Paralichthys oblongus.

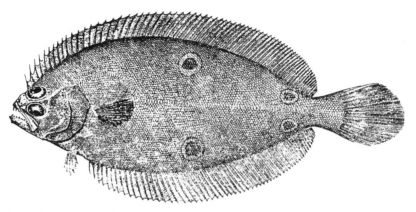

From Bigelow and Schroeder 1953.

Flatfishes are benthic, carnivorous, marine fishes (with a few freshwater derivatives) that lack a gas bladder; most live in shallow to moderate depths in arctic, temperate, and tropical locales. Some species are able to change the coloration of the eyed side to match the shading pattern of the background upon which they rest. Flatfishes constitute important fisheries for forms known as dab, flounders, halibuts, plaice, sole, tonguefishes, turbots, and whiffs. Approximately 490 of the 570 species are in five large families. Ancestral, intermediate, and closely related forms to flatfishes are not known, and the fossil record is limited.

The primitive **psettodid** flatfishes show the least movement of the eye during metamorphosis. Among the **bothid** lefteye flounders is the peacock flounder (*Bothus lunatus*) of the Caribbean. The family **Paralichthyidae** includes the summer flounder (*Paralichthys dentatus*) and California halibut (*P. californicus*), the latter species reaching a size of 1.5 m and 30 kg. The **pleuronectid** righteye flounders include the larger halibuts such as the Atlantic and Pacific halibuts. Females of the Pacific species, *Hippoglossus stenolepis*, may live 40 years and reach barndoor proportions of 3 m long and 200 kg mass, but may not mature until they are 16 years old. Several commercially important flatfish species are pleuronectids, including arrowtooth flounder, petrale sole, rex sole, winter flounder, yellowtail flounder, and English sole.

The true soles, however, are in the family **Soleidae**, which are usually right-eyed. Among the soleids is the Red Sea Moses sole, *Pardachirus marmoratus*, which exudes a toxin, pardaxin, reported to be a natural shark repellant. The common sole, *Solea solea*, of European waters is reported to raise its black-tipped pectoral fin when disturbed in an action that mimics the raising of the dorsal fin of venomous trachinid weeverfishes. The **cynoglossid** tonguefishes are the most elongate of the flatfishes and also show considerable variability in habitat choice, including shallow-water and burrowing forms (e.g., blackcheek tonguefish, *Symphurus plagiusa*), several species that occur as deep as 1900 m, species that invade rivers (e.g., the hogchoker, *Trinectes maculatus*), as well as purely freshwater forms (three species from Indonesia).

ORDER TETRAODONTIFORMES

The pinnacle of teleostean evolution is reached among the highly derived fishes of the order Tetraodontiformes. The name refers to the common pattern of four teeth in the outer jaws of puffers (an alternative name, Plectognathi, means twisted jaw). Tetraodontiforms are characterized by a high degree of fusion or loss of numerous bones in both the head and body. In the head region, such bones as the parietals, nasal, infraorbital, and posttemporal are commonly missing; both hyomandibular and palatine bones may be firmly attached to the skull; and the maxilla is fused to the premaxilla. The pelvic fins and lower vertebral ribs are often missing, and vertebral number is reduced from the common acanthopterygian condition of 26 to as few as 16. The skin has a thick, leathery feel and is covered by scales that are modified into spines, bony plates, or ossicles, some of rather spectacular proportions.

Tetraodontiforms tend to eat animals that are generally unavailable to most other reef fishes, such as sponges, sea urchins, hard corals, and jellyfishes. Some are predators on sessile benthic invertebrates (triggerfishes, puffers); others are water-column swimmers above the reef that feed on zooplankton (black durgon triggerfish); still others are large, offshore planktivorous species (gray triggerfish) or jellyfish feeders (ocean sunfish). All but 20 species are marine. Even though tetraodontiforms are recognized as the most advanced of the teleosts, the fossil record for the group goes back at least to the early Eocene and quite likely the late Cretaceous, once again pointing out that modern bony fishes have a long evolutionary history.

The superfamily of leatherjackets (Balistoidea) contains the triggerfishes (**Balistidae**) and the filefishes (**Monacanthidae**) (Fig. 15.17). In many species, the first dorsal spine is particularly long and stout and can be locked in the erect position via an interaction with the second spine. The base of the smaller second spine protrudes forward and fits into a groove on the posterior edge of the first spine, locking the first spine into position. Depressing the second spine releases the lock, hence the name triggerfish (see Fig. 15.7).

FIGURE 15.17. *Tetraodontiforms. (A) A balistid, the gray triggerfish,* Balistes capris-cus-mis maculata. *(B) The spine-locking mechanism of triggerfishes, showing how the base of the second dorsal spine fits into and helps lock the first spine in the erect position. Pushing posteriorly on the second spine releases the first spine. (C) The ocean sunfish,* Mola mola, *a member of the family considered the evolutionarily most advanced of all teleosts.*

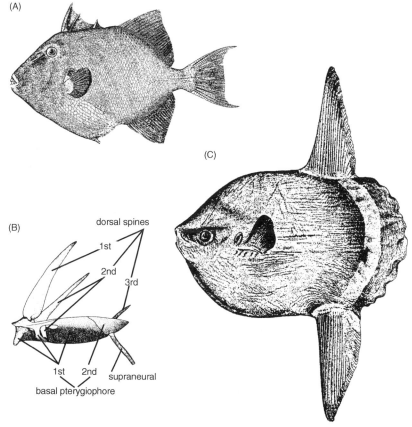

(A)

(C)

(B)

dorsal spines

1st

2nd

3rd

1st   2nd   supraneural

basal pterygiophore

(A) From Bigelow and Schroeder 1953; (B), (C) from Tyler 1980.

Sound production in this group is common, produced by grinding of teeth or vibration of the gas bladder via the pectoral spine; the legendary Hawaiian name for the triggerfishes *Rhinecanthus aculeatus* and *R. rectangulus* is *humuhumu nukunuku apua'a*, which means "the fish that sews with a needle and grunts like a pig." These fishes can also rotate their two eyes independently. In the **ostraciid** boxfishes, the entire body except the fins and caudal peduncle are encased in a bony box, which is triangular or rectangular in cross section. Stout spines sometimes protrude anteriorly just above the eyes and posteriorly just ahead of the anal fin. Swimming is accomplished via undulations of median fins (see Chap. 8, Locomotory Types).

The superfamily Tetraodontoidea contains the puffers and ocean sunfishes. Ironically, and despite family names focused on dentition, these fishes lack true teeth. Instead the jaw bone itself has a cutting edge that differentiates into separated teeth or is fused as a parrotlike beak. Puffers are able to inflate their body by filling the stomach with water (see Box 19.1). Three families of puffers are generally recognized based on the number of toothlike structures in each jaw: a single three-toothed species (**Triodontidae**); the smooth and sharpnose puffers with four teeth (**Tetra-**

odontidae); and the spiny puffers, burrfishes, and porcupinefishes with two fused teeth (**Diodontidae**). Tetraodontids have prickly skin. Because of bone loss and fusion, tetraodontids produce tasty, boneless fillets, but many species concentrate a powerful toxin, tetraodotoxin, in their viscera that can cause death in humans. Specially licensed and supervised *Fugu* restaurants in Japan serve the meat of puffers of the genus *Takifugu*, which contains small amounts of the toxin and provides a narcotic high. Despite such chemical protection, pufferfishes are commonly eaten by sea snakes. The few freshwater members in the order are tetraodontids. Diodontids have spines of varying length that are erected when the fish inflates, creating a large, round, and essentially inedible pincushion.

The most advanced tetraodontiforms and teleosts are the three species of very unfishlike-appearing, temperate and tropical molas (**Molidae**) (see Fig. 15.17). The body is essentially rectangular in side view with very tall, thin dorsal and anal fins that propel the fish. They lack a true tail but instead have a "pseudocaudal" tail fin made up primarily of dorsal and anal fin rays. The ocean sunfish, *Mola mola*, gets to be 2 m long and weighs as much as 1000 kg, with fish twice this size suspected. Fecundities of 300 million eggs have been reported, an apparent record among fishes. Although usually pelagic feeders on jelly-fishes, ocean sunfishes periodically move inshore to kelp beds off central California, apparently to have external parasites removed by cleanerfishes in the nearshore region.

Interestingly, molas have a large number of cartilaginous elements or cartilage-lined bones in their skull and in their fin supports (Tyler 1980). It is somewhat ironic that such a highly derived group returns anatomically to a starting point in fish evolution, as represented by primitive hagfishes, lampreys, sharks, and chondrosteans with their cartilaginous skeletons. Molas can therefore serve to remind us of several important aspects of the evolutionary process and our attempts to understand it. The task of determining primitive versus advanced traits can be complicated by secondarily derived characteristics, such as a very advanced species with a gular bone (e.g., the luciocephalid pikehead) or with a cartilaginous skeleton (see also the account above for the icosteid ragfish). And also important is that phylogenetically primitive fishes are not necessarily poorly adapted or somehow inferior to more advanced groups; the mola's rediscovery of the utility of cartilage underscores the observation that all living fishes are the successful result of the trial and error processes of mutation and natural selection.

SUMMARY

1. Most fishes (13,500 species, 251 families) belong to the superorder Acanthopterygii and have highly protrusible jaws, a complex pharyngeal apparatus, two dorsal fins, and spines in the first dorsal, anal, and pelvic fins. Three series are recognized: the Mugilomorpha, Atherinomorpha, and Percomorpha. Mugilomorph mullets are marine and freshwater fishes with unconnected pectoral and pelvic girdles. Atherinomorphs (silversides, needlefishes, flyingfishes, halfbeaks, killifishes, and livebearers) are shallow-water, marine or freshwater fishes that live near the surface and have a unique jaw protrusion mechanism.

2. The remaining seven orders are percomorphs. Stephanoberyciforms (whalefishes) and beryciforms (flashlight fishes, roughies, squirrelfishes) are moderate to deep-sea fishes, except for the reef-dwelling squirrelfishes; most are nocturnal.

3. Gasterosteiforms are small fishes with dermal armor plates, small mouths, and unorthodox propulsion. Pipefish and seahorse males become pregnant, carrying developing embryos in a ventral pouch. Scorpaeniforms are predominantly marine, benthic fishes, except for freshwater cottid sculpins; many have head spines and venomous fin spines (e.g., turkeyfishes, stonefishes, scorpionfishes).

4. The largest percomorph order is the Perciformes, with 150 families and 9000 species, including most marine and freshwater fishes of littoral zones. Perciforms have abdominal pelvic fins, lateral pectoral fins, and fewer than 18 caudal rays. Basal families are in the suborder Percoidei (71 families, 2900 species), including snooks, temperate basses, and the diverse sea basses. Centrarchid sunfishes and percids (darters, perches, pikeperches) are important freshwater percoids of North America. Other percoids include predatory carangoids (e.g., carangid jacks, pompanos), cobia, dolphinfishes, and the shark-sucking remoras. Heavy-bodied, tropical, benthic predatory families (snappers, sparoids) as well as fishes with barbels or feelers (croakers, threadfins, goatfishes) are placed here. Other percoid families include archerfishes, which shoot insects out of overhanging vegetation, and the colorful coral reef butterflyfishes and angelfishes.

5. The suborder Labroidei has several speciose families. Cichlids are primarily freshwater fishes that have undergone explosive speciation, forming species flocks in central African lakes. Other labroid families are the primarily tropical damselfishes, wrasses, and parrotfishes, and the temperate surfperches and odacids. Zooarchoids are tide pool and deep-ocean, benthic fishes (eelpouts, gunnels, wolf-eels). Notothenioid icefishes dominate the Antarctic and have many adaptations to very cold water. Blennioids (clinids, blennies) and gobioids (sleepers, gobies) are large suborders of generally small, benthic, marine species.

6. The suborder Acanthuroidei includes the spadefishes, rabbitfishes, and surgeonfishes, the latter two being important families of coral reef herbivores. Scombroids (barracudas, mackerels, tunas, billfishes) are the fastest and largest predatory bony fishes in the sea. Tunas and billfishes have independently evolved endothermy and heat conservation. Stromateoids are also largely pelagic marine fishes that associate with floating objects. The most advanced perciform suborder is the Anabantoidei (gouramis), which are African and Asian freshwater fishes with specialized gill structures for breathing atmospheric oxygen.

7. Pleuronectiform flatfishes begin life as pelagic, symmetric larvae but metamorphose into adults that lie on the bottom on one side and displace many body organs, most noticeably the eyes. Tetraodontiforms are primarily tropical reef dwellers, including triggerfishes, boxfishes, puffers, and porcupinefishes. They often have beaklike jaws, many fused skull and axial bones, and spiny, leathery skin. The most advanced tetraodontiforms are the giant ocean sunfishes, which have a surprising amount of cartilage in their skeletons.

SUPPLEMENTAL READING

Berra 1981; Carroll 1988; Nelson 1994; Paxton and Eschmeyer 1995; Wheeler 1975.

PART 4

Zoogeography, Habitats, and Adaptations

16 Zoogeography

Fishes occur almost everywhere that there is water, but even a casual glance at species lists from different localities demonstrates that few places have the same kinds of fishes. The question of interest then becomes one of discerning patterns in the present distribution of different species, genera, families, and higher taxa, and then trying to understand how these patterns are related to the evolution of the different groups. Basically, we are asking how and why fish faunas differ and how different fishes got where they are today. These questions form the basis of the science of zoogeography.

MAJOR ZOOGEOGRAPHIC REGIONS

Historically, the study of fish distribution has been divided into marine and freshwater components. Marine fishes comprise about 58% of the approximately 25,000 species of fishes, whereas freshwater fishes make up about 41% (Fig. 16.1).

Marine Fishes

Although terrestrial humans refer to our planet as Earth, it is really Planet Ocean. Not only is 71% of the planet's surface covered with water, but because water is dense enough to support life from the surface down to 11,000 m, the total oceanic living volume is 300 times greater than in the terrestrial part. Anyone first observing our planet from outer space would be struck by this and surely would name the planet for its water cover, unique in the solar system. Much recent attention has been directed at **biodiversity** (the numbers of species present) in tropical rainforests. The huge biodiversity in tropical areas is accounted for largely by radiation of one group, the insects. If we turn to the sea, we may not find as many total species, in part because insects have not diversified there. However, at the level of **phyletic diversity** (numbers of different phyla present), the seas support a greater biodiversity than does land. Of 33 animal phyla, 32 occur in the sea, and 15 of these are exclusively marine (Norse 1993).

Four main ecological divisions of marine fishes can be recognized:

1. **Epipelagic** fishes, which dwell from the surface down to 200 m, make up 1.3% of the total, or about 325 species.
2. **Deep pelagic** fishes include about 1250 species, or about 5% of the total. These water-column-dwelling fishes can be further subdivided into **mesopelagic** fishes, which live between 200 and 1000 m, and deeper-dwelling **bathypelagic** fishes.
3. **Deep benthic** fishes comprise about 1500 species, or 6.4% of the total.
4. **Littoral** or **continental shelf** species, shallow-dwelling fishes that inhabit the shore and shelf above 200 m, are the largest group, constituting 45% of the total, or about 11,250 species.

Epipelagic fishes. The diversity and adaptations of surface-dwelling fishes are treated in Chapter 17 (Pelagic Fishes), so attention here will be focused on their zoogeography. Many epipelagic species are worldwide in distribution. However, inshore epipelagic species frequently have more restricted distributions. One member of a family may be confined to one side of an ocean and be represented by another, **allopatric** species (a closely related species not occurring in the same area), living on the other side of the ocean. As examples, consider the distribution patterns of some tunas (Scombridae), halfbeaks (Hemiramphidae), and needlefishes (Belonidae). Most species of tunas of the genus *Thunnus* are widespread offshore. Several species, including albacore (*T. alalunga*), yellowfin (*T. albacares*), and bigeye (*T. obesus*), have continuous distributions, so there is inferred genetic interchange among populations in the Atlantic, Indian, and Pacific Oceans. Little tunas of the genus *Euthynnus* have a different distribution pattern, more closely associated with the shore. There is one Atlantic species (*E. alletteratus*), one Indo-West Pacific species (*E. affinis*), and one eastern Pacific species (*E. lineatus*). Among Spanish mackerels, *Scomberomorus*, distributions of species are even more shore-associated (Collette and Russo 1985b), with species not crossing from one side of the Atlantic to the other or from the Indo-West Pacific to the eastern Pacific. Distinct species occur in the western Atlantic, eastern Atlantic, parts of the Indo-West Pacific, and eastern Pacific.

Turning to the halfbeaks, species of the genus *Hemiramphus* are more widespread than species of the more inshore genus *Hyporhamphus*. For example, two species of *Hemiramphus* are found on both sides of the Atlantic, and one species (*He. far*) is widespread throughout the Indo-West Pacific. In contrast, all species of *Hyporhamphus* in the western Atlantic differ from those in the eastern Atlantic, Indo-West Pacific, and eastern Pacific.

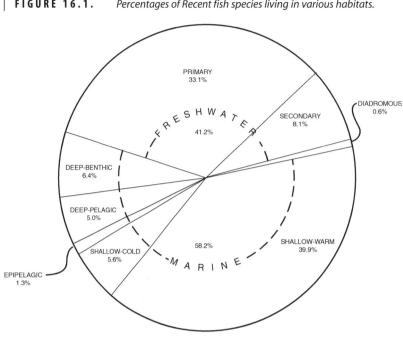

FIGURE 16.1.   *Percentages of Recent fish species living in various habitats.*

From Cohen 1970.

Similarly, needlefishes of the genera *Ablennes* and *Tylosurus* are much more widespread than species of the genus *Strongylura* (Cressey and Collette 1974). *Ablennes* is worldwide, and two species of *Tylosurus* (*T. acus* and *T. crocodilus*) are worldwide, with different subspecies recognized in parts of their ranges. Species of *Strongylura* are more numerous and have more restricted distributions, like species of *Hyporhamphus*.

Distributions of epipelagic inshore fishes may be limited by temperature, either directly or indirectly. For example, for needlefishes (Fig. 16.2), a clear correlation exists between temperature and the northernmost and southernmost distribution records of species of *Strongylura*.

Bathypelagic and mesopelagic fishes.   Many species of deepwater fishes (see Chap. 17, The Deep Sea) are also widespread. In looking at their distributions, we cannot use just surface maps, because ocean basins have underwater **sills**, ridges that act as barriers to the distribution of deep-water fishes. Sills act as barriers partly physically and partly through restricting the mixing of waters. For example, the Mediterranean Sea is continuous with the Atlantic Ocean at the surface via the Straits of Gibraltar. However, at 1200 m, the Mediterranean Sea is 14°C, whereas the adjacent Atlantic Ocean is 2.5°C at the same depth. Similarly, the Red Sea is 23°C at 125 m, whereas the Indian Ocean is 2.5°C. Fishes adapted to the cool temperatures of the Atlantic and Indian Oceans may not be able to penetrate the Mediterranean and Red Seas because they cannot tolerate the warm temperatures at the sill that separates the ocean from the adjacent sea.

Littoral fishes.   Temperature is also a major limiting factor for the distribution of shallow-water fishes. The greatest abundance of marine fish species is in tropical waters. Part of this great biodiversity is associated with the coral reefs that provide habitat for the fishes and their prey. Coral reefs are restricted to depths above 50 m in clear waters warmer than 18°C; corals reach their maximum extent at 23 to 25°C. Common groups of coral reef fishes include the moray eels (Muraenidae); squirrelfishes (Holocentridae); several families of percoids such as the sea basses (Serranidae), grunts (Pomadasyidae and Haemulidae), snappers (Lutjanidae), cardinalfishes (Apogonidae), and butterflyfishes (Chaetodontidae); and some more highly evolved families such as damselfishes (Pomacentridae), wrasses (Labridae), parrotfishes (Scaridae), surgeonfishes (Acanthuridae), triggerfishes (Balistidae), and boxfishes (Ostraciidae).

Major marine regions.   Distributions of inshore marine fishes indicate four major marine regions, in order of decreasing biodiversity: 1) Indo-West Pacific, 2) western Atlantic, 3) eastern Pacific, and 4) eastern Atlantic. The regions are separated from each other either by continents or by large expanses of open ocean. The **Indo-West Pacific Region**—from South Africa and the Red Sea east through Indonesia and Australia to Hawaii and the South Pacific Islands all the way to Easter Island—contains about one-third of the species of shallow marine fishes, which amounts to about 3000 species, compared to no more than 1200 in any other region. Biodiversity is also high in other marine taxa in this region (Briggs 1974). There are approximately 70 genera and about 500 species of hermatypic (reef-building) corals in this region, which is 10 times the number of species present in the western Atlantic (Rosen 1988). Among other groups, the Indo-West Pacific contains about 1000 species of bivalve mollusks (including all the giant clams, Tridac-

FIGURE 16.2.   *Distributions of marine populations of 11 species of needlefishes of the genus* Strongylura *in relation to the 23.9°C isotherm.*

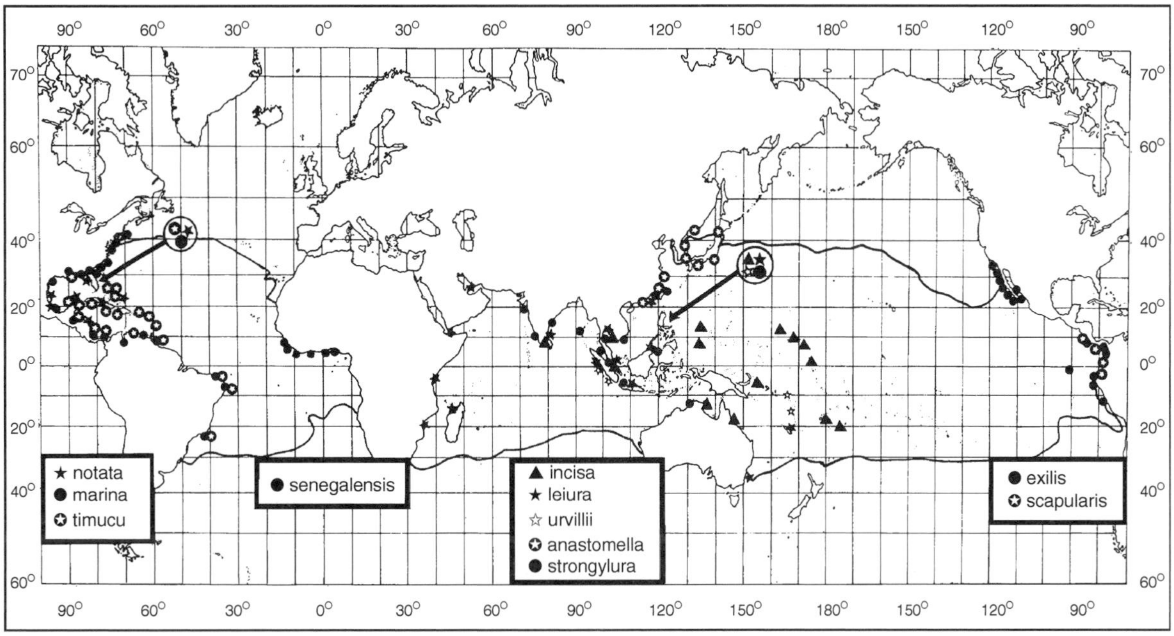

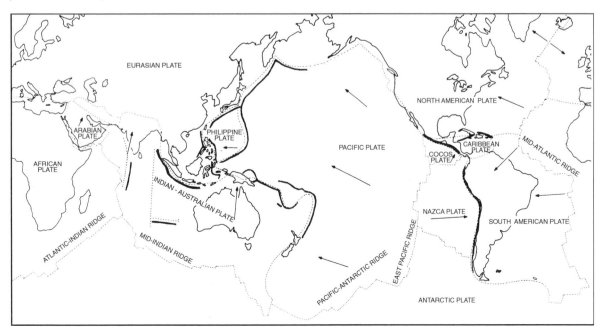

Adapted from Cressey and Collette 1970.

nidae), twice that in the western Atlantic; and 49 of the 50 species of sea snakes (Hydrophiidae), compared to 1 species in the eastern Pacific. Families of fishes such as the whitings (Sillaginidae) and rabbitfishes (Siganidae) are endemic to the Indo-West Pacific.

The Pacific Plate (Fig. 16.3) forms a major biogeographic unit of the Indo-West Pacific. The Pacific Plate is the largest of the earth's lithospheric plates (Box 16.1) and occupies most of the area that has been referred to as the Pacific Basin (Springer 1982). The number of taxa de-

FIGURE 16.3.   *The earth's important tectonic features. Dashed lines denote margins of major lithospheric plates. Arrows indicate direction of plate movements.*

Adapted from Springer 1982.

BOX
16.1 ***Continental Drift, Tectonic Plates, and Fish Distributions***

Continents and ocean basins have changed dramatically in location and size during the earth's history. Continents are blocks of largely granitic and sedimentary rocks. Continents "drift" because they literally float on top of the earth's denser basaltic crust. New crust develops at midoceanic ridges as basalt upwells from the earth's mantle. This basalt flows outward, causing spreading of the basaltic seafloor **plates** and widening of ocean basins. The moving plates carry the overlying continents with them. Crust finally is subducted (dives under) at oceanic trenches and plate margins. The notion of **continental drift** was first seriously proposed by Alfred Wegener in 1915 to explain the fit that the west coast of Africa makes with the east coast of South America. Wegener's concept was ridiculed by most scientists for many years but has gained acceptance as geophysical and paleontologic evidence accumulated.

Understanding the present distributions of fishes and many other taxa requires understanding that the present arrangement of the continents is very different from past arrangements. The present continents were at one time all part of a single landmass, **Pangaea**, that had coalesced by the Silurian (430 million years ago). About 180 million years ago, during the Mesozoic, Pangaea split into a northern portion, **Laurasia** (Eurasia and North America), and a southern portion, **Gondwana**. Gondwana later split into South America, Africa, Australia, and Antarctica about 90 million years ago. Present distributions of several fish taxa, such as lungfishes, osteoglossomorphs, and ostariophysans (see below), may well have taken place when the southern continents were still connected, before the breakup of Gondwana.

creases sharply as one proceeds eastward across the western margin of the plate. In addition, there is a high degree of endemism on the plate.

An instructive example of the kinds of distributions one finds associated with specific regions occurs with Spanish mackerels of the genus *Scomberomorus* (Collette and Russo 1985b). Of the 18 species in the genus, 10 occur in the Indo-West Pacific, but they are noticeably absent from the Pacific Plate (Springer 1982, Fig. 40). One species, *Scomberomorus commerson*, is widespread throughout much of the Indo-West Pacific. This distributional pattern cannot tell us much because, as with plesiomorphic characters (see Chap. 2), widespread species are not as informative about the causes of the distribution patterns as those with more restricted distributions. The ranges of three species, *S. guttatus*, *S. koreanus*, and *S. lineolatus* (Fig. 16.4), stop at the continental margin, at what is known as Wallace's Line (see below under Oriental Region). Australia and southern New Guinea, east of Wallace's Line, have a different Spanish mackerel fauna that consists of four species: *S. multiradiatus*, *S. semifasciatus*, *S. queenslandicus*, and *S. munroi*. These four species do not extend into the East Indies or even to the north coast of New Guinea, although they easily could swim that far. This distribution pattern is obviously not simply a result of present ecological factors but instead must be historical, related to the earlier evolution and dispersal of the genus. The present island of New Guinea resulted from the collision of two plates that may have contained two different fish faunas.

The **Western Atlantic Region** includes the temperate shores of North America, the Gulf of Mexico, the tropical shores of the Caribbean Sea, and the tropical and temper-

ate shores of South America. Radiation of tropical shore fishes in the western Atlantic is associated with the presence of the habitat created by West Indian coral reefs; 16 to 24 genera of zooxanthellate (containing symbiotic algae) corals occur there (Rosen 1988). The fish fauna of the Western Atlantic Region comprises about 1200 species. This coral reef fish fauna was once thought to be divided into northern and southern parts by the freshwater outflow of the Amazon River. Bottom trawling below the freshwater outflow of the Amazon River in 1975 filled in part of the supposed "gap" in our knowledge (Collette and Rützler 1977). At 14 benthic stations off the mouth of the Amazon under the superficial freshwater layer, a typical reef fish fauna of 45 species was found (Fig. 16.5), but these "coral reef" species were associated with 35 species of sponges in water too turbid for coral growth. The sponges provide the necessary structural habitat in which the fishes can live, which in turn allows genetic continuity between the two supposedly separated populations.

Although many groups show their maximum diversity in the Indo-West Pacific, a few show maximum diversity in the Americas. One example is the toadfishes, Batrachoididae, in which two-thirds of the species occur in New World waters (Collette and Russo 1981). The most generalized subfamily, the Batrachoidinae, is worldwide. However, the two most specialized subfamilies, the luminous midshipmen (Porichthyinae) and the venomous toadfishes (Thalassophryninae), are restricted to the western Atlantic and eastern Pacific (plus a few freshwater species derived from Atlantic or Pacific marine species).

The **Eastern Pacific Region** contains another radiation related to and only recently separated from the west-

FIGURE 16.4. *Ranges of seven Indo-West Pacific species of* Scomberomorus, *three continental and four Australian, with reference to Wallace's Line delimiting the continental margin.*

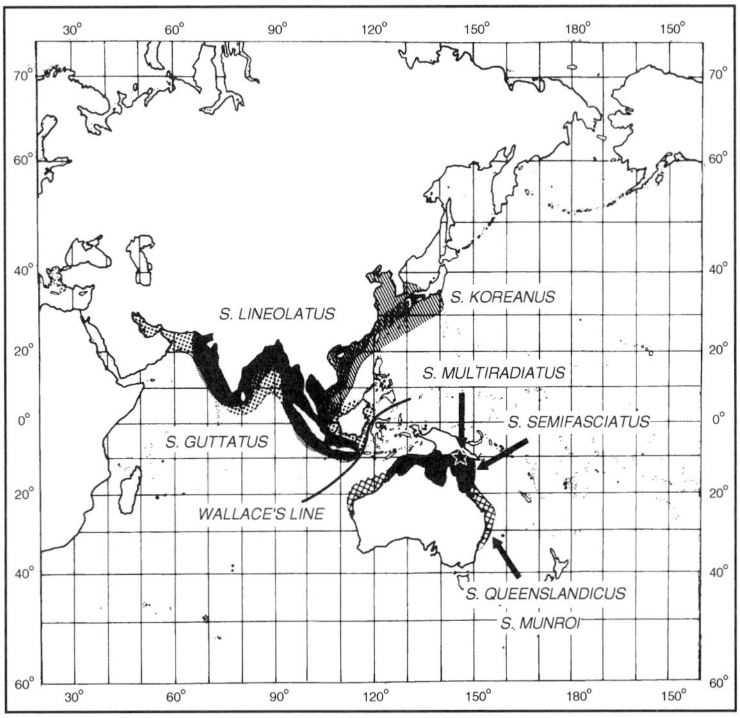

Modified from Collette and Russo 1985b.

ern Atlantic. The region contains only four to eight genera of zooxanthellate corals (Rosen 1988) and fewer species of fishes than are present in the western Atlantic. Some widespread taxa, such as the bluefish (*Pomatomus saltatrix*) and cobia (*Rachycentron canadum*), are absent (Springer 1982, Figs. 33, 35). The **Eastern Pacific Barrier**, the huge expanse of open water between the central and South Pacific Islands and the American mainland, acts as a distance barrier limiting movement of 86% of shore species from the central Pacific (Briggs 1974). Elevation of the Panamanian Isthmus approximately 3 million years ago separated the continuous distribution of species into eastern Pacific and western Atlantic populations. David Starr Jordan, pioneer American ichthyologist (see Chap. 1), referred to such pairs of species as **geminate species**, related species divided by the isthmus, such as the Spanish mackerels, *Scomberomorus sierra,* in the eastern Pacific, and *S. brasiliensis,* in the Caribbean Sea. Some geminate species have clearly differentiated into what can be called good species from a morphological and sometimes from a genetic point of view as well, such as Spanish mackerels and toadfishes of the genus *Batrachoides.* Others, such as the halfbeak, *Hyporhamphus unifasciatus,* are less well differentiated morphologically, making molecular methods useful to reach decisions on the status of the populations on either side of the isthmus.

The completeness of the Eastern Pacific Barrier is not quite as distinct as Briggs (1974) implied. Several Indo-West Pacific shore fishes actually cross the Eastern Pacific Barrier and are found at offshore islands such as the Revilligedos off the coast of Mexico, and Clipperton and Cocos off the coast of Costa Rica. Distributions of a species of mackerel (*Scomber australasicus*) and a needlefish (*Tylosurus acus melanotus*) extend from the western Pacific through the Hawaiian Islands to these islands, but these species are replaced by related forms (*S. japonicus* and *T. acus pacificus*) along the eastern Pacific coast of Middle America. These exceptions to the completeness of the Eastern Pacific Barrier may be related to habitat differences between the offshore islands and the mainland.

The Panama Canal connects the eastern Pacific with the western Atlantic. However, unlike the Suez Canal, the Panama Canal is not at sea level. It contains a freshwater lake, Lake Gatun, in its middle, and freshwater is used to raise the water level in a series of locks to lift ships up to the lake and then down to the ocean on the other side. This freshwater barrier prevents marine species from moving between the two oceans. A proposed sea level canal would allow mixing of the two different faunas and could have grave effects on the fishes and marine invertebrates on both sides of the isthmus. Diseases, parasites, and aggressive Indo-West Pacific species that pose little current danger in the eastern Pacific, such as the crown-of-thorns starfish and a sea snake, might do severe damage to coral reefs and the fish fauna of the western Atlantic (Briggs 1974; see Chap. 24).

Eastern Atlantic Region.   In the eastern Atlantic Ocean, tropical shore fishes are restricted to the Gulf of Guinea, a

FIGURE 16.5. *Distribution of coral reef fishes (shaded area) in the tropical western Atlantic. Black dots indicate 14 stations where coral reef fishes were caught in association with sponges.*

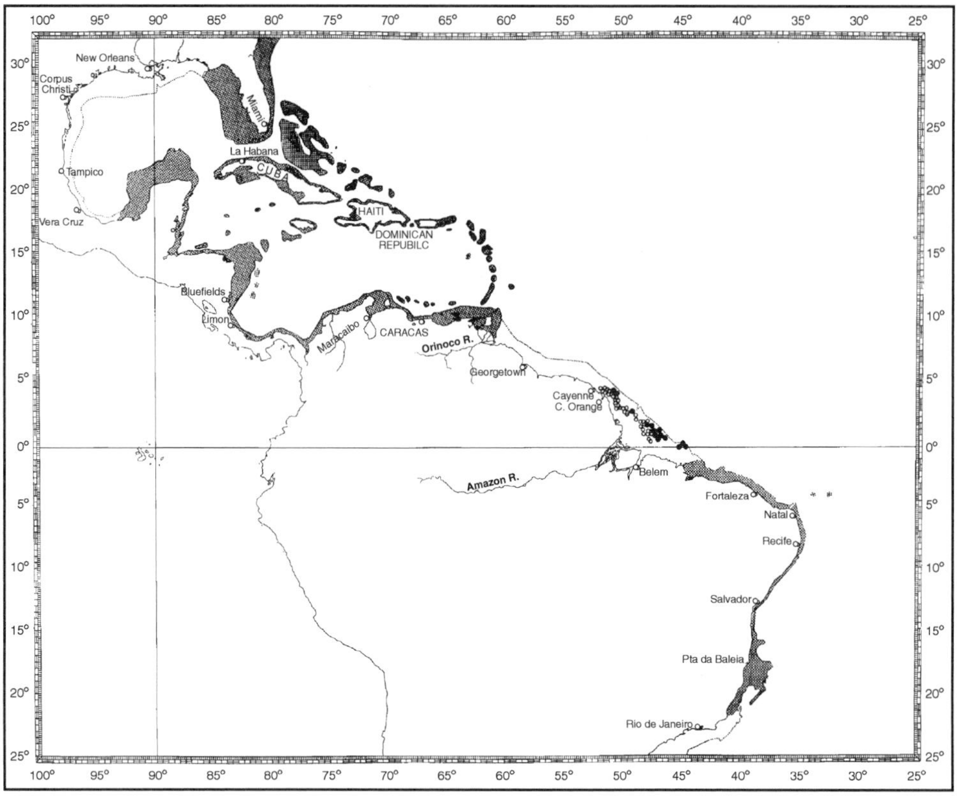

Adapted from Collette and Rützler 1977.

small area that extends from Dakar, Senegal, to Angola, and includes offshore islands such as the Cape Verde Islands, Annobon, Fernando Po, Ascension, and St. Helena. Coral cover is sparse in the tropical part of the eastern Atlantic, partly due to the large amount of freshwater runoff and accompanying sediment that flows out of such rivers as the Congo, Niger, and Volta. Only a few eastern Atlantic localities have as many as four to eight genera of zooxanthellate corals, other localities having only one to three genera (Rosen 1988). The eastern Atlantic is depauperate in many fish and invertebrate groups and contains only about 500 species of shore fishes. A few families, such as the porgies (Sparidae), have radiated in the eastern Atlantic.

Comparisons among genera of shore fishes of the western Atlantic, eastern Pacific, and eastern Atlantic demonstrate the relative depauperate nature of many eastern Atlantic groups. For example, four genera that contain two to four species each in the western Atlantic and eastern Pacific have only a single species in the eastern Atlantic (Table 16.1). Such comparison of patterns of diversity among different families can tell us much about not only the zoogeography of a group but also its probable phylogenetic history, particularly if we apply modern approaches to both zoogeography and phylogeny (Box 16.2).

The **Mediterranean Sea** is a somewhat depauperate part of the eastern Atlantic Ocean, with about 540 species

Table 16.1. Numbers of species in selected genera of inshore fishes from the western Atlantic, eastern Pacific, and Gulf of Guinea (numbers in parentheses indicate freshwater species of marine origin).

Family	Genus	Western Atlantic	Eastern Pacific	Eastern Atlantic
Scombridae	*Scomberomorus*	4	2	1
Belonidae	*Strongylura*	3 (+1)	2 (+1)	1
Hemiramphidae	*Hyporhamphus*	3 (+2)	4	1
Batrachoididae	*Batrachoides*	3 (+1)	3	1
Totals		13 (+4)	11 (+1)	4

Vicariance Biogeography and Spanish Mackerels of the Scomberomorus regalis *Species Group*

Historically, the first step in biogeography, the study of the distribution of all the organisms on our planet, has been to map the distributions of plants and animals (Ekman 1953; Darlington 1957; Briggs 1974, 1995). The next step was to ask, Why are these species distributed in this manner? Until recently, the answers were usually couched in terms of "areas of endemism," which refers to regions to which certain species are restricted. Sometimes such areas were equated with "centers of origin" where species evolved and from which species dispersed to wherever they are found today. There is little doubt that such dispersal occurs, but we are still left with the question, Is this the *primary* explanation for the present distribution of most species?

A potentially more complete answer can be given by a relatively recent concept known as **vicariance biogeography** (Nelson and Rosen 1981; Nelson and Platnick 1981). This method requires good phylogenetic information on the relationships of the species that comprise a group, as well as their distributions. Present distributions can then be compared with phylogenetic relationships to assess relative movements of biota or pieces of real estate containing the biota. This comparison is accomplished by replacing the terminal taxa in a cladogram with their geographic distributions to form an **area cladogram**. The search is for repeated distribution patterns that may be explained by **vicariant events**, such as movements of continents, collisions of islands on different tectonic plates, outcroppings of mountains or peninsulas to divide populations, or capture of a stream by headwater erosion of another drainage.

The process by which cladistics and zoogeography complement each other and inform us about a group's phylogeny can be demonstrated by the *Scomberomorus regalis* species group of Spanish mackerels. This group is defined as monophyletic based on the unique presence of nasal denticles, toothlike structures within the nasal cavity (Collette and Russo 1985a,b). There are six Atlantic and eastern Pacific species relevant to this discussion: *tritor, maculatus, concolor, sierra, brasiliensis,* and *regalis* (Fig. 16.6).

The five most advanced species (all except *tritor*) have a branch arising from the fourth left epibranchial artery (Fig. 16.7). The four most advanced species (all except *tritor* and *maculatus*) have developed a long posterior process on the pelvic girdle. The three most advanced species (*sierra, brasiliensis,* and *regalis*) have a coeliaco-mesenteric shunt connecting the fourth right epibranchial artery with the coeliaco-mesenteric artery. The two most advanced species (*brasiliensis* and *regalis*) have lost the pterotic spine.

Comparing the distribution of the six species (see Fig. 16.6) with the phylogeny (see Fig. 16.7) indicates that the eastern Atlantic species (*tritor*) is the plesiomorphic sister species of the rest of the species group. Next comes the western Atlantic *maculatus* and then the two eastern Pacific species, *concolor* and *sierra*, suggesting speciation following elevation of the Isthmus of Panama, as discussed earlier for geminate species. The two most advanced species, *brasiliensis* and *regalis*, are both found in the western Atlantic, with *regalis* occupying an unusual habitat for Spanish mackerels, namely, coral reefs.

These patterns among Spanish mackerels can then be compared with patterns of other species in Table 16.1 to see if there are commonalities among distributions. Are the single eastern Atlantic species of the halfbeak *Hyporhamphus*, of the needlefish *Strongylura*, and of the toadfish *Batrachoides* the plesiomorphic sister species of the western Atlantic and eastern Pacific species in these genera? Did these patterns arise from the widening of the Atlantic Ocean as the plates containing the Americas and the Old World (see Fig. 16.3) moved farther apart? Molecular genetic studies are needed to provide additional information on relationships of such taxa and estimates of the timing of evolution of the species involved.

of fishes. Drying out in the past eliminated many fishes from the Mediterranean, and cooler temperatures in the Straits of Gibraltar prevented warm-water fishes found in the Gulf of Guinea from moving into warm waters of the eastern Mediterranean. In 1869, a sea level route, the Suez Canal, was opened, connecting the warm but depauperate eastern Mediterranean with the Red Sea, the latter being part of the rich Indo-West Pacific Region. For some time after construction, faunal transfers between the Red Sea and the Mediterranean were inhibited by the saline waters of the Bitter Lakes in the middle of the canal. In the first edition of *A History of Fishes* in 1931, Norman reported that 16 species of Red Sea marine fishes had moved through the Suez Canal and established themselves in the eastern Mediterranean Sea. In the 1960s, 1970s, and 1980s, Adam Ben-Tuvia and others raised the number to 24, 27, 31, 46, and most recently (Golani 1993) to 52. All but one of the species (a sea bass, *Dicentrarchus*) are what have been termed **Lessepsian migrants**, having moved in one direction, from the Red Sea into the eastern

FIGURE 16.6.   *Ranges of the* regalis *group of* Scomberomorus.

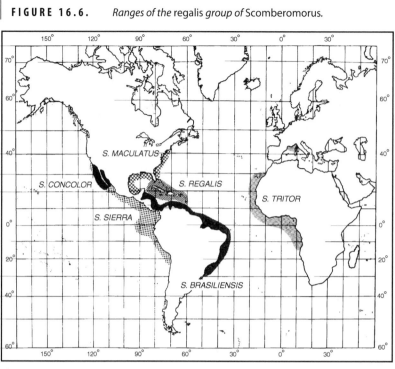

Adapted from Collette and Russo 1985a.

Mediterranean. As an example of how successful a migrant can be, the brushtooth lizardfish (*Saurida undosquamis*) was first taken in the Mediterranean in 1952. By 1955, 266 tons of this lizardfish were landed by local trawlers, constituting close to 20% of the trawler catch in Israeli waters (Golani 1993).

Why have these movements been virtually one-directional? First, the diversity of inshore fishes is greater in the Red Sea, part of the Indo-West Pacific fish fauna, than in the Mediterranean, suggesting that niches are more completely filled in the Red Sea, which means fewer ecological opportunities for new immigrants. Second, there appears to be an "empty niche" in the eastern Mediterranean, associated with water temperature, with temperatures again being suitable for warm-water fishes.

Finally, many of the species that penetrated the canal are widespread species, adapted to a wide variety of living conditions. Consider the distributions of three of the invading species, a halfbeak and two mackerels. *Hemiramphus far* is the most widespread member of its genus, known from South Africa across the Indian Ocean north to Okinawa, south to Australia, and east to Tonga and Fiji. It, rather than the Red Sea-Persian Gulf *He. marginatus*, successfully moved through the canal and established populations that have now spread west and north as far as Albania. The two mackerels, the narrow-barred Spanish mackerel (*Scomberomorus commerson*) and the Indian mackerel (*Rastrelliger kanagurta*), are the most widespread members of their genera, occurring from South Africa north to the Red Sea, east to China and Japan, and south

FIGURE 16.7.   *Cladogram of the* Scomberomorus regalis *group.*

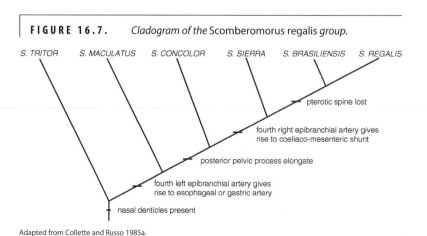

Adapted from Collette and Russo 1985a.

to Australia and Fiji (Collette and Nauen 1983, maps on pp. 49 and 63). Such generally successful colonist species could have been predicted as the most likely taxa to take advantage of the opportunities in the eastern Mediterranean because they are adapted to a wide range of ecological conditions.

Arctic and antarctic fishes. Diversity and adaptations of fishes of the far north and south are treated in Chapter 17 (Polar Regions). Marine shore and continental shelf species down to 200 m from arctic and antarctic waters account for 5.6% of the total fish fauna.

The arctic region, north of 60° in the Pacific (approximately Nunivak Island, Alaska) to Newfoundland and northern Norway in the Atlantic has 20% to 25% endemism (Briggs 1974). The most speciose families include the skates, herrings, salmons, smelts, cods, eelpouts, greenlings, sculpins, poachers, snailfishes, pricklebacks, wolffishes, gunnels, and righteye flounders. Most of these groups have higher species diversity in the Pacific than in the Atlantic portions of the region (Briggs 1974, 1995).

Antarctica and the surrounding Southern Ocean contain 274 species of fishes from 49 families. The immediate antarctic region has 174 species in 13 families, 88% of which are endemic. Antarctica has a higher level of endemism of fishes and invertebrates than the Arctic Ocean, which contains 1.5 times the fish species and twice as many families (Briggs 1974; Eastman and Devries 1986; Stromberg 1991). Of the fishes in the immediate antarctic region, six families in the suborder Notothenioidei account for 55% of the species and more than 90% of the individuals. Primitive notothenioids, such as the Bovichthyidae, occur in Southern Hemisphere habitats of Australia, New Zealand, and South America. Some families occur in both Antarctica and surrounding continents (Nototheniidae and Channichthyidae), some occur in Antarctica and nearby oceanic islands such as the Falklands (Harpagiferidae), and one family (Bathydraconidae) is restricted to Antarctica. Of the notothenioids, 97% are antarctic endemics; even 70% of the nonnotothenioids are endemic. Six other families that contribute several species to the region are, in order of species diversity: snailfishes, eelpouts, skates, eel cods, deep-sea cods, and southern flounders (Gon and Heemstra 1990; Kock 1992; Eastman 1993; Miller 1993).

Some cool-water species, such as the mackerel *Scomber japonicus*, whose distributions are interrupted by low-latitude regions are considered to have **antitropical distributions**, in that they are present in temperate waters on either side of the equator (Hubbs 1952). Other cold-water species show **tropical submergence**, that is, they continue their ranges into tropical regions by submerging, moving into deeper waters that are the same temperature as the cold waters of arctic and antarctic regions.

Freshwater Fishes

Freshwater fishes make a much larger contribution to biodiversity than might be expected based on area alone. Cohen (1970) was surprised to find that about 41% of the world's fish species live in freshwaters (see Fig. 16.1). But Horn (1972) demonstrated the real significance of this by pointing out that freshwaters comprise only 0.0093% of the water on the planet, which means that nearly half of all fish species live in less than 1% of the world's water supply. Another way of looking at this is to compare the mean volume of water per species of marine versus freshwater fish. The average marine species has 113,000 km^3 available to it, whereas the average freshwater species has only 15 km^3!

The probable causes of this 7500-fold disparity in biodiversity in the two major habitat types are undoubtedly very complex, involving ecological as well as historical (phylogenetic and geologic) factors. Two likely influences are productivity and isolation. Shallow waters receive significant sunlight, allowing photosynthesis, which forms the base of food webs. Most freshwaters are shallow and relatively productive, whereas most water in the world's oceans lies well below the euphotic zone where primary productivity occurs (see Fig. 17.1). Shallow marine waters are productive and support a diverse fauna, as is evident in the coastal zones that support 45% of all fish species (see Fig. 16.1). The potential for isolation is a historical factor that differs greatly between marine and freshwater habitats. Marine habitats are broadly continuous; significant faunal breaks occur primarily where continental landmasses, large rivers or sills, and major oceanic currents act as geographic boundaries. Freshwaters, in contrast, are frequently and readily broken up into isolated water bodies. Drought, volcanoes, landslides, tectonic uplifting, and beavers are some of the agents that can lead to a body of water losing its connections with other bodies, which in turn isolates the fishes in that body from gene flow with other areas. Genetic isolation is a driving force of evolution, leading to such dramatic events as explosive speciation and formation of species flocks (see Box 15.1). Isolating events are, therefore, much more common in freshwaters, and it is therefore not so surprising that so many species of freshwater fishes have arisen in such little space.

Freshwater fishes versus fishes in freshwaters. Up to this point in our discussion, we have used the term *freshwater fish* a little carelessly. Historically, there was also much confusion surrounding the term until George Myers (1938) clarified the problem. He distinguished between **primary** freshwater fishes, whose members are very strictly confined to freshwater, and **secondary** freshwater fishes, whose members are generally restricted to freshwater but may occasionally enter salt water. Most families of primary freshwater fishes have had a long evolutionary history of physiological inability to survive in the sea. The term **peripheral** has been used for a number of species and genera of marine families that have taken up more or less permanent residence in freshwater or that spend part of their life cycle in freshwater and another part in marine habitats (such **diadromous** fishes are discussed in Chap. 22, Diadromy). There are about 85 families of primary freshwater fishes, 11 of secondary, and more than 30 of peripheral freshwater fishes (Table 16.2).

The origins of freshwater fishes, and the importance of distinguishing among the different types, become

Table 16.2. Primary, secondary, and selected peripheral freshwater fish families and the geographic areas where they occur.

Family*	Division	Nearctic	Neotropical	Palearctic	Ethiopian	Oriental	Australian
Petromyzontidae	per	x		x			
Geotriidae	per		x				x
Mordaciidae	per		x				x
Potamotrygonidae	per		x				
Ceratodontidae	1st						x
Lepidosirenidae	1st		x				
Protopteridae	1st				x		
Polypteridae	1st				x		
Acipenseridae	per	x		x		x	
Polyodontidae	1st	x		x			
Lepisosteidae	2nd	x	x				
Amiidae	1st	x					
Denticipitidae	1st				x		
Osteoglossidae	1st		x		x	x	x
Pantodontidae	1st				x		
Hiodontidae	1st	x					
Notopteridae	1st				x	x	
Mormyridae	1st				x		
Gymnarchidae	1st				x		
Salmonidae	per	x		x			
Plecoglossidae	per			x			
Osmeridae	per	x		x			
Salangidae	per			x		x	
Retropinnidae	per						x
Prototroctidae	per						x
Galaxiidae	per		x		x		x
Aplochitonidae	per		x				x
Lepidogalaxiidae	?						x
Esocidae	1st	x		x			
Umbridae	1st	x		x			
Kneriidae	1st				x		
Phractolaemidae	1st				x		
Characidae	1st	x	x		x		
Erythrinidae	1st		x				
Ctenoluciidae	1st		x				
Hepsetidae	1st				x		
Cynodontidae	1st		x				
Lebiasinidae	1st		x				
Parodontidae	1st		x				
Gasteropelecidae	1st		x				
Prochilodontidae	1st		x				
Curimatidae	1st		x				
Anostomidae	1st		x				
Hemiodontidae	1st		x				
Chilodontidae	1st		x				
Distichodontidae	1st				x		
Citharinidae	1st				x		
Ichthyboridae	1st				x		

Table 16.2. (Continued)

Family*	Division	Nearctic	Neotropical	Palearctic	Ethiopian	Oriental	Australian
Gymnotidae	1st		x				
Electrophoridae	1st		x				
Apteronotidae	1st		x				
Rhamphichthyidae	1st		x				
Cyprinidae	1st	x		x	x	x	
Gyrinocheilidae	1st					x	
Psilorhynchidae	1st					x	
Catostomidae	1st	x		x			
Homalopteridae	1st					x	
Cobitidae	1st			x	x	x	
Diplomystidae	1st		x				
Ictaluridae	1st	x					
Bagridae	1st			x	x	x	
Cranoglanididae	1st					x	
Siluridae	1st			x		x	
Schilbeidae	1st				x	x	
Pangasiidae	1st					x	
Amblycipitidae	1st					x	
Amphiliidae	1st				x		
Akysidae	1st					x	
Sisoridae	1st			x		x	
Clariidae	1st			x	x	x	
Heteropneustidae	1st					x	
Chacidae	1st					x	
Olyridae	1st					x	
Malapteruridae	1st				x		
Mochokidae	1st				x		
Doradidae	1st		x				
Auchenipteridae	1st		x				
Aspredinidae	1st		x				
Pimelodontidae	1st		x				
Ageneiosidae	1st		x				
Hypophthalmidae	1st		x				
Helogeneidae	1st		x				
Cetopsidae	1st		x				
Trichomycteridae	1st		x				
Callichthyidae	1st		x				
Loricariidae	1st		x				
Astroblepidae	1st		x				
Amblyopsidae	1st	x					
Aphredoderidae	1st	x					
Percopsidae	1st	x					
Oryziatidae	2nd			x		x	
Adrianichthyidae	2nd					x	
Horaichthyidae	2nd					x	
Cyprinodontidae	2nd	x	x	x	x	x	
Goodeidae	2nd	x					
Anablepidae	2nd		x				

Table 16.2.   (Continued)

Family*	Division	Nearctic	Neotropical	Palearctic	Ethiopian	Oriental	Australian
Jenynsiidae	2nd		x				
Poeciliidae	2nd	x	x				
Melanotaeniidae	2nd						x
Neostethidae	per					x	
Phallostethidae	per					x	
Gasterosteidae	per	x		x			
Indostomidae	per					x	
Channidae	1st			x	x	x	
Synbranchidae	per		x	x	x	x	x
Cottidae	per	x		x			
Cottocomephoridae	per			x			
Comephoridae	per			x			
Percicthyidae	per	x	x	x			x
Centrarchidae	1st	x					
Percidae	1st	x		x			
Toxotidae	per					x	x
Scatophagidae	per				x	x	x
Enoplosidae	per						x
Nandidae	1st		x		x	x	
Embiotocidae	per	x		x			
Cichlidae	2nd	x	x		x	x	
Gadopsidae	per						x
Bovichthyidae	per		x				x
Rhyacichthyidae	per					x	x
Kurtidae	per					x	x
Anabantidae	1st				x	x	
Belontiidae	1st			x		x	
Helostomatidae	1st					x	
Osphronemidae	1st					x	
Luciocephalidae	1st					x	
Mastacembelidae	1st			x	x	x	
Chaudhuriidae	1st					x	

*The following widely distributed peripheral families (mostly marine) are omitted from this analysis. Carcharhinidae, Elopidae, Megalopidae, Anguillidae, Clupeidae, Engraulidae, Chanidae, Ariidae, Plotosidae, Batrachoididae, Gadidae, Ophidiidae, Hemiramphidae, Belonidae, Atherinidae, Syngathidae, Alabetidae, Centropomidae, Ambassidae, Teraponidae, Kuhliidae, Sparidae, Sciaenidae, Monodactylidae, Mugilidae, Polynemidae, Gobiidae, Soleidae, Tetraodontidae.
1st = primary; 2nd = secondary; per = peripheral.
From Berra 1981.

particularly clear when the composition of fishes in freshwaters of islands is considered. Continental islands that have been connected to the adjacent mainland, such as Trinidad, have the same kinds of fishes as are present on the adjacent mainland of South America. Oceanic islands such as the West Indies and Hawaii, which have never been connected with continents, have no native primary freshwater fishes. All the fishes in their freshwaters are secondary or peripheral species.

Freshwater zoogeographic regions.   An effective way to understand the distribution of freshwater fishes is to recognize six regions or realms, as Alfred Russel Wallace first proposed in 1876: 1)Nearctic (North America except tropical Mexico); 2) Neotropical (Middle and South America, including tropical Mexico); 3) Palearctic (Europe and Asia north of the Himalaya Mountains); 4) African (or Ethiopian); 5) Oriental (Indian subcontinent, Southeast Asia, the Philippines, and most of Indonesia); and 6) Australian (Australia, New Guinea, and New Zealand). A detailed discussion of the fauna of these regions was provided by Darlington (1957), and maps of the distributions of freshwater fish families are presented by Berra (1981).

The **Nearctic Region** consists of North America south to the Mexican plateau. The North American freshwater fish fauna is the best known and has been mapped by Lee et al. (1980) and discussed thoroughly by Hocutt and Wiley (1986) and Mayden (1993). There are 14 families of primary freshwater fishes (see Table 16.2) and a total of about 950 species of fishes in the region. The most speciose families include three families of ostariophysans—Cyprinidae, Catostomidae, and Ictaluridae (the only North American family of Recent catfishes), plus two percoid families—the Percidae (especially the darters) and the Centrarchidae.

This region can be divided into three subregions: 1) the Arctic-Atlantic; 2) the Pacific; and 3) the Mexican Transition, consisting of most of Mexico (Moyle and Cech 1996). The **Arctic-Atlantic Subregion** contains six provinces.

1. The **Mississippi Province** is the largest Nearctic province, comprising the area drained by the Mississippi and Missouri Rivers, and contains the most species, about 300.
2. The **Great Lakes-St. Lawrence Province** is largely derived from the Mississippi Province.
3. The **Atlantic Coastal Province** contains the rivers that drain into the Atlantic Ocean from New Brunswick to Florida and continues through Florida's gulf-draining rivers. The northern part of the province has a relatively high proportion of anadromous fishes, while the southern portion has numbers of secondary fishes that have invaded from marine waters.
4. The **Hudson Bay Province** includes most of central Canada and part of the United States. Its fish fauna of about 100 species (Crossman and McAllister 1986) is most similar to that of the Mississippi Province. Minnows are an important component in the southernmost part of the province but are largely replaced by trouts, sculpins, suckers, and pickerels in the northern part of the province.
5. The **Arctic Province** includes Canadian and Alaskan rivers that drain into the Arctic Ocean. More than half of the 66 freshwater species (Lindsey and McPhail 1986) belong to diadromous families, one-third are primary freshwater species, and 11% belong to marine families. This province shares some species, such as the Alaskan blackfish (*Dallia pectoralis*), with Siberia, which belongs in the Palearctic Region.
6. The **Rio Grande Province** is essentially the Rio Grande and its tributaries. It contains 154 species, including the northernmost species of two Neotropical families, the Characidae (*Astyanax mexicanus*) and the Cichlidae (*Cichlasoma cyanoguttatum*).

The **Pacific Subregion** contains the Pacific drainages from the Yukon River to Mexico and also the interior drainages west of the Rocky Mountains. It can be subdivided into seven provinces (Moyle and Cech 1996).

1. The **Alaskan Coastal Province** includes the coastal drainages from the Aleutian peninsula to the Canadian border with Alaska. This province contains only 34 freshwater species and could be incorporated into the Arctic Province. Three-quarters of the freshwater species are from diadromous families.
2. The **Columbia Province** is the Columbia River drainage and contains 61 freshwater species. Important families are the Cyprinidae, Salmonidae, and Cottidae. There are two endemic relict species, a mudminnow (Umbridae) and a troutperch (Percopsidae).
3. The **Klamath Province**, the Klamath and Rogue Rivers along the border of California and Oregon, contains about 30 freshwater species, with more than half being representatives of diadromous families such as trouts and lampreys.
4. Small coastal streams entering the Pacific Ocean between Monterey Bay and the tip of Baja California are recognized as a separate province, the **South Coastal Province**, by Moyle and Cech (1996). The province contains only 12 species of freshwater fishes.
5. The **Great Basin Province** contains 150 internal drainage systems that are now very arid but contained large lakes during periods of the Pleistocene. There are about 50 species, mostly minnows, suckers, killifishes, and whitefishes, with about 80% endemism (Miller 1958).
6. The **Sacramento Province** consists mostly of the Sacramento-San Joaquin drainage system of California. It contains 43 species of freshwater fishes, of which 42% are endemic, including 10 species of minnows and the Sacramento perch (*Archoplites interruptus*), the only centrarchid occurring natively outside of eastern North America.
7. The **Colorado Province** contains about 32 species of freshwater fishes.

The **Neotropical Region** consists of South America and Middle America, into which some North American species have moved. South America has the largest freshwater fish fauna in the world, with 32 families of primary freshwater fishes (see Table 16.2) and a total of more than 2500 species in freshwaters. There are no minnows or suckers in South America, but their ecological equivalents may be in the 8 families of characins with over 1200 species. Other ostariophysans include 13 families of catfishes with about 1300 species, and 6 families of Gymnotiformes. Cichlids (about 150 species) are the most speciose percoid group. Representatives of many marine families have also invaded South America: freshwater stingrays (Potamotrygoninae), endemic to South America; herrings (Clupeidae); toadfishes (Batrachoididae); needlefishes (Belonidae, three endemic genera and 7 species); croakers (Sciaenidae); and soles (Achiridae). Gery (1969) recognized eight faunistic regions (Fig. 16.8): 1) Orinoco-Venezuelan, with 325 species; 2) Magdalenean, with 150 species; 3) Trans-Andean, with 390 species; 4) Andean, including the endemic fauna of the cyprinodontoid *Orestias*; 5) Paranean; 6) Patagonian; 7) Guianean-Amazonian, the richest basin, with more than 1300 species; and 8) East Brazilian, comprising the smaller coastal provinces of Brazil.

The **Palearctic Region** of northern Europe and Asia

FIGURE 16.8.   *South American faunistic regions. Numbers refer to regions in the text, dotted lines delimit ancient shields, and arrows suggest movement of ancestral stocks within the region.*

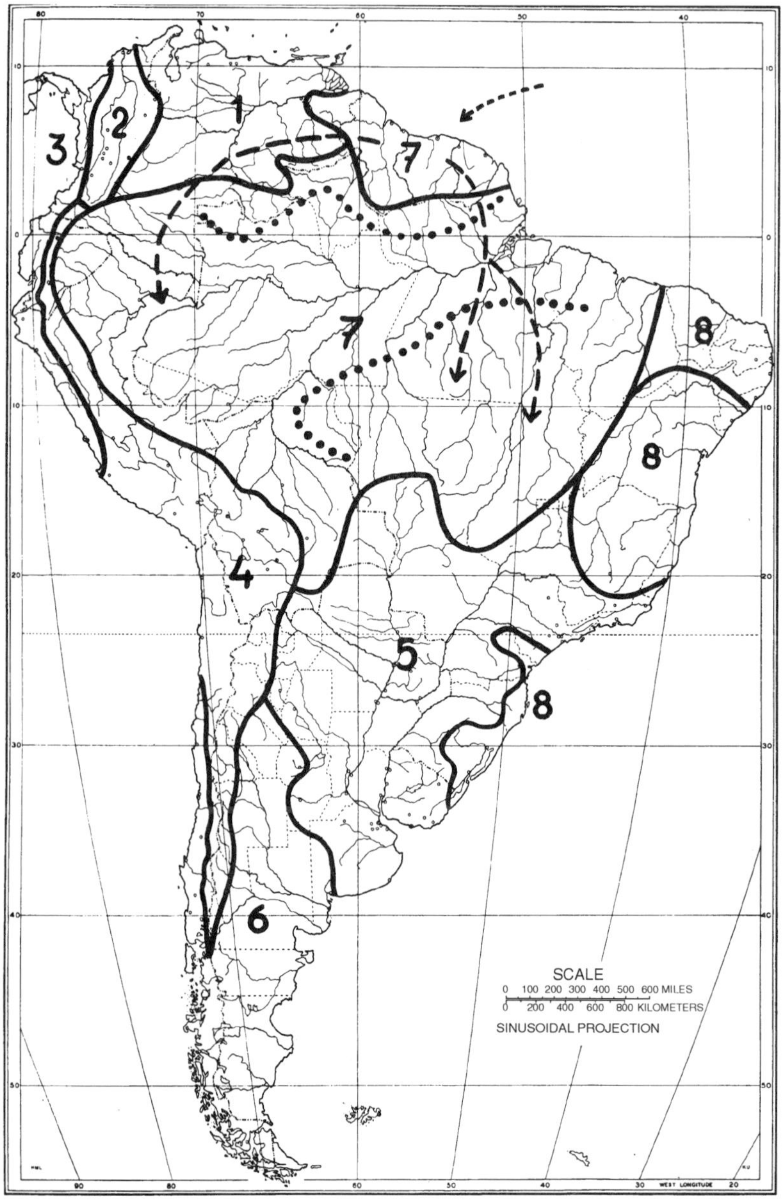

Adapted from Gery 1969.

contains only 14 families of primary freshwater fishes (see Table 16.2) with many minnows and loaches but only about 10 species of catfishes from four families. Nonostariophysans include perches (Percidae), pickerels (Esocidae), and a mudminnow (Umbridae, *Umbra krameri*).

The **African Region** has a diverse freshwater fish fauna that includes 27 families of primary freshwater fishes (see Table 16.2) and a number of primitive species as discussed below under archaic distributions. The region contains a total of about 2000 species of primary and secondary fishes belonging to about 280 genera and 47 families

(Roberts 1976). Almost half the species are ostariophysans (300 species of minnows, 190 characins, and more than 360 catfishes from 6 families). Roberts recognized 10 ichthyofaunal provinces (from north to south): Maghreb, Abyssinian Highlands, Upper Guinea, Nilo-Sudan, Lower Guinea, Zaire, East Coast, Quanza, Zambezi, and Cape of Good Hope.

The **Oriental Region** includes India, southern China, Southeast Asia, the Philippines, and the East Indies out to Borneo and Bali (Fig. 16.9). Alfred Russel Wallace (1860, 1876) proposed a boundary between the Oriental and Australian faunas that Thomas Huxley named for him as

FIGURE 16.9.   *Wallace's Line and Weber's Line separating the Oriental and Australian Regions. The 200-m contour line is drawn in showing how the major landmasses would be connected if the sea retreated to the 200-m line.*

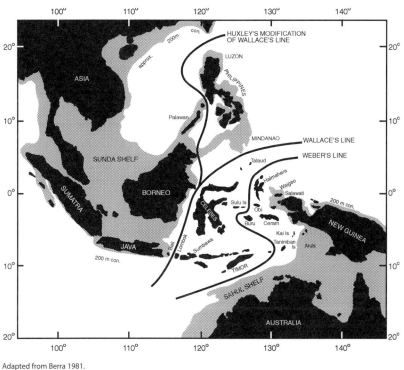

Adapted from Berra 1981.

Wallace's Line. Some authors extend the line even farther to the east (Weber's Line) to also include the Celebes (now Sulawesi) and some other Indonesian Islands in the Oriental Region. The region contains 28 families of primary freshwater fishes (see Table 16.2) with 12 families of catfishes and 4 families of cypriniform ostariophysans: minnows (Cyprinidae), loaches (Cobitidae), algae eaters (Gyrinocheilidae), and river loaches (Balitoridae), which are endemic to the region. Nonostariophysan families include snakeheads (Channidae), spiny eels (Mastacembelidae), labyrinth fishes (Anabantoidei), and a few cichlids. Only two species of primary freshwater fishes occur to the east of Wallace's Line. All other fishes in freshwaters east of the line have been derived from marine groups.

The **Australian Region** has only two species of primary freshwater fishes, both ancient relicts of a much wider archaic distribution pattern (see next section). Two families of secondary and 16 families of peripheral freshwater fishes of marine origin (see Table 16.2) include freshwater species of catfishes (two marine families, Ariidae and Plotosidae), silversides (Atherinidae), rainbowfishes (Melanotaeniidae), halfbeaks (Hemiramphidae), needlefishes (Belonidae), Teraponidae, Centropomidae, Percichthyidae, and gobies (Gobiidae).

Archaic freshwater fish distributions.   The distribution of six groups of primitive primary freshwater fishes (see Chap. 13) dates back long enough that their present distribution may be based on a different arrangement of the conti-

nents. These groups include the lungfishes (Dipnoi), Polypteriformes, Polyodontidae, Lepisosteidae, Amiidae, and Osteoglossomorpha.

There are three living genera of lungfishes (see Fig. 13.8): South American *Lepidosiren*, African *Protopterus*, and Australian *Neoceratodus*. Placement of *Lepidosiren* and *Protopterus* together in a single family, Lepidosirenidae, instead of in separate families, emphasizes their close relationships (Lundberg 1993). *Neoceratodus* is the most different lungfish, morphologically and physiologically, and has a relict distribution, restricted to portions of the Burnett and Mary Rivers in southeastern Queensland (see Chap. 13, Subclass Dipnoi).

Other archaic groups include the bichirs, Polypteridae, which consist of two living African genera—*Polypterus* with 10 species and the monotypic *Erpetoichthys* (previously *Calamoichthys*)—and a fossil genus (*Dajetella*) from the late Cretaceous and Paleocene of Bolivia (Lundberg 1993). There are 2 species of paddlefishes, family Polyodontidae: one (*Polyodon spathula*) from the Mississippi River of North America and the other (*Psephurus gladius*) from the Yangtze River of China. The gars, Lepisosteidae, are usually considered as secondary freshwater fishes and comprise 7 species of *Lepisosteus* in North America, Central America, and Cuba, with fossils known from India and Europe. Only one Recent species of bowfin, Amiidae, remains: *Amia calva* of the United States; fossil species have been found on all the continents except Australia (Nelson 1994). Fossils show that the present-day distribution of these groups is a relict of their original, much wider distribution.

Another archaic group is the Osteoglossomorpha, the most primitive subdivision of the Teleostei. It is more speciose and more widespread than the Dipnoi. It includes six families (Fig. 16.10). There are four genera in the Osteoglossidae, two in each of two subfamilies (or families, Lundberg 1993): Heterotinae, *Heterotis niloticus* in the Nilo-Sudan Province of Africa and *Arapaima gigas* from the Amazonian lowlands and Guianas of South America; and Osteoglossinae, two species of *Osteoglossum* from South America and three species of *Scleropages* from Queensland, New Guinea, and Southeast Asia. The lungfish *Neoceratodus forsteri* and *Scleropages* are the only native primary freshwater fishes found in Australia.

Other osteoglossomorphs include the African freshwater butterflyfish *Pantodon* (Pantodontidae), sister group of the Osteoglossidae; the North American Hiodontidae, the goldeye *Hiodon alosoides* and the mooneye *H. tergisus*; the African knifefishes, Notopteridae, and the African Mormy-

iformes, elephantfishes, Mormyridae, with 150 species; and the monotypic Gymnarchidae (*Gymnarchus niloticus*). Archaic fish distributions are summarized in Figure 16.11.

More recent distributions. Five groups of primary freshwater fishes have more recent distributions. These include the pickerels and relatives, the darters and perches, the sunfishes, the cichlids, and the Ostariophysi.

The suborder Esocoidei contains two families: Esocidae, the pickerels from North America and Eurasia, and Umbridae, the mudminnows from eastern and western United States and one relict species, *Umbra krameri*, from the Danube River in Europe (Fig. 16.12). The northern pike, *Esox lucius*, ranges across northern North America, Europe, and Asia, giving it the broadest natural distribution of any fish in the Northern Hemisphere.

Three families of acanthopterygian fishes are of major importance in freshwaters. The Percidae, the perches and darters, includes about 166 species, 14 of which are European and 152 of which are American, including all the

FIGURE 16.10. *Osteoglossidae and their distribution.* (A) Heterotis. (B) Osteoglossum. (C) Arapaima. (D) Scleropages.

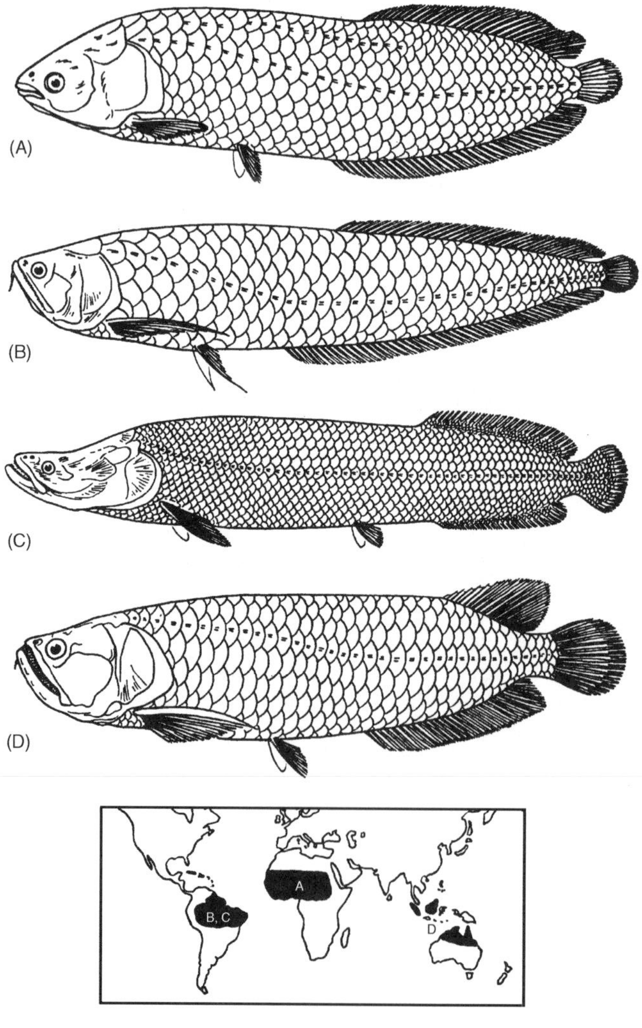

From Norman and Greenwood 1975.

FIGURE 16.11.   *Summary of archaic freshwater fish distributions.*

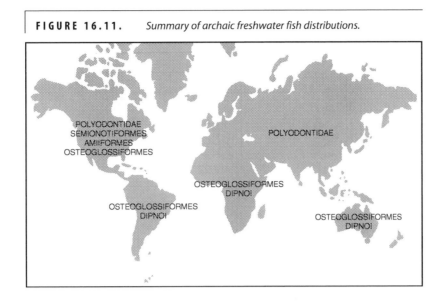

darters, tribe Etheostomatini (Fig. 16.13). The black basses and sunfishes (Centrarchidae) include 25 species, 24 from eastern North America and 1 relict species, *Archoplites interruptus*, from California (Fig. 16.14). The Cichlidae ecologically replaces the Centrarchidae and Percidae in the southern continents of South America, Africa, Madagascar, and southern India (see Fig. 16.14). This large family, with more than 700 species (see Box 15.1), is usually considered a secondary freshwater family because some species show salinity tolerance. Distributions of the first four groups are summarized in Figure 16.15.

The largest group of freshwater fishes is the series Otophysi of the superorder Ostariophysi, which includes four orders: Cypriniformes, Characiformes, Siluriformes, and Gymnotiformes.

The Cypriniformes includes three large and two small families found primarily in the northern continents. The Cyprinidae, the carps and minnows, is the largest family of freshwater fishes, with about 2000 species. It is found in North America, Africa, Europe, and Asia (Fig. 16.16). The highest diversity of cyprinids is found in Asia. The 75 species of suckers, Catostomidae, are confined to North America, except for a relict genus in China, *Myxocyprinus*, and a recent reinvasion of Siberia by *Catostomus catostomus* (Fig. 16.17). The loaches, Cobitidae, are found in Eurasia. The two smaller families, the Gyrinocheilidae and the Balitoridae, occur in Southeast Asia.

The Characiformes, the characins, comprises 10 to 16 families of tetras and relatives. The greatest diversity of the order is in South America, with 200 genera and over

FIGURE 16.12.   *Distribution of the pickerels, family Esocidae.*

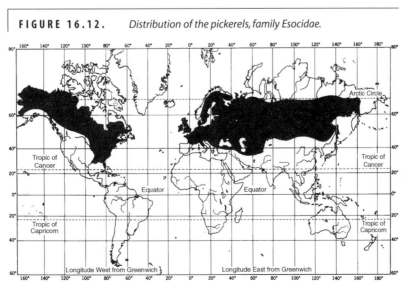

Adapted from Lagler et al. 1977.

FIGURE 16.13.   *Distribution of the darters and perches, Percidae.*

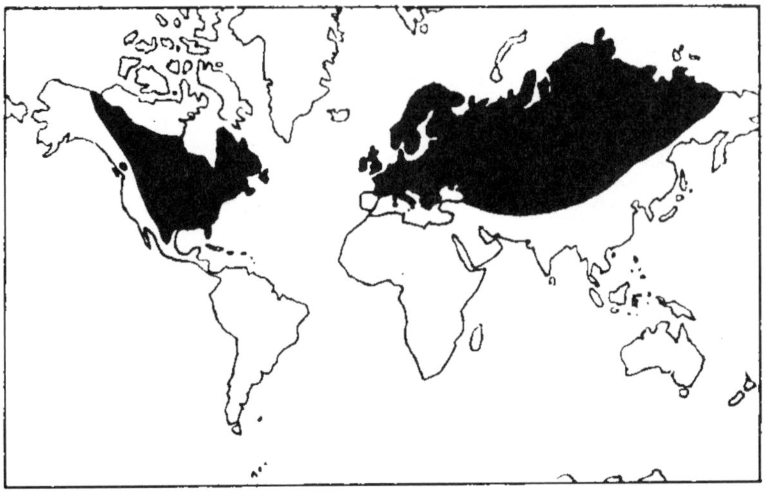

Adapted from Norman and Greenwood 1975.

1000 species. Characins are also widespread in Africa, with 23 genera and 150 species, currently placed in 4 families.

The Gymnotiformes comprises six families of electric fishes restricted to South America.

The catfishes, Siluriformes, include about 34 families of diverse fishes and about 2000 species. There is only 1 family in North America, the Ictaluridae, including *Ictalurus*, *Ameiurus*, and the madtoms of the genus *Noturus*. There are 14 endemic families and more than 1200 species in South America (Lundberg 1993). Some of the families are the suckermouth catfishes, Loricariidae and Astroblepidae, the popular aquarium fishes in the Callichthyidae, and the parasitic catfishes in the Trichomycteri-

dae. Africa has 6 freshwater families with about 400 species. Europe has only the Siluridae with *Parasiluris* and the huge *Siluris glanis*. There are several families in Asia. The catfishes also contain 2 marine families, Ariidae and Plotosidae, making them exceptions to the primary freshwater fish nature of the Siluriformes. To further complicate the issue, the Plotosidae has secondarily invaded freshwaters of Australia and New Guinea. The dominance of otophysans among primary freshwater fishes is summarized in Figure 16.18.

Similarities between South America and Africa.   The primary freshwater fishes of South America and Africa are remarkably similar. The Dipnoi and Osteoglossomorpha among ar-

FIGURE 16.14.   *Distribution of the sunfishes, Centrarchidae, and cichlids, Cichlidae.*

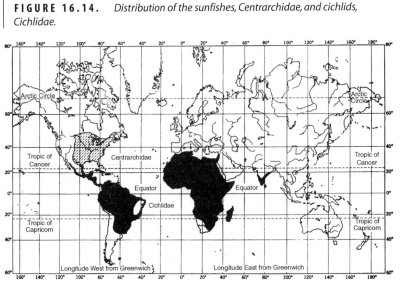

Adapted from Lagler et al. 1977.

FIGURE 16.15. *Summary of recent primary fish distributions (other than Ostariophysi).*

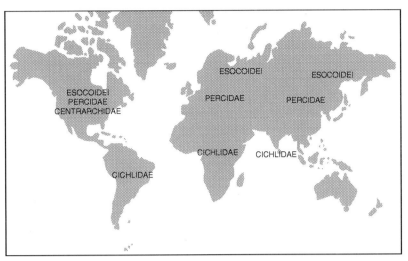

chaic fishes link South America, Africa, and Australia. The more recent distributions of the Characiformes and Cichlidae also link South America and Africa. A question arising from these parallels is whether this similarity is due to dispersal or vicariance. Some researchers believe that the Ostariophysi originated in Southeast Asia and dispersed to South America from Africa across a direct land bridge. However, land bridges work particularly poorly for freshwater fishes because rivers seldom run lengthwise along such bridges. Other researchers favor dispersal of Ostariophysi through North America, but this is not supported by recent distributions or by fossil evidence. Among 13 putative African-South American clades, only 3 clades—lepidosirenid lungfishes, polypterid bichirs, and doradoid

catfishes—clearly fit the simple drift-vicariance model, with a common ancestor inhabiting freshwaters of the African-South American landmass before the opening of the Atlantic Ocean (Lundberg 1993). Distributions of other groups, such as cichlids and characins, are more difficult to explain this way because they have evolved more recently, so perhaps some dispersal has taken place in addition to vicariant events. Explanation for the distribution of a freshwater galaxiid in South America and Australia (Box 16.3) is a good example of the argument over dispersal versus vicariance.

Middle America. The freshwater fish faunas of North and South America are very different. Minnows, suckers,

FIGURE 16.16. *Distribution of the minnows and carps, Cyprinidae.*

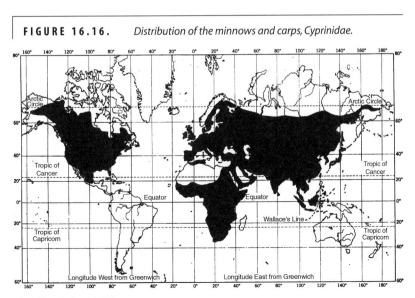

Adapted from Lagler et al. 1977.

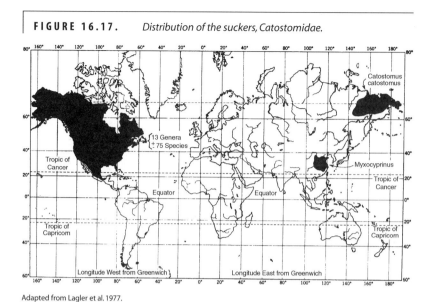

FIGURE 16.17.   *Distribution of the suckers, Catostomidae.*

Adapted from Lagler et al. 1977.

ictalurid catfishes, darters, and sunfishes predominate in the north, whereas cichlids, characins, gymnotoids, and a wide array of different catfish families are common in the south. What happens in the region between North and South America, in Middle America?

Middle America acts partly as a **filter barrier** slowing the movement of North American fishes south and South American fishes north, thus allowing invasion by marine groups. A few representatives of North American families such as Ictaluridae, Catostomidae, and *Lepisosteus* extend south to Costa Rica (Fig. 16.19). A few representatives of South American families such as Cichlidae (*Cichlasoma*) and Characidae (*Astyanax*) extend north to the Rio Grande. Of the approximately 456 species of freshwater fishes in Central America, over 75%

comprise secondary freshwater fishes such as Cyprinodontidae, Poeciliidae, Cichlidae, and marine invaders, and only 104 species are primary freshwater fishes (Miller 1966).

The general pattern of contributions from both major regions and of many secondary freshwater fishes also holds for more localized faunas in the region. For example, the **Usumacinta Province**, comprising the Grijalva and Usumacinta Rivers of Guatemala and Mexico, contains a mix of North American and South American fishes totaling over 200 species (Miller 1966). Two secondary groups, the cyprinodontoids and the cichlids, comprise about 90 species. There are also a large number of marine derivatives (18 species in nine families). Among these are a few endemics, including species from

FIGURE 16.18.   *Summary of otophysan distribution.*

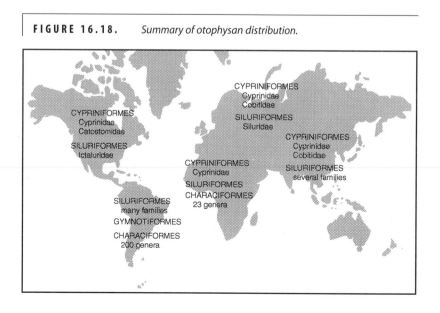

Vicariance Versus Dispersal: Galaxias maculatus

Vicariance and dispersal have been postulated by different sets of authors to explain the distribution of *Galaxias maculatus*, a small diadromous fish with a highly disjunct distribution in streams in eastern and western Australia, New Zealand, South America, and some oceanic islands (Berra et al. in press). The argument over whether vicariance or dispersal best explains the distribution of *Galaxias maculatus* started over 100 years ago and was particularly vociferous from the late 1970s through the mid-1990s (see summary in Berra et al., 1996).

Rosen (1978) considered galaxiid fishes to be part of a pan-austral Gondwanan biota that was fragmented by the movement of the southern continents during the Mesozoic. Rosen concluded that the present distribution resulted from vicariant events. *Galaxias maculatus* is the only galaxiid that breeds in brackish water, and its transparent whitebait larvae grow in the ocean before returning to freshwater. McDowell (1978), therefore, found it simpler to accept a relatively recent oceanic dispersal by a fish with a marine juvenile stage. Recent allozyme electrophoresis shows that *G. maculatus* is a single species with surprisingly little genetic variation across its entire range (Berra et al. in press). These data tend to support a dispersal hypothesis in this case, but mitochondrial DNA analysis is needed to further test the hypothesis.

---

three marine families, Belonidae (*Strongylura hubbsi*), Hemiramphidae (*Hyporhamphus mexicanus*), and Batrachoididae (*Batrachoides goldmani*). Freshwater species of these three families occur elsewhere, but what factors have led to this remarkable parallel derivation of endemic freshwater species? Part of the explanation is due to the depauperate nature of the primary freshwater fishes of the region, but this is true of many other rivers south through Panama. More information is needed on the nature of the water in the Usumacinta Province. Is it high in ions, as is true of southern Florida freshwaters that also contain a number of marine species (but not endemics derived from marine species)? Have there been historical factors or vicariant events involved? Here is an interesting problem involving phylogeny, biogeography, and physiology that awaits solution.

FIGURE 16.19. *Distributional limits of certain primary and secondary freshwater fishes in Central America. A single characin* (Astyanax mexicanus) *and a single cichlid* (Cichlasoma cyanoguttatum) *reach north to the Rio Grande. (CR = Costa Rica, P = Panama)*

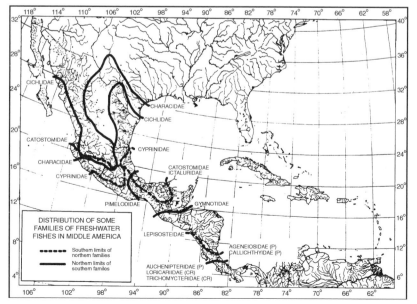

Adapted from Miller 1966, Fig. 1.

SUMMARY

1. Zoogeography is the study of the distribution of animals. About 58% of the 25,000 species of fishes are marine, 41% are freshwater species, and 1% move between the two habitats.

2. Four main ecological divisions of marine fishes are 1) epipelagic, surface-dwelling species (about 1% of all fishes); 2) deep pelagic species (5%); 3) deep benthic species (6%); and 4) inshore, littoral or continental shelf species (46%).

3. Inshore marine fishes occur in four major regions in order of decreasing biodiversity: 1) Indo-West Pacific, 2) western Atlantic, 3) eastern Pacific, and 4) eastern Atlantic.

4. Vicariance biogeography combines phylogenetic information with distribution patterns to assess relative movements of biota or habitat. Vicariant events (historical events that divided populations) often explain present distributions.

5. Freshwater fishes are surprisingly diverse, accounting for 41% of the world's fish species, although freshwater constitutes only 0.0093% of the planet's water.

6. Six freshwater regions are the 1) Nearctic Region (North America except tropical Mexico), with 14 families and about 950 species; 2) Neotropical Region (South and Middle America), 32 families and more than 2500 species; 3) Palearctic Region (Europe and Asia north of the Himalayas), 14 families, with many minnows and loaches; 4) African (or Ethiopian) Region, 47 families and 2000 species, including many primitive species; 5) Oriental Region (India, southern China, Southeast Asia, the Philippines, and the East Indies to Wallace's Line), 28 families, including 12 families of catfishes and 4 cypriniform families; and 6) Australian Region (Australia, New Guinea, and New Zealand), with only 2 primary freshwater species, both ancient relicts of a much wider archaic distribution pattern.

7. Six groups of primitive primary freshwater fishes have ancient distributions best explained by continental drift: lungfishes, bichirs, paddlefishes, gars, bowfin, and bonytongues.

8. Most primary freshwater fishes belong to the Ostariophysi, which includes Cypriniformes, Characiformes, Siluriformes, and Gymnotiformes. Cypriniform carps, minnows, loaches, and suckers are found primarily on the northern continents. Characiform characins comprise 10–16 families, with greatest diversity in South America (200 genera, more than 1000 species) but with 23 genera and 150 species in Africa. Siluriform catfishes, which include 34 families and about 2000 species, occur on all continents. Gymnotiform electric fishes comprise 6 families restricted to South America.

SUPPLEMENTAL READING

Berra 1981; Briggs 1974, 1995; Darlington 1957; Ekman 1953; Nelson and Platnick 1981; Nelson and Rosen 1981.

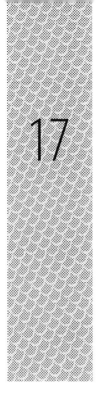

17

Special Habitats and Special Adaptations

Given our themes of diversity and adaptation, it seems appropriate to explore habitats and geographic regions that have led to spectacular evolutionary events among fishes. Certain climatic regimes and regions appear unusually harsh for successful invasion by complex vertebrate life forms. But fishes have been able to occupy almost all naturally occurring aquatic ecosystems that have any degree of permanence or at least predictability. It is often quite easy to determine the major selective pressures impinging on fishes in these habitats, and it is also often obvious what physiological, anatomic, and ecological adaptations have evolved in response to specific environmental pressures.

An axiom of evolutionary biology is that animals exposed to similar selection pressures are likely to evolve similar adaptations. This axiom, formalized as **the principle of convergence**, states that the stronger the selection pressures, the more similar unrelated animals will appear. In other words, where selection pressures are particularly extreme, animals will converge in morphology, physiology, behavior, and ecology, approaching an optimal design for that particular set of environmental forces. The special habitats discussed below—the deep sea, the open sea, polar regions, deserts, and caves—show this principle in operation.

THE DEEP SEA

The most diverse deep-sea fish assemblages occur between 40°N and 40°S latitudes, roughly between San Francisco and Melbourne, Australia, in the Pacific basin and between New York City and the Cape of Good Hope in the Atlantic basin. Separation of deep-sea fishes occurs more on a vertical than on a latitudinal basis (Fig. 17.1). The three major regions of open water are **mesopelagic** (200 to 1000 m), **bathypelagic** (1000 to 4000 m), and **abyssal** (4000 to 6000 m); deep-sea regions below 6000 m are referred to as **hadal** depths. A second group of **benthal**, or bottom-associated, species swims just above the bottom (= **benthopelagic**) or lives in contact with it (= **benthic**), usually along the upper continental slope at depths of less than 1000 m; corresponding ecological zones of benthal species are referred to as bathyal, abyssal, and hadal. The upper 200 m of the open sea, termed the **epipelagic** or **euphotic** zone, has its own distinctive subset of fishes (see below, Pelagic Fishes). This is the region where the photosynthetic activity of phytoplankton exceeds the respiration of the plants and animals living there, in other words, where the ratio of production to

respiration is greater than 1. Deeper waters are dependent on the euphotic zone for energy, which is provided in the form of sinking detritus such as feces, molted exoskeletons, and dead bodies (Marshall 1971; Wheeler 1975; Nelson 1994).

The deep-sea fishes of the mesopelagic and bathypelagic regions are readily recognized by just about anyone with a passing interest in fishes or marine biology. Deep-sea fishes often have light-emitting organs, termed **photophores**; large or long mouths studded with daggerlike teeth; chin barbels or dorsal fin rays modified as lures; long, thin bones; and greatly enlarged, tubular eyes or greatly reduced eyes (Marshall 1954, 1971). Such familiar appearances could result from a relative scarcity of forms. For example, widespread familiarity with deep-sea fishes could occur if we were exposed to many illustrations of the same strange animals. However, as the taxonomic listing in Table 17.1 reveals, the recognizability of deep-sea fishes is not a function of scarcity or a depauperate fauna. More than 1000 species of fishes inhabit the open waters of the deep sea, and another 1000 species are benthal, with good representation across orders of cartilaginous fishes and superorders of bony fishes. Similarities among unrelated fishes are therefore not due to phylogenetic relationships but to convergent adaptations.

The Physical Environment of the Deep Sea

Deep-sea fishes look alike because different ancestors invaded the deep sea from shallow regions and evolved similar anatomic and physiological solutions to an extreme environment. Understanding the convergent adaptations of deep-sea fishes requires that we first understand the physical environment of the deep sea and its influences on biota. Five physical factors—pressure, temperature, space, light, and food—contrast markedly between the surface and the deep sea and appear to have been

FIGURE 17.1.   *Regions and physical features of the deep-sea environment relative to depth. Representative species are a mesopelagic lanternfish, bathypelagic ceratioid anglerfish, benthopelagic rattail and halosaur, and benthic snailfish and greeneye. Many mesopelagic species undergo a **diurnal vertical migration** (DVM) to shallower waters at dusk, returning to deeper water at dawn. Total biomass of living organisms, available light, and temperature all decline with depth in the deep sea.*

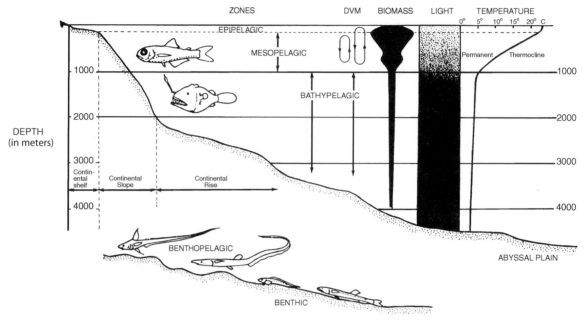

From Marshall 1971; used with permission.

strong selective forces on fishes (Marshall 1971; Hochachka and Somero 1984).

Pressure.   The weight of the overlying column of water, measured in atmospheres (atm), increases constantly with depth at a rate of 1 atm/10 m of descent (1 atm = 1.03 kg/cm^2, or 14.7 lbs/in^2). Thus between the top of the mesopelagic region at 200 m and the lower bathypelagic region at 4000 m, pressure increases 20-fold, from 20 to 400 atm. The deepest-living fishes, the neobythitine cuskeels, *Bassogigas profundissimus* and *Abyssobrotula galatheae*, have been collected at 7160 m and 8370 m respectively, where they would experience pressures of 700 to 800 atm, or approximately 12,000 lbs/in^2 (Nielsen and Munk 1964; Nielsen 1977). Below the surface, pressure at any given depth is constant and predictable, whereas at the surface it can change rapidly and significantly with each passing wave.

The tremendous pressures of the deep sea do not create problems for most biological structures because fishes are made up primarily of water and dissolved minerals, which are relatively incompressible. However, pressure has an influence on the volume of water molecules, water-containing compounds, and proteins, which affects the rates of chemical reactions. Several deep mesopelagic and bathypelagic species have evolved proteins that are much less sensitive to the effects of pressure than are those of their shallow-water relatives (Hochachka and Somero 1984; Somero et al. 1991). Gas-containing structures are particularly affected, because both volume relationships and gas solubility are sensitive to pressure. The organ most affected is the gas bladder, because it is difficult to secrete gas into a gas-filled bladder under high pressure. Three trends occur in the gas bladders of deep-sea fishes that reflect the constraints of pressure.

1. The efficiency of gas secretion depends on the interchange surface of the capillaries of the rete mirabile, the main gas-secreting organ (see Chap. 5, Buoyancy Regulation). Whereas the retes of epipelagic fishes are usually less than 1 mm long, retes of upper mesopelagic fishes are 1 to 2 mm long, those of lower mesopelagic fishes are 3 to 7 mm long, and those of some bathypelagic fishes are 15 to 20 mm long.

2. Although mesopelagic fishes have large gas-filled bladders, most bathypelagic fishes have lost their gas bladders. Flotation might therefore be a problem for these fishes, but their body musculature and skeletons are reduced as energy-saving mechanisms, and they consequently approach neutral buoyancy. As long as a fish remains at relatively constant depths, it has minimal need for buoyancy control. However, many mesopelagic fishes undergo diurnal vertical migrations, have a greater need to adjust their buoyancy, and have retained their gas bladders.

Deep benthopelagic fishes are able to hover just above the bottom with minimal energy expenditure via a different mechanism. Instead of

Table 17.1. Representative teleostean taxa from the three major deep-sea habitat types. Approximate number of deep-sea families is given in parentheses the first time a group is listed.

Mesopelagic (≈750 spp.)
Superorder Elopomorpha
 Notacanthiformes (3 families): Notacanthidae–spiny eels
 Anguilliformes (6): Nemichthyidae–snipe eels; Synaphobranchidae–cutthroat eels
Superorder Protacanthopterygii
 Osmeriformes (5): Bathylagidae–deep-sea smelts; Opisthoproctidae–barreleyes; Alepocephalidae–slickheads; Platytroctidae
Superorder Stenopterygii
 Stomiiformes (5): Gonostomatidae–bristlemouths; Sternoptychidae–hatchetfishes; Stomiidae–barbeled dragonfishes
Superorder Cyclosquamata
 Aulopiformes (11): Giganturidae–giganturids; Synodontidae–lizardfishes; Paralepididae–barracudinas;
  Evermannellidae–sabertooth fishes; Alepisauridae–lancetfishes
Superorder Scopelomorpha
 Myctophiformes (2): Neoscopelidae; Myctophidae–lanternfishes
Superorder Lampridiomorpha
 Lampridiformes (4): Stylephoridae–tube-eyes
Superorder Acanthopterygii
 Stephanoberyciformes: Mirapinnidae–hairyfish
 Perciformes: Chiasmodontidae–swallowers; Gempylidae–snake mackerels

hatchetfish
giganturid

Bathypelagic (≈200 spp.)
Superorder Elopomorpha
 Anguilliformes: Nemichthyidae–snipe eels; Serrivomeridae–sawtooth eels; Saccopharyngidae–gulpers; Eurypharyngidae–swallowers
Superorder Protacanthopterygii
 Osmeriformes: Alepocephalidae–slickheads
Superorder Stenopterygii
 Stomiiformes: Gonostomatidae–bristlemouths
Superorder Paracanthopterygii
 Gadiformes: Melanonidae–melanonids; Macrouridae–grenadiers and rattails
 Ophidiiformes: Ophidiidae–cusk-eels; Bythitidae–brotulas
 Lophiiformes (12): Ceratioidei–deep-sea anglerfishes, seadevils (11 families)
Superorder Acanthopterygii
 Stephanoberyciformes: Melamphaidae–bigscale fishes; Stephanoberycidae–pricklefishes; Cetomimoidei–whalefishes (3 families)
 Beryciformes (9): Anoplogastridae–fangtooth
 Perciformes: Chiasmodontidae–swallowers

gulper

Benthal (≈1000 benthopelagic and benthic spp.)*
Superorder Elopomorpha
 Notacanthiformes: Notacanthidae–spiny eels; Halosauridae–halosaurs
 Anguilliformes: Synaphobranchidae–cutthroat eels
Superorder Cyclosquamata
 Aulopiformes: Chlorophthalmidae–greeneyes; Ipnopidae–spiderfishes and tripodfishes
Superorder Paracanthopterygii
 Gadiformes: Merlucciidae–merlucciid hakes; Moridae–morid cods; Macrouridae–grenadiers
 Ophidiiformes: Ophidiidae–cusk-eels; Bythitidae–brotulas; Aphyonidae
 Lophiiformes: Ogcocephalidae–batfishes
Superorder Acanthopterygii
 Beryciformes: Caproidae–boarfishes
 Scorpaeniformes: Liparidae–snailfishes
 Perciformes: Zoarcidae–eelpouts; Bathydraconidae–antarctic dragonfishes

whalefish
brotula

*Chimaeras and many squaloid sharks are benthopelagic. Most benthal fishes live above 1000 m, although some grenadiers and rattails live between 1000 and 4000 m, macruronid southern hakes live somewhat deeper, tripodfish live to 6000 m, snailfishes live to 7000 m, and neobythitine cusk-eels live down to 8000 m.
Based on Marshall 1971, 1980; Wheeler 1975; Gage and Tyler 1991; Nelson 1994. Figures from Marshall 1971; used with permission.

trying to secrete gases against incredible pressure gradients, they have evolved lipid-filled gas bladders. Because lipids are relatively incompressible and are lighter than seawater, they provide flotation. Interestingly, the larvae of these fishes have gas-filled bladders, but these larvae, and the larvae of nearly all deep-sea fishes, are epipelagic, where the costs of gas secretion and buoyancy adjustment are much less. Benthopela-gic squaloid sharks such as *Centroscymnus* and *Etmopterus* show parallel evolution. These deep-sea sharks have exceptionally large livers that account for 25% of their total body mass. Their livers contain large quantities of the low-density lipid **squalene**. Deep-water holocephalans also achieve neutral buoyancy via squalene and by reduced calcification of their cartilaginous skeletons (Bone et al. 1994).

3. Most deep-sea fishes belong to the relatively primitive teleostean superorders Protacanthopterygii, Stenopterygii, Cyclosquamata, and Scopelomorpha. These taxa typically have a direct, physostomous connection between the gas bladder and the gut. Deep-sea fishes are, however, "secondarily" physoclistous, having closed the pneumatic duct, thus preventing gas from escaping out the mouth.

Temperature.   At the surface, temperature is highly discontinuous, changing markedly both seasonally and daily. In the deep sea, temperature is a predictable function of depth. Surface waters are warmer than deeper waters. Water temperature declines with depth through the mesopelagic region across a permanent thermocline until one reaches the bathypelagic region, where temperature remains a relatively constant 2 to 5°C, depending on depth.

Temperature is a strong predictor of distribution for different taxa of deep-sea fishes. Ceratioid anglerfishes and darkly colored species of the bristlemouths (*Cyclothone*) are restricted to the deeper region. Even within the mesopelagic zone, species sort out by temperature. Hatchetfishes, pale *Cyclothone*, and malacosteine loosejaws are restricted to the lower half at temperatures between 5 and 10°C, whereas lanternfishes and astronesthine and melanostomiatine stomiiforms occur in the upper half at 10 to 20°C. Latitudinal differences in temperature-depth relationships lead to distributional differences within species. Some species such as ceratioid anglers that are mesopelagic at high latitudes occur in bathypelagic waters at lower latitudes, a phenomenon known as **tropical submergence**, which results from the warmer surface temperatures in the tropics.

Since temperature remains fairly constant at any given depth, absolute temperature is a minimal constraint on a fish that does not move vertically. But vertically migrating mesopelagic species must swim through and function across a temperature range of as much as 20°C (see Fig. 17.1). Lanternfish species that migrate vertically have larger amounts of DNA per cell than do species that are nonmigratory. Increased DNA could potentially allow for multiple enzyme systems that function at the different temperatures encountered by the fishes (Ebeling et al. 1971).

Space.   The volume occupied by the deep sea is immense. Approximately 70% of the earth's surface is covered by ocean, and 90% of the surface of the ocean overlies water deeper than 1000 m. The bathypelagic region, which makes up 75% of the ocean, is therefore the largest habitat type on earth. This large volume creates problems of finding food, conspecifics, and mates, because bathypelagic fishes are never abundant. Life in the bathypelagos is extremely dilute. For example, female ceratioid anglerfishes are distributed at a density of about 1 per 800,000 m³, which means a male anglerfish is searching for an object the size of a football in a space about the size of a large, totally darkened football stadium.

Deep-sea fishes show numerous adaptations that reflect the difficulties of finding potential mates that are widely distributed in a dark expanse. Unlike most shallow-water forms, many deep-sea fishes are sexually dimorphic in ways directly associated with mate localization. Mesopelagic fishes, such as lanternfishes and stomiiforms, have species-specific and sex-specific patterns and sizes of light organs, structures that first assure that individuals associate with the right species and then that the sexes can tell one another apart. Among benthopelagic taxa, such as macrourids, brotulids, and morids, males often have larger muscles attached to their gas bladders that are likely used to vibrate the bladder and produce sounds that can attract females from a considerable distance.

Some of the most bizarre sexual dimorphisms occur among bathypelagic species, where problems of mate localization are acute. The most speciose group of bathypelagic fishes is the ceratioid anglerfishes, of which there are 11 families and about 110 species (Bertelsen 1951; Pietsch 1976; Nelson 1994). In many of these families, the males are dwarfed, reaching only 20 to 40 mm long, whereas females attain lengths 10 or more times that size, up to 1.2 m in one species. The males are entirely parasitic on the females in 5 families (Fig. 17.2). Once a male finds a female, he attaches to her by his mouth, his mouth tissue fuses with her skin, and he becomes parasitically dependent on her for nutrition. His internal organs degenerate, with the exception of his testes, which can take up more than half of his coelom. The premium placed on locating a female is reflected throughout the anatomy and physiology of searching males. During this phase, males have highly lamellated olfactory organs and well-developed olfactory tracts, bulbs, and forebrains, whereas females have almost entirely degenerate olfactory systems. Males also have extensive red muscle fibers, the kind used for sustained swimming. Females have predominantly white muscle fibers, which usually function for short bursts of swimming. Males of some species possess enlarged, tubular eyes that are extremely sensitive to light (see next section), whereas females have small, relatively insensitive eyes. Males also have high lipid reserves in their livers, which they need because their jaw teeth become replaced by beaklike denticles that are useless for feeding but are apparently specialized for holding onto a female.

All this comparative evidence indicates that male ceratioid anglerfishes are adapted for swimming over large expanses of ocean, searching for the luminescent glow and some olfactory cue emitted by females, whereas females are floating relatively passively, using their bioluminescent lures to attract prey at which they make sudden lunges, and trailing pheromones through the still waters. The coevolved nature of these traits is evident from the dependence of both sexes on locating each other. Neither sex matures until the male attaches to the female.

Convergence occurs in the unrelated bathypelagic bristlemouths, which are probably the most abundant vertebrates on earth. Again, males are smaller than females and have a well-developed olfactory apparatus, extensive red muscle fibers, and larger livers and fat reserves. Although the males are not parasitic on the females, they are un-

FIGURE 17.2. *A female ceratioid anglerfish,* Borophryne apogon. *Object projecting from the female's snout is her illicial lure. Arrows indicate two dwarf, 15- to 20-mm-long, parasitic males attached ventrally to the 100-mm-long female.*

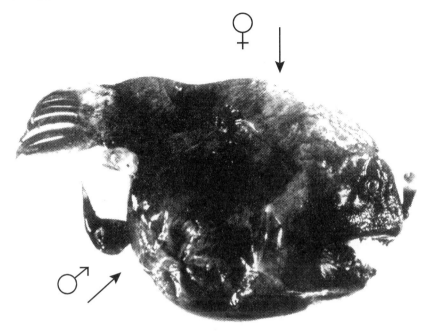

Photo by W. T. O'Day in Pietsch 1976; used with permission.

usual in that they are protandrous hermaphrodites, which means that an individual matures first as a male and then later switches sex and becomes a female. Sex change theory predicts just such a switch because relative fitness favors being a male when small and a female when large (see Chap. 10, Determination, Differentiation, and Maturation; Chap. 20, Gender Roles in Fishes).

Light. Below the euphotic zone, light is insufficiently strong to promote significant plant growth. Light visible to the human eye is extinguished by 200 to 800 m depth, even in the uniformly clear water of the mesopelagic and bathypelagic regions. Deep-sea fishes are 15 to 30 times more sensitive to light than humans and can detect light down to between 700 and 1300 m, depending on surface clarity. The mesopelagic region is often termed the twilight zone, whereas the bathypelagic region is continually dark. What little light that passes into the mesopelagic region has been differentially absorbed and scattered by water molecules and turbidity and is limited to relatively short, blue-green wavelengths centered on 470 nm.

The greatly reduced illumination of the mesopelagic region, and the missing light of the bathypelagic region, have produced obvious adaptations among both the eyes and photophores of fishes living there. Bathypelagic fishes live in permanent darkness and, with the exception of male ceratioid anglers, have greatly reduced eyes that probably function primarily for detecting nearby bioluminescence. Mesopelagic fishes have modifications to their eyes that generally increase their ability to capture what little ambient light is available. They have very large eyes,

often measuring 50% of head length; most North American freshwater fishes have eye diameters that are only 10% to 20% of head length.

Mesopelagic fishes also have comparatively large pupils and lenses and lengthened eyes. Elongation results either from a space between the pupil and lens, termed the **aphakic** (or **lensless**) **space**, or from lengthening of the retina-containing portion of the eye posterior to the lens. Aphakic spaces have evolved convergently in protacanthopterygian platytroctids and deep-sea smelts, stenopterygian loosejaws, cyclosquamate waryfishes, and scopelomorph lanternfishes, most of which live in the upper region of the mesopelagic zone. **Tubular eyes**, more characteristic of deeper mesopelagic species, have evolved convergently in four superorders and five orders of mesopelagic fishes, including protacanthopterygian barreleyes, stenopterygian hatchetfishes, paracanthopterygian anglerfishes, and acanthopterygian whalefishes. Eye elongation provides two visual benefits: increasing the sensitivity of the eye to light by about 10% and also increasing binocular overlap, which aids depth perception (Marshall 1971; Lockett 1977).

Mesopelagic fishes have pure-rod retinas with visual pigments that are maximally sensitive at about 470 nm, which is a good match to the light environment at mesopelagic depths and also matches the light output from photophores, which are much more common among mesopelagic than bathypelagic fishes. Bioluminescence has evolved independently in at least five superorders of deep-sea teleosts—protacanthopterygians, stenopterygians, scopelomorphs, paracanthopterygians, and acanthopterygians—as well as in dogfish sharks, squids, crustaceans, and other invertebrates. Light organs, in addition to identifying the species and sex of the emitter, may also illuminate nearby prey. The structure that bioluminesces may be a simple luminescent gland backed by black skin that emits on its own or contains bioluminescent bacteria. More complex circular photophores may be backed by silvery reflective material with a lens through which light passes.

In highly derived photophores, the lens may be pigmented and hence the light transmitted is of a different wavelength, as in the malacosteine loosejaws, which have a red filter over the subocular photophores and also have retinal reflectors and receptors sensitive to red wavelengths. This unique combination of luminescent emission and spectral sensitivity could give loosejaws a private channel over which they could communicate without being detected by potential predators or prey. It could also serve to maximize illumination of red mesopelagic crustaceans (Lockett 1977; Denton et al. 1985). Photophores tend to flash on for 0.2 to 4.0 seconds, depending on species. Different species of lanternfishes may have similar photophore patterns but different flash rates, suggesting a convergence in communication tactics between deep-sea fishes and fireflies (Meinsinger and Case 1990).

Food.   Limited light and huge volume mean that food is extremely scarce in the deep sea. All marine food chains, except at thermal vents, originate in the euphotic zone, which makes up only 3% of the ocean. Food for bathypelagic fishes must therefore first pass through the filter of vertebrates, invertebrates, and bacteria in the mesopelagic zone; much of this food rains down weakly, unpredictably, and patchily in the form of carcasses, sinking sargassum weed, detritus, and feces. All deep-sea fishes are carnivorous, feeding on zooplankton, larger invertebrates, or other fishes. Zooplankton biomass at the top of the bathypelagos is only about 1% of what it is at the surface, and densities of benthic invertebrates decrease with depth and distance from continental shores. High densities, diversities, and productivity of invertebrates at thermal vents on the deep-sea floor do not support a similar abundance or diversity of fishes. Only three species, a bythitid brotula and two zoarcid eelpouts, are endemic to and frequent vent areas (Grassle 1986; Cohen et al. 1990). A general scarcity of food in the deep sea puts a premium on both saving and obtaining energy. Convergent traits in both categories are readily apparent.

Foraging adaptations.   Deep-sea fishes show a number of convergent foraging traits. In general, zooplanktivores have small mouths and numerous, relatively fine gill rakers, whereas predators on larger animals have larger mouths and fewer, coarser gill rakers. Daggerlike teeth or some other form of long, sharp dentition is so characteristic of deep-sea forms that their family names often refer directly or indirectly to this trait, including such colorfully named groups as dragonfishes, daggertooths, bristlemouths, snaggletooths, viperfishes, sabertooths, and fangtooths. Large, expandable mouths,

hinged jaws, or distensible stomachs are also reflected in such names as gulpers, swallowers, and loosejaws. Saccopharyngoid gulper and swallower eels (Fig. 17.3) have enormous mouths that can expand to more than 10 times the volume of the animal's entire body, the largest mouthbody volume of any known vertebrate (Nielsen et al. 1989). Black dragonfishes, viperfishes, ceratioid anglerfishes, and sabertooth fishes can swallow prey larger than themselves, as much as three times so in the case of the anglerfishes. Their swallowing abilities are increased because the pectoral girdle is disconnected from the skull, enlarging the intercleithral space of the throat (see Chap. 8, Pharyngeal Jaws). All of these anatomic specializations point to a strategy of taking advantage of any feeding opportunity that may come along, regardless of the size of the prey.

A small number of shallow-water paracanthopterygian species, notably the goosefishes, frogfishes, batfishes, and anglerfishes, possess modified dorsal spines that are waved in front of prey species to lure them within striking distance. Such lures reach their greatest and most diverse development among mesopelagic and bathypelagic fishes, where they occur on viperfishes, various dragonfishes, astronesthine snaggletooths, and most ceratioid anglerfishes, and arguably as luminescent organs in the mouths of hatchetfishes and on the illuminated tail tip of the gulper eels. The typical anglerfish lure consists of an elongate dorsal spine, the **illicium**, tipped by an expanded structure called the **esca**. Escae tend to have species-specific shapes, can regenerate if damaged, and are moved in a variety of motions that imitate the swimming of a small fish or shrimp (see Pietsch 1974).

Most mesopelagic fishes undertake evening migrations from the relatively unproductive mesopelagic region to the richer epipelagic zone to feed; they then return to the mesopelagic region at dawn (see Fig. 17.1). The migration involves movements to near the surface from as deep as 700 m, can take an hour or more, and may entail considerable energy expenditure. This movement is so characteristic of mesopelagic fishes, crustaceans, and mollusks that the community of organisms that migrates is referred to as the **deep-scattering layer** (DSL), whose presence is discernible on sonar screens because of reflection of sonar signals off the gas bladders of the fishes. Hypotheses about the adaptiveness of the migration include 1) a net energy gain from feeding in warm water and metabolizing in cold water and 2) exploitation of surface currents that bring new food into the water column above the migrator. It is apparent that the migration serves a foraging purpose, given the 100-fold difference in plankton biomass between the two regions and also given that stomachs of migrators are empty in the evening before migration and full in the morning after migration.

It is in the deeper region of the bathypelagos that we find the most extreme adaptations for opportunistic prey capture and energy conservation. Bathypelagic fishes remain in place, perhaps because external cues of changing daylight are lacking or the energetic costs of migrating are too high. They instead lure prey with bioluminescent lures. Observations from submersibles suggest that bathypelagic forms adapt a float-and-wait foraging mode,

FIGURE 17.3.   *Extreme movements of the head and mouth during swallowing in the viperfish,* Chauliodus *sloani. (A) Mouth at rest, showing the premaxillary and mandibular teeth that sit outside the jaw when the mouth is closed. The maxillary and palatine teeth are small and slant backward. (B) Mouth opened maximally as prey is captured and impaled on the palatine teeth prior to swallowing. The anterior vertebrae and neurocranium are raised, the mandibulo-quadrate joint at the back corner of the mouth is pushed forward, and the gill covers are pushed forward and separated from the gills and gill arches. The heart, ventral aorta, and branchial arteries are also displaced backward and downward. Such wide expansion of the mouth accommodates very large prey and is in part necessary for prey to pass between the large fangs.*

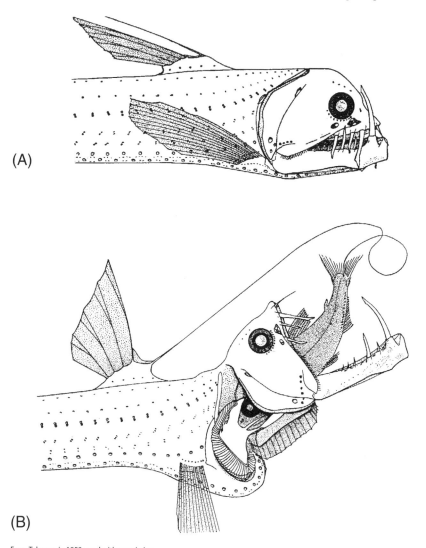

(A)

(B)

From Tchernavin 1953; used with permission.

hovering relatively motionless in the water column and making quick lunges at prey.

Energy conservation.   Deep-sea fishes minimize their daily and long-term expenditure of calories in many ways. Biochemically, rates of enzymatic and metabolic activity and even levels of ATP-generating enzymes are lower in deep-sea fishes than in shallow-water relatives, which conserves energy used in locomotion, osmotic regulation, and protein synthesis (Somero et al. 1991). Energy savings are also accomplished via elimination or replacement of heavy components. Structurally, bathy-pelagic fishes are fragile compared with shallow-water, mesopelagic, and even deep-sea benthic fishes. Many of the heavy bony elements of shallow-water relatives have been eliminated. Pelvic fins are often missing or reduced to rudiments, bones of the head are reduced to thin strands, and many species are scaleless. Spines are rare among deep-sea fishes; even the few acanthopterygian groups that have managed to invade the deep sea, such as melamphid bigscale fishes and chiasmodontid swallowers, have very feeble fin spines. Body musculature is also greatly reduced, by as much as 95% in the trunk and caudal regions compared with shallow-water forms.

Without trunk musculature, predator evasion becomes a problem. Most deep-sea fishes are colored in ways that should minimize their detection by potential predators. Mesopelagic fishes tend to be silvery or brown with ventral photophores that point downward. Silvery fishes disappear in open water (see Chap. 19, Invisible Fishes). Ventral photophores may aid in breaking up the silhouette of the fish when viewed from below against the backdrop of weak downwelling light. Bathypelagic fishes are generally dark brown or black, as would be expected where the background is black. Additional energy savings are attained by replacing heavy structural components with less-dense substances. Where glycerol lipids occur in shallow-water fishes, deep-sea forms have less-dense waxy esters. These structural changes save energy because metabolic costs of both construction and maintenance are reduced. In addition, elimination and replacement of heavy elements reduce the mass of the fish, making it closer to neutral buoyancy and eliminating costs associated with fighting gravity.

Bathypelagic fishes as a group tend to have free neuromasts in their lateral lines, rather than having lateral line organs contained in canals, as in mesopelagic and benthic groups. Free neuromasts in shallow-water fishes, such as goosefishes, cavefishes, and many gobies, are usually associated with a very sedentary lifestyle, again suggesting a premium on energy-conserving tactics and an ability to detect minor water disturbances among bathypelagic species.

Convergence in the Deep Sea

The deep sea offers numerous striking examples of the principle of convergence. Benthopelagic fishes from at least 12 different families have evolved an eel-like body that tapers to a pointed tail, often involving fusion of elongated dorsal and anal fins with the tail fin (Gage and Tyler 1991). Another aspect of convergence exemplified in the deep sea is that selection pressures can override phylogenetic patterns, producing closely related fishes that are biologically very different because they live in different habitats (Marshall 1971).

Gonostoma denudatum and *G. bathyphilum* are Atlantic bristlemouths in the stenopterygian family Gonostomatidae. *G. denudatum* is a mesopelagic fish, whereas *G. bathyphilum*, as its name implies, is a bathypelagic species. *G. denudatum* is silvery in color and has prominent photophores, well-developed olfactory and optic organs and body musculature, a well-ossified skeleton, a large gas bladder, large gill surface per unit weight, large kidneys, and well-developed brain regions associated with these various structures. *G. bathyphilum*, in contrast, is black, has small photophores and small eyes, small olfactory organs (except in males), weak lateral muscles, a poorly ossified skeleton, no gas bladder, small gills and kidneys, and smaller brain regions. Only the jaws of *G. bathyphilum* are larger than its mesopelagic congener. Similar comparisons can be drawn between other mesopelagic and bathypelagic gonostomatids, and between mesopelagic

and bathypelagic fishes in general. Even bathypelagic forms derived from benthopelagic lineages, such as the macrourids and brotulids, have converged on bathypelagic traits (Marshall 1971).

The extreme demands of the deep-sea habitat have also led to convergence in nonteleostean lineages. The mesopelagic cookie cutter sharks, *Isistius* spp., have a high squalene content in their livers, which increases buoyancy. They also possess photophores and migrate vertically with the biota of the DSL (the widespread nature of bioluminescence, with some fish producing their own light and others using symbiotic bacteria, is in itself a remarkable convergence). Deep-sea sharks and holocephalans possess visual pigments that absorb light maximally at the wavelengths that penetrate to mesopelagic depths, as is the case for another mesopelagic nonteleost, the living coelacanth (*Latimeria chalumnae*). Deep-sea crustaceans and mollusks have also evolved anatomic and physiological traits similar to those of fishes, including the emission of luminous ink (e.g., platytroctids, ceratioids, and squids) (Marshall 1980; Hochachka and Somero 1984).

PELAGIC FISHES

The epipelagic region is technically the upper 200 m of the ocean off the continental shelves (see Fig. 17.1), but the terms **epipelagic** and **pelagic** are often used synonymously to describe fishes that swim in the upper 100 to 200 m of coastal and open-sea areas. Common pelagic groups include many species of elasmobranchs (mako, whitetip, silky, and whale sharks), clupeoids (herrings, sardines, sprats, shads, pilchards, menhadens, anchovies), atherinomorphs (flyingfishes, halfbeaks, needlefishes, sauries, silversides), opahs, oarfishes, bluefish, carangids (scads, jacks, pilotfishes), dolphinfishes, remoras, pomfrets, barracudas, scombroids (cutlassfishes, mackerels, Spanish mackerels, tunas, swordfish, billfishes), butterfishes, and tetraodontiforms (triggerfishes, molas). The pelagic realm is unquestionably the most important and productive region of the sea as far as human consumption is concerned. Pelagic fishes constitute nearly half of the 70 to 80 million tons of fish captured annually worldwide. Coastal pelagics, particularly clupeoids, make up about one-third of the total, and offshore pelagics such as tunas and billfishes make up an additional 15% (Blaxter and Hunter 1982; Groombridge 1992).

Characteristics of the pelagic region are high solar insolation, variable production that can be very high in regions of upwelling or convergence of major currents, large volume, and a lack of physical structure. The abundance and diversity of fishes in the open sea is made possible by the periodic high productivity that occurs as nutrient-rich cold water upwells to the surface, promoting the bloom of algal plankton species and creating a trophic cascade, at least until the nutrients are used up. The greatest concentrations of fishes in the sea, and the largest fisheries, occur in such areas of upwelling. Upwelling areas may account for 70% of the world fisheries catch

(Cushing 1975). The anchovy fisheries of South America and Africa, and the sardine fisheries of North America and Japan have been direct results of pelagic fishes accumulating in areas of upwelling. Several of these fisheries have collapsed through a combination of overexploitation and shifts in oceanographic conditions that reduced the magnitude of the upwelling (see Chap. 25, Commercial Exploitation). The boom-and-bust cycles of temperate pelagics result from a patchy distribution of food in both time and space interacting with life history patterns of high-latitude pelagic species, which puts a premium on an ability to travel long distances and locate blooms.

Adaptations

Many common threads run through the biology of pelagic fishes, suggesting convergent adaptation to pronounced and predictable selection pressures. In general, pelagic fishes are countershaded and silvery, round or slightly compressed, and streamlined with forked or lunate tails; are schooling; have efficient respiration and food conversion capabilities and a high percentage of red muscle and lipids; are migratory; and account for all fish examples of endothermy. Differences in most of these characters correspond to how pelagic a species is; extreme examples are found among the open-water, migratory tunas, which have the fastest digestion rates, the highest metabolic rates, and the most extreme specializations for sustained levels of rapid locomotion of any fish (Magnuson 1978) and are among the most advanced of the teleost fishes.

Several superlatives apply to pelagic fishes and reflect adaptations to life in open water and an emphasis on continual swimming, often associated with long-distance migrations. Large sharks, salmons, tunas, and billfishes move thousands of kilometers annually (see Chap. 22, Annual and Supra-annual Patterns: Migrations), but even smaller coastal pelagics can make annual migrations of 150 km (sprats) and as far as 2000 km (herring) (Cushing 1975). To sustain continual swimming, pelagics have the highest proportion of red muscle among ecological groups of fishes. Within the mackerels and tunas, the amount of red muscle increases in the more advanced groups, which are also increasingly pelagic and inhabit colder water during their seasonal migrations. In primitive mackerels, the red muscle is limited to a peripheral, lateral band of the body, whereas in advanced tunas the red muscle is more extensive, occurs deeper in the body musculature, and is kept warm by the countercurrent heat exchangers that are also more developed in advanced scombrids (see Chap. 7, Endothermic Fishes) (Sharp and Pirages 1978). Countercurrent exchangers have evolved convergently in tunas and mackerel sharks, both pelagic fishes that range into cold temperate and deep waters. This convergence suggests that endothermy and heat conservation arose independently in these groups and allowed otherwise tropical fishes to expand their ranges into colder regions (Block et al. 1993).

Body shapes and composition in pelagics reflect the demands of continual swimming. Unlike benthic fishes with depressed bodies and littoral zone fishes with deep, circular, compressed bodies, pelagic fishes tend to have fusiform shapes that minimize drag. This is accomplished with a rounder cross section and by placing the maximum circumference of the body two-fifths of the way back from the head, an ideal streamlined shape also evolved convergently by pelagic sharks, whales, dolphins, and extinct ichthyosaurs (see Chap. 8, Locomotion: Movement and Shape). Streamlining is enhanced by having relatively small fins or having depressions or grooves on the body surface into which the fins can fit during swimming (e.g., tunas, billfishes).

In high-speed fishes such as sauries, mackerels, and tunas, a series of small finlets occur both dorsally and ventrally anterior to the tail. These finlets may prevent vortices from developing in water moving from the median fins and body surfaces toward the tail, which would allow the tail to push against less turbulence. The extremely small second dorsal and anal fins of mackerel sharks, swordfishes, and billfishes could function analogously. Tunas add a corselet of large scales around the anterior region of maximum girth, which may reduce drag and thus create more favorable water flow conditions posteriorly where actual propulsion occurs. In the region of the caudal peduncle and tail, sharks, jacks, tunas, swordfishes, and billfishes have a single or multiple keels that extend laterally. In the tunas, a single peduncular keel is supplemented by a pair of smaller caudal keels that angle toward each other posteriorly (Fig. 17.4). Peduncular keels reduce drag as the narrow peduncle is swept through the water, whereas caudal keels may act as a nozzle that accelerates water moving across the tail, adding to its propulsive force (Collette 1978). Peduncular keels have evolved convergently in cetaceans, but the keels are oriented vertically, as would be expected from their mode of swimming.

Many pelagic fishes swim continuously. In the bluefish, jacks, tunas, swordfish, and billfishes, this constant activity is linked to a respiratory mode known as ram gill ventilation (see Chap. 5, Water as a Respiratory Environment). Instead of pumping water via a muscular buccal pump, pelagic fishes swim with their mouths open while water flows across the gill surfaces. Ram gill ventilation requires that a fish swim continuously at speeds of at least 65 cm/sec, which is easily attained by any but the smallest tunas at their cruising speed of 1 body length per second. The more common buccal pump mechanism accounts for 15% of the total energy expended by a fish, suggesting that ram ventilation conserves energy. A trade-off arises because tunas and billfishes have minimal branchiostegal development and have lost the ability to pump water across their gills. They must therefore move continuously to breathe. However, these fishes are negatively buoyant and must move to keep from sinking anyway (Roberts 1978).

The high levels of activity of pelagics are fueled by an efficient circulatory system. Pelagics have an enhanced capacity for supplying oxygen to their muscles. For example, menhadens, bluefish, and tunas have two to three times the hemoglobin concentration of typical inshore, sedentary forms; hemoglobin concentration in tunas is more like that of a homeothermic mammal than that of

FIGURE 17.4.   *Keels and tails in scombrid fishes. The evolution of mackerels and tunas has involved increasing degrees of pelagic activity. The more primitive mackerels and Spanish mackerels live inshore and swim more slowly and less continuously. More advanced high-seas tunas swim continuously and faster and are more migratory. These ecological differences are reflected in tail shape and accessories, with more-efficient, high aspect ratio tails and more-elaborate keels characterizing the more-pelagic tuna. (A) Mackerels have forked tails with one pair of fleshy caudal keels. (B) Spanish mackerels have a semilunate tail, caudal keels, and a median peduncular keel, but the peduncular keel is external only, lacking internal bony supports (B', dorsal view of peduncle skeleton). (C) Tunas have lunate tails and multiple keels, with (C') lateral extensions of the peduncular vertebrae supporting the keels. Lunate tails and peduncular keels have also evolved in mackerel sharks, jacks, and billfishes.*

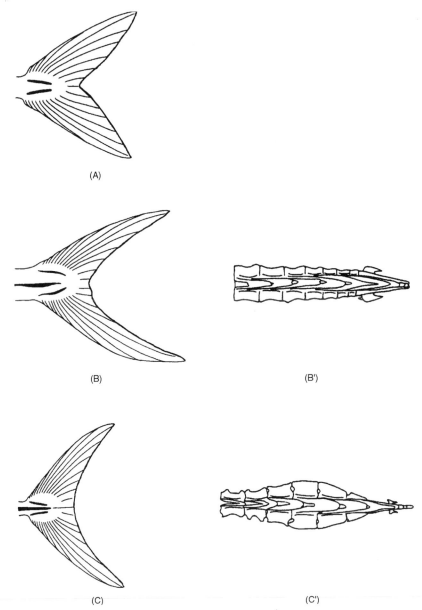

(A)

(B)   (B')

(C)   (C')

From Collette and Chao 1975 and Collette 1978; used with permission.

a fish. Tunas have large hearts that account for 2% of body mass and have concomitantly large blood volumes. The uptake of oxygen and release of carbon dioxide at the gills in herrings and mackerels are facilitated by exceedingly thin lamellar walls (5 to 7 μm thick) and numerous lamellae (> 30/mm); comparable values for less-active, inshore species are 10 to 25 μm and 15 to 25 lamellae per millimeter. The surface area of the gill lamellae relative to body weight is very high in mackerel sharks, menhadens, bluefishes, dolphinfishes, and tunas. The efficiency of the lamellae is enhanced by fusion of adjacent lamellae and elaboration of the leading and trailing edges of the gill filaments. These modifications have occurred convergently in tunas, swordfish, and billfishes but not in the less pelagic mackerels. Tunas remove more oxygen from the water as it passes over their gills than any other fish. This highly efficient oxygen uptake system is necessary to fuel their extremely high metabolic rates (Steen and Berg 1966; Collette 1978; Blaxter and Hunter 1982).

Foraging

An open-water existence limits the foraging options open to pelagic fishes. As a result, the fishes feed on phytoplankton, zooplankton, or each other. Many clupeoids utilize phytoplankton directly by swimming through plankton concentrations with an open mouth, thereby filtering the particles out of the water in a pharyngeal basket that has densely packed gill rakers (100 to 300 per centimeter) and includes an epibranchial organ that releases digestive enzymes while the food is still in the oral region. The digestive tract is long and has numerous pyloric caeca. Food passes very rapidly through this system, often taking less than an hour, but these fish can utilize a broad array of food types and are very efficient at converting food into protein.

The foraging and migratory patterns of such pelagics as tunas and billfishes become clearer when the nature of food availability in open tropical seas is considered. Estimates of zooplankton resources in the central Pacific indicate average densities on the order of 25 parts per *billion*. Large pelagic predators are feeding at even higher trophic levels, so their food is scarcer by one or two orders of magnitude. Since no animal is going to survive on food distributed evenly at such low densities, the success and rapid growth rates of many tunas attest to the extreme patchiness of food on the high seas. A nomadic lifestyle, driven by high metabolism and rapid swimming, makes sense when vast expanses must be covered in search of such patchily distributed resources (Kitchell et al. 1978).

Life History Patterns in Pelagic Fishes

Pelagics are by definition open-ocean fishes throughout their lives. Two general patterns characterize the overall life histories of pelagic fishes, brought on by the relationship of parental versus larval food requirements, lifespan, spawning frequency, oceanic currents, and fish mobility. These patterns are referred to as **cyclonic** and **anticyclonic**.

Cyclonic patterns characterize higher-latitude species such as Atlantic herring, in which the adults and larvae live in different parts of the ocean. Adults have a seasonal feeding area and tend to spawn once per year. Before they spawn, they migrate upcurrent to a region where food for larvae and juveniles will be particularly abundant. Larvae and later juveniles drift with the currents to the adult feeding region. These fish invest considerable energy into each spawning episode, both in terms of the costs of the migration and also in egg production. Because of the spatial separation of adult and larval habitats, adults may not have reliable cues for predicting conditions at the spawning grounds, which leads to highly variable spawning success and large fluctuations in year class strength (see Chap. 23, Population Dynamics and Regulation).

Anticyclonic patterns are more characteristic of low-latitude species such as tropical tunas and scads. The comparative aseasonality of tropical waters leads to less temporal fluctuation but extreme spatial variation in productivity. Adults move in a roughly annual loop through a major ocean basin, during which time they spawn repeatedly (with the exception of bluefin tunas) rather than only in particular locales. Larvae and juveniles develop and feed along with adults, carried by the same current system in their relatively nomadic existence. The energy put into reproduction is spread out among several spawning episodes. Adults can use local environmental cues to determine the appropriateness of conditions for larvae, which is critical given the low productivity and patchiness of tropical open oceans. Hence anticyclonic species often show weaker fluctuations in year class strength. Within families, tropical species mature more quickly and live shorter lives.

Interestingly, tunas evolved in the tropics, but some species such as giant bluefin spend a large part of the year feeding in productive temperate locales. Bluefin show the phylogenetic constraint of their tropical history by returning to the tropical waters of the Gulf of Mexico to spawn, forcing them into what is more of a cyclonic than an anticyclonic pattern (Rivas 1978). The same historical factors constrain anguillid eels such as the American, European, and Japanese species, which also return from temperate feeding locales to tropical breeding locales; however, several years pass between the two life history stages (see Chap. 22, Representative Life Histories of Migratory Fishes).

The high but periodic productivity of small planktonic animals in the open sea and the presence of major ocean currents have been contributing factors in the evolution of dispersive, planktonic larvae in most marine fishes, regardless of whether the adults are planktonic, pelagic, demersal, deep sea, or inshore (see Chap. 9).

Flotsam

A special open-ocean fauna occurs around what little structure is found in the open sea. Floating bits of seaweed (usually sargassum), jellyfishes, siphonophores, and driftwood almost always have fishes associated with them. Many flotsam-associated fishes such as filefishes and jacks are the juveniles of inshore or pelagic species; others such as sargassumfishes and driftfishes are found nowhere else, attesting to the reliability of occurrence of such objects. Flotsam also serves as an attractor for large predators, such as sharks, dolphinfishes, tunas, and billfishes (Gooding and Magnuson 1967); a single log will commonly have more than four hundred 5-kg tuna associated with it, often involving several species (Sharp 1978). It has been suggested that concentrations of flotsam are indicators of regions of high productivity in the open sea because the flotsam accumulates at the tops of vertical circulation patterns (Langmuir cells) that also concentrate nutrients and zooplankton (Maser and Sedell 1994). The mechanisms by which pelagics locate floating objects and their importance to fishes that do not feed around them remain a matter of conjecture (see Fig. 19.5).

Evolution and Convergence

The greatest development of a pelagic fish fauna is in the ocean. However, most major lakes have an open-water fauna that consists partly of members typically associated with open waters as well as species whose ancestors were obviously inhabitants of nearshore regions. These **limnetic** fishes include osteoglossomorphs (goldeye, mooneye), clupeids (shads), characins, cyprinids (golden shiner, rudd), salmonids (whitefishes, trouts, chars), smelts, silversides, moronid basses, and cichlids. Many of these fishes live at the air-water interface and show specializations that are apparently influenced by this habitat, including upturned mouths, ventrally positioned lateral lines, and convergent fin placement and body proportions. These surface-dwelling traits occur in both marine and freshwater families, including characins, minnows, silversides, marine and freshwater flyingfishes (exocoetids and gasteropelicids), halfbeaks, and killifishes (Marshall 1971).

Regardless of ancestry, the same anatomic and behavioral themes that are seen in the ocean recur in freshwater limnetic species, including silvery color, compressed bodies, forked tails, schooling, high lipid content, and planktivorous feeding adaptations. Analogously, *Pleuragramma antarcticum*, a pelagic nototheniid in antarctic waters, shows many traits characteristic of epipelagic fishes worldwide. Although derived from stocky, dark-colored, benthic ancestors, *Pleuragramma* has deciduous scales, a silvery body, forked tail, high lipid content for buoyancy, and is compressed in cross section. The pelagic larvae of many benthic antarctic fishes are also silvery and compressed, and have forked tails (Eastman 1993; see below, Antarctic Fishes). These examples of convergence suggest that fairly uniform and continuous selection pressures characterize the open-water habitat.

With the exception of the clupeoids, most successful taxa of adult marine pelagic fishes are acanthopterygians. Missing among otherwise successful marine groups are elopiforms and paracanthopterygians, although both groups have done well in deep-sea mesopelagic and bathypelagic regions. These two groups may be phylogenetically constrained from inhabiting shallow, open-water regions, not the least because of their tendency to be nocturnal in habit. Other strongly nocturnal taxa are also largely missing from pelagic and limnetic habitats, including the otherwise successful catfishes, sea basses, croakers, grunts, and snappers, to name a few.

POLAR REGIONS

The far north (**arctic**) and south (**antarctic**) polar regions are roughly the areas above 60° latitude. They have much in common, primarily related to cold water temperatures and short growing seasons, but they differ geologically and environmentally and support very different biotas, including fishes. The Arctic is a frozen oceanic region surrounded almost entirely by land, whereas the Antarctic is a frozen continent surrounded by ocean (Fig. 17.5). Freshwater fishes are lacking from the Antarctic because most water bodies have permanent ice cover and many freeze to the bottom during the winter. High arctic lakes and rivers have a limited fish fauna; 55 species occur in the Canadian Arctic, but most of these are primarily temperate species at the northern edge of their range (Scott and Crossman 1973). Freshwater fishes at high latitudes show interesting behavioral adjustments to the strong effects that seasonality has on light levels, day length, and growing season (see Box 17.1). Polar oceans are in a liquid state below the first few meters and have more fishes, but

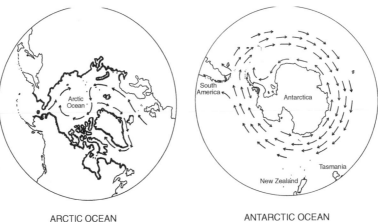

FIGURE 17.5.   *North and south polar regions. General oceanic circulation patterns are shown by arrows. The Arctic Ocean centers on the North Pole; the southern limits of the region are indicated by the dark continental borders. The Southern Ocean surrounds Antarctica.*

ARCTIC OCEAN

ANTARCTIC OCEAN

From Briggs 1995; used with permission.

the superabundance of ice at the surface, plus scouring by ice or ice anchored to shallow bottoms, limits the distribution and behavior of polar fishes, which have developed remarkable adaptations to avoid freezing to death.

Antarctic Fishes

Antarctica is surrounded by at least 900 km of the open, deep Southern Ocean that flows around and away from the continent. Strong circumpolar currents and distinct temperature differences occur between the polar and subpolar regions, delimited by a region known as the Antarctic Convergence at 50° to 60°S. This region creates a distance, depth, and thermal barrier to interchange between the cold-adapted species of the antarctic region and warm-adapted species to the north. Antarctic fishes have also had sufficient time to adapt and speciate; the antarctic region has been at its present locale with its present climate for about 20 to 25 million years, having separated from Australia during the early Cenozoic (Hubold 1991; Eastman 1993). Spatial and temporal seclusion and climatic extremes have resulted in a diverse fish fauna dominated by endemic notothenioid thornfishes, cod icefishes, channichthyid crocodile icefishes, plunderfishes, and dragonfishes, as well as several nonnotothenioid groups (see Chap. 16, Major Marine Regions).

Notothenioids as a group are benthic fishes, and fully half of all species still live on the bottom in less than 1000 m of water (Fig. 17.6). As is general among benthic fishes, they lack gas bladders, are dark in coloration and round or depressed in cross section, and have a square or rounded tail. Benthic forms often seek cover inside sponges, either as a refuge from predatory mammals or as a spawning substrate. Eggs placed inside hard sponges such as hexactinellid glass sponges are probably protected from most predators (Dayton et al. 1974; Konecki and Targett 1989). Larvae are pelagic and show adaptations specific to shallow, open-water existence, including silvery coloration, relatively compressed bodies, and forked tails (see above, Pelagic Fishes). Notothenioids have also radiated into most nonbenthic niches and consequently show substantial variation in body form and behavioral tactics, starting with a common body plan. A few species, including the abundant cod icefish, *Pleuragramma antarcticum*, are pelagic zooplanktivores. So-called cryopelagic fishes live in open water just below the ice. The food chain for these fishes starts with ice algae, which is eaten by amphipods and euphausiids, which are in turn eaten by the fishes. Cryopelagic fishes have a uniform light coloration, which may help them blend in with the icy background against which they would be viewed. They also possess better chemical defenses against freezing and have greater buoyancy than benthic relatives.

Notothenioids are interesting reproductively because they produce a small number of relatively large 2- to 5-mm eggs during a short, 1- to 2-month spawning season. The unhatched larvae have developmental periods of 2 to 6 months, followed by a long, slow-growing pelagic stage that lasts a few months to 1 year. Many benthic species exhibit parental and biparental guarding (Daniels 1979; Kellerman and North 1991).

Notothenioids are opportunistic feeders, taking a wide range of prey types, with many pelagic and mesopelagic juveniles and adults feeding on the ubiquitous krill, *Euphausia superba*, which is also the major prey of whales, penguins, and other seabirds. Although the annual temperature variation in Antarctica is seldom more than 4°C (−2 to 2°C), and in some locales as little as 0.1°C, fishes show marked variation in feeding rate during summer versus winter. Rates are still relatively high during winter (e.g., 65% of summertime intake in *Harpagifer antarcticus*; Targett et al. 1987), unlike temperate locales, where many fishes cease feeding during winter.

Mesopelagic fishes are particularly abundant throughout the water column of the Southern Ocean. Lanternfishes are the most diverse group of mesopelagic fishes at lower latitudes but are epipelagic in the Antarctic. The lanternfish, *Electrona antarctica*, is the most common fish above 200 m. It feeds heavily during the day, in contrast to the typical mesopelagic pattern of nocturnal foraging that characterizes lanternfishes at lower latitudes. "Mesopelagic" species are also an important component of the community living near the ice edges, or oceanic marginal ice zone. Large numbers of myctophid lanternfishes are eaten in the open sea and at the edge of the pack ice by seabirds, whales, and seals. A commercial midwater

FIGURE 17.6.   *Body form and habitat types of common antarctic nototheniid fishes. The dots show the preferred depths and habitats.*

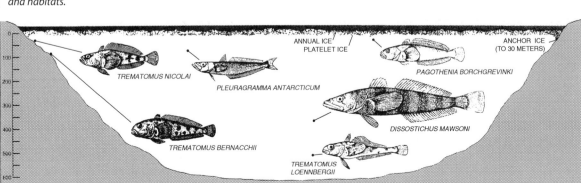

trawl fishery even exists for mesopelagic species, with annual catches of the lanternfish, *Electrona carlsbergi*, exceeding 78,000 tons (> 70 million kg) from the South Georgia Island region. As with their more northerly, low-latitude relatives, deep-living mesopelagic fishes in the Antarctic show lower enzyme activity and slower metabolic rates than shallow-water forms, which is interpreted as an adaptation to low food availability at depth (see above, The Deep Sea) (Kellerman and North 1994).

Harpagiferid plunderfishes, which are advanced perciform fishes, are remarkably similar in morphology and behavior to the relatively primitive scorpaeniform sculpins of northern temperate waters. Similarities may represent adaptations to a predominantly benthic existence, including a relatively depressed, elongate, tapering body; large, spiny head with large eyes and a large, terminal mouth; long dorsal and anal fins; large pectoral fins; rounded caudal fin; and a dorsally located lateral line. Both groups show ecological and behavioral similarities as well, feeding by a sit-and-wait mode on relatively large, mobile benthic invertebrates. In essence, plunderfishes and sculpins have converged to fill similar niches in their respective communities (Wyanski and Targett 1981).

Adaptations and Constraints of Antarctic Fishes

Notothenioids are best known for two adaptations related to existence in the cold, often energy-limited waters of the area, where water temperatures average −1.87°C and total darkness prevails for 4 months each year. First, their blood contains remarkably effective antifreeze compounds that depress the freezing point of their body fluids and make it possible for them to live in water that is colder than the freezing point of most fish blood, including, remarkably, their own. Second, some have evolved neutral buoyancy, which has permitted these species to move off the crowded bottom where most notothenioids live and into the water column.

No known species of fish can actually tolerate having its tissues freeze. The major threat to fishes in the Antarctic is ice, which floats at the surface in the form of bergs, sheets, and platelet ice but also attaches to the bottom in water less than 30 m deep in a form called anchor ice. The greatest danger comes from ice crystals penetrating or propagating across the body and seeding the formation of ice inside the fish, which would cause cell rupture. Many antarctic fishes live in water that is colder than their blood's freezing point. Fishes from lower latitudes typically freeze when placed in water colder than −0.8°C, whereas antarctic fishes can live in water as cold as −2.19°C. They accomplish this because their blood contains not only the salts normally found in fish blood but also as many as eight different glycopeptide antifreeze compounds. The glycopeptides apparently function by keeping the ice from propagating across the fish's skin. A notothenioid can be cooled as low as −6°C without freezing, as long as free ice is not in the water.

Several other adaptations accompany the production of antifreeze compounds. Notothenioids are relatively unusual among teleosts in that their kidneys lack glomeruli, which are the structures that remove small molecules from body fluids and transfer them to the urine for excretion. Glomeruli would remove the antifreeze glycopeptides, which would be energetically expensive to replace continually. A fairly strong correlation exists between antifreeze effectiveness and the frequency with which a species encounters free ice. For example, the shallow-water bathydraconid dragonfishes frequently come in contact with ice and have the highest levels of antifreeze compounds. Within the cod icefish genus *Trematomus*, shallow-water species that live in the coldest water and rest in ice holes or on anchor ice have freezing points of −1.98 to −2.07°C, whereas deeper-living species that seldom encounter ice crystals freeze at −1.83 to −1.92°C. Even within species, shallow-water populations have significantly more freezing resistance than deeper-water populations (DeVries 1970). The primitive bovichthyid thornfishes of New Zealand live in temperate waters and do not produce antifreeze. Bovichthyids possess glomeruli, indicating that the aglomerular condition of antarctic species evolved along with other adaptations to the colder antarctic environment (Eastman 1993).

Neutral buoyancy has developed in at least two water-column-dwelling members of the family Nototheniidae, the cod icefish, *Pleuragramma antarcticum,* and its giant predator, the antarctic toothfish, *Dissostichus mawsoni.* Whereas most antarctic fishes are 15 to 30 cm long, toothfish reach lengths of 1.6 m and weights of more than 70 kg. Neutral buoyancy allows these fishes to occupy the comparatively underutilized water-column zone, thus taking them away from threatening anchor ice crystals and into a region of seasonally abundant food sources such as fish larvae and krill. Both species have evolved from benthic ancestors and have retained what can only be viewed as a phylogenetic constraint on living in open water: They are similar to benthic notothenioids in that they lack a gas bladder.

As fish muscle and bone are relatively dense, a fish without a gas bladder would constantly have to fight gravity to stay in the water column. Neutral buoyancy in these two nototheniids is achieved via several mechanisms. Toothfish have cartilaginous skulls, caudal skeletons, and pectoral girdles, which reduces their mass because cartilage is less dense than bone. The skeleton itself is less mineralized than in benthic relatives, by a factor of 6 in the toothfish and 12 in *Pleuragramma*. Bone is also reduced in the vertebral column, which is essentially hollow except for the notochord. Additional buoyancy is achieved by lipid deposits dispersed around the body, including a blubber layer under the skin and fat cells or sacs located between muscle fibers or muscle bundles (Eastman and DeVries 1986; Eastman 1993). Weightlessness via analogous routes of weight reduction and replacement is also seen convergently in bathypelagic fishes, another water-column-dwelling group in which evolution has placed a strong premium on energy-saving tactics.

A unique trait of channichthyid icefishes may represent an evolutionary adjustment to polar conditions. These fishes are sometimes referred to as "white-blooded" or "bloodless" because their blood contains no hemoglobin and their muscles contain no myoglobin, giving them a very pale appearance. The highly oxygenated, cold waters of Antarctica may have been responsible for the evolutionary loss of respiratory pigments, perhaps via a "regressive" evolutionary process similar to the one that led to pigmentless, eyeless cave fishes (see below, Caves). Channichthyids possess a number of other characteristics that have evolved in conjunction with a lack of hemoglobin, including relatively low metabolic requirements (reduced protein synthesis, reduced activity, slow growth); increased vascularization of skin and fins to increase gas exchange; and an increase in cardiac size, output, and blood volume (Hemmingsen 1991). Some nototheniids have increased blood volumes and reduced hemoglobin concentrations, perhaps reflecting an intermediate stage in the response to respiratory conditions in the Antarctic that have led to the hemoglobin-free condition of the channichthyid icefishes (Wells et al. 1980).

Arctic Fishes

The Arctic has fewer endemic fishes due to the combined effects of less geographic isolation and younger age. The oceanic environment between subarctic (or boreal) and arctic areas is fairly continuous. On the western, Pacific side, the Bering Sea flows into the Arctic Ocean and has done so since the Bering Strait opened up 3.5 million years ago. Similarly, on the eastern, Atlantic side, the Arctic Ocean is directly connected to the Greenland Sea. Hence, arctic fishes are either species that evolved there since the current climate developed or are cold-tolerant Pacific or Atlantic species that experience gene flow from source areas rather than being endemic to the Arctic itself. The Arctic has undergone repeated warming and cooling until about 3 million years ago, when the present cold conditions stabilized, leaving less time for organisms to adapt to current conditions (Briggs 1995). Consequently, fishes in the northern polar region have had less time to speciate.

Adaptations to cold are evident in arctic fishes, where species have converged with antarctic fishes in the production of antifreeze compounds. Glycoprotein antifreeze occurs in arctic and Greenland cod, whereas warty sculpin, Canadian eelpout, and Alaska plaice possess peptide antifreezes (Clarke 1983). Arctic cod are frequently observed resting in contact with ice and taking refuge inside holes in ice, so their potential for encountering seed crystals is very high. In some of these fishes, kidney glomeruli are convergently reduced to help retain antifreeze compounds in the body (Eastman 1993). Several boreal cods, sculpins, eelpouts, and flatfishes whose ranges extend into arctic waters also have antifreeze compounds in their blood.

Water temperatures show greater annual and latitudinal variation in the Arctic than in the Antarctic, which means that fishes are likely to encounter extreme winter cold but also relatively high summer temperatures. Winter temperatures do commonly drop to $-1.8°C$ as in the Antarctic, but water can reach 7° or 8°C during the summer. The greater seasonal range is reflected in the tolerance of different species to warm temperatures, as well as differences in seasonal production of antifreeze. Few antarctic fishes can tolerate water temperatures above 7° or 8°C, regardless of acclimation temperature, whereas arctic species have upper lethal temperatures of 10° to 20°C, depending on species and acclimation temperature (DeVries 1977). Several north polar species produce less antifreeze during the summer, particularly among boreal fishes that may encounter temperatures well above freezing in the summer. Winter flounder, *Pleuronectes americanus*, have a blood volume of 3% antifreeze in winter and 0% in summer. Reduced antifreeze production during warmer months probably saves energy and may also increase the blood's capacity to carry oxygen or nutrients.

DESERTS AND OTHER SEASONALLY ARID HABITATS

Deserts seem improbable places for fishes to occur. However, algae and many invertebrates capitalize on the periodic availability of water in arid regions. It is not surprising then to find a small number of fishes capable of surviving under conditions of periodic **dewatering** in desert regions around the world, presenting dramatic examples of adaptation and convergent evolution.

Deserts are difficult to define because they differ in altitude, temperature range, amount of rainfall, and seasonality of water availability, among other traits. Many treatments consider a desert as an area that receives less than 30 cm of rainfall annually. A more general definition is that a desert is an area where "biological potentialities are severely limited by lack of water" (Goodall 1976), a definition that stresses the common thread of water scarcity as the significant selection factor and can therefore apply to areas with seasonal droughts, such as swamplands that dry up periodically.

For fishes, the disappearance of water is only the most extreme state in a continuum of conditions that occur during dewatering. As water evaporates, temperatures generally rise, dissolved substances such as salts become more concentrated, oxygen tension drops, carbon dioxide increases, and competition and predation intensify. Desert fishes must therefore be tolerant of widely varying and extreme salinity, alkalinity, temperature, and depleted oxygen (see Box 17.2). They may also have to be able to outcompete other fishes and avoid predators despite physiological stress. Desert stream fishes also have to withstand periodic flash flooding.

Desert-adapted fishes, not counting species that migrate to more permanent habitats when waters recede, often show three general adaptations: 1) an annual life history involving egg deposition in mud during the wet season, an egg resting period (**diapause**) during the dry season, death of the adults, and egg hatching when habitats are reinundated the next year; 2) **accessory respiratory structures** for using atmospheric oxygen (lungs, gill and mouth chambers, cutaneous respiration); and 3) **estivation**, in which adults pass the dry season in some sort of resting state.

Deserts occur on all major continents, and many of these deserts contain fishes. Africa has many habitats that

---

BOX 17.1 *The Effects of High Latitude on Activity Cycles and Predator-Prey Interactions*

As discussed in Chapter 22 (Diel Patterns), most fishes have particular periods of activity, feeding either during daylight or darkness, with a small number primarily active during crepuscular periods of dawn and dusk. These cycles of activity have a strong endogenous basis and are maintained for some time under laboratory conditions of constant light or darkness. However, in nature, the activity cycles are cued by the rising and setting of the sun.

The situation at high latitudes presents a very different set of environmental influences and selective pressures. Above the arctic circle, light levels never reach "nighttime" values during midsummer, and growing seasons are short and intense. Winter brings a time of continuous relative darkness and low food availability. Summer and winter therefore present extreme and opposite light conditions. Do fishes maintain strict diurnality or nocturnality under such variable and extreme conditions, or do they adjust their activity patterns to the changing seasons?

Laboratory studies with European species whose natural ranges extend beyond the arctic circle have produced some striking and seemingly adaptive departures from the standard picture developed at lower latitudes. The burbot, *Lota lota*, belongs to a family of strongly nocturnal fishes, the cods (Gadidae). At intermediate latitudes below the arctic circle, burbot are nocturnal throughout the year. However, at higher latitudes, a peculiar pattern occurs. During the summer the fish are continually active, whereas during the winter they shift to diurnal behavior. During spring and fall they are primarily nocturnal. Similar activity cycles have been observed in other nocturnal or crepuscular species, including sculpins and brown trout, and can be induced experimentally in brown bullheads.

Interpreting these patterns is not immediately easy. The best explanation, however, is that the change to arrhythmic, continual behavior in summer is a means of taking advantage of high, continuous, and aperiodic levels of algal and aquatic insect production during the short growing season of summer. Limiting activity to the short nighttime period each day during summer would severely restrict an animal's intake. Nocturnality during spring and fall may represent a return to the normal, evolved response of the species as day length and twilight length closely approximate the more usual and widespread conditions at lower latitudes. The switch to diurnality during winter in an animal well adapted to function in the dark remains puzzling. Regardless, changes in the length of and light intensity during twilight provide the apparent cues that lead to the phase shifts observed in these fishes (Müller 1978a,b).

The influence of twilight length at high latitudes is also shown in the predator-prey relations of marine fishes. Dawn and dusk at low latitudes are the times when fish switch between feeding and resting and are often times of maximal predator activity. If twilight is a dangerous time for prey fishes at low latitudes where twilight lasts for a relatively short time, we might expect the prolonged twilight that occurs at higher latitudes to be even more dangerous.

Conducting extensive underwater observations at high latitudes can be uncomfortable, and few such studies have been attempted. In the one instance in which the question of twilight interactions was addressed, observers found that extended twilight meant extended periods of predation. Hobson (1986) watched sculpins, greenlings, and flatfishes preying on Pacific sand lances, *Ammodytes hexapterus*, in Alaska. Sand lances school and feed on zooplankton during the day and bury in the sand at night. Schools of sand lances are relatively immune to these benthic predators during daylight, and the predators do not occur at night in the limited resting areas that the sand lances use. However, during twilight, the predators aggregate in the resting area under the schools as they break up. The predators are particularly effective at capturing sand lances that have just entered the sand or that re-emerge shortly after burying because of apparent dissatisfaction with their initial choice of resting site. The twilight transition from schooling to resting appears to be the most dangerous time for the sand lances.

Twilight conditions at the date and latitude of observation (May, 57°N) were very long, lasting about 2 hours. This is about twice as long as at tropical latitudes, where similar observations have been made with different predators and prey. The period of intense predation in Alaska is also about twice as long as that observed at tropical locales. The longer days of spring and summer at high latitudes mean that diurnal fishes experience a much longer foraging period, but this increase is bought at the high price of increased predation during the lengthened twilight periods.

dry up seasonally and that contain fishes with desert adaptations. Among the most successful groups in Africa are cyprinodontoid killifishes and rivulines, which are popular aquarium species. Many of these fishes (e.g., *Fundulosoma, Notobranchius, Aphyosemion* spp.) are **annual**, living for 8 months in mud holes, swamps, and puddles (Fig. 17.7). They mature after only 4 to 8 weeks, spawning daily and burying eggs as much as 15 cm deep in muddy bottoms, a remarkable feat for a 5-cm-long fish. The adults die, and the eggs spend the dry season in a state of arrested development until the next rains come. Some eggs can remain in such a state of diapause for up to 5.5 years. An annual life history effectively maintains a permanent population in a temporary habitat (Wourms 1972; Simpson 1979).

African and Asian clariid or walking catfishes are capable of leaving drying water bodies and moving across up to 200 m of moist grass in search of water. They will also bury themselves as deep as 3 m in sandy sediments as water levels drop. They can survive by employing aerial respiration via treelike suprabranchial organs over the second and fourth gill arches, although they cannot sur-

vive if the sand dries up (Bruton 1979). The African lungfishes (Protopteridae) are true estivators. During a drought, they burrow into mud, secrete a cocoon, and enter a torpid condition in dry mud until the next rains, an event for which they can wait 4 years (see Chap. 13, Subclass Dipnoi: The Lungfishes).

Many other fishes in African swamps are adapted to the deoxygenation that accompanies seasonal dry periods, using a variety of air-breathing mechanisms (see Table 5.1). Killifishes and *Hepsetus odoe*, the pike characin, are surface dwellers, taking advantage of higher oxygen tensions near the air-water interface. Lungfishes and bichirs use lungs, clariid catfishes have gill chamber organs, anabantids have labyrinth organs, snakeheads have pharyngeal diverticula, and featherfin knifefishes and phractolaemids have alveolar gas bladders (Table 5.1).

In South America, drought resistance has evolved in parallel to the African examples. Many fishes of the Amazon region have evolved means of using atmospheric oxygen when drought or vegetative decay lower oxygen levels (Kramer et al. 1978). Surface swimmers, such as arowanas and some characids (pacus, *Brycon*), have vascu-

FIGURE 17.7.   *Life cycle of annual cyprinodontoids, as shown by the Venezuelan* Austrofundulus myersi. *1) Spawning occurs over a protracted period; 2) shelled eggs are deposited in the mud; 3) as water dries up, adults die but eggs remain viable in an arrested developmental stage; 4) with the return of the rains, eggs hatch; 5) larvae and juveniles grow rapidly; 6) maturation occurs after only 1 or 2 months, followed by spawning.*

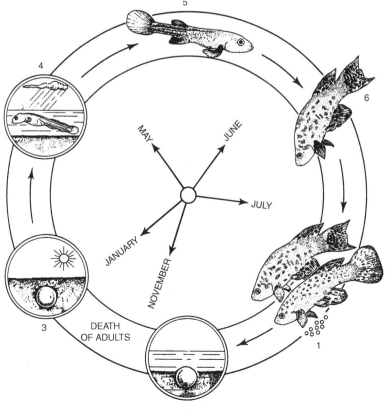

larized lips. Modifications of the alimentary tract to absorb oxygen are common, including the mouth region of electric eels and swamp eels, air-filled stomachs in loricariid catfishes, a vascularized hindgut in callichthyid armored catfishes, and a vascularized gas bladder in lungfish, arapaima, and erythrinid trahiras. As with the walking catfishes, South American species are reported to abandon drying pools and cross small stretches of wet vegetation or mud in an apparent search for new and wetter habitats (e.g., erythrinids such as *Hoplias* and *Hoplerythrinus*, callichthyid catfishes, some rivulines) (Lowe-McConnell 1987).

Conventional desert areas also exist in South America. The Chaco region of northwestern Paraguay receives less than 30 cm of water annually, with a normal 3-month winter drought period that can last as long as a year (Smith 1981b). During the annual drought, aquatic habitats become isolated and dry up. During the rainy season, these habitats are often repopulated by fishes from overflowing portions of the Paraguay River. The ichthyofauna of the Chaco consists of both drought-adapted and nonadapted species. Adaptations to drought include estivation in mud by juvenile and adult lungfish (*Lepidosiren*); accessory respiratory structures for using atmospheric oxygen (lungfishes; catfishes, *Hoplosternum*, *Pterygoplicthys*; characins, *Hoplias*); and annual life histories and diapausing eggs among cyprinodontoids, which are also successful throughout much of tropical South America (e.g., *Cynolebias*, *Rivulus*, *Austrofundulus*). Localized extirpation occurs annually in species that invade the Chaco region during the wet season but that lack the abilities to overcome drought conditions.

Australia is largely a desert continent. Its freshwater fish fauna is dominated by marine derivatives, such as river eels, plotosid catfishes, rainbowfishes, barramundi, temperate basses, grunters, pygmy perches, gobies, and sleepers. Several Australian fishes show distinct adaptations to periodic drought. The endemic, monotypic salamanderfish, *Lepidogalaxias salamandroides*, occurs commonly in southwestern Australian habitats that dry up during the annual summer drought. As waters recede, the fish burrows into bottom sediments and surrounds itself with a thick mucous coat (Berra and Allen 1989; Pusey 1990). Two major problems faced by estivating fishes are water loss and a concomitant buildup of toxic nitrogenous wastes such as urea, which normally must be transported away with water-wasting urine. Salamanderfish conserve water by absorbing it from the surrounding soil until soil moisture content approaches zero. They avoid production of nitrogenous wastes in part by metabolizing lipids rather than proteins; the end point of lipid metabolism is carbon dioxide, not nitrogen compounds (Pusey 1989). Several species in the related family Galaxiidae in Australia and New Zealand occur in similar temporary habitats and may also estivate during dry periods.

Some gobies and hardyhead silversides that live in desert springs in central Australia are exceptionally tolerant to high temperatures, high salinities, and low dissolved oxygen. The desert goby, *Chlamydogobius eremius*, typifies desert-adapted species in its ability to survive an extreme range of conditions, including ionic concentra-tions ranging from distilled water to water more saline than seawater, temperatures between 5° and 40°C, and oxygen concentrations below 1 part per million (ppm). To avoid lethal conditions in thermal springs or high summer temperatures, it seeks cooler vertical or lateral portions of springs, buries itself in cooler silt, and even emerges from the water to capitalize on evaporative cooling and aerial respiration (Glover 1982).

North American Deserts

Additional examples of desert adaptations could be presented from almost any continent (except Antarctica), but some of the best-studied desert fishes occur in the southwestern United States. The Basin and Range Province of North America contains four different deserts—the Great Basin, Mojave, Sonoran, and Chihuahuan Deserts (Naiman and Soltz 1981). The province, which includes such seemingly inhospitable areas for fishes as Death Valley, Ash Meadows, Salt Creek, and Devil's Hole, constitutes less than 10% of the total land area of North America. Although desert conditions have existed periodically in the region for approximately 70 million years, the southwestern deserts as they exist today are relatively young, with no more than 12,000 years having passed since the last wetter, pluvial period when the area contained abundant, interconnected standing and running water. Despite their relative youth and small size, southwestern deserts contain 182 native species, 149 of which are endemic to the basin and many of which are endemic to single locales. Endemicity in the fishes of the desert Southwest is the highest of any place in North America.

Two major types of desert habitat are occupied by fishes: 1) isolated pools and basins supplied by underground springs that have fairly regular flow and 2) intermittent marsh and arroyo habitats along flowing water courses that originate in wetter areas such as mountainous highlands and flow into arid regions. The native fishes of the region belong to five principal families and segregate according to fish size, habitat size, and environmental extremes. Small livebearers (Poeciliidae) and even smaller desert pupfishes of the family Cyprinodontidae live in the most extreme or isolated habitats, such as intermittent streams and spring basins; these fishes include 20 desert-adapted species in the genus *Cyprinodon*. Small streams contain small minnows (Cyprinidae) that are less than 6 cm long; larger streams and small rivers support medium-sized suckers (Catostomidae) and trout (Salmonidae). The largest fishes, such as large suckers and the Colorado squawfish (*Ptychocheilus lucius*, Cyprinidae, up to 2 m), live in large rivers. Body size is intimately tied to habitat size (Smith 1981). The smallest pupfish, the 2-cm-long Devil's Hole pupfish, *Cyprinodon diabolis*, lives on an 18-m^2 shelf in a spring basin in the smallest habitat of any known vertebrate. In contrast, the Colorado squawfish is the largest minnow in North America and lives in the area's largest habitat, the Colorado River.

The fishes in marshes and small streams experience the harshest conditions and show the strongest adaptations to desert existence. Desert pupfishes show extraordinary tolerances to environmental extremes. They can live in water with as little as 0.13 mg of oxygen per liter (0.13 ppm dissolved oxygen), which is a record for fishes that do not supplement gill respiration with some accessory breathing apparatus. Most fishes show stress at less than 5 ppm, depending on water temperature. Although these are freshwater fishes, some desert pupfishes can tolerate salinities over 100 parts per thousand (ppt) and as high as 140 ppt, three to four times that of seawater. Pupfishes experience water temperatures that vary from freezing in winter to 44°C in summer, the highest recorded for a habitat containing live fishes; the Cottonball Marsh pupfish tolerates higher temperatures than any other known teleost (Feldmeth 1981). Many of the spring-dwelling pupfishes have lost their lateral lines and pelvic fins, which may be energy-saving responses in isolated habitats that lack predators.

Other taxa show physiological and behavioral adjustments to drought conditions, such as the longfin dace of the Sonoran Desert, adults of which move into moist algae during hot days and emerge during cooler nights to forage in only a few millimeters of water (Minckley and Barber 1971). Although most emphasis is given to periods of low water in deserts, a major influence on stream- and river-dwelling fishes is the periodic occurrence of flash floods, when waters can change from low-flow, nearly stagnant conditions to raging torrents in a matter of seconds (Naiman 1981). Colorado River endemics, such as the humpback chub (*Gila cypha*), bonytail (*G. elegans*), and razorback sucker (*Xyrauchen texanus*), have anterior humps, flattened heads, keeled napes, cylindrical bodies, small scales, and elongate, narrow caudal peduncles that may provide hydrodynamic stability during periods of high or turbulent flow (see Fig. 25.2).

It is not only large fishes that show high-flow adaptations. The threatened 5-cm Gila topminnow, *Poeciliopsis occidentalis*, has been extirpated through much of its range due to predation by introduced mosquitofish, *Gambusia affinis*. However, topminnows are able to coexist with mosquitofish in streams that experience periodic flash floods because the topminnows show instinctive behavioral adaptations to high discharge, including rapid movement to shoreline areas as waters rise and proper orientation to strong currents. Mosquitofish, which evolved in southeastern regions that lack flash floods, behave inappropriately and are flushed out of rivers when floods occur (Meffe 1984).

Although the desert pupfishes and other fishes survive and reproduce in the extreme conditions of the desert Southwest, these fishes do not exhibit several other traits common to many desert forms, such as estivation, air breathing, or diapausing eggs. Adaptations of the desert pupfishes are most likely extensions of capabilities possessed by ancestral lineages rather than being newly evolved. Cyprinodontids are small fishes that frequently inhabit estuaries where temperature, salinity, and oxygen availability vary widely. Adaptation to such estuarine conditions would constitute preadaptation for desert conditions. Given the superlatives accompanying the above descriptions of thermal, salinity, and oxygen tolerance in pupfishes, additional adaptation may have been unnecessary. Working against the evolution of desert-specific adaptations are both the comparative youth of the region and the periodic connection of desert water courses and pools with each other and with estuarine and riverine areas that serve as sources of new immigrants. Selection for desert adaptations would be relaxed during wetter periods, and dilution of such adaptations would also occur due to gene flow from source areas.

Given the limited extent, isolation, and small populations characteristic of desert habitats, it is not surprising that southwestern fishes are very sensitive to environmental degradation (Soltz and Naiman 1981; Miller 1981). A variety of human activities have led to declines and extinctions, including pumping of springs and groundwater, pollution by humans and livestock, draining of marshes, damming of streams, introductions of exotic competitors and predators, and hybridization (see Chap. 25). Approximately 15 species and numerous localized populations of southwestern fishes are extinct. Desert species account for nearly two-thirds of the federally listed endangered and threatened fishes in North America. The fishes of this region have adapted well to the environmental challenges of extreme desert conditions, but nothing in their history allows them to handle the kinds of insults that often result from careless or callous humans (Pister 1981; Minckley and Deacon 1991; Rinne and Minckley 1991).

CAVES

Among the more extreme aquatic environments imaginable are underground water systems where no light penetrates and where food availability depends on infrequent replenishment from surface regions. However, cave living has advantages, including a scarcity of competitors and predators and a constant, relatively moderate climate. Fishes have evolved independently in caves around the world, and not surprisingly, similar adaptations to cave life have evolved repeatedly despite phylogenetic differences. The darkness, low productivity, and even high atmospheric pressure of cave environments have also led to some surprisingly strong convergences between cave and deep-sea fishes.

Caves usually develop in limestone formations (karst) because of the solubility of carbonaceous rock, although caves exist in other rock types, such as lava tubes on volcanic slopes. Caves include places where water dives underground and resurfaces after a short distance, or where springs upwell near the surface and are illuminated by dim but daily fluctuating daylight (technically a **cavern**). The classic cave environment is a continuously dark, subterranean system where fluctuations in temperature, oxygen, and energy availability are minimal and where little interchange occurs with other areas. The biota of caves are especially interesting because a continuum of habitats exists between the surface, caverns, and deep caves. We can consequently often identify closely related and even ancestral organisms from which cave populations and species evolved. This allows comparison of cave

BOX 17.2   *Acidity, Alkalinity, and Salinity*

The acidity or alkalinity of a water body strongly determines the existence and types of fishes that occur there. Seawater is naturally buffered against abnormal shifts in hydrogen ion content (pH), and hence pH is seldom a concern for marine fishes. Freshwater, in contrast, is easily affected by substances that alter pH. Changes in acidity in turn affect the activity of metals and other potential toxins in the water. Fishes normally live in water with a pH range between 6 and 8, a pH of 7 being neutral. Acidic conditions (pH $\ll$ 7) often result from the decay of organic matter that is not filtered through soil or further broken down. Examples of naturally occurring, low-pH water (pH 3.8 to 4.9) are seen in the black or tea-stained coloration of many swamps, the black-water rivers of the southeastern United States, and the black waters of major tributaries of the Amazon (e.g., the Rio Negro) and many African rivers. Such soft waters are also low in dissolved substances and inorganic ions but high in organic acids such as humic and fovic acids (Lowe-McConnell 1987).

Some fishes have evolved under conditions of low pH and do best in slightly acidic waters (e.g., many tetras), whereas other groups are intolerant of acidic waters. Minnows, which are so widespread throughout North America, are often missing from river systems where the pH falls below 4.5 (Laerm and Freeman 1986), although cyprinids do well in Southeast Asian waters with low pH. Acid rain, a lowering of pH that results from industrial pollution, causes reproductive failure in many fishes and has eliminated fishes from the poorly buffered lakes of the Adirondack Mountains in New York and in many lakes throughout Scandinavia (Baker and Schofield 1985).

High pH is caused by an abundance of hydroxyl (OH) groups, producing alkaline conditions. High alkalinity occurs naturally in waters that run through limestone rocks or where extensive evaporation occurs. Some fish have adapted to alkaline conditions that are lethal to most other animals. A small (< 8 cm) African cichlid, *Oreochromis grahami*, is the only fish that can live in Lake Magadi in Kenya under conditions of extreme alkalinity, salinity, and temperature. Water flows into the lake from hot springs at a pH of 10.5, a salinity of 40 ppt, and a temperature of 45°C. The water has a high load of sodium bicarbonate, sodium chloride, sodium sulphate, and sodium fluoride and has a conductivity of 160,000 μmho/cm (most African lakes have a pH of 7 to 9 and a conductivity of 100 to 1000 μmho/cm). The fish occupy pools and graze on algae at temperatures below 41°C. Their upper temperature limit creates a distinctive browse line where inflowing spring water has cooled sufficiently to allow fish activity. Algae in regions above 40°C are safe from fish grazing (Coe 1966; Fryer and Iles 1972).

Salinity determines the distribution of many if not most fish families. Biogeographic categories of freshwater fishes focus on whether taxa can tolerate salinities greater than a few parts per thousand. In one approach (see Berra 1981; Briggs 1995), freshwater fishes are classified as **primary** (those that cannot cross saltwater boundaries, such as minnows, characins, most catfishes, pike), **secondary** (those that can cross at least short saltwater regions, such as cyprinodontoids, cichlids), and **peripheral** (those derived from marine families or that spend part of their lives in the ocean, such as salmons, sculpins) (see Table 16.2).

The actual barriers to free movement between regions of high and low salt concentration are physiological in nature. At the simplest, freshwater fishes have a need to conserve salts and eliminate water, whereas saltwater fishes must conserve water and stem the influx of salts (see Chap. 7, Osmoregulation and Ion Balance). Extremes of and rapid changes in ionic concentration can cause osmotic stress. Although pure distilled water is stressful, it is also unusual in nature and hence an uncommon limitation. Hypersalinity occurs in many areas, either as a result of heated water flowing through easily soluble rocks or due to daily or seasonal evaporation and concentration of salts as water courses dry up during low tides or droughts.

Some of the most widely distributed families in freshwater turn out to be those that show a high tolerance to both rapid fluctuations and extreme conditions of salinity. Many cyprinodontoid killifishes and pupfishes can tolerate ranges of salinity from 0% to 100% of seawater (100% $\approx$ 35 ppt) and appear to tolerate rapid shifts in salinity from high to low concentration, such as those brought on by rainstorms. Some, such as the Mediterranean *Aphanius* and several North American *Cyprinodon*, live in water two to three times saltier than seawater. These capabilities have preadapted them for life in isolated habitats such as desert springs and pools (Roberts 1975). Similar abilities characterize cichlids and gobies, two of the world's largest families of fishes. *Tilapia amphimelas*, a cichlid, inhabits Lake Manyara in Africa, where the sodium content is twice that of seawater and is increased ionically by abundant potassium salts (Fryer and Isles 1972).

Certain large inland water bodies are too saline to support even the most osmotically tolerant species, including the Dead Sea of the Middle East and Great Salt Lake in Utah, where salinities exceed 200 ppt. Water withdrawal due to human activities can cause salinization of a lake and threaten the fishes there, as is occurring in the Aral Sea of the former Soviet Union (see Chap. 25).

and surface forms and analysis of the processes and selection pressures that have produced cave adaptations.

Approximately 90 species and 14 families of teleostean fishes have colonized caves. With the exception of some ophidiid cusk-eels and gobies, the families are restricted to freshwater. Most cave fishes are ostariophysans (characins, loaches, minnows, 5 catfish families), which is not surprising given the overwhelming success of this superorder in freshwater habitats. The remaining 4 families are either paracanthopterygian (ambloypsid cavefishes) or acanthopterygian (poeciliid livebearers, synbranchid swamp eels, and mastacembelid spiny eels). Only 1 family, the amblyopsid cavefishes, consists primarily (4 of 6 species) of cave-dwelling forms (Fig. 17.8).

Adaptations to Cave Living

Typical cave-adapted fishes are characterized by a lack of pigmentation, reduced squamation, a reduction or loss of light receptors (involving eyes and the pineal gland), greatly expanded lateral line and external chemosensory receptors, and relative decreases versus increases in brain areas associated with vision versus hearing and chemoreception respectively. Behaviors typically mediated by vision are lost, such as schooling, the dorsal light reaction, and circadian rhythms (Wilkens 1988) (see Chaps. 6, 22). Taste buds in surface-dwelling *Astyanax fasciatus*, a characin, are generally restricted to the mouth region, whereas in cave-adapted populations of the same species they cover the lower jaw and ventral areas of the head. Chemosensory capabilities are better in cave forms; cave-adapted *A. fasciatus* are about four times more effective in finding meat on the bottom of a darkened aquarium than are the surface forms. Adaptations to unpredictable or irregularly occurring food supplies also exist. When fed ad libitum (as much as they can consume), cave *Astyanax* build up larger fat reserves than surface forms, again by a factor of four (37% of body mass vs. 9%).

Parallel comparisons can be made within the family of cavefishes (Amblyopsidae). The cave genera (*Amblyopsis, Typhlichthys, Speoplatyrhinus*) swim more efficiently, have lower metabolic rates, and find prey quicker and at greater distances in the dark than surface forms (*Chologaster*). The cave forms are also better at avoiding obstacles and at memorizing the locations of objects than are the surface fish. Cave catfishes (blindcats) in the North American

family Ictaluridae show parallel changes with respect to eye loss, absence of pigmentation, pineal reduction, enlarged lateral line pores and canals, and brain modifications. Many analogous adaptations have also been observed in other cave-adapted taxa, including beetles, amphipods, crickets, crayfishes, shrimps, and salamanders (Poulson 1963; Poulson and White 1969; Culver 1982; Langecker and Longley 1993; Parzefall 1993).

Adjustments to cave existence also occur in the reproductive biology and life history traits of cave-dwelling fishes. Not surprisingly, visual displays are generally lacking during courtship of cave species, even in families such as livebearers and characins, whose surface forms commonly have such displays (Parzefall 1993). With respect to life history traits, cave-adapted amblyopsids produce fewer but larger eggs with greater yolk supplies, have larvae that spend more time before hatching, and have a later age at maturation and longer lifespans (Bechler 1983). Reproductive rates of cave populations are surprisingly low. Only about 10% of the mature fish in a population of cavefishes may breed in any one year, each female producing 40 to 60 large eggs. These eggs are incubated in the mother's gill cavity for 4 to 5 months, long after the young are free-swimming. This may be the longest period of parental care for an externally fertilized fish species. Many of these characteristics are what one would expect in a habitat where adult mortality and interspecific competition were low, environmental conditions were stable, and food was scarce (Culver 1982; see Chap. 23).

The degree of anatomic and behavioral change in a cave population is often correlated with the length of time since the cave was colonized. Eye loss, characteristic of cave-adapted forms, shows some responsiveness to light availability. When young *Astyanax fasciatus* from caves of different presumed ages are raised in the presence of light, individuals from old cave populations do not develop eyes, surface populations develop eyes, and populations thought to have invaded caves more recently vary in eye size (Parzefall 1993).

Food sources in caves are rather limited. Since no photosynthesis can occur in the sunless cave environment, food can only arrive if brought in by other animals or carried in by percolation through the rock or by water currents, such as during occasional floods. Common food types differ among families, but bat and cricket guano,

FIGURE 17.8. *An amblyopsid cavefish,* Typhlichthys subterraneus, *about 6 cm long. Note regressed eyes and the development of sensory canals and pores on the head and body, and even on the tail.*

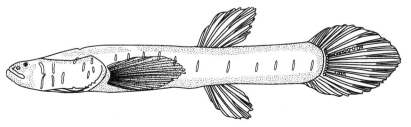

Drawing by J. Tomelleri; used with permission.

bacteria, algae, small invertebrates (isopods, amphipods, copepods), and conspecifics are the common food types of most groups (Parzefall 1993). In Mexican caves containing the livebearer *Poecilia mexicana*, bat guano is supplemented by bacteria associated with sulfur springs in the cave, an interesting analog to deep-sea vent communities (see above, The Deep Sea). Cave fishes respond to chemical or mechanical cues given off by the food; a clay ball dropped into water containing cave fishes will induce active swimming and searching by fish within a meter of the ball.

Cave fishes usually live at low densities, particularly those in isolated deep caves. Population density is strongly correlated with food availability, which again correlates with degree of isolation. Typical population densities of such fishes as the amblyopsid cavefishes are low, ranging from 0.005 to 0.15 fish per square meter. The blind cave fish, *Astyanax fasciatus*, can reach densities of 15/m², and *Poecilia mexicana* can reach densities of 200/m² where sulfur springs occur. Near-surface caves that contain bats as an energy source host even higher densities of cave-dwelling fishes.

Preadaptation, Evolution, and Convergence

Adaptation to the cave environment often involves two contrasting trends in the development of structures. Organs that may have been useful to surface ancestors but are of limited use in the cave, such as eyes and pigment, are gradually lost, a process known as **regressive evolution**. They are replaced by hypertrophied (overdeveloped) structures, such as widely distributed and enlarged lateral line and chemosensory receptors and their neural correlates. The mechanisms and agents of selection leading to regressive evolution—namely, the relative importance of neutral or directional selection, pleiotropy, energy economy, population size, time since isolation, and gene flow—remain a matter of active debate (Culver 1982).

Some groups possess preadaptations that may have made the transition to cave life quicker. Surface-dwelling Mexican characins show reduced eye development when raised in the dark, and blinded surface fish are as effective at avoiding obstacles as are cave-adapted fish. Ten of the 14 cave families commonly contain nocturnal species; nocturnality and its attendant emphasis on nonvisual sensory modes would be an important preadaptation for cave living. Some cave-dwelling characins develop taste buds outside the mouth. This pattern also exists in surface-dwelling ictalurid catfishes; in fact, taste buds are more numerous on the barbels and general body surface than in the mouth of ictalurids, which could make transition to a cave environment easier. Seven of the eight spiny-rayed cave forms, including the solitary perciform, are eel-like. An elongate body and other eel-like features occur in nearly one-third of cave forms, such as the synbranchid swamp eels, mastacembelid spiny eels, cuskeels, clariid catfishes, loaches, trichomycterid catfishes, and arguably the amblyopsid cavefishes themselves. Anguilliform swimming may be advantageous in the narrow confines of many caves (see Chap. 8, Locomotory Types).

Several authors have noted the similarities in traits between cave fishes and bathypelagic deep-sea forms, referring to the similarities as the "deep-sea syndrome." Similar adaptations in the two habitat types include loss of pigmentation, squamation, and light receptors; expanded lateral line and chemosensory receptors; and attendant modifications in the brain. In the blind catfishes, which live deeper than most other cave fishes (400 to 500 m), additional convergences occur in terms of reduced body size, gas bladder regression, large lipid deposits, and reduction of body musculature and skeletal ossification. These changes can be viewed as adaptations to overcome problems associated with energy conservation in an environment with limited food availability (Langecker and Longley 1993). These parallels underscore once again the descriptive power of the principle of convergence: If selection pressures and processes are strong and analogous, convergence can occur not just among species within a habitat but also between habitats.

SUMMARY

1. The principle of convergence states that strong selection pressures tend to produce strong similarities in unrelated animals. Several aquatic habitats offer examples. Mesopelagic ocean depths between 200 and 1000 m contain 750 species of fishes that are typically dark in color, with photophores, large mouths, slender teeth, reduced skeletons and squamation, long rete mirabilia, low enzyme activity, and daily vertical migrations. Bathypelagic fishes (1000 to 4000 m; 200 species) show stronger and more bizarre convergences, including sex reversal, extreme skeletal and musculature reduction, eye loss, longer retes, marked sexual dimorphism, and behavioral energy conservation. These characteristics are apparent adaptations to low energy availability.

2. Oceanic, pelagic fishes swim in the upper 100 to 200 m of water. This is the primary region for commercial fish production and is the habitat of herringlike fishes, sauries, carangoids, dolphinfishes, mackerels, tunas, and billfishes. Pelagic fishes are typically streamlined, silvery, and migratory, with a high proportion of red muscle for sustained swimming. They respire efficiently and save energy by using ram gill ventilation. Life history differences between temperate and tropical species are influenced by seasonal and spatial food availability, and lead to dramatic differences in year class fluctuations. Freshwater pelagics have converged on many traits with oceanic species.

3. The polar arctic and antarctic regions lie above 60° latitude. The Antarctic has more endemic, specialized fishes, half of which are in the icefish suborder Notothenioidei. Antarctic fishes avoid freezing because their blood contains antifreeze compounds. Channichthyids are unusually pale because they lack hemoglobin and myoglobin. Some notothenioids have evolved neutral buoyancy via reduced skeletal miner-

alization and increased lipid deposition. Arctic fishes have converged on similar traits.

4. Desert freshwater fishes live on almost all continents in regions where water scarcity creates extreme conditions. Desert fishes often possess accessory respiratory structures for using atmospheric oxygen and have a life cycle that includes a resting stage during droughts, either involving a diapausing egg or an estivating adult. In addition to low oxygen, desert fishes often encounter extremes of salinity and alkalinity. The southwestern deserts of the United States have a surprising diversity of endemic fishes, many of which are threatened.

5. Cave fishes live in lightless, freshwater environments where food is scarce. Cave-adapted forms typically have reduced eyes, pigmentation, and squamation; low metabolic activity and reproductive rates; low population densities; and increased chemosensory and lateral line development. Cave-dwelling fishes have converged on many of the traits evolved by deep-sea fishes, probably in response to food and light scarcity.

SUPPLEMENTAL READING

Briggs 1995; Culver 1982; Eastman 1993; Gage and Tyler 1991; Marshall 1971; Marshall 1980; Naiman and Soltz 1981; Sharp and Dizon 1978.

PART 5

Behavior and Ecology

18 *Fishes as Predators*

Predation has an overriding influence on the morphology, behavior, and ecology of fishes. The selection forces in operation are obvious and strong: Fish with a relative feeding advantage will prosper, whereas fish that get caught are eliminated from the gene pool. The next two chapters explore the behavior and ecology of feeding in fishes, with emphasis on the evolutionary interplay between predatory and escape tactics, the so-called predator-prey arms race. For organizational purposes, adaptations are classified by where they appear to fit in the **predation cycle**, which is the sequence of events involving searching, pursuing, attacking, capturing, and handling prey (Curio 1976). Often the distinction between phases is blurred; for example, pursuit and attack may occur simultaneously, as can attack and capture. Structures employed in feeding and their functions are detailed in Chapter 8 (Feeding: Biting, Sucking, Chewing, and Swallowing).

An oft-cited adaptation in foraging contexts is the formation of groups, termed **shoals** when swimming is unorganized, **schools** when individuals are polarized, that is, swimming parallel and in the same direction (Pitcher and Parrish 1993). Groups function both to increase feeding success and to deter predators (see Chaps. 18, 19). Group function also changes at different phases of the predation cycle (see also Chap. 21, Aggregations). We close this chapter with a brief discussion of foraging theory, which attempts to understand foraging in the general context of costs, benefits, and relative fitness.

SEARCH

Predators can search for prey actively or passively. Active search implies locomotion while the predator scans the environment with any of the six sensory modes discussed in Chapter 6. Water-column searchers, such as herrings, anchovies, minnows, tunas, and billfishes, rely heavily on vision, as do nocturnal plankton feeders. Olfaction, gustation, and hearing are also important for some water-column searchers, particularly sharks. Low-frequency sounds of 20 to 300 Hz are especially attractive to sharks, whereas amino acids elicit feeding responses in many predatory fishes. Smell, taste, touch, or electrolocalization (passive or active) is employed extensively by benthic and nocturnal foragers such as eels, catfishes, gymnotid knifefishes, sea robins (triglids), goatfishes (mullids), and threadfins (polynemids); chemoreception and touch are used by other groups that possess barbels, such as sturgeons, minnows, cods, and croakers.

Some fishes search by speculation, much as chickens scratch where buried prey are likely to occur. Goatfishes move along the bottom probing into sediments with their muscular barbels, which are equipped with abundant taste receptors; some goatfishes flush prey by inserting their mobile barbels into refuge holes where prey have sought shelter (Hobson 1974). Boxfishes (Ostraciidae) expel jets of water from their mouths to blast sand away from potential buried prey. Logperch (Percidae) roll stones with their snouts in search of hidden insect larvae. These speculating foragers frequently have attendant species that follow them and snap up prey disturbed by the forager's activity.

The energy expended in active search can be saved by camouflaged predators that lie in wait on the bottom or in other structure. Such camouflage is often termed **protective resemblance** when hiding from predators, or **aggressive resemblance** when lying in wait (the latter usage is inaccurate behaviorally since the term *aggression* should be reserved for combat situations between animals, not for predatory activities). Benthic, camouflaged predators lie on rocks or soft bottoms or can be slightly (or greatly) buried by sediment. Their skin is colored to resemble algae-covered rocks, tunicates, sponges, and other bottom types. Wartlike and other fleshy outgrowths of skin and fins are common. These fish rush explosively from the bottom to capture prey or open their typically large mouths rapidly and inhale prey. Many scorpionfishes (Scorpaenidae), flatheads (Platycephalidae), sea basses (Serranidae), and hawkfishes (Cirrhitidae) rest exposed on the bottom, whereas lizardfishes (Synodontidae), stonefishes (Synanceiidae), stargazers (Uranoscopidae), and flatfishes (Pleuronectiformes) lie with only their eyes exposed above the sediment. For such lie-in-wait predators, vision is the primary sense mode by which prey are detected, except for the elasmobranchs, which may also use electrical cues. Many benthic, immobile ambushers appear surprisingly conspicuous, at least to a human observer. They may rely on prey habituating to their presence and thus growing careless.

Some water-column predators, including countershaded or silvery-sided fishes such as gars (Lepisosteidae), pikes (Esocidae), and barracuda (Sphyraenidae), also lie in wait, floating motionless near or below the surface and darting at prey that fail to recognize them. This group also includes substrate- and leaf-mimicking species such as trumpetfishes (Aulostomidae) and leaffishes (Nandidae). Many predators shift among search patterns. Trumpetfish lie in wait among gorgonian corals to ambush roving prey, hide behind swimming herbivores such as parrotfishes, or swim actively in the water column and attack relatively stationary schools of zooplanktivores. By day, torpedo rays erupt from the sand at prey that have wandered over them, whereas at night they swim actively above the bottom in search of swimming prey. Prey behavior and density often determine which search mode will be employed. For example, young lumpfish (Cyclopteridae) cling to rocks with their modified pelvic fins and make short excursions to feed on nearby zooplankton when prey densities are high. At low prey densities, the larvae swim through the water column searching for and feeding on plankters, thereby incurring the greater costs of active search but avoiding starvation (Brown 1986; Helfman 1990).

Considerable attention has been paid to the search tactics and detection capabilities of zooplanktivorous fishes. Fish swim through the water column scanning an area ahead of them that is shaped approximately like a hemisphere, the widest part being closest to the fish. The volume of this search space, the distance from objects at which fish react, and the size object that a fish is capable of detecting change with fish size, water clarity, and illumination level. Large juveniles can detect smaller objects than can small juveniles, and most fishes react farther away in clearer water or after light levels exceed some threshold value (Hairston et al. 1982).

Reaction distance is heavily dependent on prey size, to the extent that most zooplanktivores will react to and pursue the largest-appearing prey in their visual field. This means that a small zooplankter near a fish may be taken preferentially to a larger plankter farther away because the smaller prey appears larger (the **apparent size hypothesis**). However, prey immobility and location also affect selection, with smaller prey being preferred if they are mobile or are more directly in front of the forager (O'Brien et al. 1985; O'Brien 1987). The speeds at which fish search appear to approach the optimal in terms of maximizing intake relative to energy expense. For example, the actual sustained search speed of a 40-cm salmonid is 3 body lengths per second (BLs^{-1}), which is close to the calculated optimum sustained speed of 2.9 BLs^{-1} (Ware 1978; Hart 1993). Speeds vary as a function of fish size (= metabolic rate) and food concentration.

Although group formation is most commonly viewed as an antipredator response (see Chap. 19), grouped fishes may search more successfully than individuals. Foragers in groups may locate food sooner, take food in faster, have more time available for foraging, and grow faster than solitary foragers. For example, in minnows (*Phoxinus*

phoxinus, Cyprinidae), goldfish (*Carassius auratus*, Cyprinidae), and stone loaches (*Noemacheilus barbatulus*, Cobitidae), shoal members spend less time before finding food than do solitary individuals, and the benefit increases with increasing shoal size (Pitcher and Parrish 1993). Accelerated rates arise because a fish in a shoal can search for food while simultaneously watching for signs of successful feeding in shoalmates, thus increasing the area over which it effectively searches. Also, the time each individual spends scanning for predators may decrease, leaving more time for feeding. These benefits depend on minimal competition for food, with competition increasing as group size increases.

PURSUIT

Pursuit places a predator close enough to attack prey. Two dramatically different categories of predators have developed. One evolutionary course has favored species that maximize their speed while overtaking fleeing prey; the other course requires minimal aerobic output but a proliferation of deceptive tactics.

Cursorial, chasing predators are capable of high-speed sustained chases of rapidly swimming prey. Such fishes include the apex pelagic predators (lamnid sharks, tunas, billfishes). Morphologically, these are the most streamlined fishes, having bodies that are round in cross section and taper to a thin, laterally keeled caudal peduncle, with the greatest body depth one-third of the way back from the head. The tail is narrow with a high aspect ratio (the ratio of height to depth; see Chap. 8, Locomotion: Movement and Shape), and the median and paired fins typically fit into grooves or depressions during high-speed swimming. Nearer to shore, where prey can escape into structure, streamlining is sacrificed to allow for rapid braking and small-radius turns. Bodies are more oval in cross section, fins are larger, and tails broader. Predators such as salmons, snook (Centropomidae), striped bass (Moronidae), black basses (Centrarchidae), and large-bodied cichlids (e.g., peacock bass) are included here.

In lurking or lie-in-wait predators that swim above the bottom, pursuit is synonymous with attack. These fishes, which rely on **fast-start performance**, have converged on a general body morphology that permits fast starts at a sacrifice in sustained speed and maneuverability. The group includes pikelike predators, including gars, pike, pickerel, needlefish, barracuda, and specialized fishes in such diverse families as the characins and cichlids (Fig. 18.1). These fishes have elongate, flexible bodies; long (produced) snouts with many sharp, often thin teeth; broad, symmetric median fins placed far back on the body opposite one another; and relatively large caudal fins with low aspect ratios. These fishes typically hover high in the water column or lurk motionless on the edges of vegetation beds, relying on their camouflage to gain access to prey.

A class of predators has developed around the energy savings that can be gained if prey can be **lured** within range of attack. The success of this ploy depends on the

FIGURE 18.1. *Variations on a theme: convergence in morphology among fast-start predators. Lurking predators that swim in the water column tend to be elongate with long mouths, sharp teeth, and fins set far back on the body. Examples from six different orders and eight families are shown, including one extinct form. (A) Lepisosteiformes, Lepisosteus (Lepisosteidae), gar, 1 m. (B) Characiformes, Ctenolucius (Ctenoluciidae), pike-characid, 1 m. (C) Esociformes, Esox (Esocidae), pike, 1 m. (D) Cyprinodontiformes, Ablennes (Belonidae), needlefish, 1 m. (E) Cyprinodonti-formes, Belonesox (Poeciliidae), pike killifish, 20 cm. (F) Perciformes, Sphyraena (Sphyraenidae), barracuda, 1 m. (G) Perciformes, Luciocephalus (Luciocephalidae), pikehead, 15 cm. (H) Osteolepiformes, Eusthenopteron (Eusthenopteridae), an extinct Devonian osteolepimorph, 75 cm.*

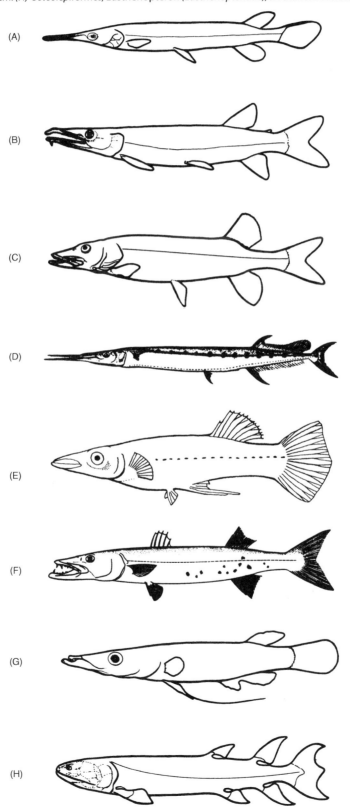

(A)-(C), (G), (H) From Nelson 1994; (D) from Collette 1995; (F) from Jordan 1905; used with permission.

prey not recognizing the predator until it is too late to flee, or on a willingness of the prey to approach the predator. Predators can induce prey to approach if the predator looks like something the prey might want to eat. This sort of deception, misnamed **aggressive mimicry**, can involve all or part of the predator's body. In goose-fishes, anglerfishes, and frogfishes (Lophiiformes) (see Chap. 17, The Deep Sea), the first dorsal spine is elongate, and its end is highly modified into a species-typical **esca**, or lure, that resembles a small fish, shrimp, or worm (Fig. 18.2). The lure is wriggled in a lifelike manner, adding to the deception; in some species, the lure secretes a chemical attractant. The body of the predator is camouflaged to resemble the bottom or, in the case of deep-sea anglers with bioluminescent lures, the dark surrounding waters. Small fishes approach the lure and are quickly inhaled by the large mouth, and their escape is often prevented by long, backward-facing teeth (Pietsch and Grobecker 1978, 1987).

Luring has evolved independently in other groups. A scorpionfish (Scorpaenidae) also uses a modified dorsal spine for a lure, whereas stargazers (Uranoscopidae) have a buccal lure, antarctic plunderfishes use chin barbels, chacid catfishes use maxillary barbels, and snake eels (Ophichthidae) have a lingual (tongue) lure (Randall and Kuiter 1989). *Haplochromis livingstoni*, a large predatory cichlid from Lake Malawi, Africa (see Box 15.2), lures prey in a unique manner. *H. livingstoni* capitalizes on the tendency of many small cichlids to scavenge on recently dead fishes. The predator lies on its side on the bottom and assumes a blotchy coloration typical of dead fish. When scavengers come to investigate and even pick at its

body, the predator erupts from the bottom and engulfs them. This is the only known example of thanatosis, or **death feigning,** in fishes (McKaye 1981a). As a final twist, stonefish lie buried on the bottom and strike at prey fishes in a narrow zone directly above their mouths. If a potential prey fish swims between the mouth and the dorsal fin, the stonefish will raise its dorsal fin, chasing the fish back into the strike zone (Grobecker 1983).

Approach to a prey fish is facilitated by camouflage. Although most predators and their prey are counter-shaded or silvery, such coloration disguises a fish only when it is seen from the side. This is not the view that a prey fish has of an approaching predator. A convergent coloration trait shared by slow-stalking predators is the **split-head color pattern** (Barlow 1967). A dark or light line that contrasts with general body coloration runs from the tip of the snout along the midline between the eyes to the top of head or dorsal fin (see Fig. 19.2). This coloration is evident in pickerel (Esocidae), some soapfishes and sea basses (Serranidae), the tripletail *Datnioides* (Lobotidae), the leaffish *Polycentrus schomburgkii* (Nandidae), and some hawkfishes, piscivores that are otherwise protectively colored and that approach their prey slowly and head-on.

The split-head pattern operates on the principle of **disruptive coloration**, dividing the head into halves and disrupting its outline. Prey may consequently require a moment to recognize the pattern as a whole, threatening head. Because predator-prey interactions occur on a time scale of tens of milliseconds, a moment's delay in recognition may be all that a predator requires to attack successfully. Prey fishes are often frightened by general,

FIGURE 18.2.   *"Aggressive" mimicry in the frogfish* Antennarius maculatus. *The fishlike lure, or esca, sits at the end of the elongate first dorsal spine, termed the illicium. The resemblance of the lure to a real fish is increased by an anterior eyespot, vertical bars and mottling on the body, and finlike appendages. The lure is waved by movements of the illicium, thereby attracting potential prey fishes.*

head-on facial characteristics of predators (e.g., Dill 1974; Karplus et al. 1982); hence disguising the face eliminates critical cues used in predator recognition (Fig. 18.3).

Group hunting is comparatively rare in fishes. Apparent cooperative feeding, involving some form of coordinated herding or driving of prey by circling or advancing predators, has been observed in blacktip reef sharks (Carcharhinidae) (W. B. Masse, personal communication), jacks, black skipjack tuna, and sailfishes (Istiophoridae), and suggested for thresher sharks (Alopiidae), a piranha (Characidae), and bluefin tuna (Bigelow and Schroeder 1948; Hiatt and Brock 1948; Voss 1956; Potts 1980; Partridge 1982; Sazima and Machado 1990). None of these fishes cooperate to the extent observed in pack-hunting mammals such as hyenas, wolves, lions, or killer whales.

ATTACK AND CAPTURE

An actual attack takes place as the predator launches itself at the prey and engulfs the prey in its mouth. The evolutionary advance of fishes is synonymous with the development of jaws (see below and Chaps. 8 and 11), which function to surround, impale, or inhale prey and then pass the prey posteriorly for processing (= handling or subduing; see below). Active prey are brought into the mouth by overtaking, extending the mouth, and suction, often in combination. Fast-start predators overtake or intercept their prey and impale them on sharp teeth. Overtaking may involve rapid swimming via body musculature, but many fishes also shoot the mouth out to surround the prey, as in largemouth bass (Nyberg 1971). The much-heralded evolution of protrusible jaws and the pipette mouth, which reach their greatest development in acanthopterygian fishes, is arguably as much for overtaking prey as for suctioning (see Chaps. 8, 11).

Many benthic, lie-in-wait predators have particularly large mouths that can be opened and closed rapidly. Frogfishes, *Antennarius* (Antennariidae), can expand their oral cavity up to 14-fold, engulfing prey in less than 6 milliseconds. Stonefish, *Synanceia* (Synanceiidae), engulf prey in 15 milliseconds. Speed is not solely characteristic of large fish feeding on large prey. Zooplanktivorous damselfishes (*Chromis*, Pomacentridae) project their mouths to capture individual plankters in 6 to 10 milliseconds; jaw protrusion followed by suction is used during capture (Coughlin and Strickler 1990). Slower movements suffice for nonevasive planktonic prey. Whale sharks, basking sharks, manta rays, herrings, anchovies, and mackerels swim through plankton concentrations with their mouths open, passively filtering prey out of the water with their fine gill rakers. Some cyprinids and cichlids filter concentrated plankton by remaining in one place and pumping water in and out of the mouth, again using the gill rakers as a sieve (Drenner et al. 1987; Ehlinger 1989).

A few specialized predators immobilize prey before engulfing them. These specialists include torpedo rays and electric eels, thresher sharks, sawsharks, sawfishes, swordfish, and billfishes (Box 18.1). In these predators, the attack involves a high degree of energy output, followed by a relatively leisurely capture. Electricity-generating predators stun individual prey with a powerful discharge; the process

FIGURE 18.3. *Is this the face of death? (Left) The general features of a predatory face that elicit fright responses in prey fishes are a broad head; a wide, downturned mouth; and ringed, broadly elliptical eyes. (Right) Head-on views of piscivorous reef fishes: (A)* Epinephelus summana, *a sea bass; (B)* Cheilinus trilobates, *a wrasse; (C)* Lutjanus kasmira, *a snapper; (D)* Cephalopholis argus, *a sea bass; (E)* Epinephelus fario, *a sea bass; and (F)* Synodus variegatus, *a lizardfish.*

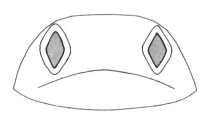

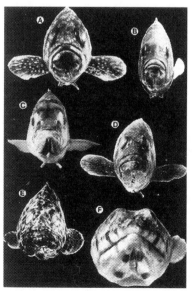

Of Spears and Hammers

Specialized structures in large predatory fishes suggest a foraging adaptation. However, verifying this suggested function is difficult because such predators are relatively rare, they are difficult or dangerous to observe, and predation is an infrequent event. Two such structures are the spears of billfishes and the hammers of hammerhead sharks.

Billfishes belong to two related families in the suborder Scombroidei. The bills in both families develop as forward growths of the upper jaw. In the swordfish (Xiphiidae), the sword is flattened and smooth, whereas in the marlins, spearfishes, and sailfishes (Istiophoridae), the bill is rough and round in cross section. Much debate has centered around whether billfish spear their prey or strike laterally to stun the prey. The occurrence of healthy billfishes with greatly reduced or even missing bills has led some workers to suggest that the bill serves no feeding role. However, good observational evidence indicates that the bill can serve both purposes. Sport fishers frequently observe marlin knocking prey such as small tuna into the air, or find their lures with characteristic scratch marks indicative of strong sideways blows of the rough spear. Tuna and dolphinfish perforated with spear holes have been found in marlin stomachs, and observers have watched marlin spear hooked tuna prior to swallowing them. A fortuitous, if not unsettling, observation was made by two spearfishermen off Durban, South Africa. Free-diving in about 20 m of water, one diver

speared an amberjack, *Seriola lalandi* (Carangidae), weighing approximately 15 kg.

The fish pulled off the spear and dashed straight for Roxburgh [at the surface] who simultaneously observed a 3–4 m marlin [probably a black marlin, *Makaira indica*] making a direct charge for the amberjack which was now hiding behind him. At the last moment the marlin halted and Roxburgh was able to push the bill aside after which the marlin circled [the] diver and amberjack several times. Seconds later the amberjack dashed off at great speed to the bottom, closely followed by the highly agitated marlin. Within an estimated 5 sec the marlin had reached its prey and impaled it on its bill. The marlin then shook the amberjack free and swallowed it. Duration of the entire incident was an estimated 30–50 sec (van der Elst and Roxburgh 1981, 215).

The swordfish, *Xiphias gladius*, evidently uses its smooth, flat bill primarily to decapitate cephalopod prey and slash them into swallowable pieces. Slashing also occurs as a swordfish enters a shoal of prey; maimed fish are picked up on subsequent passes. Spearing may occur defensively or during territorial encounters; broken swordfish (and billfish) bills have been found embedded in boat hulls and other objects. The deep-submersible *Alvin* was attacked and skewered by a 60-kg

is well studied for torpedo rays but more poorly understood for electric eels (Bray and Hixon 1977; Lowe et al. 1994). In torpedo rays, the predator encompasses the prey with its pectoral fins and discharges its electric organ. The immobilized prey is then grasped in the mouth.

Thresher sharks, sawfishes, swordfish, and billfishes are among the few predators that are minimally bothered by the confusion created by a large group of prey fishes (see Chap. 19, Responses of Aggregated Prey). These predators enter a school, slash laterally with their bill (or tail in the thresher shark), and then pick up incapacitated or decapitated prey. The actual pattern is poorly understood. Another fish that incapacitates its prey prior to capture is the archerfish (Toxotidae), which feeds on flying or resting terrestrial insects. Archerfishes have a groove in the top of the mouth along which the tongue fits. The fish shoots droplets of water directionally along the groove at its aerial prey. The insects fall into the water and are snapped up by the waiting archer. Anabantid fishes of the genus *Colisa* also spit water at insects (Dill 1977).

The actual strike of pikelike predators is short and fast, involving a prestrike S-shaped bending of the body and maximal forward propulsion driven by the combined surface area of the median and caudal fins (Webb 1986). Prey are impaled on the sharp jaw teeth, manipulated into a headfirst position, and swallowed; in barracuda, large prey can be cut into smaller pieces for swallowing. In most predators, the attack is focused on the center of mass of the prey's body, because the escape response involves a pivoting on the center of mass, and hence the center moves least relative to other prey body parts (see Chap. 19).

When attacking prey in shoals, the major obstacle to successful prey capture is the presence of other shoal members; a direct relationship exists between likelihood of escape and numbers of individuals in the shoal. Hence many predators engage in tactics that separate prey individuals from the group. For example, when pike (*Esox lucius*, Esocidae) attack minnow (*Phoxinus phoxinus*) schools, the predator's strike often leaves individual prey separated from the school. The predator then preferen-

---

BOX 18.1 *(Continued)*

---

swordfish at a depth of 600 m; the fish was still stuck when the sub was brought to the surface (Wisner 1958; Ellis 1985).

A greater mystery surrounds the expanded cephalic lobes of hammerhead sharks (Sphyrnidae). The anterior and lateral margins of the chondrocranium, particularly the olfactory and optic regions, are slightly to greatly expanded and flattened among members of this family. The eyes and nostrils sit at the ends of these cephalic lobes. It has been variously suggested that this highly modified head functions 1) as a bowplane to increase maneuverability, 2) to increase stereoscopic or binocular vision in the forward direction, 3) to increase stereo-olfaction, which would allow localization of odors, or 4) to expand the sensory area for pressure and electromagnetic detection (Compagno 1984). None of these explanations is necessarily exclusive, and all could operate to some degree, although experimental tests have failed to demonstrate exceptional capabilities with respect to hydrodynamic efficiency or olfactory localization (Johnsen and Teeter 1985; Parsons 1990). However, underwater observations suggest an additional if not primary use of the hammer during prey capture.

Stingrays figure commonly in the diets of many sphyrnids. The great hammerhead, *Sphyrna mokarran*, appears to be something of a stingray specialist: Individuals have been found with as many as 96 stingray barbs embedded in the mouth, throat, and tongue. Researchers in the Ba-

hamas observed an incident suggesting that the hammer can function specifically to facilitate feeding by sphyrnids on stingrays (Fig. 18.4). Snorkeling in 6 m of water over a sea-grass bed, the divers witnessed a 3-m-long great hammerhead pursuing a 93-cm-wide southern stingray (*Dasyatis americana*, Dasyatidae). The shark thrust its head down against the middorsal region of the ray, knocking the ray to the bottom. The shark then pinned the ray against the bottom using its head and pivoted around on the back of the ray, taking a bite from the left front margin of the ray's disk. The ray limped off, and the shark pinned it to the bottom again, pivoted, and took a bite from the right anterior margin of the disk. This incapacitated the ray, which was then consumed by the shark.

In this incident, the shark used its expanded head not only to knock the stingray to the bottom and disrupt its flight but also to hold the stingray on the bottom while the shark pivoted around to the anterior portion of the ray's disk to take bites. "It is doubtful that a shark with a conical snout could have been as effective in restricting the ray's movements (i.e., applying pressure to both pectoral fins simultaneously)" (Strong et al. 1990, 839). The researchers concluded that, while possibly serving improved sensory functions, the size and orientation of the expanded cephalic lobes of hammerhead sharks can be directly responsible for making sphyrnids efficient predators on large batoids.

---

tially chases these individuals, which may account for 89% of the predator's success (Magurran and Pitcher 1987). Similar group-separating tactics are employed by other predators, such as blue jack (*Caranx melampygus*, Carangidae) when attacking mixed-species schools of snappers, and piranha (*Serrasalmus spilopleura*, Characidae) when attacking cichlid shoals (Potts 1980; Sazima and Machado 1990). Stragglers are up to 50 times more likely to be attacked than fish within a group, and success rates for attacks on stragglers can be four times higher than for attacks on the main shoal (Parrish 1989a,b).

Aggressive mimicry also occurs in a group context, the victims either being fish and invertebrates exterior to the shoal or even unsuspecting members of the shoal itself. In both cases, shoal membership by the mimic allows the predator to get close enough to attack. Juveniles of the Indo-Pacific grouper, *Anyperodon leucogrammicus* (Serranidae), swim with and resemble adult females of four similar species of wrasses (Labridae). Small fishes have no reason to fear the wrasses, which feed on a variety of small benthic invertebrates. However, the grouper is a piscivore

and has been observed snaring small damselfishes while swimming among the shoaling wrasses.

Similarly, hamlets of the West Indian genus *Hypoplectrus* (Serranidae) resemble damselfishes and angelfishes, the models here being zooplanktivores or herbivores. The hamlets are carnivores, and their presumed mimicry could allow them to sneak up on invertebrate prey that do not distinguish them from the otherwise harmless, and usually more numerous, model species. Some scale-eating fishes attack from within shoals, their primary prey being the shoalmates that they resemble. These predators include a characid, *Probolodus heterostomus,* which feeds on schooling characids in the genus *Astyanax,* and a cichlid, *Corematodus shiranus*, which resembles and schools with its prey, a tilapiine cichlid, *Oreochromis squamipinnis* (Sazima 1977; Thresher 1978).

Prey choice is also affected by relative numbers of different kinds of individuals in the prey shoal. **Oddity** in appearance or behavior often stimulates attack. In mixed schools, the minority species is often attacked disproportionately, as happens when gafftopsail pompano (Carangi-

FIGURE 18.4.   *Use of the hammer during feeding by a great hammerhead shark. A 3-m-long hammerhead captured and consumed a 1-m-wide southern stingray by knocking it to the bottom with its hammer and then using the hammer to hold the ray against the bottom while the shark pivoted around and fed on the front margins of the pectoral fins. (A) Shark chases the ray. (B) Shark strikes downward across the back of the ray with the flat underside of its hammer. (C) Ray bounces off the bottom from the force of the blow while the shark brakes with its pectoral fins. (D) Shark delivers a second downward blow across the back of the ray. (E) Shark pivots while holding the ray against the bottom and takes a bite from the front of the ray's left pectoral fin. (F) Injured ray attempts to swim off, followed by the shark.*

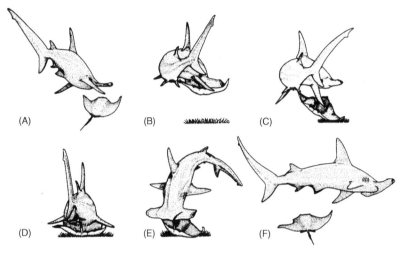

(A)   (B)   (C)

(D)   (E)   (F)

From Strong et al. 1990; used with permission.

dae) feed on anchovetas (Engraulidae) that are schooling with more numerous flatiron herring (Clupeidae) (Hobson 1968). In experimental trials, largemouth bass are much more successful predators when one or two blue-dyed minnows are added to small minnow shoals; the odd individuals are taken preferentially (Landeau and Terborgh 1986). Predation on odd individuals within a school may result from them standing out against the background of the more common type or, as suggested by Landeau and Terborgh (1986), because oddity alleviates the **confusion effect** created by a mass of similar-appearing prey animals.

Whether predatory fishes are more likely to attack injured individuals within a group remains unclear. Certainly sharks and barracudas are attracted to injured or erratically swimming prey, and the literature on mammalian predators (e.g., lions, wolves, hyenas) suggests that fishes would also feed preferentially on injured individuals. However, the only quantitative study of the subject indicates, if anything, an avoidance of injured prey (Major 1979). Regardless, predation on odd individuals indicates the strong selection that occurs for uniformity of appearance and behavior within a shoaling species.

HANDLING

Handling includes any postcapture manipulation required to subdue prey and make it ingestible and digestible. Fishes that feed on hard-bodied prey or on prey with primary external defenses such as spines, shells, bony

scales, or toxic skins must usually spend time and energy in handling, as do scavengers that feed on prey too large to be swallowed whole (Box 18.2).

Assuming prey are small enough to be swallowed and possess no exceptional defenses, such as bony armor, poisonous spines, or toxic skin secretions, piscivores face one major handling task. The prey must be manipulated into a headfirst orientation for swallowing (Reimchen 1991). Such manipulation is less important with nonspiny fishes such as clupeiforms, most ostariophysans, and salmoniforms, but even soft-rayed fishes can become lodged in the throat if they are large and swallowed tail first. Headfirst swallowing facilitates depression of the dorsal, anal, and pelvic fins, all of which can anchor themselves in the predator's mouth or throat. Additionally, headfirst swallowing reduces the likelihood of escape from the mouth, since few non-eel-like fishes can swim backward effectively.

Positioning is accomplished by jaw and head movements. Since many predators will attack fishes as long as themselves (or even longer in the case of many deep-sea species), it is not unusual to observe a piscivore swimming with the tail or more of a prey fish protruding from its mouth. Such a large meal obviously represents an energetic bonanza but is obtained at a potential cost if it hampers the predator's ability to escape its predators. On occasion, one finds a dead predator floating with a large, dead prey fish protruding from its mouth, testimony to the importance of gape limitation and the evolution of an ability to estimate accurately the size of potential prey.

BOX 18.2	*Overcoming Gape Limitation: Spinning and Knotting in Eels*

With few exceptions, fishes are gape limited. This means their diets are constrained to include only items that can be swallowed whole. Swallowing entails movement of prey down the throat, which cannot expand to a width greater than the space between the cleithral bones. Feeders on hard-bodied prey can be limited by the relatively small gape of their pharyngeal jaws (Wainwright 1988). Therefore, large, energetically valuable, even easily obtained food items must be passed up by most fishes.

The exceptions to this rule constitute some rather spectacularly endowed predators that are capable of chopping large items into smaller, swallowable pieces. Sharks as a group can take bites out of prey or cut prey items into pieces, undoubtedly a trait that has contributed substantially to their 400-million-year success story. Among bony fishes, a few species have such sharklike capabilities, notably piranhas and African tigerfish (Characidae), bluefish (Pomatomidae), and barracuda (Sphyraenidae)—all fishes with specialized cutting or chopping dentition and powerful jaw muscles. Some advanced coral reef species use powerful jaws and teeth to tear pieces out of sponges, such as various pufferfishes (Tetraodontidae, Diodontidae). Others can take small pieces of fins or flesh from prey, including some characins and cichlids, sabre-toothed blennies (Blenniidae), and at least one species of *Forcipiger* butterflyfish (Chaetodontidae).

Probably the most common tactic for overcoming gape limitation is nibbling. Many small-mouthed shallow-water marine and freshwater fishes can nibble small pieces from large prey, including centrarchid sunfishes, cichlids, damselfishes, wrasses, and surgeonfishes. But the great majority of fishes, those that utilize suction regularly during feeding, lack both the dentition and the jaw strength to nibble effectively. However, one body form among fishes that are otherwise suction feeders has allowed the development of an alternative solution to gape limitation. Eel-like fishes and other elongate, aquatic vertebrates can spin rapidly around their long body axis while holding on to food and thus can tear chunks from the larger mass of a prey item.

Eels as a group are predators and scavengers. The American eel, *Anguilla rostrata*, and other members of the family Anguillidae feed by cruising close to the bottom and poking their snouts into sediment and crevices. Typical stomach contents after a night's foraging include a variety of small invertebrates and fishes, most of which are inhaled via the standard teleostean process of inertial suction. Occasionally, an eel will be fortunate enough to encounter a dead or dying fish, or a crab that has just molted and has a soft carapace, or a clam with its siphon sticking above the surface. Anguillids lack both the dentition and jaw musculature necessary for chopping or nibbling at such items. What they do instead is grasp the item in their mouth and try to suction it. If this does not work, they give it a few shakes and tugs. If the item still does not yield, the eel will hold onto the food and rotate rapidly, up to 14 rotations per second (Fig. 18.5). This action twists the food and shears off a smaller piece. If the piece is small enough to swallow (if the width of the food is less than 85% of the eel's jaw width), it is suctioned. Otherwise the eel will shake the food some more, or will wedge it against the bottom or in a crevice and start spinning again until a small enough piece is removed.

Rotational feeding has now been documented for more than 20 fish species, including other anguillids, moray eels (Muraenidae), conger eels (Congridae), a clariid catfish, a rockling (Gadidae), a rattail (Macrouridae), rice eels (Synbranchidae), a sablefish (Anoplopomatidae), a greenling (Hexagrammidae), a sculpin (Cottidae), pricklebacks (Stichaeidae), a gunnel (Pholidae), a cod icefish (Nototheniidae), and a dab (Pleuronectidae). Amphibians that spin include tadpoles, sirens, and caecilians, most of which are eel-like in body form. Even a crocodile engaged in its infamous "death roll" while feeding folds its legs alongside its body when spinning, making it more eel-like. All species are primarily aquatic.

Moray eels add an option to their food-handling modes. Although capable of shaking and spinning, morays also tie themselves in knots. A moray will grab a live prey fish, tie an overhand knot in its tail, quickly run the knot up to its head, and lever the knot against the prey. The combined force of the strong jaws, sharp teeth, and pressure from the knot can decapitate a prey fish, thus disabling it and also making it small enough to swallow. Hagfishes also use knots when tearing chunks from dead fish, although the process is considerably slower. Hagfishes have not been observed spinning.

Both rotational feeding and knotting are closely linked to an elongate body form. In evolving eel-like forms, lineages apparently reap the additional benefit of being able to spin about their long body axis, an action not available to more conventionally shaped fishes. Without sacrificing the use of suction forces for feeding on common, small prey, rotational feeders can also overcome gape limitation and avail themselves of the occasional jackpot that a recently dead or dying fish represents (Helfman and Clark 1986; Miller 1987, 1989; Helfman 1990).

FIGURE 18.5.   *Rotational feeding in American eels. (A) Grasp: A 50-cm eel grasps the bait (a snapper fillet tied to a weight). (B) Initial torsion: The eel develops a twist in its body just prior to spinning. Note that the ventral surface of the head (light coloration) faces the camera and the ventral surface of the posterior half of the body faces upward, whereas the dorsal (dark) region between the head and midpoint of the body face the camera. (C) Spinning: After the first initial rotations, spinning continues with no apparent twisting in the body; internal forces generating the spins are not understood. (D) Withdrawal with food: The eel has removed a piece of food and is backing away from the bait.*

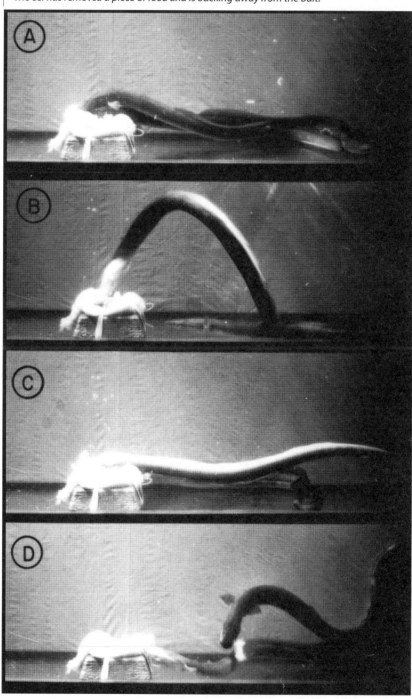

From Helfman and Clark 1986; used with permission.

Preparation for swallowing is accomplished by teeth. Dentition type is a reliable indicator of both prey type and foraging tactics (see Chap. 8, Dentition). Piscivores either hold their prey with large, sharp-pointed teeth or numerous needlelike teeth, or chop up their prey with flat, bladelike teeth. These teeth may be in the marginal jaws or on the palate and tongue. Insect feeders generally have moderately stout, conical, recurved teeth, again marginally or as part of the pharyngeal apparatus (see Chaps. 8, 14). Most fishes that feed on mollusks or echinoderms crush the shell in the mouth, using molarlike teeth. Fishes with parrotlike beaks (parrotfishes, puffers) feed on tough sponges, algae, or coral, with supplemental crushing in a pharyngeal mill in parrotfishes. Gill rakers also characterize different foraging types, functioning either for filtering or as mechanical barriers to escape. Numerous long, thin gill rakers filter out small plankters as water passes through the mouth and out the opercular openings; gill raker spacing is usually directly related to prey size. Fish eaters have harder, stouter, more widely spaced rakers that prevent escape through the gill opening (e.g., sea basses, largemouth bass).

Some armored and otherwise defended prey require special handling tactics. Sea urchins are abundant, and their internal organs are edible. However, the defensive spines must first be removed. Special methods for despining urchins include plucking individual spines off to expose the outer test (triggerfishes, Balistidae); blowing water jets to roll the urchin over, exposing the relatively spineless ventral surface (triggerfishes); or picking the urchin up by a spine and bashing it open on a rock (wrasses, Labridae). Wrasses also smash crabs against rocks to remove a leg or claw, which they then crush in their pharyngeal jaws (Wainwright 1988). Some predators apparently wash distasteful substances off the surface of prey by manipulating them in the mouth. For example, a largemouth bass fed whirligig beetles that secrete noxious chemicals or mealworms dipped in distasteful chemicals will repeatedly slosh the beetle or worm in its mouth and spit it out several times before finally swallowing it. Undipped worms are simply swallowed (T. Eisner, personal communication).

Final handling occurs in the stomach and intestines. Chemical breakdown via acids and enzymes is the rule, supplemented by mechanical grinding in the gizzards of gizzard shad (Clupeidae) and mullets (Mugilidae), and the gizzardlike stomach of milkfish (Chanidae), some characoids (Prochilodontidae, Curimatidae), butterfishes (Stromateidae), and surgeonfishes (Acanthuridae). Deep-sea fishes such as black swallowers (Chiasmodontidae) have highly distensible stomachs that expand to accommodate prey considerably longer than the body of the predator (see Chap. 17, The Deep Sea).

SCAVENGERS, DETRITIVORES, AND HERBIVORES

Scavengers and Detritivores

Many fishes scavenge on dead and dying animals. A few species obtain most of their nutrition through scavenging (e.g., hagfishes) or detritivory (e.g., some minnows and suckers, curimatids, prochilodontids, mullets, some Old World cichlids), whereas others supplement predation and omnivory with scavenging (e.g., catfishes, anguillid eels). Importantly, most predators will not pass up freshly dead prey (otherwise bait would not work in hook-and-line fisheries), and most scavengers and herbivores will take advantage of easily captured live prey. In essence, although dietary specializations certainly exist, fishes are highly opportunistic and will eat available prey of the appropriate size. At Johnston Atoll in the tropical Pacific, discarded doughnuts are eaten readily at the surface by such carnivores as snake eels, butterflyfishes, and flounders, and by such herbivores as damselfishes, parrotfishes, and surgeonfishes (D. A. Mann, personal communication).

For scavenging animals, the predation cycle is usually shortened to search, wait, manipulate, and handle (see Box 18.2), whereas for detritivores and herbivores the waiting is eliminated. The major task befalling detritivores is one of separating edible, fine particulate organic matter from any refractory, inedible sediments ingested. Ridges in the mouth and a maze of passageways associated with the gill rakers and epibranchial organs accomplish this in characoids. A winnowing process occurs in the orobranchial chambers as the fish picks up a mouthful of bottom material, sifts it in the mouth, and expels inedible sediments back out the mouth or out the gill openings. Detritivores have some of the longest or most complexly folded intestines of any fishes, attesting to the resistance of detritus to enzymatic digestion (Bowen 1983).

Herbivores

Herbivory is relatively rare among fishes compared to its widespread occurrence in mammals and birds. Nonteleostean fishes are exclusively carnivorous, with the possible exception of limited herbivory in the Australian lungfish, *Neoceratodus forsteri*. In teleosts, we find the evolution of pharyngeal mills and gizzards—mechanisms for rupturing cell walls and digesting plant matter. The most diverse freshwater fish taxa include substantial numbers of herbivorous species (characoids, minnows, catfishes, cichlids), and herbivores on coral reefs are among the most abundant fishes there (e.g., halfbeaks, parrotfishes, blennies, surgeonfishes, rabbitfishes). Temperate waters are relatively lacking in herbivores, although some marine families (porgies, sea chubs, Aplodactylidae, Odacidae, pricklebacks) feed heavily on plant matter (Horn 1989).

Herbivory requires accurate search and efficient handling. Herbivores, particularly those that browse on upright macroalgae and do not graze on finer algal turfs, appear to use visual cues for selecting edible versus inedible species. Consequently, herbivory is primarily a daytime activity. Targeted search is necessary because plants defend themselves by being tough or by producing chemicals, often in the form of halogenated terpenoids. Herbivorous fishes show strong preferences among algal types, feeding preferentially on species that lack structural and chemical defenses, while avoiding limestone-encrusted species or algae that contain deterrent chemicals. Some of these chemicals can slow growth or cause death in fishes (Horn 1989; Hay 1991).

Specializations for handling plants relate to the difficulty with which cell walls are disrupted, cellulose is digested, or defensive structures and chemicals are overcome. Herbivorous fishes typically have long guts, high ingestion rates, and rapid gut transit times. Large quantities of plant matter are passed through the gut, and relatively little nutrition is assimilated from each ingested fraction. Cell walls are broken down in pharyngeal mills or lyzed in highly acidic (pH as low as 1.5) stomachs, although conclusive evidence of enzymes capable of digesting cellulose (i.e., cellulase) is lacking. Unlike insects and many herbivorous vertebrates, fishes also generally lack endosymbiotic bacteria and other microbes that aid in digestion of plant matter. The exceptions include surgeonfishes, which contain bacteria, flagellates, and peculiar protistlike organisms, and sea chubs (Kyphosidae), which possess a unique digestive tract morphology and a hindgut microflora that aids in digestive fermentation (Fishelson et al. 1985; Rimmer and Wiebe 1987). Interestingly, some sea chubs feed heavily on brown algae that are avoided by most other herbivores (Horn 1989; Kramer and Bryant 1995).

Herbivory on coral reefs is intimately linked to both shoaling and territoriality. Most herbivores either defend exclusive territories (e.g., damselfishes, adult parrotfishes, blennies, surgeonfishes) or roam about the reef in monospecific or heterospecific shoals (sea chubs, parrotfishes, surgeonfishes, rabbitfishes). Territorial defense is very successful against solitary foragers but less so against grouped foragers. Individuals in large groups sustain fewer territorial attacks and have higher feeding rates than solitary foragers or members of small groups. Hence territoriality by some fishes promotes aggregation behavior in others (Robertson et al. 1976; Foster 1985).

FORAGING THEORY: OPTIMAL FISHES

In an attempt to identify general principles that determine what, where, and how animals feed, behavioral ecologists have developed a series of explanatory hypotheses based on the evolutionary expectation that natural selection favors animals that forage efficiently (see Hart 1993). In a foraging context, an animal is considered to behave **optimally** if it chooses among alternatives to maximize the ratio of benefits to costs. Benefits include calories and nutrients ingested, whereas costs involve energy used up, time lost to other activities, or exposure to predators or parasites. Ultimately, benefits and costs have to be measured in terms of an animal's lifetime reproductive success or **fitness**, measured as genetically related individuals produced in later generations. Successful animals are those that pass on more genes than other members of a population (Krebs and Davies 1991).

Fishes perform optimally when choosing food types, feeding locales and times, and foraging modes. One expectation of optimality theory is that foragers should be selective when presented with an overabundance of high-quality food and progressively less selective as food becomes less abundant or lower in quality. For bluegill sunfish eating water fleas (*Daphnia*), large prey are more profitable, providing the most energy relative to effort expended. When offered *Daphnia* of three different sizes and at superabundant densities (350 of each size class), only the largest prey are eaten. At low densities (20 per size class), all prey are depleted equally and completely. At intermediate prey densities (200 per size class), the largest zooplankters are consumed first, then the intermediate prey, and finally the smallest, least profitable prey. The general predictions of foraging theory apply well to food choice in this species (Werner and Hall 1974).

For a predator to exploit food patches optimally, it must assess the profitability of different patches, use the patch that gives it the best return relative to effort, and switch to new patches as resources are depleted. When South American cichlids, *Aequidens curviceps*, were presented with two food patches of different profitability, they aggregated in the more profitable patch in direct proportion to the difference in food availability. Fish moved between patches periodically, feeding most where food was most abundant, then switching as food was depleted. Similar results have been obtained in studies of minnows, guppies, and sticklebacks (Godin and Keenleyside 1984; Abrahams 1989).

Natural selection should also produce foragers that choose a method of food handling that gives them the greatest relative return for their effort. American eels employ three modes for handling food. Small pieces of food (< 85% of jaw width) can be suctioned into the mouth and swallowed. Larger pieces require dismembering. Large but soft pieces are grasped and shaken until a piece is removed, whereas large, firm foods are grasped and spun (see Fig. 18.5). In terms of net energy return and growth rate, suction is the most profitable and spinning the least profitable food-handling type, with shaking falling somewhere between. When offered food types in a two-way choice situation, eels consistently preferred suction food over shake food and spin food, and shake food over spin food, again conforming to the expectations of the cost-benefit approach (Helfman and Winkelman 1991; Helfman 1994).

SUMMARY

1. Successful predators usually search, pursue, attack, capture, and finally handle prey, using different structures and behaviors at different stages of the predation cycle.

2. Search is active or passive; detection can depend on all six senses. In active search, a fish moves through the water (zooplanktivores, tunas), whereas in passive search, a sedentary, camouflaged predator lies in wait on the bottom or in the water column (scorpionfishes, gars). Grouped fish find food faster than solitary fish.

3. Pursuit places a predator close enough to attack prey. Chasing-type predators (carangoids, billfishes) and ambushing predators (pikes, pike characins) have streamlined bodies and rely on sustained or burst speed to overtake prey. Lie-in-wait predators may use lures (anglerfishes, chacid catfishes) or may even feign death (a cichlid) to bring the prey within striking distance.

4. Attack and capture are often synonymous in fishes. Most fishes use their rapidly protrusible mouth to both overtake and suck in fleeing prey. In a few species, prey are first incapacitated prior to capture (torpedo rays, electric eels, sawfishes, thresher sharks, hammerhead sharks, billfishes). Predators on grouped prey generally try to separate an individual from the group before attacking to overcome the confusion effect.

5. Handling makes prey swallowable and digestible, such as positioning prey so it can be swallowed headfirst, or through removal of spines and shells via chewing. Final digestion is mostly chemical in action, although mechanical grinding occurs in those fishes with a gizzardlike structure.

6. Many predatory fishes supplement their diets by scavenging, and some specialists rely primarily on recently dead animals or detritus for food. Handling in scavengers requires separation of edible from inedible, either in the mouth or stomach. With few exceptions, fishes are gape limited and cannot attack food larger than they can swallow. Large prey must be dismembered, either by chopping or crushing with jaw and pharyngeal teeth or, in the case of many eel-like fishes, by twisting or spinning prey until swallowable pieces are broken off.

7. Herbivory is more common in tropical than temperate habitats. Herbivores must be able to identify inedible plants and to overcome mechanical and chemical defenses via chewing or chemical digestion; fishes generally lack endosymbiotic bacteria that break down plants. Herbivores tend to have longer guts than carnivores because of the refractory nature of plant material. Territoriality on coral reefs is common among herbivorous fishes and is often overcome by shoaling behavior in competitors.

8. Cost-benefit analyses of foraging behavior have repeatedly indicated that natural selection favors individuals that forage efficiently in terms of food types eaten, feeding locales, and methods of foraging.

SUPPLEMENTAL READING

Curio 1976; Feder and Lauder 1986; Gerking 1994; Ivlev 1961; Keenelyside 1979; Kerfoot and Sih 1987; Krebs and Davies 1991; Noakes et al. 1983; Pitcher 1993; Wootton 1990.

19

Fishes as Prey

The critical tasks facing a prey individual are to avoid detection, evade pursuit, prevent or deflect attack and capture, discourage handling, and ultimately escape from the predator. Just as predators have evolved different adaptations at different phases in the predation cycle, so have prey developed corresponding antipredator tactics. Many of these defenses are structural, involving modified body parts or adaptive use of coloration. Other defenses are behavioral, and many defenses combine actions with structures. Defenses generally function to break the predation cycle, the earlier the better: An attribute that makes it difficult for the predator to find the prey (e.g., camouflage) carries less risk of injury than an attribute that deters the predator during handling (e.g., toxic skin). As an integration of topics covered here and in the previous chapter, we end this chapter with a brief discussion of the trade-off in which fishes find themselves when their own feeding activities expose them to the threat of predation.

AVOIDING DETECTION

Camouflage

The key to lowering the probability of death during the search phase of predation is either to avoid detection by the predator or to detect the predator first. In the former case, some form of camouflage is used; in the latter case, the all-important element of surprise is eliminated. The same principles govern camouflage in both predators and prey. The task is often more difficult for prey because, unlike predators that can blend into the background and sit and wait for prey to blunder by, prey fishes must themselves search for food without being detected by predators.

It is obviously advantageous for a fish not to have to flee from a predator. The best way to accomplish this is to avoid detection by the predator in the first place. Fishes avoid detection by visually hunting predators chiefly in two ways: either 1) by appearing like something unfishlike and therefore unrecognizable or 2) by disappearing entirely. In both categories, deception is accomplished either by **reduction of photocontrast** with the background or by **disruption of the outline** of the fish. A common form of the first tactic, appearing unfishlike through photocontrast reduction, is called **protective resemblance**. Here a fish matches its background so accurately that it appears to blend in with it. Resemblance is achieved through constant or variable coloration and epidermal body growths that match surrounding objects. As most predatory animals are highly sensitive to movement, protective resemblance is usually enhanced by freezing in place.

Examples of remarkable resemblance to background structures among fishes are numerous. Sargassumfishes (Antennariidae) and leafy seadragons (Syngnathidae) hover among seaweed and mimic it. Clingfishes (Gobieso-cidae), shrimp fishes (Centriscidae), and cardinalfishes (Apogonidae) have long black stripes and hover among the spines of sea urchins. The greenish bodies of yellow-spotted gobies (Gobiidae) match both the green stalks and yellow polyps of the antipatherian sea whips on which the fish rest. Agonid sea poachers have rugose bodies covered in brown, orange, black, white, and red that match the sponge- and algae-covered bottom on which they are found. Green pipefishes (Syngnathidae) and wrasses (Labridae) live among green-stemmed sea grasses. Flatfishes are masters of camouflage, changing color and pattern to resemble a variety of bottom types (e.g., Patzner et al. 1992). Predatory fishes also employ protective resemblance and relative immobility as they lie in wait on the bottom for prey (or hide from their own predators). These predators include lizardfishes, goosefishes, stonefishes, scorpionfishes, toadfishes, flatheads, and stargazers.

Crypticity in the above examples usually involves blending into the background. An alternative tactic is to be obvious but to appear as an inedible object. This is achieved by mimicking distasteful or otherwise inedible organisms and objects. Juvenile sweetlips (Haemulidae) and batfish (Ephippidae) have the coloration and unfishlike swimming behavior of flatworms and nudibranchs. Juvenile burrfishes (Tetraodontidae) mimic opisthobranch mollusks. The invertebrates that are mimicked possess skin toxins, are brightly colored, behave conspicuously, and are avoided or rejected by most predatory fishes. Several small fishes mimic small floating sticks, blades of grass, or dead leaves, including juvenile needlefishes, halfbeaks, ephippid batfishes, and lobotid tripletails, and adult nandid leaffishes (Randall and Randall 1960; Randall and Emery 1971). Some special cases involve mimicry of dangerous fishes by otherwise harmless species. The plesiopid *Calloplesiops altivelis* has a dark body with small white spots and a white-ringed ocellus, or eyespot, at the posterior base of its dorsal fin (Fig. 19.1). When frightened, it swims into a crevice but leaves the posterior portion of its body in the open, expanding its dorsal, caudal, and anal fins. In this posture, it appears remarkably similar to the protruding head, eye, and mouth of a common moray eel, *Gymnothorax meleagris* (Muraenidae), and may thus intimidate potential predators (McCosker 1977).

Another means of appearing unfishlike is to be **disruptively colored.** Vertebrates recognize organisms by their outlines and by the gradual shading differences that exist among regions and features within an outline. The body of a disruptively colored fish has areas of contrasting color, usually black and white, that break up the outline

of the fish, making it appear unfishlike. This is one explanation for the bold coloration of some reef fishes such as humbug damselfish, rock beauty angelfish, and some croakers, as well as the vertical barring and dorsal spots of many shallow-water species (sculpins, black-banded sunfish, darters, cichlid angelfishes, jacks, barracudas, tunas) that will be viewed in the dappled light of vegetated areas or in the flickering light created by wavelets passing overhead (McFarland and Loew 1983). Even strikingly colored reef fishes, along with their less-colorful, shallow-water counterparts on reefs and in temperate lakes and kelp beds, assume a dark and light, blotchy coloration when resting at night. This disruptive pattern presumably breaks up their outlines and makes them more difficult to discern at low light levels. Disruptive coloration is also a reasonable explanation for the common occurrence of

FIGURE 19.1. *The tail region of the plesiopid reef fish* Calloplesiops *(bottom) may intimidate predators by mimicking the head of a moray eel (top), with which it occurs.*

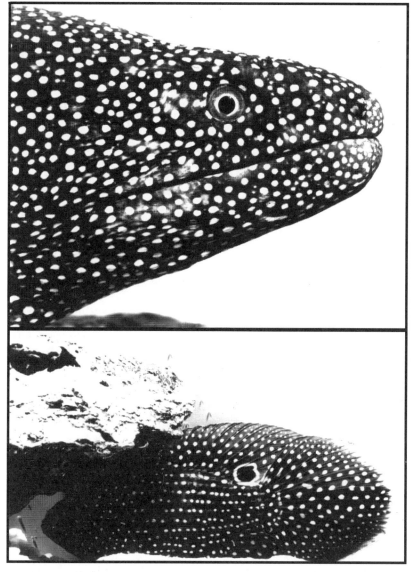

FIGURE 19.2. *Examples and functions of disruptive coloration in fishes. (A) The jacknife fish,* Eques lanceolatus, *may use boldly contrasting, dark and light regions to emphasize those parts of its outline that are not fishlike in appearance, thus momentarily confusing a potential predator. (B) Many lurking predators possess a dark or light interorbital stripe that could disrupt their head outline when viewed head-on, making recognition by prey momentarily difficult. This large (ca. 45 cm) Japanese snook (probably* Lates japonicus) *has a distinctive, light-colored interorbital stripe.*

A
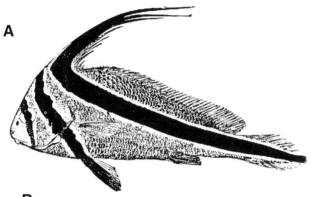

B

(A) From Cott 1940; used with permission. (B) Photo by J. DeVivo.

split-head coloration in many lurking predators (Fig. 19.2B; see Chap. 18, Pursuit).

Invisible Fishes

Protective resemblance and disruptive coloration are camouflage tactics available to all organisms, regardless of habitat (Cott 1940; Edmunds 1974; Lythgoe 1979). Coloration that makes an animal literally disappear from view, rather than blend in with its background, exploits unique features of the distribution of light underwater. In air, all portions of the visible spectrum, from deep blue to deep red (approximately 400 to 700 nm) are well represented. Hence objects of all possible colors can be found in most habitats. In addition, brightness varies substantially and irregularly as a function of sun and viewing angle. The brightest part of the sky may be the horizon during early morning and late afternoon. The ground, vegetation, or objects below an observer or viewed laterally may be as bright or brighter than objects overhead.

In water, however, light has a much more predictable distribution, particularly in open-water situations where the bottom is not visible. Sunlight is refracted at the water's surface, being bent downward even at relatively low sun angles. Hence the brightest light is consistently directly overhead, or **downwelling**. Water molecules act as a powerful filter, both absorbing and scattering light; light attenuation is even stronger if turbidity-causing dissolved or suspended particles are present. Since all light underwater, with the minor exception of bioluminescence, originates as sunlight, objects viewed from above or horizontally will reflect light that has passed through the water filter. **Upwelling** light, which consists of light photons that have passed down and then back up again through the water column, is the weakest component; upwelling light is typically only about 1% as strong as downwelling light. Horizontal **spacelight** is intermediate in strength, but again it consists of light that has first passed vertically down and then horizontally through the water; spacelight is on average about 5% as strong as downwelling light.

The attenuation of light with depth is also very symmetric around the vertical, which means that a diver measuring light at 45° from the vertical will see the same quantity and quality of light whether the meter is pointed at a 45° angle north, south, or in any other direction. Similarly, light measured 90° off vertical (i.e., horizontally) will be identical ahead of a viewer, behind a viewer, and so on. These two physical characteristics of light in water—uniform reduction with depth and uniform attenuation around the vertical—are primary influences on fish coloration, particularly in the context of camouflage tactics that render the fish invisible.

Invisibility can be accomplished via three mechanisms: countershading, silvery sides, and transparency. **Countershaded** fishes grade from dark on top to light on the bottom (Fig. 19.3). The actual color or shade of the fish is less important than 1) the strength of the light that the fish reflects, which differs at different angles from the horizontal; 2) the background light against which the fish will be compared; and 3) the viewing angle of the observer. Countershading is easiest to understand when viewing a fish from above. A fish with a dark dorsum absorbs bright downwelling light, thus presenting a dark target against the dark background of dim upwelling light. However, most fishes are viewed by their predators or prey from the side, and here the intricacies of counter-

FIGURE 19.3. *Countershaded fishes disappear in the water column because their graded coloration reflects light in a manner that makes them match natural background light. (A) A uniformly colored gray fish illuminated primarily from above (= natural lighting) would have a relatively bright dorsum and a relatively dark ventrum. (B) A countershaded fish viewed under unnatural uniform illumination is not camouflaged because it contrasts with the gradient of illumination of background light. (C) In a countershaded fish, the gradual transition from dark dorsum to light ventrum, which is opposite to the actual distribution of background light, has an averaging or canceling effect. In this way the top of the fish is seen as dark against dark, the middle as intermediate brightness against intermediate, and the light belly as light against light, which eliminates contrast between the fish and its background.*

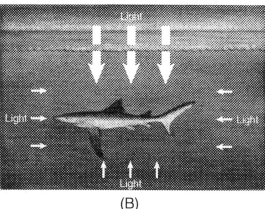
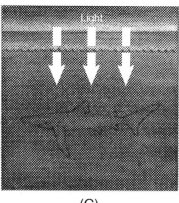

(A)   (B)   (C)

shading work best. When viewed from slightly above the horizontal, the darker dorsolateral surface of the fish absorbs relatively bright downwelling light, creating a dark target that is seen against the darkened background of slightly upwelling light. Similarly, if viewed from slightly below the horizontal, a light-colored ventrolateral surface reflects weak upwelling light, creating a relatively bright target seen against the lighter background of slightly downwelling light. A countershaded fish disappears into the background because the gradation of its color is opposite to the distribution of light in water, which creates a target that is identical to the background. The fish reflects light that is roughly equivalent to the background against which it is seen at all viewing angles: dark against dark, light against light, intermediate against intermediate.

Countershading makes even more sense if one considers how a uniformly colored fish, or one that has reverse countershading, might appear. A uniformly light-colored fish would blend into the brighter background when viewed from positions below the horizontal because it would reflect available upwelling light. However, when viewed from above, it would reflect even more downwelling light and hence become a bright target viewed against the dark background of weak upwelling light. Conversely, a uniformly dark fish would absorb downwelling light and appear dark against the relatively dark upwelling background, but it would also absorb weak upwelling light and appear as a dark object against a bright background when viewed from below.

Reverse countershading would make the fish conspicuous at all angles of viewing: light against dark when viewed from above, and dark against light when viewed from below. Reverse countershading occurs in fishes, but these exceptions prove the rule that countershading is camouflage. Many male fishes, such as sticklebacks, sunfishes, cichlids, and wrasses, take on bright dorsal or dark ventral colors during the breeding season, a time when

conspicuousness helps them attract females and repel territorial intruders. The best proof by exception comes from the reversely-countershaded mochokid upside-down catfishes, which feed on the undersides of leaves and even swim in open water in an upside-down orientation. Colorful reef fishes often superimpose their bright coloration over a countershaded body and vary the dominant color pattern depending on whether they are engaged in social interactions or avoiding predators (see Box 21.1).

It is not immediately obvious that silvery sides make a fish invisible. **Mirror-sided** fishes include some of the world's most abundant and commercially important species, including herrings, anchovies, minnows, salmons, smelts, silversides, mackerels, and tunas. These and other mirror-sided fishes are predominantly open-water, pelagic species that take advantage of the unique light conditions that prevail underwater. To understand how mirror sides work, one must imagine a piece of plate glass suspended in midwater (Fig. 19.4A). The glass is invisible because the background light passes right through it; an observer sees the water-column background, not the glass. Because light attenuates uniformly with depth and is distributed symmetrically around the vertical, a flat mirror suspended underwater achieves the same effect as clear glass. The mirror reflects light of intensity and color that is identical to the light that would be passing through a piece of glass suspended at the same locale, making it seem as if the mirror were not there. Light coming from a 45° angle above the horizontal and reflecting off the mirror and into the eyes of an observer located 45° below horizontal is identical to light that would pass through the mirror if it were clear glass (Fig. 19.4A). An observer comparing the light reflected off the fish with the background light sees no difference; the mirror, or fish, consequently disappears into the background.

Crucial to the function of mirror sides is that the fish maintains a vertical orientation at all times, since any

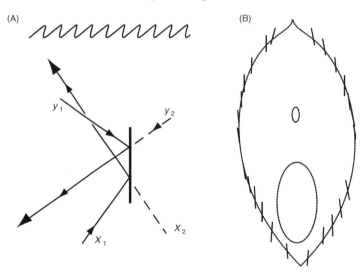

FIGURE 19.4. *The functional morphology of mirror sides in fishes. (A) A clear plate of glass suspended in water is invisible because background light passes directly through it (extensions of dashed lines X_2 and Y_2 to observers' eyes); an observer sees no difference between the glass plate and its background. A mirror suspended in water also disappears, because light reflected off the mirror (solid lines X_1 and Y_1) is identical to the background light that would pass through the object if it were clear (dashed lines). (B) Cross section through the body of a bleak (Alburnus alburnus) to show orientation of the reflective platelets in silvery fishes. Platelets are embedded in the skin and scales and are oriented vertically, even along the curved surfaces of the fish.*

After Denton and Nicol 1965.

deviation from verticality will reflect light that is either brighter or darker than the background. Anyone who has watched a school of bait fishes has witnessed periodic bright flashes as individuals deviate from vertical swimming. Mirror-sided fishes maximize verticality by being laterally compressed. The guanine and hypoxanthine crystals that actually reflect the light are embedded in the scales and skin and are stacked together in platelets. The reflecting crystals are separated by a space equal to about one-quarter of the wavelength of the light usually reflected, the theoretical optimum spacing for achieving reflectivity. Whereas the scales and skin conform to the curvature of the body, the reflecting platelets in the scales have a vertical orientation, even in regions where the body is curved (see Fig. 19.4B) (Denton and Nicol 1962, 1965; Denton 1971).

The third means of achieving invisibility is via relative **transparency**. This is a characteristic of fishes that live in very clear water immediately below the surface where the effects of sun angle are strongest and symmetric distribution around the vertical is weakest. Halfbeaks and needlefishes fall into this category, along with some specialized freshwater forms such as the x-ray tetra, *Pristella maxillaris*, and glass bloodtail, *Prionobrama filligera* (Characidae); African and Asian glass catfishes (Schilbeidae, Siluridae); bagrid catfishes in the genera *Chandramara* and *Pelteobagrus*; a gymnotid knifefish, *Eigenmannia*; and the glass fish, *Chanda* (Chandidae). Larvae and young juveniles of many fishes are pelagic and transparent, their pigmentation de-

veloping along with their later habitat preferences (see Chap. 9, Larvae). The body musculature and to some extent the bones of such fishes are translucent; consequently, these fishes are difficult to see. However, certain structures, most notably the eye, brain, and gonads, apparently cannot function in a transparent state; the gut also often contains prey that is opaque or quickly turns opaque when digested. These pigmented organs often have a silvery coating. The function of this silver film is poorly understood. It might reflect light as do the silver platelets in scales as described above, or the coating could shield delicate structures from harmful ultraviolet radiation that penetrates into clear, shallow water.

Early Detection

Predator-prey interactions often occur over a period of tens of milliseconds. With such rapid reaction times, predators generally require an element of surprise to be successful. The element of surprise can be eliminated if prey detect the predator before the predator detects the prey, or at least if the predator is seen before it gets within striking distance. Early detection can be achieved through the collective vigilance of a shoal: Fish in shoals detect an approaching predator more quickly than do solitary fish (Magurran et al. 1985; Milinski 1993). Many shoals, as well as solitary fish, gain a relative visual advantage over approaching predators by hovering under structure such as floating logs or vegetation, undercut

FIGURE 19.5.   *The advantage to fishes of hovering in shade. On a sunny day in a lake, a shaded observer has a relative visual advantage over sunlit observers. When horizontal visibility is 10 m, a shaded observer can detect sunlit objects 12 m away, which is approximately 1.2 times better visibility than experienced by a sunlit observer viewing a sunlit target (= a 20% advantage over ambient conditions). More significantly, the sunlit observer cannot see an object in the shade until it is 6 m away, which gives the shaded observer a 100% advantage over the sunlit observer. The relative visual advantage decreases on cloudy days, as does the attractiveness of overhead objects.*

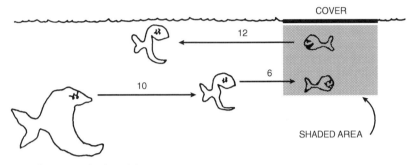

From Helfman 1979; used with permission.

banks or coral ledges, overhanging trees, or docks and bridges. Under appropriate conditions, a shaded fish can see an approaching fish in sunlight as much as 2.5 times farther away than the sunlit fish can see the shaded fish (Fig. 19.5). This phenomenon can easily be experienced by a diver approaching a ledge or dock; objects in the shadow of the ledge are difficult to discern until the observer swims into the shade of the object. The implications for predators lurking or prey hiding in shade are obvious: A predator approaching a shaded prey fish will lose all of the advantages of surprise because the predator will be spotted long before it can see the prey. Prey careless enough to pass near shaded structure are likely to be captured by predators lurking undetected within the darkened region.

The visual advantages and disadvantages around shade arise as a result of physical and physiological phenomena associated with the way in which the vertebrate eye responds to stimuli of different strengths. Vertebrate sensors adapt rapidly to the strongest background stimulus present. Sensitivity of the eye to light is determined by the brightest features of the environment: The eye becomes relatively incapable of seeing dimly lit objects against a bright background. An observer in a darkened room is hidden from the view of people in sunlight outside the room, but the sunlit people are easily seen by the shaded voyeur. In water, this effect is heightened by turbidity. As light enters water, it reflects off particles such as phytoplankton and silt. When looking horizontally underwater on a sunny day, particles are particularly obvious as bright blotches. Particles closest to the observer's eyes are the brightest because the light reflected off them travels the shortest distance. Turbidity therefore creates a bright region adjacent to the observer, a **veil of brightness** that intervenes between the eye and more-distant targets (Lythgoe 1979). The eye quickly adapts to this bright region, and objects farther away in the darker background of spacelight become more difficult to detect.

The eye's adaptation to bright background and veiling brightness combine to explain the relative visual advantage of a fish in shade. The shaded fish's eye is adapted to low illumination levels and hence can see objects in the shade as well as more brightly lit objects outside the shade. Veiling brightness is reduced because the overhead object shades out the sunlight closest to the eye, which eliminates the strongest component of the veil. Observers in the sunlight have their view obscured by reflecting particles and must see with an eye that has been desensitized to dim light because of the surrounding bright conditions.

Hovering in shade is a tactic commonly employed by resting fishes. Many nocturnally active fishes form daytime resting schools in shaded regions by day, including various herrings, silversides, squirrelfishes (Holocentridae), glasseyes (Priacanthidae), snappers, and copper sweepers (Pempheridae). Diurnally active fishes also hover in shade when resting, including various suckers (Catostomidae), centrarchid sunfishes, jacks (Carangidae), and goatfishes (Mullidae). The relative advantage accrues to predators as well as prey; it is not unusual for solitary, lurking predators to hover in shaded areas and strike out at prey that pass by (e.g., trout; pickerel; snook, Centropomidae; largemouth bass; barracuda).

Shoaling and Search

The antipredation benefits of group formation apply to all phases of the predation cycle, including search. Fish in a shoal have a lower probability of being found by a predator than the same fish distributed solitarily (Brock and Riffenberg 1960). Shoals are undoubtedly more conspicuous than solitary fish, so shoals provide no camouflage value; however, they may serve an inhibitory function, because at the edge of visibility, a shoal may be mistaken for a large fish and therefore be avoided by an approaching predator (Pitcher and Parrish 1993). Shoal formation is probably common in prey fishes because of the necessity to move and find food, particularly among herbivorous and planktivorous fishes. Highly evolved protective resemblance is not an option for such fishes; hence group formation is an alternative.

Upon detection of a predator, fish in shoals typically shift to polarized, schooling tactics. Behaviors are emphasized that preserve the integrity of the threatened group (Pitcher and Parrish 1993). Subgroups stream toward the main group (but move as coordinated units, not as individuals), interindividual distances decrease, and movements become synchronized among school members. Heterospecific shoals (those containing more than one species) sort out by species, conspecifics associating with individuals of their own species and size. If few conspecifics exist, members of the minority species may seek shelter rather than wind up as the odd members of a school (e.g., parrotfish, Scaridae; Wolf 1985). In some situations, members of the prey group will actually move away from the shoal, approach the predator, and then return to the shoal. These **predator inspection visits** have been witnessed in mosquitofish and guppies (Poeciliidae), sticklebacks (Gasterosteidae), bluegills (Centrarchidae), and gobies (Gobiidae). The behavior may 1) allow prey fish to assess the identity, motivational state, or other traits of the predator or 2) inform the predator that it has lost the element of surprise and that an attack is unlikely to be successful (Magurran 1986).

Prey can also discourage a searching predator by behaving aggressively. Several prey species actually attack potential predators and drive them from the area. This behavior, best known from bird studies and commonly called **mobbing**, has been documented for individuals or groups of snappers, goatfishes, butterflyfishes, damselfishes, wrasses, and surgeonfishes interacting with predatory moray and snake eels, lizardfish, trumpetfish, scorpionfish, stonefish, flatheads, barracuda, and flatfish; and for bluegill and longear sunfish and largemouth bass interacting with turtles and water snakes. Mobbing fish may contact the head or tail of the predator or may display in front of the predator by swimming in place and erecting dorsal spines and rolling the body. Mobbing reduces the predation rate in an area because mobbed predators take longer to return to an area than predators that are ignored (Ishihara 1987). Predators may leave an area because the physical attacks of the mobbing fish are injurious or because the actions of the mobbers notify other prey individuals to the presence of the predator, which lowers the predator's potential success in the area, analogous to the alarm calls of birds and small mammals (Helfman 1989).

Either inspection or mobbing might explain why some prey converge on or follow predators immediately after a successful attack on the group. This action has been observed in yellow perch attacked by pike, in snappers attacked by jacks, in bluegill attacked by pickerel, in territorial damselfish attacked by several predators, and in planktivorous damselfish attacked by trumpetfish (Nursall 1973; Potts 1980; Dominey 1983; Ishihara 1987; GSH personal observation).

The focus of this discussion has been on avoiding detection by visual predators. However, many nocturnal predators and those that live in turbid habitats rely heavily on acoustic, bioelectrical, and chemical cues to find prey. Little is known about mechanisms for confusing predators or avoiding detection via these channels. In terrestrial environments, both predators and prey possess attributes that function to muffle sounds, such as the serrated feathers on the leading edge of owls' wings, or the pads on the feet of felids. In contrast, sound is difficult to localize underwater. Localization requires some difference in timing or amplitude upon arrival of a sound at members of a pair of receptors. Sound travels relatively rapidly in water (4.5 times faster than in air) and hence arrives on both sides of a fish at very nearly the same time. Predators often know that prey exist in the area but cannot tell in what direction or how far away. Sharks and a few teleosts (e.g., cod, Gadidae; squirrelfishes, Holocentridae; cutlassfishes, Trichiuridae) have been shown to localize sound; this ability might encourage selection for acoustic-dampening structures or behaviors in prey. Comparatively little is known about the behavioral ecology of electrolocalization (see Chap. 21, Electric Communication), whether prey somehow insulate their electrical output or maximize their ionic similarity with their surroundings to avoid detection by passive and active electrolocators. Chemical detection of prey is well known (see Chap. 6). It has been suggested, although not demonstrated, that the mucous cocoons that many parrotfishes secrete from their gills while resting at night could seal off chemical cues used by predators such as moray eels (Winn and Bardach 1959), although tactile predators and external parasites might also be deceived or deterred.

EVADING PURSUIT

Once a predator finds and recognizes prey, pursuit is likely. Antipursuit tactics involve discouraging the predator due to real or feigned unpalatability, shelter seeking, outdistancing or outmaneuvering the predator, or disappearing into the background.

Some fishes possess stout, sharp, sometimes poison-laden spines that can be used in defense against predators. Others possess toxic chemicals in their skin and internal organs. Noxious prey typically advertise their unpalatability with coloration and movement that make them and their defenses quite evident. Such **aposematic** (warning) coloration or behavior is typical of animals that are dangerous or inedible (e.g., many bees, wasps, caterpillars, butterflies, bufonid toads, porcupines, skunks). This deliberate, slow movement aids learning of the warning signal without eliciting attack from predators that are otherwise conditioned to pursue rapidly fleeing prey. By advertising their inedibility, prey short-circuit the predation cycle at an early phase, saving the energetic costs of flight and the possible injury costs of being handled; predators in turn save time and energy and avoid potential injury or possible death.

Many fishes with obvious defenses have bold or bright coloration that could provide a warning to predators. The scalpel-like enlarged scales in the caudal peduncle of surgeonfishes (Acanthuridae) may be surrounded by a bright yellow or orange patch. Lionfish (Pteroidae) have contrasting red, black, and white fins and a stately posture that accentuates their poisonous fin spines. Weevers (Trachinidae) erect their dark-colored, highly venomous dorsal fins when disturbed. Many pufferfishes (Tetraodontidae and other tetraodontiforms), including the famous *fugu* puffers served in exclusive Japanese restaurants, contain powerful tetradotoxins in their skin, liver, and gonads. These fishes have contrasting rather than countershaded body markings and move about in exposed locations during the day. Some plotosid marine catfishes with exceedingly powerful spine venoms have contrasting dark and light coloration and shoal conspicuously during the day, a very uncatfish-like activity time.

Many zooplanktivores take shelter in bottom structure when pursued by predators. Hence many anthiine serranids, fusiliers (Lutjanidae), butterflyfishes (Chaetodontidae), damselfishes (Pomacentridae), wrasses (Labridae), and surgeonfishes (Acanthuridae) will dive toward the coral when disturbed. Morphology and behavior correlate strongly with vulnerability among these fishes. Small species and small members of large species feed closer to the bottom to compensate for their slower swimming speeds. Species that forage farther from the bottom tend to have more fusiform bodies and more deeply forked tails, both characteristics of faster-swimming fishes (see Chap. 6) (Davis and Birdsong 1973; Hobson 1991; Fig. 19.6). Many fishes are permanently associated with holes, cracks,

or tubes in the bottom to which they retreat when threatened (e.g., garden eels, Congridae; jawfishes, Opisthognathidae; tilefishes, Malacanthidae; tubeblennies, Chaenopsidae; hover gobies, Gobiidae). A few, such as the razorfishes (*Hemipteronotus*, Labridae), dive into sand with incredible speed. Fishes associated with macrophyte beds in lakes and on reefs typically use the vegetation as shelter when pursued (e.g., centrarchid sunfishes, cichlids, rabbitfishes, kelpfishes).

Evading a pursuing predator in open water requires superior speed or maneuverability. The odds here favor the predator, since predators are of necessity larger than their prey, and larger fishes can usually swim faster. However, small size does enhance maneuverability. To outdistance a predator, small pelagic prey can take advantage of the drag reduction that can be gained by becoming airborne, a tactic used most effectively by the exocoetid flyingfishes (Davenport 1992, 1994). Flyingfishes commonly double their speed after emerging into the air, accelerating from about 36 km/h in water to as much as 72 km/h while airborne. They typically take off into the wind and travel for 30 seconds and as far as 400 m in a series of up to 12 flights. Multiple flights are interspersed with periods of rapid taxiing when only the beating, elongate lower lobe of the tail fin contacts the water surface. Fish may reach an altitude of 8 m. Refraction at the water surface makes them undetectable by predators except when a calm sea and bright sun create a visible

FIGURE 19.6. *Predator avoidance has shaped body morphology in zooplanktivorous reef fishes. Fishes that feed close to protective structure tend to be more deep-bodied, with square or rounded tails; those that forage higher in the water column are more streamlined, with forked tails. All these fishes dive for the coral when threatened by predators. Streamlining may also facilitate holding position in the stronger currents higher above the reef. Lettering on the photograph of the reef indicates the zones where the five different fishes typically feed: (A)–(C) damselfishes (pomacentrids), (D)* Anthias *(a serranid), (E)* Pterocaesio *(a lutjanid).*

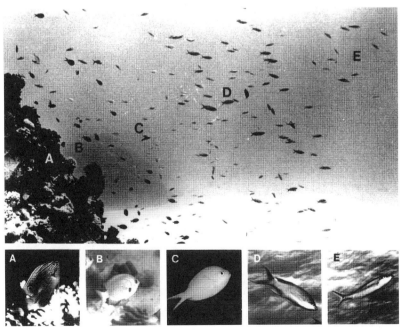

From Hobson 1991; used with permission.

shadow. Hence the flyingfish's reentry point would be largely unpredictable, particularly if it followed a curving flight path. Flyingfishes glide rather than fly, with gliding efficiency depending largely on wing surface area divided by body mass. Flyingfishes come in two-winged varieties with enlarged, cambered (curved) pectoral fins and four-winged varieties, which also have enlarged pelvic fins. These fishes are of necessity small, never exceeding 50 cm in length. A 100-g flyingfish has a total pectoral fin surface area of about 200 cm². For a 100-kg tuna to have proportionally large lifting surfaces, its pectoral fins would each have to be about 100,000 cm² in area.

Other "flying" fishes include the African freshwater butterflyfish, *Pantodon* (Pantodontidae), and the South American freshwater hatchetfishes (Gasteropelecidae), the latter species generating flying forces by vibrating its pectoral fins via pectoral muscles that may account for 25% of its body weight. The greatly enlarged pectoral fins of adult dactylopterid flying gurnards are expanded during cruising over the bottom; these fish have never been observed airborne. Many fishes leap into the air when escaping predators, including minnows (Cyprinidae), halfbeaks (Hemiramphidae), needlefishes (Belonidae), sauries (Scomberesocidae), cyprinodontids, atherinids, bluefish (Pomatomidae), mullets (Mugillidae), and tunas and mackerels (Scombridae). And some predators (dolphinfishes, mackerels, tunas, billfishes) leap into the air in horizontal pursuit of prey, but how these leaps are timed and aimed remains unstudied. Flyingfishes do have one advantage over airborne predators: A flyingfish's cornea is flattened, which gives it the ability to focus in both

water and air. Other fishes have a curved cornea, which only allows focusing in water.

PREVENTING AND DEFLECTING ATTACKS

When actually attacked, a prey fish can make a quick evasive move, employ an active defense, rely on passive structural defenses to deflect the predator, or use a combination of actions. Rapid, fast-start escape movements that lead to maximal acceleration away from the attacking predator are almost universal among fishes, developing early in the ontogeny of larvae and continuing to function into adulthood relatively unchanged (Webb 1986). Fast-start escape movements occur in response to visual, acoustic, tactile, electrical, and water displacement stimuli. The reaction to water displacement is significant, since many predators of small fishes use suction feeding, which means that a larva would experience the water around it suddenly moving toward a predator. Anatomic and behavioral features of this escape response indicate that it operates near the physical and temporal limits of nerve conduction and muscle contraction, emphasizing the importance of predation during early life as well as later in most fishes (see Box 6.1).

Responses of Aggregated Prey

Shoals under attack perform a series of identifiable maneuvers, elements of which have been noted for groups of minnows and shiners (Cyprinidae), yellow perch (Perci-

FIGURE 19.7.   *The graded responses of minnows under attack. Responses increase in intensity as the predator's actions become more threatening. The hierarchy of responses begins at the top (compact) and proceeds clockwise to flash expansion.*

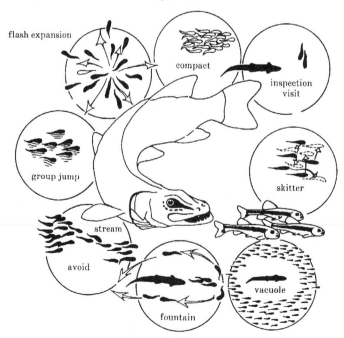

dae), snappers (Lutjanidae), and sand lances (Ammodytidae), among others (Pitcher and Parrish 1993). The tactic employed is dependent on the type of predator and the intensity of its attack (Fig. 19.7). Shoals generally avoid a slowly approaching predator by maintaining about a 5 to 15 prey body length space between the predator and the group. Often predators will swim slowly through a prey school, at which time the school separates ahead of the predator and then closes back together behind it, basically creating a prey-free vacuole around the predator. The function of the predator's slow maneuver is unclear; possibly the predator is testing for injured prey or hoping to catch an inattentive individual off guard. In more concerted attacks, prey expand rapidly out from the point of attack, scattering in different directions and fleeing the scene or seeking refuge in a nearby structure.

Group membership may reduce the statistical likelihood of an individual being attacked by a predator, producing a **dilution** or **attack abatement effect**. If a predator only consumes one or a few prey once it encounters a group, then the likelihood of any one individual being eaten decreases as group size increases. Additional benefits of grouping include passage of information about a predator to individuals unexposed to the predator. This is termed the **Trafalgar effect**, after the Battle of Trafalgar, when Admiral Nelson sent information through his fleet using flag codes, informing ships on the far side of the fleet about the enemy's actions (Pitcher and Parrish 1993).

The most widely demonstrated cause of a decrease in predator success is the **confusion effect**. In studies with largemouth bass feeding on minnows, pike feeding on perch, sticklebacks feeding on water fleas, and jacks feeding on anchovies (Godin 1986; Landeau and Terborgh 1986), it has been shown that predators catch fewer prey from large than from small schools. Success declines because the predator switches targets as it moves through the school, apparently confused by the number of multiple, edible objects moving across its field of vision.

Specific behaviors seemingly function to increase confusion. In **skittering**, as displayed by minnows, an individual accelerates rapidly, rises in the water column, and then quickly rejoins the group. **Protean** behavior, seen in anchovies and silversides (Engraulidae, Atherinidae), involves quick, uncoordinated up-and-down movements by several adjacent individuals just prior to resumption of polarized schooling. **Roll-and-flash**, often seen in herring (Clupeidae) schools, occurs when an individual rotates on its long body axis and reflects bright sunlight; it then returns to a normal upright position. The eye is quickly drawn to the point of the flash, but the fish seemingly disappears when upright orientation is resumed (a similar distractive function has been proposed for ink squirting in octopods and squid under attack).

The physiological basis of the confusion effect is poorly understood, although it may relate to an information overload problem whereby the predator's capacity for processing information is exhausted by the sheer number of objects in the visual field (Milinski 1990). Both invertebrate and vertebrate predators are subject to the confusion effect, as anyone who has ever attempted to net individual fish from a school in a large aquarium can attest. Not surprisingly, prey fish choose to join the larger of two schools when given an opportunity, and the speed with which this decision is made increases when predators are present (Hager and Helfman 1991).

DISCOURAGING CAPTURE AND HANDLING

Capture refers to initial ingestion of the prey; in fishes this involves taking the prey into the mouth. Defenses against capture exploit the gape limitation that constrains most predators to feeding on prey small enough to be swallowed whole (see Box 18.2). Hence many anticapture adaptations involve permanent or temporary increases in prey body size and elaboration of body armor that make it difficult to 1) bring prey into the mouth, 2) close the mouth once the prey are there, or 3) swallow captured prey. Many fishes have dorsal and anal fins that appear out of proportion to their bodies, or bodies with exaggerated depth, such as citharinids, silver dollars (Characidae), veliferids, snipefishes (Macrorhamphosidae), crappies (Centrarchidae), fanfishes (Bramidae), manefish (Caristiidae), butterflyfishes (Chaetodontidae), tangs (Acanthuridae), and Moorish idols (Zanclidae). Greatly elongate dorsal, pelvic, and anal fins in many larval fishes (e.g., ribbonfishes, sea basses) may also function to reduce predation. Spiny pufferfishes increase their body depth and volume by inflating their stomachs with water and erecting their spines; the spines are modified scales with three-pronged, interlocking bases embedded in the puffer's skin that prevent their depression (Box 19.1). The evolutionary development of spines that defines the Acanthopterygii accomplishes a similar defense. Predators focus their attacks on the center of mass of the body, which is often where the prey's body is deepest; this depth is increased by erectable spines. A temporary increase in depth can be achieved by erecting the dorsal, pelvic, and anal fins with their stiff armament. A bluegill sunfish increases its body depth by about 40% by erecting its fins, making it a larger and hence less desirable food item for most predators. Erecting fins as a predator approaches may be a way of discouraging the predator before it attacks.

The effectiveness of erected spines in preventing passage of prey toward the predator's throat can be enhanced by additional structures. Sticklebacks lodge themselves in the mouths of predators such as pike by locking their dorsal and pelvic spines, forcing the predator to break the spines before swallowing can occur. Triggerfishes link the first two dorsal spines to prevent depression of the dorsal fin. Triggerfishes can wedge themselves into a crevice or a predator's mouth and lock the spines; the second spine (the trigger) has to be pushed posteriorly to depress the dorsal fin (see Chap. 15, Order Tetraodontiformes). Indo-Pacific rabbitfishes (Siganidae) possess several unusual spine adaptations. The first dorsal spine points forward (retrorse) instead of up, which could inhibit headfirst swallowing by predators, and each pelvic fin has hard spines at the leading and trailing edge of the fin. These fish are difficult to handle without getting punctured, and

| BOX 19.1 | *The Functional Morphology of Pufferfish Inflation* |

Camouflage, due to countershading, disruptive coloration, or invisibility, functions during the search and detection phases of predatory behavior. Many fishes have placed less emphasis on crypticity and instead discourage attacks using traits that make them less desirable as prey. The scales and spines of squirrelfishes, lionfishes, rabbitfishes, and surgeonfishes, among many others, provide such mechanical defenses. Antihandling adaptations reach an extreme of specialization in the porcupinefish, balloonfish, and spiny pufferfishes (Diodontidae), where jaw musculature and bones, skin, stomach, scales, body musculature, and peritoneal cavity have all been modified to contribute to a coordinated defensive tactic in which a threatened fish turns itself into an inedible, spiny sphere (Brainerd 1994; Wainwright et al. 1995).

Balloonfish, in response to attacks by fishes or birds, increase their volume threefold by pumping water or air into their stomach (Fig. 19.8). The pumping mechanism involves rhythmic mouth cavity expansions and compressions that are driven by specialized muscles and unique couplings of the mouth floor, branchiostegals, and pectoral girdle. Pumping is rapid, taking about 15 seconds and 40 separate puffs to complete. Stomach pressure and the force exerted by each puff increase as the stomach fills, analogous to a human inflating a balloon. The stomach, which has lost its digestive function, rests in a folded state and then expands 50- to 100-fold when filled with water. When a balloonfish is fully inflated, the unique peritoneal cavity, now filled by the expanded stomach, extends anteriorly to the tip of the dentary, dorsally around the body musculature, and posteriorly around the dorsal and anal fins. In related tetraodontiforms that do not puff up, such as triggerfishes, the peritoneal cavity is limited to the anterior ventral quarter of the body.

The skin of the balloonfish plays an integral role during inflation. Both the dermis and epidermis contain numerous internal microfolds that make a puffer's skin highly stretchable. The skin can stretch about eight times more than other types of fish skin. When the skin is finally stretched to the point of stiffening, this rigidity aids in the erection and anchoring of the spines. The spines themselves, which are modified scales, are among the longest and most elaborate of all teleost scales. Collagen fibers in the skin attach to a forward-pointing process at the base of each spine. As the skin is stretched, a posterior-pulling force is applied to the forward-pointing process, levering the spine into an erect position.

In most fishes, the skin is tightly bound to the body musculature by a cross helix of collagen fibers and may serve as an external tendon that contributes to locomotion (see Chap. 8, Locomotion: Movement and Shape). In conjunction with high stretchability, the skin of the balloonfish lacks such tight attachment to the underlying musculature, and the special orientation of the collagen fibers in the dermis prevents the transmission of force from muscles to skin. Locomotory capability has been traded off in favor of inflation; puffers consequently swim via a unique combination of pectoral and median fin undulations referred to as diodontiform swimming, which is a modified form of tetraodontiform swimming (see Table 8.1). In an inflated balloonfish, which is more spherical than any other vertebrate, a "stiff skin surrounding a ball of incompressible water provides a rigid framework for the support of the spines," creating an object too large and too difficult for most predators to swallow (Brainerd 1994, 17a).

the spines are covered with a toxic slime that causes painful wounds, at least in humans.

Dermal and epidermal defenses play an important role in resisting capture and in complicating handling. Many fishes exude mucus upon capture. This slime may make the fish slippery and harder to hold (hagfishes, anguillid eels), but in many the slime or other skin secretions contain distasteful substances that cause rejection by the predator (some moray eels, Muraenidae; marine catfishes, Ariidae; toadfishes, Batrachoididae; clingfishes, Gobiesocidae; soapfishes, Grammistidae; gobies, Gobiidae; trunkfishes, Ostraciidae) (Hori et al. 1979; Smith 1992). In certain soleid flatfishes, toxic steroid aminoglycosides secreted from glands at the base of the dorsal and anal fins have a repellent effect on predators such as sharks (Primor et al. 1978; Tachibana et al. 1984). External defenses in some species include a thickened or hardened dermis (e.g., the ganoid scales of lepisosteid gars and polypterid bichirs, the carapace made up of scale plates in ostraciid boxfishes, and the toughened skin of balistid leatherjackets). Populations of threespine sticklebacks, *Gasterosteus aculeatus* (Gasterosteidae), that co-occur with predators have more lateral bony scutes and longer dorsal spines than do comparatively predator-free populations (Reim-

FIGURE 19.8.   *Balloonfish inflate themselves with water in response to being handled by potential predators. They undergo a threefold increase in volume, which turns them into a sphere with projecting spines. Extremities that might offer a predator a grasping point, such as the caudal (C), pectoral (P), and other fins, sit largely within the protective framework of the spines when the fish is inflated. Inflation occurs as water is pumped into the stomach, which expands up to 100-fold to fill an unusually large peritoneal space. The spines are embedded in a highly derived, stretchable skin.*

From Brainerd 1994; used with permission.

chen 1983; FitzGerald and Wootton 1994). Many shoaling species have easily dislodged, deciduous scales, which may allow them to slip away from predators, analogous to the easily shed wing scales of moths and butterflies.

A special case of a handling-induced, antipredator response in shoaling fishes and a few other species is the production of and reaction to alarm chemicals. Alarm reactions are best known in ostariophysans, where they were first discovered (see Chap. 14, Superorder Ostariophysi). Substances and reactions also occur in some sculpins, dart-

ers, and gobies, and are suspected in a few galaxiids, killifishes, livebearers, and silversides (Smith 1986, 1992). The alarm substance is released when the skin of a fish is broken, such as during a predatory attack. Reactions depend on the species and the situation. Shoaling fishes often react by schooling tightly and moving away from the area where alarm substance is released. Some solitary cyprinids sink to the bottom, whereas benthic species (gudgeon, Cyprinidae; loach, Cobitidae; suckers, Catostomidae) freeze in place, utilizing their cryptic coloration to avoid detection.

When alarm substances are in the water, overhead predators such as birds cause shiners (Cyprinidae) to hide in vegetation, whereas fish predators elicit a strong schooling response. The alarm reaction spreads as additional individuals detect the alarm substance or as they react visually to schoolmates. Many fishes show an alarm reaction to water in which predators have been kept, indicating again the probable importance of chemical interactions among fishes (see Chap. 21, Chemoreception).

Some fishes use nonchemical channels to transmit alarm signals. Visual signals induced by predators include increased fin-flicking rates in schooling characins and in parental cichlids guarding young, head bobbing by gobies, and inspection visits and mobbing as discussed above. Many fishes emit distress sounds when held, prodded, or speared (e.g., catfishes, grunts, drums, triggerfishes). At least three families of fishes (cods, squirrelfishes, groupers) produce distinctive sounds when confronted with predators. Squirrelfishes produce a staccato sound that causes conspecifics to take refuge or inspect the predator (Myrberg 1981; Smith 1992).

The adaptive significance of responding to an alarm signal is obvious: It is advantageous to know that a predator is active in an area and to take appropriate action. Coordinated flight behavior within a school lessens a predator's chance of additional success. Evolution of an ability to generate an alarm substance or signal is more problematic. Unless there is a high probability that schoolmates are close genetic relatives (e.g., kin selection), little benefit accrues to an altruistic, injured individual that produces an alarm substance and is consequently deserted by its schoolmates. One possible advantage to producing a rapidly diffusing alarm chemical is that it might attract other predators, including predators larger than the one that caused the initial injury. Such larger predators could frighten the initial attacker into leaving the area, thus allowing the injured prey to escape (Mathis et al. 1995; see also Chap. 21, Intraspecific Communication: Sound).

FORAGING THEORY: BALANCING FORAGING AGAINST PREDATORY THREAT

It is important to realize that most predators are also prey and that foraging decisions often must be made in the context of danger to the feeder. These conflicting demands create a **foraging-predation risk trade-off**. What evidence exists to suggest that foraging fishes take into account risk to their own survival when choosing among food types, locales, and methods? Sticklebacks feed more slowly and gobies eat less in the presence of predators, even when the predators are behind a transparent partition (Magnhagen 1988; Milinski 1993). Juvenile coho salmon are less willing to travel long distances to intercept floating prey when presented with photographs of large predatory trout (Dill and Fraser 1984). Juvenile black surfperch (Embiotocidae) shift from feeding in low-growing, exposed algae when predatory kelp bass (Serranidae) are absent to feeding in tall, bushy algae when predators

are present (Holbrook and Schmitt 1988a). Herbivorous minnows and loricariid catfishes abandon shallow areas of high algal productivity for deeper, less-productive areas to avoid both predatory birds and fishes (Power 1987). Juvenile bluegill grow fastest when feeding in open water on zooplankton. When largemouth bass are released in the open-water areas, smaller, more vulnerable bluegill move into vegetated areas where they have lower mortality rates from predation but also grow more slowly because of lower food intake. Bluegill too large to be swallowed by the bass remain in the open-water areas (Werner et al. 1983).

The balance between foraging and predation risk is shown by studies that vary the degree of threat and the strength of the reward. Minnows and surfperch will risk feeding in patches where predators are present if food densities are high; otherwise they avoid patches with predators (Gilliam and Fraser 1987; Holbrook and Schmitt 1988b). Additional factors can affect the decision process. Hunger or parasite loads cause sticklebacks to resume feeding sooner after exposure to predators and also cause fishes to feed closer to potential predators (Godin and Sproul 1988; Milinski 1993). When offered food in the presence of a cichlid predator, female guppies will accept more risk for more reward, whereas male guppies avoid the predator regardless of food availability, implying that reproductive output is more dependent on food intake in females than in males (see also Chaps. 20, 23) (Abrahams and Dill 1989).

Finally, prey fish do not feed or avoid predators in an all-or-none fashion. They appear to weigh the potential threat from a predator and to take action appropriate to the degree of threat. This **threat sensitivity** is evident in individuals and groups. Territorial damselfishes respond more strongly to larger than to smaller predators and to predators in a strike pose than to searching predators. The strength of the flight behavior elicited also increases as the predator draws closer (Helfman 1989). Minnows in shoals employ a graded series of escape responses that increase in strength and effectiveness as a pike escalates its attack (see Fig. 19.7), and sticklebacks spend progressively less time foraging as predatory trout increase in number (Fraser and Huntingford 1986; Magurran and Pitcher 1987). Threat sensitivity makes good Darwinian sense. A prey individual that is capable of assessing just how threatening a predator is and that is able to devote an appropriate amount of time and energy avoiding the predator will have more time for other fitness-influencing activities (feeding, breeding, defending a territory or young) than will an individual that flees or hides any time a predator arrives in the area.

SUMMARY

1. To escape predators, prey must avoid detection, evade pursuit, prevent or deflect attack and capture, and discourage handling. Different behaviors and structures achieve these different functions. Detection is avoided by various types of camouflage that make a

prey appear unfishlike or inedible, blend into its background, or just disappear. Tactics employed include protective resemblance, mimicry, disruptive coloration, countershading, mirror sides, and transparency. Shoaling may increase the ability of prey to detect and assess approaching predators, as does hovering under shade-producing objects.

2. To discourage pursuit, a prey fish can demonstrate or feign unpalatability, outdistance or outmaneuver the predator, or seek shelter in the bottom, in vegetation, or in the water column. Some fishes with toxic skin or spines advertise these structures via color or behavior (surgeonfishes, lionfishes, weeverfishes, pufferfishes). Many fishes dive into holes in coral or rocks or into the sand when pursued. Open-water species must outdistance a predator; flyingfishes leap from the water and glide at twice their swimming speed.

3. An attack can be prevented by quick evasive action or by active or passive defense. Rapid, fast-start moves are almost universal among fishes from early ontogeny on. Schooling fishes undergo a series of maneuvers when attacked; aggregating is particularly effective during attack because multiple targets presented by the prey confuse the predator.

4. Capture is commonly discouraged by exploiting the gape limitation of most predators. Prey fishes increase their body size by erecting their spines. Balloonfishes combine an inflatable stomach, stretchable skin, and scales modified into long, stout spines to create an inedible object. Easily displaced scales may allow prey to slip out of the mouth of a predator. Slime and toxic or distasteful skin secretions also discourage capture (moray eels, toadfishes, soapfishes, soleid soles, trunkfishes). Specialized skin cells in ostariophysan and a few other fishes secrete an alarm substance that warns schoolmates of an attack. Alarm substances and alarm calls may also attract secondary predators, which could frighten the first predator into releasing a victim.

5. Foraging puts fishes at risk of becoming prey themselves. Thus many fishes must trade off their own foraging against the risk of predation. Experimental studies have shown that fishes often give up foraging opportunities when threatened by predators but will risk greater threats when exceptionally hungry or when the rewards are high enough. Prey are sensitive to the degree of threat presented by a predator and are able to take evasive action proportional to the degree of threat.

SUPPLEMENTARY READING

Cott 1940; Curio 1976; Edmunds 1974; Feder and Lauder 1986; Keenleyside 1979; Lythgoe 1979; Pitcher 1993; Wootton 1990.

20 Fishes as Social Animals: Reproduction

Evolutionary success is determined by an individual's ability to place genes in future generations, relative to the success of conspecifics. To transmit genes, individuals must mate with conspecifics. However, many fish species are relatively solitary as adults. During breeding seasons, fish must overcome individualistic habits and seek out potential mating partners. Suitable spawning habitats and substrates must be found or even modified into nests, activities of males and females must be synchronized, and tactics for avoiding hybridization (**species-isolating mechanisms**) must be employed. Aggregations of reproductively active individuals create potential competition for spawning sites and partners, eliciting territorial and mate choice behavior. Courting and spawning may distract the attention of participants and thus make them more vulnerable to predators. After spawning, many species engage in varying degrees of parental care. All these activities and characteristics constitute the diverse mating systems of fishes. Our emphasis here will be on the diversity of mating systems, with focus on patterns and adaptations where they can be identified.

REPRODUCTIVE PATTERNS AMONG FISHES

Important components of a breeding system include frequency of mating, number of partners, and gender role of average individuals (Table 20.1). Fishes show greater diversity in these traits than do other vertebrates. Most fishes follow the mammalian and avian norm of remaining one gender throughout adult life, but many fish species change sex, and some are parthenogenetic, producing young from unfertilized eggs (see Chap. 10, Determination, Differentiation, and Maturation; Chap. 23, Life Histories and Reproductive Ecology). Some fishes retain a single mate, perhaps for life, others mate promiscuously, and a few are haremic. And in some fishes, a single breeding system may not characterize the entire species.

Lifetime Reproductive Opportunities

Most fishes are **iteroparous**, spawning more than once during their lives (e.g., sharks, lungfishes, sturgeons, gars, tarpons, minnows, trouts, codfishes, sea basses). However, some well-known species are **semelparous**, spawning one time and dying. Semelparity characterizes most salmon of the genus *Oncorhynchus* (e.g., pink, chum, chinook, coho, and sockeye salmon). These fishes hatch in freshwater, migrate to the sea for a period of 1 to 4 years, and then return to their natal (birth) stream, where they spawn and die. Although the life cycle of females appears to be relatively fixed across a species, intrapopulational differences

in male cycles are common (see below under Alternative Mating Systems and Tactics).

Other semelparous fishes include lampreys, anguillid (freshwater) eels, and the salmoniform southern smelts (Retropinnidae) and galaxiids of Australia and New Zealand (McDowall 1987). American shad (*Alosa sapidissima*) are semelparous in southern locations (30° to 33°N), largely iteroparous at northern latitudes (41° to 47°N), and variably iteroparous at intermediate latitudes (Leggett and Carscadden 1978). With the exception of such annual fishes as aplocheilid rivulines, semelparous fishes are diadromous or include at least one major migratory phase in their life cycle. Anguillid eels show sex-based differences in tactics within an overall semelparous strategy. Males often mature rapidly (ca. 3 to 6 years) and at a uniformly small size (30 to 45 cm), regardless of locale, whereas females are consistently longer (35 to 100 cm) and may mature quickly (4 to 13 years) at low latitudes or slowly (6 to 43 years) at high latitudes. Slow-maturing females grow larger and produce more eggs than smaller, faster-maturing females. In American eels, males have a relatively restricted geographic distribution, occurring primarily in estuaries of the southeastern United States, whereas females are found throughout the North American range of the species and in all habitats. As far as is known, all members of an anguillid species migrate to the same oceanic region to spawn and die (Sargasso Sea in the western Atlantic for American and European eels, the Philippine Sea for Japanese eels; see Chap. 22, Representative Life Histories of Migratory Fishes) (Helfman et al. 1987; Jessop 1987).

Table 20.1. A summary of components of breeding systems in fishes, with representative taxa. Accurate categorization is often hampered by the difficulty of following individual fish over extended periods in the wild. Although families are listed for some components, exceptions are common within a family.

I. Number of breeding opportunities

    A. Semelparous (spawn once and die): lampreys, river eels, some South American knifefishes, Pacific salmons, capelin

    B. Iteroparous (multiple spawnings)

        1. A single, extended spawning season: most annuals (rivulines)

        2. Multiple spawning seasons: most species (elasmobranchs, lungfishes, perciforms)

II. Mating system

    A. Promiscuous (both sexes with multiple partners during breeding season): herrings, livebearers, sticklebacks, greenlings, epinepheline sea basses, damselfishes, wrasses, surgeonfishes

    B. Polygamous

        1. Polygyny (male has multiple partners each breeding season): sculpins, sunfishes, darters, most cichlids; (haremic): serranine sea basses, angelfishes, hawkfishes, humbug damselfishes, wrasses, parrotfishes, surgeonfishes, trunkfishes, triggerfishes

        2. Polyandry (female has multiple partners each breeding season): anemonefishes (in some circumstances)

        3. Monogamy (mating partners remain together for extended period or the same pair reforms to spawn repeatedly): bullheads, some pipefishes, *Serranus,* hamlets, jawfishes, damselfishes, tilefishes, butterflyfishes, hawkfishes, cichlids, blennies

III. Gender system

    A. Gonochoristic (sex fixed at maturation): most species (e.g., elasmobranchs, lungfishes, sturgeons, bichirs, bonytongues, clupeiforms, cypriniforms, salmoniforms, beryciforms, scombroids)

    B. Hermaphroditic (sex may change after maturation)

        1. Simultaneous (both sexes in one individual): *Rivulus*, hamlets, *Serranus*

        2. Sequential (individual is first one sex and then changes to the other)

            a) Protandrous (male first, change to female): anemonefishes, some moray eels, *Lates calcarifer* (Centropomidae)

            b) Protogynous (female first, change to male): *Anthias*, humbug damselfishes, angelfishes, wrasses, parrotfishes, gobies

    C. Parthenogenetic (egg development occurs without fertilization)

        1. Gynogenetic: *Poeciliopsis, Poecilia formosa* (no male contribution, only egg activation)

        2. Hybridogenetic: *Poeciliopsis* (male contribution discarded each generation)

IV. Secondary sexual characteristics (traits not associated with fertilization or parental care)

    A. Monomorphic (no distinguishable external difference between sexes): most species (clupeiforms, carp, most catfishes, frogfishes, mullets, snappers, butterflyfishes)

    B. Sexually dimorphic

        1. Permanently dimorphic (sexes usually distinguishable in mature individuals): *Poecilia,* anthiine sea basses, dolphinfishes, *Cichlasoma*, some angelfishes, wrasses, parrotfishes, chaenopsid blennies, dragonets, Siamese fighting fishes

        2. Seasonally dimorphic (including color change only during spawning act): many cypriniforms, Pacific salmons, sticklebacks, lionfishes, epinepheline sea basses, some cardinalfishes (female), darters, some angelfishes, damselfishes, wrasses, blennies, surgeonfishes, porcupinefishes (female)

        3. Polymorphic (either sex has more than one form): precocial and adult male salmon; 10 and 20 males in wrasses, parrotfishes

V. Spawning site preparation (see Table 20.2)

    A. No preparation: most species of broadcast spawners (e.g., herring)

    B. Site prepared and defended: sticklebacks, damselfishes, sunfishes, cichlids, blennies, gobies

VI. Place of fertilization

    A. External: most species (lampreys, lungfishes, bowfin, tarpons, eels, herrings, minnows, characins, salmons, pickerels, codfishes, anglerfishes, sunfishes, marlins, flatfishes, pufferfishes, porcupinefishes)

    B. Internal: elasmobranchs, coelacanth, livebearers, freshwater halfbeaks, scorpionfishes, surfperches, eelpouts, clinids

    C. Buccal (in the mouth): some cichlids

VII. Parental care (see Table 20.2)

    A. No parental care: most species

    B. Male parental care: sea catfishes, sticklebacks, pipefishes, greenlings

    C. Female parental care

        1. Oviparity with postspawning care: *Oreochromis*

        2. Ovoviviparity without postspawning care: rockfishes (*Sebastes*)

        3. Viviparity without postspawning care: elasmobranchs, *Poecilia*, surfperches

    D. Biparental care: bullheads, discus, *Cichlasoma*

    E. Juvenile helpers: some African cichlids (*Lamprologus*)

Modified from Wootton 1990; used with permission.

Mating Systems

Mating systems are defined by the number of mating partners an individual has during a breeding season (see Table 20.1). The three most common categories are promiscuous, polygamous, and monogamous. **Promiscuous** breeders are those in which little or no obvious mate choice occurs and in which both males and females spawn with multiple partners, either at one time or over a short period. Such spawning has been documented for the Baltic herring (Clupeidae), guppies (Poeciliidae), Nassau groupers (Serranidae), humbug damselfish colonies (Pomacentridae), cichlids, and the Creole wrasse (Labridae) (Thresher 1984; Barlow 1991; Turner 1994).

Polygamy, in which only one sex has multiple partners, takes multiple forms. **Polyandry**, in which one female mates with several males (and presumably not vice versa) is relatively uncommon, so far documented only in an anemonefish (Pomacentridae) (Moyer and Sawyers 1973). Polyandry might also be descriptive of female ceratioid anglerfishes that have more than one male attached (see Chap. 17, The Deep Sea). **Polygyny** is the most common form, involving males as the polygamous sex. Territorial males that care for eggs and young are frequently visited by several females, as in sculpins, sunfishes, darters, damselfishes, and two cichlids. Polygyny can also develop into harem formation, in which a male has exclusive breeding rights to a number of females that he may guard. Harems have been observed in numerous cichlids and in several coral reef families (e.g., tilefishes, damselfishes, wrasses, parrotfishes, surgeonfishes, triggerfishes).

Many polygynous animals form leks, which are traditional areas where several males congregate for the sole purpose of displaying to females (Emlen and Oring 1977). Females are often attracted to a male in response to his central position within the lekking ground or to the vigor of his display and bright plumage. Lekking is common in birds and mammals, in which only the female provides parental care. Some African cichlids come closest to forming true leks. Large numbers (ca. 50,000) of male *Cyrtocara eucinostomus* congregate along a shallow 4-km-long shelf in Lake Malawi and build sand nests and display to passing females each morning. Females spawn and then mouth brood eggs elsewhere. The male aggregations break up each afternoon, when fish feed (McKaye 1983, 1991). Some fishes form leklike aggregations of males (e.g., damselfishes, wrasses, parrotfishes, surgeonfishes), but the display ground is also an appropriate place for launching or caring for eggs, which stretches the definition of lekking (Loiselle and Barlow 1978; Moyer and Yogo 1982). In a unique variation on leklike behavior, female triggerfish (*Odonus niger*, Balistidae) form a communal display ground for 1 day before spawning, after which they all mate with a single, nearby male (Fricke 1980).

In **monogamous** systems, fish live in pairs that stay together and mate, or mate with the same individual repeatedly and exclusively, regardless of pairing at nonmating times. Strongly pairing species include North American freshwater catfishes, many butterflyfishes, most substrate-guarding and some mouth-brooding cichlids, and anemonefishes; in the butterflyfishes, pairs may remain together for several years and probably mate for life (Reese 1975). Monogamous coral reef fishes commonly spawn with the same partner on a daily basis over an extended period without ensuing care of young, whereas freshwater species such as cichlids spawn over a limited time and then both parents typically care for the young. Monogamy is also known in freshwater bonytongues, bagrid and airsac catfishes, and snakeheads, and among marine pipefishes, sea horses, jawfishes, damselfishes, blennies, gobies, surgeonfishes, triggerfishes, filefishes, pufferfishes, and hermaphroditic hamlets (Barlow 1984, 1986; Thresher 1984; Turner 1986).

Gender Roles in Fishes

Although the vast majority of fishes are gonochoristic, with sex determined at an early age and remaining fixed as male or female, a significant number of fishes can function as males and/or females simultaneously or sequentially (see also Chap. 10, Determination, Differentiation, and Maturation). The environmental correlates and evolutionary causes of sex change in fishes have been the subject of considerable study and speculation (Box 20.1).

Sex reversal has evolved, apparently independently, in at least 23 families belonging to seven teleostean orders, including moray eels (Anguilliformes), loaches (Cypriniformes), lightfishes (Stomiiformes), killifishes (Atheriniformes), swamp eels (Synbranchiformes), flatheads (Scorpaeniformes), and 14 perciform families (snooks, sea basses, soapfishes, emperors, rovers, porgies, threadfins, angelfishes, bandfishes, cichlids, damselfishes, wrasses, parrotfishes, and gobies). Sex changers can be either 1) **simultaneous hermaphrodites**, capable of releasing viable eggs or sperm during the same spawning, or 2) **sequential hermaphrodites**, functioning as males during one life phase, and as females during another. Among sequential hermaphrodites, **protandrous** fishes develop first as males and then later change to females, whereas **protogynous** fishes mature first as females and then later become males. Variations on these patterns exist, such as protogynous populations with some males that develop directly from juveniles, or simultaneous hermaphrodites that lose the ability to function as one sex (Smith 1975; Warner 1978; Sadovy and Shapiro 1987).

Protogyny is by far the most common form of hermaphroditism. In a classic study, Robertson (1972) found that the Indo-Pacific cleaner wrasse, *Labroides dimidiatus*, formed harems of one large male and up to 10 females. Breeding access to the male was determined by a behavioral dominance hierarchy, or peck order, with the largest female dominating the next smallest and so on. If the top (alpha) female was removed, the next largest female assumed her role, and everyone else moved up a step. If the male was removed, the alpha female began courting females within an hour and developed functional testes within 2 weeks (see also Kuwamura 1984).

Protogyny in wrasses can take other forms. In the Caribbean bluehead wrasse, *Thalassoma bifasciatum*, fish usually begin life as predominantly yellow females or

---

BOX 20.1 *"When the Going Gets Tough, the Tough Change Sex"*[1]: *The Evolution of Sex Change in Fishes*

The topic of sex or gender change and its relationship to hermaphroditism has sparked a great deal of debate among biologists. Two questions dominate the discussion of sex-changing fishes (and of sex allocation in organisms in general). The questions are "Why change sex?" and "When to change?" Answers lie primarily in the ecologies of individual species, greatly influenced by the relative reproductive success of males versus females at different sizes. Animals change gender when, at a given size, the reproductive success of one gender becomes higher than if the individual remained the other gender (Ghiselin 1969; Warner 1975). The value of changing gender is reduced by the costs of changing, such as lost reproductive opportunities while undergoing the change and metabolic costs of altering gonads.

This **size advantage model** assumes indeterminate growth and increased fecundity with increasing size. Reproductive success in females is generally limited by gamete production, whereas males are limited by the number of mates they can acquire (Bateman's principle). Males, including small males, generally produce a surplus of sperm, most of which never encounter an egg. In contrast, egg production increases with growth in females, and each egg is likely to be fertilized. These circumstances would dictate that sex changers be protandrous: Small females produce very few eggs, but small males can fertilize many eggs. Such conditions would select for fish that began life as small but functional males and changed to female when they were large enough to produce more eggs than a small male could fertilize. For example, in pair-spawning, monogamous anemonefish, lifetime egg production of the pair is maximized by having the larger fish a female.

However, male fish often compete for females (see below), and the outcome of such competition is frequently determined by body size, with larger males winning. Hence one large, behaviorally dominant male can monopolize many females and fertilize their eggs, as in the bluehead wrasse and *Anthias* examples mentioned earlier. Under these circumstances, the greatest advantage accrues to the largest males, and the tactic to follow

is to be a female first (since small males have such limited competitive and therefore reproductive success) and then change to male because of the advantage conferred upon large males. The age or size at which an individual should change sex is probably determined by an interaction between body size and social structure (numbers of males and females, dominance hierarchies) of the population (Shapiro 1987). Social control also explains protogynous sex change in the Midas cichlid, *Cichlasoma citrinellum*, one of the few freshwater changers. Large fish tend to be male, smaller fish are female. Growth rate depends on behavioral interactions of juveniles, with dominant fish growing faster. The size advantage model again explains the course of change, since female Midas cichlids mate preferentially with larger males. This preference is adaptive, because the male has primary responsibility for defending the breeding site and larger males are better defenders (Francis and Barlow 1993).

Left unanswered is why more species of fishes and other vertebrates do not change sex. Ideas on this subject focus on the relative costs of changing sex in different taxa, the existence of dimorphic sex chromosomes (which are generally lacking in fishes), and differences in sex determination mechanisms (Warner 1978). Add to these the realization that evolution is predominantly a conservative process. Biological systems are complex, which is certainly a description of the reproductive systems of fishes, given the behavioral, ecological, physiological, and anatomic components involved. Alterations to complex systems are likely to destroy the homeostasis that has evolved among the components. Hence the advantages of sex change would have to be very large to overcome fitness losses due to disruption of the coevolved gene complexes that code for the systems. Sex change then becomes an alternative to gonochorism, but one that does not offer a sufficiently large advantage to overcome the costs of refitting the reproductive systems of most fishes. Gonochorism obviously works well for most species; since it is not broken, there is little selective advantage in fixing it.

[1]Warner 1982, 43.

---

similarly colored males (initial-phase coloration). Any of the initial-phase fish can change into larger, terminal-phase males, which develop a blue head, a black-and-white midbody saddle, and a green posterior region. Large males set up territories over coral heads that females pre-

fer as spawning sites. Some females are intercepted by and spawn with groups of up to 15 smaller males, but the largest, pair-spawning males have the highest spawning success. A territory-holding male may receive 40 to 100 spawnings per day, whereas a nearby group-spawning male may receive only 1 or 2 matings, and his sperm will often be diluted by the gamete output of other males in the group (Warner et al. 1975; Warner 1991).

Other well-studied protogynous species include the anthiine serranid, *Anthias squamipinnis*, a pair-spawning species that forms large aggregations in which females may outnumber males by 36:1. The precision of social control of sex change in this species is remarkable: If nine males are removed from a large group, nine females change sex to replace them. Sex change to male in *Anthias* also occurs if the female-male ratio exceeds a threshold value (Shapiro 1979, 1987). The commonness of protogyny probably reflects the fact that most teleosts, including gonochoristic species, differentiate first as nonfunctional females.

Protandry has been reported in moray eels, loaches, lightfishes, platycephalids, snooks, porgies, threadfins, and damselfishes. The popular clown- or anemonefishes (*Amphiprion* spp., Pomacentridae) live in groups of two large and several small individuals in an anemone. Only the two largest fish in an anemone are sexually mature, the largest individual being female and the next largest being male. Although smaller fish may be as old as the spawning individuals, the behavioral dominance of the mature pair keeps these smaller males from maturing and growing, and a dominance hierarchy exists among the smaller males. In essence, "low ranking males are psychophysiologically castrated" (Fricke and Fricke 1977, 830). If the female dies, the male changes sex to female, and the next largest fish in the group takes over his former role and grows rapidly (Allen 1975; Moyer and Nakazano 1978).

Simultaneous hermaphroditism (cosexuality) is known from only 3 shallow-water families and 11 of the 12 families in the deep-sea order Aulopiformes (lizardfishes, Synodontidae, are the exception) (Smith 1975; Warner 1978). Three species of New World cyprinodontiform rivulines (Aplocheilidae) are capable of self-fertilization (*Cynolebias* spp. of South America and the mangrove rivuline, *Rivulus marmoratus*, of North and Central America). Self-fertilization in *Rivulus* is internal, producing clonal populations of homozygous, genetically identical hermaphroditic fish. Functional males can be produced, depending on temperature and day length (Harrington 1971, 1975). Cyprinodontiform fishes are often colonists of small streams on islands and other seasonally adverse habitats (see Chap. 17). Self-fertilization may be one means of assuring mates in low-density populations that frequently become isolated, a scenario that could also be applied to the deep-sea aulopiforms.

The other species of simultaneous hermaphrodites occur among the small hamlets (*Hypoplectrus*, *Serranus*). Each individual is physiologically capable of producing sperm and eggs at the same time, but behaviorally these fishes function as only one sex at a time during a spawning bout. In Caribbean hamlets (*Hypoplectrus*), spawning bouts can last for several hours, during which time members of a pair alternate sex roles, one fish first behaving as the female and releasing eggs and then behaving as the male and releasing sperm (Pressley 1981; Fischer and Petersen 1987). The eastern Pacific *Serranus fasciatus* is haremic, sex-changing, and simultaneous: One male guards and spawns with several hermaphrodites that act as females. If the male is

removed, the largest hermaphrodite changes into a male (Fischer and Petersen 1987). Serranines have separate external openings for the release of eggs and sperm (in addition to an anus), which may prevent internal or accidental self-fertilization. Self-fertilization may occur in some serranines, but only in aquaria (Thresher 1984).

One additional group of fishes departs from normal gonochoristic gender roles. Livebearers in Mexico and Texas include parthenogenetic "species" that are all-female but require sperm from males of other species to activate cell division in their eggs. Parthenogenesis in livebearers takes two forms: gynogenesis and hybridogenesis (Fig. 20.1). **Gynogenetic** females are usually triploid and produce eggs that are also 3n. These eggs are activated by sperm from other species, but no sperm material is incorporated; hence daughters are genetically identical to their mothers. **Hybridogenetic** females, in contrast, are diploid and produce haploid eggs that, during the reduction division of meiosis, keep maternal genes and discard paternal genes. Upon mating, these eggs unite with sperm from males of another species, forming a new, diploid hybrid daughter (no sons are produced). When the daughter mates, she again produces eggs that are haploid and "female." Hence the maternal lineage is conserved, and the male's genetic contribution is lost after one generation. These parthenogenetic "species" are thought to have arisen originally as hybrids between *Poeciliopsis monacha* females and males of four congeners, *P. lucida*, *P. occidentalis*, *P. latidens*, and *P. viriosa*. The males of the four species are the usual sperm donors during mating. An additional species, the Amazon molly, *Poecilia formosa*, is diploid and gynogenetic. Sperm from two other species (*P. mexicana* and *P. latipinna*) activate the eggs but contribute no genetic material (Schultz 1971, 1977; Vrijenhoek 1984). Natural gynogenesis has also been reported for the cyprinid *Cyprinus auratus gibelio* (Price 1984).

An immediate question that arises is how natural selection maintains males that waste gametes so wantonly. Apparently, dominance hierarchies among donor-male populations of livebearers exclude many males from mating with conspecific females. These are often the males that participate in the parasitized, heterospecific spawnings. Satellite or peripheral males that have very low reproductive output are characteristic of many vertebrate species (these are often the sneakers and streakers discussed below), providing an abundance of otherwise unused sperm (Moore 1984). Additionally, laboratory tests of mate preferences in sexual females show that sexual females are more attracted to males that the females observed courting gynogenetic females. Apparently, a male can increase his chances of mating with a sexual female if he spends time courting asexual females, because sexual females copy the choices made by female gynogens. It is not known whether sexual females prefer males that mate with sexual females over those that mate with gynogenetic females (Schlupp et al. 1994).

Certain generalities arise from surveys of sex change in shallow-water fishes, as do exceptions. Sex change is largely a tropical and subtropical, marine phenomenon (Policansky 1982c; Warner 1982). Cool temperate and freshwater sex changers are known (e.g., loaches, swamp

FIGURE 20.1. *Parthenogenesis in Mexican livebearers. (A) In gynogenesis, a triploid female (designated MLL, shorthand for* Poeciliopsis monacha-lucida-lucida) *produces 3n eggs that are activated but not fertilized by sperm from a male P. lucida (L'). A daughter identical to the mother is produced. (B) In hybridogenesis, a diploid mother (ML, for P. monacha-lucida) produces haploid eggs (M) that contain only the maternal genome. Sperm from P. lucida (L') combine to form a diploid daughter (ML'), but this male component will be discarded again during the next round of gamete production, and all future eggs will continue to have solely* monacha *genes.*

(A) GYNOGENESIS   (B) HYBRIDOGENESIS

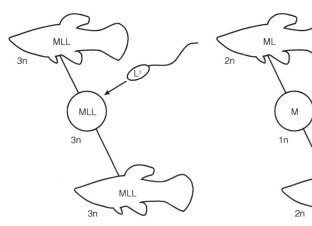

After Vrijenhoek 1984 and Allendorf and Ferguson 1990.

eels, wrasses, gobies, cichlids) but are relatively uncommon compared with tropical marine hermaphrodites. Patterns often follow familial lines, all members of a family being either protandrous or protogynous. Exceptions include the protogynous or simultaneously hermaphroditic sea basses (a family that is probably polyphyletic) and the protandrous or protogynous porgies. However, population differences are becoming increasingly well known in sex-changing fishes. The cleaner wrasse, *Labroides dimidiatus*, is haremic under some conditions but forms pairs under others. Bluehead wrasse, *Thalassoma bifasciatum*, are dominated by territorial-spawning males on small reefs with small populations, but by group-spawning males on large reefs with dense populations. Resource limitation—either food availability or reef size—and population size are frequent determinants of variation in mating systems. Clearly, sex change and mating systems respond to environmental variability (see Thresher 1984; Shapiro 1991; Warner 1991).

COURTSHIP AND SPAWNING

Sexual Selection, Dimorphism, and Mate Choice

Some traits of an animal function primarily to attract mates or to aid in battles between members of one sex for access to the other sex. Such **sexually selected** traits confer a mating advantage on an individual; they are a subset of natural selection, which usually involves traits that confer a survival advantage. Sexually selected traits may serve no other purpose than mating and may even handicap the possessor with respect to other, fitness-influencing activities. However, sexually selected traits can also confer a positive survival advantage, such as large size in males, which provides a physical defense from predators and is also favored by females during mating (Box 20.2). Sexually selected traits are often referred to as **secondary** sexual characteristics. **Primary** characteristics include ovipositors, genitalia, and other copulatory structures such as claspers in elasmobranchs, gonopodia or priapia in livebearers and phallostethids, or brood patches or pouches and other structures used in parental care. In some instances, a character may be both secondary and primary, serving both in mate attraction and in copulation or parental care. In sticklebacks, males are attracted to females with swollen bellies, but the swelling results from the female's ripe ovaries (Wootton 1976).

Secondary sexual characteristics have four general attributes: 1) They are restricted to or are expressed differentially in one sex (usually the male), 2) they do not appear until maturation, 3) they often develop during a breeding season and then regress, and 4) they generally do not enhance survival. Secondary characteristics take the form of **sexual dimorphisms** (differences in body parts between sexes), such as differences in body size, head shape, fin shape, dentition, and body ornamentation, or as **dichromatisms** (differences in coloration). We know comparatively little about electrical, chemical, and acoustic differences between the sexes, although differences in anatomy and physiology associated with these sensory modes are common (e.g., in elephantfishes, salmons, minnows, gymnotid knifefishes, toadfishes, croakers, damselfishes, and gobies). For example, males of the plainfin

BOX
20.2
Sexual Selection in Fishes

The major feature of sexually selected traits is that one sex bases its mating preferences on a character or set of characters in the other sex. The basis of mate choice is complex, involving a combination of factors related to male-male competition and female attraction, but sexually dimorphic traits serve as immediate (proximate) cues to a potential mate's ultimate reproductive quality. In nest-guarding or otherwise territorial species, males typically compete for spawning sites, and then females choose males based on male size and coloration, territory size and location, and quality of the oviposition substrate.

It is often difficult to determine which is more important: male attributes or the qualities of the site he possesses (Kodric-Brown 1990). Relevant male attributes include size, which is directly correlated with intrasexual dominance in many families (e.g., minnows, salmons, sticklebacks, sculpins, sunfishes, darters, wrasses, blennies, gobies). Females actively choose larger males in many nesting species in which males guard eggs and fry. Such a size-based preference is an adaptive choice by females: Larger males are generally more effective nest guarders, and hence choosing a larger male is a means of ensuring better protection from egg predators (Downhower et al. 1983; Fitzgerald and Wootton 1994). Bright coloration can serve as both a dominance signal and a mate attractant (minnows, livebearers, killifishes, sticklebacks, greenlings, darters); larger males are also often the most colorful. Female choice based in part on territory or spawning site characteristics has been demonstrated in damselfishes, wrasses, tripterygiids, blennies, and gobies. Given the positive correlations among the many traits that may indicate male quality (e.g., size, color, health, courtship intensity, dominance, territory size, quality of paternal care), it is not surprising that females often rely on a variety of characteristics rather than a single male trait in making their choices (Kodric-Brown 1990).

It is widely held that the sex that expends the greatest energy in gamete production will be in relatively limited supply, that the other sex will compete for the limiting sex, and that the limiting sex will protect its large investment by being choosiest about mating partners. Dimorphic males and female choice are generally the rule in fishes, reflecting the greater cost of producing eggs than sperm. Evidence of the generalization can also be found in role-reversed species in which males become "pregnant" and carry eggs internally, such as the pipefishes and sea horses. In pipefishes, females are the more colorful sex, and males select mates based on female size and the intensity of courtship displays (Berglund et al. 1986). The female is also the dimorphic sex in a limited number of species, notably in some cardinalfishes,

damselfishes, and porcupinefishes. In the first two families, males mouth brood or defend the eggs.

Several well-studied instances of sexual dimorphism and sexual selection have given us insight into how such traits evolve and function. Male swordtails (*Xiphophorus* spp., Poeciliidae) develop a colorful, elongate lower caudal fin extension, the sword, when they mature. The sword continues to grow as the fish ages and may exceed the length of the rest of the body in older individuals. The sword serves no role in male-male interactions and may be a liability in predator avoidance. The only known function of this "elaborate, highly conspicuous ornament" is during courtship, when the male displays the sword to females (Basolo 1990a, 333). Females prefer to mate with males with longer swords; when given a choice between two males with different length swords, the longer sword is preferred, and the strength of the preference increases as the difference between sword length increases. Hence female choice for long swords has selected for increased sword length in males.

Interestingly, swordlessness is the probable ancestral condition in the genus. Platyfish are primitive congeners of swordtails, but platyfish do not develop swords. When plastic swords are surgically attached to the tails of male platyfish, female platyfish prefer males with the artificial swords over normal, swordless males. Swords in male swordtails apparently evolved in response to a preexisting preference in the female's information-processing system, which, when combined with natural variation in tail length, selected for males with progressively longer swords (Basolo 1990b).

As is often the case, strong selection for sexually dimorphic traits is achieved at a cost. The trade-off operating here is sexual selection versus conspicuousness to predators. For example, male guppies attract females via a series of displays that are enhanced by brightly colored spots on the male's side. The largest, brightest, and most diverse spots are most attractive to females and predators alike. In Trinidad, predators are more common and diverse in lowland areas than in upland areas, and males in these predator-dense streams tend to have fewer, less-intense, smaller, and less-colorful spots. Hence a compromise is struck between attracting mates and attracting predators (Endler 1991). It makes little evolutionary sense that females in predator-dense areas would prefer to mate with males that were also very attractive to predators, if for no other reason than the female's sons would be relatively vulnerable. Not surprisingly, females from predator-dense areas generally do not show a preference for more-colorful males (Houde and Endler 1990; J. Endler, personal communication).

midshipman, *Porichthys notatus* (Batrachoididae), attract females by "humming," a sound produced by contracting large muscles attached to the gas bladder walls. The sound-producing ability and its importance in courtship are reflected in numerous differences between the sexes. Males differ from females in having larger body size, different color, larger sonic muscles, and differing neural circuitry—involving larger cell bodies, dendrites, and axons in the brain—than are found in the females (Bass 1996).

Although the vast majority of fish species show no obvious sexual dimorphisms, many species are distinctly dimorphic. Males are often the larger sex in salmon, sauries, wrasses, and clingfishes, whereas females are larger in mackerel sharks, sturgeons, true eels, ceratioid anglers, sticklebacks, halfbeaks, silversides, livebearers, blennies, and billfishes. Male dolphinfish (*Coryphaena hippurus*, Coryphaenidae) are larger and have a distinctly blunter head than females. During spawning migrations, male anguillid eels develop larger eyes than females, and male salmon develop distinctly concave upper and lower jaws, called **kype**, that preclude feeding. Males may have trailing filaments at the ends of their dorsal, anal, or caudal fins (characins, African rivulines, rainbowfishes, anthiine sea basses, cichlids, wrasses); enlarged median or paired fins (lampreys, bichirs, freshwater flyingfishes, minnows, characins, killifishes, livebearers, dragonets, gobies, climbing gouramis); elongate tail fins (livebearers); or elongate, lobule-tipped gill covers or pelvic fins that are displayed in front of females during courtship (*Corynopoma*, Characidae; *Ophthalmochromis*, Cichlidae). In elasmobranchs, males may have longer or sharper teeth than females, which serve to grasp the female during courtship and copulation; females in turn may have thicker skin than males.

Differences in body ornamentation include small bumps on the body, scales, and fins of mostly male fishes in 25 different families of primarily soft-rayed teleosts. These bumps are called nuptial or breeding tubercles when of epidermal origin, or contact organs when of dermal origin (Fig. 20.2) (Collette 1977). Scale differences also include scale type: Male cyprinodontids have ctenoid scales, whereas females have cycloid scales (Berra 1981). Other dimorphic ornaments include small hooks on the anal fins of some male characins (a kind of contact organ), rostral papillae in male blind cavefishes, photophore patterns in lanternfishes, and pigmented egg dummies on the anal fins of cichlids. Some male minnows, cichlids, wrasses, and parrotfishes develop bony or fatty humps on the front (nuchal region) of their head. Coloration differences are widespread and are usually expressed as more brightly colored males, either permanently (bowfin, livebearers, killifishes, rainbowfishes, cichlids, wrasses, anabantids) or seasonally (minnows, sticklebacks, darters, sunfishes, cichlids). Male color change often involves development of bright or dark patches where they are most conspicuous because they break the rules of crypticity. Hence dark and light patches exist adjacent to one another, color transitions become sudden rather than gradual, or reverse countershading develops (e.g., sticklebacks, sunfishes, temperate wrasses). Conspicuousness in the breeding season reinforces the premise that animals are willing to risk increased predatory threat for a chance to reproduce (Breder and Rosen 1966; Fryer and Iles 1972).

Site Selection and Preparation

Many fishes spawn in nests. Nests may be no more elaborate than a simple depression that either or both sexes excavate in the bottom by fanning vigorously with the fins or by swimming rapidly in place (e.g., lampreys, bowfin,

FIGURE 20.2. *Breeding tubercles in fishes. Males of at least 25 different families develop keratinized bumps on their fins and body during the breeding season that may help maintain contact between spawning fish and stimulate females to spawn. (A) Tubercles on the head of an approximately 15-cm-long male river chub,* Nocomis micropogon. *The swollen region on the top of the head is also characteristic of male river chubs during the breeding season. (B) Internal structure of a tubercle from the snout of a gyrinocheilid algae eater. The tubercle consists of an outer cap of epidermal keratin, with concentrations of replacement keratin lying in a pit at the base of the tubercle that will replace the tubercle in the event of its loss.*

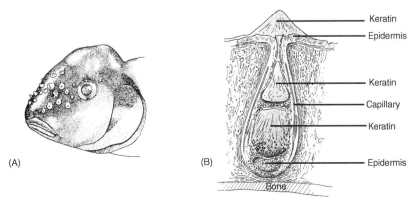

(A) (B)

(A) From Jenkins and Burkhead 1994; used with permission. (B) From Wiley and Collette 1970; used with permission.

trahiras, sunfishes). In lampreys and salmon, the eggs are covered with additional sand or gravel and then abandoned, whereas in other species the male remains to guard the exposed eggs. Many species spawn in a rock crevice or space on or under a shell or rock or in a hollowed log that has been excavated or picked clear of growth and debris (e.g., ictalurid catfishes, sculpins, poachers, darters, cichlids, damselfishes, clingfishes, sleepers, gobies). Some damselfishes "sandblast" a surface clean by spitting sand against it and then fanning it with their fins.

The European wrasse, *Crenilabrus melops*, lines a rock crevice with different types of algae. Softer algae at the back of the crevice serve as a substrate for spawning, whereas tougher, stickier coralline algae are packed into the outer portion of the crevice to protect the eggs. Burrows are excavated and later guarded by lepidosirenid lungfishes, ictalurid catfishes, jawfishes, tilefishes, and gobies. Males of some stream minnows construct nest mounds by piling as many as 14,500 small stones 6 to 12 mm in diameter that are carried in the mouth to the nest site from as far as 4 m away (e.g., *Nocomis, Semotilus, Exoglossum*). Eggs deposited by the female fall into the interstices between stones and are covered by additional stones; the nest is guarded and kept free of silt by the male (Breder and Rosen 1966; Keenleyside 1979; Potts 1984; Thresher 1984; Wallin 1989). In African cichlids, males build nests in sand, some of elaborate design (Fig. 20.3). These structures have been likened to the bowers of bowerbirds because they are primarily display, courtship, and spawning stations from which the female picks the eggs up in her mouth almost as quickly as they are laid (Fryer and Iles 1972; McKaye 1991).

Nests can also be constructed of intrinsically produced materials, sometimes combined with extrinsically gathered objects. Gouramis and Siamese fighting fishes (Anabantoidei) and pike characins produce bubble or froth nests, consisting of mucus-covered bubbles that stick together in a mass. Some anabantoids add plant fragments, detritus, sand, and even fecal particles to the bubble mass. Male sticklebacks pass a mucoidal, threadlike substance manufactured in the kidneys out through the cloaca, using it to bind together pieces of leaves, grasses, and algal filaments that create a cup or tunnel nest in which a female deposits eggs. All of these structures are tended by the male following spawning (Wootton 1976; Keenleyside 1979).

Courtship Patterns

Courtship, the series of behavioral actions performed by one or both members of a mating pair just prior to spawning, has several functions that maximize the efficiency of the spawning act. Courtship aids in species recognition (a premating species-isolating mechanism), pair bonding, orientation to the spawning site, and synchronization of gamete release. Courtship is often necessary to overcome territorial aggression by the male, who might otherwise drive the female away from the site (in many species, males already have eggs in their territories and must guard against predation by conspecifics of both sexes). Courtship may be relatively simple (as in herring) or may involve a large number or a complex progression of displays and signals by one or both members (e.g., *Corynopoma*, Characidae; guppies; sticklebacks).

During courtship, individuals frequently change color

FIGURE 20.3.   *Spawning nests or bowers of African cichlids. Male cichlids construct sand structures that vary from simple pits to complex structures, where females deposit eggs just prior to picking them up in their mouths for brooding: (A) Tilapia (Oreochromis?) andersonii and T. nilotica; (B) T. variabilis; (C) T. cf. macrochir from Lake Bangweulu; (D) T. cf. macrochir from Lake Mweru; radiating spokes are created by the male plowing through sand with his open mouth from the focal point to the edge. Structures vary from 15 to 150 cm across.*

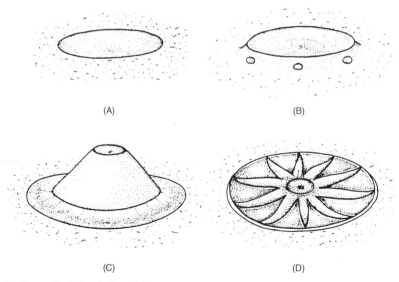

(A)   (B)

(C)   (D)

From Fryer and Iles 1972; used with permission.

from their normal, countershaded patterns to bolder, contrasting color patterns. In many species (e.g., minnows, silversides, cichlids), existing body coloration intensifies, or the head becomes dark relative to the remainder of the body. Sound production during courtship, usually by the male, occurs in many fish families (codfishes, toadfishes, sunfishes, grunts, sciaenids, damselfishes, gobies), often in accompaniment with visual displays involving exaggerated or rapid swimming patterns, erection of fins, and jumping out of the water (Fine et al. 1977; Myrberg 1981). Chemical stimulants are also involved. Male goldfish, zebra danios (Cyprinidae), and gobies begin courtship activities when exposed to water that previously held a gravid female, and gravid female gobies are attracted to male-produced androgynous substances (Hara 1986). Species and sex recognition during courtship in cichlids occurs more quickly when individuals receive both visual and chemical cues from potential mates (Barlow 1992).

Some appreciation of the evolutionary premium placed on successful courtship can be gained by realizing that the gas bladder muscles that produce the boatwhistle mating call of the male oyster toadfish, *Opsanus tau*, contract at a rate of 200 Hz. This makes them the fastest contracting vertebrate muscle known, the next closest being the shaker muscles at the base of the tail of the rattlesnakes, which contract at only half that rate (Rome et al. 1996).

The Spawning Act

The final act of spawning may take place in the water column, above the bottom, in contact with plants and rocks, and in some special cases, out of water. In external fertilizers, behaviors associated with spawning often involve rapid swimming, quivering, vibrating, fin spreading, and enfolding of the female with the male's fins or body. The breeding tubercles and contact organs common in many fishes (see above) may help maintain contact between members of a pair and may also stimulate the female. Internally fertilizing species also engage in elaborate courtship sequences. Male guppies perform a variety of actions involving following, luring, biting, and sigmoid swimming that display their fins and body coloration until a female allows them to approach and copulate. The sequence and types of displays by the male serve as a species-isolating mechanism, in that females reject males of the wrong species after viewing their courtship displays (Keenleyside 1979).

Species-specific sounds may also be produced during the spawning act itself. For example, in the simultaneously hermaphroditic hamlets, the "male" emits a courtship call and the "female" a spawning call. As individuals switch roles during a prolonged spawning bout, they also switch the sounds they produce (Lobel 1992).

Although the great majority of fishes spawn as part of large groups, pairing of individual males and females within these groups is common. Short-term pair formation probably assures efficient gamete release and fertilization; haphazard release of gametes could result in a large proportion of eggs going unfertilized, because sperm become inviable and are rapidly diluted in open water and eggs become unfertilizable within minutes after release (Hubbs 1967; Petersen et al. 1992). Codfishes spawn in large aggregations, but males establish small territories and actively court individual females using visual, tactile, and acoustic signals. The pair then moves synchronously to the surface, where gametes are released while the genital openings of both fish are in close contact (Brawn 1961). Aggressive defense of females and pair spawning also occurs in schooling tunas (Magnuson and Prescott 1966). Pair spawning characterizes most epipheline sea basses, which also form large breeding aggregations.

Group spawning, involving more than two fish, usually involves one female accompanied by several males. This is the pattern in group-spawning minnows, salmon, smelt, wrasses, and surgeonfishes. In groups of bluehead wrasse, males release sperm in direct proportion to the number of eggs released by the female and the number of competing males in the group (Shapiro et al. 1994). In fishes with alternative mating systems, such as wrasses and parrotfishes on coral reefs, some individuals spawn as pairs whereas others spawn in multiple-male groups (see above). Truly random spawning associations, as described for promiscuous species, occur most frequently in water-column spawners or in such benthic spawners as herring (Keenleyside 1979).

Water-column spawners on coral reefs often rush rapidly upward and release their gametes at the top of the rush, sometimes near the surface. Speeds approach 40 km/h in the striped parrotfish, *Scarus croicensis* (Colin 1978). This pattern has been observed in more than 50 species in at least 18 families. Its function(s) are debated. Movement up in the water column places the eggs out of reach of many benthic or near-benthic invertebrate and vertebrate zooplanktivores and into currents that promote dispersal. However, by moving away from the reef, spawning adults face the conflicting threat of piscivores; not surprisingly, well-defended spawners (e.g., larger sea basses, trunkfishes, porcupinefishes) move higher in the water or spawn more slowly than smaller, more vulnerable species (Thresher 1984). For many species, the **spawning rush** may serve as a final synchronizing event in the courtship sequence and may also help evade the sneakers and streakers that abound close to the reef. Left unanswered is the question of why surface rushes are uncommon in other habitat types.

Substrate-spawning fishes are less likely to form large groups than water-column spawners; they also release fewer eggs at each spawning. Males typically set up territories over appropriate spawning substrate, chase away intruding males, and court passing females. Females enter the territory and deposit one or a few adhesive eggs while the male folds his body or fins around her and presses her against the substrate. Sperm release occurs almost immediately, again in part because sneaker and streaker males are always nearby. Paternity can be assured, from the male's point of view, if females facilitate fertilization by taking sperm into the mouth, as happens in many mouth-brooding cichlids (Fig. 20.4).

An interesting variation on oral fertilization occurs in callichthyid catfishes of the genus *Corydoras*, the popular armored catfishes of the aquarium trade. In these catfishes, the female places her mouth over the genital

FIGURE 20.4.   *Fertilization occurs in the mouth of some female African cichlids. (A) Males of many African cichlids have round spots, termed egg dummies, on their anal fins. During spawning, the female repeatedly deposits a few eggs on the spawning site and then immediately takes them up in her mouth. The male spreads his anal fin against the bottom, and the female mouths the egg dummies as the male ejaculates. (B) In some species, females instead mouth genital tassels, which are elongate, orange lobules that grow from the genital region. (C) Other males have greatly elongate pelvic fins with enlarged, conspicuous tips that reach to the cloaca. All such structures and behaviors may facilitate fertilization, assure paternity, and minimize predation on newly laid eggs.*

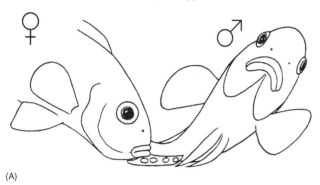

(A)

(B)

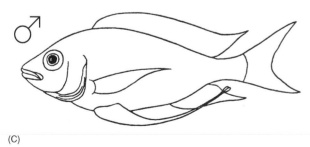

(C)

From Fryer and Iles 1972; used with permission.

opening of the male and drinks his sperm. She then passes the sperm rapidly through her digestive system, extrudes eggs that are held between her pelvic fins, and releases the male's sperm to fertilize the eggs, which are then deposited on the substrate. **Sperm drinking** could be one way for a female to maximize control over which male fertilized her eggs. Sperm viability in the female's gut may be facilitated by the specialized nature of the callichthyid intestine, which is modified for air breathing (see Chap. 5, Aerial Respiration). Callichthyids pass air bubbles rapidly from their mouths to their intestines, perhaps preadapting them for passing sperm through quickly and unharmed (Kohda et al. 1995).

Fishes that spawn on the bottom generally use available structure to protect their eggs. Many eggs are adhesive and stick to plants, rocks, woody debris, shells, or other hard substrates (e.g., herring, silversides, cichlids). Some eggs have tendrils or projections that wrap around plants and debris (e.g., skates, halfbeaks, flyingfishes). Eggs of the Port Jackson shark (Heterodontidae) have an augerlike whorl around their exterior. Females lay these eggs in cracks, and water motion apparently serves to screw the egg deeper into the substrate. California grunion (*Leuresthes tenuis*, Atherinidae) spawn on sandy beaches after high tides on dark nights following a full or new moon (see Chap. 22, Semilunar and Lunar Patterns). Capelin (*Mallotus villosus*, Osmeridae) also ride waves up beaches and deposit their eggs in the sand, although subtidal spawning is more common. A few fish species use live invertebrates as a spawning site. The marine snailfish, *Careproctus* (Liparidae), lays its eggs inside the gill chambers of a box crab. Species of bitterling, *Rhodeus* (Cyprinidae), use freshwater mussels as a spawning site. The male first defends and displays over a particular mussel. The female deposits eggs into the gill chamber of the bivalve using her long ovipositor, after which the male ejaculates over the incurrent siphon of the mussel. Eggs develop inside the mussel and emerge as free-swimming young (Breder and Rosen 1966).

The act of spawning brings together fish that may normally be solitary, territorial, or extremely sensitive to predators, often at locales they seldom frequent. As a possible mechanism to help overcome behaviors that would be counterproductive at this critical moment in a fish's life, many fishes exhibit **spawning stupor**. When in a spawning aggregation or mode, species that are normally difficult to approach or are very active instead move slowly, in an almost trancelike state. They take little or no evasive action when approached by predators or divers. Spawning stupor has been observed in minnows, suckers, mullets, silversides, sea basses, pompanos, and snappers (Johannes 1981; Helfman 1986). Most observations are anecdotal, leaving open an excellent opportunity for quantified or manipulative investigations.

PARENTAL CARE

Extent and Diversity of Care

Parental care is surprisingly common, widespread, and diverse in fishes (Breder and Rosen 1966; Blumer 1979, 1982; Baylis 1981; Potts 1984; Keenleyside 1991b; Sargent and Gross 1994). Although most species scatter or abandon eggs upon or after fertilization, about 90 of the approxi-

Table 20.2. A classification of reproductive guilds in teleost fishes, based largely on spawning site and parental care patterns. Specific examples of many groups are given in Table 20.1.

I. Nonguarding species

    A. Open substrate spawners

        1. Pelagic spawners

        2. Benthic spawners

            a) Spawners on coarse bottoms (rocks, gravel)

                (1) Pelagic free-embryo and larvae

                (2) Benthic free-embryo and larvae

            b) Spawners on plants

                (1) Nonobligatory

                (2) Obligatory

            c) Spawners on sandy bottoms

    B. Brood hiders

        1. Benthic spawners

        2. Cave spawners

        3. Spawners on/in invertebrates

        4. Beach spawners

        5. Annual fishes

II. Guarders

    A. Substrate choosers

        1. Rock spawners

        2. Plant spawners

        3. Terrestrial spawners

        4. Pelagic spawners

    B. Nest spawners

        1. Rock and gravel nesters

        2. Sand nesters

        3. Plant material nesters

            a) Gluemakers

            b) Nongluemakers

        4. Bubble nesters

        5. Hole nesters

        6. Miscellaneous materials nesters

        7. Anemone nesters

III. Bearers

    A. External bearers

        1. Transfer brooders

        2. Forehead brooders

        3. Mouth brooders

        4. Gill chamber brooders

        5. Skin brooders

        6. Pouch brooders

    B. Internal bearers

        1. Ovi-ovoviviparous

        2. Ovoviviparous

        3. Viviparous

From Moyle and Cech 1988 and Wootton 1990; based on Balon 1975, 1981 with modifications.

mately 420 families of bony fishes include species that engage in some form of defense or manipulation of eggs and young (Table 20.2). Parental care is more necessary for demersal or adhesive eggs, which are likely to be found by predators searching along the bottom or among plants or other structure. Hence it is not surprising that few if any species with pelagic, floating eggs provide care for them.

Parental care includes construction and maintenance of a nest; burying eggs once deposited in the nest; chasing potential egg and fry predators from the nest; fanning or splashing eggs or young with the mouth, fins, or body to provide oxygen and to flush away sediments and metabolic wastes; removal of dead or diseased eggs; cleaning eggs by taking them into the mouth and then returning them to the nest; carrying eggs or young in the mouth or gill chambers, in a ventral brood pouch, or externally on the head, back, or belly; coiling around the egg mass to prevent desiccation; retrieving eggs or young that wander from the nest or aggregation; accompanying foraging young and providing refuge or defense when predators approach; secreting specialized mucus upon which free-swimming young feed; and provisioning young or aiding them in the capture of food. Internal gestation of young is a special type of parental care shown by many Chondrichthyes, but it occurs in only 14 families of bony fishes. Placental connections, common in Chondrichthyes, occur in only one osteichthyan family, the Mexican livebearers (Goodeidae; Fig. 20.5) (Blumer 1979; Wourms 1988).

Some parental activities represent surprisingly unique specializations. For example, in many seahorses and pipefishes, females lay their eggs at the entry to brood pouches on the male's belly. The pregnant male fertilizes the eggs and retains and nourishes the eggs and young inside the pouch until they reach a relatively advanced stage of development. "Birth" involves contractions and contortions by the male that expel the young from the pouch. In the kurtoid nurseryfishes, an advanced perciform suborder of the southwestern Pacific, males develop a unique, downward-bent, occipital hook on their foreheads to which the eggs are attached and carried until hatching. Just how the eggs get there is a matter of conjecture.

One of the most unusual patterns of parental activity is shown by the spraying characin (*Copella*, Lebiasinidae), which deposits its eggs out of water. Male and female line up under a leaf and leap together as much as 10 cm into the air, turning upside down and adhering momentarily to the leaf's underside. In this manner, a dozen or so fertilized eggs are stuck repeatedly to the leaf. Over the next 2 to 3 days, the male moistens the egg mass by splashing it at 1-minute intervals with flips of his tail, correcting for refraction at the water's surface (Fig. 20.6). Newly hatched young fall into the water (Krekorian and Dunham 1972).

Preventing desiccation of eggs exposed to air also explains unusual parental care in intertidal species. Several small, elongate intertidal fishes coil their bodies around the egg mass as the tide goes out, thus trapping a small pool of water in which the eggs sit. This behavior has been recorded for pricklebacks, gunnels, and wolf-eels (Blumer 1982). Other fishes that spawn in the intertidal, such as temperate wrasses and sculpins, cover the eggs with algae, thus controlling desiccation during low tide (Potts 1984).

FIGURE 20.5.   *Well-developed (near-term) embryos in the ovary of a Mexican goodeid,* Ameca splendens; *13 embryos are visible. The anterior third of the ovary is not shown. Fingerlike extensions projecting forward from the ovary are trophotaenia, which are epithelial structures that grow from the embryo's anal region and serve to take up nutrients provided by the mother. Trophotaenia have evolved convergently in goodeids, ophidioids, and embiotocid surfperches. of = oocytes; om = ovarian mesentery; os = ovarian septum; ow = ovarian wall. Scale bar = 1 cm.*

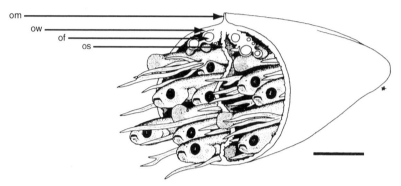

Illustration by Julian Lombardi; from Wourms et al. 1988; used with permission.

A few fishes provide food for their young via epidermal secretions. Such **trophic provisioning** has been observed in bagrid catfishes and in several cichlids, including the discus fish (*Symphysodon*), the Midas cichlid, and members of the genera *Aequidens*, *Etroplus*, and *Oreochromis* (Fig. 20.7). This form of parental care is suspected in numerous other cichlids, as well as in a bonytongue and a damselfish. Thickened scales and increased mucus production have been identified in the adults of provisioning species. The importance of provisioning relative to other food sources of the young is unclear (Noakes 1979). Many sharks produce trophic eggs that are eaten by developing embryos prior to hatching (Wourms 1981; Heemstra and Greenwood 1992; see Chap. 12). One bagrid catfish in Lake Malawi, Africa, feeds trophic eggs to free-living juveniles. *Bagrus meridionalis* young position themselves under the vent of the guarding female and apparently ingest eggs as they are extruded by the mother; 40% of the young in a nest may have such eggs in their stomachs. Circumstantial evidence indicates that the parental male also helps feed the young by uncovering invertebrates present in the nest or even by spitting invertebrates captured elsewhere into the nest (LoVullo et al. 1992).

Many species carry the eggs rather than leaving them deposited on the substrate. **Mouth** or **oral brooding** is the most common form of egg carrying, having been documented in at least six families (sea catfishes, lumpfishes, cardinalfishes, cichlids, jawfishes, gouramis); gill chamber brooding occurs in North American blind cavefishes. Eggs are picked up, usually by the male, shortly after fertilization. In the case of some cichlids, eggs are fertilized in the female's mouth, where they are retained (see Fig. 20.4). In cichlids, oral brooding extends well beyond hatching. Free-swimming young forage as part of a shoal near the female. When predators approach, the female signals the young by backing up slowly with the head down. The young swim toward her head, and she sucks them into her mouth. Some predatory cichlids will ram the head region of females that are carrying young, forcing them to spit a few out, which are then engulfed by the predator (McKaye and Kocher 1983). Other forms of external carrying include attachment of eggs to the male's lower lip (suckermouth armored catfishes) or head (nurseryfishes), or the belly of either parent (bagrid and banjo catfishes) (Breder and Rosen 1966; Balon 1975; Blumer 1982).

FIGURE 20.6.   *Parental care in the spraying characin,* Copella *sp. Eggs in this species are deposited on the undersides of overhanging vegetation, out of the water. The male guards the eggs, splashing them periodically with his tail to keep them moist.*

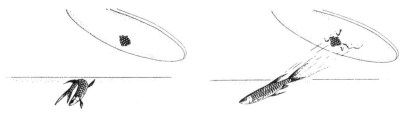

From Krekorian and Dunham 1972; used with permission.

FIGURE 20.7.   *Parental care in cichlid fishes. Two of the more striking forms of parental care exhibited by members of the diverse cichlid family are shown. (A) Mouth brooding. A female opens her mouth after signaling danger to a shoal of young. Mouth brooding of eggs is fairly widespread in fishes, but brooding of free-swimming young is relatively rare. (B) Provisioning young. Few fishes with external fertilization actually provide nutrition for their young. Many cichlids are suspected of provisioning, but the behavior is best known in the discus fish, Symphysodon. A pair of discus is shown, with the young feeding on the mucous secretions of one member.*

(A)

(B)

(A) From Fryer and Iles 1972; used with permission. (B) From Herald 1961; used with permission.

The Gender of Caregivers

The most common caregiver in fishes is the male. Males alone or in combination with females (= biparental care) account for approximately 80% of 77 families in which the sex of the caregiver is known; males alone care for young in 36 families (Blumer 1979, 1982). The predominance of male parental care in fishes contrasts markedly with its occurrence in other vertebrates, in which care by females (amphibians, mammals) or both parents (birds) is more common (posthatching care is uncommon in reptiles). Male guarding may be explainable as an evolutionary result of external fertilization and a male's way of assuring that he alone fertilizes a batch of eggs (= **paternity assurance**). To accomplish this, a male should 1) provide a suitable locale where females will lay eggs to be fertilized and 2) guard the eggs so that no other male can fertilize them. Paternity assurance was the likely driving force behind the evolution of brood pouches in pipefishes and seahorses. The female deposits eggs in the male's abdominal pouch, where only his sperm are likely to reach the eggs. In most species, extending care beyond the fertilization stage greatly increases the probability of successful hatching and dispersal, thereby increasing the likelihood that offspring will live to reproduce. Egg and larval predators are common in all environments, as are fungal infections. A guarding male can chase off fishes and invertebrates that might eat the eggs, and can remove diseased or dead eggs, thus slowing the spread of fungi and other infectious pathogens.

Males may care for young longer because females are more likely to spawn with males that already have eggs or young in the nest (e.g., fathead minnow, threespined stickleback, painted greenling, river bullhead, tessellated darter, browncheek blenny). An unexpected outcome of a female preference for males with eggs is nest and egg usurpation (also known as **allopaternal** care). Male sticklebacks will raid other males' nests, steal eggs, and deposit these eggs in their own nest. Male fathead minnows evict males from existing nests and then guard the acquired eggs. In **brood piracy**, a large male may usurp the nest of another male, spawn, and then abandon the nest to be guarded by the original territory holder (van den Berghe 1988; Magnhagen 1992).

A related phenomenon is interspecific **brood parasitism**, in which one species spawns in a nest constructed by and guarded by another species. Several species, including gars and minnows, spawn in nests guarded by male sunfishes. Small minnows also spawn over the mound nests built by larger minnows, such as bluehead chub (see above); the eggs are guarded by the large male chub. The chub may benefit by a dilution effect whereby predators are likely to eat the more numerous minnow eggs, whereas the minnow eggs receive the protection of a nest guarded by a large male. A dilution effect probably explains why bagrid catfish tolerate and guard cichlid young in their nests. Mistaken identity cannot be invoked, since the guarding catfish parents selectively chase cichlid young to the periphery of the nest, exposing the cichlids to higher predation rates and decreasing mortality in the catfish young. The young cichlids benefit from the protection of two large catfish plus the mother cichlid, which remains nearby. In addition, young cichlids are allowed to feed on the catfishes' bodies (McKaye 1981; Unger and Sargent 1988; Wallin 1992; McKaye et al. 1992).

If male care evolved to ensure that no other males fertilized the eggs, then males would not be expected to provide care in the 14 teleostean families with internal fertilization. This is almost universally true and even applies to species that are exceptional relative to the familial norm. For example, most sculpins have external fertilization and male parental care. In *Clinocottus analis* and *Oligocottus* spp., fertilization is internal and male parental care is absent (Perrone and Zaret 1979). A cardinalfish, *Apogon imberbis*, is the only known species with internal fertilization and male care. Spawning occurs repeatedly over an extended 5-day period, the male chases off other males, and he also picks up eggs in his mouth immediately after they are laid—all actions that would minimize the opportunity for other males to fertilize the eggs (Blumer 1979).

A final category of care deserving attention is the phenomenon of **cooperative breeding**, or **helpers at the nest**. Nonparental caregivers, usually young from a previous breeding episode, remain with the parents and feed and defend new young or defend and maintain the territory. Such helpers occur in over 150 bird and 25 mammal species and in at least 6 species of African cichlids (Taborsky 1984). In the best-studied species, *Lamprologus*

brichardi, of Lake Tanganyika, helpers remain for about a year through two to three subsequent breeding cycles. They clean and fan eggs, larvae, and fry; remove sand and snails from the breeding hole; and defend the parental territory. Helpers suffer slower growth rates than nonhelping individuals but receive protection from predators due to territorial shelters and the protective activities of larger family members. Females with helpers produce more fry. Helping generally imposes a cost, because helpers do not reproduce directly while remaining with their parents. However, helpers may promote their fitness (contribution of genes to future generations) more by raising siblings to whom they are closely related than by attempting to breed on their own. Helping is thus an example where kin selection explains an apparently altruistic activity.

The Costs of Care

As with foraging (see Chap. 18), caring for offspring by parents creates a series of potential trade-offs. A guarding parent often has reduced opportunity to feed, which may reduce later gamete production (e.g., in salmons, ricefishes, and livebearers; Blumer 1979). These costs can be overcome in part if the male eats some of the eggs, a phenomenon known as **filial cannibalism** (Fitzgerald 1992). Feeding is strongly curtailed in mouth-brooding species. Parental tasks may preclude continued spawning, which compromises future reproductive output. The decision of when to abandon current progeny will therefore be influenced by how much a parent's guarding activities can reduce mortality in the current clutch versus what opportunities for breeding exist in the near future (Perrone and Zaret 1979; Sargent and Gross 1994). Short breeding seasons, scarce additional mates, and short lifetimes would favor parental care of existing offspring over searching for additional spawning opportunities. Females can exploit this dilemma by preferring males that are already guarding eggs (see above). Caring for young also carries predatory risks. Brood defense may reduce predation on the young but simultaneously increases the parent's exposure to predators. Guarding parental sticklebacks, pumpkinseed sunfishes, and gobies take more risks as their offspring grow, indicating that the value of the brood can increase relative to parental survival during the parental care phase (Colgan and Gross 1977; Pressley 1981; Magnhagen and Vestergaard 1991).

Finally, an inverse relationship often exists between degree of care and number of eggs produced. Pelagic egg scatterers produce hundreds of thousands or millions of tiny eggs that they abandon, whereas species that participate in extensive parental care characteristically produce relatively small clutches of dozens to a few hundred larger eggs. High-quality care may only be possible when small numbers of young are produced. However, the ultimate evolutionary product is how many offspring make it into the next breeding generation. The existence of alternative tactics within and among species attests to the fact that no single reproductive system is universally correct.

ALTERNATIVE MATING SYSTEMS AND TACTICS

The literature on social and reproductive behavior in fishes has increasingly focused on the variety of tactics that fishes use, both among and within species. That interspecific differences should arise is not surprising given the different ecologies and evolutionary histories of different lineages. Intraspecific variation is more puzzling, since we tend to think in terms of species characteristics and species-typical behavior. In addition to initial and terminal males in sex-changing wrasses, small alternative males also exist among gonochoristic fishes such as minnows, salmons, livebearers, gobies, and sunfishes (Fig. 20.8). In the bluegill sunfish, *Lepomis macrochirus*, larger, older **parental** males (17 cm long, 8.5 years old) construct

nests, court females, and then guard the eggs that they fertilize. Two forms of cuckolder males parasitize the parental males. **Satellite** males are intermediate in size and age (9 cm, 4 years old); they mimic female coloration and behavior and hence gain access to a nest, interposing themselves between the parental male and the female during spawning. Smaller, **sneaker** males (7 cm, 3 years old) lurk in nearby vegetation and dart through nests during spawnings, depositing sperm literally on the run. These three options arise from two discrete alternative life histories: parental males that delay maturation, grow large, and begin spawning when they are older than 7 years old versus cuckolder males that mature as small 2-year-olds, acting first as sneakers and then later (when they achieve the size of reproductive females) as satellite

FIGURE 20.8. *Variation in life history and behavior results in alternative mating tactics in male bluegill sunfish and Pacific salmon. Both species are characterized by large, territorial males that court females and fight other males versus smaller males that interject themselves during spawnings by larger males. (I) Bluegill males occur as A) large males that dig nests, spawn, and guard eggs; B) small sneaker males that hide in vegetation and dart past spawning pairs, quickly depositing sperm; and C) intermediate satellite males that mimic female coloration and behavior and thus gain access to spawning pairs. (II) The life history alternatives for male bluegill are to mature at a young age and small size and adopt a sneaker role and then later a female mimic role, or to mature later at a larger size and adopt a courting and parental role. All females mature at an intermediate size and age. (III) Coho salmon males occur as large, hooknosed males that fight for access to females in midstream, or as smaller jacks that hide near structure or in shallow water and sneak copulations. Life history alternatives for salmon are to mature at a young age and small size and adopt a sneaker role, or to mature later at a larger size and adopt a fighting role. All females mature at a relatively large size and old age.*

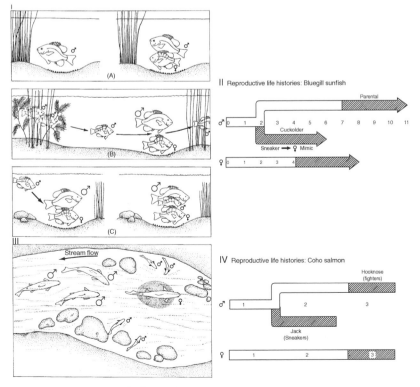

males. Cuckoldry becomes a viable alternative only when parental males are abundant, because parental males provide the breeding opportunity (Gross 1984).

Pacific salmons of the genus *Oncorhynchus* demonstrate an analogous pattern (see Fig. 20.8). Male coho salmon, *O. kisutch,* occur as two types in spawning streams. Large (52 cm long, 2.5 years old), colorful hooknose males court the females that have dug linear nests (redds) in the gravel bottom. Breeding success is directly related to male size and proximity to females; larger fish fight successfully to be closest and thereby spawn most. A second group of males, called jacks, are smaller and younger (34 cm long, 1.5 years old). These males hide in nearby stream debris and dash onto the redd as the hooknoses are spawning with females. Intermediate-size fish are relatively rare, probably because they are too small to fight successfully and too large to hide successfully. Again, sneaking provides a spawning opportunity for small males that otherwise could not compete with large males for access to females (Gross 1984).

In both examples, it is still unknown whether the alternative tactics are brought about by genetic or environmental influences or a combination of the two. Are cuckolders and jacks genetically programmed to behave as such; do they develop in response to immediate environmental conditions, including the density of larger, parental or hooknose fish (in theoretical terms, are these mixed or conditional evolutionarily stable strategies); or do genes and environment combine to determine proportions of males? A combination of influences is indicated by the salmon data. Jacks develop from fish that grow faster when young. Clearcutting around streams typically raises stream temperatures and increases the amount of debris, promoting faster growth and thereby producing more jacks and more habitat favorable to jacks. Hatcheries favor and produce faster-growing fish, and intense fishing pressure also targets the larger fish. Ironically, human activity and exploitation appear to be selecting for less-desirable, smaller, "alternative" fish (Gross 1991).

Alternative mating tactics representing variations on the above patterns exist in other fishes (e.g., in fallfish minnows, cichlids, peacock wrasses, and gobies; Ross 1983; van den Berghe 1988; Barlow 1991; Magnhagen 1992). Why fishes are so labile in mating and life history patterns is problematic. In many instances, the existence of one pattern creates the conditions that favor the development of the other: Sneakers depend on territorial males to provide them with breeding opportunities. However, sneaking is negatively correlated with frequency in that, because of competition, the advantages of sneaking decrease as the density of sneakers increases. Alternatively, no single strategy may confer a consistently greater selective advantage over others, so each is favored at different times and is consequently maintained at least at low frequencies in a population. Finally, differing reproductive modes may represent nothing more than alternative, equally valuable adaptations to similar environmental forces (Fischer and Petersen 1987).

SUMMARY

1. Reproductive success is the ultimate determinant of adaptations. Factors that characterize the breeding systems of fishes include frequency of mating, number of partners, and gender role of individuals. Not surprisingly, fishes show considerable inter- and intrataxonomic variation in all factors. Most fishes spawn repeatedly, but some (lampreys, eels, salmons) spawn only once during their lives. Most fishes have multiple breeding partners (polygyny, polyandry), but examples of monogamy are not uncommon.

2. Gender roles in fishes are labile. Although most fishes remain one functional sex throughout their adult lives, sex reversal in either direction is fairly common. Strong influences on the occurrence, timing, and direction of sex reversal include the social environment, in terms of the number of members of each sex, their relative positions in dominance hierarchies, and the relative reproductive contributions of the sexes at different sizes. All-female species of parthenogenetic livebearers use sperm from donor males to activate cell division in eggs, but the male's genes do not contribute to future generations.

3. Sexually selected traits are those that result from intrasexual competition and from the breeding preferences of the opposite sex. Anatomic differences between sexes, called sexual dimorphisms, include differences in body size, shape, color, dentition, or ornamentation. In most fishes, the male is the dimorphic sex (except that females are generally larger). Many fishes develop breeding tubercles or contact organs, again primarily in the males. A major cost of dimorphisms is that the dimorphic sex is more conspicuous to predators.

4. Many fishes spawn in nests that may be simple depressions in the bottom or more elaborate structures made of rocks or vegetation and constructed (usually) by the male, occasionally glued together with body secretions. Truly random mating is rare; some choice of mates is the norm. The spawning act itself often involves elaborate body and fin movements, color change, and chemical and sound production.

5. Males are more likely than females to guard eggs or young. Internal gestation by females occurs in most elasmobranchs and in a few bony fishes. Mouth brooding is practiced by many cichlids, some catfishes, cardinalfishes, and a few other families. Eggs may also be carried in male brood pouches or attached to the head (seahorses, nurseryfishes). Some cichlids provide food for young in the form of epidermal secretions. Parental care obviously increases survival of the young but limits the future spawning activities of the parents and exposes them to increased predation. Helpers at the nest, in the form of young from previous broods, are known from a few cichlids.

6. Not all members of a population use the same reproductive tactics. Sneaking is known from many species (salmons, sunfishes, wrasses, gobies), in which generally smaller males lurk on the edge of a spawning area and dash in rapidly while a territorial male is spawning. Whether such alternative tactics are genetically fixed in individuals or represent modifiable responses is unknown.

SUPPLEMENTAL READING

Breder and Rosen 1966; Fryer and Iles 1972; Keenleyside 1979; Keenleyside 1991a; Pitcher 1994; Potts and Wootton 1984; Reese and Lighter 1978; Reinboth 1975; Thresher 1984; Wootton 1990.

21

Fishes as Social Animals: Aggregation, Aggression, and Cooperation

Fishes associate during nonmating periods to help or hinder one another. Social interactions, involving aggression or cooperation, occur between individuals of the same species (**intraspecific** interactions) as well as between different species (**interspecific** interactions). Nonreproductive social patterns in fishes involve solitary or territorial individuals, pairs, loose aggregations, and relatively permanent schools or colonies that may change daily, seasonally, and ontogenetically. Fishes keep apart or together through highly evolved transfers of information (communication) that may involve several sensory modes. Our object in this chapter is to review examples of nonmating social interactions in fishes, particularly the influence that communication has on patterns of aggregation, spacing, aggression, and cooperation, to show the diversity of evolved solutions to problems of survival and, ultimately, reproduction. Reproduction and predator-prey interactions are sufficiently influential in fishes to warrant separate chapters (see Chaps. 18, 19, 20).

COMMUNICATION

Communication involves the transfer of information between individuals during which at least the signal sender derives some adaptive benefit (Myrberg 1981). To send information, the signal sender must **contrast** with (stand out from) the background. Although this is most obvious (to us) in a visual context—where bright objects are most easily seen against dark backgrounds and dark objects against bright backgrounds—the contrast principle applies to all sensory modes. Background noise—be it visual, acoustic, chemical, tactile, or electrical—will mask a signal. Information is transmitted when the signal exceeds the noise; conversely, an animal becomes cryptic if it blends in with the background. The message sent usually results in repulsion or attraction, or may inform the signal receiver about the physiological state or behavioral motivation of the sender. Frequently, signals from several modes are combined to enhance the message and reduce ambiguity.

Vision

Vision plays a critical role in fish communication in most environments (see Chap. 6). Coloration is dependent on hue (wavelength mixtures), saturation (wavelength purity), and brightness (light intensity) (Hailman 1977; Levine et al. 1980). Coloration is incorporated into scales, skin, fins, and eyes as the product of pigments, achro-

matic elements, or structural colors. Pigmented cells (**chromatophores**) in the dermis contain carotenoids and other compounds, and reflect yellow, orange, and red. Achromatics are black and white. Black coloration results from movement of melanin granules within melanophores; dispersed melanin darkens a fish, whereas melanin concentrated in the melanophores makes the fish appear lighter in color. White coloration comes from light reflected by guanine crystals in leucophores and iridophores. Greens, blues, and violets are generally structural colors produced by light refracted and reflected by layers of skin and scales; color depends on the thickness of the layers relative to the wavelength of the light (Lythgoe 1979; Levine et al. 1980).

The diversity of color in fishes is essentially unlimited, ranging from uniformly dark black or red in many deep-sea forms, to silvery in pelagic and water-column fishes, to countershaded in nearshore fishes of most littoral communities, to the strikingly contrasted colors of tropical freshwater and marine fishes (see Box 21.1). Visibility (and invisibility) depend on a combination of fish color and the transmission qualities of water in specific habitats. Melanin and guanine reflect light across the entire visible spectrum and are therefore available for use in almost all habitats. Black and white are among the most commonly used colors in fishes (e.g., in minnows, characins, catfishes, sunfishes, damselfishes, butterflyfishes, grunts, drums, cichlids, gobies, triggerfishes). In the clear waters of a coral reef or tropical lake, yellow and its complement indigo blue are most visible; these are the colors commonly found on butterflyfishes, angelfishes, grunts, damselfishes, parrotfishes, and wrasses on reefs and on characins, minnows, guppies, rainbowfishes, and cichlids in tropical waters. Nearshore temperate habitats, particularly in freshwater, tend to be stained with organic compounds that give them a yellowish tinge. Red and its complement blue-green are more visible under these conditions, so it is not surprising that the breeding colors of darters, sunfishes, minnows, sticklebacks, and salmonids often incorporate these (Lythgoe 1979).

Colors on a fish's body may be used in static or dynamic displays. **Static** coloration is generally an identification badge that informs about the species, sex, reproductive condition, or age of a fish. Species are identified through a combination of body form and color; ichthyologists as well as fishes use this combination in determining a species' identification (Thresher 1976). In the myctophid lanternfishes, the number and pattern of photophores (light organs) is species specific and probably aids in schooling and as a sexual isolating mechanism. The taxonomic skills of many fishes are quite good; the Beau Gregory damselfish can apparently distinguish among approximately 50 different species of reef fishes that intrude on its territory (Ebersole 1977). Sexual dimorphism in coloration and body morphology is common in fishes, occurring as a permanent distinction in many tropical species or more seasonally in temperate fishes; generally males are the dimorphic or more distinctive sex (see Chap. 20). Ontogenetically distinctive coloration may aid in identification of potential schoolmates, augmenting the tendency of fishes to aggregate with members of equal size (see below). In at least 18 coral reef families, juveniles differ from adults in color pattern (Thresher 1984; Box 21.1). As French grunts (Haemulidae) settle from the plankton and take up residence on a coral reef, they develop at least four distinctive color phases associated with changes in habitat and behavior (McFarland 1980).

Dynamic displays involve either rapid exposure of colored, previously hidden structures or changes in color. Dynamic displays include movements of the body, fins, operculae, and mouth. Often fin erection or gill cover flaring exposes patches of color that contrast sharply with surrounding structures. Grunts open their mouths in head-to-head encounters to expose a bright red mouth lining. Many fishes flare their gill covers during aggressive, head-on encounters; gills and gill margins often contrast with the rest of the body (salmonids, centrarchid sunfishes, cichlids, labrids, Siamese fighting fishes). Fin erection and coloration play a dominant role in visual displays, probably because the movement associated with their erection is particularly eye-catching. As a result, differential coloration of fins (inclusion of spots and stripes) is common. Fin flicking serves in calling young to parents, as a schooling signal, and during **agonistic** interactions ("agonistic" refers to aggressive and submissive activities, as in the verb "to agonize").

Changeable colors serve primarily to advertise alterations in the behavioral state of a fish or to conceal a fish from aggressors or predators. During agonistic, predator-prey, and breeding interactions, individuals will blanch or darken and develop bars or spots on a moment-to-moment basis (minnows, dolphinfishes, rudderfishes, cichlids, damselfishes, surgeonfishes, tunas, flatfishes). One can often predict the winner of a territorial encounter by observing differences in body shading. On a circadian basis (see Chap. 22), even the most colorful fishes by day turn relatively dull or blotchy at night. For example, neon and cardinal tetras (*Paracheirodon*, Characidae), which are brilliant blue-green and red by day, assume an inconspicuous pinkish tinge as they rest on the bottom at night (Lythgoe and Shand 1983). Such changes suggest that many visually mediated agonistic interactions cease with nightfall but that many piscivorous fishes are still capable of locating prey at night using visual cues (Helfman 1993).

Short-term color change is primarily under immediate control of the nervous system, whereas longer-term, ontogenetic and seasonal change is more likely controlled by hormone levels. Seasonal color change is most often associated with the onset of breeding activity, when territorial males develop bright, contrasting coloration (see Chap. 20). In the spring, North American minnows and darters assume color patterns that rival those in any tropical reef or river assemblage. Females in many of these species undergo less-dramatic seasonal changes. Interesting ontogenetic changes occur in migratory salmonids and anguillids. Many juvenile salmonids live in streams and combine countershading with vertically oblong, dark "parr" marks that may be disruptive in function. These fish migrate to the open ocean as smolts and develop a silvery coloration that is more effective camouflage in open-water, pelagic situations. Upon returning to their natal (birth) stream, many species assume a bright, boldly contrasting breeding coloration that is the antithesis of camouflage. Anguillid eels change from transparent pelagic larvae, to countershaded stream and lake-dwelling juveniles, to bronze or silvery oceanic, reproducing adults.

Visual agonistic displays often involve highly stereotyped movements. Combat may involve lateral displays, where two fish swim in place with fins spread, oriented either parallel or antiparallel (head to tail) (Fig. 21.1). As an interaction escalates, fish may begin body beating, a vigorous swimming in place that pushes water at an opponent and that may indicate relative strengths of the combatants. Hence tactile and acoustic near-field information may be added to the visual display. Antiparallel fish may strike one another with the pectoral fins (as in the anemonefish, *Amphiprion*) or may "carousel," swimming in tight circles around one another. Carouseling can lead to biting of caudal fins or chasing. Color changes frequently accompany lateral displays, and "color fights" occur in some species as different color phases indicate different levels of aggression (e.g., the nandid, *Badis badis*; Barlow 1963). Frontal displays, sometimes with fish facing each other head-on and even grabbing each other's mouth, are also common (e.g., in grunts, corkwing wrasse, kissing gouramis).

Ritualized combat can decide the outcome of an interaction without actual physical fighting. It is in the best

BOX
21.1
The Function(s) of Coloration in Coral Reef Fishes

No single topic has dominated the literature on coloration in fishes more than the question of why many coral reef species are so brightly and boldly ornamented. Numerous coral reef families contain spectacularly colored members, including sea basses, grunts, drums, butterflyfishes, angelfishes, Moorish idols, hawkfishes, damselfishes, wrasses, parrotfishes, dragonets, gobies, surgeonfishes, rabbitfishes, triggerfishes, and filefishes. Even normally drab families such as nurse sharks and moray eels have coral reef representatives with bold coloration.

Discounting speculation that bright coloration in reef fishes is nonadaptive, most hypotheses have focused on coloration serving an informational or anti-informational function. Anti-informational, camouflage functions were favored by earlier workers who felt that bright coloration served as protective resemblance when a fish was compared with the brightly colored background of corals, sponges, tunicates, and algae on the reef (Longley 1917). However, background matching (aside from countershading) more often characterizes benthic animals that are relatively immobile. Active reef fishes would continually change the background against which they were viewed, often making them a contrasting, conspicuous target.

Crypsis may be more effectively achieved via **disruptive coloration** (Cott 1957; see Chap. 19, Camouflage). Many reef fishes contain large patches of dark and light or adjacent contrasting colors. Boldly contrasting dark and light areas that do not follow the outlines of the body tend to break up that profile. Visually hunting predators recognize objects in part by their shape, and a disrupted body outline is more difficult to identify. Barlow (1967) suggested that the split-head coloration of many reef piscivores that lurk for or sneak up on prey (e.g., groupers, triplefins) was a disruptive pattern.

Another visual cue used by predators and prey alike is the eye itself. Animals notice and watch the eyes of other animals. When spearfishing, experienced hunters avoid looking directly at their prey and can thereby move closer. In many reef fishes, real eyes are often concealed, and false eyes are created. This is accomplished via dark head coloration around a dark eye, or reticulated or mottled body coloration that may include numerous circular patterns (Fig. 21.3). These patterns make the eye blend into the head. Additionally, a prominent dark line that conforms to the profile of the head may pass through the eye. This line is usually a lateral stripe in elongate fishes or a vertical bar in deep-bodied fishes (Barlow 1972). False eyespots, consisting of concentric, contrasting colors and termed **ocelli**, may also occur on or near the caudal peduncle (butterflyfishes, snappers, juvenile damselfishes and angelfishes, bothid flounders; also bowfin, red drum, several New World cichlids) or on the soft dorsal fin (damselfishes, plesiopids; also notopterid featherfins, centrarchid fliers, mastacembelid eels). Such ocelli may draw attention away from the animal's real and presumably vulnerable eye, may intimidate or disorient a predator that is about to strike, or may function as a head mimic and reduce damage from fin-biting predators (Neudecker 1989; Winemiller 1990; Meadows 1993). Ocelli may also serve as shoaling signals to maintain group cohesion.

The alternative interpretation of reef fish coloration is that it serves an information-providing function. Many reef fishes are highly social, interacting both intra- and interspecifically. The clear waters of the reef offer an opportunity for visual signals to evolve. Reef fishes have been referred to as **poster colored** to emphasize their conspicuousness and the possible advertising function of their color patterns (Lorenz 1962; see also Breder 1949). The species-specific nature of color patterns also argues for their role in helping individuals tell species apart during mating, aggregating, or territorial encounters (e.g., Harrington 1993). Color patterns often differ among individuals, allowing individual recognition of territorial neighbors or of partners as a pair moves across the reef (Reese 1981). The placement of yellow, red, and black patches posteriorly may help the trailing member of a pair maintain visual contact in the complex reef environment (Kelley

---

interests of both opponents to settle a dispute without incurring injury. The potential for such injury obviously varies among species but can be considerable, as has been discovered by scuba divers who ignored the distinctive head-swinging displays of apparently territorial gray reef sharks (Johnson and Nelson 1973; Fig. 21.2).

A particularly nice example of the multiple functions of visual displays involves the flashlight fish, *Photoblepharon palpebratus* (Anomalopidae; Morin et al. 1975). This 6-cm-long, nocturnally active fish lives in shallow waters of the Red Sea. It possesses a semicircular luminous organ just below its eye that contains continuously emitting bioluminescent bacteria. The light can be turned on and off by means of a muscular lid. The flashlight fish is

BOX 21.1 *(Continued)*

and Hourigan 1983). Some angelfishes and surgeonfishes are colorful and territorial as juveniles, but both color and aggressiveness fade later in life. Such an ontogenetic correlation between agonism and poster coloration is additional support for an informational function of reef fish coloration (Thresher 1984). Also, the eye of many reef fishes is, if anything, exaggerated and highlighted, sitting at the convergence of radiating lines or outlined in bright, contrasting colors (e.g., in some sea basses, angelfishes, damselfishes, wrasses, jawfishes, clinids, surgeonfishes; also some centrarchid basses and cichlids).

Undisturbed reef systems contain an abundance of large, visually hunting predators (lizardfishes, trumpetfishes, cornetfishes, scorpionfishes, flatheads, groupers, hawkfishes, jacks, snappers, emperors, barracuda, flatfishes; Hobson 1994). How can small reef fishes afford to be conspicuous? The answer may lie partly in the reef structure itself. Few other habitats contain the variety and number of hiding places of a healthy coral reef. With adequate refuge sites available, the coloration of fishes that live close to the reef is less constrained by predators than in related, nonreef species. Such correlations hold well for families like damselfishes, wrasses, parrotfishes, and rabbitfishes in which water-column- or grassbed-dwelling species are often countershaded or drab, whereas near-reef species are more boldly colored.

Water clarity can also work against predators, whose activities are conspicuous to potential prey at considerable distances. Diurnally active prey fishes typically have eyes containing dense arrays of small cones that are ideal motion detectors during bright illumination (see Chap. 6, Vision). Any predatory movements will be detected at great distances, allowing prey to take flight or cover long before an attack occurs. Not surprisingly, nocturnally active fishes are not typically poster colored but instead possess relatively uniform coloration. Diurnal fishes seek shelter at night and assume subdued colors that include blotchy, presumably disruptive-camouflage hues.

Bright illumination, clear water, and abundant

refuges have apparently served to liberate coloration in diurnal reef fishes from its usual anti-informational, cryptic function to an informational, communicative function (Thresher 1977). An analogous pattern holds for African cichlids, where small species that live over and take refuge in complex substrates (rocks, snail shells) tend to be much more boldly colored than larger relatives that live over sand or in the water column (Barlow 1991).

Are reef fishes cryptically or conspicuously colored? In all likelihood, they are both. Many early treatments attempted to explain reef fish coloration from dead specimens or from photographs of live individuals taken with unnaturally powerful lights striking fish at unnatural angles. Behavioral observations may provide the best answers. Agonistic encounters occur between neighboring fishes, whereas predatory encounters occur over larger distances. The bright and contrasting colors that many reef species use in their signals and displays may not be visible over the distances at which predator-prey interactions occur because red, orange, and yellow wavelengths are attenuated much more quickly in clear water than are blues and greens. With the brighter colors gone, such patterns as countershading may conceal a potential prey individual from a searching predator.

In addition, reef fish color is not static. Butterflyfishes are among the most colorful of the reef fishes. They and many other boldly colored reef fishes are also countershaded. During aggressive intraspecific encounters in *Chaetodon lunula*, the countershading fades and the yellow coloration intensifies. Intensification of species-specific coloration occurs in many other reef fishes as well as in temperate marine and freshwater species during mating and agonistic encounters (Thresher 1984). A reef fish can mask its poster colors to hide from a predator or to appease a competitor, or it can intensify coloration to intimidate the competition. One can only conclude that reef fish coloration is dynamic and multifunctional (Hamilton and Peterman 1971; Ehrlich et al. 1977).

---

unique in that it forms shoals at night and uses its light for feeding and predator avoidance, and in behavioral interactions. The light is turned on to attract zooplankton prey and then to illuminate prey. If approached by a predator, the flashlight fish swims with the light on and then turns it off and changes direction. The fish thus moves to a place that was unpredictable from its former

direction of movement. In a social context, shoals form at night when small groups swim close enough to see each other's lights. Male-female pairs hold territories over the reef. If an intruding *Photoblepharon* approaches, the female swims up to it with her light off and then turns the light on literally in the face of the intruder, causing it to depart (Morin et al. 1975).

FIGURE 21.1.   *Lateral and frontal displays in fishes. During agonistic interactions, fish may line up parallel, antiparallel, or head to head; they may remain stationary, spread fins or operculae, change colors, swim in place, or circle one another. (A) Typical swimming-in-place lateral display when water currents (arrows) are directed at the head of the opponent, as happens in many cichlids. (B) Lateral display in the clownfish, Amphiprion, during which individuals strike each other with their pectoral fins. (C) Head-to-head pushing in the butterflyfish,* Chelmon rostratus.

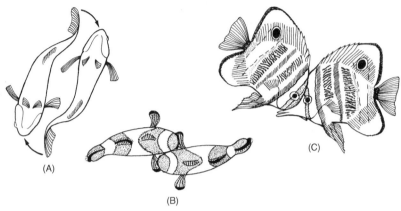

(A)   (B)   (C)

(A) After Chiszar 1978; (B) after Eibl-Eibesfeldt 1960; (C) after Eibl-Eibesfeldt 1970; used with permission.

FIGURE 21.2.   *Exaggerated swimming display of the gray reef shark,* Carcharhinus menissorah. *When approached by a diver, another shark, or a small submarine, or when competing for food, gray sharks lift the snout, arch the back, lower the pectoral fins, and swim in a tense, exaggerated manner (exaggerated postures shown on left, comparatively normal swimming postures on right). If the intrusion continues, the displaying shark may attack the intruder. Similar displays, without attacks, have been observed in Galapagos, silky, lemon, and bonnethead sharks.*

DISPLAY   NON-DISPLAY

From Johnson and Nelson 1973; used with permission.

FIGURE 21.3. *Color patterns that camouflage the eye in reef fishes. Predators and prey alike focus on the eyes of other fishes, and many fishes have color patterns that tend to mask the eye or call attention away from it. (A) A blackcap basslet (Serranidae): The dark eye is contained in a dark area. (B) A peacock flounder (Bothidae) and (C) a wrasse (Labridae): Numerous false eyes call attention away from the real eye. (D) A frogfish (Antennariidae): A small eye is subsumed in a series of radiating patterns that converge on different points.*

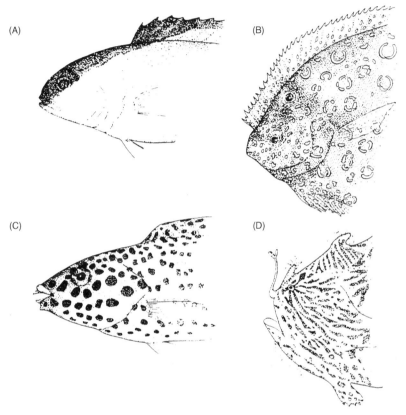

From Barlow 1972; used with permission.

Sound

Sound production occurs in at least 50 families of cartilaginous and bony fishes (Myrberg 1981; Hawkins and Myrberg 1983; Hawkins 1993). Sound production most commonly involves 1) prey responses to being startled or handled by predators ("stay away" and "release" signals); 2) mate attraction, arousal, approach, or coordination sounds; 3) agonistic interactions with competitors for mates and resources ("stay away" signals); and 4) attraction of shoalmates.

Startle and release calls occur in families as different as eagle rays, herring, characins, catfishes (of many families), cods, squirrelfishes, sea robins, and porcupinefishes. They are elicited when a fish is grabbed, poked, or even surprised. A sudden grunt, croak, or drumbeat might distract a predator, perhaps causing it to release its grip on the prey or hesitate in its attack long enough for the prey to escape. A release call could also attract additional predators, including predators on the individual holding the signaler. A small predator with prey in its mouth could be handicapped in its own efforts to evade a larger predator and might abandon its meal rather than risk becoming one (Mathis et al. 1995). Release sounds could also function as alarm calls (see Chap. 19, Discouraging Capture and Handling) that notify conspecifics of a predator's presence and activity. The caller would have to have close relatives nearby that could benefit from the sound to offset fitness losses to the signaler from being eaten.

Sound is an integral part of the courtship and spawning behavior of many fishes (see Chap. 20). Some sounds produced by male damselfishes (Pomacentridae) and European croakers (Sciaenidae) drive off intruding males. Territorial males also produce vocalizations to bring females closer during courtship (e.g., toadfishes, centrarchid sunfishes, gobies). Signaling rate frequently increases as a female draws nearer or during the spawning act itself (cods, serranids), suggesting that acoustic communication synchronizes activities between members of a pair. In at least one species of African mouth-brooding cichlid, male vocalizations stimulate gonadal activity in females, paralleling a widely observed phenomenon in seasonally breeding birds (Myrberg 1981; Lobel 1992).

During agonistic encounters associated with territorial behavior, sounds are usually produced by an aggressive or dominant animal; the response of the submissive animal is usually to retreat from the signal sender. Sound production during agonistic interactions occurs in many teleosts, including sea catfishes (Ariidae), loaches (Cobitidae), gouramis (Belontiidae), squirrelfishes (Holocentridae), damselfishes (Pomacentridae), and triggerfishes (Balistidae). Submissive animals also produce sounds that may reduce aggression in an opponent, as recorded from anemonefishes (*Amphiprion,* Pomacentridae) (Myrberg 1981; Hawkins 1993). The importance of sound production during territoriality is evident in the loach, *Botia horae* (Cobitidae), which vocalizes and displays visually to repel intruders of shelter sites. When experimentally muted, residents are unable to repel intruders, whereas sham-operated and intact animals defend their territories successfully (Valinsky and Rigley 1981).

Sound production also functions during shoal formation. Although most sounds are produced actively by vibrating the swim bladder or stridulating of teeth, bones, and fin spines, some important sounds result from water displacement by fins and bodies during swimming and are detected via the lateral line of neighboring fish. Such water displacement informs aggregating fishes of their location relative to schoolmates, serving as a minor repulsive force that combines with visual input to maintain distance between individuals. When pollack (Gadidae) are experimentally blinded, they swim slightly farther from schoolmates than when intact. In unblinded fish in which the lateral line nerve is severed and acoustic information therefore eliminated, they swim closer than normal to schoolmates (Pitcher et al. 1976).

Interestingly, more actively schooling species within a family (e.g., among cods and damselfishes) are relatively quiet, and sound production in group-spawning fishes is not as common as it is in solitary, territorial species. Whether this silence helps prevent detection by predators or results from other factors is unknown (Hawkins and Myrberg 1983; Hawkins 1993). Eavesdropping by predators may be a significant cost of sound production. Many predatory fishes (sharks, jacks, groupers, snappers, barracuda, tunas) are attracted to the incidental, low-frequency sounds produced by feeding or injured fishes.

Active sound production also has its costs. Bottlenose dolphins (*Tursiops truncatus*) include a disproportionate number of sound-producing fishes (e.g., croakers, grunts, toadfishes) in their diet (Barros and Myrberg 1987). Interception of signals, whether by predators, competitors, or potential prey, is always a potential cost affecting the evolution and use of communication signals by a species.

Chemical Communication

Exchanges involving chemicals primarily involve the release and reception of **pheromones**, which are chemicals secreted by one fish and detected by conspecifics and that produce a particular behavioral or developmental response in the receiving individual (Hara 1982, 1993; Liley 1982). Chemicals are sensed by both gustation (taste) and olfaction (smell) in fishes. Sensory receptors are often located not only in the mouth and nostrils but also on the barbels or even the body surface in many fishes (see Chap.

6, Chemoreception) or on filamentous, muscularized pelvic fins (e.g., gouramis, Belontiidae).

Chemicals play an important role in food finding and predator avoidance (see Chaps. 18, 19), mating (Chap. 20), migration (Chap. 22), parental care, species and individual recognition, aggregation, and aggression in fishes. Parents of several cichlid species recognize their young via chemical signals, and young recognize their own parents (Myrberg 1975). In salmonids, skin mucus contains species-specific amino acids that are used for individual and sexual recognition. Species recognition in other species is also mediated by chemicals in skin mucus (Hara 1993). Bullhead catfishes (Ictaluridae) and European minnows (*Phoxinus phoxinus*) can recognize individual conspecifics based on odor. Several schooling species (herring, minnows, plotosid catfish, young salmonids) show an attraction response to water that has contained conspecifics (Pfeiffer 1982).

Chemically mediated agonistic interactions include scent marking of territories or shelters, which takes advantage of the persistence of chemical signals relative to other sensory modes (Atema 1979). Members of sexual pairs of blind gobies, *Typhlogobius californiensis*, defend a burrow against individuals of the same sex. Recognition of burrow mates and of intruders is based on chemical cues. Yellow bullhead catfishes, *Ameiurus natalis*, develop dominant-subordinate relationships that are mediated by chemical secretions. Experimentally blinded fish can discriminate between odors produced by different individuals. If a dominant fish is removed and then returned to a tank the next day, a blinded subordinate still treats it as dominant. If the dominant is returned after losing an agonistic encounter with a third fish, the previous subordinate will attack it, again based on chemicals apparently produced in the skin mucus. Bullheads also produce an aggression-inhibiting pheromone when living in groups. Fighting by aggressive individuals even decreases when they are exposed to water in which a communal group was living. In Siamese fighting fishes, *Betta splendens* (Belontiidae), males display more actively in front of mirrors when placed in water that had contained another male (Todd et al. 1967; Todd 1971; Hara 1993).

Touch

Tactile information is transmitted at close range, when fish are in contact. Accurate information about relative strength of combatants can be exchanged during pushing matches or when fish lock jaws during mouth fighting. Many fights escalate into and end in biting. Anemonefish strike each other with their pectoral fins during antiparallel lateral displays (see Fig. 21.1). Fish frequently touch each other with tactile sensors, such as barbels in catfishes, loaches, and goatfishes, but also with long, filamentous pelvic fins (e.g., gouramis); sea robins (Triglidae) have touch receptors in their separated anterior pectoral fin rays. Nuptial tubercles, epidermal bumps on the body and fins of many fishes (see Chap. 20), are used to stimulate potential mates and maintain contact between a pair during breeding. Courtship and copulation in many sharks involves biting of the female by the male; the male often holds onto the female's fins or body for prolonged periods.

Touching between fishes is fairly uncommon, except during extreme fights, mating, and parent-offspring interactions. In parental species, young frequently contact the parent, usually using the mouth to feed on parental tissue or mucus. Such behavior also serves to maintain cohesion between parent and young, to promote parental behavior (perhaps by stimulating the production of parental behavior-inducing hormones), and to communicate the behavioral state of the young, such as fear or hunger. Parent-touching behavior occurs in bonytongues (Osteoglossidae), catfishes (Bagridae), damselfishes, and more than 20 species of cichlids (Noakes 1979).

Electric Communication

Fishes are unique in that some species both produce and receive electrical information based on very weak electrical output (see Chap. 6, Electroreception, Electric Communication). The electric organ discharge (EOD) is species specific and often sex specific. A fish can modify amplitude, frequency, pulse length, or interpulse length of its discharge or alter parts of its EOD, such as the fundamental frequency or peak power frequency. Fish can thus exchange information about species, sex, size, maturational and motivational state, location, distance, and individual identification. Electrical discharges are used commonly during agonistic interactions (Bullock et al. 1972; Westby 1979; Hagedorn 1986; Hopkins 1986).

AGONISTIC INTERACTIONS

Aggressive interactions usually result from competition or potential competition for valuable resources. Defendable resources include food and feeding areas, refuge and resting sites, mates and mating grounds, eggs, and young. Defense can produce dominance hierarchies in aggregating fishes or territoriality in more solitary species. In addition, a hierarchy can exist among neighboring territory holders, and dominant-subordinate relationships often exist when solitary fish meet.

Behavioral Hierarchies

Dominance hierarchies (peck orders) are either linear or despotic. In **linear** hierarchies, an alpha animal dominates all others, a beta animal is subordinate to the alpha but dominates lower-ranked individuals, and so on, down to the last, or omega, individual. Such a hierarchy exists in harems of the sex-changing cleaner wrasse, *Labroides dimidiatus* (see Chap. 20). A single male dominates up to six females, which in turn have their own linear hierarchy. Linear hierarchies also exist in salmonids, several livebearers, and centrarchid sunfishes (Gorlick 1976).

In **despotic** situations, a single individual, the despot, is dominant over all other individuals, while subordinate animals have approximately equal ranks. In captive an-

guillid eels, a single large individual can monopolize 95% of a 300-liter aquarium, relegating 25 other individuals to the remaining area, where they mass together in continual contact. Despotic hierarchies have also been observed in coho salmon, *Oncorhynchus kisutch*, and bullhead catfishes, *Ameiurus* spp. (Todd 1971; Paszkowski and Olla 1985).

Dominance can be determined by size, sex, age, prior residency, and previous experience. In general, large fish dominate over smaller, older over younger, and residents over intruders. In many species, males usually dominate females, whereas in others, such as guppies, females dominate males. Previous experience, in terms of recent wins and losses, often determines the outcome of future interactions; victorious fish tend to be aggressive and defeated fish submissive. Dominant fish typically occupy the most favorable microhabitats, relegating subordinates to suboptimal sites with respect to cover availability, current velocity, or prey densities. As a consequence, dominant individuals will have higher feeding rates, which ultimately lead to faster growth, better condition, and higher fitness (e.g., salmonids; Bachman 1984; Gotceitas and Godin 1992). Dominance hierarchies have also been observed in requiem and hammerhead sharks, minnows, ictalurid catfishes, amblyopsid cavefishes, cods, ricefishes (Oryziidae), topminnows, livebearers, centrarchid sunfishes, cichlids, labrids, blennies, and boxfishes (Ostraciidae).

Territoriality

Territoriality implies a defended space, either the **personal space** around an individual (= individual distance) or a bounded area around some resource. A territory may encompass several resources, as in male pomacentrid damselfishes in which the territory provides food (algae), a spawning site and the eggs spawned there, and refuge holes that provide both protection from predators and a place where the territory holder can rest at night. Territories are often subunits of the larger home range occupied by an individual (see next section). Territoriality is widespread in fishes, occurring in such diverse groups as anguillid eels, cyprinids, ictalurid catfishes, gymnotid knifefishes, salmonids (affecting stocking programs and the effects of introduced species), frogfishes, rockfishes, sculpins, sunfishes and black basses, butterflyfishes, cichlids, damselfishes, barracuda, blennies, gobies, surgeonfishes, and anabantids. Territoriality has not been observed in agnathans or elasmobranchs, although the threat responses of gray reef sharks (see Fig. 21.2) may represent defense of personal space.

Territorial defense often involves displays such as fin and gill spreading, lateral displays and exaggerated swimming in place, vocalizations, chasing, and, as a last resort, biting. Prolonged exchanges of displays frequently occur at territorial boundaries. When territories are being established or contested (as opposed to temporary trespassing), priority of ownership, previous experience, and individual size usually determine the outcome of a dispute. Again, territory holders win over intruders, previous winners defeat previous losers, and large fish win over small fish. Territories near one another can create **territorial mosaics** of several contiguous territories (e.g., salmonids, pomacentrids, mudskippers, blennies; Keenleyside 1979).

The costs of territorial defense—energy and time expended, exposure to predators, resource loss to competitors while defending distant portions of a territory—increase with increasing territory size. Food production affects territory size, because a territorial animal must often meet its daily energy requirements from the resources available within its territory. As would be expected, increased food density leads to a decrease in territory size (e.g., in rainbow trout; rockfishes, Scorpaenidae; surfperch, Embiotocidae; several damselfishes; Hixon 1980a). Interestingly, in Beau Gregory damselfish, males decrease territory size with increasing food, but females respond by increasing territory size. Larger females can produce more eggs, and hence increased energy intake apparently overcomes the costs of defending a larger territory (Ebersole 1980).

Territoriality is often flexible. Territorial boundaries and intensity of defense can vary as a function of the relative impacts of different intruders. Herbivorous damselfishes defend a larger space against large competitors such as parrotfishes and surgeonfishes than against small damselfishes; damselfishes also tolerate large competitors for shorter times inside the territory, attacking them more aggressively. The strongest attacks are directed at potential egg predators, which have the greatest relative impact on the reproductive success of the damselfish. Juvenile coho salmon defend a larger territory against larger conspecific intruders. Butterflyfishes chase species with which they overlap in diet but tolerate the presence of noncompetitors (Myrberg and Thresher 1974; Reese 1975; Ebersole 1977; Dill 1978).

Territoriality may also vary over time at several levels. Juvenile grunts (Haemulidae) form daytime resting shoals over coral heads. Individuals stake out small territories of about 0.04 m² within the shoal; the territories often contain refuge sites from predators. These territories are defended vigorously with open-mouth displays, chases, and biting. In the evening, the shoal becomes a polarized school that moves from the reef to adjacent grassbeds to feed. No agonism is seen during the migratory period, as such behavior would negate the antipredator function of the school when moving across the dangerous reef edge. Once in the grassbeds, the shoals break up, and the fish occur as widely spaced, foraging individuals, implying space-enforcing behaviors (McFarland and Hillis 1982). Territoriality changes with age in many species. Young Atlantic salmon are territorial in streams. As they grow and their food requirements shift to larger prey, they move into deeper water and join foraging groups that have dominance hierarchies rather than territories (Wankowski and Thorpe 1979). In some species, agonistic interactions occur during the breeding season, with fish aggregating peaceably at other times (e.g., codfishes).

Home Ranges

Territories are usually a spatial subset of the larger area that a fish uses in its daily activities. Such **home ranges** or activity spaces are common in fishes, which move over the same parts of the habitat at fairly predictable intervals, often daily but also at other time scales. Home ranges are dependent on fish size and species. Larger species generally move over larger ranges. The home range may be very restricted, as in the few square meters around a coral head (e.g., gobies and damselfishes) or contained in a tide pool (e.g., pricklebacks) (Sale 1971; Horn and Gibson 1988). Benthic stream fishes may have ranges of 50 to 100 m² (Hill and Grossman 1987). Intermediate ranges of a few hundred square meters characterize many lake, kelp bed, and reef species, whereas pelagic predators such as tunas, salmons, large sharks, and billfishes cross entire oceans seasonally or repeatedly. But even these oceanic wanderers show evidence of periodic residence in certain areas on a seasonal basis (see Chap. 22, Annual and Supra-annual Patterns: Migrations).

Although a fish may spend 90% of its time each day within its home range, it is common to encounter individuals many meters or even kilometers away from their usual activity space. These periodic excursions imply a well-developed homing ability in many species. Numerous studies, involving experimental displacements of tagged individuals, have repeatedly shown a strong tendency to return to home sites in many fishes. The tidepool sculpin, *Oligocottus maculosus*, can be displaced as far as 100 m from its home tidepool and will find its way back, using either visual or chemical cues. Older fish can still remember the way home after 6 months in captivity. Younger fish have shorter memory spans and require both visual and chemical cues to find home successfully, whereas older individuals do not require both types of information (Horn and Gibson 1988). In some species, adults find their way around the home range by identifying landmarks, creating a **cognitive map** of the locale (Reese 1989). In general, older fish have a stronger homing tendency and may occupy smaller home ranges than younger individuals. Since juveniles are the colonists that most often invade recently vacated or newly created habitat, this generalization is not surprising (Gibson 1993).

Use of a home range is affected by several components of a fish's biology. Normal ranges are often deserted during the breeding season. This may involve no more than a female damselfish having to leave her territory to lay eggs in the adjacent territory of a male, but can also involve long-distance movements of 100 km or more to traditional group-spawning areas, such as occurs in many sea basses on coral reefs. Colorado squawfish, *Ptychocheilus lucius* (Cyprinidae) make annual round-trip movements of as much as 400 river km between traditional spawning and normal home range areas.

Home range also interacts with shoaling behavior in some species and can differ among individuals within a species. Yellow perch, *Perca flavescens*, form loose shoals (see next section) of many individuals that forage in the shallow regions of North American lakes. Home range size is directly correlated with the amount of time individuals spend in shoals. Individuals with strong shoaling tendencies also have larger home ranges. As a shoal enters the residence area of an individual, the resident fish joins the shoal until the shoal moves to the boundary of the home range. Home ranges and fidelity to particular sites have probably arisen because intimate knowledge of an area increases an individual's ability to relocate productive feeding areas or effective refuge and resting sites, reducing the amount of energy expended and risk incurred while searching for such locales (Helfman 1984; Tyus 1985; Shapiro et al. 1993).

AGGREGATIONS

Shoaling

The most obvious form of social behavior in fishes is the formation of groups, either unorganized **shoals** or organized, polarized **schools** (Fig. 21.4). By convention, some social attraction among individuals is required for a group to be considered a shoal or a school, whereas fish that are mutually attracted to food or other resources are an **aggregation** (e.g., Freeman and Grossman 1992). Shoals involve social attraction, coordination, and numbers. Two fish are not a shoal, because one fish often leads and the other follows. However, when three or more fish co-occur, each fish reacts to the movements of all adjacent fish. The shoal becomes the leader and the fish become the followers. Operationally defined, a shoal is "a group of 3 or more fish in which each member constantly

adjusts its speed and direction to match those of the other members" (Partridge 1982, 115).

As many as half of all fish species may form aggregations at some time in their life. Aggregations can serve several purposes concurrently. The chief functions are to reduce the success of predators (see Chap. 19), increase foraging success (Chap. 18), synchronize breeding behavior (Chap. 20), and increase hydrodynamic efficiency. Some species shoal throughout their lives (e.g., many herrings, anchovies, minnows, silversides), others only as juveniles (bowfin, plotosid catfishes, pufferfishes). Some species aggregate when young, disband as juveniles or adults, and reaggregate to spawn, either in groups or as pairs (many salmonids, sea basses). Foraging aggregations may turn into breeding aggregations as fishes migrate to traditional spawning locations and are joined by members of other aggregations (yellow perch, grunts, rabbitfishes). Normally solitary adults may congregate during the winter, and such aggregations probably remain together through a spring spawning season (e.g., carp). Within a species, schooling tendency may change with predation intensity. Guppies in predator-dense habitats school throughout their lives, but only juveniles school where predators are rare. European minnows that co-occur with predators inherit a stronger schooling tendency than do minnows without predators (Magurran 1990). The intense predation pressure that small fishes experience and the prevalence of shoaling behavior, particularly

FIGURE 21.4.   *Types and activities of fish aggregations. Shoals contain fish attracted to one another but whose activities are only loosely coordinated. In schools, behavior is synchronized: Fish often swim parallel, in the same direction, with fairly uniform spacing (termed the "polarized school" by many authors). Foraging and spawning groups generally form shoals, whereas predator avoidance often results in highly synchronized schooling activities. In this figure, five common antipredator actions and their relationship to grouping behavior are shown in the smallest circles (see also Chap. 19, Responses of Aggregated Prey).*

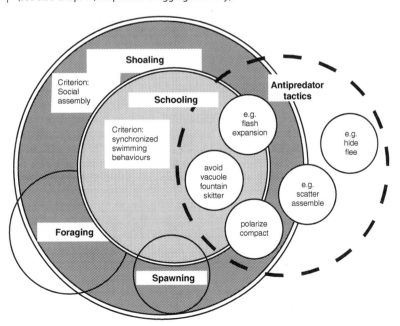

After Pitcher 1983; used by permission of the publisher Academic Press Limited, London.

among juveniles and small species, attest to the antipredator function of most aggregations (see Chap. 19) (Shaw 1970, 1978; Magurran 1990; Pitcher and Parrish 1993).

Regardless of function, most fish shoals are relatively unstable. Few fishes maintain their groups through an entire 24-hour period. Many shoals form each morning, disband at night, and reform the following morning but with different individuals. In fact, available evidence indicates little shoal fidelity in most fishes: Dace, minnows, yellow perch, surgeonfishes, parrotfishes, and bluegill sunfish join and leave foraging shoals frequently (Freeman and Grossman 1992). Fidelity may be stronger in nocturnal fishes that reaggregate each dawn and form daytime resting schools at fixed refuge locales (e.g., squirrelfishes, grunts, copper sweepers, some bullhead catfishes; Hobson 1973; Helfman 1993). In some nocturnal species, relatively complex social structures and interactions develop that rival the societies of birds and mammals (Box 21.2). Migratory schools (e.g., large tunas, bluefish) may also show strong group fidelity, but definitive information is lacking.

Fishes space themselves fairly regularly within fish schools (e.g., Partridge et al. 1983; Abrahams and Colgan 1985). A perfect crystal lattice is the theoretical ideal spatial distribution for school members: Neighbors should be 0.3 to 0.4 body lengths apart, 5 body lengths behind, and centered between preceding fish, with neighbors beating their tails in antiphase (opposite directions). Fish could gain a 65% energy savings from the wakes and vortices generated by fish around them (Weihs 1975). However, few if any groups achieve the proposed ideal lattice structure (Pitcher and Parrish 1993). Earlier authors had proposed that a hydrodynamic advantage may also

develop through drag reduction when one fish swims through the mucus produced by fish ahead of it in a school (Breder 1976). However, insufficient mucus is produced, even in schools of a billion fish, to affect drag significantly (Parrish and Kroen 1988). Predation also disrupts spacing, because individuals under attack should attempt to place schoolmates between themselves and a predator (the **selfish herd** phenomenon; Hamilton 1971). Again, however, experimental tests call into question whether central locations are in fact safer (Parrish 1989). Obviously, many factors (size, sensory input to the lateral line, visibility of neighbors, swimming speed, species composition, vulnerability to and behavior of predators, social status) contribute to the exact (and variable) structure of schools.

Colonial fishes form essentially stationary aggregations. Colonies may exist for breeding, as when male sunfishes and cichlids and female triggerfishes aggregate and construct nests or set up display sites. Some damselfishes set up contiguous territories in suitable habitat patches on a coral reef. Three-spot and bicolor damselfishes will be found in areas of a few square meters, even though adjacent, similar reef areas contain no such fishes (Schmale 1981). Garden eels (Congridae) occupy small burrows a few centimeters apart on sandy regions of coral reefs (Fig. 21.5). Jawfishes (Opisthognathidae), another burrowing coral reef form, tend to form colonies of two to nine individuals on rubble-strewn sandy bottoms (Colin 1973).

Optimal Group Size

Predator success decreases as prey group size increases (Neill and Cullen 1974; Landeau and Terborgh 1986).

FIGURE 21.5. *Garden eels live in colonies of several hundred individuals on sand bottoms near coral reefs. Individuals feed on zooplankton during the day, often extending just the anterior portion of their bodies out of burrows, the sides of which are cemented with mucus produced by the fish's skin. Withdrawal of one individual into its burrow stimulates withdrawal of all other members of the colony.*

Photo by H. Fricke.

BOX
21.2

Social Transmission of Cultural Traditions in Fishes

Traditions are social behaviors maintained across generations, either by inheritance or by learning, as when young individuals are taught by or observe and copy the actions of older individuals. Culture, to biologists, is not the refinement of tastes and artistic judgment but rather involves the behavioral transmission of information. By watching and copying the activities of other, usually older individuals, young chimpanzees learn how to use grass stems to fish termites out of mounds, young antelope learn the locations of communal display (lekking) grounds, and young oystercatchers learn how to open bivalves; fishing, lekking, and bivalve opening are hence cultural activities (Bonner 1980).

Fishes also exhibit a number of behavioral traditions. The same breeding locales are frequently used year after year in both marine and freshwater fishes. Although part of this continued use relates to site-specific, appropriate conditions for spawning, dispersing, or caring for larvae, seemingly adequate, nearby sites are ignored while the traditional site continues to be used. Traditional breeding locales have been found in numerous fish species (e.g., herring, groupers, snappers, surgeonfishes, rabbitfishes, parrotfishes, wrasses, mullets) (Loiselle and Barlow 1978; Johannes 1981; Thresher 1984; Turner 1986). The return of salmon to their natal stream to spawn is not included here, because information is obtained through individual imprinting and memory and may have a hereditary component (McIsaac and Quinn 1988).

The process by which traditions are established and maintained has been investigated with respect to breeding sites in wrasses and twilight migration routes in grunts. Bluehead wrasses, *Thalassoma bifasciatum*, have a mating system in which many females mate with solitary large males that hold territories. The locations of these territorial mating sites may remain stable for more than 12 years, encompassing four wrasse generations. Adjacent, seemingly appropriate sites (downcurrent edges of a reef that include vertical projections) are not used. Female choice of sites, rather than choice of males, determines where males establish territories. To test whether traditions were maintained by a genetic response or through social transmission, Warner (1988, 1990) removed entire populations from reefs and replaced them with naive individuals. He found that breeding sites chosen by transplanted groups

were a random sample of the available locales and that newly used territories eventually became traditional breeding sites. Hence former traditional sites were in fact maintained by social convention. Interestingly, if an additional removal/replacement manipulation was performed on the same reefs, the second group of transplants tended to prefer the same sites as the first transplanted group. This suggests that fish assessed site quality and had definite preferences, which did not necessarily include the original traditionally used spawning locales. Hence tradition is powerful enough that a breeding locale may continue to be used even though that locale may not be the best available in the habitat.

Juvenile grunts (*Haemulon*), as well as many other nocturnal reef fishes, undergo a remarkably predictable migration at dawn and dusk each day that probably thwarts the success of twilight predators (McFarland et al. 1979). Grunts feed on invertebrates at night in the grassbeds that adjoin patch and fringing reefs. By day, they form resting schools over coral heads. The locations of such schools and the routes taken by the school resident over a coral head represent traditional activities. Over more than a 3-year period, a school will take the same approximate route to and from the grassbed, even though no individual grunt in the school is more than 2 years old. How do younger fish know the correct route between resting site and grassbed?

To test for the relative influences of genetic and social transmission, Helfman and Schultz (1984) transplanted individuals between schools of juvenile French grunts. After mapping established migratory routes, members of distant groups were added to resident schools. Transplanted fish (identified by small injected paint marks) were allowed to follow residents for four twilights. Then all residents were removed, and the migration of the transplanted fish was observed. The route was similar to the one taken previously by the residents (Fig. 21.6). To test for a possibly innate response (i.e., given the terrain, any grunt at this locale would take the same route regardless of experience), new transplants were given no opportunity to observe migrating fish. These control fish migrated in a variety of directions. Hence the social traditions of resting site locale and twilight migration route in grunts are established via cultural transmission.

Larger prey groups also experience greater competitive success, foraging efficiency, and hydrodynamic efficiency. Consequently, selection should favor prey fish that join and maintain large shoals. But above some group size, benefits are offset by competition for food and mates, interference between individuals in avoiding predators (e.g., confusion among group members due to collisions, indecisiveness, or obstructed views), and increased conspicuousness of large versus small groups (Pitcher and Wyche 1983; Abrahams and Colgan 1985; Parrish 1988). Oxygen consumption in large groups may also leave trailing members in regions of depleted oxygen (McFarland and Moss 1967). Given these costs and benefits, can an optimal group size be determined? Do fish tend to form or join optimally sized shoals?

Optimal group size is complicated because antipredator functions probably favor larger optima than do feeding groups. If dominance hierarchies exist, dominant individuals, with their preferential access to resources, may have a larger optimum than would subordinate members. Subordinate animals must decide between sustaining the costs of a large group versus being alone. The conflicting needs of group members may make attainment of an optimum impossible. Minimal group size should be easier to determine than maximal size. Prey should avoid being alone or joining very small groups. When fathead minnows are allowed to choose between shoals of different sizes, they consistently choose the larger of the two shoals, particularly when one shoal is relatively small (fewer than eight fish) and a predator is present. This suggests that minnows alter their decisions when potential costs (e.g., predation) are greater (Hager and Helfman 1991).

INTERSPECIFIC RELATIONS: SYMBIOSES

Symbiosis is the living together of two unrelated organisms. In **parasitism**, one member of a pair benefits and the other suffers a reduction in fitness; in **mutualism**, both members of a pair benefit; in **commensalism**, one benefits and the other is neither harmed nor helped. Mutualistic relationships are the most interesting because they indicate a relatively long period of coevolution between pair members. Fishes form mutualistic relationships with other fishes and with a variety of invertebrate species.

Parasitism

The abundance and diversity of external and internal parasites of fishes are tremendous and beyond the scope of this discussion (e.g., Dogiel et al. 1961; Sinderman 1990; Gabda 1991). A few fish are internal parasites on other animals, including members of three families of fishes: synaphobranchid eels, trichomycterid catfishes, and carapid pearlfishes (see below). The snubnose parasitic eel, *Simenchelys parasiticus* (Synaphobranchidae), burrows into the flesh of bottom-living fishes such as halibut and has even been found in the heart of mako shark (see

Chap. 14, Subdivision Elopomorpha). The trichomycterids are particularly insidious because, in addition to feeding on tissue and blood in a host's gill cavity, the diminutive, eel-like candiru (*Vandellia*, 2.5 cm) of South America purportedly enters the urethra of human bathers and wedges itself there with its opercular spines, requiring surgical removal.

Many fishes, including marine catfish, characins, tigerperches (Theraponidae), carangids, sea chubs (Kyphosidae), cichlids, blennies, and spikefishes (Triacanthodidae) fall into a category somewhere between parasitism and predation by removing scales or taking pieces out of fins of other fishes (Losey 1978; Noakes 1979). Populations of scale-eating cichlids from Lake Tanganyika (*Perissodus*) contain even numbers of individuals whose mouths twist left or right, facilitating scale removal from the right or left sides of their prey respectively (Hori 1993). Cookie cutter sharks, *Isistius* spp. (Squalidae), remove plugs of flesh and blubber from tunas and cetaceans, and lampreys rasp through the skin of numerous fishes and feed on tissues and body fluids (see Chaps. 12, 13). At least two species—the cutlips minnow, *Exoglossum maxilingua* of North America, and the eyebiter cichlid, *Haplochromis compressiceps* of Lake Malawi—commonly remove eyeballs from unsuspecting prey (neither species' behavior has been studied adequately). Access to food sources for many of these **partial consumers** depends on deceit; small piranhas resemble and school with other characins and then bite the tails off their schoolmates; several juvenile carangids (e.g., *Scomberoides*, *Oligoplites*) resemble their silverside and anchovy schoolmates, whose scales they remove; sabretooth blennies mimic cleaner fishes (see below); and cookie cutter sharks may mimic bioluminescent invertebrates that live in the deep-scattering layer of the mesopelagic region (see Chap. 17, The Deep Sea) (Losey 1978; Sazima and Machado 1990).

Mutualism and Commensalism

Some of the best-studied mutualistic interactions involve fishes that pick external parasites from other fishes. Although **cleaning** behavior exists in almost all aquatic environments and involves dozens of shrimp and fish species (Sulak 1975; DeMartini and Coyer 1981; Lucas and Benkert 1983), coevolved relationships are more obvious on coral reefs (Limbaugh 1961; Feder 1966; Losey 1987). Juveniles of a number of wrasse, butterflyfish, damselfish, and angelfish species clean other fishes, but cleaning specialists occur only among the cleaner wrasses (*Labroides*) of the Pacific and the neon gobies (*Gobiosoma*) of the Caribbean (Fig. 21.7). Cleaners are usually territorial, occupying well-defined and often prominent coral heads or other locales referred to as **cleaning stations**. Communication between host and cleaner is obvious and stereotyped. Host fishes of numerous species approach these stations and pose, frequently assuming head-up or head-down positions while hovering in the water column, blanching in color, spreading their fins, and opening their mouths. Cleaners approach in a bouncing or tail-wagging manner and then pick over the host's body surface, frequently entering the mouth or gill covers of

FIGURE 21.6. *Testing for social learning in juvenile grunts. Twilight migration routes at four experimental sites are shown. Resident fish are those with an established migration route (dashed lines). Transplants were brought from another location and allowed to follow resident fish. Control fish were also transplanted but were released at the resident site after resident fish were removed. Hence controls had no opportunity to learn the routes. At sites 1 through 3, transplants adopted the resident route and used it even when residents were absent (dotted lines). Controls never used the resident route (wavy lines). At site 4, residents did not migrate but instead drifted away in different directions. Transplants also drifted away from the resting site, whereas controls underwent a distinctive migration.*

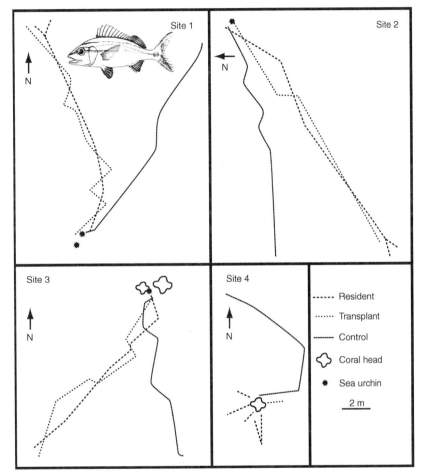

Modified from Helfman and Schultz 1984; used with permission. Grunt drawing from Gilligan 1989.

herbivores and piscivores alike. Parasites, particularly copepods, are removed, as are pieces of tissue around wounds. Since hosts without parasites or wounds will solicit cleaning, tactile stimulation alone by the cleaner must attract some fishes (Losey 1979). A cleaning bout is terminated when the cleaner leaves or the host fish shudders or snaps its mouth closed and open. Cleaning relationships have been exploited by humans, who use the European corkwing wrasse, *Crenilabrus melops*, to help reduce external parasite infestations on Atlantic salmon kept in aquaculture pens (Sayer et al. 1996).

Many cleaners have converged on coloration patterns involving bold stripes. Both cleaner fish and shrimp appear to be largely immune to predation and consequently service such predators as moray eels, sea basses, snappers,

and barracuda. In the Indo-Pacific, at least one wrasse species, *Diproctacanthus xanthurus*, cleans damselfishes that do not leave their territories and hence cannot take advantage of cleaning stations. Two species of sabretoothed blennies, *Aspidontus taeniatus* and *Plagiotremus rhinorhynchus*, mimic the coloration and behavior of *Labroides* spp. to gain access to posing hosts, from which they take pieces of fins and body tissue. The deception is most successful with young hosts. The importance of cleaner fishes in reef fish dynamics may differ at different locales. Experimental removal of cleaners from a Caribbean reef led to a decrease in host fish density and an increase in parasitic infections, whereas a similar removal in Hawaii led to no such problems (Losey 1978; Randall and Helfman 1972; Gorlick et al. 1987).

FIGURE 21.7. *Typical cleaning activities of a cleaner wrasse,* Labroides *sp., from the tropical Pacific. Cleaner fishes exist in almost all habitats, but only in tropical seas are species relatively specialized for this role. (A) A large wrasse poses in a head-up position while the smaller cleaner inspects it for parasites and necrotic tissue, contacting the host with its pelvic, anal, and caudal fins. (B) The cleaner solicits posing from a potential host by riding it while flicking the host with its pelvic fins. (C) Cleaner about to remove a parasite or necrotic tissue from the anal fin of a carangid.*

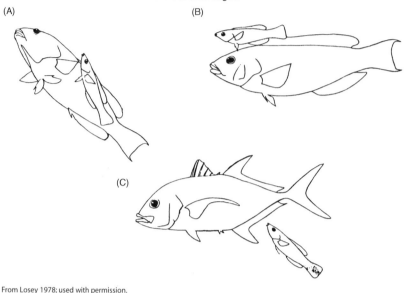

(A)   (B)   (C)

From Losey 1978; used with permission.

Some interspecific associations involve exploitation of one species' feeding habits to the benefit of another species. Fishes that dig in the substrate, such as stingrays, goatfishes (Mullidae), suckers (Catostomidae), and yellow perch, are commonly followed by other fishes that feed on invertebrates disturbed by the digger's activities (termed "**scroungers** and **producers**" in the behavioral literature). The purportedly commensal relationship between sharksucking remoras (Echeneidae) and large hosts such as sharks and rays is probably of this nature, involving feeding by the remora on leftovers following a host's meal. However, actual interactions between host and hitchhiker are seldom observed. Some remoras may clean parasites off their hosts, which would be mutualistic, whereas others may create a hydrodynamic burden, particularly when attached to relatively small hosts, creating a parasitic situation.

Symbiotic interactions with nonfish species generally involve use of invertebrates as spawning substrates, as predator refuges, to avoid extreme climatic environmental conditions (e.g., protection from desiccation or wave action), and as shoalmates. The bitterling, a European cyprinid, lays eggs in the mantle cavity of a freshwater mussel. These eggs hatch after a month and are expelled in the excurrent flow from the mussel. The mussel benefits in that its own larvae are parasitic on the gills of adult bitterlings. The proximity that bitterlings maintain to the mussels, with males establishing territories over the mussel and females inserting their ovipositors into the bivalve, undoubtedly facilitates attachment of the mussel larvae to the fish host (Breder and Rosen 1966).

Use of an invertebrate as a structural refuge against predators is common. Shrimpfish (Centriscidae), clingfishes (Gobiesocidae), cardinalfishes (Apogonidae), and juvenile grunts hover among the spines of long-spined sea urchins or rest beneath the urchins. Such fishes are usually clear or white with black stripes. Some fishes seek shelter inside living invertebrates, a habit called **inquilinism** or **endoecism**. The Caribbean conchfish, *Astrapogon stellatus* (Apogonidae), lives by day in the mantle cavity of a queen conch, a large gastropod. Individual conchfish forage at night on small crustaceans and enter the siphon canal of a live conch an hour before sunrise. Other cardinalfishes live with crown-of-thorns starfish, sea anemones, and sea urchins. Members of the elongate pearlfish family, Carapidae, similarly live by day inside mollusks and various echinoderms such as sea cucumbers and pincushion starfish, which they enter via the anus. Pearlfishes, the more primitive species of which are free-living, are also nocturnal foragers on small invertebrates but cross the line from commensalism to parasitism by consuming the viscera of their host (Thresher 1980, 1984). Many gobies live among sponges, corals, sea whips, and brain corals or share burrows created by various shrimplike crustaceans. In tropical gobies that cohabit with alpheid shrimp, the goby serves as a sentry while the prawn digs and maintains the burrow. Communication between partners is primarily tactile: The essentially blind shrimp maintains antennal contact with the goby's tail and senses tail flicks executed by the goby when predators approach (Fig. 21.8; Preston 1978; Karplus 1979).

Symbioses between fishes and cnidarians are common,

FIGURE 21.8.   *Goby-shrimp symbiosis. Several tropical goby species in the Pacific, Indian, and Atlantic Oceans live with burrowing alpheid shrimps. The goby stands guard at the burrow entrance while the shrimp excavates and repairs the burrow. The shrimp maintains contact with the goby via its antennae. If the goby is removed, shrimp will often seal up the burrow entrance and not emerge until the goby is replaced.*

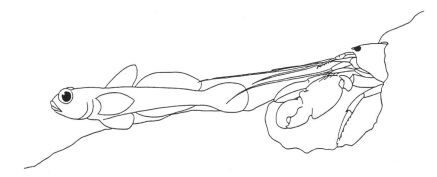

From Losey 1978, used with permission.

including fish that live on soft and hard corals (various gobies) or among the tentacles of jellyfish and Portuguese men-of-war (e.g., the man-of-war fish, *Nomeus gronovii,* Nomeidae). Although many fishes associate with sea anemones, the most highly evolved relationship is between pomacentrid anemonefishes and large sea anemones in the tropical Pacific and Indian Oceans (Fig. 21.9). Approximately 30 fish species in the genera *Amphiprion* and *Premnas* and 10 species of anemones are involved. Details differ among species of fishes and anemones, but basically any other fish that touches the anemone's tentacles is likely to be stung by nematocysts (stinging cells), paralyzed, and consumed, whereas anemonefish frequently contact the tentacles and are not stung. Although

the exact mechanism that protects anemonefishes from nematocysts remains unclear, mucous secretion, presumably by the fish, plays an integral role. An anemonefish's intimacy with its host is considerable; most individuals seldom move more than a few meters from their host, and adults remain with a single host for life. The relationship is considered mutualistic because the fish gains protection from predators on both itself and its eggs (which are laid on coral rock under the anemone) and also consumes other anemone symbionts and even the anemone itself. In turn, anemonefish chase away predatory butterflyfishes that eat anemones. The fish also may remove feces and debris from the anemone's upper surface, may drop food onto the anemone, may consume anemone

FIGURE 21.9.   *Anemonefish,* Amphiprion bicinctus, *move among tentacles of an anemone. Stinging cells in the anemone's tentacles would paralyze other fishes but are not discharged when contacted by a resident anemonefish.*

Photo by H. Fricke.

parasites, and excretes waste products that may stimulate the growth of symbiotic algae (zooxanthellae) within the anemone (Mariscal 1970; Allen 1975; Fautin 1991).

For fishes that use invertebrates as protection, such as gobies in invertebrate burrows and clownfishes in anemones, refuges may be in short supply, and territorial defense of the structure is fairly common (e.g., Grossman 1980). This is the probable explanation for the complex social system, territorial defense, and sex reversal of anemonefishes (see Chap. 20), populations of which appear to be limited by the number of available anemones.

Interspecific Shoaling

Many fish aggregations contain members of more than one species, forming **heterospecific** shoals. In monospecific aggregations, most fishes are of similar sizes (Pitcher 1986). This level of conformity is necessary, because unusual-appearing individuals contrast with the background of the more common fish and are preferentially selected by attacking predators. In a school, a perfectly healthy and normally swimming individual will be conspicuous to a predator if it is a different size or coloration from the majority of its schoolmates. Additionally, most schools cruise at efficient speeds and escape predators at high speeds that are both dependent on body length. A relatively large or small fish is likely to have different optimal swimming speeds than different-sized schoolmates; a relatively small fish will find itself trailing the school after a fast acceleration. Stragglers, like odd fish, are preferentially attacked by predators. When several abundant, morphologically similar species school together, they tend to segregate by species, either associating with conspecifics more closely or even creating horizontal layers that are relatively monospecific. Hence each fish gains the added benefit of being in a large school but avoids the risk of being the odd individual among another species (Allan 1986; Parrish 1988, 1989).

Hydrodynamics and predator avoidance dictate uniformity across a school. However, fishes that aggregate for foraging reasons are not as constrained by the need to be similar. For example, foraging schools of parrotfishes and surgeonfishes in the Caribbean frequently include other parrotfishes and surgeonfishes, as well as hamlets, butterflyfishes, wrasses, goatfishes, and trumpetfishes. Surgeonfishes and parrotfishes both feed on algae and thus benefit from the large numbers that overwhelm territorial herbivorous damselfishes. Carnivorous species may consume invertebrates flushed by the activities of the herbivores or may capitalize on territorial swamping and feed on invertebrates that live in algal mats of the territory or on the eggs of the damselfish. Larger predators, such as trumpetfish, may use the school or its members as moving blinds that conceal the predator and allow it to feed on the damselfish itself (trumpetfish will change color to match that of large or abundant school members). The presumed costs that small carnivores might suffer due to increased conspicuousness in a heterospecific shoal are apparently outweighed by gaining access to otherwise defended resources. The trade-off is underscored by the evasive maneuvers that minority fish take when a mixed-species shoal is threatened. Rather than flee with the school, odd fish abandon the school and seek nearby shelter (Robertson et al. 1976; Aronson 1983; Wolf 1985).

Finally, fishes form shoals with nonfish species. For reasons that remain puzzling, many tunas school with or below various dolphin species in the tropical Pacific. Fishing boats seek out the dolphin schools and surround them with large purse seine nets as the tuna remain below the mammals; the dolphins unfortunately become bycatch and frequently drown. On a less grand scale, postlarval French grunts school with dense clouds of mysid shrimps shortly after the grunts settle from the plankton and onto coral reefs. Both species are similar in size (8 to 13 mm) and appearance, but the mysids greatly outnumber the grunts. Grunts benefit from the antipredation function of the schools, affording them a degree of protection probably related to the number of mysids in a school. As the grunts grow, they school more on the periphery of the mysid aggregation and feed on the mysids. What began as a commensal or mutualistic relationship turns into a predator-prey interaction (McFarland and Kotchian 1982).

SUMMARY

1. Social aggregations, except where fish incidentally converge on a resource, require that communication be maintained; territorial defense similarly requires communication. Fishes use all six senses to communicate with one another. Static and dynamic visual displays—using colors and movement of fins or gill covers or turning photophores on and off—are common. The multiple functions of bright coloration in reef fishes have long been debated. They may have evolved in part because clear water makes predators detectable at a distance and the reef provides many places to hide from predators, thus eliminating a major cost of being colorful.

2. Fishes use sound when grasped by a predator, when spawning, when defending territories, and in maintaining shoals. Sounds are produced by vibrating swim bladders, by rubbing bones or teeth together, and by the movement of the fish through the water. Eavesdropping by predators may be a significant cost of sound production.

3. Chemical production and detection function during food finding, predator avoidance, mating, migration, parental care, territoriality, individual recognition, and aggregation. Pheromones are chemicals produced for intraspecific communication. Shoaling species are attracted to water that has contained conspecifics, and aggression can be reduced by production of specific chemicals in catfishes. Tactile communication is fairly uncommon and occurs primarily during mating, parent-offspring activities, and extreme fights. Electric communication is used extensively in families that have evolved the ability to produce and detect weak electrical fields (South

American knifefishes, elephantfishes); electrical output is often species, sex, and size specific.

4. Agonistic interactions involve aggression and submission, usually between conspecifics interacting in dominance hierarchies or during territorial encounters. Territoriality is common in fishes, which defend feeding, breeding, resting, and predator refuge territories.

5. Activity in fishes is often limited to a fairly defined area, termed a home range. A home range may be as small as a few square meters or as large as many square kilometers; larger species and individuals generally move over larger ranges. Individuals have an internal map of their range and a highly developed ability and strong tendency to return to their home range when experimentally displaced.

6. Fishes aggregate in loosely organized shoals or tightly organized schools. Aggregations function to increase food-finding ability, for reproduction, to save energy, and chiefly to decrease the success of predators. Most aggregations form and break up repeatedly, but some have long-term stability that exceeds the lifespan of individual members and are thus traditional.

7. Symbiotic relationships between species include parasitism, mutualism, and commensalism. Three fish families are known internal parasites (cutthroat eels, candiru catfishes, and pearlfishes). Mutualistic relationships include the many species that pick external parasites off other fishes, clownfish-anemone associations, and shrimp-goby pairs. Commensal relationships usually involve a fish living in association with an invertebrate, and may include sharksuckers attached to large elasmobranchs.

SUPPLEMENTAL READING

Ali 1980; Bullock and Heiligenberg 1986; Cott 1957; Fryer and Iles 1972; Hailman 1977; Hara 1982; Keenleyside 1979, 1991; Lythgoe 1979; Pitcher 1993; Potts and Wootton 1984; Reese and Lighter 1978; Tavolga et al. 1981; Thresher 1980, 1984; Wootton 1990.

22

Cycles of Activity and Behavior

A fundamental characteristic of biological systems is their cyclical nature. Physiological and behavioral cycles exist at numerous temporal scales. Hearts beat and nerves discharge spontaneously on a regular rhythm, producing such predictable cycles as brain "waves." Hormone production, respiration, locomotor activity, and photomechanical movements within the eye show distinct cycles. Such processes are usually driven by a neural pacemaker or are linked to external cues. Hence the rising and setting of the sun, the phases of the moon, and the annual orbit of the earth around the sun create periodic physical stimuli such as illumination, climatic, and tidal cycles that in turn determine the onset, timing, and periodicity of many activities in fishes.

The commonness of such cycles is not surprising; evolutionary adaptation is facilitated by constant or at least predictable selection pressures, and the external cues for such cycles as day length, tides, and seasonal climate are distinct and relatively predictable (Schwassmann 1980). Our object in this chapter is to review some pronounced biological cycles in fishes that are driven by internal (**endogenous**) clocks, by external (**exogenous**) cues, and by a combination of factors. We will focus on daily, semilunar (or biweekly), monthly (or lunar), seasonal, and annual patterns of activity, particularly those that involve foraging, migration, and reproduction.

DIEL PATTERNS

The 24-hour (diel or daily) periodicity of the earth's rotation creates a predictable pattern of light and darkness that has a profound effect on the biology of almost all animals and plants. Organisms are cued externally by the time of sunrise or sunset, or by the length of the day, night, or twilight; or their activities are determined by an internal clock with a roughly 24-hour period that may be reset by some external light cue.

Light-Induced Activity Patterns

Activity patterns in fishes generally represent a direct response to changing light levels but are also affected by the activity patterns of their predators and prey. In most environments, fishes are **diurnal** and tend to feed primarily during the day or are **nocturnal** and feed by night, while some feed primarily during **crepuscular** periods of twilight, and fewer show no periodicity (Table 22.1). On average, about one-half to two-thirds of the species in an assemblage will be diurnal, one-quarter to one-third are nocturnal, and about 10% are crepuscular.

These distinctions are sharpest at tropical latitudes, where families can often be characterized as diurnal, nocturnal, or crepuscular (Hobson 1991; Helfman 1993). On coral reefs, herbivorous fishes are almost exclusively diurnal, for reasons that remain unclear but may relate to the need to visually identify edible and inedible algae. Parrotfishes, surgeonfishes, rabbitfishes, and sea chubs roam the reef, often in large shoals. They feed on algae and sea grass, some of which is defended by territorial damselfishes, parrotfishes, and blennies. Fishes that feed primarily on encrusting sponges, tunicates, corals, and hydrozoans are also largely diurnal; this group includes angelfishes, butterflyfishes, pufferfishes, and triggerfishes. Many wrasses, butterflyfishes, goatfishes, mojarras, and small sea basses eat mobile or buried invertebrates (e.g., small crustaceans and polychaete worms) and are also diurnal. Zooplanktivores are particularly abundant and conspicuous during the day, including anthiine sea basses, damselfishes, wrasses, fusiliers, and butterflyfishes. Zooplanktivores also form large shoals by day; these aggregations often remain over a particular section of the reef and wait for currents to bring planktonic prey to them. A small group of piscivores, including lizardfishes, trumpetfishes, cornetfishes, scorpionfishes, jacks, hawkfishes, barracudas, and flatfishes, are active primarily during daylight. Cleaning fishes that pick external parasites from other fishes (wrasses, gobies, and juvenile angelfishes and butterflyfishes) are active by day (see Chap. 21, Interspecific Relations: Symbioses). Some of the largest shoals of fishes encountered by day in tropical waters are nonfeeding, resting shoals of nocturnal foragers. Sometimes numbering in the millions, aggregations of zooplanktivorous silversides, anchovies, and herrings frequently hover near structure or in sandy embayments

Table 22.1.   Diel activity patterns, of better-known groups and families of teleostean fishes defined according to when fishes feed. For many families, activity patterns are only known for a few species. Some large families appear under more than one heading because of intrafamilial variability.

All or most species studied diurnal	All or most species studied nocturnal	Both diurnal and nocturnal species	Several crepuscular species*	Several species without well-defined activity periods
Acanthuridae, surgeonfishes	Amiiidae (bowfin) Anguillidae (eels)	Carangidae (jacks)	Carangidae (jacks)	Aulostomidae (trumpetfishes)
Ammodytidae, sandeels	Anomalopidae (flashlightfishes)	Catostomidae (suckers) Centrarchidae (sunfishes)	Elopidae (tarpon)	Muraenidae (moray eels)
Anthiinae	Apogonidae (cardinalfishes)	Congridae (conger eels)	Fistulariidae (cornetfishes)	Scombridae (mackerels)
Characoidei (characins)	Batrachoididae (toadfishes)	Cyprinidae (minnows)	Gadoidei (cod)	Scorpaenidae (scorpionfishes)
Chaetodontidae (butterflyfishes)		Gadoidei (cods)	Lutjanidae (snappers)	Serranidae (groupers)
Cichlidae (cichlids)	Clupeidae (herring)	Leiognathidae (ponyfishes)	Serranidae (groupers)	
Cirrhitidae (hawkfishes)	Diodontidae (porcupinefishes)	Mullidae (goatfishes)		
Cyprinodontidae (killifishes)				
Embiotocidae (surfperch)	Grammistidae (soapfishes)	Pleuronectiformes (flatfishes)		
Esocidae (pikes)	Gymnotoidei (knifefishes)	Salmonidae (salmonids)		
Gobiidae (gobies)	Haemulidae (grunts)	Serranidae (groupers)		
Kyphosidae (sea chubs)	Holocentridae (squirrelfishes)	Sphyraenidae (barracudas)		
Labridae (wrasses)	Kuhliidae (aholeholes)			
Mugilidae (mullets)	Lutjanidae (snappers)			
Mullidae (goatfishes)	Mormyridae (elephantfishes)			
Percidae (perches)	Ophichthidae (snake eels)			
Pomacanthidae (angelfishes)				
Pomacentridae (damselfishes)	Ophidiidae (cusk eels)			
Scaridae (parrotfishes)	Pempheridae (sweepers)			
Siganidae (rabbitfishes)	Sciaenidae (drums)			
Synodontidae (lizardfishes)	Siluriformes (catfishes)			

*Also active at other times.
From Helfman 1993; used with permission.

(Parrish 1992). They may be subjected to attacks by roving sea basses, jacks, and tunas. Better protected are the daytime resting shoals of invertebrate-feeding squirrelfishes, copper sweepers, cardinalfishes, and grunts that occur over coral or in caves.

The prey resources available on a coral reef by day change dramatically at night (Hobson 1991). Many invertebrates bury themselves in the sand or in small holes in the coral, whereas others come out of hiding and move about the reef and into the water column. The maximum size of zooplankton increases substantially from relatively small (< 1 mm long) diurnal forms to larger (> 2 mm long) nocturnal animals; these larger invertebrates make up most of the diet of nocturnal zooplanktivores. Almost all nocturnally active fishes are carnivorous, feed on mobile invertebrates, and have relatively large eyes and large mouths. Grunts, snappers, porgies, and emperors are generally found close to the bottom, whereas zooplanktivorous anchovies, herrings, silversides, squirrelfishes, cardinalfishes, glasseye snappers, and copper sweepers forage higher in the water column. Predation pressure on fishes is evidently reduced at night; the few piscivores that roam the reef after nightfall include various eels, sharks, squirrelfishes, snappers, groupers, and jacks. A reduction in predation is reflected in the morphology of nocturnal planktivores, which tend to be less streamlined than their diurnal counterparts, and also in a general lack of shoaling behavior at night by either nocturnal or diurnal species.

Successful feeding by piscivores occurs primarily during the transitional periods of evening and morning twilight, when diurnal and nocturnal groups essentially replace one another ecologically (see Box 22.1). Crepuscular predators include tarpon, cornetfishes, groupers, snappers, and jacks. Although their activities are concentrated at this time, most of these fishes, and predatory fishes in general, are highly opportunistic and will take prey any time of the day or night.

Fishes in other habitats and latitudes vary in the predictability of their daily activity cycles. Tantalizingly little is known about tropical freshwater assemblages (Lowe-McConnell 1987). Characins and cichlids are predominantly diurnal, catfishes nocturnal, but beyond such generalizations our knowledge is relatively limited. At higher latitudes, activity cycles are similar in some respects but notably different in others. In temperate lakes, familial distinctions are weaker than on coral reefs. Diurnal zooplanktivores are abundant (minnows, sunfishes, perches) and have nocturnal counterparts among the herrings, minnows, whitefishes, and sunfishes (Helfman 1981, 1993). Herbivores are relatively rare. Diurnal invertebrate feeders include minnows, suckers, mudminnows, topminnows, sunfishes, and perches. Their nocturnal counterparts include eels, catfishes, trout, sculpins, sunfishes, and drums. Piscivores, which again have an activity peak at twilight, are represented during the day by pickerels, pike, and black basses, and at night by bowfin, salmonids, burbot, temperate basses, sunfishes, and pikeperches. Nocturnal fishes rest by day amid vegetation or other structure or form daytime resting shoals. Diurnal fishes often sink to the bottom and rest in relatively exposed locations at night.

Nearshore California kelp beds also contain diurnal, nocturnal, and crepuscular species, but again familial distinctions are blurred (Ebeling and Hixon 1991). By day, shoaling zooplanktivores are abundant (silversides, sea basses, surfperches, damselfishes, wrasses, scorpionfishes), as are diurnal invertebrate feeders (sea basses, surfperches, clinids, gobies); these groups have nocturnal equivalents among the scorpionfishes, grunts, croakers, and surfperches that feed on relatively large prey both near and above the bottom. Herbivores are relatively rare, although they are more abundant in shallower, intertidal areas (blennies, pricklebacks, gobies; Horn 1989). Piscivores (sea basses, scorpionfishes, greenlings) are active primarily during twilight or nighttime. Nocturnal fishes form daytime resting aggregations, and diurnal fishes rest at night either in holes or in exposed locales.

At temperate latitudes, twilight changeover patterns (Box 22.1) are more variable compared to coral reef species. Activity patterns of temperate lake and kelp bed fishes are less precise in that 1) many species feed both diurnally and nocturnally, 2) reputedly diurnal and nocturnal species overlap in activity times, 3) species within a family vary in major periods of activity, and 4) individuals within a species vary in twilight changeover activities. Such variation could result from longer twilight lengths at higher latitudes, where dark adaptation of a fish's eye keeps pace with rate of light change, and hence diurnal species can maintain activity well into twilight and nocturnal species can commence activity before twilight ends. A latitudinal gradient in twilight length could interact with reduced predation pressure, reduced species diversity, or greater climatic instability to produce the relatively "unstructured" temporal patterns at higher latitudes (Helfman 1993).

The above discussion focuses on active animals. However, inactivity accounts for half the diel cycle in most fishes, which often prompts the question of whether or not fish sleep. **Sleep** occurs when a fish assumes a typical resting posture for a prolonged period, uses some form of shelter, and is relatively insensitive to disturbance (Reebs 1992). By this definition, many species sleep, including dasyatid stingrays, minnows, bullhead catfishes, butterflyfishes, cichlids, mullet, wrasses, and surgeonfishes. Croakers apparently do not sleep. Some parrotfishes and wrasses secrete a mucous envelope around themselves at night while sleeping, a behavior that has been subjected to much speculation but little actual investigation. The mucous cocoon may thwart roving nocturnal predators such as moray eels, or it may be an incidental by-product of mucus production that occurs throughout the diel cycle but does not accumulate during the day because the fish are continually active (J. E. Randall, personal communication). The adaptive significance of sleep in fishes remains a matter of debate. One likely function is immobilization during a period when an animal is relatively inefficient at both foraging and predator avoidance. Hence energy is conserved and predators avoided, but only if a refuge is found before a quiescent state is assumed (Reebs 1992).

Vertical migrations. An entirely different daily rhythm of migration that appears largely dependent on light levels is the vertical migration undertaken by numerous fish species in both marine and freshwater habitats (see Chap. 17, The Deep Sea). In most fishes, this movement involves an upward migration at dusk to feed and a downward migration at dawn. For example, alewives (*Alosa pseudoharengus*, Clupeidae) migrate upward in lakes in the evening at a rate that parallels the migration of their prey, a mysid shrimp. Zooplankton often migrate to the surface at night to take advantage of reduced visual acuity in visually hunting zooplanktivorous fishes. Predator avoidance could also explain vertical movements in many larval and juvenile fishes (e.g., sockeye salmon, walleye pollock); by remaining in dark, deep waters by day, vertical migrators can avoid visually orienting, diurnal predators.

Vertical migration could also increase a fish's encounter rate with plankton if currents differ at the surface

BOX 22.1 — *Death in the Children's Hour: Changeover at Twilight on Coral Reefs*

"Between the dark and the daylight,
When the night is beginning to lower,
Comes a pause in the day's occupations,
That is known as the Children's Hour."

(H. W. Longfellow; Cody 1927)

If one enters the water on a coral reef slightly before sunset and follows the events that occur over the next half hour, a striking sequence takes place in the activity and composition of the fishes. This sequence has been observed at several Pacific and Caribbean locales and probably occurs in most coral reef assemblages. Similarities among sites, despite dissimilar species, suggest that common selection pressures are operating, leading to convergence in the behavior of the fishes. The **twilight changeover period** involves approximately four different phases of activity (Hobson 1968, 1972, 1991; Collette and Talbot 1972; Helfman 1993).

1. Migrations of diurnal fishes. Beginning about an hour before sunset, zooplanktivorous fishes (e.g., anthiine serranids, butterflyfishes, damselfishes) descend from the water column to the reef, and large herbivores (e.g., parrotfishes, surgeonfishes) migrate from daytime feeding locales to nighttime resting locales along predictable paths.
2. Cover seeking of diurnal fishes. From just before sunset until about 20 minutes after sunset, diurnal fishes seek shelter in the reef. Small individuals enter holes and cracks in the reef, whereas larger fish nestle under overhangs and in depressions. A species sequence exists; wrasses are among the first to seek shelter, followed by zooplanktivorous damselfishes, butterflyfishes, larger damselfishes, and parrotfishes. The time at which a species seeks cover is constant, with only a few minutes' variation from one evening to the next.
3. The quiet period: evacuation of the water column. Beginning about 10 to 15 minutes after sunset, the level of activity and number of fishes above the reef drops precipitously. For the next 15 to 20 minutes, activity by small fishes in the water column comes to a standstill. Whereas minutes earlier the reef was alive with migrating and feeding fishes, the water column is now empty, leaving an observer with an uneasy feeling of abandonment. Hobson (1972) termed this phase the **quiet period**, when neither diurnal nor nocturnal fishes are moving about. All activity does not cease, however. Predatory fishes, such as groupers, jacks, and snappers, are active at this time, generally swimming close to the bottom and striking up at prey fishes still in the water column. This predatory tactic undoubtedly capitalizes on the difficulty prey have seeing dark-colored predators below them against the background of the darkened reef, whereas the predators are striking up at targets that are silhouetted against the lighter evening sky.
4. Emergence and migration of nocturnal fishes. The end of the quiet period is marked by the movement of nocturnal fishes up into the water column and along the reef face. Bigeyes, cardinalfishes, and croakers appear over the reef about a half hour after sunset and begin

versus at depth. Many oceanic regions are characterized by surface currents that flow in one direction and deeper waters that flow in a different direction. By swimming down, a fish can remain in relatively stationary, deep waters by day as the surface currents replenish the food supply in the waters above, a scenario analogous to feeding off a moving conveyor belt. It has also been postulated but not demonstrated that fishes could gain an energetic advantage by moving into warm surface waters to feed actively and then returning to cooler, deeper waters to metabolize and grow (McLaren 1974; Janssen and Brandt 1980; McKeown 1984; Neilson and Perry 1990).

Not all diel activity cycles relate only to feeding and predator avoidance. The timing of spawning is quite predictable for many species and even families. Diurnal spawners include many minnows, sunfishes, darters, cichlids, and wrasses. Twilight spawning characterizes some damselfishes (dawn) and butterflyfishes, wrasses, parrotfishes, and bothid flounders (dusk). Nocturnal spawning, not surprisingly, is difficult to observe but is known in the yellow perch, which is a strongly diurnal feeder (see below).

Circadian Rhythms

A **circadian** rhythm is a pattern of activity governed by an internal clock with a period of roughly 24 hours. The actual onset of activity may be shifted each day (the clock may be reset) by some external stimulus or *Zeitgeber* (German for "time giver") such as sunrise. The need for an external resetting mechanism becomes obvious when one realizes how much day length changes during different seasons.

BOX
22.1   *(Continued)*

feeding on invertebrates. Grunts, squirrel-fishes, and copper sweepers migrate along predictable paths from daytime resting locales to nighttime feeding areas (see Box 21.2). The water column above and around the reef is now occupied by active fishes.

The evening sequence is repeated again in reverse at dawn. Nocturnal fishes migrate back to resting locales and seek shelter, a morning quiet period occurs when predators are most active, and then diurnal fishes reoccupy the water column and migrate to their daytime feeding locales. The predictability of times and locales, cued primarily by specific light levels, is striking.

Crepuscular predators are the apparent key to understanding the predictable, convergent nature of events during twilight on coral reefs. Predatory threat results from a combination of environmental, physiological, and behavioral factors unique to crepuscular periods (Fig. 22.1). During twilight, light declines from daytime levels of about 10,000 lux to nighttime levels of about 0.0001 lux. The light-adapted, cone-dominated eyes of many diurnal species cannot dark-adapt quickly enough and thus become ineffective at capturing light in the changing, dimmer conditions of dusk. At the same time, conditions are still too bright for the sensitive, rod-dominated eyes of nocturnal fishes, which are highly effective at capturing light. Twilight is a period of intermediate conditions, when cones still function, but not with great efficiency. Many reef predators have fewer but larger cones than are found in diurnal eyes and more but smaller cones than are found in nocturnal eyes. This intermediate eye provides less visual acuity than a diurnal eye during the day and is less effective at light capture than is a nocturnal eye at night. However, the intermediate eye is relatively better during the changing conditions of twilight, when neither the diurnal nor the nocturnal eye functions well.

The light-capturing photopigments in the retinas of reef fishes also indicate an influence of twilight conditions. Diurnal, nocturnal, and crepuscular fishes have rod pigments that are most sensitive to light in the blue-green portion of the spectrum (about 490 nm), which matches prevailing wavelengths during twilight better than it matches the dominant greener nighttime light at 580 nm. Both diurnal and nocturnal fishes appear to sacrifice nocturnal vision in favor of being able to capture light during the dangerous crepuscular periods.

These anatomic and physiological differences, combined with the predatory tactic of striking up at backlit prey in the water column, help explain why the postsunset minutes are so dangerous for potential prey species. The quiet period of inactivity by small diurnal and nocturnal fishes appears to be a direct result of the threat of being eaten by predators at that time. Evening and morning twilight may account for only 5% of the 24-hour diel cycle, but conditions at twilight have an apparent influence out of proportion to the absolute time involved.

Tides and feeding events can also serve as *Zeitgebers*. Activity rhythms in many teleosts can become established (entrained) if a meal is provided at a fixed time each day. Fish then develop an activity rhythm that anticipates the time of feeding, even in the absence of food and in constant light (Spieler 1992). In the absence of a *Zeitgeber*, such as during experimental conditions of constant light or darkness, rhythms are often maintained at slightly more or less than 24 hours and are referred to as **free-running**.

Free-running rhythms, involving either diurnal activity and nocturnal inactivity or the converse, have been demonstrated in a number of fishes, including hagfishes, swell sharks, anguillid eels, minnows, suckers, South American knifefishes, burbots (Gadidae), and killifishes (Boujard and Leatherland 1992; Reebs 1992). Some species, such as horn sharks, show no distinct circadian patterns.

Normally-distinct activity cycles can be disrupted by experimental additions of predators or by removal of resting structure. Distinct cycles also often break down during the breeding season. Many strongly diurnal reef fish species spawn late into evening twilight, and normally diurnal minnows, yellow perch, and gobies spawn at night. The adaptive function of such breakdowns in periodicity is not understood. More obvious is the adaptiveness of a loss of activity rhythms in species that demonstrate parental care. Eggs and larvae must be guarded and fanned throughout the diel cycle, not just when the parents are normally active. Studies of several species, including catfishes, sticklebacks, centrarchid sunfishes, cichlids, and damselfishes, indicate that parental care is provided during the time period when adults would normally be inactive (Reebs 1992).

FIGURE 22.1.   *Light availability, dark adaptation, and predator-prey interactions at evening twilight on a coral reef. Available light (curved line) decreases maximally during the period from 13 to 33 minutes after sunset. This is the time (stippled area) when 1) diurnal eyes are dark-adapting, 2) predators are maximally active and successful, and 3) diurnal and nocturnal prey species abandon the water column, creating the quiet period. Approximate lux values for light units are 10^{14} photons = 10,000 lux, 10^8 photons = 0.0001 lux.*

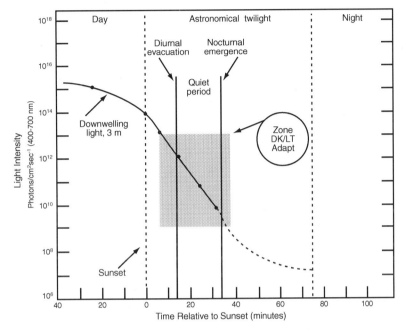

Modified from Munz and McFarland 1973; used with kind permission from Pergamon Press Ltd.

Circadian rhythms control many other aspects of fish behavior, morphology, and physiology. Many functions are under neuroendocrine control. The pineal organ on the dorsal surface of the brain secretes the hormone **melatonin**, which has a direct effect on seasonal control of reproduction, sexual maturation, development, and growth, as well as shorter-term effects on coloration, locomotor activity, and social behavior (see Chap. 7, The Endocrine System). Melatonin is secreted on a circadian basis, maximally at night and minimally during the day. This rhythm, which is entrained by light and temperature detected by the pineal, is maintained even in cultured pineal tissue removed from a fish (Zachman et al. 1992).

Secretion of hormones, such as prolactin, estradiol, progesterone, cortisol, testosterone, thyroxine, and thyroid, also follow endogenous (internally generated) circadian, semilunar, or lunar periodicities that are in turn affected by day length, temperature, and other hormone concentrations. Changing the light or temperature regime, or injecting a fish with hormones or hormone precursors, will cause changes in swimming activity and rest, temperature and salinity selection, reproduction, fat deposition, weight gain, and other aspects of growth. Hence the light-dark cycle can affect the timing of a neural pacemaker or clock, which in turn determines the timing of neural and hormonal cycles, which then entrain cellular rhythms in tissues, all governing the activity and behavior of the fish (Meier 1992).

The physical location of the clock (or clocks) in fishes remains a mystery. In mammals, a region in the brain, specifically the suprachiasmatic nucleus of the hypothalamus, serves as an endogenous oscillator (master clock). No direct analog of the suprachiasmatic nucleus has been found in fishes, although the hypothalamus has neural connections to light-receiving structures and other features that make it a candidate region for such a role, and the pineal has also been implicated in the control of many circadian rhythms in fishes (Boujard and Leatherland 1992; Holmqvist et al. 1992).

TIDAL PATTERNS

Tidal cycles are caused by the gravitational pull of the moon and to a lesser extent the sun on the oceanic water mass. Most coastlines experience a semidiurnal tidal regime that involves two high tides and two low tides each day, highs and lows being separated by about 6.2 hours. Relatively strong spring tides and relatively weak neap tides occur at biweekly intervals. Daily tidal ranges can vary from a few centimeters to several meters, depending on locale. Any marine animal that lives in the intertidal zone must either move out of the area in anticipation of or with lowering water levels, or face desiccation. Movements in general have to be synchronized with the fall and rise of tides. A drop in hydrostatic water pressure from a smaller overlying water column could serve as an external cue of a falling tide, and flooding of a pool could indicate an incoming tide. However, most intertidal animals, including fishes, appear to anticipate tidal changes via an internal clock that is reset by exogenous (external) cues (see above; Fig. 22.2).

Shallow intertidal areas—mud flats, salt marshes, sea grass beds, mangroves, reef flats, or the rocky intertidal—are among the most productive regions in the sea. Plants grow rapidly in warm, shallow water, resulting in an abundant food base. Shallow depths mean that large aquatic predators are relatively scarce. These conditions create a relative bonanza for fish species that can adapt to physiological conditions within the intertidal. Tidal regions are by definition fluctuating environments, except that most fluctuations occur with predictable periodicity. Hence animals can capitalize on the fluctuations, or at least adapt to predictable environmental constraints. Falling tides in particular create numerous problems for fishes, including desiccation; rapid changes in and exposure to extreme temperatures, pH, and salinity; and exposure to terrestrial and aerial predators. Fishes generally follow one of two courses in dealing with low tide conditions: Either they abandon shallow water with falling tides, or they remain in the intertidal and seek shelter in cracks or algae, under rocks, or in pools (Gibson 1992, 1993).

The former, termed **visiting** species, migrate in and out with the tides. This is particularly common among the many species that use intertidal areas as nursery grounds or refuges for juvenile fishes. Salt marsh creeks along the Atlantic coast of North America serve as such nurseries for numerous species of worm eels, herrings, croakers, porgies, mullets, and flatfishes, as well as housing adults of dozens of other species. Approximately 80% of the commercial landings from the U.S. Atlantic and Gulf of Mexico fisheries consists of species that spawn offshore and use salt marshes for nurseries (Shenker and Dean 1979; Weinstein 1979; Miller et al. 1985). Juvenile fishes slosh in and out of the intertidal zones, being carried out of a marsh creek with the outgoing tide and back in again with the flood tide. Analogous tidal nursery situations exist in many parts of the world, such as mangroves in most tropical areas and the tidal swamps of western and northern Australia, which are used by juveniles of 24 families of fishes (Davis 1988). Adults of coral reef species also show on-off reef movements that correspond with tides. Low tides on hot days can create hot, anoxic conditions over large sections of reef and sandy tidal flats. Such areas are commonly avoided during low tides but are reoccupied by fishes that move back onto the reef from the deeper reef face or from channels as cooler, oxygenated water floods the region during incoming tides.

Intertidal **resident** species remain in the intertidal at low tide and hide in areas insulated from complete desiccation, or make periodic visits into water or spray zones. Residents show the greatest degree of adaptation to the intertidal environment. Most are relatively small (< 20 cm), which allows them to hide in holes and cracks or under piles of vegetation (pricklebacks, gunnels, sculpins,

FIGURE 22.2. *Endogenous, circatidal activity rhythm of the shanny,* Lipophrys pholis, *held in the laboratory under continuous light. Activity is indicated by the darkened histograms; times of high tide where the fish was captured are denoted by the vertical dotted lines. The shanny is normally active at high tide. In the absence of tidal stimuli, the activity cycle "free runs" with a period of 12.4 hours, displacing it slightly from predicted high tide with each cycle. In the field, the clock would be recalibrated (reset) by the hydrostatic pressure of high tide, which would keep the fish's activity synchronized with actual tides.*

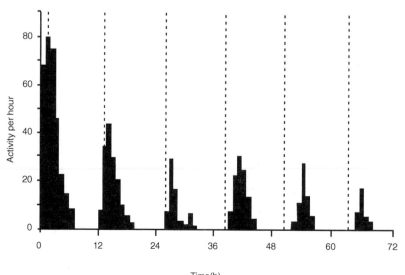

From Gibson 1993 after Northcott et al. 1990; used with permission of Cambridge University Press.

clingfishes, blennioids, gobies) and also presents less surface area to turbulence. Bodies are either thin and elongate (gunnels, pricklebacks, clinids) or depressed (sculpins, clingfishes). The intertidal is frequently exposed to breaking waves, particularly during high tides, and body morphologies of many species are convergent with those of fishes that live in other high-energy environments, such as river rapids (e.g., sculpins, darters, Asian hillstream loaches, loach gobies, New Zealand torrentfish). Suction cups, formed by fused pelvic fins, occur convergently in clingfishes, snailfishes, and gobies. Also convergent in high-energy and bottom-dwelling fishes is negative buoyancy, achieved through a missing or greatly reduced gas bladder. Some intertidal residents have evolved extreme tolerance to water loss: Clingfishes can live up to 4 days out of water if humidity exceeds 90%, and they can sustain as much as 60% loss of total water content. This tolerance exceeds that of many amphibians. Mudskippers of the tropical Indo-Pacific spend 80% to 90% of their time out of water, being submerged and inactive only during high tides (Horn and Gibson 1988; Gibson 1992).

Tidal cycles in many fishes have an apparent endogenous basis. European shanny, *Lipophrys pholis* (Blenniidae), and goby, *Gobius paganellus* (Gobiidae), held in the laboratory under constant conditions still show an activity pattern that corresponds to a semidiurnal tidal regime (see below). They rest at the local time of low tides and swim actively at the expected time of high tides (see Fig. 22.2). Similar patterns, also known from anguillid eels, tomcod (Gadidae), clingfishes, killifishes, sculpins, mudskippers, and flatfishes, appear to reflect an internal **circatidal** clock with a period of about 12 hours, but one cannot completely rule out the possibility that the fishes are sensing fluctuations in the gravitational pull of the moon and sun. The clock may be reset by changing water depth, which in the field would correlate with fluctuating hydrostatic pressure caused by alterations in the weight of the water column above the fish (Gibson 1982, 1992).

Activity cycles in many intertidal species are constrained by the cyclical nature of oxygen availability. Photosynthesis during the day oxygenates the water, but both plants and animals consume oxygen at night. Hence many intertidal fishes reduce their activity, and their oxygen consumption, at night (Horn and Gibson 1988). Tides may override normal activity patterns for species that occur in a variety of habitats. American eels, which are strongly nocturnal in nontidal habitats, travel with tidal currents by day and swim against them while foraging at night. Both eels and killifish capitalize on tidal flooding to gain access to the food resources of the salt marsh surface, eels by night and killifish by day (Weisberg et al. 1981; Helfman et al. 1983). Conversely, intertidal fishes that live in regions with very minimal tidal ranges, as in the Mediterranean, synchronize their activity patterns with day-night cycles instead of with tides (Gibson 1993).

Utilization of inshore areas by larval and juvenile fishes creates a particular logistic problem related to tidal cycles. Water flow is favorable for entry into such areas during only half of the tidal cycle; for much of the time, small fishes must fight outflowing currents of several knots, impeding or reversing any progress they may have made. Fishes overcome this problem by engaging in **selective tidal stream transport,** or modulated drift (e.g., McCleave and Wippelhauser 1987). Typically, immigrant fishes move up into the water column on incoming tides but move down close to the bottom on outgoing tides. They are consequently carried inshore with incoming currents but minimize slipping back offshore by taking advantage of reduced ebb currents as water is slowed by bottom topography and friction. They thus "ratchet" themselves into the estuary. Selective tidal stream transport has been observed in postlarval American eels, spot (Sciaenidae), and flounders. Adult anguillid eels, cod, and flatfishes on spawning migrations use similar transport mechanisms to move along shore or in open water. Selective tidal stream transport could be adaptive as a directional aid and also reduces the energy and time required to reach a particular locale; the response might be driven by an endogenous circatidal clock, as discussed in the previous section (Miller 1988; Gibson 1992; see also Box 9.1).

SEMILUNAR AND LUNAR PATTERNS

One postulated function of cyclical behavior is the opportunity that it affords individuals to synchronize their behavior with that of conspecifics. Nowhere is such synchronization more obvious and necessary than in reproduction. Not only must both sexes aggregate at the same locale to release gametes, but preparatory events of gametogenesis (gamete production) and secondary sex character development must also occur with similar timetables that converge on the same small time window. Predictable external cues that occur over biologically appropriate time periods are prime candidates as drivers of such cycles. The monthly orbit of the moon is a particularly common *Zeitgeber* because of its predictability, but also because of the direct links among lunar phase, nocturnal illumination, and tidal and current strength.

An illustrative example of this interrelationship involves the grunion, *Leuresthes tenuis* (Atherinidae), which spawn literally on the beaches of southern and Baja California (Fig. 22.3; see also Chap. 20). Grunion spawn after high tides on the three or four dark nights following a full or new moon during the spring and summer, the time when highest tides occur at night. These lunar periods correspond to spring tides, when water is pushed to its maximum height up the beaches. The same wave cover will not occur again for at least 2 weeks. Females ride waves up the beach, dig with their tails into the wet sand, and deposit eggs. Males deposit sperm around the females. The eggs normally develop in the moist sand for about 10 days. They will not hatch until the next spring tides, when waves once again cover the higher portions of the beach and the agitation of the breaking waves stimulates hatching. If waves do not reach the eggs, they can delay hatching for an additional 2 weeks (however, gonad maturation is on an 18-day, not 14-day, cycle). Spring tides promote synchronization of male and female behavior, allow a sufficient

FIGURE 22.3. *Semilunar spawning cycle of the grunion. Grunion are small (15 cm) atherinids that frequent the nearshore waters of Baja and southern California. They spawn every 2 weeks in the summer during spring tides, when waves sweep farthest up the beaches. Adults ride waves up the beach, dig in the sand, and deposit and fertilize eggs. Spawning occurs shortly after peak tides, ensuring that eggs will not be inundated again until the next spring tide, 10 to 14 days later, when they are ready to hatch. Fish symbols indicate nights of spawning.*

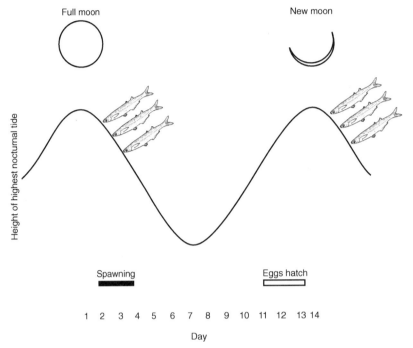

Redrawn from Alcock 1989; used with permission.

2-week interval for embryogenesis, and allow spawning under the cover of darkness, which may lessen predation on adults. The eggs are deposited where further wave action will not expose them for 2 weeks, again potentially reducing predation. A second grunion species, *L. sardina*, engages in similar behavior but spawns both day and night. In its Gulf of California locale, spawning coincides with the period of highest tides (Clark 1925; Gibson 1978).

Grunion reproduction is an example of **semilunar** periodicity; such cycles usually involve a 14.7-day interval. Synchronization with a period of maximum spring tides is common in fishes with semilunar, lunar, or longer cycles. Two groups with such rhythms are intertidal spawners and coral reef spawners. Intertidal spawners include the grunions as well as a Southern Hemisphere whitebait, *Galaxias attenuatus*, surf smelt (*Hypomesus*), silversides (*Menidia*), killifishes (*Fundulus*), four-eyed fish (*Anableps*), and a Japanese pufferfish (*Fugu*). Intertidal spawners deposit eggs during spring high tides on sand or pebble beaches or amid algal and root mats, leaf axils, and bivalve shells. Spawning often occurs on the days or nights following full or new moons. The eggs are exposed to air sometime during the next 2 weeks before the next spring tides, at which time they hatch. Smelt and killifish eggs actually require aerial incubation and will die if continuously immersed, apparently because of relatively low oxygen concentrations in water. Reduction of aquatic predation on the eggs is often postulated as the primary selection pressure favoring intertidal spawning, although predation on puffer eggs is minimal due to their toxicity (Gibson 1978; Taylor 1984, 1990; Leatherland et al. 1992).

Many coral reef fishes exhibit lunar or semilunar synchronization in their spawning activities (Johannes 1978; Gladstone and Westoby 1988; Robertson 1991). Most larger coral reef species that spawn in the water column do so at twilight or at night during the new moon, full moon, or both, often on high or ebbing tides. Larger species move to deep water, often to form spawning aggregations at predictable locales. Smaller species rush momentarily into the water column above the reef. Both groups have pelagic eggs and larvae, and larvae return to reefs primarily during spring tides. Johannes (1978) postulated that such behavior serves to move eggs and larvae out of the range of the abundant benthic and demersal predators on the reef, basically exporting the eggs out of the adult habitat but not necessarily out to sea.

Other explanations for the periodicity and timing of coral reef fish spawning focus on either larval or adult biology. Larval hypotheses, in addition to those proposed by Johannes, include maximized dispersal of larvae to distant habitats, swamping or saturation of predators, synchronization of larval production with production of invertebrate larvae on which they feed, reduction of competition among different larval cohorts, nocturnal spawning to minimize ultraviolet damage to floating eggs, and optimization of the timing of larval settlement in appropriate reef habitats. Adult biology hypotheses focus on synchronization of activities among spawners, optimization of the conditions under which adults spawn, and improvement of conditions for egg guarding in species that show parental care. No single hypothesis explains reproductive timing in all species (Barlow 1981; Taylor 1984; Robertson et al. 1990; Gibson 1992), but larval biology hypotheses may pertain more to water-column spawners, and adult biology hypotheses may better explain selective forces acting on benthic spawners that guard their eggs. Proof by exception comes from studies of species with unpalatable eggs, larvae, and adults. The sharpnose puffer, *Canthigaster valentini*, is protected by noxious chemicals at all life history stages (Gladstone and Westoby 1988). It spawns near the bottom during the day throughout the year, shows no definitive cycling, has an unhurried courtship display, and exhibits no parental care; embryos hatch on incoming tides. Apparent liberation from predation has also removed the selection pressures that induce periodicity and other more usual spawning behaviors in reef fishes. Few direct tests of most hypotheses have been attempted, and much exciting work remains to be done on this topic.

A variety of adaptive scenarios can be postulated for lunar cycles synchronized with spring tides in nearshore marine and estuarine fishes. More puzzling are lunar spawning cycles in some freshwater and high-seas fishes. For example, two Lake Tanganyika cichlids spawn during full moons, which might minimize diurnal predation on eggs while enabling adults to monitor the activities of their own, nocturnal, catfish predators. An apparent semilunar spawning cycle in an offshore gadoid, *Enchelyopus cimbrius*, is even more puzzling, aside from synchronization of reproductive behavior among adults (Gibson 1978; Taylor 1984; Leatherland et al. 1992).

Lunar cycles have been found in other aspects of the biology of fishes. White suckers show an endogenous lunar rhythm of temperature preference, selecting relatively high temperatures during the new moon and lower temperatures during the full moon. Guppies show a change in the spectral sensitivity of their retinas that has a lunar periodicity. Semilunar cycles have been found in feeding rate, body mass, body length, scale growth, otolith deposition, RNA/DNA content, and in concentrations of various plasma constituents in fishes. These phenomena may all be interrelated; feeding rate could affect all the other growth and condition parameters.

The downstream migration of young salmonids is also cued by lunar events. Newly emerged coho salmon fry (*Oncorhynchus kisutch*) move downstream from spawning redds at night during the new moon. Cover of darkness may reduce predation, whereas synchronization could aid in forming shoals and also swamp any predators that were present (see Chaps. 19, 21). Older smolts move downstream by day during the new moon, which delivers the migrants at the river mouth during spring low tides, which in turn aids movement out of the river and into the sea. Models explaining smoltification and migration in several salmonids (see Chap. 10) suggest that the full moon initiates a process of morphologic, behavioral, and physiological changes (perhaps mediated by thyroid hormone) that prime the animal for eventual downstream migration during the new moon. The new moon is also the primary time of upstream migration by elvers and downstream migration of maturing adults of anguillid eels. Again, the underlying mechanisms and clocks driving these periodicities remain a mystery (Leatherland et al. 1992).

SEASONAL PATTERNS

Activity and Distribution

The ectothermic nature of fishes makes them highly responsive to seasonal fluctuations in temperature. At temperate and polar latitudes, food availability, vegetative cover, turbulence, oxygen availability, and water clarity all vary greatly among seasons. Ice cover on high-latitude lakes leads to oxygen depletion and **winterkill** conditions; thermal stratification at lower latitudes creates analogous **summerkill** conditions. Hence fishes in these habitats characteristically move into and out of shallow, nearshore zones with the progression of the seasons.

For example, a typical pattern in a lake in the northeastern United States finds the shallow regions devoid of vegetation and fishes in early spring after ice melt and until surface water warms above about 10°C. With warming water, minnows, catfishes, pickerel, sunfishes, black basses, killifishes, and yellow perch move into nearshore regions. At water temperatures of around 15°C, sunfishes and many others spawn and vegetation growth is apparent. In late spring and early summer, as temperatures exceed 20°C, vegetation is well established and fish are distributed throughout the littoral zone, including deeper portions such as drop-offs down to the thermocline. In late summer and early fall, as temperatures fall below about 15°C and plants begin to die back, fishes first move from the deeper littoral zones to the shallower regions. As temperatures fall below 10°C and vegetation becomes sparse, fishes abandon nearshore regions, presumably for deeper water. If periods of warm weather occur during fall, fish will reoccupy and abandon the shallows as the water warms and recools (Hall and Werner 1977; Keast and Harker 1977; Emery 1978; Helfman 1979).

We know much about fish distribution and activity during spring, summer, and autumn. However, winter biology remains poorly understood. In North American temperate and arctic lakes, many fishes feed actively despite thick ice cover; smelt, salmonids (landlocked salmon, lake trout, charr), esocids (northern pike, chain pickerel), percids (yellow perch, walleye, sauger), and centrarchid sunfishes all feed actively. Where many lake fishes go in winter remains a mystery. Deeper water is a logical choice, because vegetation, which provides shelter for both fishes and their prey during warm months, disappears from shallows in winter. Also, winter storms make shallow regions unstable before ice cover develops and cause ice grinding when ice breaks up. Small and large temperate North American lakes experience a net decrease in both diversity and abundance of fishes in shallow waters during the winter. Centrarchids as a group move into deeper water. Large carp and bigmouth buffalo (*Ictiobus cyprinellus*) in Lake Mendota, Wisconsin, move to two traditional winter aggregation areas in relatively deep (5 to 7 m) water. Gill netting in those areas during one winter caught 43,000 kg of carp and 3000 kg of bigmouth buffalo, whereas nets set at other locales caught nothing. Northern pike show a tendency to occupy deeper water and to swim farther offshore under ice.

Some species remain in shallows or move into them from deeper water. Minnows remain in the littoral zone and occupy piles of twigs or small cracks in rocks and logs, or are even buried 0.5 m down in gravelly bottoms. Salmon and trout, whitefishes, burbot, and sculpin occupy deeper water by summer but move into shallower water to feed under the ice. In the Laurentian Great Lakes, fishes abandon the shallows in fall and early winter, but some (whitefishes, herrings, salmonids, troutperch, sculpins, and suckers) return after ice cover develops and filamentous algae appear. Again, early winter storms make the shallows down to about 10 m depth a turbulent and unstable habitat. At very high, arctic latitudes, little day/night or summer/winter differences are shown by the relatively depauperate fish faunas of lakes (Diana et al. 1977; Johnsen and Hasler 1977; Emery 1978; Helfman 1979).

The extreme conditions that develop during winter in ponds and small lakes lead to different behavior patterns. As ice and snow cover develop, deoxygenation occurs, beginning at the lake bottom and moving up through the water column. At very low oxygen tension levels (e.g., < 0.5 mg/L), lake species that are most resistant to winterkill mortality (e.g., mudminnows, Umbridae; pike; yellow perch) engage in behaviors that enhance their survival. They move up in the water column and take up positions immediately under the ice where it is thinnest and where oxygen concentrations are greatest, with their noses in contact with the ice. They seek out gas bubbles and inhale water from around the bubbles. Mudminnows will even engulf bubbles (air bubbles are squeezed out of ice as it freezes or are exhaled by aquatic mammals such as beavers and muskrats). Sunfishes, such as bluegill, swim throughout the water column and frequently encounter deoxygenated water; they are the first to die under winterkill conditions (Petrosky and Magnuson 1973; Klinger et al. 1982).

Temperate marine fishes also exhibit cycles of small-scale, seasonal movements that relate to temperature and climatic changes (longer migrations are discussed below). Many fishes abandon shallower waters when large algae die back in winter. In central California kelp beds, juvenile fishes inhabit understory kelp during spring and summer, using it for shelter and eating the invertebrates that live there. Juveniles disappear each fall and winter as the understory dies back or is reduced by periodic storms. Adults of resident species, particularly surfperches (Embiotocidae) and predatory kelp bass (*Paralabrax*, Serranidae) tend to remain in the area year-round but undergo changes in diet and foraging locale as the resource base shifts. Southern California bays and estuaries undergo a marked cycle of species richness and individual abundance, both of which peak in summer and are lowest in winter. The fauna contains resident (topsmelts, surfperches, gobies, flatfishes) and seasonal (anchovies, mullets) species. Seasonal movements in and out of the bays are strongly linked to changes in temperature, salinity, and the productivity of macroalgae. In Puget Sound, Washington, which is relatively protected from winter storms, rockfishes (*Sebastes*, Scorpaenidae) school in midwater and move down a few meters to slightly deeper water in winter. Benthic species remain in kelp bed and reef areas year-round (Ebeling and Laur 1985; Horn and Allen 1985; Ebeling and Hixon 1991).

On the Atlantic coast of North America, common names imply seasonal cycles of movement and abundance. Summer flounder (*Paralichthys dentatus*, Bothidae) spend warmer months nearshore along the coastline and in bays. They migrate offshore in the fall to deeper (30 to 200 m) water to spawn. In contrast, winter flounder (*Pleuronectes americanus*, Pleuronectidae) migrate to deeper water in the summer and then return to bays as the water cools; they also spawn in winter. Other species undergo seasonal movements that differ by individual age. Adult tautog (*Tautoga onitis*, Labridae) move offshore in fall as water temperatures drop below about 10°C, while young tautog and cunner (*Tautogolabrus adspersus*) move from grass and algal beds that are dying back to other shallow habitats that provide greater shelter before these fishes enter a winter torpid state. A pattern in many temperate marine environments is a dependence on algae as a refuge or as an indirect or direct food source. As colder months approach and algal beds cease productivity and lose their "above-ground" parts, many species abandon these regions for deeper waters or waters that will provide cover during months of low food production (Bigelow and Schroeder 1953; Olla et al. 1979; Rogers and Van Den Avyle 1982).

Reproductive Seasonality

The most notably seasonal activity in fishes and most other organisms is reproduction. Successful reproduction requires careful synchrony in physiology, anatomy, and behavior of both sexes. Spawning occurs when both sexes have completed gametogenesis, gamete maturation, secondary sex character development, and spawning readiness and arrive at the proper spawning locale at the same time. A series of environmental cues are likely to trigger each stage of a reproductive cycle. Seasonally dependable cues, particularly those that may ensure survival of larvae (plankton blooms, sea temperature changes, alterations in currents), are the most likely ones to be used and are usually associated with seasonal, cyclical climatic events, such as monsoonal rains, oceanic surface and upwelling currents (e.g., El Niño), and temperature cycles. Although environmental cues influence timing, flatfishes and sea bass held under constant laboratory conditions still show predictable seasonality in their gonadal cycles, indicating an endogenous basis to reproductive cycles (Bye 1990).

Seasonal cycles occur in most families, but seasons are defined differently in temperate versus tropical habitats. Most species in temperate latitudes spawn in spring or summer, a few in autumn and winter. Conditions favoring larval growth and survival appear to be primary determinants of the phasing of reproduction. In temperate locales, spring and summer are not only times of maximal food productivity but also periods when protective vegetation is maximally available. Although many fishes in tropical and even subtropical regions breed year-round (e.g., livebearers, numerous reef species), even these species show periods of peak reproductive activity that occur at relatively predictable times of the year.

Temperate freshwater fishes undergo reproductive cycles that are influenced strongly by changing photoperiod (day length) and temperature. Because gametogenesis is a complicated and lengthy process (see Chap. 9), environmental conditions at the time of initiation of gametogenesis will be different from those in effect when spawning occurs. Hence different cues are used at different phases in the cycle. In salmonids, spawning time is heritable, occurring at the same time each year over a period of 2 to 6 weeks in a particular genetic strain (Scott 1990). The rhythm is circannual, endogenous, and entrained by environmental cues, primarily photoperiod, but can be modified by temperature. Salmonids are generally divided into fall (September to December) and winter (January to March) spawners. Most species are fall spawners, including brown trout, brook trout, lake trout, and Atlantic salmon. Rainbow trout are generally late winter spawners. In both groups, the reproductive cycle is initiated during the previous springtime in response to increasing day length.

Temperate cyprinids, such as golden shiner (*Notemigonus crysoleucas*), goldfish (*Carassius auratus*), humpback chub (*Gila cypha*), and lake chub (*Couesius plumbeus*), all spawn in late spring and early summer. Gametogenesis begins in the fall in response to decreasing temperatures and shortening day length, advances slowly during the winter, and then accelerates and is completed in spring in response to increasing day length and rising temperatures. Sticklebacks have a similar cycle, as do most spring/early summer spawning fishes in temperate locales. Carp (*Cyprinus carpio*) show a variant cycle, involving gonad development in late summer, quiescence in winter, and then final maturation of oocytes and spawning in the spring. The European tench, *Tinca tinca*, is unusual in that it spawns in the fall. Late fall, winter, and early spring in temperate lakes are too unproductive for the small larvae of most species, and consequently spawning does not normally occur during those seasons. The exceptionally large size of eggs and physiological tolerance of cold temperatures in salmonids may explain their fall/winter spawning and success at high latitudes (Baggerman 1990; Hontela and Stacey 1990).

Temperate fishes, as well as many other animals and plants, use photoperiod as a proximate environmental indicator of current and future climate. Typically, long days (e.g., > 13 hours of light) and warm temperature cause gonadal recrudescence (resumption of gametogenic activity), whereas short days inhibit recrudescence, regardless of temperature. Available evidence suggests that many temperate fish species have an endogenous, circannual clock that drives reproductive activities, and that this clock is affected by another, circadian clock of photosensitivity. A critical piece of information is day length; days shorter than some minimum cause both initiation and cessation of reproductive behavior. The circadian clock can tell a fish if day length is increasing, but the fish must be most sensitive to light at a time of the day when daylight would indicate increasing day length (Fig. 22.4). Daylight during the first 8 to 10 hours after sunrise could occur at just about any time of year, but daylight 10 to 12 hours or more after sunrise will not occur during winter. Hence fish have a clock that tells them how many hours have passed since sunrise, and they tend to be insensitive to light during the first 10 hours or so after sunrise. Light encountered after that period, during the **photoinducible** phase of photosensitivity, has a strong influence on gonad development. The existence of a photoinducible phase and a photosensitive circadian rhythm was discovered by exposing fish to 2-hour pulses of light at different times of the day. Sticklebacks exposed to light 14 to 16 hours after sunrise showed greater rates of sexual maturation than fish experiencing light at other times of a light-dark cycle. The position and length of the photoinducible phase change with season and temperature (Baggerman 1990; Taylor 1990).

Seasonality among freshwater fishes at tropical latitudes (between 30°N and 30°S latitude) is defined more by rainfall than by temperature (Goulding 1980; Lowe-McConnell 1987; Munro 1990a). Regions between 15° north and south of the equator generally have two rainy seasons per year, whereas higher tropical latitudes have one rainy and one dry season. The floodplains or seasonal swamps created by a rising river are common spawning and nursery grounds in many locales, and many riverine fishes have reproductive cycles that coincide with seasonal inundation of gallery forests and swamps. Newly inundated areas are advantageous spawning locales because 1) accumulated nutrients are released, which creates plankton blooms and food for progeny; and 2) predation is minimized by abundant vegetation for refuging and because large flooded expanses minimize contact with predators. In contrast, receding water and reduced habitat space during the dry season mean that predators and competitors are concentrated, which also leads to deoxygenated water. Dry season and aseasonal spawners often provide extensive parental care, including provisioning of young, and/or possess secondary breathing structures (e.g., bagrid catfishes, cichlids, anabantoids).

Adults of many tropical freshwater species, including mormyrids, large characins, gymnotid knifefishes, and cyprinids, migrate up tributaries and onto flooded plains to spawn (Munro 1990a; Fig. 22.5). Lake-dwelling species and populations in these families move into tributary streams, whereas lacustrine herrings, silversides, and percomorphs often spawn within the lake itself. Seasonality in other riverine species involves migrations upriver to headwater regions in anticipation of seasonal rains (e.g., large characins). Many small characins, killifishes, livebearers, and cichlids reproduce year-round, although

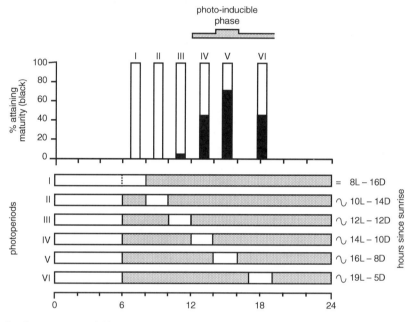

FIGURE 22.4. *Sticklebacks have a daily rhythm of photosensitivity that is maximal about 14 to 16 hours after sunrise. During this maximally sensitive period, exposure to light induces sexual maturation. The photoinducible phase helps the fish determine if day length is increasing, as would happen during spring and summer. Experimental manipulations of daily light/dark cycles can pinpoint the existence and position of the photoinducible phase by providing a 2-hour pulse of light at different times after sunrise. The open bars at the bottom indicate when lights were on, the darkened portion when lights were off (0 = sunrise). Light 14 to 16 hours after sunrise would be naturally experienced when days consisted of 16 hours of light and 8 hours of dark, as would happen during summer.*

From Baggerman 1985; used with permission.

peaks in recruitment often correspond with high water. Worldwide, predatory species often spawn earlier than their prey, thus assuring a food source for young predators. For example, the South American characin, *Hoplias malabaricus*, is predatory throughout life and breeds earlier than most other species. In contrast, juvenile piranhas are omnivorous, and adults breed along with other nonpredatory species. Regardless, spawning migrations upriver or onto flooded areas must be preceded by gonadal recrudescence that anticipates seasonal rainfall by several months. The cues that stimulate gonadal growth and gametogenesis are poorly understood but may include photoperiod and temperature changes (particularly at higher latitudes in the tropics), social interactions, food availability, and energy stores, as well as endogenously controlled rhythms, perhaps entrained by previous spawning itself (Munro 1990a).

Temperate marine teleosts have restricted spawning periods that vary by species, locale, and genetic stock (Bye 1990). At any particular locale, however, a stock is likely to have a fairly short and predictable time period during which most spawning occurs. Typically, pelagic species spawn over a 4-month period, with a shorter period of maximal activity. For example, cod in the North Sea off the northeast coast of England spawn between January and May, with 70% of eggs produced during 6 weeks of that pe-

riod. Peak spawning of European plaice (*Pleuronectes platessa*) in the Southern Bight of the North Sea occurred within 1 week of January 19 over the 39-year period between 1911 and 1950. The actual locale of spawning is also fairly predictable and defined by oceanic phenomena such as thermoclines or frontal areas, which are transition regions between differing water masses. American and European eels migrate to spawn in a region of the Sargasso Sea where a persistent frontal zone, defined by marked horizontal differences in temperature, salinity, and water density, exists every spring (McCleave et al. 1987). Inshore, temperate marine species are strongly seasonal. Coastal California species spawn mostly in spring and summer (e.g., grunion, surfperches, halibut, sheephead, blacksmith, croakers, sea basses, pricklebacks), some spawn in winter or early spring (rockfishes, starry flounder, lingcod), and a few spawn in the fall (greenlings, cabezon). Predominantly spring and summer spawning also typifies Atlantic temperate fishes (halibut, killifishes, most flatfishes, sea basses, porgies, many croakers, wrasses), with a few winter spawners (summer and winter flounder, sculpins, some croakers) (Ferraro 1980; Holt et al. 1985).

Coral reefs experience less extreme seasonal variation than temperate habitats and are less subject to the vagaries of rainfall than tropical freshwater systems. As a consequence, many coral reef fishes spawn through most or all

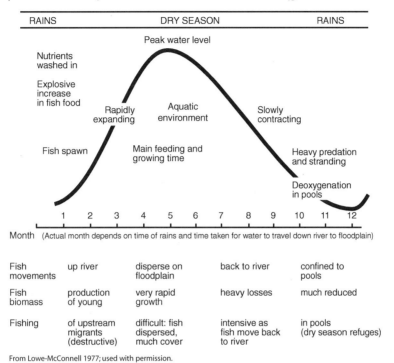

FIGURE 22.5. *The seasonal progression of events for many fishes in large tropical rivers. Seasonal flooding of highly productive gallery forests and swamps opens these areas up to lateral migration, feeding, and spawning by fishes. Many regions show two rainy seasons, and both result in lateral migrations. Fat stores increase as fishes capitalize on the food abundance in flooded regions.*

From Lowe-McConnell 1977; used with permission.

of the year, particularly at low latitudes (e.g., many damselfishes, wrasses, parrotfishes, grunts, surgeonfishes). Nevertheless, seasonal reproduction is also common among many families, including groupers, snappers, damselfishes, rabbitfishes, gobies, and pufferfishes. Seasonal spawning peaks have been found in most tropical locales, including sites in the Indian Ocean, South China Sea, and tropical Atlantic and Pacific Oceans. Spring peaks are most common, followed next by two periods of major spawning activity in both spring and fall. The most common environmental correlate of these peaks is that they occur when major currents around islands are weakest. Spawning during slack current periods would minimize the long-distance dispersal of larvae away from home reefs (Johannes 1978; Sale 1978; Robertson 1991).

Recruitment of larvae to the reef is also cyclical and seasonal. Even species that breed throughout the year show seasonal peaks in the arrival of larvae and episodic pulses of larval arrival. French grunts, *Haemulon flavolineatum*, in the Caribbean breed year-round. Larvae arrive in semilunar pulses over an 8-month period, with greatest recruitment in May, June, and November. Damselfishes in the southern Great Barrier Reef breed during a 5-month, summertime period, and larvae are recruited in pulses, with one or a few major pulses accounting for most arrivals. Larvae arrive on a lunar cycle, the major feature being that little recruitment occurs around the time of the full moon. Most settlement of fish larvae on

coral reefs occurs at night, suggesting a strong influence of visual predators. Avoidance of full moon periods would have a similar function. From the above, it is evident that spawning and recruitment do not necessarily follow the same timetables. Larvae can be produced, but conditions for settling after the 1- to 2-month larval period may not be favorable. In fact, abundance of larvae offshore and settlement of larvae on the reef do not necessarily coincide. For example, larvae of Nassau grouper, *Epinephelus striatus*, can be found along the Bahamas bank of the western Atlantic over a 2- to 3-month period, but actual larval settlement occurs almost entirely over a 4- to 6-night period when storm-driven currents push water and larvae into shore (McFarland et al. 1985; Doherty and Williams 1988; Doherty 1991; Shenker et al. 1993).

Seasonal Reproduction: Proximate and Ultimate Factors

Munro (1990b) has proposed a classification of the proximate cues that determine the occurrence of different portions of the reproductive cycle. He recognizes four factors that control the development and synchrony of breeding cycles.

1. **Predictive cues** are general periodic environmental events that a fish can use to predict that the spawning season is approaching. Changing day length and temperature are predictive cues

that are likely to trigger the onset of gametogenesis and secondary sex character development. Gametogenesis may have an endogenous circannual rhythm that is entrained by some predictive environmental cue (e.g., heteropneustid catfish, rainbow trout, sticklebacks).

2. **Synchronizing cues** signal the arrival of spawning conditions. Typically, the presence of a suitably appearing and behaving mate, perhaps releasing pheromones, may serve as such a cue, causing final gamete maturation and release. Presence of vegetation or other spawning substrate plays a role in some species. Synchrony is important to ensure contact between sperm and eggs and to prevent hybridization. In many species, gametes decline in fertility rapidly after ovulation and spermiation. Hence a small temporal window of spawning receptivity and opportunity exists.

3. **Terminating cues** signal the end of the spawning period. Because breeding conditions remain optimal for a short period, including the above-mentioned changes in gamete viability, breeding seasons are typically short. Gonad regression occurs after breeding in response to environmental cues (i.e., changes in predictive cues), exhaustion of gametes, or departure or changes in behavior of conspecifics. Nest-guarding species may respond to the presence of eggs in a nest, causing hormonal changes that inhibit spawning and encourage egg care and aggression.

4. The first three categories of cues can all be modified by secondary factors such as water quality, lunar cycle, adult nutrition, predator presence, and social interactions. These **modifying factors** are the causes of intraspecific variation in breeding at different latitudes or in different habitats.

Evolutionarily, why is seasonal breeding so prevalent in fishes? Gamete production, particularly in females, is energetically expensive. Gametes are usually released in batches; time and energy are required to replenish gametic products, even in males (Nakatsuru and Kramer 1982; Shapiro et al. 1994). Because courtship and spawning, and parental care where it occurs, require time and energy and expose participants to predators, few fishes can afford to reproduce year-round. Hence a decision in evolutionary terms must be made as to the optimal time to reproduce, optimality being defined in terms of the relative costs and benefits of current versus future reproduction (see Chap. 23, Life Histories and Reproductive Ecology). Conditions for egg dispersal, larval survival and growth, and larval recruitment vary through the year and are dependent on seasonally driven climatic variation. In most species, spawning appears to be synchronized with periods most favorable for the survival of young. In temperate marine fishes with pelagic larvae, food availability is one critical determinant. Spawning coincides with seasonal blooms of zooplankton, thus maximizing the chances that larvae will encounter prey during the critical period shortly after they use up the energy stores of their yolk supply (the match-mismatch hypothesis of Cushing

1973; see Chap. 9, Larval Feeding and Survival). Individuals that spawn at times when the probabilities of egg, larval, and their own survival are higher will be more successful than individuals that spawn at less suitable times (Munro 1990a).

ANNUAL AND SUPRA-ANNUAL PATTERNS: MIGRATIONS

Many fishes engage in periodic long-distance movements. A vast literature exists on various aspects of migratory behavior (e.g., Harden-Jones 1968; Leggett 1977; Baker 1978; Northcote 1978; McCleave et al. 1984; McKeown 1984). Our focus will be on species that undergo fairly large-scale migratory cycles with an annual or greater period, either in the ocean or between the ocean and freshwater (we will neglect the many species that undergo reproductive migrations within freshwater, the so-called potamodromous fishes, although some of these are dealt with above).

Migrations take several general forms. Reproductive migrations take animals from a feeding locale to a spawning locale, moving the animal from a habitat that is optimal for adult survival to one that is better for larval or juvenile survival. Fish that spawn several times in their lives (the iteroparous condition) may undergo this migration more than once (e.g., Atlantic sturgeon, American shad, Atlantic salmon, and the world's largest salmon, the taimen salmon of Siberia, *Hucho taimen*, which may weigh 70 kg). Semelparous fishes, those that spawn once and die, undergo the migration only once (e.g., sea lampreys, anguillid eels, Pacific salmons, some galaxiids). Inherent in reproductively migrating species is the complementary migration that juveniles take to juvenile and adult feeding areas. In some species, nonspawning juveniles and adults also migrate between feeding and spawning areas along with reproductively active individuals (e.g., sturgeon). Reproductive migrations may involve movement between the sea and freshwater (diadromy, see below) or may entail movements within ocean basins in a roughly circular or back-and-forth pattern (bluefish, tunas). Other species engage in transoceanic, seasonal migrations that do not appear linked directly to reproduction but instead probably place adult fish in optimal locales to intercept seasonally available food sources (pelagic sharks, billfishes) or may move individuals away from climatically unfavorable areas to regions that are less harsh (e.g., summer and winter flounder).

Diadromy

Many species of migratory fishes move predictably between fresh- and saltwater at relatively fixed times in their lives. These **diadromous** (running between two places) fishes include about 160 species, or a little less than 1% of all fish species. However, many of them are very important commercially, and their complex life histories are fascinating (Table 22.2). Diadromy takes three different forms: anadromy, catadromy, and amphidromy (Fig.

Table 22.2. Families of known diadromous fishes.

Anadromous	Catadromous	Amphidromous
Petromyzontidae, lampreys	Anguillidae, true eels	Plecoglossidae, ayu
Geotriidae, southern lampreys	Galaxiidae, galaxiids	Prototroctidae, southern graylings
Mordaciidae, southern lampreys	Scorpaenidae, scorpionfishes	Aplochitonidae, whitebaits
Acipenseridae, sturgeons	Moronidae, temperate basses	Galaxiidae, galaxiids
Clupeidae, herrings	Centropomidae, snooks	Clupeidae, herrings
Salmonidae, salmons	Kuhliidae, aholeholes	Cottidae, sculpins
Osmeridae, smelts	Mugilidae, mullets	Mugiloididae, sandperches
Salangidae, icefishes	Bovichthyidae, bovichthyids	Eleotridae, sleepers
Retropinnidae, New Zealand smelts	Pleuronectidae, righteye flounders	Syngnathidae, pipefishes
Gobiidae, gobies		
Aplochitonidae, whitebaits		
Ariidae, sea catfishes		
Soleidae, soles		
Gasterosteidae, sticklebacks		
Gadidae, cods		
Moronidae, temperate basses		
Cottidae, sculpins		
Gobiidae, gobies		

Modified from McDowall 1987.

FIGURE 22.6. *Diadromy takes three general forms: anadromy, catadromy, and amphidromy. In anadromy, adults spawn in freshwater; juveniles move to saltwater for several years of feeding and growth, and then migrate back to freshwater to spawn. In catadromy, adults spawn at sea; juveniles migrate to freshwater for several years to feed and then return to the sea to spawn. In amphidromy, spawning can occur in either fresh- or saltwater; larvae migrate to the other habitat for an initial feeding and growth period, then migrate to the original habitat as juveniles or adults, where they remain for additional feeding and growth prior to spawning. B = birth, G = growth, R = reproduction.*

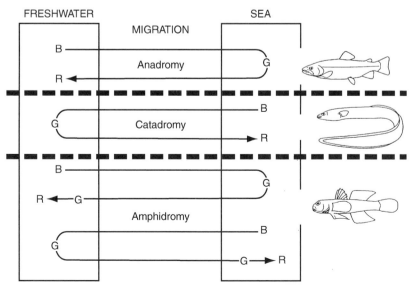

Modified from Gross 1987; used with permission.

22.6). **Anadromous** fishes, such as lampreys, sturgeons, shads, Pacific salmons, smelts, and striped bass, spend most of their lives in the ocean and then migrate to freshwater to spawn. Many anadromous species (sea lampreys, alewives, blueback herring, Atlantic salmon, rainbow trout) develop **landlocked** populations that never migrate to the sea but instead spawn in inlet streams to large lakes. **Catadromous** fishes, such as anguillid eels, mullets, temperate basses, and some sculpins, spend most of their lives in freshwater and then return to the ocean to spawn. **Amphidromous** fishes (ayu, galaxiids, southern graylings, sandperches, sleepers, gobies) move between marine and freshwater at certain phases of their lives, but the final migration occurs long before maturation and spawning occur. Of diadromous fishes, anadromous fishes make up about 54%, catadromous 25%, and amphidromous 21% (McDowall 1987, 1988).

The geographic distribution of the different forms of diadromy is interesting because it provides insight into the evolution of the behavior (Fig. 22.7). Anadromy is largely a Northern Hemisphere, high-latitude phenomenon; catadromy is more common at low latitudes and in the Southern Hemisphere; and amphidromy has a bimodal distribution at middle latitudes in both hemispheres, with greater representation in the Southern Hemisphere (McDowall 1987).

Interpretations of why different forms of diadromy prevail at different latitudes are confounded by phylogenetic histories, but one appealing analysis views diadromous migrations as complex adaptations that function to place both larvae and adults in environments where food is most abundant (Gross 1987; Gross et al. 1988). For a long-distance migration to evolve, the gains in fitness from moving must exceed the fitness an individual would have achieved had it remained in its original habitat. Gains have to be sufficiently large to also overcome losses—including osmotic costs, energy and time lost, and predation risk—incurred while migrating. If an individual migrates during its life, it should ideally 1) spawn in a place of low predator density to minimize egg mortality but where 2) larvae can drift passively to locations of higher productivity appropriate to their growth needs, and it should 3) place juveniles and young adults in areas where they can maximize their feeding, thus allowing them to build up energy stores necessary for 4) the long migration back to the optimal (low-productivity) spawning locale.

Available evidence indicates very different growth and reproduction rates in the different habitats. Juvenile Pacific salmon may increase their daily growth rate by 50% during their first week in the ocean. When landlocked, sockeye salmon, *Oncorhynchus nerka*, are referred to as kokanee and seldom attain 25% of the body size of anadromous individuals. Comparisons between diadromous and nondiadromous stocks of the same salmonid species (cutthroat trout, rainbow trout, Atlantic salmon, brown trout, and others) indicate that diadromous stocks produce on average three times as many eggs as nondiadromous conspecifics, probably because diadromous fish grow to larger size as a result of increased feeding in the ocean. Mortality rates are difficult to estimate; however, the threefold reproductive advantage of diadromy

FIGURE 22.7. *The latitudinal distribution and frequency of different forms of diadromy among major fish groups. The number of species employing each tactic is plotted as a function of latitude, showing anadromy to be largely a northern, temperate and polar phenomenon; catadromy to be more tropical and subtropical in distribution; and amphidromy to be more bipolar and temperate.*

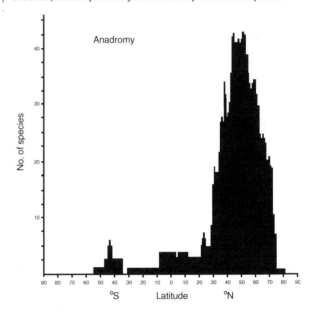

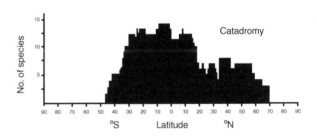

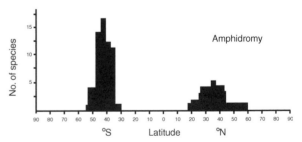

Modified from McDowall 1987; used with permission.

would permit up to a threefold increase in mortality at sea before the benefits of diadromy were negated.

Combining such growth and reproduction differences with information about relative productivity of freshwaters and oceans at different latitudes suggests that the temperate prevalence of anadromy and more tropical occurrence of catadromy are evolutionarily logical. In temperate regions, oceans tend to be more productive than freshwaters, whereas freshwaters are more productive than the ocean in tropical regions. Hence the primary feeding habitat for a migratory species should be the ocean at high latitudes versus rivers and lakes at lower latitudes. Survival of young may be enhanced if spawning occurs where productivity, and hence presumably predation, are least: streams and rivers at high latitudes, the ocean at lower latitudes. As long as the costs of movement between spawning and growth habitats are not excessive, the existence of anadromy in colder climates and catadromy in warmer climates appears adaptive. Amphidromy may serve as an intermediate or stepping-stone condition in the evolution of anadromy or catadromy (the catadromous nature of American and European eels, discussed below, may reflect the evolutionary history of the family, which is primarily tropical and which still entails spawning at tropical latitudes) (Gross 1987; Gross et al. 1988).

Mechanisms of Migration

Fishes may move thousands of kilometers through the open and seemingly landmark-free ocean. A great deal of research has focused on the means by which fish undertake long-distance migrations, specifically how they orient toward and locate their ultimate destinations. Research has identified numerous possible cues used in orientation, including sun and polarized light, geomagnetic and geoelectric fields, currents, olfaction, and temperature discontinuities and isolines (Leggett 1977; McCleave et al. 1984; McKeown 1984).

Birds use a sun compass and internal clock to orient. An animal must be able to sense the time of day, the altitude, azimuth (angle with the horizontal), and compass direction of the sun at a given time and date, correcting for the 15°/hr movement of the sun across the sky. Experimental evidence suggests that some fishes use such a mechanism. Swordfish (*Xiphias gladius*) can maintain a constant compass heading in the open sea for several days. Displaced parrotfish return relatively directly to their home locations on sunny days. When the sun is obscured, when fitted with eyecaps, or when held in darkness such that their internal clocks have been shifted 6 hours, displaced fish are disoriented or move in a direction appropriate for a 6-hour clock shift.

Juvenile sockeye salmon (*Oncorhynchus nerka*) have a sun compass that they complement with a magnetic compass at night or during overcast conditions. Polarized light can also provide directional cues, and sockeye salmon are able to detect and discriminate between vertically and horizontally polarized light, which could aid them particularly during dawn and dusk migrations toward the sea, when light is maximally polarized. Minnows, other salmonids, halfbeaks (Hemiramphidae), and cichlids can also sense polarized light, which often involves detection of ultraviolet radiation undiscernible to the human eye (Quinn and Brannon 1982; McKeown 1984; Hawryshyn 1992).

A magnetic compass implies a sensitivity to the earth's magnetic fields. Such a sensitivity has been demonstrated in elasmobranchs, anguillid eels, sockeye salmon, and tunas (see Chap. 6, Magnetic Reception; Chap. 12, Sen-

sory Physiology). Sharks are theoretically capable of navigating using geomagnetic cues, since they can detect fields 10 to 100 times weaker than the earth's magnetic field, as well as fields created by ocean currents moving through the earth's magnetic field or fields induced by their own movement. An induced field would change as the animal's compass heading changed, being strongest when moving east or west and weakest when heading north or south, thus giving it directional information. A magnetic compass could be useful in transoceanic migrations undertaken by large pelagic sharks (e.g., blue and tiger sharks; see Chap. 12). Orientation abilities are also needed for homing, as happens when scalloped hammerhead sharks, *Sphyrna lewini*, return daily to small seamounts in the Sea of Cortez after foraging offshore at night. Scalloped hammerheads may use a combination of directional cues, including visual landmarks, auditory cues produced by fishes and invertebrates, electrical cues induced by site-specific currents, and geomagnetic fields at seamounts. The use of multiple cues and redundant systems is a general feature of migratory animals. Redundant information increases the accuracy of the information, and backup systems provide information when conditions interfere with or negate the use of other cues (Kalmijn 1982; Klimley 1995; Klimley et al. 1988).

Water currents serve to transport fish eggs, larvae, and adults but may also provide orientational information. Where currents border on other water masses, differences in water density, turbulence, turbidity, temperature, salinity, chemical composition, oxygen content, and color could all act as landmarks to a migrating fish (once inside a current and out of sight of or contact with the bottom or other stationary objects, it is difficult to imagine that a fish could sense the water's movement, unless the fish could detect induced magnetic fields as discussed above). In shallow waters, many fishes show a positive or negative rheotactic response that causes them to move up- or downstream respectively. The strength and direction of response may change with season and ontogeny. Selective tidal stream transport (see above) is such a response, whereby a fish moving upriver in an estuary swims actively against an ebbing tide and drifts passively with a flooding tide.

Olfactory cues are often carried on currents. Homing of salmon to chemicals in the streams in which they were spawned (see below) probably applies to many stream and intertidal fishes (e.g., minnows, sculpins, blennies), although the age at which a fish learns the chemical fingerprint of a water body will vary. Sensitivities to familiar chemicals are extreme, on the order of $1:1 \times 10^{-10}$ or 10^{-19}, depending on species, suggesting that just a few molecules of a substance are necessary for detection (Hara 1993).

Seasonal movement is induced or directed by temperature changes in several migratory species. American shad, *Alosa sapidissima*, move north along the Atlantic seaboard in the spring, staying in their preferred water temperatures of 13° to 18°C. Individuals may winter as far south as Florida and spawn in Nova Scotia, 3000 km away. Some oceanic species follow specific isotherms during seasonal migrations. Albacore tuna, *Thunnus alalunga*, move north during the summer along the Pacific Coast of North America, staying within a fairly narrow temperature zone of 14.4° to 16.1°C; east-west movements are contained within a temperature range of 14° to 20°C. Onshore arrival of water masses of the preferred temperature serve as predictors of arrival of the fish. Many other tuna species also migrate to stay within fairly narrow temperature ranges.

Many pelagic fisheries, which rely on oceanic migrations to bring fish into regions on a seasonal basis, are highly dependent on water masses of the correct temperatures moving into specific areas. Cod and capelin (*Mallotus villosus*) in the Barents Sea of northern Europe are available to Finnish fisheries in cold years when fish migrate farther west to warmer waters. In warm years, fish restrict their movements to the eastern side of the basin and are then exploited in the Murmansk area. The response to temperature may be a direct, behavioral one involving thermal preferences, or an indirect response related to food abundance. Often, plankton blooms are associated with changing water temperatures, and hence fish may be tracking food availability, which responds to temperature. Herring in the Norwegian and Greenland Seas migrate in response to the inflow of warm Atlantic water, which in turn stimulates plankton growth and food availability (Leggett 1977; McKeown 1984; Dadswell et al. 1987).

Representative Life Histories of Migratory Fishes

Among vertebrates, fishes stand out in terms of the complexity of their life histories, and migratory fishes have among the most complex life histories. Details of a few of the better-known and more interesting species are highlighted below.

Anadromy.   Some of the most spectacular examples of highly evolved, complex migrations involve fishes that spawn in freshwater but spend most of their lives at sea. Included among anadromous fishes are lampreys, sturgeons, shads and herrings, salmons and trouts, and striped bass (see Table 22.2). The classic case involves Pacific salmons. Chinook salmon, *Oncorhynchus tshawytscha*, can serve as an example. Chinook salmon spawn in streams of the Pacific northwestern coast of North America during the spring, summer, and fall. Eggs are buried in gravel nests and hatch into alevins (yolk-sac larvae), which emerge and make their way downstream, eventually transforming into silvery smolts after a few months to 2 years, depending on when they were spawned. Smolts move out into the ocean, grow into juveniles and adults, and move in a series of counterclockwise ellipses through the northeastern Pacific that may carry them as far north and west as the Aleutian Islands of Alaska or as far south as northern California, covering distances of several thousand kilometers (sockeye salmon, *O. nerka*, migrate even farther from land and in larger circles in the open sea, and may cover tens of thousands of kilometers). After 1 to 8 years, these adults mature and return to the nearshore area. A coastal migration eventually carries them to the mouth of the river from which they migrated as smolts. They enter this river and work their way up, bypassing hundreds of potentially usable streams and innumerable, seemingly insurmountable barriers such as

rapids and waterfalls. In this final stage, they cease feeding and change to a reddish color; the males develop the characteristic hooknosed appearance known as kipe. They ultimately find the natal stream in which they were hatched, where they spawn and die (Netboy 1980; Healey and Groot 1987; Brown 1990; Groot and Margolis 1991).

At each juncture in this complicated journey, directional decisions must be made by migrating fish. Numerous mechanisms, which vary depending on life history stage and habitat, have been proposed to provide directional information for a migrating fry, smolt, juvenile, and adult (Fig. 22.8). Movement by young fish from spawning sites and natal rivers and lakes to the ocean involves a combination of responses to light (including a sun compass and discrimination of polarized light), geomagnetic cues, and water currents. The fish must also imprint on (learn) the chemical fingerprint of its home stream and river. Open-ocean migration and eventual home-stream selection offer very different problems that probably require different orientation systems. Quinn (1982) has proposed a combined map-compass-calendar system to explain movements on the high seas. The map would involve learned or genetic knowledge of the distribution of the earth's magnetic field (which has also been mapped by oceanographers and is predictable). Compass directions, provided by celestial and magnetic cues, can be used to maintain directional headings. The calendar would require an assessment of day length or change in day length, with input from an endogenous circannual clock. Integration of all this information would tell a fish where it was, where it was going, and how long it would take to get there, forming the basis of a navigational system.

Once a maturing adult arrived in the coastline region of its home river, it would shift to an olfactorily guided response to natural chemicals contained in different rivers. Having remembered the chemical fingerprint of its natal stream, it would move upriver and reject any stream mouth it passed that did not have the appropriate bouquet. Upon encountering the correct chemical cues, it would move upcurrent in that system until it arrived at the appropriate spawning site. The **olfactory hypothesis** has received experimental confirmation in studies with coho salmon, *O. kisutch*, transplanted into Lake Michigan. Fish were imprinted on synthetic chemicals in hatchery water and released. Eighteen months later, most chemically imprinted salmon that entered streams chose streams containing the synthetic chemicals (Hasler and Scholz 1983; Quinn and Dittman 1990; see Box 4.2).

Home-stream return, perhaps involving olfactory guidance, also occurs in striped bass, *Morone saxatilis*, which forms stocks along the U.S. Atlantic coast that are associated with major river systems. Fish migrate north in the spring and south in the fall along the Atlantic coast but return each spring to spawn in their natal rivers. American shad, alewives, and blueback herring also home to their natal rivers to spawn (Boreman and Lewis 1987; Loesch 1987; Quinn and Leggett 1987).

Despite this homing sense, mistakes sometimes do occur. Although as many as 98.6% of chinook salmon may home correctly to the Cowlitz River in Washington, the same species may show 10% to 13% straying rates in Cali-

FIGURE 22.8. *Characteristic life history of a Pacific salmon, as seen in the sockeye salmon,* Oncorhynchus nerka. *At different stages, different orientation mechanisms are likely to come into play to aid the developing, growing, or maturing fish find its way to, through, and away from the sea and back to its natal river to spawn. Responses to light, gravity, and current are initially important for recent hatchlings. Later, sun compass and magnetic detection, backed up by other cues, aid a fish moving downstream and into the ocean. Finally, the memory of home-stream chemicals on which the juvenile imprinted leads the maturing adult back to its spawning grounds.*

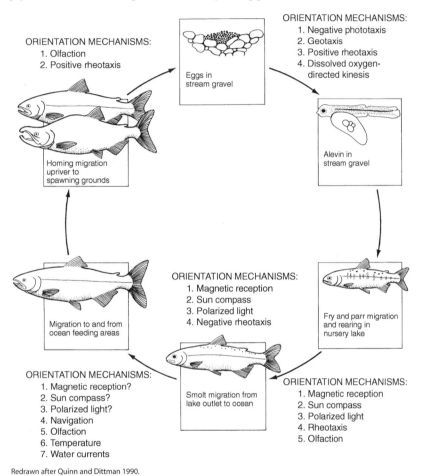

ORIENTATION MECHANISMS:
1. Olfaction
2. Positive rheotaxis

Eggs in
stream gravel

ORIENTATION MECHANISMS:
1. Negative phototaxis
2. Geotaxis
3. Positive rheotaxis
4. Dissolved oxygen-
   directed kinesis

Alevin in
stream gravel

Homing migration
upriver to
spawning grounds

ORIENTATION MECHANISMS:
1. Magnetic reception
2. Sun compass
3. Polarized light
4. Negative rheotaxis

Fry and parr migration
and rearing in
nursery lake

Migration to and from
ocean feeding areas

Smolt migration from
lake outlet to ocean

ORIENTATION MECHANISMS:
1. Magnetic reception?
2. Sun compass?
3. Polarized light?
4. Navigation
5. Olfaction
6. Temperature
7. Water currents

ORIENTATION MECHANISMS:
1. Magnetic reception
2. Sun compass
3. Polarized light
4. Rheotaxis
5. Olfaction

Redrawn after Quinn and Dittman 1990.

fornia rivers. In fact, tagging studies show that as many as 47% of fish may wind up in the wrong, or at least in a nonnatal, stream. However, the pattern of straying is adaptive. High fidelity (low straying) rates characterize species and populations that spawn in large, stable rivers, whereas straying is more common in fish that come primarily from small, unstable rivers with variable flow characteristics, where juvenile survival is also more variable. Straying can then be viewed as an alternative life history trait that functions as a bet-hedging tactic to ensure survival of some offspring in situations where the natal river may become uninhabitable (Quinn 1984; see Chap. 20 for other examples of alternative reproductive tactics in salmonids).

Catadromy. Life history variation occurs among salmons and sturgeons, with some species and populations being landlocked or seldom entering the sea. In contrast, all 15 species of the eel family Anguillidae are thought to spawn

in the sea but grow up in freshwater. The best-known species are the European, Japanese, and American eels, all of which undergo larval and adult migrations of truly epic proportions. The American eel, *Anguilla rostrata*, can serve as an example.

American eels spawn in the Sargasso Sea, an unproductive region of the western Atlantic northeast of Hispaniola and the Bahamas. The exact locale of spawning remained a mystery until the 1920s, when Danish biologist Johannes Schmidt analyzed 25 years of oceanic plankton tows and determined that the smallest eel larvae of both American and European species (known as **leptocephali** and once thought to be a different species of fish altogether) were captured in this area. Schmidt's results have been subsequently confirmed by captures of even smaller larvae (< 7 mm long) of both species at the same time from the same locale; American and European eels spawn in overlapping areas during early spring and then drift northward with major ocean currents. European larvae apparently do not metamorphose until they are 2 to 3 years old; hence they float past the North American continent. American eel leptocephali in contrast metamorphose after about 1 year and, using transport mechanisms that remain unresolved, move westward to inshore waters. Interestingly, hybrids between the European and American species stop halfway, in Iceland. Mysteriously, leptocephali are not thought to feed, or at least they have nonfunctional guts during most of their larval phase. Leptocephali are next attracted by the mixture of organic materials dissolved in outflowing freshwaters and migrate upriver, moving by selective tidal stream transport (see above) and transforming into transparent, miniature (50 mm) eels known as **glass eels**. As they move upriver, they become pigmented and are called **elvers**.

Elvers grow into juvenile **yellow eels**, which take up residence in freshwater for periods that range from 3 to 40 years, the time depending on sex and latitude. Males are more abundant in southerly latitudes and in estuaries. They never grow larger than 44 cm and usually mature after 3 to 10 years. Females are likely to be found throughout a river system, from the estuary all the way up to headwaters. In fact, female American eels probably have the widest geographic and environmental range of any nonintroduced freshwater fish anywhere in the world. Their habitats include rapidly flowing, clear, headwater streams, large lakes and rivers, underground cave springs, lowland rivers and swamps, down to estuarine salt marshes. They are found from Iceland to Venezuela, including most Caribbean Islands and Bermuda, and range up the Mississippi River to its headwaters and as far west as the Yucatan Peninsula.

Maturation varies as a function of this range. In general, more northerly populations and those farther from the Sargasso Sea contain older, larger (and usually female) animals. Maturation may take from 4 to 13 years at southerly locales and as much as 43 years in Nova Scotia (Jessop 1987). As the animals mature, they turn a silvery bronze color, the pectoral fins become pointed, the eyes enlarge (particularly in males), and fat stores are accumulated. These nonfeeding, **silver eels** then migrate back to the Sargasso Sea, with migrations beginning earlier for animals traveling farther, which apparently synchronizes the time of arrival at the spawning grounds. Silver eels travel as much as 5000 km to spawn and then apparently die. Conjecture surrounds this stage, because no one has either seen a fully mature anguillid eel or located an adult eel at the presumed spawning grounds (Tesch 1977; McCleave et al. 1987; Helfman et al. 1987; Avise et al. 1990).

Oceanodromy.   Oceanodromous fishes migrate within ocean basins, usually in a circuit and usually traveling with major ocean currents. The migration serves to place different life history stages in seasonally appropriate locales and hence may include an area for spawning from which eggs and larvae float to a nursery area, winter and summer feeding areas for juveniles and adults, as well as

migratory zones through which a stock moves. Juveniles may move between seasonal feeding areas for several years before maturing and migrating to spawning grounds. The great tunas (*Thunnus*, Scombridae), particularly those living in temperate waters, are representative. Subspecies or stocks have been suggested for different ocean basins in the past, but movement across ocean basins and probable mixing of stocks tends to eliminate genetic differences (e.g., for the highly migratory albacore tuna, *Thunnus alalunga*; Graves and Dizon 1989). Bluefin tuna are subdivided into northern and southern species (Atlantic bluefin, *Thunnus thynnus,* in the northern Atlantic and Pacific, and southern bluefin, *T. maccoyii,* off Australia, New Zealand, South Africa, etc.). Bluefin tagged off Florida have been recaptured in Norway, involving a minimum migration distance of 10,000 km. In addition, fish of different sizes have different migratory patterns, and adults of different sizes may spawn at different times and places. In the western North Atlantic, the largest bluefin (120 to 900 kg) have a migratory cycle that begins on summer feeding grounds (May to September) over the Continental Shelf from Cape Hatteras to Nova Scotia. This is followed by fall and winter movements offshore and south to wintering grounds that include the Bahamas, Greater Antilles, Caribbean, and Gulf of Mexico. In the spring, the giants move northward in oceanic waters, then onto the Continental Shelf in late spring, and finally back to the summer feeding grounds. Spawning occurs in southern waters (Gulf of Mexico, Straits of Florida) in May and June, and in the Mediterranean and Black Seas during warm summer months. How transatlantic movements fit into this picture and whether each individual actually makes the migration annually remain unclear (McClane 1974; Richards 1976; Rivas 1979).

Differentiation into stocks, some that mix and some that do not, appears common among oceanodromous fishes. Atlantic herring (*Clupea harengus*) are subdivided into several spawning groups or stocks (six alone in the northeast Atlantic) which may be further subdivided into isolated stocks in various estuaries and inlets. Migrations carry different stocks to overlapping feeding areas, but spawning occurs at separate times and places. Atlantic cod (*Gadus morhua,* Gadidae) and plaice (*Pleuronectes platessa,* Pleuronectidae) are also differentiated into several migratory stocks with distinct spawning grounds. Bluefish (*Pomatomus saltatrix,* Pomatomidae) occur worldwide in warmer oceans except for the eastern Pacific. Schools apparently migrate onshore and offshore with the seasons, perhaps following baitfish. Along the U.S. Atlantic coast, this migration involves an inshore migration in spring and summer and a return to offshore locales in fall and winter, which corresponds with movements of a primary prey species, menhaden (*Brevoortia tyrannus*). Bluefish movements occur progressively later as one travels north, and the pattern is complicated by a degree of north-south migration (McKeown 1984; Hersey 1988). Other, larger predators, such as blue marlin (*Makaira nigricans,* Istiophoridae), also are oceanodromous, move seasonally, and form local but wide-ranging stocks (see also Chap. 12 on pelagic sharks).

Summary

1. Biological systems are cyclical in nature. On a 24-hour, diel cycle, most fishes are diurnal, being active by day and resting at night. Fewer species are nocturnal, and some predators are crepuscular, being active primarily during dusk and dawn twilight periods. The external cue for activity and inactivity appears to be the setting and rising of the sun. Distinctive activity cycles are most pronounced in tropical environments, less so at higher latitudes.

2. Circadian rhythms have a period of approximately 24 hours, driven by an internal (endogenous) clock. Activity patterns in many fishes have an underlying circadian rhythm, as does hormone secretion. Other endogenous cycles include circatidal patterns in intertidal fishes that correspond to twice-daily high and low tides.

3. Intertidal fishes either move in and out with tides (juveniles using the shallows as a nursery area) or remain in the intertidal at low tide and seek shelter to avoid heat, desiccation, and oxygen stress (resident families such as gunnels, sculpins, blennies, gobies). Nursery species gain access to tidal areas by moving up into the water column with flooding currents and by hugging the bottom on ebbing currents.

4. Many fishes spawn on a biweekly (semilunar) or monthly (lunar) cycle. For example, grunion spawn every 2 weeks during the summer, laying their eggs in the sand high on the beach just as the tide turns. Their eggs then hatch 2 weeks later when the next spring tides cover them. Many coral reef species spawn biweekly or monthly when tides and therefore currents are maximal, which may help move the eggs away from abundant, reef-dwelling zooplanktivores. Nonreproductive migrations in many fishes (eels, salmons) are also tied to lunar cycles.

5. Fishes at nontropical latitudes spawn more seasonally, usually during spring, probably to allow larvae and juveniles to take advantage of spring and summer blooms of plankton. Cycles are set in motion by changing day length. Winter activities of many temperate species are poorly understood; burying and aggregating in deeper water have been documented. Problems of low oxygen under ice can be overcome by inhaling bubbles that form directly under ice. Freshwater tropical fishes spawn seasonally in response to changing rainfall regimes, often migrating into flooded forests and swamps to reproduce.

6. The life cycle of many fishes includes a migration over long distances, either as part of reproduction or to take advantage of seasonal changes in food availability. Movement between fresh- and saltwater is called diadromy; anadromous species spawn in freshwater but grow in the ocean (lampreys, sturgeons, salmons), catadromous species spawn at sea and grow in freshwater (eels, mullets, temperate basses), and amphidromous species move between habitats more than once (galaxiids, southern graylings, sleepers). Anadromy is more common at temperate, northern locales, whereas catadromy occurs more at southern locales and at low latitudes. These patterns place early life history stages in habitats most favorable for growth.

7. Fishes navigate across distances by orienting to cues of light, geomagnetism, currents, odors, and temperature. Sun compasses are used by many species, as is polarized light. Elasmobranchs, eels, salmons, and tunas are sensitive to the earth's magnetic field. Chemical sensitivities in fishes are extreme, on the order of $1:1 \times 10^{-10}$ or 10^{-19}. Pacific salmons learn the odor fingerprint of the stream in which they are born and then, using olfactory cues, return to their natal stream after years at sea. Catadromous eels begin life in tropical seas, float to continental areas on currents as larvae, grow in freshwater, and then migrate thousands of kilometers back to their open-ocean spawning region using orientation cues that are a mystery.

Supplemental Reading

Ali 1980, 1992; American Fisheries Society 1984; Baker 1978; Brown 1990; Dadswell et al. 1987; Gauthreaux 1980; Goulding 1980; Groot and Margolis 1991; Harden-Jones 1968; Hersey 1988; Lowe-McConnell 1987; McCleave et al. 1984; McDowall 1988; McKeown 1984; Munro et al. 1990; Thorpe 1978.

23

Individuals, Populations, and Assemblages

Fishes interact with their own and other fish species, with other animals and plants, and with their physical surroundings. These interactions affect the birth, growth, reproduction, death, and movement of individuals; the distribution and abundance of populations; the transport and exchange of energy and nutrients among members of communities; and the flow of matter and energy into, out of, and between ecosystems. The science of ecology is concerned with the relationship between organisms and the biotic and abiotic environment, and more specifically with how environmental variation influences the distribution, abundance, and function of organisms. Ecology is a vast subject, and the literature on fish ecology is voluminous. The bibliography of one relatively recent and comprehensive textbook (Wootton 1990) runs for over 40 pages; such a large topic can obviously receive only cursory treatment in two chapters.

Our focus will be at different levels of ecological organization: individuals, populations, and assemblages in this chapter, communities and ecosystems in the next. These distinctions are recognizably artificial; one cannot understand ecological attributes of individuals of one species in an area (a **population**) without also considering the other species of fishes (the **assemblage**), other taxa with which the individuals interact (the **community**), and abiotic influences on individuals, including patterns of nutrient and energy transfer (the **ecosystem**). Our emphasis once again is on the diversity of adaptations shown by fishes in an ecological context. Certain topics that commonly fall under the heading of ecology have been treated elsewhere and will only be examined briefly here (e.g., feeding and predator-prey relations, Chaps. 18, 19; growth and reproduction, Chaps. 9, 10, 20; energetics, Chap. 5; symbioses, Chap. 21).

INDIVIDUALS

Ecological adaptations are traits of an individual that ensure its survival and reproduction in response to selection pressures from the biotic and abiotic environment. It is important to emphasize the individual as the basic unit of adaptation, since natural selection operates primarily at the level of the individual, favoring individuals of one genotype while selecting against individuals with less-favorable genotypes. We can then ask if survival and reproduction are enhanced by how an individual selects an appropriate habitat in which to live (discussed in the section below, Assemblages), how it budgets its time and energy among the activities and conflicting demands pre-

sented to it on a daily basis, and how it eventually partitions energy into growth versus reproduction.

Life Histories and Reproductive Ecology

A **life history** can be viewed as how an individual divides up its time and resources among the often-conflicting demands associated with maintenance, growth, reproduction, mortality, and migration. Life history characteristics, or **traits**, are measurable aspects of an individual's life history and include age- and size-specific birth rates (and associated characteristics such as clutch size, egg size, offspring provisioning, and clutch frequency) and the probabilities of death and migration (Congdon et al. 1982; Dunham et al. 1989). These traits vary among species, among populations within a species, and among individuals and sexes within a population in ways that make evolutionary sense, indicating their adaptiveness. The challenge to biologists is to identify trends in life history traits, identify the likely selection pressures causing variation, and interpret the adaptiveness of the variation (Potts and Wootton 1984; Stearns 1992; Winemiller and Rose 1992). Many life history traits are correlated, which means they are inherited together and change in direct relationship with one another, making it somewhat difficult to isolate the exact interaction between environment and adaptation. Nonetheless, their importance in determining reproductive success is obvious.

Analyses of life history traits focus on females in part because female reproductive effort produces eggs, each of which has a much greater likelihood of becoming a new individual than is the case for the millions of sperm produced by a male. About one dozen life history characteristics (termed traits by some but not all authors) have direct links with reproduction and have been identified and quantified in many fishes. Background detail on reproductive biology and anatomy is presented in Chapters 9 and 10.

1. Age and size at maturation. A complex but fascinating trade-off exists between early versus late maturation; the trade-off depends on the probability of successful reproduction versus the risk of death. A female that delays maturation until she is larger and older will produce more eggs at each spawning but runs the risk of dying before she ever reproduces (Fig. 23.1). A fish that spawns at an earlier age stands a greater chance of getting some genes into the next generation at least once. However, younger fish are smaller and hence produce fewer and often smaller eggs, which lessens the chance that any will make it past egg and larval predators and starvation. Also, by allocating energy to reproduction, the earlier-spawning fish has slower somatic (body) growth and is then more subject to predation because of smaller body size (Werner et al. 1983). Additionally, reproduction uses up much energy, potentially placing a smaller fish with lesser energy stores in a weakened condition, which reduces the chances of future reproduction. Theoretically, females in populations where adult survival is poor should reproduce at an earlier age than in populations where survival is better. Female guppies in downstream locales in Trinidad where predators are abundant do mature earlier than upstream populations with fewer predators (Reznick and Endler 1982). Similarly, individuals in commercially exploited fish populations, particularly those in which adults are targeted by the fishery, often reproduce at earlier ages than do fish in unexploited populations (e.g., O'Brien et al. 1993).

2. Body size. Even very large predators eat relatively small prey, but only large predators can eat large prey. Therefore, larger fishes are susceptible to predation by fewer predators. Larger fishes are also able to store more energy, to swim faster and farther, and to better overcome harsh abiotic conditions such as strong currents (Karr et al. 1992). Size determines territorial interactions as well as male mating success in many fishes. This premium on large size exists at a cost, since energy allocated to somatic growth is unavailable for immediate reproduction, as discussed above.

3. Longevity. The longer an individual lives, the more reproductive opportunities it should have, discounted by how long it waits until first reproduction (see point 1 above) and how long an interval exists between reproductive periods (see point 10 below).

4. Clutch size. How many eggs a female produces at each spawning varies as a function of body condition and size, age, egg size, and number of spawnings per season (egg number is referred to as **fecundity** and can be subdivided into batch, breeding season, or lifetime fecundity, although fecundity commonly refers to number of eggs or young produced per year; see also Chap. 9, Eggs and Sperm). Combining clutch size with egg size (see point 5 below) gives a measure of **reproductive allotment**, which is the percentage of a female's weight devoted to eggs or embryos. Theoretically, females in populations where adult survival is poor should devote more energy to repro-

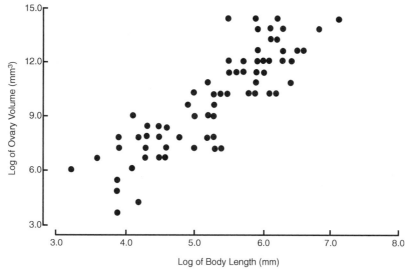

FIGURE 23.1. *The larger a fish is, the more eggs she produces. Since fish grow throughout their lives (= indeterminate growth), older fish are usually larger. Within a species, larger individuals generally produce more eggs and often larger eggs. Values are plotted for a variety of Canadian freshwater species.*

Redrawn after Wootton (1990).

duction than in populations where survival is better. Reproductive allotment in female guppies in predator-dense populations is 30% greater than in females subject to less predation (Reznick and Endler 1982). Commercially exploited species (e.g., pike, halibut) show increased fecundities as compared to unexploited populations, a change in a life history trait that could compensate for high levels of predation or exploitation (Policansky 1993).

5. Egg size and size at birth.   Both mean and range of egg size vary within species and within individuals (Wootton 1990). Eggs spawned early in a season by multiple spawners tend to be larger. In the least killifish, *Heterandria formosa* (Poeciliidae), the more broods a female produces, the smaller the young from each brood will be. In the orangethroat darter, *Etheostoma spectabile* (Percidae), offspring hatched from larger eggs are larger and are less likely to starve (Marsh 1986). Greater investment in each egg, largely in terms of amount of yolk, increases the chances of survival for that offspring. A larger larva is better able to avoid predators and feed independently (e.g., Richards and Lindeman 1987). The volume of the ovarian space in a female determines fecundity, producing an inverse relationship between egg size and number of eggs. Fish that produce larger eggs have lower batch fecundities.

6. Time until hatching and exogenous feeding.   For egg-laying species, which is most fishes, eggs are deposited on substrates or in the water column and are essentially defenseless, either immobile or floating. After hatching, yolk-sac fry are inefficient swimmers. The longer a larva spends growing inside the egg or absorbing yolk resources, the larger it will become before having to obtain its own food. A larva trades off the increased vulnerability it experiences while being passive against the advantages it will have in finding food and avoiding predators once it achieves independence from the egg shell and yolk sac.

7. Larval growth rate and interval length.   Rapid growth provides a larva with the same advantages as large egg size or yolk supply, namely, achieving a larger size earlier. However, rapid growth requires more energy and higher metabolism, which in turn demands more-efficient or faster feeding and an increased likelihood of starvation. A short larval period means larvae can transform quickly into juveniles and settle from the plankton and into the generally safer juvenile habitat. But if a larva finds itself in inappropriate habitat at the end of its larval period—such as far out at sea for a species adapted to shallow-water existence as a juvenile and adult—then a short larval life provides little advantage. Conversely, a long larval interval can permit long-distance dispersal. However, extended planktonic life exposes the larva for a longer time to the extreme hazards of planktonic existence, when more than 99% of larvae are eaten or starve. Growth rates of juveniles and adults are subject to advantages and constraints as discussed in point 2.

8. Spawning bouts per year and duration of spawning season.   The number of times an individual, particularly a female, spawns each year tells much about the allocation of energy to reproduction. Duration of the spawning season is more a population than an individual characteristic and is useful in assessing potential recruitment into that population. Usually, individuals have a shorter spawning season than the population as a whole.

9. Number of spawnings per lifetime.   Most fishes are **iteroparous** (itero = "to repeat"; parous = "to give birth"), spawning repeatedly throughout their lives. One-time spawners, termed **semelparous**, devote all their energy to a single, massive spawning event, after which they die (anguillid eels, Pacific salmons, some gobies).

10. Reproductive interval.   The time spent between reproductive bouts for iteroparous species varies greatly, from daily spawners that reproduce year-round in some low-latitude coral reef fishes (e.g., wrasses), to fishes that spawn every few weeks during a protracted season (e.g., grunion, darters), to seasonal spawners that may spawn only once or a few times during a limited season (snappers, groupers, larger percids, centrarchid basses), to internal bearers with long gestation periods of a year or more (some sharks), to species that may wait several years between spawnings (sturgeon). For iteroparous species that spawn repeatedly each year, reproductive interval can theoretically be adjusted in response to expected mortality levels. Where the probability of mortality is high, reproductive intervals should be short. As was the case for variable age at first reproduction and reproductive allotment (above), female guppies exposed to high levels of predation have relatively short reproductive intervals (Reznick and Endler 1982).

11. Parental care.   The degree of care given has an overwhelming influence on the mortality rate of the young and is generally inversely proportional to fecundity (see Chap. 20, Parental Care). Parental care is often distinguished as **prezygotic** (e.g., nest preparation) and **postzygotic** (e.g., internal brooding, guarding young). Care may be nonexistent to elaborate. Broadcast spawners release large numbers of eggs into the water column and provide no further care (tarpon, cods, tunas). Moderate care occurs in fishes that spawn intermediate numbers of eggs on substrates and may involve some substrate preparation such as nest digging or egg covering (salmons, grunion). More extensive care occurs in fishes that prepare a nest and then guard relatively few eggs until they hatch and perhaps a little later (sticklebacks, sunfishes, some cichlids). Intensive care is usually associated with relatively low numbers of large eggs, such as fishes that gestate young internally (livebearers, embiotocid surfperches) or incubate the eggs orally (cardinalfishes); some oral incubators continue to protect the young after hatching (some catfishes) and some cichlids even feed the young with external body secretions. Parental care increases the survival of young but occurs at a cost to the parents, because extended care increases the interval between spawnings.

12. Gender change and sex ratio variation.   Fishes in several families change sex, beginning as males and changing to females (**protandry**) or vice versa (**protogyny**) (see Chaps. 10, 20). The timing of the change is largely determined by the relative reproductive success males or females experience at the same body size. Sex change occurs at a cost in immediate reproductive output, since weeks or months may be required to convert the actual machinery of gamete production from one sex to another.

In many vertebrates (crocodilians, turtles, lizards, possums, monkeys), the sex of offspring may be determined by conditions such as the temperature at which eggs or embryos develop; this is termed **environmental sex determination** (ESD). Extreme temperatures affect sex determination in a few fishes, mostly atheriniforms such as rivulins, ricefish, and livebearers; pH can influence sex determination in some cichlids and a livebearer (Rubin 1985; Francis 1992). Naturally occurring variation in temperature determines the sex of developing Atlantic silversides, *Menidia menidia* (Atherinidae). Offspring produced early in the year at relatively low temperatures tend to be female; young produced later at higher temperatures tend to be male (see Chap. 10, Determination, Differentiation, and Maturation). Females could theoretically manipulate the sex ratio of their offspring to take advantage of disproportionate numbers of one sex or of environmental conditions that favor one sex over another. Whether and why a female silverside actively manipulates the sex ratio of her young are matters of conjecture. It is difficult to imagine how a female would assess current sex ratios, as larvae are dispersed widely and are thus unlikely to be recruited into the same population and environment as the mother (Conover and Kynard 1981; Conover and Van Voorhees 1990; D. Conover, personal communication).

13. Geographic patterns and phylogenetic constraints. In many families, related species living in different habitats often adopt life history patterns appropriate for that habitat, and unrelated fishes converge on suites of life history adaptations. Mouth-brooding fishes worldwide have converged on small clutches of large eggs, slow growth rates, and protracted breeding seasons (bonytongues, marine catfishes, cichlids). Such convergence is evidence of the importance of environmental selection factors promoting one life history over another and can be found at relatively large geographic scales. Among freshwater fishes in North America, species that mature relatively late in life tend to have larger body sizes, longer lifespans, higher fecundities, smaller eggs, few multiple spawnings, and a short spawning season (sturgeons, paddlefish, shads, muskellunge, chars, burbot). Fishes with extended spawning seasons tend to have larger eggs, multiple spawning bouts, and exhibit more parental care (cavefishes, madtom catfishes). Marine fishes that have extensive geographic ranges (tarpon, cods) also tend to have high fecundity. Anadromous species, such as salmons, striped bass, and sturgeons, mature late, grow fast as adults, live long, and have large eggs (Winemiller and Rose 1992).

As with any evolved characteristic, life history traits are influenced by the evolutionary history of the lineage to which an animal belongs. Consequently, an animal may not have life history characteristics that we expect given current conditions. Unless selection pressures have been relatively stable for many generations, an animal's adaptations will not necessarily reflect present conditions but will instead reflect past adaptive scenarios and selection pressures. For relatively conservative traits that are shared among many members of a lineage (the symplesiomorphies of cladistic analysis; see Chap. 2), historical constraints may be difficult to overcome, and species will retain seemingly nonadaptive or nonoptimal characteristics. Regardless of latitude and habitat, percopsiforms (troutperches, pirate perch, cavefishes) tend to be small, produce small clutches of large eggs, exhibit extensive parental care, and have protracted spawning seasons and slow growth rates. Within the cypriniforms, suckers in the genus *Ictiobus* are large with large clutches and few spawning bouts, whereas minnows in the genus *Notropis* are small with small clutches and frequent spawning bouts. Flatfishes as a group mature at a large size, produce large clutches of small eggs during short spawning seasons, and grow rapidly when young (Winemiller and Rose 1992).

POPULATIONS

Simply defined, a **population** consists of all individuals of a particular species in a given area. Because populations form the matrix in which individual survival and reproduction occur, expanded definitions recognize the importance of genetic structure; a population is therefore "a gene pool that has continuity through time because of the reproductive activities of the individuals in the population" (Wootton 1990, 280). Populations grow and shrink in numbers as a result of the actions and interactions of their individuals, which can change relative gene frequencies (the **genetic structure**) of the population. Much of ecology has been devoted to describing, understanding, and predicting the nature and causes of population numerical growth and decline and of genetic structure (the effects of competition and predation on population size are discussed in the section below, Assemblages).

Population Dynamics and Regulation

Population size changes as a function of four major processes: birth, death, immigration, and emigration. **Birth** and **death** rates (**age-specific reproduction** and **survivorship** rates of individuals) can be used to calculate the approximate rate at which a population will change in size (Table 23.1). Such **life table** statistics are usually calculated only for females in a population; relatively few have been constructed for fishes.

Migration, both in and out of a population, greatly complicates any attempt at predicting future population size. Fish populations increase in size due to migration as a result of either recruitment or colonization. **Recruitment** usually refers to addition to the population through reproduction, as when larvae settle out of the plankton and into the population. In fisheries terminology, recruitment generally refers to the addition of potentially catchable individuals to the **stock** in question, stock being

Table 23.1.   A life table for a cohort of brook trout, *Salvelinus fontinalis*, in Hunt Creek, Michigan, for the year 1952. Survivorship (l_x) is the probability that an individual female will live to age x; reproductive output (m_x) is the mean number of daughters produced by a female of that age (estimated as half the number of eggs produced). The reproductive rate of the population (net reproductive rate, R_0) is the sum of the survivorship and reproductive output columns (= $\Sigma l_x m_x$), which equals the average number of females being produced per female in the cohort. When R_0 is greater than 1, the population is growing; when it is less than 1, the population is shrinking.

Age Class, in Years (x)	Survivorship (l_x)	Reproductive Output per Female (m_x)	Reproductive Rate of Population $l_x m_x$
0	1.0000	0	0
1	0.0528	0	0
2	0.0206	33.7	0.6952
3	0.0039	125.6	0.4898
4	0.00051	326.9	0.1667
			$R_0 = 1.352$

From Wootton 1990, based on data of McFadden et al. 1967; used with permission.

synonymous with population. **Colonization** is addition by movement of established individuals between habitats, such as when juveniles move from a nursery habitat to an adult habitat. Because many fishes export their reproductive products in the form of pelagic larvae that are dispersed widely, reproductive events in a particular population may have little effect on later population size. In fact, most marine populations show little correspondence between stock and recruitment, which means that current population size may not predict future population size, justifying regional rather than local approaches to managing stocks (Rothschild 1986; Wootton 1990). Life tables are therefore more useful for relatively closed populations, such as in ponds and lakes, than for marine species.

Regardless of locale, fish populations vary widely and notoriously in size. The concept of the **year class** or **cohort** is important in understanding these dynamics. High population density may not necessarily indicate a sustainably reproducing population, because many of the individuals in that population may come from a single year class, whereas most other years have seen little successful reproduction (Fig. 23.2). If the successful year class is approaching the usual maximum age for that species and no younger year class is abundant, overexploitation of the dominant year class can lead to very rapid population collapse. Year class strength then becomes a critical statistic in determining management schemes for exploited populations.

Variation in numbers among year classes points out another important feature of fish populations, which is that they are **size structured**. Indeterminate growth and overlapping generations create a situation where a population may include individuals of very different sizes,

differing in body mass by as much as four or more orders of magnitude (e.g., consider bluefin tuna that weigh a fraction of a gram at birth and grow to an adult size exceeding 500 kg in mass, a range of seven orders of magnitude). Size structuring can affect population regulation, because multiple size classes provide the potential for intraspecific competition and cannibalism, which in turn may lead to differences in habitat and other resource use because of avoidance of one group by another. Such intraspecific variation has led to the concept of the **ontogenetic niche**, which recognizes the very different ecological roles that different-age and -size conspecifics are likely to play in an assemblage (Werner and Gilliam 1984; Osenberg et al. 1992). For example, pinfish (*Lagodon rhomboides*, Sparidae) start off as a carnivore and progressively shift to increasing herbivory in five distinct phases (Stoner and Livingston 1984). Largemouth bass (*Micropterus salmoides*, Centrarchidae) initially feed on zooplankton, then on littoral invertebrates, and then finally on fish, including conspecifics. As juveniles, they compete with adult bluegill (*Lepomis macrochirus*) for zooplankton. Young largemouth are, however, miniature adults and are morphologically constructed as piscivores. They are consequently less-efficient planktivores than are adult bluegill, and this morphologic constraint makes them inferior competitors (they get even later when bluegill become their major prey) (Werner and Gilliam 1984). At each size, a fish is likely to have a different set of competitors and predators, some overlapping with the previous set, producing an incredibly complex set of interactions within a community containing even a small number of species (see Fig. 24.7).

Death can come at any time, but certain life history stages are more dangerous than others. Eggs and larvae are by far the most vulnerable periods (see Chap. 9, Larval Feeding and Survival). Estimates of mortality for marine fish populations range from about 10% to 85% per day for eggs, from 5% to 70% per day for yolk-sac larvae, and from 5% to 55% per day for feeding larvae (Bailey and Houde 1989). These are daily rates. When compounded over the larval life of a species, the magnitude of the loss is more striking. For example, jack mackerel, *Trachurus symmetricus*, require 8 days from hatching until they resorb their yolk sac and begin independent feeding. During this time, when mortality falls from 80% to 50% per day, 99.5% to 99.9% of larvae are lost to predation. Exogenous feeding adds the hazard of starvation; during the first week after yolk-sac absorption, larvae die at a rate of 45% per day from starvation alone, to which can be added predation losses (Hewitt et al. 1985).

Most workers agree that predation is the major source of mortality for eggs and larvae and that mortality is strongest on eggs and small larvae. The list of predators on larvae is long and includes numerous invertebrates (ctenophores, siphonophores, jellyfishes, copepods, chaetognaths, euphausiids, shrimps, amphipods) as well as fishes. As fish grow, their strength, swimming speed, food-getting ability, and general escape ability increase. Estimates of mortality of juveniles and adults are diametrically different from the rates experienced by eggs and larvae; for example, 99.9% daily survival for juvenile or

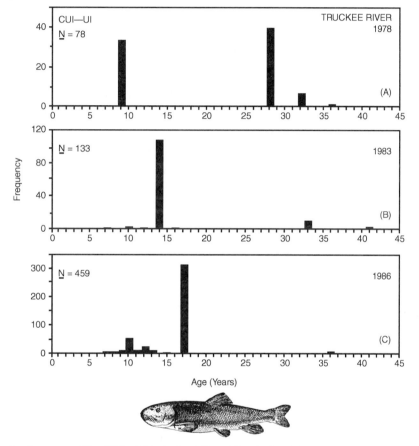

FIGURE 23.2.   *Year class strength in an endangered sucker. The cui-ui,* Chasmistes cujus *(Catostomidae), occurs presently only in Pyramid Lake, Nevada. Its long lifespan, approaching 45 years, has probably saved it from extinction. Reproductive failure in most years has resulted from drought and human diversion of water from its spawning habitat in the lower Truckee River. (A) Samples of spawning fish in 1978 indicate that the entire species has been maintained by two year classes, one born in 1950 and the other in 1969. (B) Mortality had all but eliminated the 1950 year class by 1983. (C) A bypass channel built in 1976 gave fish access to the river even at low water levels, and some successful reproduction has occurred in subsequent years (no fish occur in the 0 to 6 year classes because reproduction does not start until age 6).*

From Scoppetone and Vinyard 1991; cui-ui drawing from La Rivers 1962; used with permission.

adult nototheniid icefishes, English sole, winter flounder, cutlassfishes (Trichiuridae), mackerel, and tuna (McGurk 1986; Richards and Lindeman 1987). Freshwater salmonids (brown trout, brook trout, rainbow trout, coho salmon) sustain relatively high annual mortality rates of 60% to 90% of the adult population, which is still less than the values experienced by eggs and larvae (Alexander 1979).

Cannibalism (intraspecific predation) is widespread in fishes and may play a dominant role in population regulation in some species (Dominey and Blumer 1984; Smith and Reay 1991; Elgar and Crespi 1992). Cannibalism occurs in many chondrichthyans and in at least 36 families of teleost fishes, including herbivores, scavengers, planktivores, and piscivores. Cannibalism can have a significant impact on population dynamics. Between 30% and 70% of egg consumption is caused by conspecifics among anchovies (Engraulidae) and whitefishes (Salmonidae). In addition, adults may eat larvae and juveniles, including their own offspring, and young fish may eat siblings as well as unrelated individuals. Sixty percent of annual mortality in walleye pollock, *Theragra chalcogramma* (Gadidae) and 25% of mortality in yellow perch, *Perca flavescens*, have been attributed to cannibalism of juveniles. Year class strength is thought to be strongly dependent on cannibalism rates in pike (Esocidae); cod, haddock, and whiting (Gadidae); walleye and perch (Percidae); and Nile perch (Centropomidae). In lakes where one or only a few species occur, cannibalism may be *the* major population regulatory mechanism. In such situations, giant cannibal morphs that are specialized to feed on conspecifics may develop (e.g., landlocked arctic char,

Salvelinus alpinus; Sparholt 1985; Riget et al. 1986); such cannibalistic **polyphenism** is also known among larval salamanders and frogs.

At first glance, cannibalism might be considered counterproductive evolutionarily. However, as long as the cannibal is not eating close relatives, no fitness costs are incurred, aside from possible transmission of host-specific pathogens and parasites, whereas potential competitors are eliminated. More importantly, conspecifics represent a highly nutritious protein meal made up of optimum proportions of vitamins, minerals, and amino acids for the species in question, producing high growth rates (e.g., walleye, walleye pollock) and enhanced reproductive output (e.g., mosquitofish, Poeciliidae). Even when kin are consumed, the benefits to the cannibal of reduced competition, increased growth, and enhanced reproduction could outweigh the current costs of losing a few relatives (Dominey and Blumer 1984; Smith and Reay 1991; Sogard and Olla 1994).

Production

A topic of general interest to population ecologists and of particular interest to fisheries managers is the concept of production. How much biomass (or fish flesh) is a population producing, and how much of this is available to predators, including humans, without causing the population to crash? Can production be predicted from such measurable population traits as the birth and death schedules of different age classes (i.e., from calculations using the life table characteristics discussed above)?

Production is calculated as the growth rate of individuals over a time period multiplied by the biomass of the age class, corrected for mortality occurring during the time period (Ricker 1975; Gulland 1983; Wootton 1990). Natural production values for different populations of temperate freshwater fishes vary widely, from less than 0.1 g/m²/yr for sockeye salmon in an Oregon lake to 155 g/m²/yr for desert pupfish (*Cyprinodon nevadensis*, Cyprinodontidae) in a desert stream in California. Most populations fall near the lower end of these values, in the 1 to 10 g/m²/yr range. Tropical and fertilized ponds often show higher values (Chapman 1978). Knowing production also allows one to calculate **annual turnover**, which is the ratio of production to biomass (P/B). Turnover is an index of how productive populations and subpopulations are and can be quite useful in understanding ecosystem processes. Among different age classes, very young fishes, although constituting relatively little of the biomass of the overall population, contribute 60% to 80% of population production because of their high P/B ratios. Young fish have very high growth rates relative to their sizes and hence have high turnover rates. Yields to predators are therefore relatively higher when predators feed on young fish rather than eating older, slower-growing fish. Whether high rates of exploitation of young age classes by predators reflect some form of optimal exploitation due to relative P/B ratios, or whether they just reflect ease of capture, would make an interesting study.

Genetic Structure of Populations

Determining the physical boundaries of a population can be easy (as in isolated ponds and lakes) or difficult (large rivers and lakes, oceanic regions). **Gene flow**, the exchange of genes across boundaries and between populations, can blur the distinction between populations. Coregonine whitefishes in many lakes in Canada and sticklebacks in lakes, ponds, and rivers on the Pacific coast of North America are reproductively isolated from each other and consequently form distinct genetic groupings (**demes**) (Hagen and McPhail 1970; Bell and Richkind 1981; Smith 1981). In contrast, American eels, although distributed in ponds, lakes, streams, and rivers from Iceland to Venezuela, all return to a single spawning locale in the Sargasso Sea, remixing their genes at each reproductive episode. Such **panmictic spawning** means that the entire species consists of only one population (Avise et al. 1986). Nearshore marine species generally show levels of gene flow that are strongly related to dispersal capability and distance between populations; good dispersers that are close together are very similar genetically (Waples 1987).

Genetic analysis of high-seas species often indicates a lack of differentiation among widely spread populations that were once thought of as several species. Albacore tuna (*Thunnus alalunga*) in the North Pacific and South Atlantic, and skipjack tuna (*Katsuwonus pelamis*) and yellowfin tuna in the Atlantic and Pacific are two examples (Graves and Dizon 1988; Scoles and Graves 1993). Migration over long distances or close proximity in a nearshore marine area do not, however, guarantee gene flow. Salmon of Pacific northwestern rivers (*Oncorhynchus* spp.) migrate and intermix in the ocean during much of their lives, but genetically discrete **stocks** separate and return to their natal streams to reproduce, conserving the genetic identity of the more than 200 stocks (Nehlsen et al. 1991). Populations of live-bearing black surfperch, *Embiotoca jacksoni*, separated by only 40 to 80 km in California and Mexico, show genetic differences of a magnitude that is normally found between different species within a genus (Waples 1987; Utter and Ryman 1993).

Even an isolated lake may have more than one genetically distinct population of a species. Arctic char differentiate into different forms in a number of Scandinavian lakes. In Thingvallavatn, Iceland (the suffix *vatn* means "lake" in Icelandic), char far outnumber the other two species present, a stickleback and a trout. The char exist as four distinct forms that occupy different habitats and feed on different food types (Fig. 23.3). A large and a small morph remain near the bottom and feed on benthic invertebrates in the shallow littoral zone, whereas two other morphs frequent the water column, where one feeds on zooplankton in the limnetic (pelagic) region and the other feeds on fishes both inshore and offshore. Morphologically (and appropriately), the benthic-feeding morphs have subterminal mouths and relatively dark coloration, whereas the more pelagic forms have terminal mouths and silvery, countershaded coloration. Spawning times vary among the morphs, and the morphologic differences show up shortly after hatching and hence are

FIGURE 23.3.   *Four morphs of arctic char that differ anatomically, behaviorally, and ecologically occur in some Scandinavian lakes. Shown here are adults of the four morphs from an Icelandic lake. They are, from top to bottom, the large, benthic feeding morph (33 cm long), the small benthic feeding morph (8 cm), the piscivorous morph (35 cm), and the planktivorous morph (19 cm).*

Photo by S. Skulason from Skulason and Smith (1995); used with permission from the Food and Agriculture Organization of the United Nations.

not **ecophenotypic**, that is, they do not result from environmental influences experienced by different individuals.

The morphologic and behavioral differences among morphs have a strong genetic basis, as offspring of the different morphs retain their trophic specializations even when raised in a common laboratory environment (Skulason et al. 1993). Genetic differences can also be demonstrated biochemically. The small benthivorous morph differs from the other three morphs at one esterase enzyme locus out of 36 loci tested. The morphs may have evolved because of the availability of habitats (**adaptive zones** or open niches) brought on by the absence of other species in the lake (Magnusson and Ferguson 1987; Sandlund et al. 1988).

Populations have genetic structure. The distribution of genotypes can be characterized, either by assaying directly for relative frequencies of alleles of genes or by studying the distribution of phenotypes. Genetic analyses of populations can be a powerful technique for solving systematic questions. For example, five species and eight subspecies of kelpfishes (*Gibbonsia*, Clinidae) were recognized from western North America, based on traditional meristic and morphometric analyses. Reanalysis using allozyme data obtained by electrophoresis of enzymes extracted from various tissues indicated very little genetic differentiation among subspecies and even between some species, suggesting a high degree of gene flow among populations (Stepien and Rosenblatt 1991).

Three nominal species—the scarlet kelpfish (*G. erythra*), the crevice kelpfish (*G. montereyensis*), and an offshore Mexican endemic, *G. norae*—showed no significant differences in the frequencies of different alleles at 40 gene loci (the term **nominal** refers to a population that has been described as a separate species). Reanalysis of morphometric data showed that many anatomic differences were instead sexual dimorphisms or anatomic trends that changed continually as a function of water temperature or with depth (the latter are ecophenotypic differences, because they represent phenotypes that differ consistently and result from depth of occurrence). All scarlet kelpfish, distinguished by more caudal peduncle scales and a higher dorsal spine, were in fact males. All nominally crevice kelpfish were females. *G. norae* differed only in lower counts of scale rows and fin rays, but it is generally observed that fish that develop in warmer water lay down fewer meristic elements (Barlow 1961; see Chap. 9, Meristic Variation). Hence *G. erythra*, *G. norae*, and *G. montereyensis* represent different populations of a single species, *G. montereyensis*. Subspecies of kelpfishes also showed little genetic differentiation. Anatomic differences occurred in populations that occupied different depth zones, but characters such as dorsal spine height generally increased with depth of occurrence and were probably ecophenotypic, not genetic, traits. The combined analysis—genetic, morphologic, distributional, and ecological—indicated that three rather than five species and no subspecies of kelpfish existed. Hence genetic analysis showed the true taxonomic relationships among the different species.

Understanding the genetic makeup of a population has become increasingly important as environmental degradation and overexploitation place many populations and species at risk. A key character is the degree of genetic variation in a population. Genetic variation results from selection, mutation, dispersal (emigration out of and gene flow into populations), nonrandom mating, and genetic drift (random changes in gene frequencies, particularly in small, isolated populations). Genetic variation is a chief driving force of evolution; natural selection acts on such variation, favoring genotypes that are adapted to current conditions. Measures of genotypic variation and frequencies can tell us whether a population has become dangerously inbred and lacks the genetic diversity necessary to allow for adaptation to changing environmental conditions, whether gene flow is occurring between populations, and whether hybridization with introduced species is occurring (and hence if the genetic identity of a species is threatened) (Schonewald-Cox et al. 1983; Meffe 1986).

Hybridization

An individual contains combinations of genes that have evolved together over millions of years. Different species contain different gene combinations, which means that a hybrid individual brings together genes that have not undergone such fine-tuned coevolution. The result is that hybrid individuals are typically aberrant in some aspect of their biology, which can be expressed as faster growth or more vigorous mating behavior, a phenomenon referred to as **hybrid vigor** or **heterosis**. However, most natural hybrids are inefficient reproductively, ecologically, biochemically, physiologically, or behaviorally. They are therefore likely to be reproductive failures because of sterility, relatively infertility, or an inability to find mates, or will be ecological failures because they will be outcompeted for resources or be more prone to capture by predators than are individuals from single-species matings. Natural selection will obviously favor spawning individuals that avoid mating with members of other species. This separation of species during mating is accomplished via **species-isolating mechanisms**, which are usually anatomic or behavioral traits that keep individuals of different species from breeding with one another. Species-isolating mechanisms include genitalia that do not match up correctly, as in internally fertilized livebearers or elasmobranchs, or they may result from incompatibility between sperm and eggs, or differences in courtship patterns, timing, or location of spawning. Any inappropriate cues given by one or the other member of a spawning pair can lead to termination of the spawning act.

Hybridization in fishes often occurs when one species experiences a substantial reduction in abundance, so that rare fish tend to mate with closely related but much more abundant members of another species (Hubbs 1955). Hybridization is also common in disturbed habitats where the preferred spawning sites of one species are lacking, forcing them to spawn in another habitat and hence with another species. When exotic but related species are introduced into a region, isolating mechanisms between the introduced and the established species may not have evolved, and hybridization occurs.

Among fishes, instances of hybridization are most common in freshwater fishes and less well documented among marine species (e.g., Schwartz 1981). Freshwater families among which natural hybrids are often found include the minnows (Cyprinidae), sunfishes (Centrarchidae), and darters (Percidae). Artificial hybrids produced in aquaculture are also common, as in the sunshine bass (a cross between a male striped bass, *Morone saxatilis*, and a female white bass, *M. chrysops*), the splake (a cross between a lake trout, *Salvelinus namaycush*, and a brook trout, *S. fontinalis*), and the tiger muskellunge (a cross between a northern pike, *Esox lucius*, and a muskellunge, *E. masquinongy*). Hybrid marine fishes are much less common. The most common marine examples occur in the reef fish families of butterflyfishes (Chaetodontidae) and angelfishes (Pomacanthidae), although distinctive, complex color patterns and popularity among aquarium keepers make hybrids in these families more likely to be detected (Pyle and Randall 1994). The disproportionate numbers of hybrids among freshwater species could reflect the greater degree of physical disturbance and of species introductions in freshwater habitats (see Chap. 25).

Hybridization is usually a dead-end for the individuals produced by the hybrid cross, although successful new species may be produced via hybridization, as in the case of the parthenogenetic, unisexual livebearers of the genus *Poecilia* in Mexico (see Chap. 20, Gender Roles in Fishes).

ASSEMBLAGES

An **assemblage** consists of the various species populations of a larger taxon in a defined area. Ecological interactions occur within a fish assemblage, within a spider assemblage, and so on. Focusing on assemblages is admittedly myopic, since fishes interact with invertebrate prey and parasites, with plants as food and shelter, with reptiles, birds, and mammals as predators, and so forth. However, fish-fish interactions are particularly obvious; it is often logistically difficult to deal with all components of an ecosystem; and researchers tend to specialize and develop expertise in certain taxonomic groups (hence the rationale for producing an ichthyology or any other taxon-oriented textbook). For the purposes of the present discussion, we will look at competitive and predator-prey interactions that tend to involve fishes of different species, and will discuss prevalent ideas on how assemblages are structured and ordered, that is, how interactions between fishes affect species composition and maintenance of assemblages.

Niches and Guilds: The Ecological Role of a Species

Most discussions of the composition of natural assemblages focus on the functional roles of the different species. The ecological function of a species is synonymous with its **niche**, a broadly defined term that essentially describes what an animal eats and what eats it, its environmental and microhabitat requirements (temperature, oxygen concentration, pH, salinity, substrate type), and symbiotic associations in which it participates. Characterizing and measuring niche components and dimensions allow us to compare niches among species and measure changes in niche use that occur when species are added to or subtracted from an assemblage.

An additional, useful concept for understanding the ecological roles of different species within an assemblage is the **guild**. The guild concept emphasizes ecological rather than taxonomic similarities (Gerking 1994). A guild consists of the different species in an area that exploit similar resources in a similar way. Hence the many fishes that hover in the water column along the face of a coral reef and feed on zooplankton make up the zooplanktivore guild, as do the large predatory invertebrates feeding in the same area on the same resource. The diurnal zooplanktivore guild on a reef face includes anthine sea basses, snappers, fusiliers, butterflyfishes, damselfishes, wrasses, surgeonfishes, and triggerfishes, among others. At night, a different set of species exploits the larger (and hence different) zooplankton resource, constituting the nocturnal zooplanktivore guild: anchovies, herrings, silversides, squirrelfishes, copper sweepers, cardinalfishes, grunts, and glasseye snappers (Hobson 1975). A species may be a member of different guilds at different times in its life. Angelfishes, wrasses, and leatherjackets (*Scomberoides*, Carangidae) may belong to the cleaner fish guild as juveniles, but as adults they change to feeding on sessile invertebrates, mobile invertebrates, or small fishes respectively.

In streams, fishes can be classified by their habitat preferences as members of a benthic guild (e.g., some minnows, suckers, sculpins, darters) that feed largely on benthic invertebrates living among rocks or buried in sediments, or as members of a water-column guild feeding largely on drifting insects or on insects that fall onto the water's surface (trouts, several minnows) (Grossman and Freeman 1987). In tropical streams, guilds can include algivores, aquatic and general insectivores, piscivores, scale and fin eaters, terrestrial herbivores, and omnivores (Angermeier and Karr 1983). We can also recognize tide pool guilds, kelp bed water-column guilds, pelagic predator guilds, wavezone sand-dwelling guilds, rock-dwelling lake guilds, buried benthic predatory guilds, and so on, depending largely on habitat and foraging activities.

Habitat Use and Choice

An important component of a species' niche, and one that can easily differ among species, is its habitat. Habitat can often be described and quantified in detail. Among stream fishes, species often differ in height above bottom, preferred current strength, bottom type (particle size and type), structure, distance from shore, amount of vegetation, and type and amount of food resources (Fig. 23.4). A survey of major habitats in an eastern North American stream (rapids, riffles, runs, pools, overhangs) shows that species often segregate along vertical dimensions, with certain species typically found in contact with the bottom (catfishes, darters, sculpins, eels, grazing minnows), some just above the bottom (suckers), some species low and others higher in the water column (planktivorous minnows, trout), some close to the surface (silversides, livebearers, topminnows), some in swift water (darters, trout), and others in moderate flow or in slow flow near more rapid-flowing water (catfishes, minnows, pickerel, sunfishes). The existence of particular species in particular habitats implies an active choice by individuals (e.g., Gorman 1988). Experimental studies, usually of juveniles, generally show that individuals actively choose habitats, that the ones they prefer are the ones in which the species is most often found, and that preferred habitats are ones in which a species can successfully feed, avoid predators, and reproduce; in other words, habitat choice is an evolved aspect of a species' niche (e.g., Sale 1969). Habitat choice is, however, dynamic within a species, varying on the basis of age, size, sex, reproductive condition, geographic area, and environmental conditions (e.g., Karr et al. 1982).

Similar descriptions, based on habitat characteristics, apply to assemblages in most major habitat types. Different faunas in different geographic locales often occupy similar habitats and, in essence, converge on many niche characteristics. The Chinese algae eater, *Gyrinocheilus* (Gyrinocheilidae), belongs to a Southeast Asian family related to loaches. It lives in rivers, where it feeds on algae attached to rocks. It has a sucking mouth with which it clings to rocks even while feeding in high-flow situations. A South American catfish, *Otocinclus* (Loricariidae), similarly feeds on algae-covered rocks in flowing water. It is convergent with the algae eater in body form and size, suctorial mouth, and even coloration. Stream-dwelling galaxiids of the Southern Hemisphere, especially in New Zealand and Australia, occupy niches very similar to those of salmonid trouts in the Northern Hemisphere. They are convergent in body morphology, habitat, and foraging habits, and have been rapidly exterminated by the introduction of trout in many locales due to competition and predation.

Approximately 60 families of fishes include species that fill the eel niche in their assemblage. Only 22 of these families belong to the order Anguilliformes and are "true"

FIGURE 23.4.   *Habitat choice in stream fishes, as demonstrated by vertical segregation among cypriniform fishes in a Borneo stream. Habitat choice is one aspect of the niche of a species, species choose habitats according to specific characteristics, and species often differ in one or more quantifiable characteristics of habitat.*

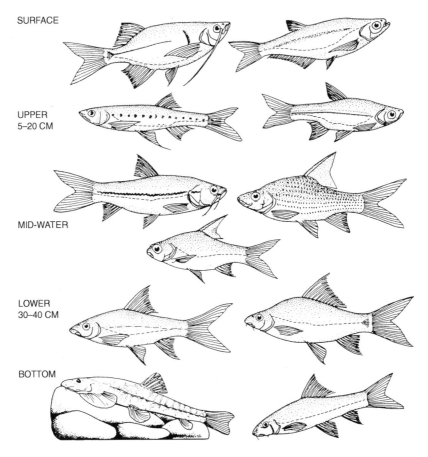

SURFACE

UPPER
5–20 CM

MID-WATER

LOWER
30–40 CM

BOTTOM

From Welcomme 1985; used with permission.

eels. Many of these fishes are convergent in habitat choice, occurring on and often in the bottom in soft sediments or among the interstices of rocks and other structure. They share other characteristics: elongate dorsal and anal fins that often lack hard spines, increased vertebral counts, reduced opercula, missing pelvic or pectoral fins, missing or embedded scales, and a pointed tail. Behaviorally they are carnivores and scavengers (some are parasitic), can swim backward and forward with equal facility (**palindromic locomotion**), and tear pieces off their prey by holding on and rotating rapidly along their long body axis (Helfman 1995; see Box 18.2).

Habitat choice changes with the seasons (see Chap. 22), with the size and age of fish (see Chaps. 9, 10), and also with the presence of other species, particularly predators and competitors. Many stream species, such as darters and benthic minnows and suckers, occupy increasingly swift water as they grow older. Often the distribution results from spawning habits: Adults migrate upstream to spawn in headwaters, and young move progressively downstream with growth (e.g., Hall 1972).

Such an ontogenetic habitat shift could reflect a body-size constraint related to the current speed at which an individual can hold position without expending excess energy (see below and Fig. 5.8). A general and somewhat unexplained pattern of increasing depth with age occurs in a number of freshwater and marine species. In many species, this trend reflects the use of inshore, shallow, productive nursery areas that are reputed to be relatively predator-free (e.g., salt marshes, mangroves); these fishes move offshore as they grow. In other species (minnows, copper sweepers, damselfishes, surgeonfishes, sunfishes, croakers, percids, clinids, wrasses, great barracuda), an actual size-depth correlation has been found (Helfman 1978; Power 1987). That the relationship exists in a diversity of habitats and involves a number of unrelated species (e.g., Polloni et al. 1979) implies a convergent, adaptive trait, but the details of this adaptation remain to be worked out in most cases.

"Bigger-deeper" distributions within species and other types of habitat shift may result from avoidance of predators, influenced by differences in vulnerability between different size classes of prey. When predatory sunfishes (*Lepomis* spp.) are present and largemouth bass are absent, young-of-the-year stoneroller minnows (*Campostoma anomalum*) occupy shallows of pools, whereas larger stonerollers prefer deeper portions. Sunfishes occupy deeper portions of pools and can eat young minnows but are too small to eat larger minnows. When bass, which can eat all minnows and also prefer deep sections, are added, all size classes of stoneroller are confined to shallow margins or emigrate from pools.

A combination of fish, avian, and mammalian predators affects the depth distributions of loricariid catfishes (including the *Plecostomus* of the aquarium trade) in Panamanian streams. These herbivores feed on algae that grows most abundantly in the shallows of pools. Small catfishes live in the more productive shallows, whereas large catfishes occur in deeper water where algal production is minimal and where the fish lose fat reserves and cease growing. The habitat differences of the two size groups are enforced by the balanced impact of terrestrial predators and piscivorous fishes. Birds and mammals can capture any size prey, whereas piscivorous characins are gape limited and cannot swallow the large catfishes. Large catfishes are therefore forced into deeper, less-productive regions by terrestrial predation. The small catfishes avoid predatory fishes that live in deep water but can hide from the birds and mammals in shallow water by seeking refuge among rocks; however, these refuges are too small for larger catfishes (Power 1987; Power et al. 1989; see also Chap. 24, Predation).

Cannibalism can also influence habitat shifts, forcing smaller fish into suboptimal habitats where their growth rates suffer. Smaller sculpins (*Cottus*) use a diversity of habitats in some streams and prefer deeper water in others when large conspecifics are missing. If larger fish are present or are added experimentally, the smaller fish shift to shallower habitats. Similar habitat shifts, resulting in occupation of suboptimal feeding habitats by small prey, have been demonstrated in other minnow species and bluegill sunfish (Werner et al. 1983; Gilliam and Fraser 1987; Schlosser 1987; Freeman and Stouder 1989).

Habitat Choice and Spatial Structure in Fish Assemblages: Zonation

Researchers have compared the fish species that occur in different habitats along environmental gradients, such as from headwaters to mouths of streams and rivers, or from shore to reef face, continental shelf, or sublittoral regions of oceans and lakes. These investigations have led to generalizations about zonation in various habitats.

Almost every fish habitat that has been studied can be divided into more-or-less distinctive zones. Early work of this type focused on stream and river (riverine, lotic, or fluviatile) fishes in the British Isles and Europe, where analogous **longitudinal zones** (habitat types along the length of a river) were identified in different systems. The zones were often named based on the common fish species present, which in turn reflected species' preferences with respect to gradient (slope), water velocity,

stream width, stream depth, temperature variation, oxygenation, and sediment type. These habitat characteristics also influenced the presence and type of vegetation both in and along a river, bottom characteristics, and invertebrate fauna. A popular classification recognized four basic zones, beginning with the headwaters and moving to the lowlands (Huet 1959; Hawkes 1975).

1. The trout zone: the narrow, shallow, cold, steep (often torrential), highly oxygenated headwater region with large rocks or gravel; common fish species show morphologic or behavioral adaptions to high flow and include brown trout, Atlantic salmon fry, bullhead sculpin (*Cottus gobio*), and minnow (*Phoxinus phoxinus*).
2. The grayling zone: deeper, less steep, with alternating riffles and pools, relatively strong currents, less rocky, more gravelly bottom, cool, slightly less oxygenated, with salmonids in the rapids and rheophilic (current-loving) minnows in the pools; common fishes are grayling (*Thymallus thymallus*), species of the trout zone, and rheophilic minnows (barbel, chub, hotu, gudgeon).
3. The barbel zone: riverine conditions of moderate gradient and current, greater depth, alternating rapids and runs (quieter flowing water), fluctuating temperatures; common fishes are rheophilic cyprinids, other cyprinids (roach, rudd, dace), and predators (pike, perch, European eel).
4. The bream zone: lowland reaches that include rivers, canals, and ditches, little current, high summer temperatures with oxygen depletion, and turbid water; common fishes are roach, rudd, dace, predators of the barbel zone, and slow-water cyprinids (carp, tench, bream).

Alternative classifications emphasize different species in different areas, but the recognition of longitudinal succession in habitats and species, often involving three to eight basic zones that differ in a few essential physical features, is common to most classificatory schemes (e.g., head stream, trout beck, minnow reach, lowland course; spring zone, upper, middle, and lower salmonid region, barbel region). Underlying causes of zonation have often been explained in terms of the physical factors listed above, which often correlate with **stream order**. Order is determined by tributary number: Small, headwater streams are first order, two first-order streams join to form a second-order stream, two second-order streams join to form a third-order stream, and so on (Kuehne 1962). The world's largest rivers are seldom much more than tenth order. For example, the Mississippi from its confluence with the Ohio to its mouth, a distance of almost 1000 km, is about an eleventh-order river. To be larger, another eleventh-order river would have to flow into it (Fremling et al. 1989). Faunal breaks, where species type and number change in direct correspondence with stream order, have been identified in several systems (Lotrich 1973; Horwitz 1978; Evans and Noble 1979), but other studies have shown that elevation, gradient, drainage area, and historical factors are also likely correlates of species richness and faunal change (Hughes and Omernik 1981; Matthews 1986; Beecher et al. 1988).

Longitudinal zonation among fishes has been described in numerous tropical and temperate riverine systems (Moyle and Nichols 1974; Horwitz 1978; Balon and Stewart 1983; Welcomme 1985). In eastern North America, headwater streams may contain on¹y a single species, such as brook trout, sculpin, or creek chub. Downstream, where streams increase in both size and velocity, habitat diversity increases, and species adapted to high-flow conditions (e.g., darters, hogsuckers, longnose dace, rheophilic minnows, sculpin, smallmouth bass) are added. Farther downstream, as the system widens and deepens, sediments are deposited, and primary and secondary production increases. Swift-water species characteristic of upstream areas may drop out, but many more species are added. Food webs become less dependent on inputs of food from the surrounding watershed (**allochthonous inputs** such as falling leaves and falling insects) and more dependent on production from the stream itself (**autochthonous inputs** from periphyton, macrophytes, and associated animals). Deep pools and slow-flowing conditions alternate with shoals and rapids. Fish diversity increases to include suckers, herbivorous minnows, catfishes, and largemouth bass. In large, slow-flowing rivers with little gradient, planktonic organisms and planktivores are added (shad, herring, silversides, paddlefish, and predators such as striped bass and pickerel, all species that also inhabit lakes) as are larger carnivores such as sturgeon. Production of food is largely from the river itself, much of it in the form of detritus-eating insects that live on snags formed by fallen trees from the gallery forest (Benke et al. 1985). The floodplain and its forest contribute substantially, both as sources of nutrients and as nursery areas during seasonal flooding. A fish fauna of backwaters, oxbows, and sloughs is added, including some that are adapted to swampy, periodically deoxygenated conditions (many sunfishes, small pickerels, livebearers, killifishes, swampfish, pirate perch, mudminnows, gars, bowfin). Finally, as the river enters the coastal zone and is subject to both the tidal and salinity influence of the ocean, a very diverse assemblage that includes freshwater species from the lower river reaches and marine species from nearshore zones inhabit the salt marsh or other estuarine regions. In all zones, fairly characteristic species are associated with relatively definable habitats.

Since the original formulations of the idea of longitudinal zonation, many corrections, modifications, and exceptions have been introduced. Zones were originally thought to be distinct; border lines were even drawn between them on maps. In reality, zones usually grade into one another, with **border zones** that are intermediate in physical nature and species composition connecting the zones. Sometimes these border zones may be longer than the fish zones they presumably separate. Also, the faunas of different zones are not necessarily distinct from one another. In fact, a commonality in flowing water is that species diversity increases as one goes downstream, and this increase is generally due to additions, not replacements, of species (see the listing

above of European river zones for an example). Rather than species dropping out from faunal lists as one progresses downstream, additional species are found: Headwater fishes are often found in downstream habitat patches (e.g., a swift shoal), although downstream species seldom occur in headwaters. Other complications include missing zones, as where rivers appear suddenly due to emergence of major springs (e.g., northern Florida) or disappear suddenly (western deserts). Reversals in zones may occur in rivers that have a stair-step course, with repeated slow-flowing flat sections that become steeper, or in rivers that flow through lakes that are cooler than the incoming river water. Seasonal migrations among zones and out of rivers and into lakes and the ocean change assemblage composition dramatically. Benthic invertebrates also occur in zones, but these zones may or may not correspond with fish zones.

Most of these exceptions do not contradict the idea of zonation, because they can be anticipated from the specific environments involved. The basic idea of a relatively few, generalizable zones, with characteristic fish species living under characteristic flow and temperature regimes, has remained a viable descriptor and predictor of fish assemblages in a surprisingly large number of lotic situations (Hawkes 1975) and can serve to indicate disruptions due to human causes (e.g., Fausch et al. 1984). The major shortcoming of viewing a river, or any aquatic habitat, as isolated sections or zones containing relatively independent and distinct assemblages is that it ignores the critical, functional linkages among sections and subsets of the biota. In this regard, the **river continuum concept** (RCC) (Vannote et al. 1980; Minshall et al. 1985) can be usefully applied to fish assemblages in rivers. The RCC views a river as an orderly progression of predictably intergrading, dependent regions containing organisms whose ecological roles reflect changes in river basin geomorphology, current speed, gradient, sediment and organic matter composition, and allochthonous versus autochthonous production (among aquatic insects, shredders and gathering collectors predominate in headwaters, shredders are replaced by scrapers in middle reaches, and filter feeders predominate in higher-order sections). Integration of fishes into the RCC remains an ongoing challenge in fish (and riverine) ecology.

Competition

Competition occurs when two consumers require a resource that is in insufficient abundance to meet the needs of both. Members of the same species can compete (**intraspecific competition**) as can members of different species (**interspecific competition**). Although intraspecific competition can affect an individual's ability to acquire resources, interspecific competition has received more attention because of the insight it provides into the coexistence of different species in an assemblage, which addresses the more general question of how biodiversity is created and maintained. In general, individuals can compete for food, feeding and resting sites, and refuges from predators and the elements (competition for mates and breeding sites is viewed as part of the reproductive biology of a species rather than as traditional competition).

To avoid or reduce competition, organisms may change the way they exploit a resource. Competition may lead to differences in resource use (**resource partitioning**), such as when two **sympatric** (living together) species feed on different sizes of a prey type or eat similar prey but in different microhabitats. Competition is more strongly implicated if these same predators feed on identical prey when they are **allopatric** (living separately). Competitive interactions can also be suspected if potential competitors shift their resource use when resources become seasonally limiting, or if population reductions of one species occur when a likely competitor is introduced into an area. However, ecological differences among species can also be caused by differences in nutritional requirements, foraging or locomotory capabilities, predator vulnerability, and phylogeny. In addition, introduced species can alter predator-prey relationships or serve as vectors for parasites and diseases, which would affect population densities of previous residents. Consequently, in order to prove that competition is in fact the cause of the dissimilarities, it is generally necessary to perform experimental manipulations of resource abundance, resource distribution, or population densities of suspected competitors. In such experiments, competition can be invoked if an inferior competitor or less-aggressive species that occupies suboptimal regions in sympatry expands its habitat or feeding habits when the superior competitor is eliminated. Reciprocal removal of the inferior competitor should have little effect on the habits of the superior species. Such experiments may be conducted fortuitously or deliberately.

A fortuitous manipulation of resource partitioning, mediated by both competition and predation, is conducted annually in Lake Tjeukemeer, The Netherlands (Lammens et al. 1985). Bream (*Abramis brama*, Cyprinidae) and European eels (*Anguilla anguilla*, Anguillidae) occur year-round in fairly stable numbers in the lake, where their chief foods are water fleas and juvenile midges respectively. Smelt (*Osmerus eperlanus*, Osmeridae), a zooplanktivore, enter the lake each spring as juveniles when water is pumped from a nearby lake as part of a water stabilization program. Smelt do not persist in the lake, because the adults are almost all consumed by predatory pikeperch (*Stizostedion lucioperca*, Percidae). When large numbers of juvenile smelt are recruited into the lake, they depress zooplankton populations, having their strongest effects on the size classes of zooplankton most used by bream. Bream respond to reductions in zooplankton resources by switching to benthic invertebrates such as midge larvae, thereby depressing that resource. Eels then respond to depletion of their primary food by switching to piscivory. When smelt are abundant, both bream and eels suffer reductions in condition (the ratio of weight to length), and bream show poor gonad development. In years when smelt recruitment is low, bream and eels switch back to their water flea and midge diets, and their growth and reproductive development improve.

A well-studied example that includes the experiments necessary to establish the causes of shifts in resource use involves sunfishes in North America. As many as eight species of centrarchid sunfishes and basses may co-occur in a single lake. Many of these species are very similar morphologically. How do they coexist without competing? When stocked separately in ponds as year-old fish, three species—bluegill sunfish (*Lepomis macrochirus*), green sunfish (*L. cyanellus*), and pumpkinseed sunfish (*L. gibbosus*)—use similar habitats and feed on similar food types. All three concentrate their time and effort on vegetation-associated invertebrates. When stocked together, bluegill and pumpkinseed shift their habitats and diet in apparent avoidance of the competitively superior green sunfish. Bluegill shift to feeding on zooplankton in open water, and pumpkinseed include more benthic prey in their diet. Green sunfish maintain a diet of vegetation-associated insects. All three species show reduced growth rates, indicating competitive reduction of resources for each species, but bluegill show the greatest declines. When bluegill and green sunfish are stocked together in ponds with little open-water habitat (i.e., no alternative habitat for bluegill), green sunfish show better growth, fuller stomachs, and a higher survival rate than bluegill.

Competition among the three species has an ontogenetic component that is felt most strongly by young fish. As the fish grow older, they begin to specialize more on different habitats. Bluegill become more adept at maneuvering in open water and suction feeding on zooplankton, and pumpkinseed develop pharyngeal dentition with which they can crush mollusks that live in sediments. Hence the potential for competition is reduced in older fish under natural conditions (Werner and Hall 1979; Werner 1984; Mittelbach 1988; Osenberg et al. 1988; Wootton 1990).

Many other investigations have demonstrated strong competitive interactions among fishes in various habitat types (e.g., tropical streams, Zaret and Rand 1971; temperate streams, Schlosser 1982; temperate marine nearshore, Hixon 1980; Holbrook and Schmitt 1989; coral reefs, Hixon and Beets 1989; see reviews in Ross 1986; Ebeling and Hixon 1991). In general, of the kinds of resources for which fishes can compete, competition for food resources (or at least differences in trophic resource use) is more common among fishes than are interspecific differences in habitat use; the reverse is true in terrestrial communities (Ross 1986). Some traits that reflect apparent adjustments to present-day competition may result from historical interactions between species, the so-called "ghosts of competition past." The influence of historical competition is frustratingly difficult to determine: Are two species different today because of their current impacts on one another or because of past interactions? Experimental manipulations of the resource in question are almost always needed to prove competition, but obviously one cannot manipulate the history of two species, and hence we can only speculate on, but not demonstrate, historical competition (Connell 1980).

Historical factors must also be considered when comparing ecological characteristics of species from unrelated taxonomic groups. The more distantly related two fish

---

| BOX 23.1 | *Habitat Choice and Environmental Physiology: A Cautionary Note* |

Species differences in habitat choice can be attributed to interactions among competitors or between predators and prey. However, one should not immediately assume that such biotic interactions explain habitat choice in general. An alternative, and often simpler, explanation is that a species occurs where it does because the fish functions best there, which is to say that choice of habitat reflects use of physiologically optimal environments rather than being the result of interactions with other species over limiting resources such as food or shelter.

Headwater streams often contain small assemblages of ecologically distinct species. In the southern Appalachian mountains of the United States, the assemblage typically consists of fewer than 10 species that are often found at different heights above the bottom and at different water velocities. Four common species are rosyside dace (*Clinostomus funduloides*) and introduced rainbow trout (*Oncorhynchus mykiss*) in the water column, and longnose dace (*Rhinichthys cataractae*) and mottled sculpin (*Cottus bairdi*) on the bottom. Facey and Grossman (1990, 1992) tested whether distribution differences among these four fishes could be explained not by **interactive segregation** (competition) but instead by energetic efficiencies, that is, whether fishes were using physiologically optimal habitats. They found that longnose dace showed no real preference but were distributed in proportion to the available velocities in the stream (interactions with other species are unlikely to produce such a statistically random distribution). Rain-

bow trout, rosyside dace, and mottled sculpin chose velocities that were lower than would be expected if they were distributed at random. Tests of energy use in flowing-water respirometers indicated that trout and dace used more oxygen as water velocity increased, that is, they occupied low-velocity regions because it was energetically expensive to live in high-flow areas. Also, trout and dace preferred velocities in which they were most efficient at capturing drifting invertebrate prey (Hill and Grossman 1993). Hence physiological costs associated with holding position at high velocities and optimal velocities for capturing prey, not interactions with other species, are the most likely determinants of where in a stream these two species are found.

Sculpin, however, occupied low-flow regions even though the respirometry data indicated they incurred little added cost at higher velocities. Their microhabitat preferences might therefore be explained by food availability, predator avoidance, or competitive interactions. Experimental manipulations involving sculpin and its most likely competitor, longnose dace, indicated minimal effects on sculpin (Barrett 1989; Stouder 1990). In headwater streams, a highly variable environment characterized by large fluctuations in water level and velocity (droughts and floods), combined with physiological constraints, prey-capture abilities, and intraspecific competition (Freeman and Stouder 1989), appears to have a greater influence on occurrence, distribution, abundance, foraging, and habitat choice than do interspecific interactions.

---

species are, the less similar they tend to be ecologically (Ross 1986). For example, generalist predators on coral reefs tend to be active at twilight or at night, have large mouths, and feed on fishes, whereas specialists are diurnal, have small mouths, and feed on sessile or small invertebrates (Hobson 1974, 1975). Resource partitioning along both trophic and temporal resource dimensions could be invoked here. However, generalist reef species tend to belong to more primitive acanthopterygian (spiny-rayed) groups (squirrelfishes, scorpionfishes, groupers), whereas specialists belong to more advanced groups (butterflyfishes, wrasses, triggerfishes). Feeding habits, morphology, and activity times in one lineage are likely to have evolved independently of what happened in a later-evolving lineage. Differing ecologies may therefore simply reflect differing phylogenetic histories. Interpreting differences in resource use as a result of competition

may also be erroneous because of physiological differences among species (Box 23.1).

Predation

Predator-prey interactions among species in an assemblage can have direct and indirect effects on prey population size and distribution. **Direct effects** include immediate mortality or delayed mortality due to injury. **Indirect effects** involve habitat shifts caused by a predator's presence that force potential prey to use suboptimal habitats, which can affect individual growth and reproduction (see above). Population-level responses associated with predation are usually density dependent and vary with the age of the prey. **Density-dependent** changes occur when the size of the prey population determines the impact of the predator. Direct density dependence is referred to as **compensatory**, meaning that predation increases to compensate for increases in prey population size. Predation by seabirds on schooling pelagic fishes is often compensatory in both the short and long term. The feeding activities of one bird draw the attention of other birds, and the number of predators arriving at the site increases in direct relation to the size of the fish school on which they are feeding. Successful feeding by the birds in turn increases survivorship of their young, which means an increase in predators in the next generation, all dependent on the size of the fish resource. Intercohort cannibalism, in which older fish eat younger age classes, can have a strong density-dependent impact on year class strength. Consider a population with three year classes. A large cannibalistic cohort can depress the numbers of the next, younger cohort. When the younger cohort reaches a size at which it is a threat to the third, youngest age class, reduced numbers in the second age group mean it will have relatively little impact, which translates into high survivorship in the third group. In this way, population cycles can be established through the density-dependent effects of cannibalism. Just this type of scenario has been invoked to explain 2-year cycles of abundance in pink salmon, *Oncorhynchus gorbuscha*, in the Pacific Northwest (Ricker 1962).

Inverse density dependence is considered **depensatory**, because relative predation risk and impact decrease as prey numbers increase. Depensatory predation occurs when a fixed number of predators become swamped or saturated by large numbers of prey. Under such conditions, the proportion of the prey captured decreases as prey numbers increase. For example, an individual salmon smolt that is migrating to the sea reduces its risk of death if it can time its downstream migration to coincide with that of other smolts, since predators take only a small number of migrants. By extension, the proportion of the prey population killed decreases as the population increases (Wootton 1990).

Regardless of the nature of the relationship between predator and prey densities, predation can have dramatic effects on prey population size. It is generally held that most mortality in eggs and larvae of species with planktonic young is due to predation, with predators taking more than 99% of the individuals (Bailey and Houde 1989; see above). When older, fish are still subject to predation, but the threat falls off progressively with increasing age and size, forming what is described as **exponentially declining mortality** (Fig. 23.5). Refuge site availability may influence the impact that predators have on later life history stages. Observations and experimental manipulations on coral reefs indicate that prey population density is directly related to the number and availability of holes, which also implies that competition for refuge sites could interact with predation to determine population density and diversity. Several experimental studies on reef fishes have in fact shown that removing predators leads to increased density of prey. The effects of predator density on prey diversity may follow a similar pattern, but surprisingly few investigations have been conducted (Hixon 1991; Hixon and Beets 1993).

Predation can also affect gene frequencies in populations through the evolution of antipredator adaptations.

Guppies in small streams in Venezuela and on the island of Trinidad occur in pools that differ in levels of predation. Upstream areas tend to have few if any predators, often limited to a single topminnow species, *Rivulus marmoratus* (Cyprinodontidae). Further downstream, more predators occur, including a cichlid (*Crenicichla*), a characin (*Hoplias*), and freshwater prawns. In areas of low predation, males tend to have many bright, colorful spots that are attractive to females but are also conspicuous to predators. Spot number, size, and brightness are inherited; the offspring of brightly colored males are brightly colored. In a series of experiments, fish from several populations were exposed to different levels of predation over several generations or were transferred between areas with high and low predation intensity. There was a regular decline in spot number, spot size, and spot brightness in populations subjected to more predation in both the field and the lab (Endler 1980, 1983).

Synthesis: What Determines Assemblage Structure Among Coral Reef Fishes?

Coral reefs contain more species of fishes than any other habitat. Almost 700 shallow-water species occur in the Caribbean, and Indo-Pacific reefs are home to more than 3000 species (see Chap. 16, Major Marine Regions). This incredible diversity has understandably drawn the attention of fish ecologists and has led to some surprisingly emotional controversies. Probably the most divisive debate concerns the maintenance of this high diversity. How does a coral reef support so many different fishes? What determines the number of species and individuals that will occur on a reef? Can one predict what kinds of fishes (both taxonomically and ecologically) will occur on different reefs or on the same reef at different times? Do the processes that determine species abundance and diversity in one tropical ocean apply to reefs in another ocean? These questions have more than theoretical value. Understanding spatial and temporal patterns of recruitment and the factors that determine the success of recruits can influence fisheries management practices such as reserve design, seasonal closures, and artificial reef design and placement (Beets 1989; Bohnsack et al. 1991).

These and related topics have nurtured what has become known as the **stochastic-deterministic debate**. The terms refer to the two general processes affecting the maintenance of diversity. **Stochastic** processes are largely random in operation. An extreme adherent of the stochastic school would argue that chance events affecting planktonic larvae and newly recruited juveniles play too large a role for us to be able to predict species composition. The ocean is a huge place with very little shallow-water habitat. A larval fish that was not eaten or did not starve to death must also be lucky enough to encounter a reef during the brief period when it is competent to settle. The larva is also likely to get eaten by the zooplanktivorous predators that abound on reefs, and it must finally find a suitable, unoccupied site in which to settle. These chance events, which are further influenced by unpredictable storms (see Chap. 24, Extreme Weather), reduce the accuracy with which we can predict the actual species and abundances of

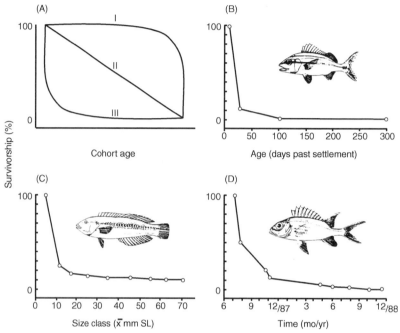

FIGURE 23.5. *Survivorship curves in theory and practice. (A) Three general forms of survivorship (percent surviving at the end of each year or within each age class) are found in natural populations. Because predation tends to be most heavily focused on young and small fishes, populations that display type III survivorship curves are most likely to be regulated by predation. (B) A grunt population (Haemulidae) in the Virgin Islands; (C) a wrasse population (Labridae) in Panama; (D) a squirrelfish population (Holocentridae) in the Virgin Islands.*

After Hixon 1991; used with permission. Fish drawings from Gilligan 1989.

fishes that will occur on a specific reef, beyond knowing what occurs in a general geographic region.

An extreme **determinist**, in contrast, would argue that biological interactions such as competition, predation, and symbiosis have led to an evolutionarily fine-tuned assemblage of species. Each species has a well-defined ecological niche that includes competitive, cooperative, and predator-prey interactions with other species. Hence new recruits will become established depending on which residents already occur at a site and which niches are unfilled at a given time (a similar debate has developed around the question of assemblage stability in temperate stream fishes; see, for example, Grossman et al. 1982, 1985 "versus" Herbold 1984; Rahel et al. 1984; Yant et al. 1984; reviewed by Strange et al. 1993).

Few fish ecologists would adhere to either extreme view. The debate, after accounting for differences in methodology and study site, really boils down to an argument as to which phenomenon—chance events or biological interactions—plays the larger role in determining species compositions at different locales. Observational and experimental studies in the Caribbean and Pacific, beginning largely with the work of Sale (1978) and Smith (1978), have led to different conclusions. Observational studies emphasize comparisons of underwater counts of individuals at several sites in one reef area, or at the same locale at several different times or after a hurricane strikes an area (in the

latter situation, one must be "fortunate" enough to have been conducting work at the site before the hurricane struck). Experimental studies usually involve removal of individuals from small coral heads or patch reefs, or addition of small patches of reef habitat followed by monitoring of recruitment and recolonization.

Reef fish assemblages are usually defined in terms of the adults present, but replacement of adults is seldom by other adults. When an adult is removed, either experimentally by a researcher or naturally by a predator, it is usually replaced by a newly settled (recruited) juvenile or a slightly older colonist. The question of interest then becomes one of whether adult population dynamics are determined by events before settlement (i.e., during the planktonic phase), during recruitment (i.e., by settling larvae), or after recruitment (i.e., due to interactions between juveniles, adults, and their competitors and predators) (Fig. 23.6).

Some workers emphasize the importance of events and interactions in the plankton in determining which species populate a reef (Wellington and Victor 1988; Doherty and Williams 1988). This view developed from observations of similar reefs in close proximity to one another that contained dissimilar species assemblages. Also, experimentally increased food and refuge availability on a reef does not necessarily lead to an increase in fish numbers at a site. Hence reefs may contain fewer individuals

FIGURE 23.6. *The various processes that operate to determine the diversity and abundance of fishes on a coral reef. Solid arrows indicate known interactions, dashed arrows possible interactions; the broken arrow between reproductive output and plank-tonic larvae refers to the uncertainty that reproduction on a reef may influence the number of recruits returning to a reef.*

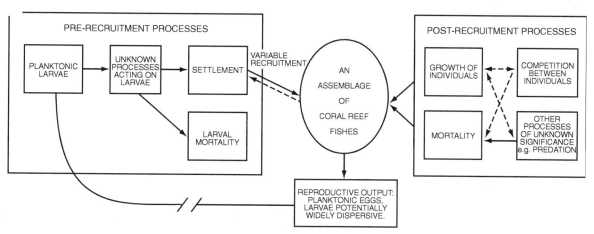

From Mapstone and Fowler 1988; used with permission.

than they can theoretically support, that is, they often exist below their carrying capacity. These findings suggest that adult populations may be limited by the number of larvae available to settle in the area, the so-called **recruitment limitation hypothesis**.

Factors that affect larval abundance include predation on larvae by vertebrate and invertebrate planktivores, food availability for larvae, and dispersal away from appropriate settling areas. It is well known that mortality is the most likely fate awaiting a planktonic larva; most studies estimate that more than 99% of larvae die or are eaten before they settle (see above). Thus antipredator, food-getting, and active dispersal adaptations of larvae themselves may be critical. Adults can also improve their offspring's chances of making it through the planktonic filter by spawning at times and places that minimize dispersal away from home reefs, which reduces the area over which larvae must search for appropriate settling habitat. Carefully chosen spawning locales can also place larvae where planktonic food tends to be concentrated (but also where predators on planktonic fish larvae also abound) (Johannes 1978).

Periodically, vast numbers of larvae that are ready to settle can be found around reefs, indicating that recruitment limitation does not apply at all times (Victor 1991; Kaufman et al. 1992). Under such circumstances, conditions that prevail in the settlement area may determine the success of recruits and the ultimate species composition in an area. Several studies have shown that small artificial reefs placed in shallow water attract large numbers of recently settled larvae to areas that were previously devoid of larvae, suggesting that appropriate, unoccupied habitat can limit recruitment. Previous occupation, also known as **priority effects**, may be crucial in determining whether larvae settle successfully. If the first occupants of a coral patch are herbivores or small planktivores, a variety of larvae will follow and take up residence. However, if the first settlers are predators such

as moray eels, squirrelfishes, snappers, or groupers, later recruitment will be greatly reduced as these small predators eat incoming fish larvae. Which larvae settle first is governed largely by chance because of the unpredictable nature of planktonic existence. Once larvae settle, their impact on later settlers is fairly predictable and depends on the ecological role of the species in question. Hence both stochastic and deterministic forces are in operation (Shulman et al. 1983; Hixon and Beets 1989; Beets 1991).

A larva that settles successfully onto a patch of reef and transforms into a juvenile is by no means guaranteed a long and productive life. Biological interactions involving predation, competition, and cooperation can have a strong impact on individual success. Mortality rates remain high once larvae have settled; 25% of recruits may die in the first 5 days after settlement, but this rate falls to less than 10% after 6 days and continues to decrease thereafter (Doherty and Sale 1986). Although the rate may slow, the numbers killed remain high, varying between 65% and 99.9% during the first year after settlement (Sweatman 1984; Shulman and Ogden 1987). Larvae may settle more successfully in isolated habitats away from major concentrations of larger fishes and then move later to the more extensive reef habitats. The move to the reef itself is also full of perils, as predators tend to patrol the edges of reefs and catch prey moving across refugeless zones (Shulman 1985). For species that have symbiotic relationships with invertebrates, such as anemonefishes, carapid pearlfishes, cardinalfishes, and many gobies, successful location of an unoccupied, species-specific host is probably a good guarantor of survival, but such hosts may be in limited supply.

One possibly influential difference between the Caribbean and the Indo-Pacific is the size of the eggs produced by residents (Thresher 1982). On average, egg sizes are smaller in the western Atlantic than the western Pacific. Smaller eggs imply greater fecundity among Atlantic species, which could lead to greater reproductive output in the western Atlantic. Assuming comparable larval mortality rates, more larvae mean more potential competition for space among recruits, tipping the scales in favor of a stronger role for deterministic interactions among Atlantic coral reef fishes than their Pacific relatives.

The bottom line to this discussion is that arguments about the relative importance of stochastic and deterministic factors and their influences on reef fish assemblage structure and dynamics oversimplify the situation. Both types of factor come into play and have different levels of influence at different times and in different places. Random events in the plankton undoubtedly influence which larvae will survive, and whether the larva settles successfully depends on whether space is available for it on the reef. Space becomes available in part as a chance result of predation and storm disturbance. But larvae, juveniles, and adults often have specific habitat preferences, as shown by the fairly distinctive zones that occur on most coral reefs (lagoonal, patch reef, back reef, reef crest, shallow and deep reef front) and the fairly predictable assemblages of species found in each zone. Nonrandom competitive, predatory, and mutualistic interactions affect the suitability of a site, the survivorship of its inhabitants, and ultimately its availability for new recruits and colonists. Space availability may depend on the guild of a fish that has been removed from the reef. We can predict that certain guilds are likely to be present on a reef, but we cannot predict which members of that guild will be present. Therefore, the relative importance of the different factors will vary from time to time and place to place.

SUMMARY

1. Ecology focuses on organism-environment interactions at the level of individuals, populations, assemblages, communities, ecosystems, and landscapes. An individual's life history results from differences in allocation of energy and resources to the often-conflicting demands of maintenance, growth, reproduction, survival, and migration.

2. Large size is advantageous in fishes; larger fishes produce more eggs and escape more predators. Reproduction at an early age and small size incurs a substantial cost in future reproduction; delayed reproduction means more eggs spawned per occasion but occurs at the risk of dying before ever spawning. Theory accurately predicts the effects of mortality on reproductive age, size, interval, and allotment; populations with high adult mortality reproduce earlier and have higher fecundity and shorter reproductive intervals.

3. Populations grow and decline as a result of age-specific reproduction and survivorship rates. Because of dispersing larvae, migration into populations (recruit-

ment and colonization) has a strong influence on year class strength. Different-age fish differ substantially in size and feeding habits, making cannibalism a frequent cause of mortality. Production is a measure of how much biomass a population produces yearly and is important in determining sustainable exploitation rates for commercial fishes. Most fish populations produce less than 10 g/m²/yr, most of which occurs in younger age classes.

4. Most populations are relatively isolated from other populations, which allows for genetic differentiation. Genetically distinct populations can exist in neighboring lakes (e.g., whitefishes, sticklebacks), and Pacific salmons occur as genetically isolated stocks in adjacent rivers. Genetic differences without apparent geographic separation occur within populations of arctic char in Scandinavia. In contrast, hybridization between species results when species-specific spawning habitat is unavailable or disproportionate numbers of one species exist. Hybridization is most common among freshwater fishes.

5. Species generally have relatively predictable habitat use patterns and predator-prey and competitive interactions (= niche). Species that utilize similar resources in similar ways are members of a guild, as in the zooplanktivore and cleaner fish guilds on coral reefs. Niches and their guild memberships change as fish grow and their food and habitat preferences change. Two general aspects of habitat use in fishes is that bigger individuals within a species occur in deeper habitats and that habitats in rivers and the species occupying them differ as one moves downstream.

6. Competition within and between species results when two consumers use a limiting resource. Shifts in resource use due to competition result in resource partitioning. Competition for food resources is most common in fishes, which can lead to dramatic habitat shifts and can also influence predator-prey interactions. Differences in resource use do not automatically imply that competition is occurring; physiological requirements, phylogenetic constraints, and differential susceptibility to predation can also produce species differences in resource use.

8. Predation can directly affect prey density through predator-caused mortality or can have indirect effects through predator avoidance that places prey in suboptimal environments, thereby slowing individual growth and reproductive output. Predation can also cause genetic differences in coloration, habitat use, and schooling and breeding behavior.

9. The incredible diversity of coral reef fishes has fueled a debate over the relative importance of the physical environment versus biological interactions as determinants of how many and what kinds of fishes occur on any one reef. Much of the debate focuses on whether adult populations are determined by larval mortality (recruitment limitation) or by events

occurring after recruitment, such as predator-prey and competitive interactions among juveniles and adults. It is likely that all factors contribute and that their relative importance differs temporally and spatially.

Supplemental Reading

Gerking 1978; Goulding 1980; Lowe-McConnell 1987; Potts and Wootton 1984; Ricker 1975; Rothschild 1986; Sale 1991; Schonewald-Cox et al. 1983; Stearns 1992; Wootton 1990.

24

Communities, Ecosystems, and the Functional Role of Fishes

Ecology textbooks differ in their approach to communities versus ecosystems, either treating them as separate entities or dealing with one as a subset of the other. A **community** is by definition the plant and animal assemblages that live together in an area. This area may have arbitrarily determined boundaries based on geopolitics but not recognized by organisms (e.g., the marine communities of Florida or Kaneohe Bay, Hawaii) or may involve biologically relevant boundaries across which few organisms pass (Lake Tahoe, the Chattahoochee River). An **ecosystem** consists of the biotic community and the abiotic environment with which the community interacts. Hence one can talk about a stream ecosystem, a lake ecosystem, or an intertidal or offshore reef ecosystem. At a larger spatial scale, one can think in terms of **watersheds**, which take into account the land from which water flows into a series of streams and eventually into a lake or river, and the hydrologic, geologic, and biological forces at work there. The next level of organization is the **landscape**, which recognizes interactions and linkages among ecosystems and the influence of human activities on these interactions (e.g., Schlosser 1991).

Traditionally, community ecology has focused on biotic interactions among different taxonomic groups and the effects such interactions have on distribution and abundance. Ecosystem ecology has focused more on the flow of energy, nutrients, and materials between components of the ecosystem and the functional roles that plants and animals play in this exchange. Our goals in this chapter are to show how fishes interact with other taxa within various communities, and to show how fishes function in the exchange of components within ecosystems.

COMMUNITY-LEVEL INTERACTIONS BETWEEN FISHES AND OTHER TAXONOMIC GROUPS

Fishes do not restrict their ecological interactions—chiefly competition, predation, parasitism, and symbiosis—to other fishes. Evolved, fitness-enhancing or -reducing interactions are also common between fishes and other taxonomic groups, ranging from plants and invertebrates to other vertebrates. Often these interactions and their influences are relatively direct, that is, through eating and being eaten. But some dramatic effects, such as changes in habitat use, food types, and life history traits, take place through fairly indirect means,

either incidental to or several steps removed from the activities of the fishes involved.

Competition

Fishes compete with a variety of organisms for food and space. Bluefish (*Pomatomus saltatrix*) and common terns (*Sterna hirundo*) have a complex feeding relationship that involves both commensalism and competition and that may be applicable to many other fish-bird interactions (Safina and Burger 1985). Both species feed on anchovies and sand lances. The seabirds are particularly dependent on prey fish during the early-summer breeding season when they must meet their own energetic demands as well as those of their growing chicks. Off Long Island, New York, bluefish arrive in large numbers around mid-July each year as part of their annual migration (see Chap. 22, Oceanodromy). A commensal relationship exists between the bluefish and terns in that the feeding activities of bluefish drive prey up in the water column, concentrate them in space, and indicate the whereabouts of prey to the seabirds, all of which facilitate prey capture by the terns. However, newly arrived bluefish consume large numbers of prey fish, which rapidly depresses the prey resource. Any birds that initiate breeding after the arrival of the bluefish tend to be unsuccessful. Hence bluefish may be significant competitors with terns and may have been a strong selective force in determining the timing of reproduction by the terns.

Predation

A wide variety of nonpiscine predators are dependent on fishes as a major component of their diets. Numerous invertebrate predators capture fishes. Predaceous water-bugs (Belostomatidae) and dragonfly larvae prey on small freshwater fishes. Marine invertebrate predators are common, including jellyfish, anemones, siphonophores, squids, cone shells, and crabs (e.g., Laughlin 1982). Among vertebrates, reptilian predators include turtles, crocodilians, varanid monitor lizards, a few iguanid lizards, and sea and water snakes; mammals include mink, raccoons, otters, seals, sea lions, bears, dolphins, whales, and bats. Amphibian predation on fishes is poorly documented, although sirens, bullfrogs, and a few other large frogs (*Pipa*, *Xenopus*) are considered to be fish predators. A host of marine bird predators concentrate on fishes, including terns, petrels, albatrosses, gannets, auks, murres, cormorants, skimmers, spoonbills, pelicans, penguins, and gulls. In freshwaters, osprey, eagles, loons, mergansers, goldeneye ducks, kingfishers, herons, egrets, and storks are a few of the birds that eat fish.

The impact of nonfish predators on fish populations and behavior can be substantial. In the Au Sable River, Michigan, mortality of adult brook and brown trout averaged 70% to 90% annually, most of which was from predation. Nonfish and nonhuman predators (mergansers, heron, kingfisher, mink, otter) accounted for 28% to 35% of this mortality (Alexander 1979). When the potential threat of predatory birds is combined with that from piscivorous fishes, it is not surprising that the distribution of many prey species reflects the foraging locales of their predators (Fig. 24.1). This combined threat can include a third dimension during the breeding season. Male dollar sunfishes (*Lepomis marginatus*) construct nests in shallow water to avoid predatory fishes that are typically deeper. The males, however, must repeatedly abandon their nests to avoid being captured by herons and kingfishers. Each time the male flees, eggs and young in the nest are subject to predation by small fishes, forcing the male into a trade-off between the conflicting demands of protecting himself and protecting his offspring (Winkelman 1993).

Fishes fall prey to less obvious but more insidious predators. Massive fish kills have long been attributed to dinoflagellate blooms (red tides) in many nearshore marine areas, but fish death has usually been considered an incidental by-product of a bloom, via such effects as insufficient oxygen concentrations. Recent studies indicate an evolved, predatory response involving a dinoflagellate, *Pfiesteria piscidida*, that has 19 different life history stages depending on environmental conditions, including the presence or absence of fish in the water (Rensberger 1992; Boyle 1996). Resting cysts of the dinoflagellate are stimulated to break open in the presence of chemicals exuded by fish. The toxic vegetative cells released from the cysts produce a neurotoxin and other substances that can kill fish in a matter of hours. Skin sloughs off the dying fish and is attacked by the dinoflagellates, which then reproduce rapidly, leading to a massive fish kill; 1 billion menhaden died during an episode in North Carolina's Neuse River estuary in 1991. When a fish population declines to some level where the cyst-breaking chemical trigger is no longer sufficiently concentrated, the dinoflagellates return to the encysted form. Hence density-dependent population regulation of fishes could occur through the population responses of a microscopic predator. Increasing frequencies of *Pfiesteria*-caused fish kills correspond to increasing concentrations of human and agricultural wastewater. Not only fish can be affected by *Pfiesteria piscida*; lesions and other pathologic symptoms—including blurred vision, erratic heart beat, and memory loss—can also occur in humans.

Parasitism

The previous example points out the difficulty of distinguishing predators, which consume most of their prey, from parasites, which consume only a small portion of their prey. Parasite-host relationships often involve co-evolved responses (see Chap. 21, Interspecific Relations: Symbioses). One particularly bizarre relationship exists between fishes and parasitic isopods in the family Cymothoidae. These isopods are frequently observed on the heads and in the gills of numerous reef fishes. However, under some circumstances, the isopod attaches to the tongue of its host, causing the tongue to degenerate to a small stub (either through direct consumption or constriction of blood supply). Fish tongues lack skeletal musculature and are not protrusible, unlike their function in other vertebrates. Instead they fill the lower part of the mouth, covering the basibranchial and basihyal bones and serving the mechanical function of holding food against the vomerine and palatine teeth during processing. When the isopod *Cymothoa exigua* attaches to and destroys the tongue of spotted rose snapper (*Lutjanus guttatus*), it resembles the shape and size of the fish's tongue and occludes with the vomerine teeth when the fish feeds. Snappers with the isopod in place for extended periods are typically in good condition, with full stomachs and accumulated fat, which is often not the case for fish that have cymothoid isopods attached to their gills. In essence, the parasite functions as a prosthetic replacement tongue, allowing its host to survive normally and hence continue to provide the parasite the nourishment it needs for its own successful reproduction (Brusca and Gilligan 1983).

THE EFFECTS OF FISHES ON PLANTS

The activities of herbivores are usually divided into browsing and grazing, which are different feeding modes that affect plants differently. **Browsing** involves removing parts of the plant on which the fish is feeding, such as tips of leaves or the leaves themselves (e.g., silver dollar characins, many cichlids, damselfishes). **Grazing** involves biting the plant off at the substrate and even taking in some of the substrate itself, as in the case of parrotfishes

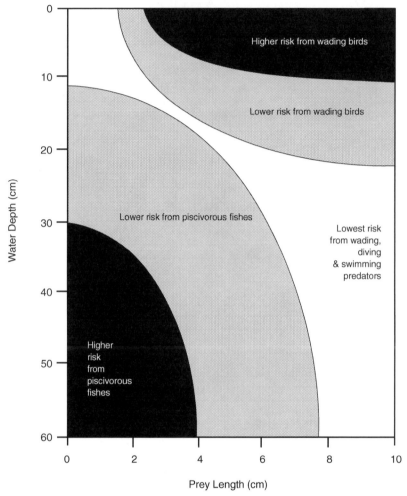

FIGURE 24.1. *The distribution of prey fishes in shallow streams reflects the risk of predation from various sources. Piscivorous fishes, which are gape limited, present the greatest threat in deeper water. Wading birds, which can dismember prey and are therefore not gape limited, present the greatest threat in shallow water. Small prey fishes are safest in shallow water because they can hide from birds among structure, whereas larger prey fishes cannot fit into small spaces. However, larger prey are safer in deeper water because many predators cannot swallow them whole.*

Modified slightly from Power 1987; used with permission.

that scrape coral surfaces in the process of eating algae. The ecological impacts of fishes on plants also vary among habitats, affecting plants by altering biomass, productivity, growth form, and species composition; by dispersing seeds; and by causing changes in the allocation of energy to vegetative versus reproductive structures.

Herbivory is variably developed in different communities. Latitude appears to correlate strongest with herbivore diversity. Above 40°N or S, herbivores are rare or lacking in marine and freshwater fish assemblages (herbivory is variously defined by different authors but generally means that plants constitute at least 25% or 50% of a fish's diet). Herbivorous fishes are most diverse, are usually most dense, and make up a larger percentage of the assemblage in tropical habitats than in temperate habitats. Fewer than 25% of the species in a temperate stream

are herbivorous, whereas 25% to 100% of the species in a tropical stream may be herbivorous. In temperate marine habitats, 5% to 15% of the species are herbivorous, whereas 30% to 50% of the species in coral reef assemblages are herbivorous (Fig. 24.2) (Horn 1989; Wootton and Oemke 1992). At least one antarctic icefish (*Notothenia neglecta*) eats macroalgae and diatoms during spring and summer, and switches to carnivory in autumn and winter (Daniels 1982).

Tropical Communities

In tropical streams, the most common families containing herbivores are minnows, characins (particularly several piranha relatives), and cichlids, and to a lesser extent catfishes, livebearers, and gouramis (Goulding

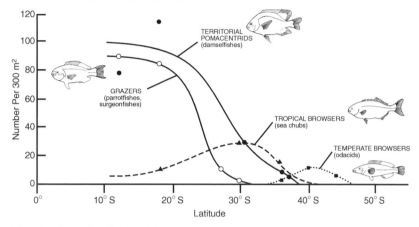

FIGURE 24.2. *Relative numbers and types of herbivores as a function of latitude. More species and types of herbivores inhabit coral reefs than occur at higher latitudes, and browsing prevails at higher latitudes. Values shown are based on Southern Hemisphere comparisons, particularly Australia and New Zealand. However, the same or ecologically similar groups and numerical trends hold for Northern Hemisphere assemblages; for example, sea chubs have temperate representatives in California (halfmoon, opaleye), and pricklebacks function as temperate browsers. The temperate east coasts of North America and southern Africa may be exceptions because of relatively more browsers among the porgies (e.g., Hay 1986).*

Redrawn from Choat 1991; used with permission.

1980; Lowe-McConnell 1987). In Panama, experimental manipulations have shown that algal biomass is reduced by the feeding activity of loricariid catfishes in both shallow (< 20 cm) and deep habitats. When algae-covered rocks are transplanted from shallow regions where biomass is usually higher to deeper regions, the algae are quickly cropped by catfishes. When the catfishes are removed, algal growth is similar at both depths. Higher-standing stocks of algae are maintained in shallows because predatory birds limit the feeding activity of the catfishes there (Power et al. 1989; see above).

Work in Costa Rican streams with a fish assemblage of at least 13 species that eat plants suggests that fishes consume a significant fraction of the macrophytes (a general term for rooted aquatic plants), algae, and grasses, as well as the leaves that fall into the stream (Wootton and Oemke 1992) (see below for a discussion of seed consumption and dispersal in tropical streams). Continuous consumption of periphyton (the algal covering on rocks) and leaves may have a strong effect on the development of an aquatic insect fauna. Temperate streams typically have a much more diverse assemblage of aquatic insects than do tropical streams; many of these insects live in and feed on periphyton and leaves. The diversity and activities of herbivorous fishes in tropical streams may keep these resources at levels too low to permit the development of a diverse fauna of herbivorous aquatic insects, despite the incredible diversity of terrestrial insects in the tropics (e.g., Flecker and Allan 1984). Why herbivorous fishes are more common in the tropics remains a matter of conjecture; perhaps comparative studies from other tropical freshwater locales will help provide the answer.

It is in tropical river systems that the role of fishes in

dispersing seeds is best known. The fruiting of many trees coincides with the annual or semiannual flooding of the rivers (see Chap. 22, Seasonal Cycles). Hence fishes gain access to the base of the trees, or trees gain access to the water. In South and Central America, fruits and seeds that fall into the water are consumed by several characoid fishes, such as pacu (*Colossoma* spp.) and *Brycon guatemalensis,* and constitute the major part of the diets of these fishes during those periods. Fruits and seeds of at least 40 different tree species are eaten by fishes in the Rio Machado region of the Amazon. Although some fruits and seeds may be killed by digestive processes, many seeds make it through the gut unharmed; germination may even be aided by the time spent in a fish's gut. In the case of *Brycon* feeding on the seeds of a common riparian fig tree, no loss in germination occurred. Importantly, seeds remained in the fish's gut for 18 to 36 hours, during which time some fish moved several kilometers. Hence consumption by fishes may aid dispersal of the tree's seeds, including dispersal in the otherwise unobtainable upstream direction (Goulding 1980; Agami and Waisel 1988; Horn 1993).

Tropical lakes have a large number of herbivorous fishes, particularly among the minnows, characins, catfishes, and cichlids. Phytoplanktivorous fishes can affect the relative abundances of phytoplankton species, even though a large fraction of the ingested phytoplankton cells may pass through a fish's gut unharmed (Miura and Wang 1985). Some cichlid species engage in suspension feeding, whereby they pump water into their mouth and out their gill openings, filtering out different size prey. In blue tilapia, *Tilapia aurea,* particles larger than 25 μm are retained by the gill rakers and by mucus-covered microbranchiospines on the gill arches; smaller particles pass through. In experimental ponds, large phytoplankton species are filtered out of the water, and smaller species come to dominate and even increase in numbers. Small phytoplankters may thrive because of nutrient enrichment due to fish excretion and also because fish filter out zooplankton (rotifers, water fleas) that prey on the phytoplankton (see below).

Interactions between plants and fishes are probably best known on coral reefs, where herbivory has led to antiherbivore adaptations in plants and in turn to adaptations by fishes to overcome plant defenses. Herbivore densities on reefs average 0.5 fish/m², as compared to 0.1/m² in most temperate marine habitats. These values do not include the abundant sea urchins, snails, and microcrustaceans that also feed on reef plants (Horn 1989). Reef fishes affect plant biomass, productivity, species composition, distribution, and growth form; whether coral reef fishes distribute reproductive products is apparently unknown. Resident herbivorous fishes can consume most of the daily productivity of the algae on a reef and in the process stimulate higher production rates in cropped algae than in uncropped algal turfs (Carpenter 1986; Klumpp and Polunin 1989). Algae may respond to foraging by changing their growth form. Algae subjected to cropping assume lower, spreading shapes, whereas an absence of herbivores leads to upright, foliose growth forms of the same species (e.g., *Lithophyllum* and *Padina*).

Defensive responses also commonly involve mechanical or chemical adjustments. Plants may be leathery or rubbery (e.g., *Sargassum*), or even hard, as in the case of coralline algae. Seaweeds also produce a variety of so-called secondary compounds, usually halogenated terpenoids, that are distasteful to fishes; more than 20 such compounds have been isolated from marine algae (e.g., caulerpenyne from *Caulerpa*, halimedatrial from *Halimeda*, several dictyols from *Dictyota*). Freshwater fishes also avoid plants with abundant phenolic compounds (Lodge 1991). Some tropical plants respond both mechanically and chemically, such as the green alga *Halimeda*, which deposits calcium carbonate in its tissues and also produces distasteful chemicals. Some plant species always possess such defenses, whereas others produce secondary compounds in direct response to recent herbivore activity. Most algae with such noxious properties are generally avoided by herbivorous fishes, although some fish species appear to specialize on particularly tough or chemically defended plant species. Sea chubs prefer brown seaweeds such as *Dictyota* that contain dictyols and that are generally avoided by most other reef fishes (Horn 1989; Hay 1991).

A possible antiherbivore adaptation may be the production of **ciguatera** toxins, a class of substances that was first noticed because of the effects it has on humans. When a person eats a ciguatoxic fish, adverse reactions include gastrointestinal distress, reversal of sensations (e.g., ice cream feeling hot), and possibly death from respiratory failure. Over 400 species of reef fishes have been implicated in ciguatera poisoning, but the most lethal sources have been large predators such as moray eels, snappers, and barracuda. These circumstances suggested that the ciguatera toxin was magnified as it passed

through the reef food chain. Extensive research has verified that the toxin or toxins originate in unicellular dinoflagellates, primarily in the genera *Gambierdiscus*, *Ostreopsis*, and *Prorocentrum*; these dinoflagellates grow as epiphytes on common reef macroalgae or on newly exposed coral surfaces. Herbivores ingest the dinoflagellates directly or indirectly when consuming plants. The toxin is apparently not broken down by the herbivore, and hence predators high in the food chain obtain a prey fish's lifetime dosage of toxin each time a prey fish is eaten. Transmission of the toxin up the food chain may be enhanced because many fish species exhibit loss of equilibrium and erratic swimming after feeding on ciguatoxic prey, which makes such fish more susceptible to predation. Herbivorous fishes appear to avoid foraging in areas where epiphytic dinoflagellates are particularly dense, suggesting that the toxin can be an effective chemical defense against herbivory (Randall 1958; Davin et al. 1988; Kohler and Kohler 1994).

One by-product of differences in plant palatability, and a factor that can affect algal species composition on reefs, is the gardening behavior of damselfishes. The algal assemblage within a damselfish territory has been justly termed a lawn because it frequently consists of a few highly palatable algal species (e.g., the red alga, *Polysiphonia*) cropped down to a fairly even level; less-desirable species are actively weeded out (Lassuy 1980; Irvine 1982). The lawn contrasts with the surrounding area, where abundant roving herbivores such as parrotfishes and surgeonfishes may graze most surfaces down to relatively bare rock or leave only crustose coralline algae. If heavy fishing pressure has removed large herbivores, surfaces outside territories may have a higher biomass of algae than inside damselfish territories. Regardless, damselfishes, whose territories may cover 40% to 50% of a reef's surface, and other territorial herbivores (parrotfishes, blennies, surgeonfishes) can have a substantial effect on overall algal species diversity and distribution (Hixon and Brostoff 1983; Klumpp et al. 1987; Horn 1989; Hay 1991).

Hixon and Brostoff (1983) found that damselfish territoriality led to higher levels of algal species diversity (Fig. 24.3). The feeding rates of damselfish inside their territories were less intense than the levels of herbivory in unguarded areas outside the territories where parrotfishes and surgeonfishes were abundant. Roving herbivores kept algal diversity low outside territories because they grazed many surfaces bare. In caged enclosures that excluded both damselfishes and other herbivores, algal diversity was again lower than in territories, because certain algal species were able to overgrow and eliminate other species. Hence the damselfish served as a **keystone** species whose activities increase algal diversity not only by decreasing the disturbance created by roving herbivores but also by decreasing a competitive dominant alga's ability to monopolize the space resources of an area.

Temperate Communities

In temperate streams and rivers, herbivory occurs primarily among minnows, catfishes, suckers, and pupfishes. Although only a few species are exclusively or even

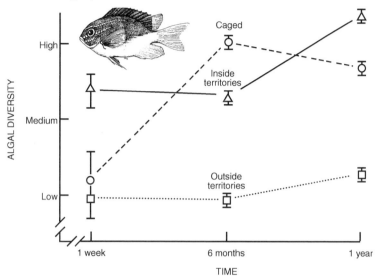

FIGURE 24.3.   *Damselfish territoriality leads to high algal species diversity. The diversity of algal species was measured over 1 year outside territories where roving herbivores were abundant, inside cages that excluded damselfish and other herbivores, and inside damselfish territories. Diversity remained low outside of cages and territories. Inside cages, diversity first increased then decreased as a red algal species became dominant. Inside territories, diversity was highest after 1 year because damselfish excluded roving herbivores that often stripped surfaces bare, while feeding by damselfish controlled the competitively dominant algal species.*

After Hixon and Brostoff 1983; damselfish drawing from Gilligan 1989.

strongly herbivorous, some of these can dominate locally in numbers or biomass and have a strong effect on algal growth (e.g., several pupfishes, *Cyprinodon* spp.; some suckers, *Catostomus* spp.). The stoneroller minnow, *Campostoma anomalum*, can attain densities of 10 fish/m² in midwestern U.S. streams, each fish taking 10 to 15 bites per minute at the algal substrate and consuming 27% of its mass in algae daily. This activity can reduce biomass, alter species composition, and increase growth rates of algal communities. The thick overstory of filamentous green algae and diatoms is removed, leaving behind a thin layer of blue-green algae (cyanobacteria), which receives more light and more nutrients. Biomass-specific primary production is increased in areas grazed by minnows, which means that the plants left behind are more productive than those removed, probably because the algal types that are removed tend to be slow-growing forms and those left behind are faster-growing species (Matthews et al. 1987; Gelwick and Matthews 1992).

Some omnivorous fishes consume large quantities of algae, particularly when preferred animal food is unavailable. It is becoming increasingly evident that many if not most herbivorous fishes readily digest and probably actively seek out animal matter, either when taken in with plants or when opportunistically available. Also, many carnivorous fishes can utilize algae as a supplemental food source when animal matter is unavailable (e.g., Kitchell and Windell 1970; Gunn et al. 1977). A Sonoran Desert minnow, the longfin dace (*Agosia chrysogaster*), prefers mayflies in the spring but takes in large quantities of algae

in the fall when mayflies are no longer available. By increasing its feeding rate during the fall to allow for the lower nitrogen content of plant material, longfin dace maintain a relatively constant uptake rate of nitrogen. Nitrogen is then egested via feces and excreted across the gills into the stream in forms that are rapidly taken up by plants and may account for 10% of the nitrogen used by an otherwise nitrogen-limited stream ecosystem (Grimm 1988).

Herbivorous fishes in temperate lakes often belong to the same families as those that occur in rivers. Their impact on plant biomass can be considerable. Roach (*Rutilus rutilus*) and rudd (*Scardinius erythrophthalmus*) in Lake Mikolajskie in Poland consume 1700 kg of macrophytes and 3800 kg of filamentous algae per hectare per year; the macrophyte values amount to about 15% of the annual total biomass (Prejs 1984). Strong selectivity is shown; roach and rudd prefer *Elodea* over *Potamogeton*, consuming 34% of the former and only 0.1% of the biomass of the latter, despite low relative abundance of *Elodea*. An optimality analysis of benefits (amount of plant obtainable, nutritive value) and costs (plant toughness, availability of leaves and plants) indicates that the preference for *Elodea* exists because it provides a very high benefit-cost ratio as compared to other plant species.

The relative effectiveness of herbivorous lake fishes in consuming plant biomass is attested to by the widespread popularity and success of introductions of the grass carp, *Ctenopharyngodon idella*. Grass carp, a native of China, have been widely introduced in the United States, Europe, and elsewhere to control unwanted aquatic plants, many of which are also introductions (e.g., Eurasian watermilfoil, *Hydrilla*, water hyacinth). Whereas most herbivorous fishes browse only the leaves of a plant and may in fact stimulate later growth, grass carp uproot and eat the entire plant. Grass carp grow to a large size (to 30 kg) and consume 70% to 80% of their body weight daily; they can consequently eliminate all macrophytes in a lake, as happened in a Texas lake where 3650 hectares of vegetation were eradicated within 2 years (Martyn et al. 1986). Total elimination of macrophytes is usually undesirable because, among other effects, it results in destruction of critical habitat for invertebrates and juvenile fishes (Allen and Wattendorf 1987). Feeding preferences by grass carp can also lead to shifts in relative abundances of different species, altering species compositions within the plant assemblage. In experimental ponds, grass carp reduced total plant biomass by feeding preferentially on *Chara*, *Elodea*, and *Potamogeton pectinatus*. Later, total plant biomass increased over original conditions because those species avoided by the grass carp (*Myriophyllum* and *P. natans*) occupied the space vacated by the preferred plants. When grass carp consume submerged vegetation, floating leafed plants can come to dominate (Fowler and Robson 1978; Shireman et al. 1986).

Phytoplanktivory also occurs in temperate lakes and can lead to reduction in phytoplankton abundance. Gizzard shad, *Dorosoma cepedianum*, selectively remove larger phytoplankton species (Drenner et al. 1984a,b). One form of phytoplanktivory, suspension feeding (see above), has been documented in numerous temperate as well as tropical families, including many commercially important species (e.g., herrings, anchovies, whitefishes, minnows, silversides, mullets, cichlids, mackerels; Lazzaro 1987). Particles can be captured on structures other than the gill rakers. A cyprinid, the blackfish (*Orthodon microlepidotus*), captures particles on its palatal organ, a mucus-covered region in the roof of the mouth (Sanderson et al. 1991).

Herbivores are often abundant in temperate marine habitats, although species diversity is lower than on tropical reefs (see Fig. 24.2). Many porgies are seasonal herbivores that take advantage of plant growth during warmer months. Temperate porgies can become particularly abundant, achieving densities of more than 7 fish/m² (Hay 1986). Temperate herbivores are usually browsers, eating the ends of algal fronds and other parts of seaweeds. Despite high seasonal abundance, available evidence indicates that temperate herbivores do not exercise the strong influence on algal ecology that is so evident in tropical marine environments. The strongest effects may be on the establishment and growth of young plants, as in the case of three California sea chub species that feed on young giant kelp, *Macrocystis*. Territorial herbivores are generally lacking in temperate habitats, perhaps because a seasonally limited and variable plant resource makes territoriality impractical for most of the year (Horn 1989). Herbivory in temperate marine and freshwaters may be important, but invertebrates probably exercise a greater influence on algal ecology than do fishes.

FISH EFFECTS ON INVERTEBRATE ACTIVITY, DISTRIBUTION, AND ABUNDANCE

Effects on Zooplankton

The influence that fish have on invertebrate prey populations in freshwater differs according to habitat (Sih et al. 1985; Northcote 1988; Walls et al. 1990). Although exceptions occur, fishes in general have a strong and direct influence on water-column prey such as zooplankton, and a lesser and sometimes undetectable influence on benthic invertebrate populations. Fishes generally crop no more than 5% to 10% of zooplankton production annually, although the impact can be much greater, as when alewives and yellow perch consumed 97% of the zooplankton in Lake Michigan in 1984 (Evans 1986). Fish predation on planktonic organisms normally causes a shift in size and species composition in freshwater zooplankton communities. **Size-selective predation** is a general phenomenon because fish prefer larger zooplankton. Larger zooplankton are caught preferentially by particulate-feeding fish that detect individual prey visually. Large zooplankton are themselves predators on smaller zooplankters. In freshwater ponds and lakes that contain zooplanktivorous fishes, the zooplankton community will be dominated numerically by small-bodied cladocerans and rotifers (*Bosmina*, *Scapholebris*, *Ceriodaphnia*), because they are not eaten by the fishes and their chief predators have been eliminated. When zooplanktivorous fishes are absent, larger zooplankton (e.g., large copepods and cladocerans, *Daphnia* spp., *Simocephalus*) abound and feed on the smaller zooplankton and phytoplankton (Fig. 24.4) (Brooks and Dodson 1965; Janssen 1980; Zaret 1980; Newman and Waters 1984; Northcote 1988).

For size-selective predation to occur, a fish must be able to assess prey size and correct for distance in the three-dimensional, forward-projecting hemisphere in which it normally searches. How this is accomplished is an area of active research. Fish size comes into play, with young fishes being apparently unable to detect small prey, whereas larger individuals have better acuity. In bluegill sunfish, 6-cm-long fish can see objects half the size that 3.5-cm fish can detect. Some evidence indicates that apparent rather than absolute prey size may be important. When offered prey of different sizes at different distances, bluegill, white crappie, and sticklebacks chose the apparently larger prey, that is, they preferred a smaller but nearer prey over a larger prey that was farther away and therefore appeared to be smaller (Hairston et al. 1982; O'Brien 1987).

Zooplanktivorous fishes cause other shifts in the ecology of their prey. Diel vertical migrations are probably a means of avoiding visually hunting fishes; plankters occur in deep, dark, cold, and relatively deoxygenated regions of lakes by day and rise into surface waters at night to feed. Prey migrations are less extensive when predation pressure from fishes is reduced. Adjustments in life history traits of prey include reductions in age at sexual maturity and average size of offspring. Predation also affects prey morphology and coloration. Some cladocerans (e.g., *Daphnia*) develop neck spines when exposed to

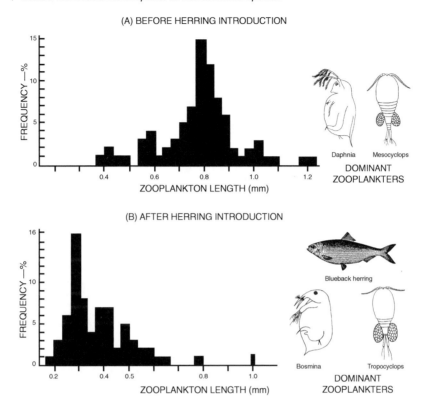

FIGURE 24.4.   *Effects of fish predation on zooplankton assemblage structure. When blueback herring were introduced into a small Connecticut lake, size-selective predation by the fish caused the average size of zooplankters to decline. Predation favored smaller individuals within species as well as smaller species.*

After Brooks and Dodson 1965 and Bigelow and Schroeder 1953.

water that has contained bluegill. Some cladocerans are relatively dark in fish-free ponds but are essentially transparent when co-occurring with fish in lakes; copepods are red in the absence of fish but take on a less vulnerable pale green color in ponds containing fishes. The costs of coloration are obvious from studies that show that some fishes will take the smaller of two forms if it is more pigmented, or will take prey with large, dark eyes over similar-size prey with smaller eyes (Zaret 1980; Lazzaro 1987; O'Brien 1987; Walls et al. 1990; Hobson 1991).

Studies in temperate and tropical nearshore marine environments also suggest a strong influence of zooplanktivorous fishes on species composition and size structure of prey populations. Small (< 1 mm), transparent zooplankters that are part of the pelagic community carried by currents (e.g., dispersing fish eggs and larvae, small copepods, larvaceans) characterize the water column and fish diets during the day. The fishes tend to aggregate at the upcurrent edges of reefs and other structures, where they intercept the incoming pelagic forms. Plankton densities are lower in downcurrent areas, indicating that fishes remove a significant number of plankters from the water column. At night, different species and sizes of zooplankters occur than are seen during the day. Larger, opaque animals (1 to 10 mm), such as mysids, large copepods, polychaete worms, amphipods, ostracods,

and crustacean larvae, emerge from the substrate or migrate vertically upward and join the small, transient zooplankters. Both groups encounter a different set of predators at night, generally large-mouthed fishes with large eyes (e.g., squirrelfishes, cardinalfishes, sweepers). Vision is the most widely used detection mode both by day and by night, but nocturnal zooplanktivores are constrained by the lack of light, making it safer for larger zooplankton that would be easily detected during the day (small plankters are seldom found in the stomachs of nocturnal fishes). Verification of the influence of predators on oceanic plankton communities would require the kinds of predator removal or enclosure experiments that have been performed in freshwater, but the open nature of oceanic coastlines and the long-distance dispersal of the prey make these kinds of experiments virtually impossible (Bray 1981; Hobson et al. 1981; Hobson 1991).

Effects on Benthic Invertebrates

Populations of benthic invertebrates in streams appear to be minimally influenced by predatory fishes, although results vary in different habitats. In Colorado, brook trout (*Salvelinus fontinalis*) were removed from a 1-km section of a stream. Invertebrate numbers and species composition were compared over 4 years in the removal section and in control sections both up- and downstream. No significant differences arose in experimental versus control sections, indicating that trout predation had little effect on invertebrate community dynamics. Similar results, involving fish removal or cages that excluded fish, have been found in streams in North Carolina, Kentucky, Czechoslovakia, and England, and in ponds in New York and South Carolina (Allan 1982, 1983; Holomuzki and Stevenson 1992). In contrast, benthic invertebrate numbers were significantly depressed by the activities of bluegill in lakes and by creek chub (*Semotilus atromaculatus*) in streams. Bluegill sunfish can also reduce invertebrate biomass and the density of pond invertebrates, and yellow perch as an introduced species can eliminate 50% of the benthic invertebrate biomass of small lakes. Predation in many of these experiments was size selective, with fish preferentially eating larger individuals and species and hence affecting community composition as well as numbers (Crowder and Cooper 1982; Mittelbach 1988; Northcote 1988; Gilliam et al. 1989).

Predators can affect more than absolute and relative prey numbers. Distributional changes arise as a result of behavioral adjustments to the threat of predation, which can have population consequences for prey. In the presence of predatory fishes such as sculpins, insect larvae such as mayflies avoid the surfaces of rocks where food availability is highest and instead are found under rocks (Kohler and McPeek 1989; Culp et al. 1991). When smallmouth bass (*Micropterus dolomieui*) are present, crayfish reduce both locomotory and foraging activities and select bottom types that provide more protection (Stein 1979). Drift—the movement of invertebrate larvae and adults downstream in the water column of lotic systems (streams and rivers)—is also affected by fish activity. In the presence of longnose dace, *Rhinichthys cataractae*, mayfly drift first increases and then decreases compared with situations when predators are absent. Hence invertebrates may first move out of an area where predators occur and then later settle and take refuge. Amphipods respond to sculpin by reducing general activity, including drift (Andersson et al. 1986; Culp et al. 1991). Cooper (1984) showed that predation by trout can have both negative and positive effects on water striders. By preying on water striders, trout increase intraspecific competition by increasing the relative density of striders in refuges away from predation risk. However, trout predation benefits water striders in that trout prey on both competitors and predators of the water striders.

Disagreement about the responses of benthic invertebrates to fish predation may reflect differences in prey mobility in different areas (Cooper et al. 1990). In streams containing highly mobile or drifting prey, or in cage studies where mesh size is large enough to allow invertebrates to recolonize, prey that are eliminated by predators will be replaced by immigration. Where minimal exchange of prey occurs between habitats, or where cage mesh sizes are too small to allow recolonization, predators have a stronger depressing effect on prey populations. Not surprisingly then, drift-feeding fishes (e.g., salmonids) have lesser effects on prey, whereas benthic-feeding fishes (e.g., some minnows, sculpins, sunfishes) tend to reduce prey populations. Lakes, ponds, and pool communities with their reduced flow regimes will be more affected by predation than will be riffles and runs. Current speed in lotic systems will operate analogously, with fast currents serving to transport immigrants and slow currents tending to delay replacement. Predatory fishes therefore play different roles depending on whether the habitat is relatively open or closed.

The apparent minimal effect that fishes have on stream invertebrate populations may be interpreted as a lack of influence of fishes on their prey. However, the flip side of this situation is that stream fishes are therefore unlikely to limit their own food supply and are thus unlikely to find themselves competing for food. Insect abundance does, however, vary significantly and can be limiting. Seasonal cycles of insect abundance, driven more by insect life histories than by fish predation, include a midsummer low after many adults have emerged from the stream and before the next cohort of prey has reached edible size. Growth rates of fishes decrease during this midsummer period. Experimental additions of food to streams during midsummer can increase the growth, survival, and energy stores of juvenile stream fishes, indicating that food was in fact limiting (Schlosser 1982; Mason 1976; Karr et al. 1992).

Fishes can also affect the numbers, species composition, and distribution of sessile and mobile marine invertebrates. Some corals appear to be restricted to shallow-water habitats by the grazing activity of fishes. Triggerfish (*Balistapus undulatus*) occur on reef slopes and deep reef areas, and eat corals such as *Pocillopora damicornis*. *Pocillopora* can be successfully transplanted to deeper water on reefs if placed in cages that exclude the triggerfish. Corals transplanted outside of cages are consumed (Neudecker 1979). Damselfish actively kill corals within their territories, which creates a growth surface for the algae on which the fish feed (Kaufman 1977). Since damselfish territories can cover a significant proportion of some shallow reefs, habitat modification by damselfish can influence the amount of coral cover. At the same time, damselfish algal turfs are important habitats for small crustaceans, polychaetes, and mollusks, and the territoriality of the damselfish protects these invertebrates from predatory fishes (Lobel 1980). Fish predation can affect the diversity of encrusting species. Tunicates in caves on the Great Barrier Reef are heavily grazed by fishes. Inside enclosures, tunicates dominate other forms, whereas outside enclosures, removal of tunicates by fishes allows for growth of bryozoans. Tunicates grow successfully only where they have natural protection, such as at the base of stinging hydroids (Day 1985).

Fish predation can limit the distribution of such mobile invertebrates as polychaetes, burrowing sea urchins, snails, stomatopods, and hermit crabs on and near reefs. Predation by stingrays may reduce snail abundance on sandy substrates near reefs. The reef itself often provides a refuge from and for fishes. Herbivorous sea urchins shelter in the reef by day and venture into surrounding grassbeds at night. Patch reefs in the Caribbean typically have a denuded ring of sand 2 to 10 m wide at their base that results from the nocturnal forays of the urchins. The width of this "halo" apparently reflects the distance that urchins can move from the reef and still return safely at dawn before diurnal predaceous fishes become active. The region around the reef, including the halo, is also relatively devoid of infaunal invertebrates. Invertebrate-feeding fishes venture off the reef and forage in the reef's vicinity, returning to the reef when threatened by predators and when resting. Invertebrate densities increase as one travels farther from the reef, unless another reef is close enough that the foraging regions of fishes from both reefs overlap (Ogden 1973; Randall 1974; Ambrose and Anderson 1990; Jones et al. 1991; Posey and Ambrose 1994).

Turnover Rates and the Inverted Food Pyramid

The **biomass** (mass per unit area) of fishes or other animals in a habitat indicates something about the nature of the community. However, biomass is a static depiction, basically a snapshot, of a very dynamic situation for which a moving picture would tell us more. **Turnover**, the ratio of production to standing crop biomass (P/B), provides the added information. Turnover, expressed in units of mass per unit area per unit time (e.g., $g/m^2/yr$), is a measure of how productive a population is over time and takes into account life table schedules of birth and death, population density, individual growth rate, and development time (Benke 1993; see Chap. 23, Production). For example, a seeming paradox occurs in many freshwater habitats when prey consumption rates of fishes are investigated. Trout in the Horokiwi Stream of New Zealand consume about 20 times the standing crop biomass of invertebrates annually; trout and stonefly consumption of prey in a Colorado stream is about 10 times greater than the standing crop biomass of prey. In some streams, biomass of predators exceeds that of prey, which would seem to violate laws of ecology and thermodynamics. Obviously, just looking at biomass tells us little about ecosystem dynamics in such a situation. The paradox of how fish can consume more prey than exist, and how more predators than prey can be maintained in a habitat, is solved when one looks at turnover rates, namely, how quickly animals reach maturity, how many times they reproduce, and how many young they produce. If benthic invertebrates go through several generations per year (which they do), then their annual production can greatly exceed biomass at any one moment, and the invertebrate community can support a much larger fish assemblage than if the fishes were solely dependent on standing crop biomass. Production values 3 to 10 times greater than biomass are not unusual (Allan 1983; Benke 1976, 1993).

FISHES IN THE ECOSYSTEM

The preceding discussions have focused on the effects that fishes as predators have on trophic levels lower in the food web of the community. Such **top-down** regulation of community dynamics can be contrasted with **bottom-up** factors affecting plant growth and subsequent plant or animal prey availability, which ultimately determines fish

abundance and diversity. From the fish's perspective, top-down processes involve the ways that fishes affect the structure and function of an ecosystem (Table 24.1), whereas bottom-up processes involve the physical and chemical factors that affect food availability for fishes. In addition to these direct interactions, ecosystem function is affected by indirect effects of different trophic levels on one another that are separated by several steps or levels in the trophic organization of a community. Fishes function in such interactions as agents that transfer and cycle nutrients, energy, and matter and that link different parts of the ecosystem together.

Indirect Effects and Trophic Cascades

The impact of fish predation on both plants and animals extends beyond the direct effects of reduction in biomass and shifts in species and size composition. Indirect effects on other components of the food web have received considerable attention in recent years. Only about half of the variation in annual primary production in lakes can be explained by changes in amounts and types of nutrients that occur there. The other half results from the indirect but important role that fishes, both piscivores and zooplanktivores, play in determining plant production in lakes. This effect can be described as a cascade of influences down through the food web of a lake from secondary consumers to primary producers (Fig. 24.5).

A typical trophic cascade involves piscivorous fishes (salmons, pike, basses) feeding on zooplanktivorous fishes (herrings, minnows), which feed on herbivorous zooplankton, which eat phytoplankton. Increasing the number of piscivores reduces the number of zooplanktivores, which increases zooplankton abundance, which leads to more removal of phytoplankton from the water column. Hence experimental additions of top predators lead to greater water clarity. The previously mentioned increase in the average size of zooplankton due to size-selective feeding of fishes occurs as an incidental result of such a manipulation. Fish predators on zooplankton are reduced, favoring larger zooplankton that would otherwise be taken by fishes and that may be too large for invertebrate predators to handle. A reversed chain of events occurs if piscivore abundance is reduced, either experimentally, through overfishing, or from fish kills such as occur during winter deoxygenation. Fewer piscivores mean more phytoplankton because of the reduction in herbivorous zooplankters. In reality, herbivorous zooplankton are never entirely eliminated (Carpenter et al. 1985; Carpenter and Kitchell 1988).

Trophic cascades are not limited to the open-water communities of lakes. Benthic communities in shallow water may also be structured by fish-mediated cascades. A fish-snail-epiphyte-macrophyte cascade was found in a Tennessee lake where redear sunfish (*Lepomis microlophus*) ate snails that grazed on epiphytes (filamentous bluegreens and diatoms) that normally infest lake weeds (Martin et al. 1992). When fish were excluded, snail abundance increased, snails depressed the epiphyte populations, and macrophyte growth increased. Strong trophic cascades similar to those found in lakes can also develop

Table 24.1.   Top-down effects of fishes in temperate lakes and streams. Whether the same suite of effects and relationships occurs in tropical freshwater systems or marine systems remains largely unstudied.

Activity	Effect on	Mechanism and consequence
Direct feeding	Water transparency	(1) Food searching stirs up bottom sediments and lowers transparency
		(2) Intense herbivory may increase transparency through removal of phytoplankton; turbidity may increase grazed and extent of fertilization effects
	Nutrient release, cycling	(1) Benthic food searching increases mud–water nutrient cycling
		(2) Littoral vegetation grazing, processing increases nutrient cycling
	Phytoplankton	(1) As in (2), water transparency
		(2) Heavy grazing commonly increases production
	Periphyton	(1) Strong cropping effect leads to increase in biomass
	Macrophytes	(2) As for periphyton
	Zooplankton	(1) Strong cropping effect on abundance especially of larger forms
		(2) Some evidence of increased production
	Zoobenthos	(1) Strong cropping effect on abundance common but not invariable in lakes and streams
		(2) Distribution and size of fish feeding cause marked seasonality in effects
		(3) Production often increased in lakes but not so in streams
Selective predation (size, visibility, motility)	Phytoplankton	(1) Shifts in relative abundance of algal size and species composition
	Zooplankton	(1) Shifts in relative abundance of species reduces algal grazing efficiency and water transparency
		(2) Changes in clutch size and timing of maturation
	Zoobenthos	(1) Heaviest predation on large body-size forms affects their cover selection, activity patterns, and reproductive behavior
	Nutrient release	(1) Shift to smaller body size of zooplankton increases nutrient release
Excretion	Nutrient release	(1) Liquid release provides quick, patchy availability
		(2) Feces release provides slower patchy availability after remineralization
		(3) Epidermal mucous release increases Fe availability to algae via chelation
Decomposition	Nutrient release	(1) Carcass remineralization provides slow, patchy releases
Migration with excretion/ decomposition	Nutrient enrichment	(1) Transport of excreta of body decomposition products from high nutrient to low nutrient regions (sea to inland waters, stream lower to upper reaches, lake layers)

Slightly modified from Northcote 1988; used with permission.

in lotic ecosystems. During the summer, low-flow conditions create a series of pools in seasonally flowing rivers. In the Eel River, California, turfs of filamentous algae up to several meters long and covered with diatoms and blue-green algae cover most of the bottom (Fig. 24.6). Through the summer, these turfs are grazed down to small tufts by midges. The midges are the major prey of predatory damselfly nymphs and the fry of California roach (*Hesperoleucas symmetricus*) and three-spined stickleback (*Gasterosteus aculeatus*). Fish fry and damselfly nymphs are in turn eaten intensively by large roach and steelhead trout (*Oncorhynchus mykiss*). When algae are allowed to grow in cages that exclude the trout and large roach, fry and damselfly nymphs abound and crop down the midges, which allows the algae to maintain fairly luxuriant growth through the summer. Hence the feeding activities of predaceous fishes cascade down through the food web and eventually determine the growth form and extent of primary producers in the river (Power 1990).

Indirect effects driven by fish predation can cause unexpected physical changes in lakes. Lake Michigan and other high-pH, hard-water lakes experience milky water during summer months. These "whiting events" result from the precipitation of limestone crystals (calcite = calcium carbonate, $CaCO_3$). Whitings can inhibit

zooplankton feeding, increase sinking rates and loss of precipitated nutrients to deeper water, and reduce light penetration and primary production. Whitings result from increased photosynthetic activity of algae at elevated summer temperatures, which removes CO_2 from the water and causes an increase in pH. $CaCO_3$ is less soluble at high pH and thus precipitates out of the water, causing the milkiness. Intensive stocking of salmonids (coho and chinook salmon; lake, rainbow, and brown trout) in Lake Michigan during the 1970s led to high salmonid populations in 1983. Salmonids ate huge numbers of zooplanktivorous alewives, allowing phytoplanktivorous cladocerans to develop large populations, which in turn ate phytoplankton. Lack of phytoplankton kept pH low, and hence no whiting event occurred that year (Stewart et al. 1981; Vanderploeg et al. 1987).

Another physical result of trophic cascades involving fishes is the effect that plankton biomass has on temperatures, thermocline placement, and seasonal mixing depths in a lake. Lakes typically have an upper epilimnetic region of warmer water, a lower hypolimnetic region of cold water, and an intermediate metalimnion or thermocline where temperatures change from warm to cold. In experimental enclosures and small (< 20 km²) lakes that lack zooplanktivorous fishes (and hence have zooplankton

FIGURE 24.5. *A trophic cascade as postulated for a lake or pond over a growing season. Solid lines represent changes in biomass or density resulting from a strong year class or experimental addition of piscivores; dashed lines show effects of winterkill or overfishing of piscivores. In North America, piscivores would include salmonids, pike, black basses, and walleye; vertebrate planktivores would include herrings, minnows, whitefishes, bluegill, and perch; invertebrate planktivores would be copepods and many insect larvae; most herivores are crustacean zooplankton.*

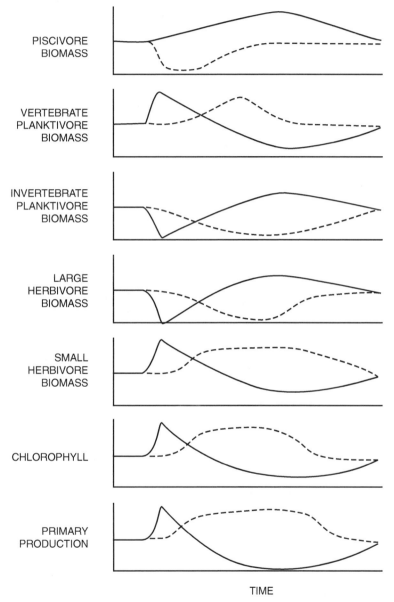

TIME

From Carpenter et al. 1985; used with permission.

that remove phytoplankton), water is clearer, temperatures are 3 to 13°C higher in the metalimnion, and the greater penetration of light and heat leads to deeper thermoclines and a deeper mixed layer of epilimnetic water. Hence the heat content of a lake and all the biological and physical processes and interactions dependent on that heat may be strongly influenced by the top-down effects of fish predation (Mazumder et al. 1990).

Trophic cascades are of interest beyond the insight they give us into the function of aquatic ecosystems.

They have been directly applied to problems associated with eutrophication in lakes, where excessive nutrient input, such as from fertilizers, leads to blooms of undesirable phytoplankton (Kitchell 1992; Carpenter and Kitchell 1993). Primary production and turbidity can be reduced by stocking piscivores that eat zooplanktivores or by selectively removing zooplanktivorous fishes. Both practices have the same theoretical result at the end of the trophic cascade: Fewer zooplanktivores mean more zooplankton, which means more consumption of phy-

FIGURE 24.6.   *Trophic cascades in rivers mediated by fish. (A) During early summertime low-flow conditions, fila-mentous algae develop long turfs on cobble bottoms. (B) By late summer, midges have consumed most of the algae and woven the remainder into small tufts. These midges are eaten by invertebrates and fish fry, which are in turn consumed by larger fish. When large fish are excluded, predators on the midges abound, the midges decline in numbers, and the algae reach abundances more like (A) than (B). Hence algal abundance is determined indirectly by the activities of large preda-tory fishes.*

From Power 1990; used with permission.   (A)   (B)

toplankton and clearer water, assuming that planktivores are not too big to eat and that phytoplankters are edible.

Finally, trophic cascades can involve humans in unex-pected ways (Fig. 24.7). Concern over shark attacks along South African beaches led to an extensive gill netting effort targeted at large sharks. This netting effort was very successful and reduced shark attack frequency, but had a deleterious, cascade effect on sport fishes. Large sharks are predators on smaller sharks, and smaller sharks compete with humans for sport fishes. Hence by reducing the abundance of large predatory sharks, smaller shark species and the young of large sharks experienced a population boom, which in turn strongly depressed the abundance of fin fishes in the region (Van der Elst 1979).

Nutrient Cycling and Transport by Fishes

Phosphorus is often the nutrient that limits primary pro-duction in lakes. Fishes excrete soluble reactive phospho-rus (SRP), a form that is readily taken up by algae. Phosphorus excretion by fishes may be the major source

of SRP in many lakes, enhancing phytoplankton produc-tion and altering algal community composition (Schin-dler et al. 1993). Fishes also function to move nutrients between different compartments of a lake ecosystem. Benthic-feeding fishes disturb sediments, accelerating the rate of exchange of nutrients between the water and bottom muds. The concentration of dissolved nitrogen and phosphorus is greater in the water and reduced in sediments when more bottom-feeding fishes are active. Vertically migrating fishes serve as transporters of impor-tant nutrients between colder, deeper waters and surface layers where most primary productivity occurs. Peamouth chub (*Mylocheilus caurinus*) feed on benthic invertebrates at depths greater than 20 m and then migrate to the surface at night during the summer and fall, where excre-tion and defecation can release nutrients in forms usable by plants. Movement of nutrients across the physical boundary of the thermocline is otherwise limited mostly to periods when the lake turns over, which may occur only once or twice annually in many temperate lakes and rarely in permanently stratified tropical lakes (Northcote et al. 1964; Northcote 1988).

Fish can represent a major reservoir of nutrients that are essential for primary production and can therefore become part of the bottom-up pathway affecting ecosystem function. In some lentic habitats, 90% of the phosphorus in the water column may be bound up in bluegill. These nutrients are released through excretion from the gills, through defecation, and through decomposition after death. Approximately 20% of the internal phosphorus entering a large Quebec lake during the spring could have come from decomposing fish that died after spawning (Nakashima and Leggett 1980). Excretion and defecation by roach (*Rutilus rutilus*) contributed about 30% of the total phosphorus in the epilimnion of a deep Norwegian lake during the growing season (Brabrand et al. 1985). Fish feces and mucus could be important sources of iron in lakes where algal growth is iron limited (Box 24.1). Fish removal experiments often lead to reductions in phosphorus and nitrogen levels in lake waters, whereas fish additions generally lead to increases in nitrogen compounds, except where herbivorous fishes are involved. When macrophytes are eaten, nutrients become available for uptake by phytoplankton (Northcote 1988).

Fishes can act to link different ecosystems together. This role is illustrated nicely by the life cycles of Pacific salmons and their impacts on the energy and nutrient budgets of the different systems they inhabit (see Chap. 22) (Brett 1986). Most salmon hatch in low-order (relatively small) headwater streams. After some time in the river or a lake, the fish move downstream and out to sea. Although abundant in numbers, their major impact on the stream or lake ecosystem while en route to the sea is as food for birds and other fishes, including larger salmon. As they move to sea, they represent a relatively minor loss of nutrients and energy from the river; about three times more phosphorus is gained during runs of adults into lakes than is lost when smolts emigrate (Stockner 1987). In the year or more they spend at sea, growth accelerates substantially. The 5-g sockeye salmon smolt that left its home river 2 to 4 years earlier may return as a 3-kg adult. In Babine Lake, British Columbia, 160 tons of smolts leave the lake each year. Despite 95% mortality while at sea, 3400 tons of maturing adults return to the lake. As thousands (historically millions) of these now mature salmon move back to their natal streams, they constitute a large proportion of the animal biomass present. They also constitute a crucial food source for predators, including seals, sea lions, and killer whales in the nearshore zone, and later eagles, bears, and other piscivores in the river itself. But the major impact of these spawning adults is not as upstream migrants, but as spent carcasses floating downstream when spawning is completed. As carcasses decompose, they are attacked by microscopic (bacteria, fungi) and macroscopic (invertebrates, ravens) scavengers and release carbon, nitrogen, phosphorus, and other nutrients. This release leads to significant increases in primary and secondary production. Primary productivity, bacterial activity, and periphyton biomass may reach their annual peaks in a stream during carcass decomposition, even though spawning runs occur during the winter (Krokhin 1975; Richey et al. 1975).

Net accumulations of nutrients also result from spawning migrations and the death and decomposition of spawning adults of other species, such as clupeids and many lake and stream fishes; these inputs can be linked to increased microbial and invertebrate production (Hall 1972; Durbin et al. 1979). Catadromous fish migrations, in contrast, can lead to substantial losses of organic matter and nutrients. An interaction between physical processes and human disturbance can influence retention of nutrients deposited by fish carcasses. Logs and debris dams create fish habitat in the form of pools just upstream and crevices and shade that provide refuge. If such large woody debris is removed from a stream, this habitat is destroyed, and retention of nutrients by the stream drops dramatically as a result of flushing of material from the stream during floods.

The potentially important role of fishes in linking different ecosystems is becoming obvious as ecologists focus attention on other migratory species. Gulf menhaden (*Brevoortia patronus*) are the numerically dominant fish species in estuaries throughout the Gulf of Mexico (Deegan 1993). These fish spawn offshore in December and January, enter estuaries as larvae in February, spend about 9 months feeding and growing in the estuary, and then migrate offshore in the fall as juveniles, having increased in body mass 80-fold (from 0.15 g to 13 g) over that time. Their emigration represents a net loss of energy, carbon, nitrogen, and phosphorus from the estuary, totaling 38 g of biomass, 930 kilojoules of energy, 22 g of carbon, 3 g of nitrogen, and 1 g of phosphorus for every square meter of estuary. These numbers indicate that between 5% and 10% of the primary productivity of the Louisiana salt marshes is exported in the form of emigrating menhaden. Menhaden may transport half of the nitrogen and phosphorus that leaves these estuaries, the remaining half leaving passively in tidal currents in the form of detritus and dissolved substances. However, the high lipid and protein content of fish makes these nutrients much more available to offshore food chains than is the case for nitrogen, phosphorus, and carbon tied up in detritus. In fact, the estuary's loss is the nearshore environment's gain, as menhaden are the major prey of many predatory fishes. The carbon contained in menhaden represents 25% to 50% of offshore production (Madden et al. 1988). Many other species of fishes, particularly various croakers, have life histories that include larval growth in estuaries followed by offshore migration of juveniles or adults (see Fig. 9.11). Hence fishes link inshore and offshore ecosystems via their role in the exchange of energy, matter, and nutrients. Additionally, nutrient cycles and food webs are seldom limited to the particular habitat in which we find a species at any given moment. Ecosystems are connected to and intimately dependent on one another.

Fishes in Food Webs

Discussing top-down and bottom-up interactions can imply that many aquatic ecosystems are characterized by linear food chains. However, this is a simplification of a much more dynamic situation, because few straight-line chains ("A eats B who eats C who eats D") exist. Instead, trophic

| BOX 24.1 | *Fish Feces in the Ecosystem: The Coprophage Connection* |

Although some fish are converted directly into other fish, squid, jellyfish, birds, and mammals via predation, a major path that energy takes from the organisms on which fish feed to other ecosystem components is through feces (Hobson 1991). For example, on coral reefs, many zooplanktivorous fishes feed on oceanic plankton and produce prodigious quantities of feces during the day that rain down on the reef; the bulk of these by-products are eaten by scatophagous or coprophagous (feces-eating) organisms. Consumers of feces include fishes and various invertebrates such as crustaceans, snails, brittlestars, and corals. Trophically, these organisms are generally considered to be herbivores, omnivores, and detritivores, but a significant proportion of their diet comes from coprophagy. When zooplankters are particularly abundant, they pass through planktivore's guts so rapidly that they appear basically undigested in the feces that fall to the reef. These conditions raise the possibility that coprophagous consumers gain more energy indirectly from zooplankton than do the planktivores that initially captured the zooplankton.

At one Indo-Pacific locale over a 4-month period, 45 different species of reef fishes from 8 families (primarily sea chubs, damselfishes, wrasses, rabbitfishes, surgeonfishes, and triggerfishes) consumed 5975 feces from 64 species in 11 families (Robertson 1982; Bailey and Robertson 1982). These coprophages were not indiscriminate consumers. Fecal material from zooplanktivores and other carnivores was preferred, whereas feces from herbivores that ate brown algae (e.g., rabbitfishes) or from corallivores that consumed carbonate skeletal material along with coral polyps (e.g., parrotfishes) were generally avoided. The higher the caloric, protein, and lipid content of the feces, the more likely it was to be consumed.

A typical fecal food chain might involve a zooplanktivorous damselfish producing feces that were eaten by a surgeonfish that normally eats red algae and whose feces are in turn eaten by a parrotfish that normally eats small algae and coral. Often, the nutritional value of feces eaten exceeded the value of the usual food of a species. Some fishes actively followed others and fed on their feces, indicating that coprophagy is not just incidental to normal feeding behavior. Interestingly, fish never ate feces from their own species, perhaps because most of the usable nutrition for that species had already been extracted, or due to the risk of parasite transfer.

A time delay may occur in the transfer of fecal material from fishes to other reef components. Following a day's (or night's) foraging, many fishes defecate or otherwise excrete at resting sites that are frequently quite distant from the original feeding site. In this way, fishes help exchange energy and nutrients between different parts of the reef. Deposition of feces and other excretory products, particularly nitrogen and phosphorus, can lead to enhanced growth by corals that live under resting schools of grunts in the Caribbean (Meyer et al. 1983).

Linkages between oceanic and nearshore ecosystems are not restricted to tropical waters. In temperate marine habitats, organic carbon was traditionally thought either to be produced in situ by reef algae or to be transported unpredictably to

interactions often involve herbivores that eat animals, carnivores that eat plants, cannibalism, reversals in energy transfer (one species eats juveniles of another species but winds up as the prey of the adults of the same species), and fishes that parasitize their predators by eating their fins, scales, and mucus. Most individuals enter a detrital food loop, either directly during decomposition or indirectly after falling prey and being processed into feces (see Box 24.1). Hence trophic interactions are more accurately described as food webs than as food chains (Fig. 24.8).

A particularly thorough analysis of the role of fishes in a community food web has been conducted for lowland streams in Venezuela and Costa Rica (Winemiller 1990). The fish assemblages at four different sites included between 20 and 83 species that fed on detritus, plants, seeds, flowers, protozoans, aquatic and terrestrial insects, numerous aquatic invertebrates (worms, crabs, shrimps,

clams, snails), fishes, larval amphibians, turtles, lizards, birds, and mammals. Food web analysis of relationships only between fish and their prey indicated between 50 and 100 interacting taxa and 200 to 1200 trophic links (i.e., one fish species might eat a dozen different food types, each one constituting a link) (see Fig. 24.8). The pimelodid catfish *Rhamdia* fed at three different trophic levels: seeds, prawns, and fishes. More than a dozen species (mostly characins, catfishes, and cichlids) were omnivores, feeding extensively on both plant and animal matter. Detritus, formed from decaying aquatic and terrestrial vegetation, was a particularly important component, accounting for 30% to 50% of the food eaten by all species. Species as diverse as characins, catfishes, livebearers, cichlids, and sleepers fed directly on detritus. Reciprocal food loops were common. The trichomycterid catfish *Ochmacanthus* fed on the external mucus of oscars

BOX 24.1 *(Continued)*

the reef by currents in the form of plankton and detritus. Kelp bed zooplanktivores, such as the abundant blacksmith (*Chromis punctipinnis*), feed on oceanic plankton during the day and return repeatedly to the same resting crevices at night, where they defecate. Blacksmith produce an average of 180 mg of feces per square meter of resting habitat each night. These feces are taken up by a variety of detritivores (gobies, clinids, shrimps, hermit crabs, amphipods, snails, brittlestars), which are in turn eaten by larger fishes. Blacksmith also excrete ammonium (NH_4^+) through their gills, which is taken up readily by growing kelp plants, thus aiding the production of the habitat on which the entire kelp bed community depends. Rather than random arrival with currents, fecal and excretory inputs by fishes are a constant and reliable source of energy, nutrients, and trace metals for detritivores and plants, thus adding an additional pathway for the active capture and transfer of potentially limiting substances in nearshore habitats (Bray et al. 1981, 1986; Rothans and Miller 1991).

The importance of fish feces in the nutrient dynamics of other aquatic ecosystems has been less extensively studied. Little is known about this topic in temperate streams, although large amounts of feces accumulate in pools in some streams and are reworked by minnows (Matthews et al. 1987). Underwater photographs in African lakes that contain large numbers of cichlid fishes often show fecal material intact on the rocks (e.g., photographs on pp. 35, 53–70 of Axelrod and Burgess 1976; pp. 15, 22, 31–42 of Lewis et al. 1986).

One would be unlikely to make such an observation on a coral reef, implying that use of fecal material by fishes and invertebrates has not evolved as extensively in tropical lakes (the name of one Indo-Pacific riverine family, the Scatophagidae, indicates that tropical coprophages are not unknown in freshwater).

Fecal material may nonetheless contribute substantially to the nutrient budget of lakes. Prejs (1984; see above) calculated that two minnow species, roach and rudd, contribute approximately 133 kg of nitrogen and 12 kg of phosphorus during the 3-month growing season in a Polish lake as a result of the consumption and processing of macrophytes; additional nutrients are released from digestion of benthic algae. Fishes are efficient consumers of plants but inefficient assimilators of the nutrients contained in the plants; 70% to 80% of ingested plant material leaves the gut as feces. In situations where high macrophyte biomass supports dense populations of herbivorous fishes, the digestive activities of the fishes can lead to undesirable plankton blooms (Prejs 1984). Where fishes use phytoplankton as a food source, as do many cichlids in African lakes, the redistribution of nutrients that fertilize phytoplankton from macrophytes through herbivores and into the water column may be important in the maintenance of a diverse and abundant fish assemblage. Overharvest of benthic-feeding herbivores could lead, indirectly, to reductions in populations of phytoplanktivores. Food webs are maintained through a complex series of linkages (Lowe-McConnell 1987).

(*Astronotus ocellaris*); oscars in turn ate the catfish. A predatory cichlid (*Cichlasoma dovii*) ate juvenile sleepers (*Gobiomorus dormitor*), whereas adult sleepers ate juvenile cichlids.

As complex as the interactions in Figure 24.8 appear, the actual food web is even more intricate. Different ontogenetic stages of a species were not separated in the analysis; nine dominant piscivorous species fed initially on zooplankton, switching later to invertebrates and eventually to fishes as they grew larger. The food webs at the four sites differed considerably due to varying diversities and species compositions, but differences also occurred within sites during the wet versus the dry season. Even this seemingly complex description of interactions simplifies the true complexity of feeding relationships in a natural community, because only the interactions involving fishes are listed; other food items of the prey of

the fishes were not considered (i.e., links between shrimp and snails or between aquatic insects and their prey). The web of feeding interactions in any community is indeed tangled.

Most descriptions of trophic relationships within a community tend to characterize species as either relative specialists or generalists, referring to whether a species feeds predominantly on one or a few food types as compared to a species that feeds on many food types or even at several trophic levels. Specialist-generalist characterizations are used in general community descriptions and have also been invoked to explain the relatively high diversity of fishes and other taxa in tropical as compared to temperate communities. Tropical species are thought to be relatively specialized. The relatively narrow niches of specialists makes it theoretically possible to fit more species into the resource spectrum of a habitat.

FIGURE 24.7.   *A trophic cascade with unexpected results. Extensive gill-netting of large sharks in the Natal region of South Africa removed large sharks, such as the bull shark,* Carcharhinus leucas, *a predator on humans. Primary prey of large sharks, such as juvenile dusky sharks,* C. obscurus, *increased in numbers. Abundant small sharks consumed sport fishes (e.g., bluefish,* Pomatomus saltatrix*). (+) = increasing population, (−) = declining population.*

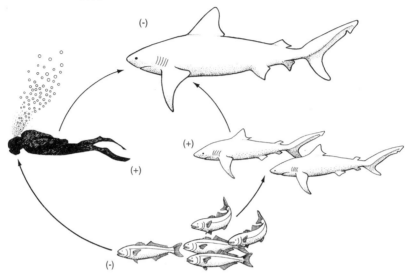

Based on data in Van der Elst 1979; drawings after Bigelow and Schroeder 1948, 1959.

FIGURE 24.8.   *Food webs involving fishes. (A) A relatively simple food web in a temperate North American lake involving humans, predatory fishes (pike, walleye), planktivorous fishes (cisco, yellow perch), invertebrate plankton, and algae. Thickness of lines reflects importance of a food item in a species' diet. (B) A small, lowland forest stream in Costa Rica, and (C) a swamp creek in Venezuela. Each numbered point in the web represents a fish species or a prey taxon eaten by fishes. The base of the food web is at the bottom and includes types of detritus, plants, and plant parts. Intermediate levels in the web represent primary consumers (herbivorous fishes and invertebrates), with predatory fishes at the top. Eleven fish species are involved in food web (B) and 51 species are involved in web (C).*

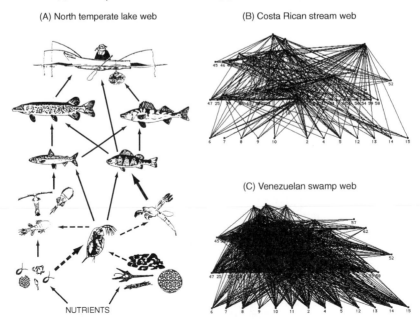

(A) North temperate lake web

(B) Costa Rican stream web

(C) Venezuelan swamp web

NUTRIENTS

(A) After Rudstam et al. 1992. (B)–(C) From Winemiller 1990; used with permission.

Although usually an accurate description, characterizing a species as either a specialist or a generalist may only apply to its habits under the feeding regime that exists in that habitat at the time of the description. An excellent example of this comes from a study of feeding habits of two Panamanian toadfishes. Both species fed almost entirely (85% to 100%) on long-spined sea urchins, *Diadema antillarum* (Robertson 1987). In early 1983, *Diadema* underwent a massive die-off that wiped out 95% to 99% of the individuals across the Caribbean region. One would have predicted that such feeding specialists as the toadfishes would suffer population declines because of their relatively invariant feeding habits. However, populations of both species changed little, reproduction continued, and food habits shifted to a variety of mobile benthic invertebrates in one species and to fishes and mobile invertebrates in the other species. Hence these classic trophic specialists became relative generalists, sounding a cautionary note for anyone attempting to characterize community feeding relationships.

Fishes as Producers and Transporters of Sand, Coral, and Rocks

Fishes not only move energy and nutrients around in aquatic ecosystems but may also contribute to the geologic dynamics of an area. Anyone who has snorkeled on a coral reef has probably witnessed the trails of white material vented from the guts of parrotfishes. This excretory material is in part sand produced in the pharyngeal mill of the parrotfish. Live and dead coral and coralline algae are ground up to separate the skeletal rock from algae growing on or in it. Parrotfish also ingest sand trapped in algal turfs, so some sand in their stomachs is newly produced and some sand is recycled sediment. As parrotfish move about the reef, they therefore create new sand and redistribute old sand. Estimates of the former (bioerosion) range between 50 and 500 g/m²/yr, of the latter (sediment turnover) two to five times those amounts (Ogden 1977; Frydl and Stearn 1978; Choat 1991).

Sand is also produced in the jaws and stomachs of other fishes that feed on corals and coralline algae, as well as those that crush mollusks and echinoderms (e.g., stingrays, emperors, wrasses, surgeonfishes, triggerfishes, puffers) (Randall 1974). Coral is moved around the reef in larger chunks by sand tilefish as they dig burrows and then construct piles of coral fragments over them. The mounds, which can measure 2 m across and several centimeters high, contain hundreds of coral and shell fragments. Frequently, tilefish mounds will be the only accumulations of hard substrate in large expanses of sand. Once abandoned by the tilefish, the mounds may be colonized by large groups of newly recruited damselfishes as well as small drums, butterflyfishes, angelfishes, and surgeonfishes (Clifton and Hunter 1972; Baird 1988). A similar role is played by nest-building minnows in many North American streams (e.g., *Nocomis, Semotilus, Exoglossum*). A chub nest can contain thousands of stones as large as 1 cm across brought from a distance of several meters (Wallin 1989; see Chap. 20, Site Selection and Preparation). The nest again constitutes hard substrate on an otherwise unstable bottom and is used for spawning by other species and is later colonized by various aquatic insects.

INFLUENCE OF PHYSICAL FACTORS AND DISTURBANCE

Ecosystem ecology is concerned not only with biological interactions but also with the effects of physical and climatic factors on ecosystem components. Often these effects are most obvious when extreme climatic or other disturbances occur. The **structure** of a community, broadly defined to include species composition, abundance, distribution, and ecological interactions, changes in response to variation in climate and other forms of disturbance. Disturbance can lead to short-term changes in the physiology, behavior, or ecology of individuals (e.g., acclimatization, movement, trophic and reproductive adjustments), which result in alterations in community structure, which in turn affect the pathways and rates at which energy and nutrients flow through an ecosystem. Abiotic factors that function as disturbances to fishes and cause alterations in community structure and ecosystem function include but are not limited to reductions in dissolved oxygen, often in concert with increased water temperature; changes in stream and river discharge as the result of storms, floods, hydroelectric plants, and drought; and cyclonic storms in coral and kelp bed habitats. Biological disturbances that have ecosystem-wide repercussions include outbreaks of disease or population explosions of destructive species that affect the food and habitat resources of a system (Karr and Freemark 1985; Karr et al. 1992).

Temperature, Oxygen, and Water Flow

Water holds relatively little oxygen, seldom more than 8 parts per million (ppm), and levels below 1 ppm are generally fatal to fishes (see Chap. 5, Water as a Respiratory Environment). Hence periods of deoxygenation—due to excessive decomposition of plants and animals, concentration of fishes in pools during drought or following floods that strand fishes in pools, high summer temperatures leading to thermal and oxygen stratification in lakes, or ice cover—are natural events that strongly affect the distribution and survival of fishes. Species with narrow ranges of tolerance for temperature and oxygen variation will obviously be most strongly affected by extreme conditions. Adult striped bass, *Morone saxatilis*, prefer temperatures between 18 and 25°C and oxygen concentrations above 3 ppm. In southern U.S. rivers such as the St. John's in Florida and Flint in Georgia, striped bass in summer avoid high water temperatures in the main river and aggregate near inflowing springs that provide cooler, preferred temperatures. By the end of summer, these narrow thermal tolerances result in emaciated fish because of limited feeding opportunities in the springs. Thermal stratification in large reservoirs during the summer can also squeeze stripers into an increasingly small region near the thermocline to the point where mortality occurs if the fish are forced into the relatively deoxygenated but cooler waters of the hypolimnion (Coutant 1985; Van Den Avyle and Evans 1990).

Extreme water flow can constrain fishes in streams and rivers. Small fishes and small species typically cannot hold position in swift water as readily as can larger individuals. Eggs and larvae may be washed out of a system or covered with silt if high-flow conditions occur during the breeding season. Juveniles of quiet-water species (e.g., many minnows, sunfishes) are frequently flushed out of headwaters during flood conditions (Schlosser 1985; Harvey 1987). Adult stream fishes can also be adversely affected if floods fill pools or riffles with debris (Minkley and Meffe 1987). However, upland streams are characterized by recurrent and dramatic fluctuations in flow regime, and the fish faunal composition of such areas can return to its original state, probably via recolonization from downstream, within 8 months following even catastrophic floods (Matthews 1986). In contrast, flooding farther downstream is often an important signal inducing spawning in many river fishes, because this is the time when riparian zones and gallery forests become inundated and create nursery habitats for juveniles (Welcomme 1985; Lowe-McConnell 1987; see Chap. 22, Reproductive Seasonality).

Many species, including relatively small fishes, are of course well adapted to high-flow conditions and have little difficulty maintaining position in surf zones (clingfishes, gobies) or swift-flowing water (e.g., homalopterid hillstream loaches, Colorado River minnows and suckers, many darters, torrent fishes of New Zealand). Marked differences in adaptation to intermittent and extreme flow can occur between species within a family. The Sonoran topminnow (*Poeciliopsis occidentalis*) and the mosquitofish (*Gambusia affinis*) are both small, morphologically similar members of the livebearer family, Poeciliidae. The topminnow has evolved in desert streams of the American Southwest that periodically experience flash floods. Mosquitofish are native to southeastern lowlands where rapidly flowing water seldom occurs. Mosquitofish have been widely introduced into southwestern regions and prey on the young of topminnows, leading to population declines of the desert species. But the speed with which mosquitofish eliminate the desert native is largely dependent on the extent of flash flooding. In locales where floods occur regularly, mosquitofish are less successful, because they lack behavioral avoidance and orientation capabilities that topminnows display when flow rate increases. The introduced species gets flushed out of the system because of its inappropriate response to floods (Meffe 1984).

The opposite conditions of low flow lead to isolated habitats, desiccation, and deoxygenation. As water levels decline, upland fishes frequently move downstream and floodplain fishes move into the main river. Isolated pools lead to an increased rate of ecological interactions as fishes crowd together. Competition may be reduced if species diverge in their resource use in response to dwindling resource availability. Seasonal shifts of just this type

have been observed among Panamanian stream fishes, where the least diet overlap among species occurs during the dry season when resources are least abundant (Zaret and Rand 1971).

Seasonal cycles of changing water levels can lead to complex interactions among community components. The Everglades region of southern Florida experiences seasonally fluctuating water levels that usually include a dry season in the spring. This dry period concentrates fishes in relatively small pools (alligator holes) where they are preyed upon heavily by herons, ibises, and storks. The birds are dependent on the fishes for successful reproduction and eat 76% of the fishes in the pools. Although population sizes of the fishes are reduced, species diversity is highest during low water, with small species of omnivores and herbivores (mostly livebearers and topminnows) dominating. If water levels are high, the fish can escape bird predation (see Fig. 24.1), and bird populations decline. Relaxed bird predation leads to overcrowding if water levels drop, resulting in 96% fish mortality from deoxygenation and also from predation by piscivorous fishes (bullhead catfish, largemouth bass, sunfishes). Overall fish diversity also declines as the predators eat the omnivores and herbivores. Hence drought conditions in the presence of predatory birds are beneficial to small fish species, but in the absence of wading birds are beneficial to larger, predatory fishes (Kushlan 1976, 1979; Karr and Freemark 1985).

Extreme Weather

Extremes of climate act as disturbances in the marine environment also, although the effects appear to differ between sites and storms. Major tropical (cyclonic) storms can generate winds in excess of 200 km/h, creating waves more than 12 m high that break on relatively shallow coral reef environments. The influence of such waves is felt far below the surface, as massive corals are broken off and tossed around at depths exceeding 15 m. In addition, literal mountains of sand are shifted around, become suspended in the water column, and scour most structure in the area. After a major cyclonic storm (such storms are generally referred to as hurricanes in the Atlantic and eastern Pacific, as typhoons in the northwestern Pacific, and as cyclones in the southwestern Pacific), few live corals remain in shallow water, and major destruction can be found down to depths of 30 to 50 m, depending on the nature and direction of the storm, tide stage when it strikes, and bottom topography.

When Hurricane Allen struck Jamaica's north coast in 1980, much of the shallow-water coral and other structure was destroyed or damaged. Damselfish algal lawns were eliminated, and the damselfish wandered around for over a week without displaying their usual territoriality. The territories they eventually set up were in deeper water and were associated with different coral types than were used before the hurricane. Parrotfish formed smaller schools and stopped reproducing for 2 weeks. Normally cryptic and nocturnal species (moray eels, squirrelfishes, hawkfish, blennies) swam out in the open by day, perhaps because their refuges were destroyed. Planktivorous fishes (damselfishes, wrasses, bogas) hugged the reef rather than foraging high in the water column. Large predators (snappers, groupers, grunts), which were previously rare, increased in number and swam conspicuously in the open, perhaps capitalizing on displaced and confused prey species. One year after the storm, species distributions and densities remained different, having shifted in favor of fishes associated with low relief (e.g., rubble vs. upright coral) habitats. Analysis of coral recovery 6 years after the storm indicated that damselfish caused a decrease in the numbers and sizes of colonies of the dominant coral species (Woodley et al. 1981; Kaufman 1983; Knowlton et al. 1990).

Even moderate storms can have strong effects on reef fishes. A series of three relatively mild (sustained winds of 60 km/h) cyclones struck the northern Great Barrier Reef over a 2-year period. These storms caused little structural damage to corals but had major behavioral and community effects on the fishes (Lassig 1983). Suspended sand, moved by strong surge and currents, forced many otherwise benthic fishes up into the water column and caused visible wounds, apparently from collisions with corals. More importantly, juveniles suffered substantial population losses, and subadults were redistributed; 60% of the species surveyed suffered density losses of juveniles following one storm. Poor recruitment in several species was attributed to injury or, more likely, settling juveniles being flushed from the system by strong currents. Hence periodic storms can have a decisive effect on the structure of some reef communities.

Analogous events follow storms in temperate marine habitats. A series of severe winter storms occurred off California in 1980, destroying much of the canopy of giant kelp in coastal kelp beds and scouring the bottom. Removal of kelp eliminated the major food of sea urchins, which switched to a diet of benthic algae and denuded the understory regions of reefs, "transforming the reef from a richly forested site to a barren area" (Stouder 1987, 74). The understory turf harbored invertebrates that were the major food types of the abundant, resident surfperches. Differences in microhabitat use and feeding patterns among surfperch species decreased as fish converged on the few areas where prey remained. Although adult surfperches remained in reef areas over the next 15 months, overall fish abundance decreased by 50% as nonresident and subadult fishes abandoned the reefs, probably because of loss of food and refuge sites and unsuccessful competition with competitively superior, resident surfperches (Ebeling et al. 1985; Stouder 1987).

Not all storms, even major storms, have such clear-cut effects on the diversity and density of reef fishes. A severe 3-day storm struck the Kona coast of Hawaii in 1980, destroying most of the shallow-water coral. Few direct fish mortalities or injuries occurred. Many shallow-water species initially fled to deeper water but returned to former areas after a few weeks or months. Sixteen months after the storm, diversity and density had returned to or surpassed prestorm levels, with the exception of a few distributional shifts involving species that remained in deeper water (Walsh 1983).

Reduction in diversity and density is reversible if the causative agents do not recur frequently. In the winter of 1977, the Florida Keys were exposed to record cold temperatures that caused an extensive fish kill. Water temperatures fell to 11°C in areas that normally do not drop much below 19°C. Dead and dying reef fishes were found throughout the area; underwater censuses showed significant decreases in both species diversity and individual densities the following summer. However, by the next year, both overall diversity and density were not different from their pre-cold-snap levels; in fact, diversity on some reefs was higher than before. These increases were largely the result of successful recruitment of many new individuals, perhaps because potential competitors and predators were eliminated by the cold weather (Bohnsack 1983; see also Thomson and Lehner 1976).

Biological Analogs of Extreme Weather

Violent storms and sudden water temperature shifts create obvious and rapid changes in the physical environment of a habitat. Fishes can experience analogous disruptive effects as a result of the action of biological processes. Major disturbance events include population explosions of animals that affect the physical structure of a habitat. One such example is the crown-of-thorns starfish, *Acanthaster planci*, a predator on live corals through much of the Indo-West Pacific Ocean. Normally, the starfish occurs at low densities (2 to 3 animals/km²), but periodically and for reasons that remain mysterious, it undergoes population explosions producing densities of several starfish per square meter, with thousands of individuals swarming over a reef. The starfish consumes the live tissue on the surface of the corals; the underlying limestone skeleton first becomes covered with algae but then collapses due to biological and physical erosion. In this manner, 95% of the coral in a large area may be killed and will require 10 to 20 years to recover (Endean 1973; Wilkinson and Macintyre 1992). Fishes that are directly or indirectly dependent on corals, either for food or shelter, suffer as a result. Coral-feeding fishes, including butterflyfishes, parrotfishes, gobies, wrasses, and triggerfishes, disappear from affected areas, leading to a 15% to 35% decline in species diversity on affected reefs. Densities of other coral-dependent fishes (many cardinalfishes, damselfishes, wrasses, gobies, blennies) also decline, leading to an overall reduction in fish density of 55% to 65% in an area (Sano et al. 1984).

SUMMARY

1. Fishes interact with nonfish taxa, competing for food and space while eating and being eaten. The distribution of many fish species represents an avoidance of piscine, mammalian, and avian predators; in many streams, fishes are squeezed out of deep water by fish predators and out of shallow water by wading birds.

2. Herbivory among fishes is more common in tropical than temperate habitats; common herbivores include minnows, characins, catfishes, and cichlids. Fishes

influence plant biomass, productivity, growth form, energy allocation, and species composition; fishes also disperse seeds. Plants have evolved mechanical and chemical defenses against herbivorous fishes. Damselfishes on reefs "garden" algae within their territories, encouraging edible species and discouraging growth of less-palatable species. Damselfish activities thus affect the diversity and distribution of algae and the many invertebrates that live in algal patches.

3. Temperate freshwater herbivores include minnows, catfishes, suckers, and pupfishes. During warm months when plants grow quickly, grazing minnows can crop most of the plant productivity. During cold months, temperate herbivores commonly shift to carnivory. Phytoplanktivory also occurs in temperate lakes, where fish such as shad can affect plankton abundance and diversity.

4. Lake fishes prefer to eat large zooplankters, which shifts the size and species constitution of the plankton to smaller zooplankton species. Marine zooplanktivores that are active by day also preferentially eat large prey. Avoidance of foraging fishes may be responsible for daily vertical migrations by zooplankters, for day-night differences in zooplankton assemblage composition, and for life history and anatomic traits of zooplankters and other invertebrates.

5. Trophic cascades describe the direct and indirect effects that predators at the top of a food web can have on trophic levels several steps below. For example, piscivorous fishes eat zooplanktivorous fishes, which feed on herbivorous zooplankton, which eat phytoplankton. Hence removing the top piscivores has the unexpected effect of increasing phytoplankton density. Complex interactions of this nature indicate that changes in fish populations can ultimately affect water chemistry, calcium carbonate deposition, the distribution of water masses of different temperatures, and ultimately the heat budget of a lake.

6. Fishes can directly affect the transport and cycling of nutrients in aquatic habitats. Phosphorus excretion by fishes is important for algal growth. Benthic fishes disturb sediments, which increases the transfer of nutrients from the mud to the water column. Fish bodies contain a large fraction of the nutrients in many ecosystems; nutrients are released through excretion from the gills, through defecation, and through decomposition after death. Vertical and horizontal migrations by fishes that feed in one area and rest in another influence coral growth on coral reefs and kelp growth in kelp beds; the long-distance migrations of salmons link oceanic ecosystems with headwater streams. Fishes can also affect the production and distribution of substrate, as when parrotfishes grind coral into sand or when tilefishes or breeding minnows pile rocks over their burrows or nests.

7. Physical factors that appear to have the greatest effects on fish assemblages include reductions in dissolved oxygen from drought and ice cover, storm-induced increases in stream and river discharge, and habitat destruction on coral reefs and kelp beds from storm-caused waves. Biological disturbances with ecosystem-wide repercussions include outbreaks of disease or population explosions of species that literally eat the food and habitat resources of a system.

Supplemental Reading

Carpenter 1988; Carpenter and Kitchell 1993; Gerking 1978; Goulding 1980; Kerfoot and Sih 1987; Kitchell 1992; Lowe-McConnell 1987; Sale 1991; Wootton 1990; Zaret 1980.

PART 6

The Future of Fishes

25

Conservation

Diversity is a major theme of this book. Fishes are objects of wonder in large part because of their incredible diversity, a diversity that makes sense when viewed in the light of evolutionary processes. We conclude this book with a chapter on what we view as a major tragedy of modern times, namely, catastrophic, human-caused reductions in the diversity of fishes and other life forms.

EXTINCTION AND BIODIVERSITY LOSS

Population declines lead to species declines and eventually to extinction. Extinction is a natural process, and natural processes can be characterized by average rates. These rates have accelerated dramatically during past periods of major environmental change. But never in the history of the earth, as we are able to read that history, have global environmental changes resulted from the actions of a single species, nor have extinction rates approached the pace established in the last decades of the twentieth century.

Historically, extinction rates for animals average 9% of existing species every million years, or one to two species per year. During the celebrated Permo-Triassic and K-T (Cretaceous-Tertiary) mass extinctions that marked the ends of the Paleozoic and Mesozoic eras respectively, extinction rates accelerated to perhaps 50% to 75% of the marine fauna over a period of 10,000 to 100,000 years (Raup 1988; Jablonski 1991). In stark contrast, extinction rates for the close of the twentieth century have been estimated at upward of 300 species per day, or 100,000 per year, about 1000 to 10,000 times background levels and 10 to 100 times greater than the major extinction catastrophes of the past (Wilson 1988; Mann 1991). While the accuracy of such estimates is difficult to verify, there is little argument that extinction rates today exceed any in the past. This astounding loss of **biodiversity**, defined as the variety of life forms and processes, can be directly linked to the activities of an overgrown and overconsumptive human population. It is the purpose of this chapter to review examples and major causes of decline in fish biodiversity worldwide and to present some of the solutions that have been suggested for slowing the rate of biodiversity loss. Fishes serve as just one example of the effects on living organisms of human-induced environmental degradation; biodiversity loss is panbiotic, cutting across and into all taxa.

Threatened and Endangered Fishes

Designation of a fish species as threatened or endangered is a complicated process influenced by political as well as biological concerns (see Wheeler and Sutcliffe 1990). The International Union for the Conservation of Nature (IUCN) publishes catalogs of such species periodically, in which it lists numbers of plants and animals worldwide that are considered to be endangered, threatened, or otherwise at risk. Its "Red List" for 1988 identified 596 fish species worldwide that were imperiled, including 24 extinct species (e.g., IUCN 1988). The accuracy of such estimates is constrained by our limited knowledge of the population status of fishes in most parts of the world; 23 of the 24 extinct taxa are from North America, which must reflect scientific effort as much as it reflects environmental degradation.

Aside from records for certain industrialized nations, we know comparatively little (Table 25.1). North America has about 1000 freshwater fish species, of which about 350 (35%) are in need of protection throughout all or part of their range. Of Europe's approximately 200 species of freshwater fishes, about 40% are in serious need of protection. Australia has a similar number of freshwater species, of which about 30% are in trouble. About 60% of South Africa's 100 species are similarly at risk (Lelek 1987; Skelton 1987; Williams et al. 1989). In tropical countries, with their thousands of endemic species and rapidly dwindling rain forests (see Chap. 16), faunal surveys are incomplete, and we have no way of estimating which species are declining and which habitats need protection. It has been estimated that, worldwide, about 1800 (20%) of the world's approximately 9000 species of freshwater fishes are already extinct or in serious decline (Moyle and Leidy 1992).

The data from species already known to be extinct are equally sobering. In North America, 40 distinct fishes (27 species and 13 subspecies) became extinct in the past century (Miller et al. 1989). Ten of these species had apparently viable populations in 1979, reflecting an increasing extinction rate (e.g., 52% of the extinctions occurred between 1900 and 1964, the remaining 48% disappeared in only the next 25 years). The causes of extinction are often discernible and underscore the environmental problems that form the focus of this chapter. Habitat alteration is the most frequently cited factor, causing 73% of extinctions. Other factors include introduced species (68%), chemical alteration or pollution (38%),

Table 25.1.   Declining freshwater fishes of the United States and Canada, as of 1989.[a]   According to the U.S. Endangered Species Act of 1973, species are designated as "endangered" if they face imminent extinction through all or a significant part of their range, as "threatened" if they are likely to become endangered in the near future, or as "of special concern" if minor disturbances to their habitat will cause them to become endangered or threatened.

Family	Species Native to United States and Canada[b]	Number of Taxa[c] Designated as		
		Endangered	Threatened	Of Special Concern
Petromyzontidae, lampreys	18	0	0	3
Acipenseridae, sturgeons	8	1	3	1
Polyodontidae, paddlefishes	1	0	0	1
Cyprinidae, minnows	221	20	25	29
Catostomidae, suckers	58	9	2	12
Ictaluridae, bullhead catfishes	39	3	8	5
Umbridae, mudminnows	4	0	0	1
Osmeridae, smelts	9	0	1	1
Salmonidae, trouts	36	6	9	19
Amblyopsidae, cavefishes	6	1	2	0
Cyprinodontidae, killifishes	51	11	9	7
Poeciliidae, livebearers	11	1	2	2
Atherinidae, silversides	13	0	2	0
Gasterosteidae, sticklebacks	6	2	2	1
Cottidae, sculpins	123	1	1	4
Centrarchidae, sunfishes	34	0	1	5
Percidae, perches	141	7	22	17
Embiotocidae, surfperches	19	0	0	1
Eleotridae, sleepers[d]	7	0	0	1
Gobiidae, gobies[e]	74	1	0	3
Totals	679	63	89	117

[a]Does not include 94 taxa listed from Mexico (4 characids, 20 cyprinids, 4 catostomids, 4 ictalurids, 4 pimelodids, 3 salmonids, 2 ophidiids, 19 cyprinodontids, 9 goodeids, 14 poeciliids, 4 atherinids, 1 synbranchid, 2 centrarchids, 3 percids, and 5 cichlids).
[b]From Robins et al. 1991, who list only described species. Excludes subspecies and introduced and extinct species.
[c]Includes undescribed species and subspecies.
[d]Includes one Hawaiian endemic.
[e]Includes three Hawaiian endemics.
From Williams et al. 1989 among other sources.

hybridization (38%), and overharvesting (15%). These factors often operate in combination. Some extinctions represent elimination of isolated populations, such as the Miller Lake lamprey, *Lampetra minima* (Fig. 25.1). This unique dwarf lamprey was endemic to one small lake in southern Oregon. Because it parasitized introduced trout, it was poisoned into extinction (Bond and Kan 1973). However, other extinct species were widespread, such as the harelip sucker, *Lagochila lacera*, which occurred commonly in large rivers of at least eight eastern states and probably succumbed to siltation of its clearwater pool habitat. The blue pike, *Stizostedion vitreum glaucum*, sustained a large fishery in Lake Erie and Lake Ontario until the mid-1950s; pollution, introduced fishes, overharvesting, and hybridization all contributed to its demise (Miller et al. 1989).

Certain obvious patterns arise from lists of species at risk. Freshwater fishes account for practically all extinct and compromised taxa, reflecting the restricted nature, human impact on, and degraded nature of freshwater habitats. Certain regions and habitat types appear most frequently on the lists. In North America, the isolated and disjunct aquatic systems of the otherwise arid Southwest, such as the spring pools and rivers of the Great Basin and of Mexico, have been centers of evolution and human-induced extinction (Minckley and Deacon 1991; see Chap. 17, North American Deserts). Specialist species endemic to small, isolated habitats make up the majority of extinct and endangered fishes, because both they and their habitats are exceedingly vulnerable to human activity. An isolated stream or pond can be easily destroyed by dumping of toxic substances, introduction of predators, habitat modification, or water withdrawal (e.g., the Devil's Hole pupfish, *Cyprinodon diabolis*, which occupies a 3 m × 6 m limestone shelf in a cave, the smallest habitat of any vertebrate). Big-river fishes with special needs for clean water, such as sturgeons, paddlefish, and some suckers and large minnows, have also been strongly affected.

FIGURE 25.1.   *The extinct Miller Lake lamprey,* Lampetra minima, *the smallest lamprey in North America and the smallest parasitic lamprey known (adult size to 129 mm). This species was probably derived from the Pacific lamprey, L.* tridentata. *Unlike other dwarf satellite lampreys (see Chap. 9), it retained predatory and parasitic feeding habits. Its predation on introduced salmonids made the Miller Lake lamprey a target of ichthyocide treatments, and it was exterminated in the mid-1950s.*

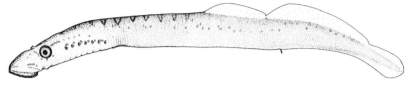

Drawing by Sara Fink. From Miller et al. 1989; used with permission.

Large rivers are primary sites of human habitation and impact; such habitats have been degraded for centuries due to pollution, siltation, water withdrawal, and damming.

Few marine fishes appear on lists of species at risk. Marine fishes have broad distributions and a greater chance for replacement by neighboring populations (except for the coelacanth, *Latimeria chalumnae*; see Chap. 13). Hence, aside from some heavily exploited coastal and pelagic species (e.g., many sharks, bluefin tuna), the most vulnerable marine fishes are estuarine species that have been affected because of their dependence on freshwater input, such as the giant totoaba, *Totoaba macdonaldi* (see below; Moyle and Leidy 1992).

GENERAL CAUSES OF BIODIVERSITY DECLINE

The close of the twentieth century has witnessed a number of well-publicized environmental problems of regional and international scale. Each problem has contributed to declines in fish biodiversity. In addition to habitat modification, species introductions, pollution, and commercial exploitation, global climate change is predicted as a future threat to aquatic ecosystems and fishes. These causes result in direct population losses due to mortality or reproductive failure, or indirect losses due to hybridization or loss of genetic diversity. At the root of each problem is human overpopulation and overconsumption. Overpopulation is particularly destructive to aquatic ecosystems and to fishes because humans concentrate along rivers and estuaries.

Habitat Loss and Modification

Human alteration of aquatic habitats is the most commonly cited cause of declines in fish populations. Habitats are altered via modification of bottom type and above-bottom structure, channelization, dam building, watershed perturbation, and competition for water.

Modification of bottom type.   Many fish species are ecologically dependent on bottom topography and above-bottom structure for successful survival. In flowing water, rocks and logs provide shelter from the current and a site of attachment for eggs, algae, and associated fauna. Undersides of rocks are a major refuge for insect larvae and other invertebrates that fishes eat. Aquatic vegetation similarly provides shelter and food attachment sites for lacustrine fishes. Human activities that disrupt, remove, or cover bottom structure will be detrimental to fishes. Such activities include dredging for navigation and to obtain construction materials, removal of logs and debris dams to aid navigation and as "habitat improvement," and watershed disruption that leads to increased erosion and silt deposition.

Woody debris in streams and rivers exemplifies the effects of habitat disruption on fishes. Woody debris, in the form of debris dams in streams and of logs (= snags) in rivers, plays a critical role in ecosystem function (Wallace and Benke 1984; Harmon et al. 1986). Debris dams retain silt, organic matter, and nutrients; offer a solid substrate for invertebrate attachment; and are a site for transformation and processing of organic matter, thus making it available for invertebrate and fish use. Woody debris also slows the flow of the water, which decreases erosion and increases the time during which nutrients are available to the food web. In coastal, low-gradient (slow-moving) rivers, many game fishes obtain more than half of their food directly from snags. Snags are the most biologically rich habitat in such rivers: Although making up only 4% of habitable surfaces, snags contain 60% of the total invertebrate biomass, provide 80% of the drifting invertebrate biomass, and produce four times more prey than mud or sand habitats (Benke et al. 1985). Government efforts at snag removal in navigable rivers of the southeastern United States began in the early 1800s. When rail transportation largely replaced river commerce in the 1850s, snag removal was less important, but the practice was continued by the U.S. Army Corps of Engineers throughout the United States until the 1950s. Many state agencies still emphasize removal of woody debris as a habitat improvement tool (Sedell et al. 1982).

In the tropics, catastrophic deforestation along rivers and streams adversely modifies both terrestrial and aquatic habitats. In tropical marine environments, coral reef destruction occurs at an equally alarming rate. Coral reefs contain the most diverse fish assemblages on earth, but reefs suffer both directly and indirectly from human activities. Habitats are destroyed by the direct mining and

BOX
25.1

The Snail Darter and the Politics of Endangerment

Darters belong to the family Percidae, which also includes such large predators as yellow perch (*Perca flavescens*) and walleye (*Stizostedion vitreum*). About 150 species and three genera (*Etheostoma*, *Percina*, and *Ammocrypta*) of darters exist; all are small (< 15 cm), benthic, often colorful, and endemic to eastern North America. The region of highest darter diversity is the mountainous areas of the southeastern United States; Tennessee alone is home to about 90 species of darters.

In this center of diversity live several darter species that are restricted to relatively small locales, including headwater streams, springs, and small lakes. Although extensive surveys of the ichthyofaunas of the areas have been made, many areas remain relatively unexplored and uncollected. Hence it was not too surprising when, in 1973, ichthyologists from the University of Tennessee discovered a previously undescribed, small (60 mm) darter in a swift-flowing, gravel shoal of the Little Tennessee River. What was surprising was the uproar the fish created.

Tellico Dam was proposed for construction on the lower Little Tennessee River by the Tennessee Valley Authority (TVA) as early as 1936. Its usefulness, beyond the jobs created during its construction, was always a matter of debate; it was the last dam proposed in the area because its construction was difficult to justify. Its environmental impact would be substantial, as the lake created behind it would flood 17,000 acres of valuable agricultural land, several important Cherokee Indian religious and ceremonial sites (including the village of Tanasi, the capital of the Cherokee Nation, from

which the state derived its name), and a renowned trout fishery. Proponents of the dam included the TVA, local land developers, and the Army Corps of Engineers. Opponents included conservationists, farmers, local landowners, fishermen, the U.S. Fish and Wildlife Service, Supreme Court Justice William O. Douglas, Tennessee Governor Winfield Dunn, and the Cherokee Indian Nation.

Plans for Tellico Dam were shelved and resurrected repeatedly until the U.S. Congress finally approved the project in 1966. Construction began the next year, only to be halted in 1971 when a federal court injunction was issued because the TVA had not filed an environmental impact statement, as required by the National Environmental Policy Act of 1969. The TVA spent 2 years preparing the impact statement, which was approved in 1973, and work recommenced. The Endangered Species Act was then passed in 1973, but no known endangered species were affected by the proposed dam. The ichthyologists chanced upon the snail darter in the region to be inundated by Tellico Dam and named it *Percina tanasi* (Etnier 1976). When extensive collections by TVA and other fish biologists failed to produce other populations of the snail darter, its endangered nature was evident: The Tellico Dam project was one of the chief threats to the species' existence. The fish was given endangered status in October 1975.

The TVA meanwhile was not idle. It undertook a massive, unauthorized 8-month transplantation program, moving 700 snail darters from the Little Tennessee River to the nearby Hiwassee River. Con-

collecting of coral, and inadvertently by harmful fishing techniques (poisons, explosives, bottom trawling), boat anchoring and diver activities, sedimentation and pollution, boat groundings, and changes in coral predator abundance as a result of fishing practices. All these phenomena lead to reductions in fish diversity and biomass, because fishes and their prey rely directly on corals for food and shelter.

Coral mining is a particularly deleterious activity. Limestone blocks are cut from the reef surface and then used in road and home building and as landfill. Massive, head-forming corals in shallow (1 to 2 m depth) water are most frequently targeted. Where heavily practiced, coral cover can change from 50% to 5%, both as a direct result of removal and as a by-product of trampling and sediment production. Recovery is slow, taking more than a decade, if it occurs at all. Fish biomass, abundance, and

diversity decline in mined areas through reduction in living coral and also through reduction in substrate rugosity (topographic complexity) (Bell and Galzin 1984; Shepherd et al. 1992).

Coral collecting for the aquarium trade also takes a significant toll on reef habitat (Derr 1992). Both live corals and algae-covered or invertebrate-encrusted dead corals are taken. Live coral and "live rock" were removed from the Florida Keys at a rate of 3 tons/day in 1989, with an annual retail value of around U.S.$10 million. Live rock, which consists of substrate built over 4000 to 7000 years, does not represent a renewable resource on the reef. Mortality rates for live corals in aquaria exceed 98% within 18 months of collection. Because of the acknowledged difficulties of keeping live coral and reef-building invertebrates in captivity, most large, commercial public aquaria use artificial corals; home aquarists should do the same.

BOX
25.1   *(Continued)*

struction on the dam accelerated in an apparent attempt to complete the dam before other complications arose. In February 1976, the TVA was sued for violating the Endangered Species Act, but the suit was not upheld. Construction continued. The court decision was appealed, and in February 1977, a U.S. court of appeals decided in favor of the fish and issued a permanent injunction against any further dam construction.

The TVA appealed on the somewhat ironic grounds that the Little Tennessee River was no longer a suitable habitat for the snail darter because of the Tellico Dam; the existing construction was blocking the upstream spawning migration of the fish. The TVA proposed transplanting all snail darters to the Hiwassee. The U.S. Supreme Court denied the TVA appeal, which was good news for the conservationists. The catch was that the Supreme Court recommended that the U.S. Congress, which had passed the Endangered Species Act in the first place, become the ultimate arbiter of the situation. Congress, amid much press coverage of the Tellico project, amended the Endangered Species Act and created an exemption committee, which consisted of Secretaries of major federal agencies and was later referred to as "The God Squad" and "The Extinction Committee." This panel had the power to exempt certain activities despite their threat to endangered species if the economic consequences of species preservation were substantial; in this circumstance, it was the darter *or* the dam. The committee met in February 1979 and voted unanimously in favor of the darter! The environmental coalition rejoiced.

The celebration was short-lived. A few months later, in a deft political maneuver, a special exemption for the Tellico project was hidden in more general energy legislation and passed Congress without debate. Many members of Congress did not even realize what they were voting on. President Carter reluctantly signed the legislation, apparently trading the snail darter for conservative votes on his Panama Canal legislation. This vote "sentenced the Little T and its snail darters to death beneath the murky waters behind Tellico Dam" (Ono et al. 1983, 185). Fifteen years after construction began, Tellico Dam was completed and closed.

Although the Endangered Species Act was weakened during the legislative battle that ensued over the snail darter, the process strengthened the species preservation movement in the United States. Never before had so much public interest and sympathy been generated for a comparatively small, economically unimportant, cold-blooded vertebrate. Fortunately, although extirpated from the Little Tennessee, the darter managed to survive the battle. The transplanted population in the Hiwassee River is viable, and additional transplants to the Holston, Elk, and French Broad Rivers are apparently successful. Additional, nontransplanted populations were later discovered in four other locales in Tennessee, Georgia, and Alabama. In fact, thanks to the concern of and efforts by the ichthyological community and an enlightened public, the snail darter's status improved from endangered to threatened as of 1984 (Etnier 1976; Ono et al. 1983; D. A. Etnier, personal communication).

Channelization.   Channelization, also referred to as "bank stabilization," involves straightening a riverine system and smoothing its sides; bends in a river are bulldozed into straight lines, and banks are covered and heightened with stones and boulders (riprap) or concrete. Rivers and streams are channelized primarily to reduce seasonal inundation of the floodplain (so called because the floodplain is the natural area that receives overflow during seasonal rains); channelization is basically the process by which a river or stream is converted into a pipe. Channelized stream segments have low habitat heterogeneity and higher velocities during higher flows; shallow-water and floodplain habitats are eliminated, both of which provide nursery areas for riverine fishes. Channelized rivers either lack fishes or are dominated by introduced species. Especially affected are big-river species, species dependent on sandy areas, and fishes that use the flood-

plain in their life cycle, including sturgeons, paddlefish, and darters of the genus *Ammocrypta*. Channelization-induced loss of the floodplain in parts of the lower Mississippi River has led to a 10-fold reduction in the standing biomass of fishes. Because channelization is often accompanied by deforestation of the floodplain to allow for agriculture and housing development, the entire hydrologic regime of a river is altered, with a result that catastrophic flooding actually increases (Simpson et al. 1982; Moyle and Leidy 1992).

The adverse effects of channelization have been so great in some areas that expensive dechannelization programs have been initiated. In southern Florida, the U.S. Army Corps of Engineers channelized the meandering, shaded, productive 103-mile-long Kissimmee River, turning it into a 56-mile-long, straight, concrete canal. Channelization resulted in drained wetlands (including desiccation of substantial portions of Everglades National Park), water pollution and eutrophication, periodic flooding, salt contamination of streams and aquifers, water table lowering and land subsidence, oxidation of peat soils, wind erosion, and marsh fires. Biological effects include a 90% decrease in wading bird populations, the deaths of 5 billion fishes and 6 billion shrimp, and extirpation of six native fishes from the Kissimmee River. Dechannelization, proposed by the State of Florida and planned in part by the Corps of Engineers, is estimated to cost over U.S.$400 million and take 10 to 15 years, or about as long as the original channelization effort required, at a price 10 times the original cost (Berger 1992).

Dam building.   Dams provide hydroelectric power, water storage capacity (although evaporation often minimizes water storage benefits in arid regions), agricultural water, recreational opportunities, and lakefront development potential. Drawbacks of dam building include flooding of agriculturally and historically valuable land. Poor watershed management, often brought on by deforestation of the land surrounding newly created lakes, leads to rapid silting-in of the lake, transforming it into a much less desirable (from a development standpoint) marsh or swamp. In tropical countries, regions around dams become uninhabitable for humans because the altered habitats favor organisms that cause such debilitating parasitic diseases as schistosomiasis and onchocerciasis (river blindness).

Not too surprisingly, fishes adapted to flowing water do not live in the impounded regions behind dams. Many productive cold-water trout fisheries have been lost behind dam walls. Stream assemblages, usually rich in native darters, minnows, suckers, and trouts, are often replaced by sunfishes and catfishes. As is the case in most disturbed habitats, introduced species come to dominate, including carp, yellow perch, mosquitofish, and lacustrine minnows. The history of the snail darter serves as a good example of the biological and political complexities of dam building (Box 25.1).

Two North American examples typify the effects of dams on aquatic faunas. The Colorado River was an ancient, warm, fast-flowing, turbid river that developed a unique fauna of streamlined fishes adapted to high flows and high temperatures. These fishes spawned in response to seasonal changes in water level and temperature. Of 32 fishes native to the Colorado River, about 75% are endemic. More than 100 dams were built along this huge desert river for water retention, flood control, and agriculture; less than 1% of the virgin flow reaches the river's mouth. The deep reservoirs formed by many dams became thermally stratified, and water released periodically from the cold, lower portions of the reservoirs chilled downstream habitats, disrupting natural spawning cycles and killing native fishes. Of the 80 fish species that now occur in the Colorado River, only about one-third are native. Of the remaining native fishes, most are threatened or endangered, including the humpback chub (*Gila cypha*), the bonytail chub (*G. elegans*), the razorback sucker (*Xyrauchen texanus*), and the Colorado squawfish (*Ptychocheilus lucius*), the largest minnow native to North

America (Fig. 25.2). The modified environment created by the dams and the success of introduced fishes are chief contributors to the decline of native fishes (Ono et al. 1983; Minckley 1991; Wydoski and Hamill 1991).

Hydroelectric dams also block movements of fishes that migrate upriver to spawn or pulverize juveniles on their downstream movements. Fishes that make it past tailwaters or through turbines often suffer from gas-bubble disease, brought on because the agitated waters below a dam can be supersaturated with gas (e.g., Raymond 1988). Habitat destruction, water flow reduction, and other dam effects are considered major factors causing the decline of salmonid stocks in western North America. The Columbia River system has a gauntlet of 28 dams that must be run by spawning adult salmonids and ocean-bound juveniles. Upstream mortality is estimated at 5% and downstream mortality at 20% *per dam* (Booth 1989). Commercial catches of salmon in the Columbia have declined dramatically (Fig. 25.3). For related reasons, approximately 106 major West Coast salmon and steelhead stocks (*Oncorhynchus* spp.) have already been extinguished, and an additional 214 native, naturally spawning stocks of Pacific salmon, steelhead, and sea-run cutthroat trout are at risk in Oregon, California, Washington, and Idaho (overfishing, hatchery introductions, and pollution also contribute to the problem) (Nehlsen et al. 1991).

Retention of sediments in reservoirs, combined with elimination of flood cycles, can have far-reaching consequences for fish production. Nutrients that would have been dispersed over many kilometers of downstream floodplain during seasonal inundation remain trapped behind a dam. Construction of three dams in northern Nigeria led to a 50% reduction in downstream fish landings. Similar effects of dams have been reported in Zambia, South Africa, Ghana, and Egypt. In Egypt, construction of the Aswan High Dam, which impounds 50% to 80% of the Nile River's flow, has caused a 96% reduction in annual landings of sardines in the eastern Mediterranean. In eastern Europe, dams along the Volga River have contributed to a 90% reduction in fish catches in the Caspian Sea. Similar, or worse, scenarios have been created in the Azov, Black, and Aral Seas (see below; Welcomme 1985; Moyle and Leidy 1992).

Watershed perturbation.   Aquatic systems include not only the water in which fishes live but also the groundwater and the surrounding terrestrial area through which water must flow. Many activities have an adverse effect on the watershed, including logging or burning of vegetation, groundwater and surface water withdrawal and contamination, overgrazing and trampling of streamside vegetation, and erosion caused by wind, water, or the movements of livestock.

Much has been written about deforestation in tropical and temperate regions, but most concern has focused on terrestrial effects. Riparian trees, those that grow along stream and river sides, interact intimately with nearby water courses. Obvious consequences of tree removal include a rise in water temperature from loss of shade; increased variation in flow rates because water uptake by

FIGURE 25.2. *Endangered fishes of the upper Colorado River. Large native Colorado River species show convergent adaptations to high-flow conditions, having long, tapered bodies with elongate caudal peduncles, small depressed skulls with "compression bows" (predorsal humps or keels), winglike fins that have hardened leading edges, and tiny or absent scales. Four of the large, endangered cypriniforms of the Colorado exemplify these traits: (A) humpback chub (*Gila cypha*), (B) razorback sucker (*Xyrauchen texanus*), (C) bonytail chub (*G. elegans*), and (D) Colorado squawfish (*Ptychocheilus lucius*). Humpback chubs occupy eddies near the swiftest waters, razorback suckers and bonytails occur in strong currents over sandy bottoms, whereas squawfish are found in riffles, eddies, and deeper pools.*

Photos from Rinne and Minkley 1991; used with permission.

plants is lost; intensified erosion leading to turbidity, siltation, and stream bank collapse (particularly where logging operations occur on steep slopes); and loss of nutrient input from falling leaves and fruit. **Siltation** is a major problem. It hinders productivity because of light reduction, eliminates refuge sites, and smothers eggs, sessile invertebrates, and plants. Siltation has been directly linked to native fish declines in many habitats, including Sri Lankan streams and South African estuaries (Moyle and Leidy 1992).

Another adverse effect of deforestation on aquatic systems involves the cessation of inputs of woody debris, in the form of branches and trunks. Such structure is crucial to the productivity of many low-gradient rivers along

FIGURE 25.3.   *Commercial catches of salmon and steelhead trout in the Columbia River over the past century. Soon after commercial exploitation began, catches rose to sustained levels of 20,000 tons annually. After dam construction, catches declined regularly and have been as low as 550 tons.*

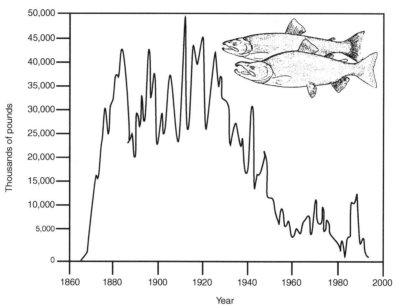

From Abramovitz 1996; used with permission.

coastal plains (see above). Many species use the exteriors and hollowed interiors of logs as spawning sites (e.g., ictalurid catfishes) or as resting sites (Lowe-McConnell 1987). The gallery forests that line lowland rivers are also major spawning sites for fishes that migrate into their flooded zones during winter or spring floods at high latitudes and during rainy seasons at low latitudes. The strong dependence of Amazonian fishes on seasonally inundated floodplains underscores the more general problem of wetlands loss through logging and filling (Goulding 1980; see Fig. 22.5).

Logging along stream courses can have quite unexpected, complicated impacts on fish populations. Clear cutting in the Carnation Creek watershed of British Columbia raised stream temperatures 1 to 3°C. Elevated temperatures caused early emergence and accelerated growth of young coho salmon, *Oncorhynchus kisutch*. Smolts migrated earlier than normal and then experienced poor ocean survival, probably because their early arrival in the ocean placed them out of synchrony with prey cycles (Holtby 1988). Logging operations high in a watershed can affect ecosystem processes at distances far removed from the actual site of disturbance, such as when increased erosion causes high levels of sediment deposition in coastal lagoons and estuaries (Moyle and Leidy 1992).

Competition for water.   Humans use water for drinking, agriculture, recreation, ornamentation, and waste disposal. All these activities have adverse effects on aquatic organisms. Consumption and irrigation necessitate water with-

drawals, leading to flow reductions in aquatic systems. Pumping of groundwater lowers water tables, which reduces the output of springs and seeps that are often necessary for maintaining year-round flow in many systems. Habitats subjected to withdrawals shrink, progressively losing heterogeneity and species. Downstream systems from which water is diverted evaporate, concentrating salts and pollutants. The universal use of waterways and bodies as dumping grounds for human waste creates environments toxic to fishes and humans.

Water withdrawal for irrigation of arid regions has created numerous ecological disasters, leading to species extinctions among fishes and other biota, and eventually producing salinated croplands and contaminated water supplies for humans. The history of species extinctions in the desert Southwest of North America, summarized briefly above, serves as one example. At a larger scale are the events surrounding desiccation of the Aral Sea in the Uzbekistan-Kazakhstan region of the former Soviet Union. In 1960, the Aral Sea was the fourth largest lake in the world, covering 68,000 km²; it supported large commercial fisheries as well as extensive hunting in its wetlands. Inputs are primarily from river flow; losses are due to evaporation. Construction of diversion canals and withdrawal of water from its two major input rivers for irrigation purposes shrank the lake to only 41,000 km² in 1987. At current withdrawal rates, the projected area of the Aral Sea in the year 2000 will be 23,400 km², or about one-third its original size. Salinity increased from 10 parts per thousand (ppt) to 27 ppt and is projected at 37 ppt, or about the salinity of seawater, by the year 2000. An original native fish fauna of 24 species has been reduced to 4 species; commercial fisheries fell from 48,000 metric tons in 1957 to zero by the early 1980s. Dust and salt storms, detectable on satellite imagery, originate on the dry lake bed and distribute 43 million metric tons of crop-destroying salt annually over a 200,000-km² area. Reduced river flow, salinization, pollution of remaining water, and lowering of the water table have led to a high incidence of intestinal illnesses and throat cancer in human populations surrounding the sea. Economic losses of approximately 2 billion rubles (= U.S. $3.2 billion) annually have been estimated for the Aral Sea region as a result of its desiccation (Micklin 1988).

Species Introductions

Movement of species into new areas is a natural zoogeographic phenomenon. When such range extension occurs as a result of human actions, it is considered an **introduction**. Natural dispersal is limited by a species' mobility and by physical barriers. Under natural conditions, species are constrained by coevolutionary processes; species have natural parasites, predators, and competitors that control population growth, and organisms typically must exploit prey taxa that have evolved defense mechanisms against their foraging tactics. When individuals of a species are introduced suddenly into an alien environment, they may find the new physical and biological factors inhospitable or even lethal. Alternatively, freedom from natural biotic control may remove all checks on population growth. It is these liberated introductions that cause the greatest problems. Many such introductions have become well-known pests: rabbits, cane toads, and prickly pear cactus in Australia; starlings, English sparrows, gypsy moths, and zebra mussels in North America; mongoose and mynah birds in Hawaii; feral goats in the Galapagos and on many other islands, to name a few. These catastrophic introductions have their counterparts in fish assemblages as well.

Introductions are called **transplants** if they occur within the country of origin but outside their native range; an **exotic** is a species introduced into a new country (Shafland and Lewis 1984). Introductions may occur through deliberate actions (game fish stocking, vegetation control, aquaculture) or inadvertent mishaps (ballast water introductions, aquaculture escapement, bait fish release). More than 160 species of fishes have been deliberately transported among different countries. As of 1990, 47 exotic species of fishes were established in the continental United States and Canada, and 63 additional species had been collected but breeding populations were not verified (Courtenay et al. 1991). Most of these fishes represent deliberate introductions by government agencies and individuals (e.g., grass carp, *Ctenopharyngodon idella*, for vegetation control; peacock cichlid, *Cichla ocellaris*, as a game fish), escapees from aquaculture facilities (*Tilapia* spp.), or inadvertent bait or aquarium releases (rudd, *Scardinius erythrophthalmus*; walking catfish, *Clarias batrachus*; redeye piranha, *Serrasalmus rhombeus*; suckermouth catfishes, *Hypostomus* spp.). Hawaii has the largest number of established exotics at 33 species, followed by southeastern and southwestern states, particularly Florida and California with their large aquarium fish-breeding industries. Depending on one's perspective, the same introduction can be claimed as a success story or reviled as a disaster (Welcomme 1984; Table 25.2).

Introduced predators. Introductions can lead to population reduction or extermination of native fishes, either directly through predation on adults, eggs, and young, or indirectly through superior competition, hybridization, or transmission of pathogens (Balon and Bruton 1986; Ross 1991; Fausch 1988). Some catastrophic introductions are inadvertent, as with the spread of the marine lamprey, *Petromyzon marinus*, into North American Great Lakes via man-made canals; lake trout, whitefishes, pikeperch, and

Table 25.2. Commonly introduced but controversial fish species. Listed are 10 of the most frequently introduced species that, although often effective in terms of the original purpose for which they were introduced, have subsequently posed serious ecological problems.

Species	Native to	Original Purpose of Introduction
Cyprinus carpio, common carp	Eurasia	food, ornamental
Carassius auratus, goldfish	eastern Asia	ornamental
Ctenopharyngodon idella, grass carp	eastern Asia	vegetation control
Oncorhynchus mykiss, rainbow trout[a]	western North America	game fish, aquaculture
Gambusia affinis, mosquitofish	eastern North America	mosquito control
Poecilia reticulata, guppy	northern South America	mosquito control
Micropterus salmoides, largemouth bass[b]	eastern North America	game fish
Cichla ocellaris, peacock cichlid	Amazon basin	game fish
Lates niloticus, Nile perch[c]	eastern Africa	food
Oreochromis spp., tilapiine cichlids[d]	eastern, central Africa	aquaculture, vegetation control

[a]Also Eurasian brown trout, *Salmo trutta*, and North American brook trout, *Salvelinus fontinalis*.
[b]Also *M. dolomieui*, smallmouth bass.
[c]*L. longispinus* or *L. macrophthalmus* may have also been introduced into Lake Victoria (see Ribbink 1987; Witte et al. 1992).
[d]*Oreochromis mossambicus, O. niloticus, Tilapia rendalli, T. zilli,* and others, including hybrids.
From Welcomme 1984, 1986 and Courtenay et al. 1984.

other species declined precipitously in the wake of the lamprey (see Chap. 13, Petromyzontiforms). Predatory species that have been widely introduced to provide sportfishing are the peacock cichlid (*Cichla ocellaris*), largemouth bass (*Micropterus salmoides*), and rainbow and brown trout (*Oncorhynchus mykiss, Salmo trutta*). Such introductions often decimate native fish faunas, including reduction of important food fishes. Peacock cichlids escaped from an impoundment and into the Chagres River, Panama. The cichlid invaded Gatun Lake and progressively eliminated seven local fish species, including an atherinid, four characins, and two poeciliids; vegetation increased and fish-eating birds were displaced (Zaret and Paine 1973; Swartzmann and Zaret 1983).

Largemouth bass have been responsible for similar community disruptions in Lago de Patzcuaro, Mexico; Lake Naivasha, Kenya; northern Italy; Zimbabwe; and Lake Lanao, Philippines. Rainbow and brown trout have led to the decline of endemic fishes in Yugoslavia, Lesotho, Colombia, Australia, New Zealand, and South Africa, and in Lake Titicaca in Bolivia and Peru (McDowell et al. 1994). In Lake Titicaca, the world's highest lake, a species flock of numerous cyprinodontids (*Orestias*) has been decimated, first through direct predation and later via competition for invertebrate prey (see Box 15.1). Brown trout in particular have been identified as an effective predator on native fishes, including other salmonids. Brown trout have contributed to the decline of several threatened salmonids, including Gila trout (*Oncorhynchus gilae*), McCloud River Dolly Varden (*Salvelinus malma*), and golden trout (*Oncorhynchus mykiss aguabonita*), the latter being the official state fish of California.

One of the most dramatic examples of the effects of an introduced predator involves the stocking of the centropomid Nile perch, *Lates niloticus*, in Lakes Victoria and Kyoga, East Africa (Ribbink 1987; Bruton 1990; Ogutu-Ohwayo 1990; Witte et al. 1992). Lake Victoria is, or was, a showcase of evolution and explosive speciation among fishes, having given rise to a species flock of perhaps 300 haplochromine cichlids, as well as three dozen other fishes. The lake is thought by many to have contained the richest lacustrine fish fauna in the world (see Box 15.1). Against the advice of ecological experts, Nile perch were stocked in the lakes in the early 1960s, "to feed on 'trash' haplochromines . . . [and convert them] . . . into more desirable table fish" (Ribbink 1987, 9). This predator, which can attain a length of 2 m and a weight of 200 kg, spread slowly through both lakes, effectively wiping out native fishes by feeding preferentially on abundant species, then shifting to other species as the density of initial prey declined. As species were eliminated, food webs in the lakes were substantially disrupted and simplified (Fig. 25.4); elimination of herbivorous cichlids led to algal blooms and attendant oxygen depletion in deep water, which caused periodic fish kills. In Lake Kyoga, the catch changed from a multispecies fishery dominated by several haplochromines to one dominated by two introduced species (Nile perch and a tilapia) and a native cyprinid (*Rastrineobola argentea*).

The contribution of haplochromines to commercial landings in portions of Lake Victoria declined from 32%

in 1977 to 1% in 1983; catch rates of haplochromines reflected this shift, declining from 27 to 0 kg/h between 1977 and 1981. Corresponding Nile perch catches increased from 1% to 68% of the catch and from 3 to 169 kg/h. Although overfishing has also been implicated in the decline of endemic cichlids in these lakes, analyses of areas subject to and free from commercial fishing or Nile perch indicate that the impact of the perch is stronger than that due to fishing (Ogutu-Ohwayo 1990; Witte et al. 1992). Just how many of the endemic cichlids have actually been exterminated is difficult to say; perhaps only 50% of the species have been described, and rare fishes are difficult to sample. However, decreasing catches indicate that populations are shrinking, and the continued threat of predation by Nile perch and commercial fishing will only exacerbate the situation. Based on comparative samples taken in 1978 and 1990, approximately 70%, or 200 species, of haplochromines are extinct or threatened with extinction (Witte et al. 1992). Given present trends, "probably more vertebrate species are at imminent risk of extinction in the African lakes than anywhere else in the world" (Ribbink 1987, 22). Events in Lake Victoria call into question recent proposals for introduction of Nile perch into Lakes Malawi and Tanganyika and point out the ecological consequences of introducing predators into any aquatic system (Witte et al. 1992).

The scenario played out in Lakes Victoria and Kyoga is one of reduced biodiversity and simplified community interactions as a cost of production of animal protein for human consumption. Successful fisheries for introduced Nile perch and tilapia have been established in those lakes, replacing the previous fisheries for smaller, native haplochromines. However, the impacts of these introductions are not limited to the aquatic ecosystems. Nile perch have a relatively high oil content. Traditional preparation methods, such as air drying, are less effective for processing Nile perch. Instead, the flesh is often smoked over wood fires; wood cutting leads to deforestation of hillsides in the Lake Victoria basin, which leads to greater silt inputs into the lake. Similar introductions, for similar purposes and with similar results, could be cited. For example, Contreras and Escalante (1984) identified nine instances in Mexico where, after the introduction of potential food fishes, the number of native, often endemic species declined by an average of 80%.

Competition.   Predation of new fishes on old is the most obvious effect of introduced species. Less well documented, but of potentially serious consequence, is the threat of competition, disease, and hybridization that can occur from introducing foreign species (Taylor et al. 1984). Competition is difficult to prove under the best-controlled experimental conditions (see Chap. 23). Evidence of competitive depression of native fishes usually takes inferential form, in terms of overlap in use of potentially limiting resources, or decline in native fishes correlated with the introduction of a nonpredator.

Diet overlap with native fishes in North America has been documented for such introduced species as brown trout, common carp, pike killifish (*Belonesox belizanus*), numerous cichlids, and two Asiatic gobies (*Acanthogobius*

FIGURE 25.4. *Effects of Nile perch introduction on the food web of Lake Victoria. (A) Food web prior to the introduction of* Lates. *Top predators included piscivorous catfishes and haplochromine cichlids, which fed on a variety of prey (including characins, cyprinids, mormyrids, catfishes, haplochromine and tilapiine cichlids, and lungfishes), which in turn fed on a variety of invertebrate prey and algae. (B) The food web after* Lates *eliminated most other fish species.* Lates *feeds on juvenile* Lates, *a cyprinid (Rastrineobola), and an introduced tilapiine cichlid.*

(A)

(B)

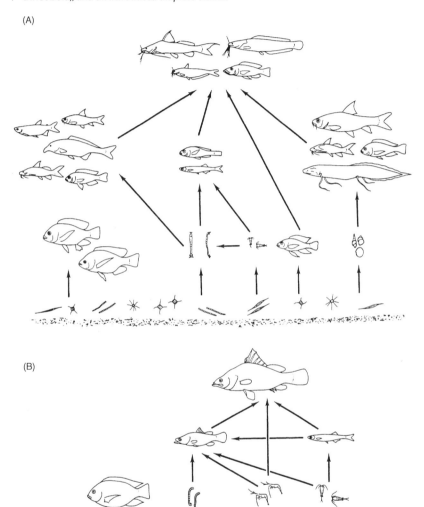

From Ligtvoet and Witte 1991; used with permission.

flavimanus and *Tridentiger trigonocephalus*). Blue tilapia (*Oreochromis aureus*) overlap extensively in diet with gizzard shad and threadfin shad (*Dorosoma cepedianum, D. petenense*, Clupeidae). Blue tilapia reproduce rapidly, forming dense populations (> 2000 kg/ha) of stunted individuals. Introductions of blue tilapia in Texas and Florida have resulted in concomitant population declines of shad, particularly of the benthic-feeding gizzard shad. Overcrowding by tilapia also inhibits largemouth bass spawning behavior, although the actual mechanisms involved (competition, chemical suppression, behavioral interference) are poorly understood (Taylor et al. 1984). Competition for food probably explains the negative impact of introduced guppies (*Poecilia reticulata*) on the endemic White River springfish (*Crenichthys b. baileyi*) in Nevada. Competition for nursery grounds led to a decline in catches of native *Tilapia variabilis* after transplantation of redbelly tilapia to Lake Victoria.

Hybridization. Hybridization and introgression (crossing of hybrid offspring with parental genotypes) have caused rapid losses of native fishes over extensive geographic areas (Echelle 1991). Hybridization can result from habitat alterations that reduce physical barriers between populations and also occurs when numbers of one species fall to the point where conspecifics are rare during mating periods, leading to interspecific matings. Rare species have fallen victim to hybridization in the U.S. Southwest, including hybridization between the threatened Clear Creek gambusia (*Gambusia heterochir*) and introduced mosquitofish (*G. affinis*), between the endangered humpback chub (*Gila cypha*) and the more common roundtail chub (*G. robusta*), and between June suckers (*Chasmistes liorus liorus*) and Utah suckers (*Catostomus ardens*). In the latter example, rarity and water drawdowns for irrigation caused enough interbreeding that June suckers became extinct, their genes existing only in a hybridized subspecies, *Chasmistes liorus mictus*.

Human-caused hybridization is particularly threatening where stocking programs bring hatchery or other strains of fishes in contact with native conspecifics. Native strains disappear as they interbreed with introduced fishes, as has happened when rainbow trout were stocked with threatened cutthroat, Gila, and Apache trouts (*Oncorhynchus clarki* ssp., *O. gilae*, *O. apache*) in western North America (Echelle 1991). Hatchery fishes often originate from a limited gene pool or from inbred lines and have reduced genetic variability compared with wild populations. Low genetic variability correlates with lower fecundity, poorer survivorship, and slower growth, as found in different populations of endangered Sonoran topminnows (*Poeciliopsis occidentalis* ssp.) (Quattro and Vrijenhoek 1989). When hatchery transplants breed with wild fish, the resulting offspring are often less diverse genetically than the wild strains. Hybrid offspring may continue to breed with and eventually eliminate native stocks, as has occurred with rainbow trout stocked widely throughout North America. Threats from transplanted and cultured fish have caused considerable concern over the genetic integrity of wild Atlantic salmon (*Salmo salar*) stocks, prompting programs to minimize the effects of sea ranching and stock enhancement programs (North Atlantic Salmon Conservation Organization 1991).

The creation of intergeneric hybrids, initially considered unlikely, has also proven troublesome. European brown trout (*Salmo trutta*) hybridize with North American brook trout (*Salvelinus fontinalis*), producing a cross known as the tiger trout (e.g., Brown 1966). The widely introduced European rudd, *Scardinius erythrophthalmus* (Cyprinidae), is a hardy, colorful bait fish cultured in the southern United States. Rudd hybridize with the native golden shiner, *Notemigonus chrysoleucas*. Rudd are known to be established in 8 states and could potentially hybridize with golden shiners in 26 states in the Mississippi River basin, with unknown consequences for the fish assemblages or ecosystems of those areas. Such risks are unnecessary given that several acceptable native bait species, including the golden shiner, already exist throughout the region (Burkhead and Williams 1991).

Parasites and diseases.   A major threat from introductions, whether exotic or transplanted, is transmission of bacterial and viral diseases and parasites to which native fishes were previously unexposed. **Furunculosis**, a fatal bacterial disease caused by *Aeromonas salmonicida*, was originally endemic to western North American strains of rainbow trout. When the trout was introduced into Europe, the disease became widespread among brown trout populations and now occurs wherever salmonids are cultured. **Whirling disease**, caused by the protozoan *Myxosoma cerebralis*, is native and originally nonpathogenic to European salmonids. It was transmitted from Europe to North America in the late 1950s, has proved extremely pathogenic to rainbow and brook trout, and has subsequently spread with exportation of North American salmonids, including back to Europe. **"Ich,"** a debilitating gill and skin infestation caused by the ciliated protozoan *Ichthyopthirius multifiliis*, originated in Asia and has spread throughout temperate regions via introductions (Hoffman and Schubert 1984; Welcomme 1984).

An interaction between genetic disease resistance and the dangers of transplantations is exemplified by fall chinook salmon (*Oncorhynchus tshawytscha*). Salmon raised from eggs taken from streams where the protozoan *Ceratomyxa shasta* is endemic show mortality rates of less than 14% when exposed to the pathogen. Salmon taken from streams where the pathogen is not native exhibit mortality rates of 88% to 100% upon exposure (Winton et al. 1983). Introduction of infected fishes into areas where specific diseases do not occur naturally, such as might occur during pen-rearing operations or a supplementation program, could have catastrophic consequences for endemic stocks of fishes.

Not only fishes are threatened by diseases carried via fish introductions; humans may also become infected. A bacterium, *Streptococcus iniae*, causes meningitis and encephalitis in fishes. Normally, such bacteria are host taxon-specific and of little concern to human health. But at least six North American cases of meningitis and related infections have been documented in human workers cleaning tilapia, which are African cichlids that are widely used in aquaculture programs (Holder 1996).

Ballast water introductions.   A little-known but significant source of introductions is the ballast water of large ships. Water is pumped into special ballast tanks or empty holds of ships to stabilize them; this water is then pumped out when cargo is taken on board at another port. Hundreds of species of fishes and invertebrates have become widely established as a result of such ballast water introductions, including such well-known pests as the zebra mussel, *Dreissena polymorpha*, and the predatory cladoceran, *Bythotrephes cederstroemi*, in the Laurentian Great Lakes. These and other invertebrates can drastically alter the food resource base for fishes via competition for or elimination of natural prey. Fish introductions via ballast water also have an impact on native fish faunas. The introduced ruffe, *Gymnocephalus cernua*, a European percid, competes with yellow perch, *Perca flavescens*, and preys on the eggs of whitefish (*Coregonus* spp.). Artificial alterations in the genetic structure of a stock can occur, such as in the case of European threespine sticklebacks, *Gasterosteus aculeatus*, which have been transported to North America, where they interbreed with North American populations.

The yellowfin goby, *Acanthogobius flavimanus*, an east Asian native, has become one of the most common benthic fishes in the San Francisco Bay-Sacramento River area (Courtenay et al. 1984; Carlton 1985; Moyle 1991).

Assessing potential impacts of introductions. Many species introductions, including those precipitating some of the worst-case ecological scenarios, occur in developing nations where the focus is on human economic and nutritional problems. In addition to the negative ecological consequences of many introductions, traditional fishing methods are frequently displaced by introduced species, requiring new harvesting technologies or replacing local artisanal fishermen with commercial or sportfishers. Although many developing nations are in desperate need of capital and of animal protein sources, simple planning measures and attention to natural distributions and local fishing techniques could often minimize results that are destructive to both the local biota and culture. Whenever an introduction is being considered, be it a transplant or an exotic, a protocol such as that outlined by Kohler and Courtenay (1986a,b) should be followed to assess the potential biological and sociological costs and benefits of the introduction.

Pollution

Pollution enters aquatic systems as sediments or in the form of dissolved or suspended substances in runoff or precipitation, while attached to sediments, or while airborne. Human-produced toxic substances number in the thousands, ranging from elemental contaminants such as chlorine and heavy metals to chemical complexes such as persistent pesticides, detergents, and petroleum products. Harmful effects on fishes occur as a result of direct toxicity or through food chain effects (e.g., eutrophication, bioaccumulation), ultimately affecting individual survival and reproduction. Food chain effects also link contaminated fishes to other endangered species, such as marine mammals and birds of prey (Lloyd 1992).

Pollution-related reductions in fish biodiversity occur worldwide. Some of the best-documented examples have occurred in North America and Europe due to acid rain and agricultural chemicals. Acid rain has a pH less than 5.6. It results when oxides of nitrogen and sulfur dioxide from internal combustion engines and coal-burning operations are further oxidized in the atmosphere to form nitric and sulfuric acid. Acid rain becomes a particularly serious problem in watersheds composed of rock types that are incapable of buffering the acids, such as the metamorphic rocks of northern North America and Europe. Acid rain has caused dramatic chemical changes in more than 100,000 lakes in Ontario and Quebec, wiping out all wild stocks of the endangered aurora trout (*Salvelinus fontinalis timagiensis*), and has reduced the range of the endangered Acadian whitefish (*Coregonus huntsmani*) by 50% (Williams et al. 1989). Similar acidification and salmonid declines have occurred in the Adirondack Mountains of New York and in many Scandinavian lakes.

Agricultural chemicals—pesticides, herbicides, and fertilizers—have been responsible for the extermination of many fishes in the American Southwest, particularly those in isolated habitats. The toxic chemicals work directly on the fishes or are ingested with food, whereas fertilizers lead to eutrophication, which changes the balance of algae from edible species to inedible blue-greens, raises lake temperature, and lowers oxygen content. The Clear Lake splittail, *Pogonichthys ciscoides*, a cyprinid endemic to Clear Lake in northern California, was extremely abundant through the 1940s. Agricultural development of the lake basin transformed the lake from a clear, cool habitat dominated by native fishes to a warm, turbid lake dominated by introduced species. The last splittail was taken from the lake in 1970. Eutrophication or toxic chemicals have been similarly implicated in the demise of such unusual fishes as the Lake Ontario kiyi, phantom shiner, stumptooth minnow, blue pike, and Utah Lake sculpin. Overall, pollution has contributed to the demise of 15 of the 40 species and subspecies of fishes that have gone extinct in North America during the past century (Williams et al. 1989).

Fishes as indicators of environmental health. "The quality of fishing reflects the quality of living." This motto of the American Sportfishing Association embodies the host of problems facing aquatic ecosystems worldwide. Lakes, rivers, and oceans with abundant, diverse fishes are reliable indicators of a healthy environment for all life forms. Quantifying the condition of aquatic habitats therefore becomes a crucial exercise in understanding and predicting potential hazards to human welfare.

Fishes can serve as indicators of the health of aquatic systems, in advance of effects on human health. At one extreme, massive fish kills indicate high levels of lethal contaminants, or low levels of oxygen. Ideally, less-acute warnings are preferable. To this end, several measures have been developed that use quantifiable aspects of fish assemblage structure, health, and behavior as a means of monitoring conditions in aquatic systems. One approach gaining increased acceptance is the **Index of Biotic Integrity** (IBI) (Karr 1981, 1991; Miller et al. 1988), which combines measurements of species composition, abundance, and trophic relationships for different habitats. The IBI provides a quantitative comparison between the habitat in question and "unimpaired" reference systems to assess relative degrees of disturbance. The IBI bases its comparisons on a number of traits that generally characterize disturbed systems, such as an increase in number of introduced species, replacement of specialist species with generalist species, decline in the number of sensitive species, impairment of reproduction, change in age structure of populations away from older age classes, and increase in disease and anatomic anomalies. The IBI was originally developed for midwestern U.S. streams but has been applied successfully in a variety of systems (Hughes and Noss 1992).

Environmental contamination is more conventionally investigated by assaying water and sediments for known toxins, correlating growth abnormalities with sediment contaminants, or observing the responses of fishes exposed to suspect water (Heath 1987; Gassman et al. 1994).

Traditionally, the concentration at which 50% of the animals die (LD_{50}) is considered a critical threshold. Lower levels of contamination can be indicated by behavioral measures, such as elevated breathing rates, coughing, chafing against the bottom, impaired locomotion and schooling, and suppressed or hyperactivity. Although such bioassays can be performed relatively rapidly, they primarily measure immediate rather than past conditions. Measurement of "body burdens" of bioaccumulated contaminants in fish tissues gives a broader picture but can vary with season, feeding habits, or metabolic activity. A more integrated, long-term picture can be obtained by measuring alterations in energetics, metabolism, growth, reproduction, and behavior (the biomarker approach of Hellawell 1983; McCarthy and Shugart 1990; see Chap. 7, Stress; and Box 7.3). At a biochemical and energetic level, stress is indicated by changes in such attributes as liver enzyme function, occurrence of DNA damage, unusual ratios of intermediate metabolites (ADP/ATP), amounts of or ability to store lipids, and growth and developmental anomalies. Histologic markers include parasite loads and damage, tissue necrosis or abnormal growth (particularly pathologies of the gills and liver), and both elevation and suppression of immune responses. At the population level, reproductive output can be monitored, whereas species richness, the presence or absence of sensitive species, and indices such as the IBI indicate assemblage- and community-level effects. These measures are useful not only for monitoring water quality as it directly affects fishes but also because fishes are effective sentinels against human health problems (Adams 1990; McCarthy and Shugart 1990).

Commercial Exploitation

Direct exploitation of fishes by humans is an obvious cause of fish population declines. However, humans are just one of many predators on most smaller fishes, and species or populations subject to predation generally possess compensatory mechanisms for sustaining predation losses (see Chap. 23, Population Dynamics and Regulation). Predation by humans, however, has extraordinary characteristics. Most "natural" predators focus their activities on young individuals, which tend to be the most abundant cohorts within a population; on sick individuals with little reproductive potential; or on old individuals that have already reproduced. Human fisheries are at best indiscriminate (e.g., trawl and purse seine fisheries); at worst they target larger individuals that have not spawned (e.g., ocean-going salmonids). As a result, many species of fresh- and saltwater fishes are in severe decline as a direct result of fishing pressure (e.g., bluefin tuna, swordfish, many sharks, striped bass, Atlantic cod, Atlantic salmon). The U.S. National Marine Fisheries Service estimated that 40% of the marine species important to commercial and recreational fisheries in the United States were exploited at unsustainable rates (NMFS 1991a,b). Many freshwater fisheries worldwide have also experienced extreme degradation and decline as a result of a variety of human-induced insults (Table 25.3).

Overfishing.   In some well-documented examples, overfishing, often in combination with climatic change, has produced dramatic crashes in seemingly inexhaustible stocks. The clupeoid fisheries of California and South America offer an interesting, interwoven example. The history of the California sardine fishery "is a classic case of the rise and fall of a fishery dependent on a pelagic species, of overcapitalization of an industry, and of too many fishing boats using new technologies to harvest a fragile, if not dwindling, resource" (Ueber and MacCall 1990, 17).

The Pacific sardine (*Sardinops sagax caeruleus*) is a 10- to 15-cm-long, schooling, epipelagic clupeid that occurs from northern Mexico to the Bering Sea. The fish were typically captured by purse seiners and canned for human consumption. The fishery off California dates to the late 1800s. By 1925, it was the largest fishery in California, with landings of about 175,000 tons. Waste from the canning process was reduced into poultry food and fertilizer. Because the value of reduced sardine soon surpassed that of the canned product, processors started reducing whole fish. Floating reduction plants, anchored outside the 3-mile limit to bypass regulative legislation, became common. Catches climbed steadily to a maximum of 790,000 tons in 1937. For the next 10 years, catches averaged 600,000 tons per year, despite state fishery calculations that the stock could only sustain a harvest of 250,000 tons annually. Catches began a steady decline, averaging 230,000 tons from 1946 to 1952, then 55,000 tons from 1953 to 1962, and finally only 24,000 tons from 1963 to 1968. Commercial fishing for Pacific sardine ended in 1968.

The collapse of the California sardine fishery was in part responsible for the later development, overexploitation, and eventual collapse of similar fisheries in South America and Africa, as well as of the king crab fishery in Alaska. Boats, gear, and processing equipment were sold below cost, or costs were underwritten by international agencies. With the influx of former sardine boats and personnel, Alaska king crab landings rose from 26,000 pounds in 1960 to 190,000 pounds in 1980, only to crash to 35,000 pounds 2 years later, despite continued activity of the imported boats (Wooster 1990). The exact causes of the decline are debated, but a likely explanation is that overexploited breeding stocks and unfavorable climatic conditions resulted in poor recruitment of young crabs, demise of the fishery, and lost jobs for most persons associated with the industry.

A similar scenario is offered for the Peruvian fishery for anchovetas (*Engraulis ringens*). The fishery became established in the 1950s, when fish were primarily used for human consumption. After 1953, reduction plants were built and boats were added to the fleet, many from the former California sardine fishery. By 1969, Peru caught more tonnage of fish than any other nation, with anchoveta accounting for up to 98% of the catch. The exploitation of anchoveta was uncontrolled: In 1970, 12.4 million metric tons (mmt) were harvested, about 5 mmt above the calculated maximum sustainable yield. The fishery collapsed soon after, falling below 1 mmt in the mid-1970s. The collapse was again probably caused by a

Table 25.3. The status of commercial fisheries in selected river systems and lakes.

Area	Known Freshwater Fish Species (number)	Extinct (percent)	Imperiled (percent)	Principal Threats
Global	9000+	{20 combined}		
Amazon River	3000+	—	—	Habitat degradation Overharvest
Asia	1500+	—	—	Habitat degradation Competition for water Overharvest
North America	950	2	37	Habitat degradation Introduced species
Mexico (arid lands)	200	8	60	Competition for water Pollution
Europe	193	—	42	Habitat degradation Pollution
South Africa	94	—	63	Habitat degradation Pollution Competition for water
Lake Victoria	350	57	43	Introduced species
Costa Rica	127	—	9	Habitat degradation
Sri Lanka	65	—	28	
Iran	159	—	22	Habitat degradation Competition for water
Australia	188	—	35	Habitat degradation Competition for water

From Abramovitz 1996; used with permission.

combination of overfished stocks and unfavorable climatic factors, including depressed upwellings associated with the El Niño–Southern Oscillation events of 1972–1973 (Caviedes and Fik 1990).

Neither the Pacific sardine nor the Peruvian anchovy has been driven close to extinction, although the term **commercial extinction** is applied to once-abundant fishes that no longer support significant fisheries. Uncontrolled exploitation of marine species, particularly those dependent on stressed estuarine systems, can lead to even more serious declines in a species' abundance. The giant totoaba, *Totoaba macdonaldi* (Sciaenidae) is endemic to the upper Gulf of California and is the largest member of its widespread family, reaching 2 m in length and weighing over 100 kg. Its numbers have been drastically reduced as a result of overfishing on the spawning grounds, dewatering of the Colorado River estuary where it spawns, and by-catch of juveniles by shrimp boats (Fig. 25.5). At one time, it ranked as the most important commercial fish species in the Gulf of California, sought chiefly for its large gas bladder, which was dried and made into soup (the remainder of the body was often discarded). Spawning fish were so abundant that they were speared from small boats. The fishery peaked in 1942 and has declined steadily since. The totoaba was declared an endangered species in 1979 (Ono et al. 1983; Cisneros-Mata et al. 1995). The clearest message from these and similar examples is that maximizing short-term profits and ignoring biological parameters, due to greed or ignorance, have long-term, dire ecological and socioeconomic consequences (Glantz and Feingold 1990).

Overfishing creates problems besides reduced availability for human exploitation. Biological responses involving genotypic and phenotypic alteration in heavily exploited fishes are not uncommon. Overfishing can create bottlenecks in the breeding biology of a species when populations reach critically small numbers, thereby reducing the genetic diversity of the species. For example, the orange roughy (*Hoplostethus atlanticus*, Trachichthyidae) fishery in New Zealand started up in the early 1980s. Within 6 years, biomass of the stocks was reduced 70%. Electrophoretic studies indicated significant reductions in genetic diversity of the three separate stocks that were monitored (Smith et al. 1991). The danger of reduced genetic diversity is that the few remaining individuals produce offspring that possess a very limited subset of the original genetic diversity of the species. Genetic adaptation to local conditions does not guarantee tolerance of new or altered environments. Altered conditions are increasingly likely due to human-caused climatic or chemi-

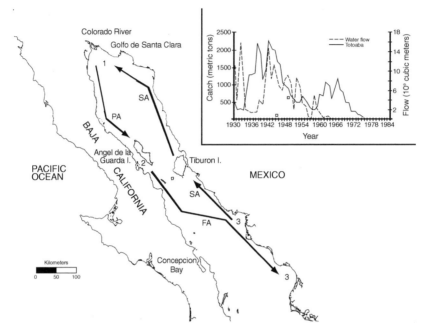

FIGURE 25.5.   *The Gulf of California and a reconstruction of the presumed seasonal migration route of the endangered totoaba, the world's largest sciaenid. SA = prespawning adults; PA = postspawning adults; FA = adults during fall migration. Numbers indicate life history zones: 1) spring spawning zone and nursery ground of juveniles; 2) summer feeding zone; and 3) autumn feeding zone. Zone 1 is now largely a biosphere reserve. Inset graph shows the relationship between water delivery from the Colorado River and totoaba population size as calculated from commercial catches.*

From Cisneros-Mata et al. 1995; used with permission of Blackwell Science.

cal changes, such as might occur from global warming or ozone depletion. Breeding bottlenecks are one step short of species extinction.

Given enough time, animals can and will adjust their life history characteristics in response to strong predation. Life history theory predicts that individuals in populations exposed to high levels of adult mortality should respond by reproducing at smaller average sizes and ages, shifting from multiple to single reproductive seasons (from iteroparity to semelparity), and having shorter lifespans. Just these kinds of changes have been observed in several exploited species, including Atlantic cod (*Gadus morhua*), haddock (*Melanogrammus aeglefinus*), other gadids, gag grouper (*Mycteroperca microlepis*), vermilion snapper (*Rhomboplites aurorubens*), Atlantic mackerel (*Scomber scombrus*), and Pacific halibut (*Hippoglossus stenolepis*). In heavily exploited chinook salmon (*Oncorhynchus tshawytscha*), average size has declined 50% and average age at maturity has declined 2 years over the past 60 years. It is unknown whether these shifts reflect 1) selection for genotypically determined differences in life history traits or 2) adjustments in the phenotype of remaining individuals. The alarming fact is that most of our marine fish stocks are overutilized and that the observed shifts in life history characteristics create fish populations that are less useful to humans (Upton 1992; O'Brien et al. 1993; Policansky 1993).

Reliance on market factors to protect declining stocks has proven illusory. A decrease in catch despite increased effort does not necessarily discourage exploitation. For example, western Atlantic stocks of bluefin tuna decreased 90% between 1972 and 1992, from 225,000 fish to 22,000 fish, and yet intensive fishing continued. Bluefin tuna meat sells in Japan for upward of U.S.$350/lb., and a single fish has brought as much as U.S.$68,000 (Williams 1992). This sort of profitability threatens other large and valuable fish species, such as sturgeons, coelacanths, swordfish, lumpfish (*Cyclopterus lumpus*, for caviar), and some tropical reef species desirable for aquaria.

By-catch.   A topic of increasing importance to the conservation of marine fishes concerns by-catch in trawl fisheries (Murray et al. 1992; Perra 1992). Few fisheries employ gear that can catch one species to the exclusion of all others. For example, dolphins, whales, turtles, and pinnipeds are frequently captured in gill nets or in purse seine nets set for tunas and billfishes. The by-catch problem is particularly acute when trawl nets with small mesh sizes are dragged along the bottom of the ocean in pursuit of groundfish or shrimp. Frequently, the incidental catch of fishes exceeds the catch of the targeted species. Although varying on a seasonal and regional basis, average fish-shrimp weight ratios of 1:1 to 3:1 have been reported for southeastern U.S. shrimp fisheries. These numbers can run as high as 130:1 (= 130 kg of "scrap" fish for each kilogram of shrimp). Overall, 105 species of finfishes are captured by shrimp trawlers in the southeastern United States. On a species basis, 5 billion Atlantic croaker (*Micropogonias undulatus*, Sciaenidae), 19 million red snapper (*Lutjanus campechanus*, Lutjanidae), and 3 million Spanish mackerel (*Scomberomorus maculatus*, Scombridae) were among nearly 10 billion individuals and 180 million kg of by-catch killed by shrimp trawlers in the Gulf of Mexico in 1989 (Nichols et al. 1990). Because of the small mesh size of the shrimp trawl nets, most of the fishes captured are 1) juveniles, 2) smaller than legal size limits, or 3) undesirable small species. In any case, these incidental captures are unmarketable and are usually shoveled back over the side of the vessel dead or dying.

The by-catch problem is complicated economically, ecologically, and sociologically. By-catch is a liability to shrimp fishermen, clogging the nets and increasing fuel costs because of increased drag on the vessel. Sorting the catch requires time, leading to spoilage of harvested shrimp and reduced time for fishing. Ecologically, high mortality rates among juvenile fishes could contribute to population declines of recreational and commercial species. Evidence to this effect exists for Gulf of Mexico red snapper and Atlantic coast weakfish (*Cynoscion regalis*, Sciaenidae). As the nearshore areas where shrimp concentrate are also important nursery grounds for many fish species, shrimp trawling could have a profound impact on stock size (e.g., Miller et al. 1990). Alternatively, by-catch is returned to the ecosystem and consumed by predators, detritivores, and decomposers, which could have a substantial positive effect on sport fish, crab, and even shrimp populations. Finally, conservation organizations decry the obvious and wanton waste associated with by-catch. Public concern over high mortality rates of endangered marine turtles captured in shrimp trawls led to the development of turtle exclusion devices (TEDs) in the 1980s. TEDs were incorporated into the shrimp net design with the purpose of directing turtles out of nets without affecting shrimp catches. Marine engineers and fishermen are developing shrimp net designs that incorporate by-catch reduction devices, although the similar sizes of shrimp and by-catch fish species make the problem technically more difficult. Other suggested solutions include prohibiting shrimping during seasons when by-catch is relatively high or where vulnerable life history stages of nontargeted species are concentrated.

Aquarium fishes. The United States imports more than 120 million tropical fishes of 300 freshwater species and 300 marine species per year, with a retail value of U.S.$300 million; worldwide, the aquarium trade is valued at more than U.S.$7 billion annually (Andrews 1990; Derr 1992). Twenty percent of the freshwater fishes are wild caught, whereas 99% of the marine fishes come directly from the sea. Marine collecting is particularly destructive because of the widespread use of poisons, such as sodium cyanide, rotenone, bleach, and quinaldine, to "anesthetize" the fishes prior to capture. Approximately 1 million kg of sodium cyanide were squirted into Philippine reefs during the 1970s and 1980s. These toxins kill targeted fishes as

well as other reef organisms. Destruction of fishes by the aquarium trade is a sordid and underpublicized fact. Mortality rates of wild-captured fishes, ignoring fishes that die during or incidental to capture, are difficult to establish and vary by locale. A figure of 66% mortality within 6 months of capture for marine tropicals has been estimated (for some species, such as coral-dependent butterflyfishes, the number is closer to 100%) (Derr 1992). A comparable estimate is unavailable for freshwater fishes, but a mortality rate of 50% between capture and arrival at the *exporter* has been estimated for South American freshwater fishes (Conroy et al. 1981). In areas of Florida, Hawaii, the Philippines, Sri Lanka, and Kenya, coveted aquarium species, especially butterflyfishes (Chaetodontidae), anemonefishes (Pomacentridae), angelfishes (Pomacanthidae), and triggerfishes (Balistidae) have become rare from overcollecting (Derr 1992).

Home aquaria are of unquestionable educational and aesthetic value. Most ichthyologists have at least one in their own home. However, these values do not justify the ecological problems created by an unregulated industry, including the detrimental effects of introduced species and diseases on native fishes, invertebrates, and plants, and the defaunation of tropical reefs and rivers. Keeping reef fishes in aquaria cannot be rationalized on the grounds of species preservation. Few reef species have been successfully bred and raised in captivity, largely because of their complex life histories and age-specific habitat and feeding requirements (see Chap. 9, Larval Feeding and Survival). Nondestructive capture methods, bag limits, restricted areas and seasons, and, most importantly, licensed captive breeding can provide a diversity of interesting fishes to meet the home aquarist's needs while protecting natural environments.

Global Climate Change

Since the industrial revolution of the late 1800s, atmospheric concentrations of greenhouse gases—mostly carbon dioxide, methane, chlorofluorocarbons, and nitrous oxide—have increased substantially as a direct result of human activity. Sunlight passes through the atmosphere and heats the planet; this heat is radiated back to space as infrared energy. The greenhouse gases act as a blanket, trapping the infrared radiation, heating the earth even further. No one questions the process; without greenhouse gases, the average temperature on earth would be about $-18°C$, or about $33°C$ colder than at present. Nor is there disagreement over the fact that greenhouse gases are increasing in the atmosphere at a rate of about 1% to 10% annually due to fossil fuel and wood burning, deforestation, cattle grazing, rice growing, and industrial pollution. What remains unknown is what effects this continued increase will have on regional and global climate. Average temperatures have increased about 0.5°C over the past century. If current trends of greenhouse gas production continue, most climate modelers predict that average temperatures will rise another 3°C over the next century, which is 10 times the rate at which the earth warmed after the last glacial advance (Ramanathan 1988; Smith 1990). Of major concern are the likely climatic effects of this temperature increase and how they will be distributed. Changes in the freeze-thaw cycle and altered wind direction and intensity have been predicted. Vagaries of ocean currents and cloud behavior will undoubtedly lead to greater warming in some regions and even cooling in others. Similarly, rainfall patterns will shift, making some regions wetter, others drier. One likely result of temperature increase will be an increase in sea level of about 0.3 to 0.7 m due to thermal expansion of oceanic water and melting of polar ice caps. The postulated consequences for fishes of such a change are potentially dramatic. Coastal wetlands, mangroves, and salt marshes, which are major nursery grounds for numerous fish species, would be flooded with seawater. Vegetation loss due to flooding has several ecological consequences. The food webs of coastal marshes depend on vegetation as both a source of and a physical trap for detritus; vegetation also provides spawning substrates, physical refugia for juvenile fishes, and substrates for prey. Marshes and their defining flora and fauna could disappear from many coastal areas (Kennedy 1990; Meier 1990).

Some global climate models indicate major shifts in ocean currents and upwelling patterns as a result of global warming. Such changes may alter or intensify the El Niño–Southern Oscillation phenomenon, which has a substantial influence on major oceanic and coastal food webs (see Chap. 17, Pelagic Fishes). Some models predict weakening of major low-latitude currents such as the Gulf Stream and Kuroshio currents, reduction in nutrient-transporting eddies of these currents, and reduced upwelling off the western coasts of South America and Africa. Shifts in ocean currents and upwellings could have far-reaching effects on the production, distribution, and year class strength of major pelagic fisheries, such as scombrids, salmonids, clupeids, and engraulids—fishes that account for nearly 70% of the world's fisheries (Bakun 1990; Francis 1990; Gucinski et al. 1990).

Variations in the frequency and severity of climatic extremes of drought, flood, and cyclonic-force storms are also predicted. Drought can have obvious, severe consequences on inland fish species, such as already imperiled desert fishes (see Chap. 17). Droughts also cause shifts in the distribution of estuarine habitats, because seawater typically intrudes farther up river basins during periods of low rainfall. Storms and attendant floods wash young fish out of appropriate habitats and can dilute high-salinity, nearshore regions with freshwater, lessening their value as nursery grounds for larvae and juveniles. Drought conditions would also exacerbate human impacts on fish habitat by reducing stream flow and increasing pollutant concentrations.

Most global climate models predict less-pronounced climatic changes at low latitudes. However, tropical animals tend to have relatively narrow climatic tolerances compared to high-latitude species and may therefore be more vulnerable to slight deviations from normal conditions (Stevens 1989). For example, abnormally long periods of average summertime temperatures and short periods of elevated temperature 1°C above normal have been linked to the phenomenon of **bleaching** in reef-building corals. Bleached corals contain fewer symbiotic algae (zooxanthellae) and less photosynthetic pigment than unbleached corals, which in turn reduces the energy stores of the corals and lessens their ability to secrete a calcium carbonate skeleton. Multiple bleaching events have been associated with coral mortality, indicating that elevated temperatures may result in coral death (Roberts 1990; Fitt et al. 1993). Given the strong dependence of coral reef fishes on living coral and the direct correlation between fish diversity and coral diversity on reefs, any reduction in coral biomass caused by elevated sea temperatures could have significant, deleterious effects on reef fishes. Of related importance to human welfare, bleached corals appear to provide an enhanced surface for the growth of dinoflagellates, including the ones responsible for ciguatera poisoning among humans that eat coral reef fishes (Kohler and Kohler 1992; see Chap. 24, The Effects of Fishes on Plants).

Significant warming trends would affect fishes on a local basis by contracting the space available for them seasonally. Higher temperatures would mean increased evaporation rates and decreased lake volumes. Fishes such as striped bass or lake trout that use deep, cool, oxygenated water as a summer refuge could be squeezed against anoxic or bottom areas as the volume of cool, oxygenated water in a lake decreased; brook trout would similarly lose access to lower portions of streams during the summer (Coutant 1990; Regier and Meisner 1990). Warm-water species could see an increase in available habitat space at northerly latitudes (Magnuson et al. 1990). Predicted elevated temperatures and drier summers at middle latitudes could reduce lake levels and increase thermal stratification and bacterial activity during the summer, which commonly results in oxygen depletion problems (McCormick 1990; Meisner et al. 1990). Warming of just a few degrees would be catastrophic for fishes already living near their critical thermal maxima (see Chap. 7, Coping with Temperatures Extremes). For example, many stream fishes of the southwestern United States die when exposed to temperatures above 38 to 40°C. Summer temperatures in rivers of this area occasionally exceed these limits, leading to heat-related deaths. A 3°C rise in temperatures could cause the extinction of 20 species of fishes endemic to this region (Matthews and Zimmerman 1990).

On a continental basis, warming would contract the geographic ranges of cold-adapted species while expanding the ranges of species that prefer warmer conditions. Warm-intolerant species near the southern edge of their range (in the Northern Hemisphere), such as races of trout and striped bass in the southeastern United States, could be eliminated from low-altitude regions, or even extirpated or driven to extinction (e.g., Coutant 1990; Meisner 1990). On an oceanic scale, major distribution patterns of many marine forms are delimited by thermal regimes and resulting current patterns (see Chap. 16). At the simplest level, any shifts in the northern or southern boundaries of defined water masses are likely to cause shifts in species distributions as well. Consequences for marine populations could be severe if, for example, prey species are less tolerant to rising temperatures than are predators, or if changes in circulation patterns move larvae too far offshore to permit migration into nearshore nursery regions.

The negative impact of climate warming will be strongest on species diversity. Cold-water species will probably be replaced by warm-water species, and warming could give some species access to high-altitude and -latitude regions that are currently too cold to inhabit. It has been estimated that 27 species currently confined to the lower Laurentian Great Lakes would gain access to Lakes Huron and Superior, which would dramatically alter assemblage relationships (Mandrak 1989). Temperature increases might also increase overall productivity, because fish production is ultimately an ectothermic process (Regier et al. 1990). But these "gains" would be offset by an overall loss of genetic and species diversity of fish faunas.

WHAT CAN BE DONE?

Biodiversity loss is a symptom of environmental deterioration on a global scale. A growing number of scientists, traditionally occupied with the descriptive and experimental pursuit of knowledge, have turned their efforts to environmental issues. These efforts are of necessity multidisciplinary, requiring integration among and knowledge from the biological and physical sciences, as well as from sociology, anthropology, and economics. Such efforts fall under the names of conservation biology, environmental science, restoration ecology, and environmental ethics, among others. Regardless, it is apparent to all concerned that the major task of conservation efforts is to reverse previous and minimize future human impacts on natural systems. These efforts fall into a number of categories beyond the crucial need for surveying, documenting, and monitoring problem areas. Four of the most commonly suggested solutions involve preservation of landscapes, development of preserves, restoration of degraded habitat, and captive breeding of endangered species. All efforts are doomed if human population growth and resource consumption cannot be checked. The impact that humans have on the earth's resources is the combined effects of large numbers of people consuming at unsustainable rates. In developed nations, achieving sustainable consumption rates will require an environmentally educated and concerned public willing to reduce its standard of living in order to maintain a high quality of life (Pister 1992).

The Level at Which Efforts Must Be Addressed

Little doubt exists as to the declining state of global fish stocks. At the four levels of biodiversity—genetic, species, communities, and landscapes—we are witnessing alarming reductions. Historically, attention and concern have been focused on threats to individual species or populations. The U.S. Endangered Species Act of 1973 focused on identifying and protecting species at risk. While that legislation was innovative and far-reaching, it is generally agreed among biologists that emphasizing species rather than habitats is at best a partial solution to biodiversity loss. Endangered species problems are re-

ally endangered habitat problems; captive-breeding programs are futile if insufficient natural habitat exists into which the species can be reintroduced. The majority of extinctions result from habitat destruction (see above). Often more than one rare species is affected by loss of a particular habitat, and organisms live in coevolved, interacting communities, the elements of which are necessary for the welfare of most of the species in question. Increasing emphasis is now being given to ecosystem and landscape preservation, which is logistically, economically, and politically more difficult to attain than species preservation.

Biological Preserves

One way to protect habitats is through the creation of biological preserves (Moyle and Leidy 1992). Few biological preserves exist today that are targeted directly at freshwater organisms (pupfishes in the Great Basin–Death Valley area of the southeastern United States are an exception; see Miller and Pister 1971). Most preserves are created as terrestrial parks that include lakes and portions of streams and rivers. Unfortunately, human activities upstream of such parks can threaten the aquatic biota in the park, and seasonal migrations by many fishes carry them beyond the protection of park and even international boundaries. Marine parks can help fishes directly, but protective laws in many such parks are logistically difficult to enforce. Multiple-use objectives often compromise the refuge quality of marine preserves. Protection is often limited; for example, spearfishing and nets may be prohibited, but not hook-and-line fishing. Again, many species move out of park waters as a normal part of their life histories, which then subjects them to commercial and recreational exploitation. Dependence on dispersed larvae for recruitment means that upstream habitats must also be protected or populations in an area may decline. Larger parks, such as much of the Great Barrier Reef in Australia and the Florida Keys National Marine Sanctuary in Florida, encompass more species and life history stages but are increasingly difficult to police. Where such preserves are established, marine life and particularly fish populations rebound dramatically (e.g., Hanauma Bay in Hawaii, Dry Tortugas in Florida). In all cases, whether the area is freshwater, estuarine, or marine, the costs of acquisition are high, and opposition to "loss" of the area to exploitation is often strong.

Restoration

Although costs are a major factor in restoration of degraded habitat, dramatic results can be achieved if restoration becomes a high priority. Reversals and partial restorations of badly polluted and degraded systems have been achieved in Puget Sound in Washington State, Kaneohe Bay in Hawaii, the upper Illinois River in Illinois, the Willamette River in Oregon, the Mattole River in northern California, and the Merrimack River in New Hampshire and Massachusetts (National Research Council 1992). Success in restoring depleted fish populations can result from such actions, as is apparently the case for lake trout (*Salvelinus namaycush*) in the Laurentian Great Lakes. However, few if any ecosystems are ever restored to their original conditions; improvement from a state of extreme degradation is often the most that can be achieved. The complexity of natural, evolved systems operates against human-induced solutions, which often strive for simplicity. In truth, restoration efforts have often demonstrated that "corrective measures to restore ecosystem function [are] obtained only at very high costs, that some attributes can be maintained only with continuous management, and that certain losses in the ecosystem [are] irreversible" (Schelske and Carpenter 1992, 383).

Captive Breeding

Captive propagation of endangered species, with eventual release into the wild, is an oft-cited solution to the extinction problem (e.g., Ribbink 1987; Johnson and Jensen 1991; but see Bruton 1990). Such efforts can be costly and time-consuming, small numbers of breeding individuals lead to genetic bottlenecks and loss of diversity, and the breeding and rearing requirements of most species are poorly known or difficult to replicate outside of the natural habitat. Conservation genetics is a relatively new discipline that focuses on the myriad problems associated with trying to resurrect endangered species from small numbers of breeding individuals (e.g., Meffe 1986). In some instances, captive propagation may be the only hope for a species.

Education and Population Control

Preservation of aquatic habitats in which fishes live requires an educated public that understands the sensitivity of, degree of human dependence on, and enormity of human impact on aquatic environments. Major changes in our attitudes—as citizens, scientists, and managers—toward how we value fishes, aquatic environments, and natural resources will have to take place (Callicott 1991; Pister 1992). Ultimately, however, the future of aquatic and terrestrial life on earth will be determined by whether the human species continues to multiply and consume at current rates. National and international programs to control human population growth are crucial to reversing global and local environmental deterioration (Becker 1992). Any refusal to take remedial action in the form of human population control only forestalls an inevitable Malthusian scenario. Predator populations cycle in step with their prey. Humans, with alternative prey resources, have managed to increase the time lag separating the cycles. But the laws and principles of natural selection and of population dynamics will eventually be upheld, and order restored.

SUMMARY

1. Extinction rates have increased dramatically in the past 50 years due to human activities; present rates are 1000 times greater than average and 10 to 100 times greater than during past periods of mass extinction. About 20% of the world's 9000 species of freshwater fishes are either extinct or nearly so; 40 fishes have gone extinct in North America in the past century, and the rate is accelerating. Marine fishes are less threatened because of their wider distributions, although many commercially important species are showing serious declines.

2. Major causes of biodiversity loss are habitat loss and modification, species introductions, pollution, commercial exploitation, and global climate change. Habitat loss occurs through modification of bottom type, as happens because of dredging, log removal, coral or gravel mining, and silt deposition due to deforestation of the surrounding watershed. Other causes of habitat loss include channelization of streams and rivers, dam building, and water withdrawal.

3. Introduced species affect native species because introduced fishes are often freed from their evolved population controls, and natives are evolutionarily unprepared for the introductions. Predation by and competition and hybridization with introduced species are common results, as is the introduction of new pathogens. The introduction of Nile perch into Lake Victoria has led to the possible extinction of hundreds of species of endemic cichlids that previously supported an important local fishery.

4. Chemical, nutrient, and sediment pollution all have adverse effects on fishes; predation on fishes by birds and mammals links aquatic and terrestrial ecosystems via such pollution. Fishes can therefore serve as valuable indicators of environmental health.

5. Approximately 40% of the commercial marine fish species important to the United States are exploited at unsustainable rates. The Pacific sardine, Peruvian anchovetta, and giant totoaba were once very abundant commercial species that no longer sustain fisheries largely due to overfishing. Some species reductions are the indirect result of other fisheries. Bycatch in the shrimp fisheries of the Gulf of Mexico greatly reduces the available stocks of red snapper and Spanish mackerel, among other species. Coral reef fishes are commercially exploited for the home aquarium trade, which has led to reef destruction and species depletion in many places. Few such fishes live more than a few months in captivity.

6. Greenhouse gases have been pumped into the atmosphere at increasing rates during the past century, raising the prospect of global warming, sea level rises, ocean current shifts, and major climatic changes such as drought, floods, and cyclonic storms. Ocean warming could destroy coral reefs and the huge diversity of fishes living there. Major shifts in freshwater fish distribution and diversity would also occur.

7. Biodiversity loss is a symptom of environmental deterioration on a global scale. Solutions to environ-

mental problems include ecosystem and landscape preservation, development of reserves, habitat restoration, and captive breeding of endangered species. None of these efforts will be successful if human population growth and overconsumption are not curtailed.

SUPPLEMENTAL READINGS

American Fisheries Society 1990a, b; Courtenay and Stauffer 1984; Fiedler and Jain 1992; Glantz and Feingold 1990; Hargrove 1989; Heath 1987; Hellawell 1983; IUCN (International Union for the Conservation of Nature) 1988; McCarthy and Shugart 1990; Minckley and Deacon 1991; National Research Council (U.S.) 1992; Ono et al. 1983; Smith and Tirpak 1989; SPA (Science and Policy Associates, Inc.) 1990; Wheeler and Sutcliffe 1990.

References

Abdel Magid, A. M. 1966. Breathing and function of the spiracles in *Polypterus senegalus*. *Anim Behav* 14:530–533.

Abdel Magid, A. M. 1967. Respiration of air by the primitive fish *Polypterus senegalus*. *Nature* (London) 215:1096–1097.

Abrahams, M. 1989. Foraging guppies and the Ideal Free Distribution: The influence of information on patch choice. *Ethology* 82:116–126.

Abrahams, M. V., and P. Colgan. 1985. Risk of predation, hydrodynamic efficiency and their influence on school structure. *Env Biol Fish* 13:195–202.

Abrahams, M., and L. M. Dill. 1989. A determination of the energetic equivalence of the risk of predation. *Ecology* 70:999–1007.

Abu-Gideiri, Y. B. 1966. The behavior and neuroanatomy of some developing teleost fishes. *J Zool* 149:215–241.

Adams S. M., ed. 1990a. Biological indicators of stress in fish. Bethesda, MD: *Amer Fish Soc Symp* 8.

Adams, S. M. 1990b. Status and use of biological indicators for evaluating the effects of stress on fish. Bethesda, MD: *Amer Fish Soc Symp* 8:1–8.

Adkins-Regan, E. Hormones and sexual differentiation. In *Hormones and reproduction in fishes, amphibians and reptiles,* ed. D. O. Norris and R. E. Jones, 1–29. New York: Plenum Press.

Agami, M., and Y. Waisel. 1988. The role of fish in distribution and germination of seeds of the submerged macrophytes *Najas marina* L. and *Ruppia maritima* L. *Oecologia* 76:83–88.

Ahlstrom, E. H., K. Amaoka, D. A. Hensley, H. G. Moser, and B. Y. Sumida. 1984. Pleuronectiformes: Development. In *Ontogeny and systematics of fishes,* ed. H. G. Moser, W. J. Richards, D. M. Cohen, M. P. Fahay, A. W. Kendall Jr., and S. L. Richardson, 640–670. Lawrence, KS: Amer Soc Ichthyol Herpetol, Spec Publ 1.

Ahlstrom, E. H., and O. P. Ball. 1954. Description of eggs and larvae of jack mackerel (*Trachurus symmetricus*) and distribution and abundance of larvae in 1950 and 1951. *Fish Bull US* 56:295–329.

Ainley, D. G., C. S. Strong, H. R. Huber, T. J. Lewis, and S. H. Morrell. 1981. Predation by sharks on pinnipeds at the Farallon Islands. *Fish Bull US* 78:941–945.

Akihito, P. 1986. Some morphological characters considered to be important in gobiid phylogeny. In *Indo-Pacific fish biology: Proc 2nd Internat Conf Indo-Pacif Fish,* ed. T. Uyeno, R. Arai, T. Taniuchi, and K. Matsuura, 629–639. Tokyo: Ichthyol Soc Japan.

Alcock, J. 1989. *Animal behavior.* 4th ed. Sunderland, MA: Sinauer.

Aldridge, R. J., and D. E. G. Briggs. 1989. A soft body of evidence. *Natur Hist* (May):6–11.

Alexander, G. R. 1979. Predators of fish in coldwater streams. In *Predator-prey systems in fisheries management,* ed. H. Clepper, 153–170. Washington, DC: Sport Fishing Institute.

Alexander, R. McN. 1983. *Animal mechanics.* 2d ed. Oxford: Blackwell Science.

Alexander, R. M. 1993. Buoyancy. In *The physiology of fishes,* ed. D. H. Evans, 75–97. Boca Raton, FL: CRC Press.

Ali, M. A., ed. 1980. *Environmental physiology of fishes.* New York: Plenum Press.

Ali, M. A., ed. 1992. *Rhythms in fishes.* New York: Plenum Press.

Ali, M. A., and H.-J. Wagner. 1975. Distribution and development of retinomotor responses. In *Vision in fishes,* ed. M. A. Ali, 369–396. New York: Plenum Press.

Allan, J. D. 1982. The effects of reduction in trout density on the invertebrate community of a mountain stream. *Ecology* 63:1444–1455.

Allan, J. D. 1983. Predator-prey relationships in streams. In *Stream ecology: Application and testing of general ecological theory,* ed. J. R. Barnes and G. W. Minshall, 191–229. New York: Plenum Press.

Allan, J. R. 1986. The influence of species composition on behaviour in mixed-species cyprinid shoals. *J Fish Biol* 29(A):97–106.

Allen, G. R. 1975. *Anemonefishes.* 2d ed. Neptune City, NJ: TFH Publ.

Allen, G. R. 1991. *Damselfishes of the world.* Melle, Germany: Mergus.

Allen, L. G. 1979. Larval development of *Gobiesox rhessodon* (Gobiesocidae) with notes on the larva of *Rimicola muscarum. Fish Bull US* 77:300–304.

Allen, S. K. Jr., and R. J. Wattendorf. 1988. Triploid grass carp: Status and management implications. *Fisheries* 12(4):20–24.

Allendorf, F. W., and M. M. Ferguson. 1990. Genetics. In *Methods for fish biology,* ed. C. B. Schreck and P. B. Moyle, 35–63. Bethesda, MD: American Fisheries Society.

Allendorf, F. W., et al. 1987. Genetics and fishery management. In *Population genetics and fishery management,* ed. N. Ryman and F. Utter, 1–19. Seattle: Washington Sea Grant Program, University of Washington Press.

Ambrose, R. F., and T. W. Anderson. 1990. Influence of an artificial reef on the surrounding infaunal community. *Mar Biol* 107:41–52.

American Fisheries Society. 1984. Rhythmicity in fishes. *Trans Amer Fish Soc* 113:411–552.

American Fisheries Society. 1990a. Effects of global climate change on our fisheries resources. *Fisheries* 15(6):2–44.

American Fisheries Society. 1990b. Symposium on effects of climate change on fish. *Trans Amer Fish Soc* 119:173–389.

Anderson, D. P. 1990. Immunological indicators: Effects of environmental stress on immune protection and disease outbreaks. *Amer Fish Soc Symp* 8:38–50.

Anderson, E. D. 1990. Fisheries models as applied to elasmobranch fisheries. In *Elasmobranchs as living resources: Advances in the biology, ecology, systematics, and the status of fisheries,* ed. H. L. Pratt Jr., S. H. Gruber, and T. Taniuchi, 473–484. Washington, DC: NOAA Tech Rept 90.

Anderson, R. O., and S. J. Gutreuter. 1984. Length, weight, and associated structural indices. In *Fisheries techniques,* ed. L. A. Nielsen and D. L. Johnson. Bethesda, MD: American Fisheries Society.

Andrews, C. 1990. The ornamental fish trade and fish conservation. *J Fish Biol* 37 (suppl A):53–59.

Andrews, S. M. 1973. Interrelationships of crossopterygians. In *Interrelationships of fishes,* ed. P. H. Greenwood, R. S. Miles, and C. Patterson. *Zool J Linn Soc* 53 (suppl 1):137–177.

Angermeier, P. L., and J. R Karr. 1983. Fish communities along environmental gradients in a system of tropical streams. *Env Biol Fish* 9:117–135.

Arkoosh, M. R., J. E. Stein, and E. Casillas. 1994. Immunotoxicology of an anadromous fish: Field and laboratory studies of B-cell mediated immunity. In *Modulators of fish immune responses*. Vol. 1, *Models for environmental toxicology, biomarkers, immunostimulators*, ed. J. S. Stolen and T. C. Fletcher, 33–48. Fair Haven, NJ: SOS Publications.

Aronson, R. B. 1983. Foraging behavior of the west Atlantic trumpetfish, *Aulostomus maculatus*: Use of large, herbivorous reef fishes as camouflage. *Bull Mar Sci* 33:166–171.

Atema, J., et al., eds. 1988. *Sensory biology of aquatic animals*. New York: Springer-Verlag.

Avise, J. C. 1994. *Molecular markers, natural history and evolution*. New York: Chapman and Hall.

Avise, J. C., G. S. Helfman, N. C. Saunders, and L. S. Hales. 1986. Mitochondrial DNA differentiation in North Atlantic eels: Population genetic consequences of an unusual life history pattern. *Proc Natl Acad Sci USA* 83:4350–4354.

Avise, J. C., W. S. Nelson, J. Arnold, R. K. Koehn, G. C. Williams, and V. Thorsteinsson. 1990. The evolutionary genetic status of Icelandic eels. *Evolution* 44:1254–1262.

Avise, J. C., and R. K. Selander. 1972. Evolutionary genetics of cave-dwelling fishes of the genus *Astyanax*. *Evolution* 26:1–19.

Axelrod, H. R., and W. E. Burgess. 1976. *African cichlids of Lakes Malawi and Tanganyika*. 5th ed. Neptune City, NJ: TFH Publ.

Bachman, R. A. 1984. Foraging behavior of free-ranging wild and hatchery brown trout in a stream. *Trans Amer Fish Soc* 113:1–32.

Baggerman, B. 1985. The role of biological rhythms in the photoperiodic regulation of seasonal breeding in the stickleback *Gasterosteus aculeatus*, L. *Neth J Zool* 35:14–31.

Baggerman, B. Sticklebacks. 1990. In *Reproductive seasonality in teleosts: Environmental influences*, ed. A. D. Munro, A. P. Scott, and T. J. Lam, 79–107. Boca Raton, FL: CRC Press.

Bailey, K. M., and E. D. Houde. 1989. Predation on eggs and larvae of marine fishes and the recruitment problem. *Adv Mar Biol* 25:1–83.

Bailey, T. G., and D. R. Robertson. 1982. Organic and caloric levels of fish feces relative to its consumption by coprophagous reef fishes. *Mar Biol* 69:45–50.

Baird, T. A. 1988. Female and male territoriality and mating system of the sand tilefish, *Malacanthus plumieri*. *Env Biol Fish* 22:101–116.

Baker, J. R., and C. L. Schofield. 1985. Acidification impacts on fish populations: A review. In *Acid deposition: Environmental, economic and policy issues*, ed. D. D. Adams and W. P. Page, 183–221. New York: Plenum Press.

Baker, R. R. 1978. *The evolutionary ecology of animal migration*. New York: Holmes and Meier.

Bakun, A. 1990. Global climate change and intensification of coastal ocean upwelling. *Science* 247:198–201.

Baldridge, H. D. 1970. Sinking factors and average densities of Florida sharks as functions of liver buoyancy. *Copeia* 1970:744–754.

Baldridge, H. D. 1972. Accumulation and function of liver oil in Florida sharks. *Copeia* 1972:306–325.

Ballantyne, J. S., M. E. Chamberlin, and T. D. Singer. 1992. Oxidative metabolism in thermogenic tissues of swordfish and mako shark. *J Exp Zool* 261:110–114.

Ballintijn, C. M., D. D. Beatty, and R. L. Saunders. 1977. Effects of pseudobranchectomy on visual pigment density and ocular PO_2 in Atlantic salmon, *Salmo salar*. *J Fish Res Board Can* 34:2185–2192.

Balon, E. K. 1975a. Reproductive guilds of fishes: A proposal and definition. *J Fish Res Board Can* 32:821–864.

Balon, E. K. 1975b. Terminology of intervals in fish development. *J Fish Res Board Can* 32:1663–1670.

Balon, E. K. 1980. Early ontogeny of the lake charr, *Salvelinus (Cristivomer) namaycush*. In *Charrs: Salmonid fishes of the genus Salvelinus. Perspectives in vertebrate science*, Vol. 1, ed. E. K. Balon, 485–562. The Hague: Dr. W. Junk.

Balon, E. K. 1981. Additions and amendments to the classification of reproductive styles in fishes. *Env Biol Fish* 6:377–389.

Balon, E. K. 1984. Reflections on some decisive events in the early life of fishes. *Trans Amer Fish Soc* 113:178–185.

Balon, E. K., and M. N. Bruton. 1986. Introduction of alien species or why scientific advice is not heeded. *Env Biol Fish* 16:225–230.

Balon, E. K., M. N. Bruton, and H. Fricke. 1988. A fiftieth anniversary reflection on the living coelacanth, *Latimeria chalumnae*: Some new interpretations of its natural history and conservation status. *Env Biol Fish* 23:241–280.

Balon, E. K., M. N. Bruton, and D. L. G. Noakes. 1994. Women in ichthyology: An anthology in honor of ET, Ro and Genie. *Env Biol Fish* 41:7–438.

Balon, E. K., and D. J. Stewart. 1983. Fish assemblages in a river with unusual gradient (Luongo, Africa—Zaire system), reflections on river zonation, and description of another new species. *Env Biol Fish* 9:225–252.

Barbour, C. D. 1973. A biogeographical history of *Chirostoma* (Pisces: Atherinidae): A species flock from the Mexican Plateau. *Copeia* 1973:533–556.

Bardack, D., and R. Zangerl. 1971. Lampreys in the fossil record. In *The biology of lampreys*, ed. M. W. Hardisty and I. C. Potter, 67–84. New York: Academic Press.

Bardack, D. 1991. First fossil hagfish (*Myxinoidea*): A record from the Pennsylvanian of Illinois. *Science* 254:701–703.

Barlow, G. W. 1961. Causes and significance of morphological variation in fishes. *System Zool* 10:105–117.

Barlow, G. W. 1963. Ethology of the Asian teleost, *Badis badis*: I. Motivation and signal value of the colour patterns. *Anim Behav* 11:97–105.

Barlow, G. W. 1967. The functional significance of the split-head color pattern as exemplified in a leaf fish, *Polycentrus schomburgkii*. *Ichthyologica* 39:57–70.

Barlow, G. W. 1972. The attitude of fish eye-lines in relation to body shape and to stripes and bars. *Copeia* 1972:4–12.

Barlow, G. W. 1981. Patterns of parental investment, dispersal and size among coral reef fishes. *Env Biol Fish* 6:65–85.

Barlow, G. W. 1984. Patterns of monogamy among teleost fishes. *Arch Fischereiwiss* 35:75–123.

Barlow, G. W. 1986. A comparison of monogamy among freshwater and coral-reef fishes. In *Indo-Pacific fish biology: Proc 2nd Internat Conf Indo-Pacific Fish*, ed. T. Uyeno, R. Arai, T. Taniuchi, and K. Matsuura, 767–775. Tokyo: Ichthyol Soc Japan.

Barlow, G. W. 1991. Mating systems among cichlid fishes. In *Cichlid fishes: Behaviour, ecology and evolution*, ed. M. H. A. Keenleyside, 173–190. London: Chapman and Hall.

Barlow, G. W. 1992. Is mating different in monogamous species? The midas cichlid fish as a case study. *Amer Zool* 32:91–99.

Barret, J. C. 1989. The effects of competition and resource availability on the behavior, microhabitat use, and diet of the mottled sculpin (*Cottus bairdi*). Ph.D. dissertation, University of Georgia, Athens, GA.

Barrington, E. W. 1957. The alimentary canal and digestion. *Physiology of Fishes* 1:109–161.

Barros, N. B., and A. A. Myrberg Jr. 1987. Prey detection by means of passive listening in bottlenose dolphins (*Tursiops truncatus*). *J Acoust Soc Amer* 82 (suppl 1):S65.

Basolo, A. L. 1990a. Female preference for male sword length in the green swordtail, *Xiphophorus helleri* (Pisces: Poeciliidae). *Anim Behav* 40:332–338.

Basolo, A. L. 1990b. Female preference predates the evolution of the sword in swordtail fish. *Science* 250:808–810.

Bass, A. H. 1996. Shaping brain sexuality. *Amer Scient* 84:352–363.

Baylis, J. R. 1981. The evolution of parental care in fishes, with reference to Darwin's rule of male sexual selection. *Env Biol Fish* 6:223–251.

Beamish, F. W. H. 1970. Oxygen consumption of largemouth bass, *Micropterus salmoides*, in relation to swimming speed and temperature. *Can J Zool* 48:1221–1228.

Beamish, F. W. H. 1978. Swimming capacity. In *Fish physiology*. Vol. 7, *Locomotion*, ed. W. S. Hoar and D. J. Randall. New York: Academic Press.

Beamish, R. J., and G. A. McFarlane. 1983. The forgotten requirement of validation of estimates of age determination. *Trans Amer Fish Soc* 112:735–743.

Beamish, R. J., and G. A. McFarlane. 1987. Current trends in age determination methodology. In *The age and growth of fish*, ed. R. C. Summerfelt and G. E. Hall, 15–42. Ames: Iowa State University Press.

Beamish, R. J., and C.-E. M. Neville. 1992. The importance of size as an isolating mechanism in lampreys. *Copeia* 1992:191–196.

Beamish, R. J., and J. H. Youson. 1987. Life history and abundance of young adult *Lampetra ayresi* in the Fraser River and their possible impact on salmon and herring stocks in the Strait of Georgia. *Can J Fish Aquat Sci* 44:525–537.

Bechler, D. L. 1983. The evolution of agonistic behavior in amblyopsid fishes. *Behav Ecol Sociobiol* 12:35–42.

Becker, C. D. 1992. Population growth versus fisheries resources. *Fisheries* 17(5):4–5.

Becker, G. C. 1983. *Fishes of Wisconsin*. Madison: University of Wisconsin Press.

Beecher, H. A., E. R. Dott, and R. F. Fernau. 1988. Fish species richness and stream order in Washington State streams. *Env Biol Fish* 22:193–209.

Beets, J. 1989. Experimental evaluation of fish recruitment to combinations of fish aggregating devices and benthic artificial reefs. *Bull Mar Sci* 44:973–983.

Beets, J. P. 1991. Aspects of recruitment and assemblage structure of Caribbean coral reef fishes. Ph.D. dissertation, University of Georgia, Athens, GA.

Bell, J. D., and R. Galzin. 1984. Influence of live coral cover on coral-reef fish communities. *Mar Ecol Prog Ser* 15:265–274.

Bell, M. A., and S. A. Foster, eds. 1994. *The evolutionary biology of the threespine stickleback*. Oxford: Oxford University Press.

Bell, M. A., and K. E. Richkind. 1981. Clinal variation of lateral plates in threespine stickleback fish. *Amer Natur* 117:113–132.

Bellwood, D. R. 1994. A phylogenetic study of the parrotfishes, family Scaridae (Pisces: Labroidei), with a revision of genera. *Rec Austral Mus Suppl* 20:1–86.

Bemis, W. E. 1987. Feeding systems of living Dipnoi: Anatomy and function. In *The biology and evolution of lungfishes*, ed. W. E. Bemis, W. W. Burggren, and N. E. Kemp, 249–275. New York: Alan R. Liss.

Bemis, W. E., W. W. Burggren, and N. E. Kemp, eds. 1987. *The biology and evolution of lungfishes*. New York: Alan R. Liss.

Bemis, W. E., and T. E. Hetherington. 1982. The rostral organ of *Latimeria chalumnae*: Morphological evidence of an electroreceptive function. *Copeia* 1982:467–471.

Benchley, P. 1974. *Jaws*. New York: Doubleday.

Benke, A. C. 1976. Dragonfly production and prey turnover. *Ecology* 57:915–927.

Benke, A. C. 1993. Concepts and patterns of invertebrate production in running waters. *Verh Internat Verein Limnol* 25:15–38.

Benke, A. C., R. L. Henry III, D. M. Gillespie, and R. J. Hunter. 1985. Importance of snag habitat for animal production in southeastern streams. *Fisheries* 10(5):8–13.

Benton, M. J., ed. 1993. *The fossil record 2*. London: Chapman and Hall.

Berg, L. S. 1947. *Classification of fishes, both recent and fossil*. Ann Arbor, MI: J. W. Edwards Brothers.

Berger, J. J. 1992. The Kissimmee riverine-floodplain system. In *Restoration of aquatic ecosystems: Science, technology, and public policy*, National Research Council (US), 477–496. Washington, DC: National Academy Press.

Berglund, A., G. Rosenqvist, and I. Svensson. 1986. Mate choice, fecundity and sexual dimorphism in two pipefish species (Syngnathidae). *Behav Ecol Sociobiol* 19:301–307.

Bernstein, J. J. 1970. Anatomy and physiology of the central nervous system. *Fish Physiol* 4:1–90.

Berra, T. M. 1981. *An atlas of distribution of the freshwater fish families of the world*. Lincoln, NE: University of Nebraska Press.

Berra, T. M., and G. R. Allen. 1989. Burrowing, emergence, behavior, and functional morphology of the Australian salamanderfish, *Lepidogalaxias salamandroides*. *Fisheries* 14(5):2–10.

Berra, T. M., and R. M. Berra. 1977. A temporal and geographical analysis of new teleost names proposed at 25 year intervals from 1869–1970. *Copeia* 1977:640–647.

Berra, T. M., L. E. L. M. Crowley, W. Ivantsoff, and P. A. Fuerst. 1996. *Galaxias maculatus*: An explanation of its biogeography. *Mar Freshwater Res* 47:845–849.

Berra, T. M., and J. B. Hutchins. 1990. A specimen of megamouth shark, *Megachasma pelagios* (Megachasmidae) from Western Australia. *Rec West Aust Mus* 14:651–656.

Berra, T. M., D. M. Sever, and G. R. Allen. 1989. Gross and histological morphology of the swimbladder and lack of accessory respiratory structures in *Lepidogalaxias salamandroides*, an aestivating fish from western Australia. *Copeia* 1989:850–856.

Bertelsen, E. 1951. The ceratioid fishes. Ontogeny, taxonomy, distribution and biology. *Dana Rep* 39:1–276.

Bertelsen, E., and J. G. Nielsen. 1987. The deep sea eel family Monognathidae (Pisces, Anguilliformes). *Steenstrupia* 13:141–198.

Beverton, R. J. 1987. Longevity in fish: Some ecological and evolutionary considerations. In *Evolution of senescence: A comparative approach*, ed. A. D. Woodhead and K. H. Thompson, 145–160. New York: Plenum Press.

Beverton, R. J., and S. J. Holt. A review of the lifespans and mortality rates of fish in nature and their relation to growth and other physical characteristics. In *The lifespan of animals*. *CIBA Fndn Colloquium on Aging* 5:142–180.

Bigelow, H. B. 1963. Family Coregonidae. In *Fishes of the western North Atlantic*. Part 3, *Soft-rayed bony fishes*, 547–597. New Haven, CT: Mem Sears Fndn Mar Res, No. 1, Part 1.

Bigelow, H. B., and I. P. Farfante. 1948a. Lancelets. In *Fishes of the western North Atlantic*, ed. J. Tee-Van, C. M. Breder, S. F. Hildebrand, A. E. Parr, and W. C. Schroeder, 1–28. New Haven, CT: Mem Sears Fndn Mar Res, No. 1, Part 1.

Bigelow, H. B., and W. C. Schroeder. 1953a. Fishes of the Gulf of Maine. Fishery Bulletin of the Fish and Wildlife Service, Vol. 53, Fish Bull 74.

Bigelow, H. B., and W. C. Schroeder. 1953b. Chimaeroids. In *Fishes of the western North Atlantic*, ed. H. B. Bigelow and W. C. Schroeder, 515–562. New Haven, CT: Mem Sears Fndn Mar Res, No. 1, Part 1.

Binkowski, F. P., and S. I. Doroshov, eds. 1985. *North American sturgeons: Biology and aquaculture potential. Developments in environmental biology of fishes*. Dordrecht: Dr. W. Junk.

Birstein, V., and W. Bemis. 1995. Will the Chinese paddlefish survive? *Sturgeon Quart* 3(2):12.

Blaber, S. J. M., D. T. Brewer, and J. P. Salini. 1994. Diet and dentition in tropical ariid catfishes from Australia. *Env Biol Fish* 40:159–174.

Black-Cleworth, P. 1970. The role of electrical discharges in the non-reproductive social behavior of *Gymnotus carapo* (Gymnotidae, Pisces). *Anim Behav Monogr* 3(1):1–77.

Blaxter, J. H. S. 1969. Development: Eggs and larvae. In *Fish physiology*, Vol. 3, ed. W. S. Hoar and D. J. Randall, 177–252. New York: Academic Press.

Blaxter, J. H. S., ed. 1974. *The early life history of fish*. New York: Springer-Verlag.

Blaxter, J. H. S. 1975. The eyes of larval fish. In *Vision in fishes*, ed. M. A. Ali, 427–444. New York: Plenum Press.

Blaxter, J. H. S. 1984. Ontogeny, systematics and fisheries. In H. G. Moser, W. J. Richards, D. M. Cohen, M. P. Fahay, A. W. Kendall Jr., and S. L. Richardson, ed. *Ontogeny and systematics of fishes*. Amer Soc Ichthyol Herpetol Spec Publ 1:1–6.

Blaxter, J. H. S., and L. A. Fuiman. 1990. The role of the sensory systems of herring larvae in evading predatory fish. *J Mar Biol Ass UK* l70:413–427.

Blaxter, J. H. S., and G. Hempel. 1963. The influence of egg size on herring larvae. *J Cons Int Explor Mer* 28:211–240.

Blaxter, J. H. S., and J. R. Hunter. 1982. The biology of clupeoid fishes. *Adv Mar Biol* 20:1–223.

Blazer, V. S. 1991. Piscine macrophage function and nutritional influences: A review. *J Aquat Anim Health* 3:77–86.

Blazer, V. S., D. E. Facey, J. W. Fournie, L. A. Courtney, and J. K. Summers. 1994. Macrophage aggregates as indicators of environmental stress. In *Modulators of fish immune responses*. Vol. 1, *Models for environmental toxicology, biomarkers, immunostimulators*, ed. J. S. Stolen and T. C. Fletcher, 169–185. Fair Haven, NJ: SOS Publications.

Blazer, V. S., R. E. Wolke, J. Brown, and C. A. Powell. 1987. Macrophage aggregates in largemouth bass: Effects of age, season, relative weight and site quality. *Aquat Toxicol* 10:199–215.

Bleckmann, H. 1993. Role of the lateral line in fish behaviour. In *The behaviour of teleost fishes*, 2d ed., ed. T. J. Pitcher, 201–246. London: Chapman and Hall.

Block, B. A. 1991. Evolutionary novelties: How fish have built a heater out of muscle. *Amer Zool* 31:726–742.

Block, B. A., and J. R. Finnerty. 1994. Endothermy in fishes: A phylogenetic analysis of constraints, predispositions, and selection pressures. *Environ Biol Fish* 40:283–302.

Block, B. A., J. R. Finnerty, A. F. R. Stewart, and J. Kidd. 1993. Evolution of endothermy in fish: Mapping physiological traits on a molecular phylogeny. *Science* 260:210–214.

Blumer, L. S. 1979. Male parental care in the bony fishes. *Quart Rev Biol* 54:149–161.

Blumer, L. S. 1982. A bibliography and categorization of bony fishes exhibiting parental care. *Zool J Linn Soc* 76:1–22.

Bodznick, D., and R. L. Boord. 1986. Electroreception in Chondrichthyes. In *Electroreception*, ed. T. H. Bullock. New York: John Wiley & Sons.

Boehlert, G. W. 1984. Scanning electron microscopy. In *Ontogeny and systematics of fishes*, ed. H. G. Moser, W. J. Richards, D. M. Cohen, M. P. Fahay, A. W. Kendall Jr., and S. L. Richardson, 43–48. Amer Soc Ichthyol Herpetol, Spec Publ 1.

Boehlert, G. W., and B. C. Mundy. 1988. Roles of behavioral and physical factors in larval and juvenile fish recruitment to estuarine nursery areas. *Amer Fish Soc Symp* 3:51–67.

Boehlert, G. W., and J. Yamada. 1991. Rockfishes of the genus *Sebastes*: Their reproduction and early life history. *Env Biol Fish* 30:7–280.

Bohlke, E. B., ed. 1989a. *Fishes of the western North Atlantic*. Part 9. Vol. 1, *Orders Anguilliformes and Saccopharyngiformes*, 1–655. New Haven, CT: Sears Fndn Mar Res, Memoir, Yale Univ.

Bohlke, E. B., ed. 1989b. *Fishes of the western North Atlantic*. Part 9. Vol. 2, *Leptocephali*, 657–1055. New Haven, CT: Sears Fndn Mar Res, Memoir, Yale Univ.

Bohlke, J. E. 1966. Order Lyomeri: Deep-sea gulpers. In *Fishes of the western North Atlantic*, ed. G. W. Mead et al., 603–625. New Haven, CT: Mem Sears Fndn Mar Res, No. 1, Part 5.

Bohnsack, J. A. 1983. Resiliency of reef fish communities in the Florida Keys following a January 1977 hypothermal fish kill. *Env Biol Fish* 9:41–53.

Bond, C. E. 1996. *Biology of fishes*. 2d ed. Philadelphia: Saunders College Publishing.

Bond, C. E., and T. T. Kan. 1973. *Lampetra* (*Entosphenus*) *minima* n. sp., a dwarfed parasitic lamprey from Oregon. *Copeia* 1973:568–574.

Bone, Q. 1972. Buoyancy and hydrodynamic functions of integument in the castor oil fish *Ruvettus pretiosus* (Pisces: Gempylidae). *Copeia* 1972:78–87.

Bone, Q. 1978. Locomotor muscle. *Fish Physiol* 7:361–424.

Bone, Q., N. B. Marshall, and J. H. S. Blaxter. 1995. *Biology of fishes*. 2d ed. London: Blackie Academic and Professional.

Bonner, J. T. 1980. *The evolution of culture in animals*. Princeton, NJ: Princeton University Press.

Booth, D. E. 1989. Hydroelectric dams and the decline of chinook salmon in the Columbia River basin. *Mar Res Econ* 6:195–211.

Boreman, J., and R. R. Lewis. 1987. Atlantic coastal migration of striped bass. In *Common strategies of anadromous and catadromous fishes*, ed. M. J. Dadswell, R. J. Klauda, C. M. Moffitt, R. L. Saunders, R. A. Rulifson, and J. E. Cooper. Amer Fish Soc Symp 1:331–339.

Bortone, S. A., and W. P. Davis. 1994. Fish intersexuality as indicators of environmental stress. *BioScience* 44:165–172.

Boschung, H. T. 1983. A new species of lancelet, *Branchiostoma longirostrum* (Order Amphioxi), from the western North Atlantic. *Northeast Gulf Sci* 6:91–97.

Boschung, H. T., and R. F. Shaw. 1988. Occurrence of planktonic lancelets from Louisiana's continental shelf, with a review of pelagic *Branchiostoma* (Order Amphioxi). *Bull Mar Sci* 43:229–240.

Boughton, D. A., B. B. Collette, and A. R. McCune. 1991. Heterochrony in jaw morphology of needlefishes (Teleostei: Belonidae). *Syst Zool* 40:329–354.

Boujard, T. 1992. Space-time organization of riverine fish communities in French Guiana. *Env Biol Fish* 34:235–246.

Boujard, T., and J. F. Leatherland. 1992. Circadian rhythms and feeding time in fishes. *Env Biol Fish* 35:109–131.

Bowen, S. H. 1983. Detritivory in neotropical fish communities. *Env Biol Fish* 9:137–144.

Boyle, R. 1996. Phantom. *Natural History* 105(3):16–19.

Brainerd, E. L. 1992. Pufferfish inflation: Functional morphology of postcranial structures in *Diodon holocanthus* (Tetraodontiformes). *J Morph* 220:1–20.

Brainerd, E. L., K. F. Liem, and C. T. Samper. 1989. Air ventilation by recoil aspiration in polypterid fishes. *Science* 246:1593–1595.

Branstetter, S. 1990. Early life-history implications of selected carcharhinoid and lamnoid sharks of the northwest Atlantic. In *Elasmobranchs as living resources: Advances in the biology, ecology, systematics, and the status of fisheries*, ed. H. L. Pratt Jr., S. H. Gruber, and T. Taniuchi, 17–28. Washington, DC: NOAA Tech Rept 90.

Branstetter, S. 1991. Shark early life history: One reason sharks are vulnerable to overfishing. In *Discovering sharks*, ed. S. H. Gruber, 29–34. Highlands, NJ: American Littoral Society, Spec Publ 14.

Branstetter, S., and J. D. McEachran. 1986. Age and growth of four carcharhinid sharks common to the Gulf of Mexico: A summary paper. In *Indo-Pacific fish biology: Proc 2nd Internat Conf Indo-Pacific Fish*, ed. T. Uyeno, R. Arai, T. Taniuchi, and K. Matsuura, 361–371. Tokyo: Ichthyol Soc Japan.

Brawn, V. M. 1961. Reproductive behaviour of the cod (*Gadus callarias* L.). *Behaviour* 18:177–198.

Brawn, V. M. 1962. Physical properties and hydrostatic function of the swimbladder of herring (*Clupea harengus* L.). *J Fish Res Board Can* 19:635–656.

Bray, R. N. 1981. Influence of water currents and zooplankton densities on daily foraging movements of blacksmith, *Chromis punctipinnis*, a planktivorous reef fish. *Fish Bull US* 78:829–841.

Bray, R. N., and M. A. Hixon. 1978. Night-shocker: Predatory behavior of the Pacific electric ray (*Torpedo californica*). *Science* 200:333–334.

Bray, R. N., A. C. Miller, and G. G. Geesey. 1981. The fish connection: A trophic link between planktonic and rocky reef communities? *Science* 214:204–205.

Bray, R. N., L. J. Purcell, and A. C. Miller. 1986. Ammonium excretion in a temperate-reef community by a planktivorous fish, *Chromis punctipinnis* (Pomacentridae), and potential uptake by young giant kelp, *Macrocystis pyrifera* (Laminariales). *Mar Biol* 90:327–334.

Breder, C. M. Jr. 1926. The locomotion of fishes. *Zoologica* (NY) 4:159–297.

Breder, C. M. Jr. 1949. On the relationship of social behavior to pigmentation in tropical shore fishes. *Bull Amer Mus Natur Hist* 94:83–106.

Breder, C. M. Jr. 1976. Fish schools as operational structures. *Fish Bull US* 74:471–502.

Breder, C. M. Jr., and D. E. Rosen. 1966. *Modes of reproduction in fishes*. Neptune City, NJ: TFH Publ.

Brett, J. R. 1957. The eye. In *The physiology of fishes*, ed. M. E. Brown, 121–154. New York: Academic Press.

Brett, J. R. 1971. Energetic responses of salmon to temperature. A study of some thermal relations in the physiology and freshwater ecology of sockeye salmon (*Oncorhynchus nerka*). *Amer Zool* 11:99–113.

Brett, J. R. 1979. Environmental factors and growth. In *Fish physiology*, Vol. 8, ed. W. S. Hoar, D. J. Randall, and J. R. Brett, 599–675. London: Academic Press.

Brett, J. R. 1986. Production energetics of a population of sockeye salmon, *Oncorhynchus nerka*. *Can J Zool* 64:555–564.

Brett, J. R., and J. M. Blackburn. 1978. Metabolic rate and energy expenditure of the spiny dogfish, *Squalus acanthias*. *J Fish Res Board Can* 35:816–821.

Brett, J. R., and T. D. D. Groves. 1979. Physiological energetics. In *Fish physiology*, ed. W. S. Hoar and D. J. Randall, 279–352. New York: Academic Press.

Briggs, D. E. G., E. N. K. Clarkson, and R. J. Aldridge. 1983. The conodont animal. *Lethaia* 16:1–14.

Briggs, J. C. 1974. *Marine zoogeography*. New York: McGraw-Hill.

Briggs, J. C. 1995. *Global biogeography*. Amsterdam: Elsevier.

Brock, V. E., and R. H. Riffenburgh. 1960. Fish schooling: A possible factor in reducing predation. *J Conseil* 25:307–317.

Brodal, A., and R. Fange. 1963. *The biology of Myxine*. Oslo: Scandinavian University Books.

Brooks, J. L., and S. I. Dodson. 1965. Predation, body size, and composition of plankton. *Science* 150:28–35.

Brothers, E. B. 1984. Otolith studies. In *Ontogeny and systematics of fishes*, ed. H. G. Moser, W. J. Richards, D. M. Cohen, M. P. Fahay, A. W. Kendall Jr., and S. L. Richardson, 50–57. Amer Soc Ichthyol Herpetol, Spec Publ 1.

Brothers, E. B., C. P. Mathews, and R. Lasker. 1976. Daily growth increments in otoliths from larval and adult fishes. *Fish Bull US* 74:1–8.

Brothers, E. B., and W. N. McFarland. 1981. Correlations between otolith microstructure, growth, and life history transitions in newly recruited French grunts [*Haemulon flavolineatum* (Desmarest), Haemulidae]. *Rapp P-V Reun Cons Int Explor Mer* 178:369–374.

Brown, B. 1990. *Mountain in the clouds: A search for the wild salmon*. New York: Collier.

Brown, J. A. 1986. The development of feeding behaviour in the lumpfish. *J Fish Biol* 29 (suppl A):171–178.

Brunori, M. 1975. Molecular adaptation to physiological requirements: The hemoglobin system of trout. *Curr Top Cell Reg* 9:1–39.

Brusca, R. C., and M. R. Gilligan. 1983. Tongue replacement in a marine fish (*Lutjanus guttatus*) by a parasitic isopod (Crustacea: Isopoda). *Copeia* 1983:813–816.

Bruton, M. N. 1979. The survival of habitat desiccation by air breathing clariid catfishes. *Env Biol Fish* 4:273–280.

Bruton, M. N., and M. J. Armstrong. 1991. The demography of the coelacanth *Latimeria chalumnae*. *Env Biol Fish* 32:301–311.

Bruton, M. N., A. J. P. Cabral, and H. Fricke. 1992. First capture of a coelacanth, *Latimeria chalumnae* (Pisces, Latimeriidae), off Mozambique. *Suid-Afrikaanse Tydskrif vir Wetenskap* 88:225–227.

Bruton, M. N., and S. E. Coutouvidis. 1991. An inventory of all known specimens of the coelacanth *Latimeria chalumnae*, with comments on trends in the catches. *Env Biol Fish* 32:371–390.

Bruton, M. N., and R. E. Stobbs. 1991. The ecology and conservation of the coelacanth *Latimeria chalumnae*. *Env Biol Fish* 32:313–339.

Bryan, P. G., and B. B. Madraisau. 1977. Larval rearing and development of *Siganus lineatus* (Pisces, Siganidae) from hatching through metamorphosis. *Aquaculture* 10:243–252.

Buckland, J. 1880. *Natural history of British fishes*. London: Unwin.

Buckley, J., and B. Kynard. 1985. Habitat use and behavior of pre-spawning and spawning shortnose sturgeon, *Acipenser brevirostrum*, in the Connecticut River. In *North American sturgeons: Biology and aquaculture potential. Developments in environmental biology of fishes 6*, ed. F. P. Binkowski and S. I. Doroshov, 111–117. Dordrecht: Dr. W. Junk.

Buddington, R. K., and J. P. Christofferson. 1985. Digestive and feeding characteristics of the chondrosteans. In *North American sturgeons: Biology and aquaculture potential. Developments in environmental biology of fishes 6*, ed. F. P. Binkowski and S. I. Doroshov, 31–41. Dordrecht: Dr. W. Junk.

Buddington, R. K., and J. M. Diamond. 1987. Pyloric ceca of fish: A "new" absorptive organ. *Amer J Physiol* 252:G65–76.

Budker, P. 1971. *The life of sharks*. New York: Columbia University Press.

Bullock, T. H., R. H. Hamstra, and H. Scheich. 1972. The jamming avoidance response of high frequency electric fish. *J Comp Physiol* 72:1–22.

Bullock, T. H., R. H. Hamstra Jr., and H. Scheich. 1972. The jamming avoidance response of high frequency electric fish. I. General features. II. Quantitative aspects. *J Comp Physiol* 77A:1–22, 23–48.

Bullock, T. H., and W. Heiligenberg, eds. 1986. *Electroreception*. New York: John Wiley & Sons.

Burgess, W. E. 1978. *Butterflyfishes of the world. A monograph of the family Chaetodontidae*. Neptune City, NJ: TFH Publ.

Burgess, W. E. 1989. *An atlas of freshwater and marine catfishes—a preliminary survey of the Siluriformes*. Neptune City, NJ: TFH Publ.

Burkhead, N. M., and J. D. Williams. 1991. An intergeneric hybrid of a native minnow, the golden shiner, and an exotic minnow, the rudd. *Trans Amer Fish Soc* 120:781–795.

Burns, J. R. 1975. Seasonal changes in the respiration of pumpkinseed, *Lepomis gibbosus*, correlated with temperature, day length, and stage of reproduction. *Physiol Zool* 48:142–149.

Busacker, G. P., I. R. Adelman, and E. M. Goolish. 1990. Growth. In *Methods for fish biology*, ed. C. B. Schreck and P. B. Moyle, 363–387. Bethesda, MD: American Fisheries Society.

Buth, D. G., T. E. Dowling, and J. R. Gold. 1991. Molecular and cytological investigations. In *Cyprinid fishes: Systematics, biology and exploitation*, ed. I. J. Winfield and J. S. Nelson, 83–126. London: Chapman and Hall.

Bye, V. J. 1990. Temperate marine teleosts. In *Reproductive seasonality in teleosts: Environmental influences*, ed. A. D. Munro, A. P. Scott, and T. J. Lam, 125–143. Boca Raton, FL: CRC Press.

Cailliet, G. M. 1990. Elasmobranch age determination and verification: An updated review. In *Elasmobranchs as living resources: Advances in the biology, ecology, systematics, and the status of fisheries*, ed. H. L. Pratt Jr., S. H. Gruber, and T. Taniuchi, 157–165. Washington, DC: NOAA Tech Rept 90.

Cailliet, G. M., M. S. Love, and A. W. Ebeling. 1986. *Fishes: A field and laboratory manual on their structure, identification, and natural history*. Belmont, CA: Wadsworth Pub.

Caira, J. N., G. W. Benz, J. Borucinska, N. E. Kohler, and J. G. Casey. 1994. Eels from the heart of a mako shark. Unpublished manuscript.

Calder, W. A. III. 1984. *Size, function, and life history*. Cambridge: Harvard University Press.

Callicott, J. B. 1991. Conservation ethics and fishery management. *Fisheries* 16(2):22–28.

Cameron, J. N. 1978. Chloride shift in fish blood. *J Exp Zool* 206:289–295.

Campbell, K. S. W., and R. E. Barwick. 1987. Paleozoic lungfishes—a review. In *The biology and evolution of lungfishes*, ed. W. E. Bemis, W. W. Burggren, and N. E. Kemp, 93–131. New York: Alan R. Liss.

Capen, R. L. 1967. *Swimbladder morphology of some mesopelagic fishes in relation to sound scattering*. San Diego, CA: US Navy Electronic Laboratory, Report 1447.

Carey, F. G., and E. Clark. 1995. Depth telemetry from the sixgill shark, *Hexanchus griseus*, at Bermuda. *Env Biol Fish* 42:7–14.

Carey, F. G., J. W. Kanwisher, O. Brazier, G. Gabrielson, J. G. Casey, and H. L. Pratt Jr. 1982. Temperature and activities of a white shark, *Carcharodon carcharias*. *Copeia* 1982:254–260.

Carey, F. G., and K. D. Lawson. 1973. Temperature regulation in free-swimming bluefin tuna. *Comp Biochem Physiol* 44A:375–392.

Carey, F. G., and J. V. Scharold. 1990. Movements of blue sharks in depth and course. *Mar Biol* 106:329–342.

Carey, F. G., and J. M. Teal. 1966. Heat conservation in tuna fish muscle. *Proc Nat Acad Sci USA* 56:1464–1469.

Carey, F. G., J. M. Teal, and J. W. Kanwisher. 1981. The visceral temperatures of mackerel sharks (Lamnidae). *Physiol Zool* 54:334–344.

Carey, F. G., J. M. Teal, J. W. Kanwisher, K. D. Lawson, and J. S. Beckett. 1971. Warm-bodied fish. *Amer Zool* 11:137–145.

Carlander, K. D. 1969. *Handbook of freshwater fishery biology*. Vol. l. Ames: Iowa State University Press.

Carline, R. F., W. E. Sharpe, and C. J. Gagen. 1992. Changes in fish communities and trout management in response to acidification of streams in Pennsylvania. *Fisheries* 17(1):33–38.

Carlton, J. T. 1985. Transoceanic and interoceanic dispersal of coastal marine organisms: The biology of ballast water. *Oceanogr Mar Biol Ann Rev* 23:313–371.

Carpenter, K. E., B. B. Collette, and J. L. Russo. 1995. Unstable and stable classifications of scombroid fishes. *Bull Mar Sci* 56:379–405.

Carpenter, R. C. 1986. Partitioning herbivory and its effects on coral reef algal communities. *Ecol Monogr* 56:345–363.

Carpenter, S. R., ed. 1988. *Complex interactions in lake communities*. New York: Springer-Verlag.

Carpenter, S. R., and J. F. Kitchell. 1988. Consumer control of lake productivity. *BioScience* 38:764–769.

Carpenter, S. R., and J. F. Kitchell, eds. 1993. *The trophic cascade in lakes*. New York: Cambridge University Press.

Carpenter, S. R., J. F. Kitchell, and J. R. Hodgson. 1985. Cascading trophic interactions and lake productivity. *BioScience* 35:634–639.

Carroll, R. L. 1988. *Vertebrate paleontology and evolution*. New York: W. H. Freeman.

Casselman, J. M. 1983. Age and growth assessment of fish from their calcified structures—techniques and tools. In *Proceedings of the international workshop on age determination of oceanic pelagic fishes: Tunas, billfishes, and sharks*, ed. E. D. Prince and L. M. Pulos, 1–17. Washington, DC: NOAA Tech Rept NMFS 8.

Casselman, J. M. 1990. Growth and relative size of calcified structures of fish. *Trans Amer Fish Soc* 119:673–688.

Casteel, R. W. 1976. *Fish remains in archaeology and paleo-environmental studies*. London: Academic Press.

Castelnau, F. L. 1879. On a new ganoid fish from Queensland. *Proc Linn Soc NSW* 3:164–165.

Castle, P. H. J. 1978. Ovigerous leptocephali of the nettastomatid eel genus *Facciolella*. *Copeia* 1978:29–33.

Castro, J. I. 1983. *The sharks of North American waters*. College Station: Texas A&M University Press.

Caviedes, C., and T. Fik. 1990. Variability in the Peruvian and Chilean fisheries. In *Climate variability, climate change and fisheries*, ed. M. H. Glantz and L. E. Feingold, 95–102. Boulder, CO: Environmental and Societal Impacts Group, Nat Ctr Atmos Res.

Chao, L. N. 1973. Digestive system and feeding habits of the cunner, *Tautogolabrus adspersus*, a stomachless fish. *Fish Bull US* 71:565–585.

Chapman, D. W. 1978. Production in fish populations. In *Ecology of freshwater fish production*, ed. S. D. Gerking, 5–25. London: Blackwell Science.

Chave, E. H., and H. A. Randall. 1971. Feeding behavior of the moray eel *Gymnothorax pictus*. *Copeia* 1971:570–574.

Chiarelli, A. B., and E. Capanna. 1973. Checklist of fish chromosomes. In *Cytotaxonomy and vertebrate evolution*, ed. A. B. Chiarelli and E. Capanna, 205–232. London: Academic Press.

Chiszar, D. 1978. Lateral displays in the lower vertebrates: Forms, functions, and origin. In *Contrasts in behavior*, ed. E. S. Reese and F. J. Lighter, 105–135. New York: Wiley-Interscience.

Choat, J. H. 1991. The biology of herbivorous fishes on coral reefs. In *The ecology of fishes on coral reefs*, ed. P. F. Sale, 120–155. San Diego, CA: Academic Press.

Cisneros-Mata, M. A., G. Montemayor-Lopez, and M. J. Roman-Rodriguez. 1995. Life history and conservation of *Totoaba macdonaldi*. *Cons Biol* 9:806–814.

Clark, D. L. 1987. Phylum Conodonta. In *Fossil invertebrates*, ed. R. S. Boardman, A. H. Cheetham, and A. J. Rowell, 636–662. Palo Alto, CA: Blackwell Science.

Clark, E., and E. Kristof. 1990. Deep sea elasmobranchs observed from submersibles in Grand Cayman, Bermuda and Bahamas. In *Elasmobranchs as living resources: Advances in the biology, ecology, systematics, and the status of fisheries*, ed. H. L. Pratt Jr., S. H. Gruber, and T. Taniuchi, 275–290. Washington, DC: NOAA Tech Rept 90.

Clark, E., and E. Kristof. 1991. How deep do sharks go? Reflections on deep sea sharks. In *Discovering sharks*, ed. S. H. Gruber, 79–84. Highlands, NJ: American Littoral Society, Spec Publ 14.

Clarke, A. 1983. Life in cold water: The physiological ecology of polar marine ectotherms. *Oceanogr Mar Biol Ann Rev* 21:341–453.

Clements, K. D., and J. H. Choat. 1995. Fermentation in tropical marine herbivorous fishes. *Physiol Zool* 68:355–378.

Clifton, H. E., and R. E. Hunter. 1972. The sand tilefish, *Malacanthus plumieri*, and the distribution of coarse debris near West Indian coral reefs. In *Results of the Tektite program: Ecology of coral reef fishes*, ed. B. B. Collette and S. A. Earle. *Natur Hist Mus Los Angeles Co Bull* 14:87–92.

Cloutier, R., and P. L. Forey. 1991. Diversity of extinct and living actinistian fishes (Sarcopterygii). *Env Biol Fish* 32:59–74.

Coad, B. W., and F. Papahn. 1988. Shark attacks in the rivers of southern Iran. *Env Biol Fish* 23:131–134.

Coburn, M. M., and J. I. Gaglione. 1992. A comparative study of percid scales (Teleostei: Perciformes). *Copeia* 1992:986–1001.

Coe, M. 1966. The biology of *Tilapia grahami* Boulenger in Lake Magadi, Kenya. *Acta Tropica* 23:146–177.

Cohen, D. E., R. H. Rosenblatt, and H. G. Moser. 1990. Biology and description of a bythitid fish from deep-sea thermal vents in the tropical eastern Pacific. *Deep-Sea Res* 37:267–283.

Cohen, D. M. 1970. How many recent fishes are there? *Proc Calif Acad Sci*, 4th ser, 38:341–346.

Cohen, D. M. 1984. Ontogeny, systematics and phylogeny. In *Ontogeny and systematics of fishes*, ed. H. G. Moser, W. J. Richards, D. M. Cohen, M. P. Fahay, A. W. Kendall Jr., and S. L. Richardson, 7–11. Amer Soc Ichthyol Herpetol, Spec Publ 1.

Colby, P. J., ed. 1977. Percid international symposium (PERCIS). *J Fish Res Board Can* 34:1445–1999.

Colgan, P. W., and M. R. Gross. 1977. Dynamics of aggression in male pumpkinseed sunfish (*Lepomis gibbosus*) over the reproductive phase. *Z Tierpsychol* 43:139–151.

Colin, P. L. 1973. Burrowing behavior of the yellowhead jawfish, *Opistognathus aurifrons*. *Copeia* 1973:84–92.

Colin, P. 1978. Daily and summer-winter variation in mass spawning of the striped parrotfish, *Scarus croicensis*. *Fish Bull US* 76:117–124.

Collette, B. B. 1962. The swamp darters of the subgenus *Hololepis* (Pisces, Percidae). *Tulane Stud Zool* 9:115–211.

Collette, B. B. 1977. Epidermal breeding tubercles and bony contact organs in fishes. *Symp Zool Soc London* 1977:225–268.

Collette, B. B. 1978. Adaptations and systematics of the mackerels and tunas. In *The physiological ecology of tunas*, ed. G. D. Sharp and A. E. Dizon, 7–39. New York: Academic Press.

Collette, B. B. 1979. Adaptations and systematics of the mackerels and tunas. In *The physiological ecology of tunas*, ed. G. D. Sharp and A. E. Dizon, 7–39. New York: Academic Press.

Collette, B. B. 1995. Hemiramphidae. In W. Fischer, F. Krupp, W. Schneider, C. Sommer, K. E. Carpenter, and V. H. Niem, eds. Guia FAO para la identificacion de especies para los fines de la pesca, Pacifico Centro-Oriental, vol. II:1175–1181.

Collette, B. B., and P. Banarescu. 1977. Systematics and zoogeography of the fishes of the family Percidae. *J Fish Res Board Can* 34:1450–1463.

Collette, B. B., and L. N. Chao. 1975. Systematics and morphology of the bonitos (*Sarda*) and their relatives (Scombridae, Sardini). *Fish Bull US* 73:516–625.

Collette, B. B., and C. E. Nauen. 1983. Scombrids of the world. *FAO Fish Synop* 125(2):137.

Collette, B. B., T. Potthoff, W. J. Richards, S. Ueyanagi, J. L. Russo, Y. Nishikawa. 1984. Scombroidei: Development and relationships. In *Ontogeny and systematics of fishes*, ed. H. G. Moser et al., 591–620. Amer Soc Ichthyol Herpetol, Spec Publ 1.

Collette, B. B., and J. L. Russo. 1981. A revision of the scaly toadfishes, genus *Batrachoides*, with descriptions of two new species from the eastern Pacific. *Bull Mar Sci* 31:197–233.

Collette, B. B., and J. L. Russo. 1985a. Interrelationships of the Spanish mackerels (Pisces: Scombridae: *Scomberomorus*) and their copepod parasites. *Cladistics* 1:141–158.

Collette, B. B., and J. L. Russo. 1985b. Morphology, systematics, and biology of the Spanish mackerels (*Scomberomorus*, Scombridae). *Fish Bull US* 82:545–692.

Collette, B. B., J. L. Russo, and L. A. Zavala-Camin. 1978. *Scomberomorus brasiliensis*, a new species of Spanish mackerel from the western Atlantic. *Fish Bull US* 76:273–280.

Collette, B. B., and K. Rutzler. 1977. Reef fishes over sponge bottoms off the mouth of the Amazon River. *Proc 3rd Int Coral Reef Symp* 305–310.

Collette, B. B., and F. H. Talbot. 1972. Activity patterns of coral reef fishes with emphasis on nocturnal-diurnal changeover. *Los Angeles Cty Natur Hist Mus Sci Bull* 14:98–124.

Collette, B. B., and M. Vecchione. 1995. Interactions between fisheries and systematics. *Fisheries* 20(1):20–25.

Compagno, L. J. V. 1981. Legend versus reality: the jaws image and shark diversity. *Oceanus* 24(4):5–16.

Compagno, L. J. V. 1984. *FAO species catalogue*. Vol. 4, *Sharks of the world. An annotated and illustrated catalogue of shark species known to date*. FAO Fisheries Synopsis No. 125. Rome: FAO.

Compagno, L. J. V. 1990a. The evolution and diversity of sharks. In *Discovering sharks*, ed. S. H. Gruber, 15–22. Highlands, NJ: American Littoral Society, Spec Publ 14.

Compagno, L. J. V. 1990b. Alternative life history styles of cartilaginous fishes in time and space. *Env Biol Fish* 28:33–75.

Compagno, L. J. V. 1990c. Relationships of the megamouth shark, *Megachasma pelagios* (Lamniformes: Megachasmidae) with comments on its feeding habits. In *Elasmobranchs as living resources: Advances in the biology, ecology, systematics, and the status of fisheries*, ed. H. L. Pratt Jr., S. H. Gruber, and T. Taniuchi, 357–380. Washington, DC: NOAA Tech Rept 90.

Conant, E. B. 1987. An historical overview of the literature of Dipnoi: Introduction to the bibliography of lungfishes. In *The biology and evolution of lungfishes*, ed. W. E. Bemis, W. W. Burggren, and N. E. Kemp, 5–13. New York: Alan R. Liss.

Cone, R. S., et al. 1990. Properties of relative weight and other condition indices. *Trans Amer Fish Soc* 119:1048–1058.

Congdon, J. D., A. E. Dunham, and D. W. Tinkle. 1982. Energy budgets and life histories of reptiles. In *Biology of the reptilia*, Vol. 13, ed. C. Gans and F. H. Pough, 233–271.

Conniff, R. 1991. The most disgusting fish in the sea. *Audubon* 93(2):100–108.

Conover, D. O., and S. W. Heins. 1987. Adaptive variation in environmental and genetic sex determination in a fish. *Nature* 326:496–498.

Conover, D. O., and B. E. Kynard. 1981. Environmental sex determination: Interaction of temperature and genotype in a fish. *Science* 213:57–59.

Conover, D. O., and D. A. Van Voorhees. 1990. Evolution of a balanced sex ratio by frequency-dependent selection in a fish. *Science* 250:1556–1558.

Conroy, D. A., J. Morales, C. Perdomo, R. A. Ruiz, and J. A. Santicana. 1981. Preliminary observations on ornamental fish diseases in northern South America. *Rivista Italiana di Piscicoltura e Ittiopatologia* 16:86–98.

Constantz, G. W. 1975. Behavioral ecology of mating in the male gila topminnow, *Poeciliopsis occidentalis* (Cyprinodontiformes: Poeciliidae). *Ecology* 56:966–973.

Contreras, S., and M. A. Escalante. 1984. Distribution and known impacts of exotic fishes in Mexico. In *Distribution, biology, and management of exotic fishes*, ed. W. R. Courtenay Jr. and J. R. Stauffer Jr., 102–130. Baltimore: Johns Hopkins University Press.

Coombs, S., P. Gorner, and H. Munz, eds. 1989. *The mechanosensory lateral line: Neurobiology and evolution*. New York: Springer-Verlag.

Cooper, J. C., and P. J. Hirsch. 1982. The role of chemoreception in salmonid homing. In *Chemoreception in fishes*, ed. T. J. Hara, 343–362. Amsterdam: Elsevier.

Cooper, S. D., S. J. Walde, and B. L. Peckarsky. 1990. Prey exchange rates and the impact of predators on prey populations in streams. *Ecology* 71:1503–1514.

Corwin, J. T. 1981. Audition in elasmobranchs. In *Hearing and sound communication in fishes*, ed. W. N. Tavolga, A. N. Popper, and R. R. Fay, 81–106. Berlin: Springer.

Cott, H. B. 1957. *Adaptive coloration in animals*. London: Methuen.

Couch, J. A., and J. W. Fournie. 1993. *Pathobiology of marine and estuarine organisms*. Boca Raton, FL: CRC Press.

Coughlin, D. J., and J. R. Strickler. 1990. Zooplankton capture by a coral reef fish: An adaptive response to evasive prey. *Env Biol Fish* 29:35–42.

Courtenay, W. R. Jr., D. A. Hensley, J. N. Taylor, and J. A. McCann. 1984. Distribution of exotic fishes in the continental United States. In *Distribution, biology, and management of exotic fishes*, ed. W. R. Courtenay Jr. and J. R. Stauffer Jr., 41–77. Baltimore: Johns Hopkins University Press.

Courtenay, W. R. Jr., D. P. Jennings, and J. D. Williams. 1991. Appendix 2, Exotic fishes. In *Common and scientific names of fishes from the United States and Canada*, 5th ed., ed. C. R. Robins et al., 97–108. Bethesda, MD: American Fisheries Society, Spec Publ 20.

Courtenay, W. R. Jr., and J. R. Stauffer Jr., eds. 1984. *Distribution, biology, and management of exotic fishes*. Baltimore: Johns Hopkins University Press.

Coutant, C. C. 1985. Striped bass, temperature, and dissolved oxygen: A speculative hypothesis of environmental risk. *Trans Amer Fish Soc* 114:31–61.

Coutant, C. C. 1990. Temperature-oxygen habitat for freshwater and coastal striped bass in a changing climate. *Trans Amer Fish Soc* 119:240–253.

Cox, D. L., and T. J. Koob. 1993. Predation on elasmobranch eggs. *Env Biol Fish* 38:117–125.

Craig, J. F. 1987. *The biology of perch and related fish*. Kent, United Kingdom: Croom Helm.

Cressey, R. F., and B. B. Collette. 1970. Copepods and needle-fishes: A study in host-parasite relationships. *Fish Bull US* 68:347–432.

Crossman, E. J., and D. E. McAllister. 1986. Zoogeography of freshwater fishes of the Hudson Bay drainage, Ungava Bay, and the Arctic archipelago. In *The zoogeography of North American freshwater fishes*, ed. C. H. Hocutt and E. O. Wiley, 53–105.

Crowder, L. B., and W. E. Cooper. 1982. Habitat structural complexity and the interaction between bluegills and their prey. *Ecology* 65:894–908.

Culp, J. M., N. E. Glozier, and G. J. Scrimgeour. 1991. Reduction of predation risk under the cover of darkness: Avoidance responses of mayfly larvae to a benthic fish. *Oecologia* 86:163–169.

Culver, D. C. 1982. *Cave life: Evolution and ecology*. Cambridge: Harvard University Press.

Cunningham, J. T., and D. M. Reid. 1932. Experimental research on the emission of oxygen by the pelvic filaments of the male *Lepidosiren* with some experiments on *Synbranchus marmoratus*. *Proc Royal Soc* 110:234–248.

Curio, E. 1976. *The ethology of predation*. Berlin: Springer-Verlag.

Cushing, D. H. 1973. *The detection of fish*. Oxford: Pergamon Press.

Cushing, D. H. 1975. *Marine ecology and fisheries*. Cambridge: Cambridge University Press.

Cuvier, G. L. C. F. D., and A. Valenciennes. 1829–1849. *Histoire naturelle de poisson*. 24 vols. Paris: Bibliotheque Centrale du Museum National d'Histoire Naturelle.

Dadswell, M. J., R. J. Klauda, C. M. Moffitt, R. L. Saunders, R. A. Rulifson, and J. E. Cooper, eds. 1987. Common strategies of anadromous and catadromous fishes. *Amer Fish Soc Symp* 1.

Dadswell, M. J., G. D. Melvin, P. J. Williams, and D. E. Themelis. 1987. Influences of origin, life history, and chance on the Atlantic Coast migration of American shad. In *Common strategies of anadromous and catadromous fishes*, ed. M. J. Dadswell,

R. J. Klauda, C. M. Moffitt, R. L. Saunders, R. A. Rulifson, and J. E. Cooper. *Amer Fish Soc Symp* 1:313–330.

Dahlgren, U. 1927. The life history of the fish *Astroscopus* (the "stargazer"). *Sci Monthly* 24:348–365.

Daniels, R. A. 1979. Nest guard replacement in the Antarctic fish *Harpagifer bispinis*: Possible altruistic behavior. *Science* 205:831–833.

Daniels, R. A. 1982. Feeding ecology of some fishes of the Antarctic Peninsula. *Fish Bull US* 80:575–588.

Darlington, P. J. 1957. *Zoogeography: The geographical distribution of animals*. New York: John Wiley.

Darwin, C. 1859. *On the origin of species . . .* London: John Murray.

Davenport, J. 1994. How and why do flying fish fly? *Rev Fish Biol Fisheries* 4:184–214.

Davenport, J., and E. Kjorsvik. 1986. Buoyancy in the lumpsucker, *Cyclopterus lumpus*. *J Mar Biol Assoc UK* 66:159–174.

Davin, W. T. Jr., C. C. Kohler, and D. R. Tindall. 1988. Ciguatera toxins adversely affect piscivorous fishes. *Trans Amer Fish Soc* 117:374–384.

Davis, T. L. O. 1988. Temporal changes in the fish fauna entering a tidal swamp system in tropical Australia. *Env Biol Fish* 21:161–172.

Davis, W. P., and R. S. Birdsong. 1973. Coral reef fishes which forage in the water column. *Helgol wiss Meeresunters* 24:292–306.

Dawson, J. A. 1963. The oral cavity, the "jaws" and the horny teeth of *Myxine glutinosa*. In *Biology of* Myxine, ed. A. Brodal and R. Fange, 231–255. Oslo: Universitetsforlaget.

Day, R. W. 1985. The effects of refuges from predators and competitors on sessile communities on a coral reef. *Proc 5th Internat Coral Reef Cong* 4:41–45.

Dayton, P. K., G. A. Robilliard, R. T. Paine, and L. B. Dayton. 1974. Biological accommodation in the benthic community at McMurdo Sound, Antarctica. *Ecol Monogr* 44:105–128.

Deegan, L. A. 1993. Nutrient and energy transport between estuaries and coastal marine ecosystems by fish migration. *Can J Fish Aquat Sci* 50:74–79.

Degnan, K. J., K. J. Karnaky Jr., and J. A. Zadunaisky. 1977. Active chloride transport in the in vitro opercular skin of a teleost (*Fundulus heteroclitus*), a gill-like epithelium rich in chloride cells. *J Physiol* 271:155–191.

de Groot, S. J. 1971. On the interrelationships between morphology of the alimentary tract, food and feeding behaviour in flatfishes (Pisces: Pleuronectiformes). *Netherlands J Sea Res* 5:121–196.

DeMartini, E. E., and J. A. Coyer. 1981. Cleaning and scale-eating in juveniles of the kyphosid fishes, *Hermosilla azurea* and *Girella nigricans*. *Copeia* 1981:785–789.

DeMartini, E. E., and R. K. Fountain. 1981. Ovarian cycling frequency and batch fecundity in the queenfish, *Seriphus politus*: Attributes representative of serial spawning fishes. *Fish Bull US* 79:547–560.

Denison, R. H. 1970. Revised classification of Pteraspididae, with descriptions of new forms from Wyoming. *Fieldiana, Geol* 20:1–41.

Denison, R. H. 1979. *Acanthodii. Handbook of paleoichthyology*. Vol. 5. Stuttgart: Gustav Fischer Verlag.

Denton, E., and J. A. C. Nicol. 1965. Studies of the reflexion of light from silvery surfaces of fishes, with special reference to the bleak, *Alburnus alburnus*. *J Mar Biol Assoc UK* 45:683–703.

Denton, E. J., P. J. Herring, E. A. Widdder, M. F. Latz, and J. F. Case. 1985. The roles of filters in the photophores of oceanic animals and their relation to vision in the oceanic environment. *Proc Roy Soc Lond, Ser B: Biol Sci* 225:63–98.

Denton, E. J., and N. B. Marshall. 1958. The buoyancy of bathypelagic fishes without a gas-filled swim bladder. *J Mar Biol Assoc UK* 37:753–767.

Denton, T. E., and W. M. Howell. 1973. Chromosomes of the African polypterid fishes, *Polypterus palmas* and *Calamoichthys calabaricus* (Pisces: Brachiopterygii). *Experientia* 29:122–124.

Derr, M. 1992. Raiders of the reef. *Audubon* 94(2):48–54.

DeSilva, C. D., S. Premawansa, and C. N. Keembiyahetty. 1986. Oxygen consumption in *Oreochromis niloticus* (L.) in relation to development, salinity, temperature and time of day. *J Fish Biol* 29:267–277.

DeVries, A. L. 1970. Freezing resistance in Antarctic fishes. In *Antarctic ecology*, Vol. 1, ed. M. W. Holdgate, 320–328. San Diego, CA: Academic Press.

DeVries, A. L. 1974. Survival at freezing temperatures. In *Biochemical and biophysical perspectives in marine biology*, Vol. 1, ed. D. C. Malins and J. R. Sargent, 289–330. New York: Academic Press.

DeVries, A. L. 1977. The physiology of cold adaptation in polar marine poikilotherms. In *Polar oceans*, ed. M. J. Dunbar, 409–417. Calgary, Alberta: Arctic Inst N Amer.

DeVries, A. L. 1988. The role of antifreeze glycopeptides and peptides in freezing avoidance of Antarctic fishes. *Comp Biochem Physiol* 90B(3):611–621.

Diana, J. S., W. C. Mackay, and M. Ehrman. 1977. Movements and habitat preference of northern pike (*Esox lucius*) in Lac Ste. Anne, Alberta. *Trans Amer Fish Soc* 106:560–565.

Dickson, I. W., and R. H. Kramer. 1971. Factors influencing scope for activity and active and standard metabolism of rainbow trout (*Salmo gairdneri*). *J Fish Res Board Can* 28:587–596.

Dill, L. M. 1974. The escape response of the zebra danio (*Brachydanio rerio*). I. The stimulus for escape. *Anim Behav* 22:711–722.

Dill, L. M. 1977. Refraction and the spitting behavior of the archerfish (*Toxotes chatareus*). *Behav Ecol Sociobiol* 2:169–184.

Dill, L. M. 1978. Aggressive distance in juvenile coho salmon (*Oncorhynchus kisutch*). *Can J Zool* 56:1441–1446.

Dill, L. M. 1987. Animal decision making and its ecological consequences: The future of aquatic ecology and behaviour. *Can J Zool* 65:803–811.

Dill, P. A. 1977. Development of behaviour in alevins of Atlantic salmon, *Salmo salar*, and rainbow trout, *S. gairdneri*. *Anim Behav* 25:116–121.

Dillard, J. G., L. K. Graham, and T. R. Russell, eds. 1986. *The paddlefish: Status, management and propagation*. Columbia, MO: North Central Division, American Fisheries Society, Spec Publ 7.

Dingerkus, G. 1979. Chordate cytogenetic studies: An analysis of their phylogenetic implications with particular reference to fishes and the living coelacanth. *Occ Pap Calif Acad Sci* 134:111–127.

Dingerkus, G., and W. M. Howell. 1976. Karyotypic analysis and evidence of tetraploidy in the North American paddlefish, *Polyodon spathula*. *Science* 194:842–844.

di Prisco, G., B. Maresca, and B. Tota, eds. 1991. *Biology of Antarctic fish*. Berlin: Springer-Verlag.

Dogiel, V. A., G. K. Petrushevski, and Y. I. Polyanski. 1961. *Parasitology of fishes*. Edinburgh: Oliver and Boyd.

Doherty, P. J. 1991. Spatial and temporal patterns in recruitment. In *The ecology of fishes on coral reefs*, ed. P. F. Sale, 261–293. San Diego, CA: Academic Press.

Doherty, P. J., and P. F. Sale. 1986. Predation on juvenile coral reef fishes: An exclusion experiment. *Coral Reefs* 4:225–234.

Doherty, P. J., and D. McB. Williams. 1988. The replenishment of coral reef fish populations. *Oceanogr Mar Biol Ann Rev* 26:487–551.

Dominey, W. J. 1983. Mobbing in colonially nesting fishes, espe-cially the bluegill, *Lepomis macrochirus*. *Copeia* 1983:1086–1088.

Dominey, W. J., and L. S. Blumer. 1984. Cannibalism of early life stages in fishes. In *Infanticide. Comparative and evolutionary perspectives*, ed. G. Hausfater and S. B. Hrdy, 43–64. New York: Aldine.

Donaldson, E. M. 1990. Reproductive indices as measures of the effects of environmental stressors in fish. *Amer Fish Soc Symp* 8:109–122.

Douglas, R. H., and M. B. A. Djamgoz, eds. 1990. *The visual system of fish*. London: Chapman and Hall.

Downhower, J. F., and L. Brown. 1981. The timing of reproduction and its behavioral consequences for mottled sculpins, *Cottus bairdi*. In *Natural selection and social behaviour*, ed. R. D. Alexander and D. W. Tinkle, 78–85. New York: Chiron Press.

Downhower, J. F., L. Brown, R. Pederson, and G. Staples. 1983. Sexual selection and sexual dimorphism in mottled sculpins. *Evolution* 37:96–103.

Drenner, R. W., K. D. Hambright, G. L. Vinyard, M. Gophen, and U. Pollingher. 1987. Experimental study of size-selective phytoplankton grazing by a filter-feeding cichlid and the cichlid's effects on plankton community structure. *Limnol Oceanogr* 32:1140–1146.

Drenner, R. W., J. R. Mummert, F. deNoyelles Jr., and D. Kettle. 1984. Selective particle ingestion by a filter-feeding fish and its impact on phytoplankton community structure. *Limnol Oceanogr* 29:941–948.

Drenner, R. W., S. B. Taylor, X. Lazzaro, and D. Kettle. 1984. Particle-grazing and plankton community impact of an omnivorous cichlid. *Trans Amer Fish Soc* 113:397–402.

Duman, J. G., and A. L. DeVries. 1974. The effects of temperature and photoperiod on antifreeze production in cold water fishes. *J Exp Zool* 190:89–98.

Duncker, H. R., and G. Fleischer, eds. 1986. *Functional morphology of vertebrates*. New York: Springer-Verlag.

Dunham, A. E., B. W. Grant, and K. L. Overall. 1989. Interfaces between biophysical and physiological ecology and the population ecology of terrestrial vertebrate ectotherms. *Physiol Zool* 62:335–355.

Dunier, M. D. 1994. Effects of environmental contaminants (pesticides and metal ions) on fish immune systems. In *Modulators of fish immune responses*. Vol. 1, *Models for environmental toxicology, biomarkers, immunostimulators*, ed. J. S. Stolen and T. C. Fletcher, 123–139. Fair Haven, NJ: SOS Publications.

Durbin, A. G., S. W. Nixon, and C. A. Oviatt. 1979. Effects of the spawning migration of the alewife, *Alosa pseudoharengus*, on freshwater ecosystems. *Ecology* 60:8–17.

Eastman, J. T. 1993. *Antarctic fish biology: Evolution in a unique environment*. San Diego, CA: Academic Press.

Eastman, J. T., and A. I. DeVries. 1986. Antarctic fishes. *Sci Amer* 255(5):106–114.

Eaton, R. C. 1991. Neuroethology of the Mauthner system. *Brain Behav Evol* 37:250–317.

Eaton, R. C., and R. DiDomenico. 1986. Role of the teleost escape response during development. *Trans Amer Fish Soc* 115:128–142.

Ebeling, A. W., N. B. Atkin, and P. Y. Setzer. 1971. Genome sizes of teleostean fishes: Increases in some deep-sea species. *Amer Natur* 105:549–562.

Ebeling, A. W., and M. A. Hixon. 1991. Tropical and temperate reef fishes: Comparison of community structures. In *The ecology of fishes on coral reefs*, ed. P. F. Sale, 509–563. San Diego, CA: Academic Press.

Ebeling, A. W., and D. R. Laur. 1985. The influence of plant cover on surfperch abundance at an offshore temperate reef. *Env Biol Fish* 12:169–179.

Ebeling, A. W., D. R. Laur, and R. J. Rowley. 1985. Severe storm disturbance and reversal of community structure in a southern California kelp forest. *Mar Biol* 84:287–294.

Ebersole, J. P. 1977. The adaptive significance of interspecific territoriality in the reef fish *Eupomacentrus leucostictus*. *Ecology* 58:914–920.

Ebersole, J. P. 1980. Food density and territory size: An alternative model and a test on the reef fish *Eupomacentrus leucostictus*. *Amer Natur* 115:492–509.

Echelle, A. A. 1991. Conservation genetics and genic diversity in freshwater fishes of western North America. In *Battle against extinction. Native fish management in the American West*, ed. W. L. Minckley and J. E. Deacon, 141–153. Tucson: University of Arizona Press.

Echelle, A. A., and A. F. Echelle. 1984. Evolutionary genetics of a "species flock": Atherinid fishes on the Mesa Central of Mexico. In *Evolution of fish species flocks*, ed. A. E. Echelle and I. Kornfield, 93–110. Orono: University of Maine Press.

Echelle, A. E., and I. Kornfield, eds. 1984. *Evolution of fish species flocks*. Orono: University of Maine Press.

Edmunds, M. 1974. *Defence in animals*. New York: Longman.

Eggington, S., and B. D. Sidell. 1989. Thermal acclimation induces adaptive changes in subcellular structure of fish skeletal muscle. *Amer J Physiol* 256:R1–R10.

Ehrlich, P. R., F. H. Talbot, P. C. Russell, and G. R. Anderson. 1977. The behaviour of chaetodontid fishes, with special reference to Lorenz's 'poster coloration' hypothesis. *J Zool Lond* 183:213–228.

Eibl-Eibesfeldt, I. 1965. *Land of a thousand atolls*. Cleveland: World Publ.

Eibl-Eibesfeldt, I. 1970. *Ethology: The biology of behavior*. Translated by E. Klinghammer. New York: Holt, Rinehart and Winston.

Eisen, J. S. 1991. Developmental neurobiology of the zebrafish. *J Neurosci* 11:311–317.

Ekman, S. 1953. *Zoogeography of the sea*. London: Sidgwick and Jackson.

Elgar, M. A., and B. J. Crespi. 1992. *Cannibalism: Ecology and evolution among diverse taxa*. New York: Oxford University Press.

Elliott, D. K. 1987. Reassessment of *Astraspis desiderata*, the oldest North American vertebrate. *Science* 237:190–191.

Ellis, A. E. 1989. The immunology of teleosts. In *Fish pathology*, 2d ed., ed. R. J. Roberts, 135–152. Philadelphia: Bailliere Tindall.

Ellis, A. E., R. J. Roberts, and P. Tytler. 1989. The anatomy and physiology of teleosts. In *Fish pathology*, 2d ed., ed. R. J. Roberts, 13–55. Philadelphia: Bailliere Tindall.

Ellis, R. 1976. *The book of sharks*. New York: Grosset and Dunlop.

Ellis, R., and J. E. McCosker. 1991. *Great white shark*. New York: Harper Collins.

Emery, A. R. 1978. The basis of fish community structure: Marine and freshwater comparisons. *Env Biol Fish* 3:33–47.

Emery, A. R., and R. E. Thresher, eds. 1980. Biology of damselfishes. *Bull Mar Sci* 30:145–328.

Emlen, S. T., and L. W. Oring. 1977. Ecology, sexual selection, and the evolution of mating systems. *Science* 197:215–223.

Endean, R. 1973. Population explosions of *Acanthaster planci* and associated destruction of hermatypic corals in the Indo-West Pacific region. In *Biology and geology of coral reefs*. Vol. 2, *Biol. I*, ed. O. A. Jones and R. Endean, 389–438. New York: Academic Press.

Endler, J. A. 1980. Natural selection on color patterns in *Poecilia reticulata*. *Evolution* 34:76–91.

Endler, J. A. 1983. Natural and sexual selection on color patterns in poeciliid fishes. *Env Biol Fish* 9:173–190.

Endler, J. A. 1991. Interactions between predators and prey. In

Behavioural ecology: An evolutionary approach, 3d ed., ed. J. R. Krebs and N. B. Davies, 169–196. Oxford: Blackwell Science.

Enger, P. S. 1967. Hearing in herring. *Comp Biochem Physiol* 22:527–538.

Erickson, D. L., J. E. Hightower, and G. D. Grossman. 1985. The relative gonadal index: An alternative index for quantification of reproductive condition. *Comp Biochem Physiol* 81A:117–120.

Eschmeyer, W. N. 1990. *Catalog of the genera of recent fishes*. San Francisco: California Academy of Sciences.

Etnier, D. A. 1976. *Percina (Imostoma) tanasi*, a new percid fish from the Little Tennessee River, Tennessee. *Proc Biol Soc Washington* 88:469–488.

Etnier, D. A., and W. C. Starnes. 1993. *The fishes of Tennessee*. Knoxville: University of Tennessee Press.

Evans, D. H., ed. 1993. *The physiology of fishes*. Boca Raton, FL: CRC Press.

Evans, D. H. 1993. Osmotic and ionic regulation. In *The physiology of fishes*, ed. D. H. Evans, 315–342. Boca Raton, FL: CRC Press.

Evans, D. H., E. Chipouras, and J. A. Payne. 1989. Immunoreactive atriopeptin in the plasma of fishes: A potential role in gill hemodynamics. *Amer J Physiol* 257:R939–R945.

Evans, D. O. 1984. Temperature independence of the annual cycle of standard metabolism in the pumpkinseed. *Trans Amer Fish Soc* 113:494–512.

Evans, H. E. 1952. The correlation of brain pattern and feeding habits in four species of cyprinid fishes. *J Comp Neur* 97:133–142.

Evans, H. M. 1940. *Brain and body of fish*. New York: McGraw-Hill.

Evans, J. W., and R. L. Noble. 1979. The longitudinal distribution of fishes in an east Texas stream. *Amer Midl Natur* 101:333–343.

Evans, M. S. 1986. Recent major declines in zooplankton populations in the inshore region of Lake Michigan: Probable causes and implications. *Can J Fish Aquat Sci* 43:154–159.

Everhart, W. H., and W. D. Youngs. 1981. *Principles of fishery science*. 2d ed. Ithaca, NY: Cornell University Press.

Facey, D. E., and G. D. Grossman. 1990. The metabolic cost of maintaining position for four North American stream fishes: Effects of season and velocity. *Physiol Zool* 63:757–776.

Facey, D. E., and G. D. Grossman. 1992. The relationship between water velocity, energetic costs, and microhabitat use in four North American stream fishes. *Hydrobiologia* 239:1–6.

Fahy, W. E. 1982. The influence of temperature change on number of pectoral fin rays developing in *Fundulus majalis* (Walbaum). *J Cons Int Explor Mer* 40:21–26.

Fange, R., and D. Grove. 1979. Digestion. In *Fish physiology*, Vol. 8, *Bioenergetics and growth*, ed. W. S. Hoar, and D. J. Randall, 3:161–260. New York: Academic Press.

Farrell, A. P. 1993. Cardiovascular system. In *The physiology of fishes*, ed. D. H. Evans, 219–250. Boca Raton, FL: CRC Press.

Farrell, A. P., and D. R. Jones. 1992. The heart. *Fish Physiol* 12A:1–88.

Farrell, A. P., and D. J. Randall. 1995. *Fish physiology*. Vol. 14. New York: Academic Press.

Farris, J. S. 1970. Methods for computing Wagner trees. *Syst Zool* 19:83–92.

Farris, J. S. 1988. Hennig86, Version 1.5. Port Jefferson Station, NY.

Fausch, K. D. 1988. Tests of competition between native and introduced salmonids in streams: What have we learned. *Can J Fish Aquat Sci* 45:2238–2246.

Fausch, K. D., J. R. Karr, and P. R. Yant. 1984. Regional application of an index of biotic integrity based on stream fish communities. *Trans Amer Fish Soc* 113:39–55.

Fautin, D. G. 1991. The anemonefish symbiosis: What is known and what is not. *Symbiosis* 10:23–46.

Feder, H. M. 1966. Cleaning symbiosis in the marine environment. In *Symbiosis*, Vol. l, ed. S. M. Henry, 327–380. New York: Academic Press.

Feder, M. E., and W. W. Burggren. 1985. Skin breathing in vertebrates. *Sci Amer* 253:126–142.

Feder, M. E., and G. V. Lauder, eds. 1986. *Predator-prey relationships: Perspectives and approaches from the study of lower vertebrates.* Chicago: University of Chicago Press.

Feldmeth, C. R. 1981. The evolution of thermal tolerance in desert pupfish (Genus *Cyprinodon*). In *Fishes in North American deserts*, ed. R. J. Naiman and D. L. Soltz, 357–384. New York: John Wiley & Sons.

Feng, A. S. 1991. Electric organs and electroreceptors. In *Comparative animal physiology*, 4th ed., ed. C. L. Prosser, 317–334. New York: John Wiley & Sons.

Fernald, R. D. 1984. Vision and behavior in an African cichlid. *Amer Sci* 72:58–65.

Fernholm, B. 1974. Diurnal variations in the behaviour of the hagfish *Eptatretus burgeri*. *Mar Biol* 27:351–356.

Fernholm, B. 1981. Thread cells from the slime glands of hagfish (Myxinidae). *Acta Zool* 62:137–145.

Ferraro, S. P. 1980. Daily time of spawning of 12 fishes in the Peconic Bays, New York. *Fish Bull US* 78:455–464.

Ferry, L. A., and G. V. Lauder. 1996. Heterocercal tail function in leopard sharks: a three-dimensional kinematic analysis of two models. *J Exp Biol* 199:2253–2268.

Fiedler, P. L., and S. K. Jain, eds. 1992. *Conservation biology: The theory and practice of nature conservation, preservation and management.* New York: Chapman and Hall.

Finch, C. E. 1990. *Longevity, senescence, and the genome.* Chicago: University of Chicago Press.

Fine, M. L., H. E. Winn, and B. Olla. 1977. Communication in fishes. In *How animals communicate*, ed. T. Sebeok, 472–518. Bloomington: Indiana University Press.

Fink, S. V., and W. L. Fink. 1981. Interrelationships of the ostariophysan fishes (Teleostei). *J Linn Soc (Zool)* 72:297–353.

Finnerty, J. R., and B. A. Block. 1995. Evolution of cytochrome *b* in the Scombroidei (Teleostei): Molecular insights into billfish (Istiophoridae and Xiphiidae) relationships. *Fish Bull US* 93:78–96.

Fischer, E. A., and C. W. Petersen. 1987. The evolution of sexual patterns in the seabasses. *BioScience* 37:482–489.

Fishelson, L., W. L. Montgomery, and A. A. Myrberg Jr. 1985. A unique symbiosis in the gut of tropical herbivorous surgeonfish (Acanthuridae: Teleostei) from the Red Sea. *Science* 229:49–51.

Fitt, W. K., H. J. Spero, J. Halas, M. W. White, and J. W. Porter. 1993. Recovery of the coral *Montastrea annularis* in the Florida Keys after the 1987 Caribbean "bleaching event." *Coral Reefs* 12:57–64.

FitzGerald, G. J. 1992. Filial cannibalism in fishes: Why do parents eat their offspring? *Trends Ecol Evol* 7:7–10.

FitzGerald, G. J., and R. J. Wootton. 1993. Behavioural ecology of sticklebacks. In *The behaviour of teleost fishes*, 2d ed., ed. T. J. Pitcher, 537–572. London: Chapman and Hall.

Flecker, A. S., and J. D. Allan. 1984. The importance of predation, substrate and spatial refugia in determining lotic insect distributions. *Oecologica* 64:306–313.

Flecker, A. S., and C. R. Townsend. 1994. Community-wide consequences of trout introduction in New Zealand streams. *Ecol Appl* 4:798–807.

Forey, P. L. 1985. *Latimeria chalumnae* and its pedigree. *Env Biol Fish* 32:75–97.

Forey, P., and P. Janvier. 1993. Agnathans and the origin of jawed vertebrates. *Nature* 361:129–134.

Fortier, L., and W. C. Leggett. 1983. Vertical migrations and transport of larval fish in a partially mixed estuary. *Can J Fish Aquat Sci* 40:154–155.

Foster, S. A. 1985. Group foraging by a coral reef fish: A mechanism for gaining access to defended resources. *Anim Behav* 33:782–792.

Fowler, M. C., and T. O. Robson. 1978. The effects of the food preferences on stocking rates of grass carp (*Ctenopharyngodon idella* Val.) on mixed plant communities. *Aquat Bot* 5:261–276.

Fox, F. Q., G. P. Richardson, and C. Kirk. 1985. Torpedo electromotor system development: Neuronal cell death and electric organ development in the fourth branchial arch. *J Comp Neur* 236:274–281.

Francis, R. C. 1990. Climate change and marine fisheries. *Fisheries* 15(6):7–9.

Francis, R. C. 1992. Sexual lability in teleosts: Developmental factors. *Quart Rev Biol* 67:1–18.

Francis, R. C., and G. W. Barlow. 1993. Social control of primary sex differentiation in the Midas cichlid. *Proc Natl Acad Sci USA* 90:10673–10675.

Fraser, D. F., and F. A. Huntingford. 1986. Feeding and avoiding predation hazard: The behavioral response of the prey. *Ethology* 73:56–68.

Frazer, T. K., W. J. Lindberg, and G. R. Stanton. 1991. Predation of sand dollars by gray triggerfish, *Balistes capriscus*, in the northern Gulf of Mexico. *Bull Mar Sci* 48:159–164.

Freeman, M. C., and G. D. Grossman. 1992. Group foraging by a stream minnow: Shoals or aggregations? *Anim Behav* 44:393–403.

Freeman, M. C., and D. J. Stouder. 1989. Intraspecific interactions influence size specific depth distribution in *Cottus bairdi*. *Env Biol Fish* 24:231–236.

Freihofer, W. C. 1963. Patterns of the ramus lateralis accessorius and their systematic significance in teleostean fishes. *Stanford Ichthy Bull* 8(2):79–169.

Fremling, C. R., J. L. Rasmussen, R. E. Sparks, S. P. Cobb, C. F. Bryan, and T. O. Claflin. 1989. Mississippi River fisheries: A case history. In *Proc Internat Large River Symp*, ed. D. P. Dodge. *Can Spec Publ Fish Aquat Sci* 106:309–351.

Fricke, H. W. 1973. Behaviour as part of ecological adaptation. *Helgolander wiss Meeresunters* 24:120–144.

Fricke, H. W. 1980. Mating systems, maternal and biparental care in triggerfish (Balistidae). *Z Tierpsychol* 53:105–122.

Fricke, H. 1986. Coelacanths. *Nat Geog* 173:825–838.

Fricke, H., and J. Frahm. 1992. Evidence for lecithotrophic viviparity in the living coelacanth. *Nuturwissenchaften* 79:476–479.

Fricke, H. W., and S. Fricke. 1977. Monogamy and sex change by aggressive dominance in coral reef fish. *Nature* 266:830–832.

Fricke, H.W., and K. Hissmann. 1994. Home range and migrations of the living coelacanth *Latimeria chalumnae*. *Mar Biol* 120:171–180.

Fricke, H., K. Hissmann, J. Schauer, O. Reinicke, L. Kasang, and R. Plante. 1991. Habitat and population size of the coelacanth *Latimeria chalumnae* at Grand Comoro. *Env Biol Fish* 32:287–300.

Fricke, H., O. Reinicke, H. Hofer, and W. Nachtigall. 1987. Locomotion of the coelacanth *Latimeria chalumnae* in its natural environment. *Nature* (London) 331–333.

Fricke, H., J. Schauer, K. Hissmann, L. Kasang, and R. Plante. 1991. Coelacanth *Latimeria chalumnae* aggregates in caves: First observations on their resting habitat and social behavior. *Env Biol Fish* 30:281–285.

Fry, F. E. J. 1971. The effect of environmental factors on the physiology of fish. In *Fish physiology*, Vol. 6, ed. W. S. Hoar and D. J. Randall, 1–98. New York: Academic Press.

Frydl, P., and C. W. Stearn. 1978. Rate of bioerosion by parrotfish in Barbados reef environments. *J Sedim Petrol* 48:1149–1158.

Fryer, G., and T. D. Iles. 1972. *The cichlid fishes of the Great Lakes of Africa.* Edinburgh: Oliver and Boyd.

Funk, V. A. 1995. Cladistic methods. In *Hawaiian biogeography,* ed. W. L. Wagner and V. A. Funk. Washington, DC: Smithsonian Institution Press.

Gabbott, S. E., R. J. Aldridge, and J. N. Theron. 1995. A giant conodont with preserved muscle tissue from the Upper Ordovician of South Africa. *Nature* (London) 374:800–802.

Gabda, J. 1991. *Marine fish parasitology.* New York: VCH Publ.

Gage, J. D., and P. A. Tyler. 1991. *Deep-sea biology: A natural history of organisms at the deep-sea floor.* Cambridge: Cambridge University Press.

Gagnier, P. Y. 1989. The oldest vertebrate: A 470-million-year-old jawless fish, *Sacabambaspis janvieri,* from the Ordovician of Bolivia. *Nat Geog Res* 5:25–253.

Gagnier, P. Y., A. R. M. Blieck, and G. S. Rodrico. 1986. First Ordovician vertebrate from South America. *Geobios* 19:629–634.

Gardiner, B. G. 1984. Devonian palaeoniscid fishes: New specimens of *Mimia* and *Moythomasia* from the Upper Devonian of Western Australia. *Bull Brit Mus Natur Hist (Geol)* 37:173–428.

Gassman, N. J., L. B. Nye, and M. C. Schmale. 1994. Distribution of abnormal biota and sediment contaminants in Biscayne Bay, Florida. *Bull Mar Sci* 54:929–943.

Gauthreaux, S. A. Jr., ed. 1980. *Animal migration, orientation and navigation.* New York: Academic Press.

Gengerke, T. W. 1986. Distribution and abundance of paddlefish in the United States. In *The paddlefish: Status, management and propagation,* ed. J. G. Dillard, L. Graham, and T. Russell, 22–35. Columbia, MO: North Central Division, American Fisheries Society, Spec Publ 7.

Gerking, S. D., ed. 1978. *Ecology of freshwater fish production.* London: Blackwell Science.

Gerking, S. D. 1994. *Feeding ecology of fish.* San Diego, CA: Academic Press.

Gery, J. 1969. The fresh-water fishes of South America. In *Biogeography and ecology in South America,* ed. E. J. Fittkau et al., 828–848. The Hague: Dr. W. Junk.

Ghiselin, M. T. 1969. The evolution of hermaphroditism among animals. *Q Rev Biol* 44:189–208.

Gibbs, R. H. Jr., and B. B. Collette. 1967. Comparative anatomy and systematics of the tunas, genus *Thunnus. US Fish Wild Serv Fish Bull* 66:65–130.

Gibbs, R. H. Jr., E. Jarosewich, and H. L. Windom. 1974. Heavy metal concentrations in museum fish specimens: Effects of preservatives and time. *Science* 184:475–477.

Gibson, R. N. 1978. Lunar and tidal rhythms in fish. In *Rhythmic activity of fishes,* ed. J. E. Thorpe, 201–213. London: Academic Press.

Gibson, R. N. 1982. Recent studies on the biology of intertidal fishes. *Oceanog Mar Biol Ann Rev* 20:363–414.

Gibson, R. N. 1992. Tidally-synchronized behaviour in marine fishes. In *Rhythms in fishes,* ed. M. A. Ali, 63–81. New York: Plenum Press.

Gibson, R. N. 1993. Intertidal teleosts: Life in a fluctuating environment. In *The behaviour of teleost fishes,* 2d ed., ed. T. J. Pitcher, 513–536. London: Chapman and Hall.

Gillett, C., and P. Rombard. 1986. Prehatching embryo survival of nine freshwater fish eggs after a pH shock during fertilization or early stages of development. *Repro Nutrit Devel* 26:1319–1333.

Gilliam, J. F., and D. F. Fraser. 1987. Habitat selection under predation hazard: Test of a model with foraging minnows. *Ecology* 68:1856–1862.

Gilliam, J. F., D. F. Fraser, and A. M. Sabat. 1989. Strong effects of foraging minnows on a stream benthic invertebrate community. *Ecology* 70:445–452.

Gilligan, M. R. 1989. An illustrated field guide to the fishes of Gray's Reef National Marine Sanctuary. Washington, DC: NOAA Tech Mem NOS MEMD 25.

Gilmore, R. G. 1991. The reproductive biology of lamnoid sharks. In *Discovering sharks,* ed. S. H. Gruber, 64–67. Highlands, NJ: American Littoral Society, Spec Publ 14.

Gilmore, R. G., J. W. Dodrill, and P. A. Linley. 1983. Embryonic development of the sand tiger shark, *Odontaspis taurus* Rafinesque. *Fish Bull US* 81:201–225.

Gladstone, W., and M. Westoby. 1988. Growth and reproduction in *Canthigaster valentini* (Pisces, Tetraodontidae): A comparison of a toxic reef fish with other reef fishes. *Env Biol Fish* 21:207–221.

Glantz, M. H., and L. E. Feingold, eds. 1990. *Climate variability, climate change and fisheries.* Boulder, CO: Environmental and Societal Impacts Group, Nat Ctr Atmos Res.

Glover, C. J. M. 1982. Adaptations of fishes in arid Australia. In *Evolution of the flora and fauna of arid Australia,* ed. W. R. Barber and P. J. M. Greenslade, 241–246. Frewville: Peacock Publications.

Godin, J.-G. J. 1986. Antipredator function of shoaling in teleost fishes: A selective review. *Naturalist Can (Rev Ecol Syst)* 113:241–250.

Godin, J.-G. J., and M. H. A. Keenleyside. 1984. Foraging on patchily distributed prey by a cichlid fish (Teleostei, Cichlidae): A test of the ideal free distribution theory. *Anim Behav* 32:1201–1213.

Godin, J.-G. J., and C. D. Sproul. Risk taking in parasitized sticklebacks under threat of predation: Effects of energetic need and food availability. *Can J Zool* 66:2360–2367.

Golani, D. 1993. The biology of the Red Sea migrant, *Saurida undosquamis* in the Mediterranean and comparison with the indigenous confamilial *Synodus saurus* (Teleostei: Synodontidae). *Hydrobiologica* 271:109–117.

Gold, J. R. 1979. Cytogenetics. In *Fish physiology,* Vol. 8, ed. W. S. Hoar, D. J. Randall, and E. M. Donaldson, 353–405. New York: Academic Press.

Gon, O., and P. C. Heemstra, eds. 1990. *Fishes of the Southern Ocean.* Grahamstown, South Africa: J. L. B. Smith Institute of Ichthyology.

Goodall, D. W. 1976. Introduction. In *Evolution of desert biota,* ed. D. W. Goodall, 3–5. Austin, TX: University of Texas Press.

Goode, G. B. 1884. *Natural history of useful aquatic animals. The fisheries and fishery industries of the United States.* Sec. l, Part 3, *The food fishes of the United States.* Washington, DC: Government Printing Office.

Gooding, R. M., and J. J. Magnuson. 1967. Ecological significance of a drifting object to pelagic fishes. *Pacif Sci* 21:486–497.

Gorbman, A. 1983. Reproduction in cyclostome fishes and its regulation. In *Fish physiology,* Vol. 9, Part A, ed. W. S. Hoar, D. J. Randall, and E. M. Donaldson, 1–29. New York: Academic Press.

Gorbman, A., H. Kobayashi, Y. Honma, and M. Matsuyama. 1990. The hagfishery of Japan. *Fisheries* 15(4):12–18.

Gorlick, D. L. 1976. Dominance hierarchies and factors influencing dominance in the guppy *Poecilia reticulata* (Peters). *Anim Behav* 24:336–346.

Gorlick, D. L., P. D. Atkins, and G. S. Losey. 1987. Effect of cleaning by *Labroides dimidiatus* (Labridae) on an ectoparasite population infecting *Pomacentrus vaiuli* (Pomacentridae) at Enewetak Atoll. *Copeia* 1987:41–45.

Gorman, G. C. 1973. The chromosomes of the Reptilia, a cytotaxonomic interpretation. In *Cytotaxonomy and vertebrate evolution,* ed. A. B. Chiarelli and E. Capanna, 349–424. London: Academic Press.

Gorman, O. T. 1988. The dynamics of habitat use in a guild of Ozark minnows. *Ecol Monogr* 58:1–18.

Gorr, T., and T. Kleinschmidt. 1993. Evolutionary relationships of the coelacanth. *Amer Sci* 81:72–82.

Gosline, W. A. 1971. *Functional morphology and classification of teleostean fishes.* Honolulu: University Press of Hawaii.

Gotceitas, V., and J.-G. J. Godin. 1992. Effects of location of food delivery and social status on foraging-site selection by juvenile Atlantic salmon. *Env Biol Fish* 35:291–300.

Gould, S. J. 1966. Allometry and size in ontogeny and phylogeny. *Biol Rev* 41:587–640.

Gould, S. J. 1977. *Ontogeny and phylogeny.* Cambridge, MA: Belknap Press.

Gould, S. J. 1990. Bully for *Brontosaurus. Natur Hist* 1990(2):16–24.

Goulding, M. 1980. *The fishes and the forest. Explorations in Amazonian natural history.* Berkeley: University of California Press.

Grande, L., and W. E. Bemis. 1991. Osteology and phylogenetic relationships of fossil and Recent paddlefishes (Polyodontidae) with comments on the interrelationships of Acipenseriformes. *J Vert Paleo* 11 (Memoir 1, suppl to No 1):1–121.

Grassle, J. F. 1986. The ecology of deep-sea hydrothermal vent communities. *Adv Mar Biol* 23:301–362.

Graves, J. E., and A. E. Dizon. 1987. Mitochondrial DNA sequence similarity of Atlantic and Pacific albacore tuna (*Thunnus alalunga*). *Can J Fish Aquat Sci* 46:870–873.

Gray, J. 1968. *Animal locomotion.* London: Weidenfeld and Nicolson.

Greenewalt, C. H. 1962. Dimensional relationships for flying animals. *Smithsonian Misc Collections* 144(2), Publ 4477:1–46.

Greenwood, P. H. 1981. *The haplochromine fishes of the east African lakes.* Ithaca, NY: Cornell University Press.

Greenwood, P. H. 1984. What *is* a species flock? In *Evolution of fish species flocks,* ed. A. E. Echelle and I. Kornfield, 13–19. Orono: University of Maine Press.

Greenwood, P. H. 1987. The natural history of African lungfishes. In *The biology and evolution of lungfishes,* ed. W. E. Bemis, W. W. Burggren, and N. E. Kemp, 163–179. New York: Alan R. Liss.

Greenwood, P. H. 1991. Speciation. In *Cichlid fishes: Behaviour, ecology and evolution,* ed. M. H. A. Keenleyside, 86–102. London: Chapman and Hall.

Greenwood, P. H., D. E. Rosen, S. H. Weitzman, and G. S. Myers. 1966. Phyletic studies of teleostean fishes, with a provisional classification of living forms. *Bull Amer Mus Nat Hist* 131:339–456.

Gregory, W. K. 1933. Fish skulls: A study of the evolution of natural mechanisms. *Trans Amer Philos Soc,* new ser, 23:71–481.

Grier, H. J. 1981. Cellular organization of the testis and spermatogenesis in fishes. *Amer Zool* 21:345–357.

Grimm, N. B. 1988. Feeding dynamics, nitrogen budgets, and ecosystem role of a desert stream omnivore, *Agosia chrysogaster* (Pisces: Cyprinidae). *Env Biol Fish* 21:143–152.

Groombridge, B., ed. 1992. *Global biodiversity: Status of the Earth's living resources.* London: Chapman and Hall.

Groot, C., and L. Margolis, eds. 1991. *Pacific salmon life histories.* Vancouver, Canada: University of British Columbia Press.

Gross, M. R. 1982. Sneakers, satellites and parentals: Polymorphic mating strategies in North American sunfishes. *Z Tierpsychol* 60:1–26.

Gross, M. R. 1984. Sunfish, salmon, and the evolution of alternative reproductive strategies and tactics in fishes. In *Fish reproduction: Strategies and tactics,* ed. G. W. Potts and R. J. Wootton, 55–75. London: Academic Press.

Gross, M. R. 1987. Evolution of diadromy in fishes. In *Common strategies of anadromous and catadromous fishes,* ed. M. J. Dadswell, R. J. Klauda, C. M. Moffitt, R. L. Saunders, R. A. Rulifson, and J. E. Cooper. *Amer Fish Soc Symp* 1:14–25.

Gross, M. R. 1991. Salmon breeding behavior and life history evolution in changing environments. *Ecology* 72:1180–1186.

Gross, M. R., R. M. Coleman, and R. M. McDowall. 1988. Aquatic productivity and the evolution of diadromous fish migration. *Science* 239:1291–1293.

Grossman, G. D. 1980. Food, fights, and burrows: The adaptive significance of intraspecific aggression in the bay goby (Pisces: Gobiidae). *Oecologia* (Berl) 45:261–266.

Grossman, G. D., and M. C. Freeman. 1987. Microhabitat use in a stream fish assemblage. *J Zool* (London) 212:151–176.

Grossman, G. D., M. C. Freeman, P. B. Moyle, and J. O. Whitaker Jr. 1985. Stochasticity and assemblage organization in an Indiana stream fish assemblage. *Amer Natur* 126:275–285.

Grossman, G. D., P. B. Moyle, and J. O. Whitaker Jr. 1982. Stochasticity in structural and functional characteristics of an Indiana stream fish assemblage: A test of community theory. *Amer Natur* 120:423–454.

Gruber, S. H., ed. 1991. *Discovering sharks.* Highlands, NJ: American Littoral Society, Spec Publ 14.

Gruber, S. H., and A. A. Myrberg Jr. 1977. Approaches to the study of the behavior of sharks. *Amer Zool* 17:471–486.

Gucinski, H., R. T. Lackey, and B. C. Spence. 1990. Global climate change: Policy implications for fisheries. *Fisheries* 15(6):33–38.

Gulland, J. A. 1983. *Fish stock assessment: A manual of basic methods.* Chichester: John Wiley & Sons.

Gunning, G. E., and R. D. Suttkus. 1991. Species dominance in the fish populations of the Pearl River at two study areas in Mississippi and Louisiana: 1966–1988. *Proc Southeast Fish Council* No. 23:7–23.

Gunther, A. C. L. G. 1859–1870. *Catalogue of the fishes in the British Museum.* 8 vols. London: Trustees of the British Museum.

Gunther, A. C. L. G. 1880. *An introduction to the study of fishes.* New Delhi: Today and Tomorrow's Book Agency.

Guthrie, D. M., and D. R. A. Muntz. 1993. Role of vision in fish behavior. In *The behaviour of teleost fishes,* 2d ed., ed. T. J. Pitcher, 87–128. Baltimore: Chapman and Hall.

Hagedorn, M. 1986. The ecology, courtship, and mating of gymnotiform electric fish. In *Electroreception,* ed. T. H. Bullock and W. Heiligenberg, 497–525. New York: John Wiley & Sons.

Hagedorn, M., M. Womble, and T. E. Finger. 1990. Synodontid catfish: A new group of weakly electric fish. *Brain Behav Evol* 35:268–277.

Hagen, D. W., and J. D. McPhail. 1970. The species problem within *Gasterosteus aculeatus* on the Pacific coast of North America. *J Fish Res Board Can* 27:147–155.

Hager, M. C., and G. S. Helfman. 1991. Safety in numbers: Shoal size choice by minnows under predatory threat. *Behav Ecol Sociobiol* 29:271–276.

Hailman, J. P. 1977. *Optical signals: Animal communication and light.* Bloomington: Indiana Press.

Hairston, N. G. Jr., K. T. Li, and S. S. Easter Jr. 1982. Fish vision and the detection of planktonic prey. *Science* 218:1240–1242.

Hales, L. S. Jr. 1987. Distribution, abundance, reproduction, food habits, age, and growth of the round scad *Decapterus punctatus,* in the South Atlantic Bight. *Fish Bull US* 85:251–268.

Hales, L. S., and M. C. Belk. 1992. Validation of otolith annuli of bluegills in a southeastern thermal reservoir. *Trans Amer Fish Soc* 121:823–830.

Hall, C. A. S. 1972. Migration and metabolism in a temperate stream ecosystem. *Ecology* 53:585–604.

Hall, D. J., and E. E. Werner. 1977. Seasonal distribution and abundance of fishes in the littoral zone of a Michigan lake. *Trans Amer Fish Soc* 106:545–555.

Hall, F. G., and F. H. McCutcheon. 1938. The affinity of hemoglobin for oxygen in marine fishes. *J Cell Comp Physiol* 11:205–212.

Halstead, L. B., Y. H. Liu, and K. P'an. 1979. Agnathans from the Devonian of China. *Nature* 282:831–833.

Hamilton, W. D. 1971. Geometry for the selfish herd. *J Theor Biol* 31:295–311.

Hamilton, W. J. III, and R. M. Peterman. 1971. Countershading in the colourful reef fish *Chaetodon lunula*: Concealment, communication or both. *Anim Behav* 19:357–364.

Hamlett, W. C. 1991. From egg to placenta: Placental reproduction in sharks. In *Discovering sharks*, ed. S. H. Gruber, 56–63. Highlands, NJ: American Littoral Society, Spec Publ 14.

Hamoir, G., and N. Geradin-Otthiers. 1980. Differentiation of the sarcoplasmic proteins of white, yellowish and cardiac muscles of an Antarctic hemoglobin-free fish, *Champsocephalus gunnari*. *Comp Biochem Physiol* 65B:199–206.

Hanson, L. H., and W. D. Swink. 1989. Downstream migration of recently metamorphosed sea lampreys in the Ocqueoc River, Michigan before and after treatment with lampricides. *N Amer J Fish Manage* 9:327–331.

Hanssens M. M., G. G. Teugels, and D. F. E. Thys Van Den Audernaerde. 1995. Subspecies in the *Polypterus palmas* complex (Brachiopterygii; Polypteridae) from West and Central Africa. *Copeia* 1995:694–705.

Hara, T. J., ed. 1982. *Chemoreception in fishes*. Amsterdam: Elsevier.

Hara, T. J., ed. 1992. *Fish chemoreception*. London: Chapman and Hall.

Hara, T. J. 1993. Role of olfaction in fish behaviour. In *The behaviour of teleost fishes*, 2d ed., ed. T. J. Pitcher, 171–199. London: Chapman and Hall.

Harden-Jones, F. R. 1968. *Fish migration*. London: Edward Arnold.

Harder, W. 1975. *Anatomy of fishes*. Stuttgart: E. Schweizerbart'sche Verlagsbuchhandlung.

Hardisty, M. W. 1979. *Biology of the cyclostomes*. London: Chapman and Hall.

Hardisty, M. W. 1982. Lampreys and hagfishes: Analysis of cyclostome relationships. In *The biology of lampreys*, Vol. 4b, ed. M. W. Hardisty and I. C. Potter, 165–259. New York: Academic Press.

Hardisty, M. W., and I. C. Potter. 1971–1982. *The biology of lampreys*. Vols. 1–4b. New York: Academic Press.

Hargrove, E. C. 1989. *Foundations of environmental ethics*. Englewood Cliffs, NJ: Prentice Hall.

Harmon, M. E., et al. 1986. Ecology of coarse woody debris in temperate ecosystems. *Adv Ecol Res* 15:133–302.

Harrington, M. E. 1993. Aggression in damselfish: Adult-juvenile interactions. *Copeia* 1993:67–74.

Harrington, R. W. Jr. 1955. The osteocranium of the American cyprinid fish, *Notropis bifrenatus*, with an annotated synonymy of teleost skull bones. *Copeia* 1956:267–290.

Harrington, R. W. Jr. 1971. How ecological and genetic factors interact to determine when self-fertilizing hermaphrodites of *Rivulus marmoratus* change into functional secondary males, with a reappraisal of the modes of intersexuality among fishes. *Copeia* 1971:389–431.

Harrington, R. W. Jr. 1975. Sex determination and differentiation among uniparental homozygotes of the hermaphroditic fish *Rivulus marmoratus* (Cyprinodontidae: Atheriniformes). In *Intersexuality in the animal kingdom*, ed. R. Reinboth, 249–262. Berlin: Springer-Verlag.

Hart, P. J. B. 1993. Teleost foraging: Facts and theories. In *The behaviour of teleost fishes*, 2d ed., ed. T. J. Pitcher, 253–284. London: Chapman and Hall.

Harvey, B. C. 1987. Susceptibility of young-of-the-year fishes to downstream displacement by flooding. *Trans Amer Fish Soc* 116:851–855.

Hasler, A. D. 1983. Synthetic chemicals and pheromones in homing salmon. In *Control processes in fish physiology*, ed. J. C. Rankin, T. J. Pitcher, and R. T. Duggan, 103–116. London: Croom Helm.

Hasler, A. D., and A. T. Scholz. 1983. *Olfactory imprinting and homing in salmon: Investigations into the mechanism of the imprinting process*. Berlin: Springer-Verlag.

Haswell, M. S., and D. J. Randall. 1978. The pattern of carbon dioxide excretion in the rainbow trout, *Salmo gairdneri*. *J Exp Biol* 72:17–24.

Hawkins, A. D. 1993. Underwater sound and fish behaviour. In *The behaviour of teleost fishes*, 2d ed., ed. T. J. Pitcher, 129–169. London: Chapman and Hall.

Hawkins, A. D. 1993. Underwater sound and fish behaviour. In *The behaviour of teleost fishes*, 2d ed., ed. T. J. Pitcher, 130–169. Baltimore: Chapman and Hall.

Hawkins, A. D., and A. A. Myrberg Jr. 1983. Hearing and sound communication under water. In *Bioacoustics, a comparative approach*, ed. B. Lewis, 347–405. London: Academic Press.

Hawryshyn, C. W. 1992. Polarization vision in fishes. *Amer Sci* 80:164–175.

Hay, M. E. 1986. Associational plant defenses and the maintenance of species diversity: Turning competitors into accomplices. *Amer Natur* 128:617–641.

Hay, M. E. 1991. Fish-seaweed interactions on coral reefs: Effects of herbivorous fishes and adaptations of their prey. In *The ecology of fishes on coral reefs*, ed. P. F. Sale, 96–119. San Diego, CA: Academic Press.

Hazel, J. R. 1993. Thermal biology. In *The physiology of fishes*, ed. D. H. Evans, 427–468. Boca Raton, FL: CRC Press.

Hazon, N., R. J. Balment, M. Perrott, and L. B. O'Toole. 1989. The renin-angiotensin and vascular and dipsogenic regulation in elasmobranchs. *Gen Comp Endocrinol* 74:230–236.

Healey, E. G. 1957. The nervous system. *Physiol Fish* 2:1–119.

Healey, M. C., and C. Groot. 1987. Marine migration and orientation of ocean-type chinook and sockeye salmon. In *Common strategies of anadromous and catadromous fishes*, ed. M. J. Dadswell, R. J. Klauda, C. M. Moffitt, R. L. Saunders, R. A. Rulifson, and J. E. Cooper. *Amer Fish Soc Symp* 1:298–312.

Heath, A. G. 1987. *Water pollution and fish physiology*. Boca Raton, FL: CRC Press.

Hebrank, M. R. 1980. Mechanical properties and locomotor functions of eel skin. *Biol Bull* 158:58–68.

Heemstra, P. C., and P. H. Greenwood. 1992. New observations on the visceral anatomy of the late-term fetuses of the living coelacanth fish and the oophagy controversy. *Proc Royal Soc London* 249B:49–55.

Heezen, B. C., and C. D. Hollister. 1971. *The face of the deep*. London: Oxford University Press.

Heiligenberg, W. 1993. Electrosensation. In *The physiology of fishes*, ed. D. H. Evans, 137–160. Boca Raton, FL: CRC Press.

Heins, D. C., and F. G. Rabito Jr. 1986. Spawning performance in North American minnows: Direct evidence of the occurrence of multiple clutches in the genus *Notropis*. *J Fish Biol* 28:343–357.

Heisler, N. 1993. Acid-base regulation. In *The physiology of fishes*, ed. D. H. Evans, 343–378. Boca Raton, FL: CRC Press.

Helfman, G. S. 1978. Patterns of community structure in fishes: Summary and overview. *Env Biol Fish* 3:129–148.

Helfman, G. S. 1979. Temporal relationships in a freshwater fish community. Ph.D. dissertation, Cornell University, Ithaca, NY.

Helfman, G. S. 1979. Fish attraction to floating objects in lakes. In *Response of fish to habitat structure in standing water*, ed. D. L. Johnson and R. A. Stein, 49–57. Columbia, MO: N Cent Div, American Fisheries Society, Spec Publ 6.

Helfman, G. S. 1981a. Twilight activities and temporal structure in a freshwater fish community. *Can J Fish Aquat Sci* 38:1405–1420.

Helfman, G. S. 1981b. The advantage to fishes of hovering in shade. *Copeia* 1981:392–400.

Helfman, G. S. 1984. School fidelity in fishes: The yellow perch pattern. *Anim Behav* 32:663–672.

Helfman, G. S. 1986. Behavioral responses of prey fishes during predator-prey interactions. In *Predator-prey relationships: Perspectives and approaches from the study of lower vertebrates*, ed. M. E. Feder and G. V. Lauder, 135–156. Chicago: University of Chicago Press.

Helfman, G. S. 1989. Threat-sensitive predator avoidance in damselfish-trumpetfish interactions. *Behav Ecol Sociobiol* 24:47–58.

Helfman, G. S. 1990. Mode selection and mode switching in foraging animals. *Adv Stud Behav* 19:249–298.

Helfman, G. S. 1993. Fish behaviour by day, night and twilight. In *The behaviour of teleost fishes*, 2d ed., ed. T. J. Pitcher, 479–512. London: Chapman and Hall.

Helfman, G. S., D. E. Facey, L. S. Hales Jr., and E. L. Bozeman Jr. 1987. Reproductive ecology of the American eel. In *Common strategies of anadromous and catadromous fishes*, ed. M. J. Dadswell, R. J. Klauda, C. M. Moffitt, R. L. Saunders, R. A. Rulifson, and J. E. Cooper. *Amer Fish Soc Symp* 1:42–56.

Helfman, G. S., and E. T. Schultz. 1984. Social transmission of behavioural traditions in a coral reef fish. *Anim Behav* 32:379–384.

Helfman, G. S., D. L. Stoneburner, E. L. Bozeman, P. A. Christian, and R. Whalen. 1983. Ultrasonic telemetry of American eel movements in a tidal creek. *Trans Amer Fish Soc* 112:105–110.

Helfman, G. S., and D. L. Winkelman. 1991. Energy trade-offs and foraging mode choice in American eels. *Ecology* 72:310–318.

Hellawell, J. M. 1983. *Biological indicators of freshwater pollution and environmental management*. New York: Elsevier Applied Science Publ.

Heller, E., and R. E. Snodgrass. 1903. Papers from the Hopkins Stanford Galapagos Expedition, 1898–1899. XV. New fishes. *Proc Wash Acad Sci* 5:189–229.

Hemmingsen, E. A. 1991. Respiratory and cardiovascular adaptations in hemoglobin-free fish: Resolved and unresolved problems. In *Biology of Antarctic fish*, ed. G. di Prisco, B. Maresca, and B. Tota, 191–203. Berlin: Springer-Verlag.

Hempel, G. 1979. *Early life history of fish: The egg stage*. Seattle: University of Washington Press.

Hennig, W. 1966. *Phylogenetic systematics*. Urbana: University of Illinois Press.

Herald, E. S. 1961. *Living fishes of the world*. Garden City, NY: Doubleday.

Herbold, B. 1984. Structure of an Indiana stream fish association: Choosing an appropriate model. *Amer Natur* 124:561–572.

Herdendorf, C. E., and T. M. Berra. 1995. A Greenland shark from the wreck of the SS *Central America* at 2,200 meters. *Copeia* 1995:950–953.

Hersey, J. 1988. *Blues*. New York: Random House.

Hertel, H. 1966. *Structure, form and movement*. New York: Reinholt.

Hewitt, R. P., G. H. Theilacker, and N. C. H. Lo. 1985. Causes of mortality in young jack mackerel. *Mar Ecol Prog Ser* 26:1–10.

Hiatt, R. W., and V. E. Brock. 1948. On the herding of prey and the schooling of the black skipjack, *Euthynnus yaito* Kishinouye. *Pac Sci* 2:297–298.

Hickman, C. P., and B. F. Trump. 1969. The kidney. In *Fish physiology*, Vol. 1, ed. W. S. Hoar and D. J. Randall, 1–239. New York: Academic Press.

Hildebrand, M. 1982. *Analysis of vertebrate structure*. 2d ed. New York: John Wiley & Sons.

Hildebrand, M. 1988. *Analysis of vertebrate structure*. 3d ed. New York: John Wiley & Sons.

Hildebrand, M., D. M. Bramble, K. F. Liem, and D. B. Wake, eds.

1985. *Functional vertebrate morphology*. Cambridge, MA: Belknap Press.

Hill, J., and G. D. Grossman. 1987. Home range estimates for three North American stream fishes. *Copeia* 1987:376–380.

Hill, J., and G. D. Grossman. 1993. An energetic model of microhabitat use for rainbow trout and rosyside dace. *Ecology* 74:685–698.

Hinton, D. E., and D. J. Lauren. 1990. Integrative histopathological approaches in detecting effects of environmental stressors on fishes. *Amer Fish Soc Symp* 8:51–66.

Hixon, M. A. 1980a. Food production and competitor density as the determinants of feeding territory size. *Amer Natur* 115:510–530.

Hixon, M. A. 1980b. Competitive interactions between California reef fishes of the genus *Embiotoca*. *Ecology* 61:918–931.

Hixon, M. A. 1986. Fish predation and local prey diversity. In *Contemporary studies on fish feeding*, ed. C. A. Simenstad and G. M. Cailliet, 235–257. Dordrecht: Dr. W. Junk.

Hixon, M. A. 1991. Predation as a process structuring coral reef fish communities. In *The ecology of fishes on coral reefs*, ed. P. F. Sale, 475–508. San Diego, CA: Academic Press.

Hixon, M. A., and J. P. Beets. 1989. Shelter characteristics and Caribbean fish assemblages: Experiments with artificial reefs. *Bull Mar Sci* 44:666–680.

Hixon, M. A., and J. P. Beets. 1993. Predation, prey refuges, and the structure of coral-reef fish assemblages. *Ecol Monogr* 63:77–101.

Hixon, M. A., and W. N. Brostoff. 1983. Damselfish as keystone species in reverse: Intermediate disturbance and diversity of reef algae. *Science* 220:511–513.

Hjort, J. 1914. Fluctuations in the great fisheries of northern Europe viewed in the light of biological research. *Rapp P-V Reun Cons Int Explor Mer* 20:1–228.

Hoar, W. S. 1957. The gonads and reproduction. *Physiol Fish* 1:287–321.

Hoar, W. S. 1969. Reproduction. In *Fish physiology*, Vol. 3, ed. W. S. Hoar and D. J. Randall, 1–72. New York: Academic Press.

Hoar, W. S., and D. J. Randall, eds. 1969–1993. *Fish physiology*. Vols. 1–13. New York: Academic Press.

Hoar, W. S., and D. J. Randall, eds. 1978. *Fish physiology*. Vol. 7, *Locomotion*. New York: Academic Press.

Hoar, W. S., and D. J. Randall, eds. 1988. *Fish physiology*. Vol. 11, *The physiology of developing fish*. Part B, *Viviparity and posthatching juveniles*. San Diego, CA: Academic Press.

Hobson, E. S. 1968. Predatory behavior of some shore fishes in the Gulf of California. *US Fish Wildl Serv Res Rept* 73:1–92.

Hobson, E. S. 1972. Activity of Hawaiian reef fishes during evening and morning transitions between daylight and darkness. *Fish Bull US* 70:715–740.

Hobson, E. S. 1973. Diel feeding migrations in tropical reef fishes. *Helgo wiss Meeresunter* 24:671–680.

Hobson, E. S. 1974. Feeding relationships of teleostean fishes on coral reefs in Kona, Hawaii. *Fish Bull US* 72:915–1031.

Hobson, E. S. 1975. Feeding patterns among tropical reef fishes. *Amer Sci* 63:382–392.

Hobson, E. S. 1986. Predation on the Pacific sand lance, *Ammodytes hexapterus* (Pisces: Ammodytidae), during the transition between day and night in southeastern Alaska. *Copeia* 1986:223–226.

Hobson, E. S. 1991. Trophic relationships of fishes specialized to feed on zooplankters above coral reefs. In *The ecology of fishes on coral reefs*, ed. P. F. Sale, 69–95. San Diego, CA: Academic Press.

Hobson, E. S. 1994. Ecological relations in the evolution of acanthopterygian fishes in warm-temperate communities of the northeastern Pacific. *Env Biol Fish* 40:49–90.

Hobson, E. S., W. N. McFarland, and J. R. Chess. 1981. Crepuscular and nocturnal activities of California nearshore fishes, with consideration of their scotopic visual pigments and the photic environment. *Fish Bull US* 79:1–30.

Hochachka, P. W., and T. P. Mommsen. 1983. Protons and anaerobiosis. *Science* 219:1391–1397.

Hochachka, P. W., and G. N. Somero. 1973. *Strategies of biochemical adaptation*. Philadelphia: Saunders.

Hochachka, P. W., and G. N. Somero. 1984. *Biochemical adaptation*. Princeton, NJ: Princeton University Press.

Hocutt, C. H., and E. O. Wiley, eds. 1986. *The zoogeography of North American freshwater fishes*. New York: John Wiley & Sons.

Hodgson, E. S., and R. F. Mathewson, eds. 1978. *Sensory biology of sharks, skates, and rays*. Arlington, VA: Office of Naval Research.

Hoese, H. D. 1985. Jumping mullet—the internal diving bell hypothesis. *Env Biol Fish* 13:309–314.

Hoffman, G. L., and G. Schubert. 1984. Some parasites of exotic fishes. In *Distribution, biology, and management of exotic fishes*, ed. W. R. Courtenay Jr. and J. R. Stauffer Jr., 233–261. Baltimore: Johns Hopkins University Press.

Holbrook, S. J., and R. J. Schmitt. 1988a. Effects of predation risk on foraging behavior: Mechanisms altering patch choice. *J Exp Mar Biol Ecol* 121:151–163.

Holbrook, S. J., and R. J. Schmitt. 1988b. The combined effects of predation risk and food reward on patch selection. *Ecology* 69:125–134.

Holbrook, S. J., and R. J. Schmitt. 1989. Resource overlap, prey dynamics, and the strength of competition. *Ecology* 70:1943–1953.

Holden, C. 1996. Fish fans beware. *Science* 273:1049.

Holland, K. N., R. W. Brill, R. K. C. Chang, J. R. Silbert, and D. A. Fournier. 1992. Physiological and behavioral thermoregulation in bigeye tuna (*Thunnus obesus*). *Nature* 358:410–412.

Holmgren, U. 1958. On the pineal organ of the tuna, *Thunnus thynnus*. *Mus Comp Zool Breviora* 100:5.

Holmqvist, B. I., T. Ostholm, and P. Ekstrom. 1992. Retinohypothalamic projections and the suprachiasmatic nucleus of the teleost brain. In *Rhythms in fishes*, ed. M. A. Ali, 292–318. New York: Plenum Press.

Holomuski, J. R., and R. J. Stevenson. 1992. Role of predatory fish in community dynamics of an ephemeral stream. *Can J Fish Aquat Sci* 49:2322–2330.

Holt, G. J., S. A. Holt, and C. R. Arnold. 1985. Diel periodicity of spawning in sciaenids. *Mar Ecol Prog Ser* 27:1–7.

Holtby, L. B. 1988. Effects of logging on stream temperature in Carnation Creek, British Columbia, and associated impacts on the coho salmon (*Oncorhynchus kisutch*). *Can J Fish Aquat Sci* 45:502–515.

Hontela, A., and N. E. Stacey. 1990. Cyprinidae. In *Reproductive seasonality in teleosts: Environmental influences*, ed. A. D. Munro, A. P. Scott, and T. J. Lam, 53–77. Boca Raton, FL: CRC Press.

Hopkins, C. D. 1972. Sex differences in electric signaling in an electric fish. *Science* 176:1035–1037.

Hopkins, C. D. 1974a. Electric communication: Functions in the social behavior of *Eigenmannia virescens*. *Behaviour* 50:270–305.

Hopkins, C. D. 1974b. Electric communication in fish. *Amer Sci* 62:426–437.

Hopkins, C. D. 1986. Behavior of Mormyridae. In *Electroreception*, ed. T. H. Bullock and W. Heiligenberg, 527–576. New York: John Wiley & Sons.

Hori, K., N. Fusetani, K. Hashimoto, K. Aida, and J. E. Randall. 1979. Occurrence of a grammistin-like mucous toxin in the clingfish *Diademichthys lineatus*. *Toxicon* 17:418–424.

Hori, M. 1993. Frequency-dependent natural selection in the handedness of scale-eating cichlid fish. *Science* 260:216–219.

Horn, M. H. 1972. The amount of space available for marine and freshwater fishes. *Fish Bull US* 70:1295–1297.

Horn, M. H. 1989. Biology of marine herbivorous fishes. *Oceanog Mar Biol Ann Rev* 27:167–272.

Horn, M. H. 1993. Feeding and movements of the fruit-eating characid fish *Brycon guatemalensis* in relation to seed dispersal in a Costa Rican rain forest. *Ann Meet Amer Soc Ichthyol Herpetol*:172 (abstract).

Horn, M. H., and L. G. Allen. 1985. Fish community ecology in southern California bays and estuaries. In *Fish community ecology in estuaries and coastal lagoons: Towards an ecosystem integration*, ed. A. Yanez-Arancibia, 169–190. Mexico City: DR (R) UNAM Press.

Horn, M. H., and R. N. Gibson. 1988. Intertidal fishes. *Sci Amer* 256(1):64–70.

Horwitz, R. J. 1978. Temporal variability patterns and the distributional patterns of stream fishes. *Ecol Monogr* 48:307–321.

Houde, A. E., and J. A. Endler. 1990. Correlated evolution of female mating preferences and male color patterns in the guppy *Poecilia reticulata*. *Science* 248:1405–1408.

Houde, E. D. 1987. Fish early life dynamics and recruitment variability. *Amer Fish Soc Symp* 2:17–29.

Houde, E. D. 1989. Comparative growth, mortality, and energetics of marine fish larvae: Temperature and implied latitudinal effects. *Fish Bull US* 87:471–495.

Hubbs, C. L. 1952. Antitropical distribution of fishes and other organisms. *Proc 7th Pacific Sci Cong* 3:324–329.

Hubbs, C. L. 1955. Hybridization between fish species in nature. *Syst Zool* 4:1–20.

Hubbs, C. L. 1964. History of ichthyology in the United States after 1850. *Copeia* 1964:42–60.

Hubbs, C. 1967. Geographic variation in survival of hybrids between etheostomatine fishes. *Texas Mem Mus Bull* 13.

Hubbs, C. L., and K. F. Lagler. 1964. *Fishes of the Great Lakes region*. Ann Arbor: University of Michigan Press.

Hubold, G. 1991. Ecology of notothenioid fish in the Weddell Sea. In *Biology of Antarctic fish*, ed. G. di Prisco, B. Maresca, and B. Tota, 3–22. Berlin: Springer-Verlag.

Huet, M. 1959. Profiles and biology of Western European streams as related to fish management. *Trans Amer Fish Soc* 88:153–163.

Hueter, R. E., and P. W. Gilbert. 1991. The sensory world of sharks. In *Discovering sharks*, ed. S. H. Gruber, 48–55. Highlands, NJ: American Littoral Society, Spec Publ 14.

Huggett, R. J., R. A. Kimerle, P. M. Mehrle Jr., and H. L. Bergman. 1992. *Biomarkers: Biochemical, physiological, and histological markers of anthropogenic stress*. Boca Raton, FL: Lewis Publishers.

Hughes, D. R. 1981. Development and organization of the posterior field of ctenoid scales in the Platycephalidae. *Copeia* 1981:596–606.

Hughes, R. M., and R. F. Noss. 1992. Biological diversity and biological integrity: Current concerns for lakes and streams. *Fisheries* 17(3):11–19.

Hughes, R. M., and J. M. Omernik. 1990. Use and misuse of the terms watershed and stream order. In *Warmwater streams symposium*, Southern Division, American Fisheries Society, ed. L. A. Krumholz, 320–326. Lawrence, KS: Allen Press.

Hughes, R. N., ed. 1990. *Behavioural mechanisms of food selection*. NATO ASI Series G 20. Berlin: Springer-Verlag.

Humphries, J. M., F. L. Bookstein, B. Chernoff, G. R. Smith, R. L. Elder, and S. G. Poss. 1981. Multivariate discrimination by shape in relation to size. *Syst Zool* 30:291–308.

Hunter, J. R. 1981. Feeding ecology and predation of marine fish larvae. In *Marine fish larvae: Morphology, ecology, and relation to fisheries*, ed. R. L. Lasker, 33–77. Seattle: Washington Sea Grant Program.

Hunter, J. R., and K. M. Coyne. 1982. The onset of schooling in northern anchovy larvae *Engraulis mordax*. *Calif Coop Oceanic Fish Invest Rep* 23:246–251.

Huntingford, F. A. 1993. Development of behaviour in fish. In *The behaviour of teleost fishes*, 2d ed., ed. T. J. Pitcher, 57–83. London: Chapman and Hall.

Huxley, J., ed. 1940. *The new systematics*. Oxford: Oxford University Press.

Ibara, R. M., L. T. Penny, A. W. Ebeling, G. van Dykhuizen, and G. Cailliet. 1983. The mating call of the plainfin midshipman fish, *Porichthys notatus*. In *Predators and prey in fishes*, ed. D. L. G. Noakes et al. The Hague: Dr. W. Junk.

Irvine, G. V. 1981. The importance of behavior in plant-herbivore interactions: A case study. In *Gutshop '81, Fish Food Habits Studies*, ed. G. M. Cailliet and C. A. Simenstad, 240–248. Seattle: University of Washington Sea Grant.

Ishihara, M. 1987. Effect of mobbing toward predators by the damselfish *Pomacentrus coelestis* (Pisces: Pomacentridae). *J Ethol* 5:43–52.

Ivlev, V. S. 1961. *Experimental ecology of the feeding of fishes*. Translated by D. Scott. New Haven, CT: Yale University Press.

Jablonski, D. 1991. Extinctions: A paleontological perspective. *Science* 253:754–757.

Jacobs, G. H. 1992. Ultraviolet vision in vertebrates. *Amer Zool* 32:544–554.

Jamieson, B. G. M. 1991. *Fish evolution and systematics: Evidence from spermatozoa*. Cambridge: Cambridge University Press.

Janssen, J. 1980. Alewives (*Alosa pseudoharengus*) and ciscoes (*Coregonus artedii*) as selective and non-selective planktivores. In *Evolution and ecology of zooplankton communities*, ed. W. C. Kerfoot, 580–586. Hanover, NH: University Press of New England.

Janssen, J., and S. B. Brandt. 1980. Feeding ecology and vertical migration of adult alewives (*Alosa pseudoharengus*) in Lake Michigan. *Can J Fish Aquat Sci* 37:177–184.

Janvier, P. 1984. The relationships of the Osteostraci and Galeaspida. *J Vert Paleo* 4:344–358.

Janvier, P. 1991. The phylogeny of the Craniata, with particular reference to the significance of fossil "agnathans." *J Vert Paleo* 1:121–159.

Janvier, P. 1995. Conodonts join the club. *Nature* (London) 374:761–762.

Janvier, P., and R. Lund. 1983. *Hardistiella montanensis* n. gen. et sp. (Petromyzontida) from the Lower Carboniferous of Montana, with remarks on the affinities of lampreys. *J Vert Paleo* 2:407–413.

Jarvik, E. 1980. *Basic structure and evolution of vertebrates*. Vol. 1. London: Academic Press.

Jenkins, R. E., and N. M. Burkhead. 1993. Freshwater fishes of Virginia. Bethesda, MD: *Amer Fish Soc*

Jensen, D. 1966. The hagfish. *Sci Amer* 214(2):82–90.

Jessen, H. 1966. Struniiformes. In *Traite de Paleontologie*, Tome 4, Vol. 3, ed. J. P. Lehman, 387–398. Paris: Masson.

Jessen, H. L. 1973. Interrelationships of actinopterygians and brachiopterygians: Evidence from pectoral anatomy. In *Interrelationships of fishes*, ed. P. H. Greenwood, R. S. Miles, and C. Patterson. *Zool J Linn Soc* 53 (suppl 1):227–232.

Jessop, B. M. 1987. Migrating American eels in Nova Scotia. *Trans Amer Fish Soc* 116:161–170.

Jimenez, B. D., and J. J. Stegeman. 1990. Detoxication enzymes as indicators of environmental stress on fish. *Amer Fish Soc Symp* 8:67–79.

Jobling, M. 1981. The influence of feeding on the metabolic rate of fishes: A short review. *J Fish Biol* 18:385–400.

Johannes, R. E. 1978. Reproductive strategies of coastal marine fishes in the tropics. *Env Biol Fish* 3:65–84.

Johannes, R. E. 1981. *Words of the lagoon: Fishing and marine lore in the Palau District of Micronesia*. Berkeley: University of California Press.

Johnsen, P. B., and A. D. Hasler. 1977. Winter aggregations of carp (*Cyprinus carpio*) as revealed by ultrasonic tracking. *Trans Amer Fish Soc* 106:556–559.

Johnsen, P. B., and J. H. Teeter. 1985. Behavioral responses of bonnethead sharks (*Sphyrna tiburo*) to controlled olfactory stimulation. *Mar Behav Physiol* 11:283–291.

Johnson, G. D. 1984. Percoidei: Development and relationships. In *Ontogeny and systematics of fishes*, ed. H. G. Moser et al., 464–498. Amer Soc Ichthyol Herpetol, Spec Publ 1.

Johnson, G. D. 1986. Scombroid phylogeny: An alternative hypothesis. *Bull Mar Sci* 39:1–41.

Johnson, G. D. 1992. Monophyly of the euteleostean clades—Neoteleostei, Eurypterygii, and Ctenosquamata. *Copeia* 1992:8–25.

Johnson, G. D. 1993. Percomorph phylogeny: Progress and problems. *Bull Mar Sci* 52:3–28.

Johnson, G. D., and E. B. Brothers. 1993. *Schindleria*: A paedomorphic goby (Teleostei; Gobioidei). *Bull Mar Sci* 52:441–471.

Johnson, G. D., and C. Patterson. 1993. Percomorph phylogeny: A survey of acanthomorphs and a new proposal. *Bull Mar Sci* 52:554–626.

Johnson, J. E., and B. L. Jensen. 1991. Hatcheries for endangered freshwater fishes. In *Battle against extinction: Native fish management in the American West*, ed. W. L. Minckley and J. E. Deacon, 199–217. Tucson: University of Arizona Press.

Johnson, R. H., and D. R. Nelson. 1973. Agonistic display in the gray reef shark, *Carcharhinus menisorrah*, and its relationship to attacks on man. *Copeia* 1973:76–84.

Johnson, T. C., et al. 1996. Late Pleistocene desiccation of Lake Victoria and rapid evolution of cichlid fishes. *Science* 273:1091–1093.

Jones, D. R., and D. J. Randall. 1978. The respiratory and circulatory systems during exercise. In *Fish physiology*. Vol. 7, *Locomotion*, ed. W. S. Hoar and D. J. Randall, 425–501. New York: Academic Press.

Jones, D. R., and T. Schwarzfeld. 1974. The oxygen cost to the metabolism and efficiency of breathing in trout (*Salmo gairdneri*). *Respir Physiol* 21:241–254.

Jones, E. C. 1971. *Isistius brasiliensis*, a squaloid shark, the probable cause of crater wounds on fishes and cetaceans. *Fish Bull US* 69:791–798.

Jones, F. R. H. 1957. The swimbladder. *Physiol Fish* 2:305–322.

Jones, G. P., D. J. Ferrell, and P. F. Sale. 1991. Fish predation and its impact on the invertebrates of coral reefs and adjacent sediments. In *The ecology of fishes on coral reefs*, ed. P. F. Sale, 156–179. San Diego, CA: Academic Press.

Jones, P. L., and B. D. Sidell. 1982. Metabolic responses of striped bass (*Morone saxatilis*) to temperature acclimation. II. Alterations in metabolic carbon sources and distributions of fiber types in locomotory muscle. *J Exp Zool* 219:163–171.

Jones, R. S. 1968. Ecological relationships in Hawaiian and Johnston Island Acanthuridae (Surgeonfishes). *Micronesica* 4:309–361.

Jones, W. R., and J. Janssen. 1992. Lateral line development and feeding behavior in the mottled sculpin, *Cottus bairdi* (Scorpaeniformes: Cottidae). *Copeia* 1992:485–492.

Jordan, D. S. 1905. *A guide to the study of fishes*. Vol. I, II. New York: Henry Holt.

Jordan, D. S. 1922. *The days of a man*. Vols. I, II. New York: World Book.

Jordan, D. S. 1923. A classification of fishes including families and genera as far as known. *Stanford Univ Pubs Biol Sci* 3:77–243.

Jordan, D. S., and B. W. Evermann. 1896–1900. The fishes of

North and Middle America. *Bull US Natl Mus* 447 (pts 1–4):1–3313.

Joung, S. J., C. T. Chen, E. Clark, S. Uchida, W.Y.P. Huang. 1996. The whale shark, *Rhincodon typus*, is a livebearer: 300 embryos found in one 'megamamma' supreme. *Env Biol Fish* 46:219–223.

Kalmijn, A. J. 1971. The electric sense of sharks and rays. *J Exp Biol* 55:371–383.

Kalmijn, A. J. 1978. Electric and magnetic sensory world of sharks, skates, and rays. In *Sensory biology of sharks, skates, and rays*, ed. E. S. Hodgson and R. F. Mathewson, 507–528. Arlington, VA: Office of Naval Research.

Kalmijn, A. J. 1982. Electric and magnetic field detection in elasmobranch fishes. *Science* 218:916–918.

Kamler, E. 1991. *Early life history of fish: An energetics approach.* London: Chapman and Hall.

Kapoor, B. G., H. Smit, and I. A. Verighina. 1975. The alimentary canal and digestion in teleosts. *Adv Mar Biol* 13:109–239.

Karnaky, K. J. Jr. 1986. Structure and function of the chloride cell of *Fundulus heteroclitus* and other teleosts. *Amer Zool* 26:209–224.

Karplus, I. 1979. The tactile communication between *Cryptocentrus steinitzi* (Pisces, Gobiidae) and *Alpheus purpurilenticularis* (Crustacea, Alpheidae). *Z Tierpsychol* 49:173–196.

Karplus, I., and D. Algom. 1981. Visual cues for predator face recognition by reef fishes. *Z Tierpsychol* 55:343–364.

Karplus, I., M. Goren, and D. Algom. 1982. A preliminary experimental analysis of predator face recognition by *Chromis caeruleus* (Pisces, Pomacentridae). *Z Tierpsychol* 58:53–65.

Karr, J. R. 1981. Assessment of biotic integrity using fish communities. *Fisheries* 6:21–27.

Karr, J. R. 1991. Biological integrity: A long-neglected aspect of water resource management. *Ecol Applic* 1:66–84.

Karr, J. R., M. Dionne, and I. J. Schlosser. 1992. Bottom-up versus top-down regulation of vertebrate populations: Lessons from birds and fish. In *Effects of resource distribution on animal-plant interactions*, ed. M. D. Hunter and T. Ohgushi, 243–286. San Diego, CA: Academic Press.

Karr, J. R., and K. E. Freemark. 1985. Disturbance and vertebrates: An integrative perspective. In *The ecology of natural disturbance and patch dynamics*, ed. S. T. A. Pickett and P. S. White, 153–168. New York: Academic Press.

Karr, J. R., L. A. Toth, and G. D. Garman. 1982. *Habitat preservation for midwest stream fishes: Principles and guidelines.* Corvallis, OR: US Environmental Protection Agency 600/3-83-006:1–120.

Kaufman, L. 1977. The threespot damselfish: Effects on benthic biota of Caribbean coral reefs. *Proc 3rd Internat Coral Reef Symp* 1:559–564.

Kaufman, L. S. 1983. Effects of Hurricane Allen on reef fish assemblages near Discovery Bay, Jamaica. *Coral Reefs* 2:43–47.

Kaufman, L., J. Ebersole, J. Beets, and C. C. McIvor. 1992. A key phase in the recruitment dynamics of coral reef fishes: Post-settlement transition. *Env Biol Fish* 34:109–118.

Keast, A., and J. Harker. 1977. Fish distribution and benthic invertebrate biomass relative to depth in an Ontario lake. *Env Biol Fish* 2:235–240.

Keast, A., and D. Webb. 1966. Mouth and body form relative to feeding ecology in the fish fauna of a small lake, Lake Opinicon, Ontario. *J Fish Res Board Can* 23:1845–1874.

Keenleyside, M. H. A. 1979. *Diversity and adaptation in fish behaviour.* Berlin: Springer-Verlag.

Keenleyside M. H. A., ed. 1991. *Cichlid fishes: Behaviour, ecology and evolution.* London: Chapman and Hall.

Kellermann, A., and A. W. North. 1994. The contribution of the BIOMASS Programme to Antarctic fish biology. In *Southern Ocean ecology: The BIOMASS perspective*, ed. S. Z. El-Sayed, 191–209. Cambridge: Cambridge University Press.

Kelley, C. D., and T. F. Hourigan. 1983. The function of conspicuous coloration in chaetodontid fishes: A new hypothesis. *Anim Behav* 31:615–617.

Kelsch, S. W., and W. H. Neill. 1990. Temperature preference versus acclimation in fishes: Selection for changing metabolic optima. *Trans Amer Fish Soc* 119:601–610.

Kemp, A. 1987. The biology of the Australian lungfish, *Neoceratodus forsteri* (Krefft 1870). In *The biology and evolution of lungfishes*, ed. W. E. Bemis, W. W. Burggren, and N. E. Kemp, 181–198. New York: Alan R. Liss.

Kendall, A. W. Jr. 1979. *Morphological comparisons of North American sea bass larvae (Pisces: Serranidae).* Washington, DC: NOAA Tech Rep NMFS Circ 428.

Kendall, A. W. Jr., E. H. Ahlstrom, and H. G. Moser. 1984. Early life history stages of fishes and their characters. In *Ontogeny and systematics of fishes*, ed. H. G. Moser, W. J. Richards, D. M. Cohen, M. P. Fahay, A. W. Kendall Jr., and S. L. Richardson, 11–22. Amer Soc Ichthyol Herpetol, Spec Publ 1.

Kendall, R. L. 1988. Taxonomic changes in North American trout names. *Trans Amer Fish Soc* 117:321.

Kennedy, V. S. 1990. Anticipated effects of climate change on estuarine and coastal fisheries. *Fisheries* 15(6):16–24.

Kerfoot, W. C., and A. Sih, eds. 1987. *Predation: Direct and indirect impacts on aquatic communities.* Hanover, NH: University Press of New England.

Keys, A. B., and E. N. Willmer. 1932. Chloride secreting cells in the gills of fishes with special references to the common eel. *J Physiol* 76:368–377.

Kirschvink, J. L., M. M. Walker, S. B. Chang, A. E. Dizon, and K. A. Peterson. 1985. Chains of single-domain magnetite particles in chinook salmon, *Oncorhynchus tshawytscha*. *J Comp Physiol A* 157:375–381.

Kitchell, J. F., ed. 1992. *Food web management. A case study of Lake Mendota.* New York: Springer-Verlag.

Kitchell, J. F., W. H. Neill, A. E. Dizon, and J. J. Magnuson. 1978. Bioenergetic spectra of skipjack and yellowfin tunas. In *The physiological ecology of tunas*, ed. G. D. Sharp and A. E. Dizon, 357–368. New York: Academic Press.

Kleckner, R. C., and W. H. Kruger. 1981. Changes in swim bladder retial morphology in *Anguilla rostrata* during premigration metamorphosis. *J Fish Biol* 18:569–577.

Klimley, A. P. 1994. The predatory behavior of the white shark. *Amer Sci* 82:122–133.

Klimley, A. P. 1995. Hammerhead city. *Natur Hist* 104(10):32–39.

Klimley, A. P., S. D. Anderson, P. Pyle, and R. H. Henderson. 1992. Spatiotemporal patterns of white shark (Carcharodon carcharias) predation at the South Farallon Islands, California. *Copeia* 1992:680–690.

Klimley, A. P., S. B. Butler, D. R. Nelson, and A. T. Stull. 1988. Diel movements of scalloped hammerhead sharks, *Sphyrna lewini* Griffith and Smith, to and from a seamount in the Gulf of California. *J Fish Biol* 33:751–761.

Klimley, A. P., and D. R. Nelson. 1981. Schooling of hammerhead sharks, *Sphyrna lewini*, in the Gulf of California. *Fish Bull US* 79:356–360.

Klinger, S. A., J. J. Magnuson, and G. W. Gallepp. 1982. Survival mechanisms of the central mudminnow (*Umbra limi*), fathead minnow (*Pimephales promelas*) and brook stickleback (*Culaea inconstans*) for low oxygen in winter. *Env Biol Fish* 7:113–120.

Klumpp, D. W., D. McKinnon, and P. Daniel. 1987. Damselfish territories: Zones of high productivity on coral reefs. *Mar Ecol Prog Ser* 40:41–51.

Klumpp, D. W., and N. V. C. Polunin. 1989. Partitioning among grazers of food resources within damselfish territories on a coral reef. *J Exp Mar Biol Ecol* 125:145–169.

Knight, C. A., C. C. Cheng, and A. L. DeVries. 1991. Adsorption of helical antifreeze peptides on specific ice crystal surface planes. *Biophys J* 59:409–418.

Knowlton, N., J. C. Lang, and B. D. Keller. 1990. Case study of natural population collapse: Post-hurricane predation on Jamaican staghorn corals. *Smithson Contrib Mar Sci* 31:1–25.

Koboyashi, H., B. Pelster, and P. Scheid. 1989. Water and lactate movement in the swimbladder of the eel, *Anguilla anguilla*. *Respir Physiol* 78:45–57.

Koboyashi, H., B. Pelster, and P. Scheid. 1990. CO_2 back-diffusion in the rete aids O_2 secretion in the swimbladder of the eel. *Respir Physiol* 79:231–242.

Kock, K.-H. 1992. *Antarctic fish and fisheries*. New York: Cambridge University Press.

Kodric-Brown, A. 1990. Mechanisms of sexual selection: Insights from fishes. *Ann Zool Fennici* 27:87–100.

Kohda, M., M. Tanimura, M. Kikue-Nakamura, and S. Yamagishi. 1995. Sperm drinking by female catfishes: A novel mode of insemination. *Env Biol Fish* 42:1–6.

Kohler, C. C., and W. R. Courtenay Jr. 1986a. Regulating introduced aquatic species: A review of past initiatives. *Fisheries* 11(2):34–38.

Kohler, C. C., and W. R. Courtenay Jr. 1986b. American Fisheries Society position on introductions of aquatic species. *Fisheries* 11(2):39–42.

Kohler, C. C., and W. A. Hubert. 1993. *Inland fisheries management in North America*. Bethesda, MD: American Fisheries Society.

Kohler, C. C., and S. T. Kohler. 1994. Ciguatera tropical fish poisoning: What's happening in the food chain? In *Proc 26th Meeting Assoc Mar Lab Carib*, ed. D. T. Gerace, 112–125, San Salvador, Bahamas.

Kohler, S. L., and M. A. McPeek. 1989. Predation risk and the foraging behavior of competing stream insects. *Ecology* 70:1811–1825.

Kohler, S. T., and C. C. Kohler. 1992. Dead bleached coral provides new surfaces for dinoflagellates implicated in ciguatera fish poisonings. *Env Biol Fish* 35:413–416.

Konecki, J. T., and T. E. Targett. 1989. Eggs and larvae of *Nototheniops larseni* from the spongocoel of a hexactinellid sponge near Hugo Island, Antarctic Peninsula. *Polar Biol* 10:197–198.

Kornfield, I., and K. E. Carpenter. 1984. Cyprinids of Lake Lanao, Philippines: Taxonomic validity, evolutionary rates and speciation scenarios. In *Evolution of fish species flocks*, ed. A. E. Echelle and I. Kornfield, 69–84. Orono: University of Maine Press.

Kornfield, I. L., and J. N. Taylor. 1983. A new species of polymorphic fish, *Cichlasoma minckleyi* from Cuatro Cienegas, Mexico (Teleostei: Cichlidae). *Proc Biol Soc Wash* 96:253–269.

Kramer, D. L., and M. J. Bryant. 1995. Intestine length in the fishes of a tropical stream: 2. Relationships to diet—the long and short of a convoluted issue. *Env Biol Fish* 42:129–141.

Kramer, D. L., C. C. Lindsey, G. E. E. Moodie, and E. D. Stevens. 1978. The fishes and the aquatic environment of the central Amazon basin, with particular reference to respiratory patterns. *Can J Zool* 56:717–729.

Kramer, D. L., and M. McClure. 1982. Aquatic surface respiration, a widespread adaptation to hypoxia in tropical freshwater fishes. *Environ Biol Fish* 7:47–55.

Krebs, J. R., and N. B. Davies. 1987. *An introduction to behavioural ecology*. 2d ed. Sunderland, MA: Sinauer Associates.

Krebs, J. R., and N. B. Davies, eds. 1991. *Behavioural ecology: An evolutionary approach*. 3d ed. London: Blackwell Science.

Krejsa, R. J., P. Bringas Jr., and H. C. Slavkin. 1990a. The cyclostome model: An interpretation of conodont element structure and function based on cyclostome tooth morphology, function, and life history. *Courier Forsch-Inst Senckenberg* 118:473–492.

Krejsa, R. J., P. Bringas Jr., and H. C. Slavkin. 1990b. A neontological interpretation of conodont elements based on agnathan cyclostome tooth structure, function, and development. *Lethaia* 23:359–378.

Krekorian, C. O., D. W. Dunham. 1972. Preliminary observations of the reproductive and parental behavior of the spraying characid *Copeina arnoldi* Regan. *Z Tierpsychol* 31:419–437.

Krogh, A. 1939. *Osmotic regulation in aquatic animals*. Cambridge: Cambridge University Press.

Krokhin, E. M. 1975. Transport of nutrients by salmon migration from the sea to lakes. In *Coupling of land and water systems*, ed. A. D. Hasler, 153–156. New York: Springer-Verlag.

Kruger, R. L., and R. W. Brocksen. 1978. Respiratory metabolism of striped bass, *Morone saxatilis* (Walbaum), in relation to temperature. *J Exp Mar Biol Ecol* 31:55–66.

Kuehne, R. A. 1962. A classification of streams illustrated by fish distribution in an eastern Kentucky creek. *Ecology* 43:608–614.

Kusher, D. I., and J. W. Crim. 1991. Immunosuppression in the bluegill (*Lepomis macrochirus*) induced by environmental exposure to cadmium. *Fish ShellF Immunol* 1:157–161.

Kushlan, J. A. 1979. Design and management of continental wildlife reserves: Lessons from the Everglades. *Biol Conserv* 15:281–290.

Lackey, R. T., and L. A. Nielsen, eds. 1980. *Fisheries management*. Oxford: Blackwell Science.

Laerm, J., and B. J. Freeman. 1986. *Fishes of the Okefenokee Swamp*. Athens: University of Georgia Press.

Lagler, K. F. 1947. Scale characters of the families of Great Lakes fishes. *Trans Amer Microscop Soc* 66:149–171.

Lagler, K. F., J. E. Bardach, R. R. Miller, and D. R. M. Passino. 1977. *Ichthyology*. 2d ed. New York: John Wiley & Sons.

Lammens, E. H. R. R., H. W. de Nie, J. Vijverberg, and W. L. T. van Dense. 1985. Resource partitioning and niche shifts of bream (*Abramis brama*) and eel (*Anguilla anguilla*) mediated by predation of smelt (*Osmerus eperlanus*) on *Daphnia hyalina*. *Can J Fish Aquat Sci* 42:1342–1351.

Landeau, L., and J. Terborgh. 1986. Oddity and the 'confusion effect' in predation. *Anim Behav* 34:1372–1380.

Langecker, T. G., and G. Longley. 1993. Morphological adaptations of the Texas blind catfishes *Trogloglanis pattersoni* and *Satan eurystomus* (Siluriformes: Ictaluridae) to their underground environment. *Copeia* 1993:976–986.

Lasker, R., ed. 1981. *Marine fish larvae*. Seattle: Washington Sea Grant Publications.

Lassig, B. R. 1983. The effects of a cyclonic storm on coral reef fish assemblages. *Env Biol Fish* 9:55–63.

Lassuy, D. R. 1980. Effects of "farming" behavior in *Eupomacentrus lividus* and *Hemiglyphidodon plagiometopon* on algal community structure. *Bull Mar Sci* 30:304–312.

Lauder, G. V. 1980. Evolution of the feeding mechanism in primitive actinopterygian fishes: A functional anatomical analysis of *Polypterus*, *Lepisosteus* and *Amia*. *J Morphology* 163:283–317.

Lauder, G. V. 1982. Patterns of evolution in the feeding mechanism of actinopterygian fishes. *Amer Zool* 22:275–285.

Lauder, G. V. Jr. 1981. Form and function: Structural analysis in evolutionary morphology. *Paleobiology* 7:430–442.

Lauder, G. Jr. 1983a. Food capture. In *Fish biomechanics*, ed. P. W. Webb and D. Weihs, 280–311. New York: Praeger.

Lauder, G. V. Jr. 1983b. Functional design and evolution of the pharyngeal jaw apparatus in euteleostean fishes. *Zool J Linn Soc* 77:1–38.

Lauder, G. V. Jr. 1985. Aquatic feeding in lower vertebrates. In *Functional vertebrate morphology*, ed. M. Hildebrand, D. M. Bramble, K. F. Liem, and D. B. Wake, 210–229, 397–399. Cambridge, MA: Belknap Press.

Lauder, G. V. Jr., and K. F. Liem. 1981. Prey capture by *Lucio-*

cephalus pulcher: Implications for models of jaw protrusion in teleost fishes. *Env Biol Fish* 6:257–268.

Lauder, G. V., and K. F. Liem. 1983. The evolution and interrelationships of the actinopterygian fishes. *Bull Mus Comp Zool* 150:95–197.

Laughlin, R. A. 1982. Feeding habits of the blue crab, *Callinectes sapidus* Rathbun, in the Apalachicola estuary, Florida. *Bull Mar Sci* 32:807–822.

Laurent, P., and S. Dunel-Erb. 1984. The pseudobranch: Morphology and function. *Fish Physiol* 10B:285–323.

Laurent, P., and S. F. Perry. 1991. Environmental effects on fish gill morphology. *Physiol Zool* 64:4–25.

Lawrence, J. M. 1957. Estimated size of various forage fishes largemouth bass can swallow. *Proc Southeastern Assoc Game Fish Comm* 11:220–226.

Lazzaro, X. 1987. A review of planktivorous fishes: Their evolution, feeding behaviors, selectivities, and impacts. *Hydrobiologia* 146:97–167.

Leaman, B. M. 1991. Reproductive styles and life history variables relative to exploitation and management of *Sebastes* stocks. *Env Biol Fish* 30:253–271.

Leatherland, J. F., K. J. Farbridge, and T. Boujard. 1992. Lunar and semi-lunar rhythms in fishes. In *Rhythms in fishes*, ed. M. A. Ali, 83–107. New York: Plenum Press.

Lee, D. S., C. R. Gilbert, C. H. Hocutt, R. E. Jenkins, D. E. McAllister, and J. R. Stauffer Jr. 1980. *Atlas of North American fresh water fishes*. Raleigh: North Carolina State Museum of Natural History.

Leggett, W. C. 1977. The ecology of fish migrations. *Ann Rev Ecol System* 8:285–308.

Leggett, W. C., and J. C. Carscadden. 1978. Latitudinal variation in reproductive characteristics of American shad (*Alosa sapidissima*): Evidence for population specific life history strategies in fish. *J Fish Res Board Can* 35:1469–1478.

Lehman, J. P. 1966. Actinopterygii. In J. Piveteau, ed. *Trait de Paleontologie* 4:1–242.

Leis, J. M. 1991. The pelagic stage of reef fishes: The larval biology of coral reef fishes. In *The ecology of fishes on coral reefs*, ed. P. F. Sale, 183–230. San Diego, CA: Academic Press.

Lelek, A. 1987. *The freshwater fishes of Europe*. Vol. 9, *Threatened fishes of Europe*. Wiesbaden, Germany: Aula Verlag.

Levine, J. S., P. S. Lobel, and E. F. MacNichol Jr. 1980. Visual communication in fishes. In *Environmental physiology of fishes*, ed. M. A. Ali, 447–475. New York: Plenum Press.

Leviton, A. E., R. H. Gibbs Jr., E. Heal, and C. E. Dawson. 1985. Standards in herpetology and ichthyology. Part I. Standard symbolic codes for institutional resource collections in herpetology and ichthyology. *Copeia* 1985:802–832.

Lewis, D., P. Reinthal, and J. Trendall. 1986. *A guide to the fishes of Lake Malawi National Park*. Gland, Switzerland: World Wildlife Federation.

Liem, K. F. 1967a. Functional morphology of the head of the anabantoid teleost fish *Helostoma temmincki. J Morph* 121:135–158.

Liem, K. F. 1967b. A morphological study of *Luciocephalus pulcher*, with notes on gular elements in other recent teleosts. *J Morph* 121:103–133.

Liem, K. F. 1973. Evolutionary strategies and morphological innovations: Cichlid pharyngeal jaws. *Syst Zool* 22:424–441.

Liem, K. F. 1978. Modulatory multiplicity in the functional repertoire of the feeding mechanism in cichlid fishes. *J Morph* 158:323–360.

Liem, K. F., and P. H. Greenwood. 1981. A functional approach to the phylogeny of the pharyngognath teleosts. *Amer Zool* 21:83–101.

Liem, K. F., and G. V. Lauder, eds. 1982. Evolutionary morphology of the actinopterygian fishes. *Amer Zool* 22:239–345.

Liem, K. F., and D. B. Wake. 1985. Morphology: Current approaches and concepts. In *Functional vertebrate morphology*, ed. M. Hildebrand, D. M. Bramble, K. F. Liem, and D. B. Wake, 366–377. Cambridge, MA: Belknap Press.

Lighthill, M. J. 1969. Hydromechanics of aquatic animal propulsion. *Ann Rev Fluid Mech* 1:413–446.

Ligtvoet, W., and F. Witte. 1991. Perturbation through predator introduction: Effects on the food web and fish yields in Lake Victoria (East Africa). In *Perturbation and recovery of terrestrial and aquatic ecosystems*, ed. O. Ravera, 263–268. Chichester: Elliss Horwood.

Liley, N. R. 1982. Chemical communication in fish. *Can J Fish Aquat Sci* 39:22–35.

Lima, S. L., and L. M. Dill. 1990. Behavioral decisions made under the risk of predation: A review and prospectus. *Can J Zool* 68:619–640.

Limbaugh, C. 1961. Cleaning symbiosis. *Sci Amer* 205:42–49.

Lin, J. J., and G. N. Somero. 1995. Temperature-dependent changes in expression of thermostable and thermolabile isozymes of cytosolic malate dehydrogenase in the eurythermal goby fish *Gillichthys mirabilis. Physiol Zool* 68:114–128.

Lindsey, C. C. 1975. Pleomerism, the widespread tendency among related fish species for vertebral number to be correlated with maximum body length. *J Fish Res Board Can* 32:2453–2469.

Lindsey, C. C. 1978. Form, function, and locomotory habits in fish. In *Fish physiology*, Vol. 7, *Locomotion*, ed. W. S. Hoar and D. J. Randall, 1–100. New York: Academic Press.

Lindsey, C. C. 1988. Factors controlling meristic variation. In *Fish physiology*. Vol. 11, *The physiology of developing fish*. Part B, *Viviparity and posthatching juveniles*, ed. W. S. Hoar and D. J. Randall, 197–274. San Diego, CA: Academic Press.

Lindsey, C. C., and J. D. McPhail. 1986. Zoogeography of fishes of the Yukon and Mackenzie basins. In *Zoogeography of North American freshwater fishes*, ed. C. H. Hocutt and E. O. Wiley, 639–674. New York: John Wiley & Sons.

Lineaweaver, T. H., and R. H. Backus. 1970. *The natural history of sharks*. Philadelphia: J. B. Lippincott.

Linnaeus, C. 1758. *Systema Naturae*. 10th ed. Stockholm: Laurentii Salvii (in Latin).

Liu, C., and Y. Zeng. 1988. Notes on the Chinese paddlefish *Psephurus gladius* (Martens). *Copeia* 1988:482–484.

Lloyd, R. 1992. *Pollution and freshwater fish*. Oxford: Fishing News Books.

Lobel, P. S. 1981. Trophic biology of herbivorous reef fishes: Alimentary pH and digestive capabilities. *J Fish Biol* 19:365–397.

Lobel, P. S. 1989. Ocean current variability and the spawning season of Hawaiian reef fishes. *Env Biol Fish* 24:161–171.

Lobel, P. S. 1992. Sounds produced by spawning fishes. *Env Biol Fish* 33:351–358.

Lockett, N. A. 1977. Adaptations to the deep-sea environment. In *Handbook of sensory physiology* VII/5, ed. F. Crescitelli, 67–192.

Lodge, D. M. 1991. Herbivory on freshwater macrophytes. *Aquat Bot* 41:195–224.

Loesch, J. G. 1987. Overview of life history aspects of anadromous alewife and blueback herring in freshwater habitats. In *Common strategies of anadromous and catadromous fishes*, ed. M. J. Dadswell, R. J. Klauda, C. M. Moffitt, R. L. Saunders, R. A. Rulifson, and J. E. Cooper. *Amer Fish Soc Symp* 1:89–103.

Loiselle, P. V. 1982. Male spawning partner preference in an arena breeding teleost, *Cyprinodon macularias californiensis* Girard (Atherinomorpha: Cyprinodontidae). *Amer Natur* 120:721–732.

Loiselle, P. V., and G. W. Barlow. 1978. Do fishes lek like birds? In *Contrasts in behavior*, ed. E. S. Reese and F. J. Lighter, 3–75. New York: Wiley-Interscience.

Long, J. A. 1995. *The rise of fishes*. Baltimore: The Johns Hopkins Press.

Longley, W. H. 1917. Studies upon the biological significance of animal coloration. I. The colors and color changes of West Indian reef-fishes. *J Exp Zool* 23:533–601.

Longwell, A. C., S. Chang, A. Hebert, J. B. Hughes, and D. Perry. 1992. Pollution and developmental abnormalities of Atlantic fishes. *Env Biol Fish* 35:1–21.

Lorenz, K. 1962. The function of color in coral reef fishes. *Proc Roy Inst Great Britain* 282–296.

Losey, G. S. 1972. Predation protection in the poison-fang blenny, *Meiacanthus atrodorsalis*, and its mimics *Ecsenius bicolor* and *Runula laudandus* (Blenniidae). *Pacif Sci* 26:129–139.

Losey, G. S. Jr. 1978. The symbiotic behavior of fishes. In *The behavior of fish and other aquatic animals*, ed. D. I. Mostofsky, 1–31. New York: Academic Press.

Losey, G. S. Jr. 1979. Fish cleaning symbiosis: Proximate causes of host behaviour. *Anim Behav* 27:669–685.

Losey, G. S. Jr. 1987. Cleaning symbiosis. *Symbiosis* 4:229–258.

Lotrich, V. A. 1973. Growth, production, and community composition of fishes inhabiting a first-, second-, and third-order stream of eastern Kentucky. *Ecol Monogr* 43:377–397.

LoVullo, T. J., J. R. Stauffer Jr., and K. R. McKaye. 1992. Diet and growth of a brood of *Bagrus meridionalis* Gunther (Siluriformes: Bagridae) in Lake Malawi, Africa. *Copeia* 1992:1084–1088.

Lowe, C. G., R. N. Bray, and D. R. Nelson. 1994. Feeding and associated electrical behavior of the Pacific electric ray *Torpedo californica* in the field. *Mar Biol* 120:161–169.

Lowe-McConnell, R. H. 1987. *Ecological studies in tropical fish communities*. London: Cambridge University Press.

Lucas, J. R., and K. A. Benkert. 1983. Variable foraging and cleaning behavior by juvenile leatherjackets, *Oligoplites saurus* (Carangidae). *Estuaries* 6:247–250.

Lund, R. 1985. The morphology of *Falcatus falcatus* (St. John and Worthen), a stethacanthid chondrichthyan from the Bear Gulch Limestone of Montana. *J Vert Paleo* 5:1–19.

Lund, R. 1986. On *Damocles serratus*, nov. gen. et sp. (Elasmobranchii: Cladodontida) from the Upper Mississippian Bear Gulch limestone of Montana. *J Vert Paleo* 6:12–19.

Lund, R. 1990. Shadows in time—a capsule history of sharks. In *Discovering sharks*, ed. S. H. Gruber, 23–28. Highlands, NJ: American Littoral Society, Spec Publ 14.

Lund, R., and W. L. Lund. 1985. Coelacanths from the Bear Gulch limestone (*Namurian*) of Montana and the evolution of the coelacanthiformes. *Bull Carnegie Mus Natur Hist* 25:1–74.

Lundberg, J. G. 1993. African-South American freshwater fish clades and continental drift: Problems with a paradigm. In *Biological relationships between Africa and South America*, ed. P. Goldblatt, 156–199. New Haven, CT: Yale University Press.

Lundberg, J. G., and L. A. McDade. 1990. Systematics. In *Methods for fish biology*, ed. C. B. Schreck and P. B. Moyle, 65–108. Bethesda, MD: American Fisheries Society.

Lythgoe, J. N. 1979. *The ecology of vision*. Oxford: Clarendon Press.

Lythgoe, J. N., and J. Shand. 1983. Diel colour changes in the neon tetra *Paracheirodon innesi*. *Env Biol Fish* 8:249–254.

Mabee, P. M. 1993. Phylogenetic interpretation of ontogenetic change: Sorting out the actual and artefactual in an empirical case study of centrarchid fishes. *Zool J Linn Soc* 107:175–291.

Macchi, G. J., L. A. Romano, and H. E. Christiansen. 1992. Melano-macrophage centres in whitemouth croaker, *Micropogonias furnieri*, as biological indicators of environmental changes. *J Fish Biol* 40:971–973.

Macleay, W. 1879. On a species of *Amphisile* from the Palau Islands. *Proc Linn Soc NSW* 3:165, pl. 19B.

Magnhagen, C. 1988. Changes in foraging as a response to predation risk in two gobiid fish species, *Pomatoschistus minutus* and *Gobius niger*. *Mar Ecol Prog Ser* 49:21–26.

Magnhagen, C. 1992. Alternative reproductive behaviour in the common goby, *Pomatoschistus microps*: An ontogenetic gradient? *Anim Behav* 44:182–184.

Magnhagen, C., and K. Vestergaard. 1991. Risk taking in relation to reproductive investments and future reproductive opportunities: Field experiments on nest-guarding common gobies, *Pomatoschistus microps*. *Behav Ecol* 2:351–359.

Magnuson, J. J. 1978. Foreword. In *The physiological ecology of tunas*, ed. G. D. Sharp and A. E. Dizon, xi–xii. New York: Academic Press.

Magnuson, J. J., and D. J. Karlen. 1970. Visual observation of fish beneath the ice in a winterkill lake. *J Fish Res Bd Can* 27:1059–1068.

Magnuson, J. J., J. D. Meisner, and D. K. Hill. 1990. Potential changes in the thermal habitat of Great Lakes fish after global climate warming. *Trans Amer Fish Soc* 119:254–264.

Magnuson, J. J., and J. H. Prescott. 1966. Courtship, locomotion, feeding, and miscellaneous behaviour of Pacific bonito (*Sarda chiliensis*). *Anim Behav* 14:54–67.

Magnusson, K. P., and M. M. Ferguson. 1987. Genetic analysis of four sympatric morphs of Arctic charr, *Salvelinus alpinus*, from Thingvallavatn, Iceland. *Env Biol Fish* 20:67–73.

Magurran, A. E. 1986. The development of shoaling behaviour in the European minnow, *Phoxinus phoxinus*. *J Fish Biol* 29 (suppl A):159–170.

Magurran, A. E. 1986. Predator inspection behaviour in minnow shoals: Differences between populations and individuals. *Behav Ecol Sociobiol* 19:267–273.

Magurran, A. E. 1990. The adaptive significance of schooling as an anti-predator defence in fish. *Ann Zool Fennici* 27:51–66.

Magurran, A. E., W. J. Oulton, and T. J. Pitcher. 1985. Vigilant behaviour and shoal size in minnows. *Z Tierpsychol* 67:167–178.

Magurran, A. E., and T. J. Pitcher. 1987. Provenance, shoal size and the sociobiology of predator-evasion behaviour in minnow shoals. *Proc R Soc Lond B* 229:439–465.

Maisey, J. G. 1980. An evaluation of jaw suspension in sharks. *Amer Mus Novitates* 2706:17.

Maisey, J. G. 1986. Heads and tails: A chordate phylogeny. *Cladistics* 2:201–256.

Major, P. F. 1979. Piscivorous predators and disabled prey. *Copeia* 1979:158–160.

Malloy, R., and F. D. Martin. 1982. Comparative development of redfin pickerel (*Esox americanus americanus*) and the eastern mudminnow (*Umbra pygmaea*). In *Proc 5th Ann Larval Fish Conf*, ed. C. F. Bryan, J. V. Conner, and F. M. Truesdale, 70–72.

Mandrak, N. E. 1989. Potential invasions of the Great Lakes by fish species associated with climate warming. *J Gt Lakes Res* 15:306–316.

Mann, C. C. 1991. Extinction: Are ecologists crying wolf? *Science* 253:736–738.

Mapstone, B. D., and A. J. Fowler. 1988. Recruitment and the structure of assemblages of fish on coral reefs. *Trends Ecol Evol* 3:72–77.

Margulies, D. 1989. Size-specific vulnerability to predation and sensory system development of white seabass, *Atractoscion nobilis*, larvae. *Fish Bull US* 87:537–552.

Mariscal, R. N. 1970. The nature of the symbiosis between Indo-Pacific anemone fishes and sea anemones. *Mar Biol* 6:58–65.

Marliave, J. B. 1986. Lack of planktonic dispersal of rocky intertidal fish larvae. *Trans Amer Fish Soc* 115:149–154.

Marshall, C. R. 1987. A list of fossil and extant dipnoans. In *The biology and evolution of lungfishes*, ed. W. E. Bemis, W. W. Burggren, and N. E. Kemp, 15–23. New York: Alan R. Liss.

Marshall, N. B. 1954. *Aspects of deep sea biology*. London: Hutchinsons.

Marshall, N. B. 1960. Swimbladder structure of deepsea fishes in relation to their systematics and biology. *Discovery Rept* 31:112.

Marshall, N. B. 1971. *Explorations in the life of fishes.* Cambridge: Harvard University Press.

Marshall, N. B. 1980. *Deep sea biology: Developments and perspectives.* New York: Garland STPM Press.

Martin, T. H., L. B. Crowder, C. F. Dumas, and J. M. Burkholder. 1992. Indirect effects of fish on macrophytes in Bays Mountain Lake: Evidence for a littoral trophic cascade. *Oecologia* 89:476–481.

Martyn, R. D., R. L. Noble, P. W. Bettoli, and R. C. Maggio. 1986. Mapping aquatic weeds with aerial color infrared photography and evaluating their control by grass carp. *J Aquat Plant Manage* 24:46–56.

Maser, C., and J. Sedell. 1994. *From the forest to the sea: The ecology of wood in streams, rivers, estuaries, and oceans.* Delray Beach, FL: St. Lucie Press.

Mason, J. C. 1976. Response of underyearling coho salmon to supplemental feeding in a natural stream. *J Wildl Manage* 40:775–788.

Matarese, A. C., and E. M. Sandknop. 1984. Identification of fish eggs. In *Ontogeny and systematics of fishes,* ed. H. G. Moser, W. J. Richards, D. M. Cohen, M. P. Fahay, A. W. Kendall Jr., and S. L. Richardson, 27–31. Amer Soc Ichthyol Herpetol, Spec Publ 1.

Matthews, W. J. 1986a. Fish faunal "breaks" and stream order in the eastern and central United States. *Env Biol Fish* 17:81–92.

Matthews, W. J. 1986b. Fish faunal structure in an Ozark stream: Stability, persistence and a catastrophic flood. *Copeia* 1986:388–397.

Matthews, W. J., and E. G. Zimmerman. 1990. Potential effects of global warming on native fishes of the Southern Great Plains and the Southwest. *Fisheries* 15(6):26–32.

Mathewson, R. F., and D. P. Rall, eds. 1967. *Sharks, skates and rays.* Baltimore: Johns Hopkins University Press.

Mathis, A., D. P. Chivers, and R. J. F. Smith. 1995. Chemical alarm signals: Predator deterrents or predator attractants? *Amer Natur* 145:994–1005.

May, R. C. 1974. Larval mortality in marine fishes and the critical period concept. In *The early life history of fish,* ed. J. H. S. Blaxter, 3–19. New York: Springer-Verlag.

Mayden, R. L., ed. 1993. *Systematics, historical ecology, and North American freshwater fishes.* Stanford: Stanford University Press.

Mayr, E. 1942. *Systematics and the origin of the species.* New York: Columbia University Press.

Mayr, E. 1974. Cladistic analysis or cladistic classification? *Zool Syst Evol-forsch* 12:94–128.

Mayr, E., and P. D. Ashlock. 1991. *Principles of systematic zoology.* New York: McGraw-Hill.

Mazumder, A., W. D. Taylor, D. J. McQueen, and D. R. S. Lean. 1990. Effects of fish and plankton on lake temperature and mixing depth. *Science* 247:312–315.

McAllister, D. E. 1968. Evolution of branchiostegals and classification of teleostome fishes. *Bull Nat Mus Canada* 221, 239.

McCarthy, J. F., and L. R. Shugart. 1990a. *Biomarkers of environmental contamination.* Boca Raton, FL: CRC Press.

McCarthy, J. F., and L. R. Shugart, eds. 1990b. Biological markers of environmental contamination. In *Biomarkers of environmental contamination.* eds. McCarthy, J. F. and L. R. Shugart, 3–14 Boca Raton, FL: CRC Press.

McClane, A. J., ed. 1974. *McClane's new standard fishing encyclopedia.* 2d ed. New York: Holt, Rinehart and Winston.

McCleave, J. D., G. P. Arnold, J. J. Dodson, and W. H. Neill, eds. 1984. *Mechanisms of migration in fishes.* New York: Plenum Press.

McCleave, J. D., R. C. Kleckner, and M. Castonguay. 1987. Reproductive sympatry of American and European eels and impli-

cations for migration and taxonomy. In *Common strategies of anadromous and catadromous fishes,* ed. M. J. Dadswell, R. J. Klauda, C. M. Moffitt, R. L. Saunders, R. A. Rulifson, and J. E. Cooper. *Amer Fish Soc Symp* 1:286–297.

McCormick, M. J. 1990. Potential changes in thermal structure and cycle of Lake Michigan due to global warming. *Trans Amer Fish Soc* 119:183–194.

McCormick, S. D., and R. L. Saunders. 1987. Preparatory physiological adaptations for marine life of salmonids: Osmoregulation, growth, and metabolism. *Amer Fish Soc Symp* 1:211–229.

McCosker, J. E. 1977. Fright posture of the plesiopid fish *Calloplesiops altivelis*: An example of Batesian mimicry. *Science* 197:400–401.

McCosker, J. E., and M. D. Lagios, eds. 1979. The biology and physiology of the living coelacanth. *Occ Pap Calif Acad Sci* 134:1–175.

McCully, H. H. 1962. The relationship of the Percidae and the Centrarchidae to the Serranidae as shown by the anatomy of their scales. *Amer Zool* 2:430 (abstract).

McCune, A. R. 1990. Evolutionary novelty and atavism in the *Semionotus* complex: Relaxed selection during colonization of an expanding lake. *Evolution* 44:71–85.

McCune, A. R., K. S. Thomson, and P. E. Olsen. 1984. Semionotid fishes from the Mesozoic Great Lakes of North America. In *Evolution of fish species flocks,* ed. A. E. Echelle and I. Kornfield, 27–44. Orono: University of Maine Press.

McDonald, D. G., V. Cavdek, and R. Ellis. 1991. Gill design in freshwater fishes: Interrelationships among gas exchange, ion regulation, and acid-base balance. *Physiol Zool* 64:103–123.

McDowall, R. D. 1978. Generalized tracks and dispersal in biogeography. *Syst Zool* 27:88–104.

McDowall, R. M. 1987. The occurrence and distribution of diadromy among fishes. In *Common strategies of anadromous and catadromous fishes,* ed. M. J. Dadswell, R. J. Klauda, C. M. Moffitt, R. L. Saunders, R. A. Rulifson, and J. E. Cooper. *Amer Fish Soc Symp* 1:1–13.

McDowall, R. M. 1988. *Diadromy in fishes.* London: Croom Helm.

McDowall, R. M. 1990. When galaxiid and salmonid fishes meet—a family reunion in New Zealand. *J Fish Biol* 37 (suppl A):35–43.

McFadden, J. T., G. R. Alexander, and D. S. Shetter. 1967. Numerical changes and population regulation in brook trout *Salvelinus fontinalis.* *J Fish Res Bd Can* 24:1425–1459.

McFarland, W. N. 1980. Observations on recruitment of haemulid fishes. *Proc Gulf Carib Fish Inst* 32:132–138.

McFarland, W. N., E. B. Brothers, J. C. Ogden, M. J. Shulman, E. L. Bermingham, and N. M. Kotchian-Prentiss. 1985. Recruitment patterns in young French grunts, *Haemulon flavolineatum* (family Haemulidae) at St. Croix, USVI. *Fish Bull US* 83:413–426.

McFarland, W. N., and Z.-M. Hillis. 1982. Observations on agonistic behavior between members of juvenile French and white grunts—family Haemulidae. *Bull Mar Sci* 32:255–268.

McFarland, W. N., and N. M. Kotchian. 1982. Interaction between schools of fish and mysids. *Behav Ecol Sociobiol* 11:71–76.

McFarland, W. N., and E. R. Loew. 1983. Wave produced changes in underwater light and their relations to vision. *Env Biol Fish* 8:11–22.

McFarland, W. N., and S. A. Moss. 1967. Internal behavior in fish schools. *Science* 156:260–262.

McFarland, W. N., J. C. Ogden, and J. N. Lythgoe. 1979. The influence of light on the twilight migrations of grunts. *Env Biol Fish* 4:9–22.

McGurk, M. D. 1986. Natural mortality of marine pelagic eggs and larvae: Role of spatial patchiness. *Mar Ecol Progr Ser* 34:227–242.

McIsaac, D. O., and T. P. Quinn. 1988. Evidence for a hereditary component in homing behavior of chinook salmon (*Oncorhynchus tshawytscha*). *Can J Fish Aquat Sci* 45:2201–2205.

McKaye, K. R. 1981. Death feigning: A unique hunting behavior by the predatory cichlid, *Haplochromis livingstoni* of Lake Malawi. *Env Biol Fish* 6:361–365.

McKaye, K. R. 1981. Natural selection and the evolution of inter-specific brood care in fishes. In *Natural selection of social behavior*, ed. R. D. Alexander and D. W. Tinkle, 173–183. New York: Chiron Press.

McKaye, K. R. 1983. Ecology and breeding behaviour of a cichlid fish, *Cyrtocara eucinostomus*, on a large lek in Lake Malawi, Africa. *Env Biol Fish* 8:81–96.

McKaye, K. R. 1991. Sexual selection and the evolution of the cichlid fishes of Lake Malawi, Africa. In *Cichlid fishes: Behaviour, ecology and evolution*, ed. M. H. A. Keenleyside, 241–257. London: Chapman and Hall.

McKaye, K. R., and T. Kocher. 1983. Head ramming behaviour by three paedophagous cichlids in Lake Malawi, Africa. *Anim Behav* 31:206–210.

McKaye, K. R., D. E. Mughogho, and T. J. Lovullo. 1992. Formation of the selfish school. *Env Biol Fish* 35:213–218.

McKenney, T. W. 1959. A contribution to the life history of the squirrel fish *Holocentrus vexillarius* Poey. *Bull Mar Sci Gulf Caribb* 9:174–221.

McKenzie, D. J., and D. J. Randall. 1990. Does *Amia calva* aestivate? *Fish Physiol Biochem* 8:147–158.

McKeown, B. A. 1984. *Fish migration*. London: Croom Helm.

McLaren, I. A. 1974. Demographic strategy of vertical migration by a marine copepod. *Amer Natur* 108:91–102.

Meadows, D. W. 1993. Morphological variation in eyespots of the foureye butterflyfish (*Chaetodon capistratus*): Implications for eyespot function. *Copeia* 1993:235–240.

Meffe, G. K. 1984. Effects of abiotic disturbance on coexistence of predator-prey fish species. *Ecology* 65:1525–1534.

Meffe, G. K. 1986. Conservation genetics and the management of endangered fishes. *Fisheries* 11:14–23.

Meffe, G. K., and F. F. Snelson Jr., eds. 1989. *Ecology and evolution of livebearing fishes (Poeciliidae)*. Engelwood Cliffs, NJ: Prentice Hall.

Meier, A. H. 1992. Circadian basis for neuroendocrine regulation. In *Rhythms in fishes*, ed. M. A. Ali, 109–126. New York: Plenum Press.

Meier, M. F. 1990. Reduced rise in sea level. *Nature* (London) 343:115–116.

Meinsinger, F., and F. Case. 1990. Luminescent properties of deep sea fish. *J Exptal Mar Biol Ecol* 144:1–16.

Meisner, J. D. 1990. Effect of climatic warming on the southern margins of the native range of brook trout, *Salvelinus fontinalis*. *Can J Fish Aquat Sci* 47:1065–1070.

Melton, W. G., and H. W. Scott. 1973. *Geol Soc Amer Spec Pap* 141:31.

Meyer, A. 1993. Phylogenetic relationships and evolutionary processes in east African cichlid fishes. *Trends Ecol Evol* 8:279–284.

Meyer, A., T. D. Kocher, P. Basasibwaki, and A. C. Wilson. 1990. Monophyletic origin of Lake Victoria cichlid fishes suggested by mitochondrial DNA sequences. *Nature* 347:550–553.

Meyer, J. L., E. T. Schultz, and G. S. Helfman. 1983. Fish schools: An asset to corals. *Science* 220:1047–1049.

Micklin, P. P. 1988. Desiccation of the Aral Sea: A water management disaster in the Soviet Union. *Science* 241:1170–1176.

Milinski, M. 1990. Information overload and food selection. In *Behavioural mechanisms of food selection*, NATO ASI Series G 20, ed. R. N. Hughes, 721–737. Berlin: Springer-Verlag.

Milinski, M. 1993. Predation risk and feeding behaviour. In *The behaviour of teleost fishes*, 2d ed., ed. T. J. Pitcher, 285–305. London: Chapman and Hall.

Miller, D. L., et al. 1988. Regional applications of an index of biotic integrity for use in water resource management. *Fisheries* 13(5):12–20.

Miller, J. M. 1988. Physical processes and the mechanisms of coastal migrations of immature marine fishes. *Amer Fish Soc Symp* 3:68–76.

Miller, J. M., L. B. Crowder, and M. L. Moser. 1985. Migration and utilization of estuarine nurseries by juvenile fishes: An evolutionary perspective. *Contrib Mar Sci* 27 (suppl):338–352.

Miller, J. M., L. J. Pietrafesa, and N. P. Smith. 1990. *Principles of hydraulic management of coastal lagoons for aquaculture and fisheries*. FAO Fish Biol Tech Pap 314.

Miller, R. G. 1993. *History and atlas of the fishes of the Antarctic Ocean*. Tucson, AZ: Foresta Inst Ocean and Mountain Studies.

Miller, R. J., and H. E. Evans. 1965. External morphology of the brain and lips in catostomid fishes. *Copeia* 1965:467–487.

Miller, R. R. 1958. Origin and affinities of the freshwater fish fauna of western North America. In *Zoogeography*, ed. C. L. Hubbs, 187–222. Washington, DC: AAAS.

Miller, R. R. 1966. Geographical distribution of Central American freshwater fishes. *Copeia* 1966:773–802.

Miller, R. R. 1981. Coevolution of deserts and pupfishes (Genus *Cyprinodon*) in the American southwest. In *Fishes in North American deserts*, ed. R. J. Naiman and D. L. Soltz, 39–94. New York: John Wiley & Sons.

Miller, R. R., and E. P. Pister. 1971. Management of the Owens pupfish, *Cyprinodon radiosus*, in Mono County, California. *Trans Amer Fish Soc* 100:502–509.

Miller, R. R., J. D. Williams, and J. E. Williams. 1989. Extinctions of North American fishes during the past century. *Fisheries* 14(6):22–38.

Miller, T. 1987. Knotting: A previously undescribed feeding behavior in muraenid eels. *Copeia* 1987:1055–1057.

Miller, T. J. 1989. Feeding behavior of *Echidna nebulosa, Enchelycore pardalis* and *Gymnomuraena zebra* (Teleostei: Muraenidae). *Copeia* 1989:662–672.

Miller, T. J., L. B. Crowder, J. A. Rice, and E. A. Marshall. 1988. Larval size and recruitment mechanisms in fishes: Toward a conceptual framework. *Can J Fish Aquat Sci* 45:1657–1670.

Milligan, C. L., D. G. McDonald, and T. Prior. 1991. Branchial acid and ammonia fluxes in response to alkalosis and acidosis in two marine teleosts: Coho salmon (*Oncorhynchus kisutch*) and starry flounder (*Platichthys stellatus*). *Physiol Zool* 64:169–192.

Minckley, W. L. 1991. Native fishes of the Grand Canyon region: An obituary? In *Committee to Review the Glen Canyon*, 124–177. Washington, DC: National Academy Press.

Minckley, W. L., and W. E. Barber. 1971. Some aspects of biology of the longfin dace, a cyprinid fish characteristic of streams in the Sonoran Desert. *Southw Natur* 15:459–464.

Minckley, W. L., and J. E. Deacon, eds. 1991. *Battle against extinction: Native fish management in the American West*. Tucson: University of Arizona Press.

Minckley, W. L., and G. K. Meffe. 1987. Differential selection by flooding in stream fish communities of the arid American Southwest. In *Community and evolutionary ecology of North American stream fishes*, ed. W. J. Matthews and D. C. Heins, 93–104. Norman: University of Oklahoma Press.

Minshall, G. W., et al. 1985. Developments in stream ecosystem theory. *Can J Fish Aquat Sci* 42:1045–1055.

Mittelbach, G. G. 1988. Competition among refuging sunfishes and effects of fish density on littoral zone invertebrates. *Ecology* 69:614–623.

Miura, T., and J. Wang. 1985. Chlorophyll *a* found in feces of phytoplanktivorous cyprinids and its photosynthetic activity. *Verh Int Ver Limnol* 22:2636–2642.

Moav, R., T. Brody, and G. Hulata. 1978. Genetic improvement of wild fish populations. *Science* 201:1090–1094.

Moller, P. 1980. Electroreception. *Oceanus* 23:44–54.

Monod, T. 1968. Le complexe urophore des poissons teleosteens. *Afrique Noire: Mem Inst Fond* 81:705.

Montgomery, J. C. 1988. Sensory physiology. In *Physiology of elasmobranch fishes*, ed. T. J. Shuttleworth, 79–98. Berlin: Springer-Verlag.

Montoya, R. V., and T. B. Thorson. 1982. The bull shark (*Carcharhinus leucas*) and largetooth sawfish (*Pristis perotteti*) in Lake Bayano, a tropical man-made impoundment in Panama. *Env Biol Fish* 7:341–347.

Moore, R. H., and D. E. Wohlschlag. 1971. Seasonal variations in the metabolism of Atlantic midshipman, *Porichthys porosissimus* (Valenciennes). *J Exp Mar Biol Ecol* 7:163–172.

Moore, W. S. 1984. Evolutionary ecology of unisexual fishes. In *Evolutionary genetics of fishes*, ed. B. J. Turner, 329–398. New York: Plenum Press.

Morin, J. G., A. Harrington, K. Nealson, N. Kriegr, T. O. Baldwin, and J. W. Hastings. 1975. Light for all reasons: Versatility in the behavioral repertoire of the flashlight fish. *Science* 190:74–76.

Moser, H. G. 1981. Morphological and functional aspects of marine fish larvae. In *Marine fish larvae: Morphology, ecology, and relation to fisheries*, ed. R. L. Lasker, 89–131. Seattle: Washington Sea Grant Program.

Moser, H. G. 1984. In *Ontogeny and systematics of fishes*, ed. H. G. Moser, W. J. Richards, D. M. Cohen, M. P. Fahay, A. W. Kendall Jr., and S. L. Richardson, 1–6. Amer Soc Ichthyol Herpetol, Spec Publ 1.

Moser, H. G., and E. H. Ahlstrom. 1970. Development of lanternfishes (family Myctophidae) in the California Current. Part I. Species with narrow-eyed larvae. *Nat Hist Mus Los Ang Cty Sci Bull* 7.

Moser, H. G., and E. H. Ahlstrom. 1974. Role of larval stages in systematic investigations of marine teleosts: The Myctophidae, a case study. *Fish Bull US* 72:391–413.

Moser, H. G., W. J. Richards, D. M. Cohen, M. P. Fahay, A. W. Kendall Jr., and S. L. Richardson, eds. 1984. *Ontogeny and systematics of fishes*. Amer Soc Ichthyol Herpetol, Spec Publ 1.

Moss, S. A. 1967. Tooth replacement in the lemon shark *Negaprion brevirostris*. In *Sharks, skates and rays*, ed. P. W. Gilbert, R. F. Mathewson, and D. P. Rall. Baltimore: Johns Hopkins Press.

Moss, S. A. 1984. *Sharks. A guide for the amateur naturalist*. Englewood Cliffs, NJ: Prentice Hall.

Moss, S. A. 1991. Anatomical features of sharks and their allies. In *Discovering sharks*, ed. S. H. Gruber, 35–40. Highlands, NJ: American Littoral Society, Spec Publ 14.

Mott, J. C. 1957. The cardiovascular system. *Physiol Fish* 1:81–108.

Motta, P. J. 1984. Mechanics and functions of jaw protrusion in teleost fishes: A review. *Copeia* 1984:1–18.

Motta, P. J. 1988. Functional morphology of the feeding apparatus of ten species of Pacific butterflyfishes (Perciformes, Chaetodontidae): An ecomorphological approach. *Env Biol Fish* 22:39–67.

Motta, P. J., ed. 1989. Butterflyfishes: Success on the coral reef. *Env Biol Fish* 25:7–246.

Moyer, J. T., and A. Nakazano. 1978. Protandrous hermaphroditism in six species of the anemonefish genus *Amphiprion* in Japan. *Jap J Ichthyol* 25:101–106.

Moyer, J. T., and C. E. Sawyers. 1973. Territorial behaviour of the anemonefish *Amphiprion xanthurus* with notes on the life history. *Jap J Ichthyol* 20:85–93.

Moyer, J. T., and Y. Yogo. 1982. The lek mating system of *Halichoeres melanochir* (Pisces: Labridae) at Miyake-jima. *Japan Z Tierpsychol* 60:209–226.

Moyle, P. B. 1991. Ballast water introductions. *Fisheries* 16(1):4–6.

Moyle, P. B., and J. J. Cech Jr. 1996. *Fishes: An introduction to ichthyology*. 3d ed. Upper Saddle River, NJ: Prentice Hall.

Moyle, P. B., and R. A. Leidy. 1992. Loss of biodiversity in aquatic ecosystems: Evidence from fish faunas. In *Conservation biology: The theory and practice of nature conservation, preservation and management*, ed. P. L. Fiedler and S. K. Jain, 127–169. New York: Chapman and Hall.

Moyle, P. B., and R. Nichols. 1974. Decline of the native fish fauna of the Sierra-Nevada foothills in central California. *Amer Midl Natur* 92:72–83.

Moy-Thomas, J. A., and R. S. Miles. 1971. *Palaeozoic fishes*. 2d ed. London: Chapman and Hall.

Müller, K. 1978a. Locomotor activity of fish and environmental oscillations. In *Rhythmic activity of fishes*, ed. J. E. Thorpe, 1–29. London: Academic Press.

Müller, K. 1978b. The flexibility of the circadian system of fish at different latitudes. In *Rhythmic activity of fishes*, ed. J. E. Thorpe, 91–104. London: Academic Press.

Munkittrick, K. R., and D. G. Dixon. 1988. Growth, fecundity, and energy stores of white sucker (*Catostomus commersoni*) from lakes containing elevated levels of copper and zinc. *Can J Fish Aquat Sci* 45:1355–1365.

Munro, A. D. 1990a. Tropical freshwater fishes. In *Reproductive seasonality in teleosts: Environmental influences*, ed. A. D. Munro, A. P. Scott, and T. J. Lam, 145–239. Boca Raton, FL: CRC Press.

Munro, A. D. 1990b. General introduction. In *Reproductive seasonality in teleosts: Environmental influences*, ed. A. D. Munro, A. P. Scott, and T. J. Lam, 1–11. Boca Raton, FL: CRC Press.

Munro, A. D., A. P. Scott, and T. J. Lam, eds. 1990. *Reproductive seasonality in teleosts: Environmental influences*. Boca Raton, FL: CRC Press.

Munz, F. W., and W. N. McFarland. 1973. The significance of spectral position in the rhodopsins of tropical marine fishes. *Vis Res* 13:1829–1874. Oxford, OX3 0BW, UK: Pergamon Press Ltd, Headington Hill Hall.

Murphy, R. C. 1971. The structure of the pineal organ of the bluefin tuna, *Thunnus thynnus*. *J Morph* 133:1–15.

Murray, J. D., J. J. Bahen, and R. A. Rulifson. 1992. Management considerations for by-catch in the North Carolina and Southeast shrimp fishery. *Fisheries* 17(1):21–26.

Musick, J. A., M. N. Bruton, and E. K. Balon, eds. 1991. The biology of *Latimeria chalumnae* and evolution of coelacanths. *Env Biol Fish* 32:1–435.

Myers, G. S. 1938. Fresh-water fishes and West Indian zoogeography. *Ann Rept Smithsonian Inst* 1937:339–364.

Myers, G. S. 1964. A brief sketch of the history of ichthyology in America to the year 1850. *Copeia* 1964:33–41.

Myrberg, A. A. Jr. 1975. The role of chemical and visual stimuli in the preferential discrimination of young by cichlid fish, *Cichlasoma nigrofasciatum* (Gunter). *Z Tierpsychol* 37:274–297.

Myrberg, A. A. Jr. 1978. Underwater sound—its effect on the behaviour of sharks. In *Sensory biology of sharks, skates, and rays*, ed. E. S. Hodgson and R. F. Mathewson. Arlington, VA: Off Nav Res, Dept Navy.

Myrberg, A. A. Jr. 1981. Sound communication and interception in fishes. In *Hearing and sound communication in fishes*, ed. W. Tavolga, A. N. Popper, and R. R. Fay, 395–452. New York: Springer-Verlag.

Myrberg, A. A. Jr. 1987. Understanding shark behavior. In *Sharks: An inquiry into biology, behavior, fisheries and use*, ed. S. Cook. Corvalis: Oregon State University Extension Service.

Myrberg, A. A. Jr., and D. R. Nelson. 1991. The behavior of sharks:

What have we learned. In *Discovering sharks*, ed. S. H. Gruber, 92–100. Highlands, NJ: American Littoral Society, Spec Publ 14.

Myrberg, A. A. Jr., and R. E. Thresher. 1974. Interspecific aggression and its relevance to the concept of territoriality in fishes. *Amer Zool* 14:81–96.

Nafpaktitis, B. G., R. H. Backus, J. E. Craddock, R. L. Haedrich, B. H. Robison, and C. Karnella. 1977. Family Myctophidae. In *Fishes of the western North Atlantic*. Part 7, *Order Iniomi (Myctophiformes)*, 13–265. New Haven, CT: Mem Sears Fndn Mar Res, No. 1, Part 7.

Nagelkerke, L. A. J., F. A. Sibbing, J. G. M. van den Boogaart, E. H. R. R. Lammens, and J. W. M. Osse. 1994. The barbs (*Barbus* spp.) of Lake Tana: A forgotten species flock. *Env Biol Fish* 39:1–22.

Naiman, R. J. 1981. An ecosystem overview: Desert fishes and their habitats. In *Fishes in North American deserts*, ed. R. J. Naiman and D. L. Soltz, 493–531. New York: John Wiley & Sons.

Naiman, R. J., and D. L. Soltz, eds. 1981. *Fishes in North American deserts*. New York: John Wiley & Sons.

Nakamura, R. 1991. *A survey of the Pacific hagfish resource off the Central California coast*. Final Report, California Marine Fisheries Impacts Program, Contract No. A-800-184. Sacramento, CA.

Nakashima, B. S., and W. C. Leggett. The role of fishes in the regulation of phosphorus availability in lakes. *Can J Fish Aquat Sci* 37:1540–1549.

Nakatsaru, K., and D. L. Kramer. 1982. Is sperm cheap? Limited male fertility and female choice in the lemon tetra (Pisces, Characidae). *Science* 216:753–755.

National Research Council (US). 1992. *Restoration of aquatic ecosystems: Science, technology, and public policy*. Washington, DC: National Academy Press.

Nehlsen, W., J. E. Williams, and J. A. Lichatowich. 1991. Pacific salmon at the crossroads: Stocks at risk from California, Oregon, Idaho, and Washington. *Fisheries* 16:4–21.

Neill, S. R. St. J., and J. M. Cullen. 1974. Experiments on whether schooling by their prey affects the hunting behaviour of cephalopod and fish predators. *J Zool Lond* 182:549–569.

Nelson, D. R. 1990. Telemetry studies of sharks: A review, with applications in resource management. In *Elasmobranchs as living resources: Advances in the biology, ecology, systematics, and the status of fisheries*, ed. H. L. Pratt Jr., S. H. Gruber, and T. Taniuchi, 239–256. Washington, DC: NOAA Tech Rept 90.

Nelson, D. R., and R. H. Johnson. 1972. Acoustic attraction of Pacific reef sharks: Effect of pulse intermittency and variability. *Comp Biochem Physiol* 42A:85–95.

Nelson, G. J. 1969. Gill arches and the phylogeny of fishes, with notes on the classification of vertebrates. *Bull Amer Mus Nat Hist* 141:475–552.

Nelson, G., and N. Platnick. 1981. *Systematics and biogeography. Cladistics and vicariance*. New York: Columbia University Press.

Nelson, G., and D. E. Rosen, eds. 1981. *Vicariance biogeography: A critique*. New York: Columbia University Press.

Nelson, J. S. 1994. *Fishes of the world*. 3d ed. New York: John Wiley & Sons.

Netboy, A. 1980. *The Columbia River salmon and steelhead trout*. Seattle: University of Washington Press.

Netsch, N. F., and A. Witt Jr. 1962. Contributions to the life history of the longnose gar (*Lepisosteus osseus*) in Missouri. *Trans Amer Fish Soc* 91:251–262.

Neudecker, S. 1979. Effects of grazing and browsing fishes on the zonation of corals in Guam. *Ecology* 60:666–672.

Neudecker, S. 1989. Eye camouflage and false eyespots: Chaetodontid responses to predators. *Env Biol Fish* 25:143–157.

Nevenzel, J. C., W. Rodegker, J. F. Mead, and M. S. Gordon. 1966. Lipids of the living coelacanth, *Latimeria chalumnae*. *Science* 152:1753–1755.

Newman, R. M., and T. F. Waters. 1984. Size-selective predation on *Gammarus pseudolimnaeus* by trout and sculpins. *Ecology* 65:1535–1545.

Nichols, S., A. Shah, G. J. Pellegrin Jr., and K. Mullin. 1990. *Updated estimates of shrimp fleet by-catch in the offshore waters of the Gulf of Mexico 1972–1989*. Pascagoula, MS: National Marine Fisheries Service.

Nico, L. G., and D. C. Taphorn. 1988. Food habits of piranhas in the low llanos of Venezuela. *Biotropica* 20:311–321.

Nielson, J. D., and R. I. Perry. 1990. Diel vertical migrations of marine fishes: An obligate or facultative process. *Adv Mar Biol* 26:115–168.

Nielsen, J. G. 1977. The deepest-living fish, *Abyssobrotula galatheae*, a new genus and species of oviparous ophidioids (Pisces, Brotulidae). *Galathea Rept* 14:41–48.

Nielsen, J. G., E. Bertelsen, and A. Jespersen. 1989. The biology of *Eurypharynx pelecanoides* (Pisces: Eurypharyngidae). *Acta Zool* (Stockholm) 70:187–197.

Nielsen, J. G., and O. Munk. 1964. A hadal fish (*Bassogigas profundissimus*) with a functional swimbladder. *Nature* 204 (4958):594–595.

Nielsen, L. A., and D. L. Johnson, eds. 1983. *Fisheries techniques*. Bethesda, MD: American Fisheries Society.

Nieuwenhuys, R., and E. Pouwels. 1983. The brain stem of actinopterygian fishes. In *Fish neurobiology*, ed. R. G. Northcutt and R. E. Davis, 1:25–87. Ann Arbor: University of Michigan Press.

Nikolsky, G. V. 1961. *Special ichthyology*. Jerusalem: Israel Program for Scientific Translations.

Nilsson, S., and S. Holmgren. 1993. Autonomic nerve functions. In *The physiology of fishes*, ed. D. H. Evans, 279–313. Boca Raton, FL: CRC Press.

NMFS (National Marine Fisheries Service). 1991a. *Our living oceans*. Washington, DC: NOAA Technical Memorandum, NMFS-F/SPO-1.

NMFS (National Marine Fisheries Service). 1991b. *Status of fishery resources off the northeastern US for 1991*. Washington, DC: NOAA Technical Memorandum, NMFS-F/NEC/86.

Noakes, D. L. G. 1979. Parent-touching behavior by young fishes: Incidence, function and causation. *Env Biol Fish* 4:389–400.

Noakes, D. L. G., and J.-G. J. Godin. 1988. Ontogeny of behaviour and concurrent development changes in sensory system in teleost fishes. In *Fish physiology*, Vol. 11, Part B, ed. W. S. Hoar and D. J. Randall. San Diego, CA: Academic Press.

Noakes, D. L. G., D. G. Lindquist, G. S. Helfman, and J. A. Ward, eds. 1983. *Predators and prey in fishes. Developments in environ biol fish 2*. The Hague: Dr. W. Junk.

Norcross, B. L., and R. F. Shaw. 1984. Oceanic and estuarine transport of fish eggs and larvae: A review. *Trans Amer Fish Soc* 113:153–165.

Norman, J. R. 1931. *A history of fishes*. New York: A. A. Wyn.

Norman, J. R. 1948. *A history of fishes*. 2d ed. New York: A. A. Wyn.

Norman, J. R., and F. C. Fraser. 1949. *Field book of giant fishes*. New York: G. P. Putnam's Sons.

Norman, J. R., and P. H. Greenwood. 1975. *A history of fishes*. 3d ed. New York: Halstead Press.

Normark, B. B., A. R. McCune, and R. G. Harrison. 1991. Phylogenetic relationships of neopterygian fishes, inferred from mitochondrial DNA sequences. *Mol Biol Evol* 8:819–834.

Norse, E. A., ed. 1993. *Global marine biological diversity*. Washington, DC: Island Press.

North Atlantic Salmon Conservation Organization. 1991. *Guidelines to minimize the threats to wild salmon stocks from salmon aquaculture*. Edinburgh: North Atlantic Salmon Conservation Organization.

Northcote, T. G. 1978. Migratory strategies and production in freshwater fishes. In *Ecology of freshwater fish populations*, ed. S. D. Gerking, 326–359. New York: John Wiley & Sons.

Northcote, T. G. 1988. Fish in the structure and function of freshwater ecosystems: A 'top-down' view. *Can J Fish Aquat Sci* 45:361–379.

Northcote, T. G., H. W. Lorz, and J. C. MacLeod. 1964. Studies on diel vertical movement of fishes in a British Columbia lake. *Verh Int Ver Limnol* 15:940–946.

Northcutt, R. G. 1977. Elasmobranch central nervous system organization and its possible evolutionary significance. *Amer Zool* 17:411–429.

Northcutt, R. G., and R. E. Davis, eds. 1983. *Fish neurobiology*. 2 vols. Ann Arbor: University of Michigan Press.

Northcutt, R. G., and C. Gans. 1983. The genesis of neural crest and epidermal placodes: A reinterpretation of vertebrate origins. *Quart Rev Biol* 58:1–28.

Northcutt, S. J., R. N. Gibson, and E. Morgan. 1990. Persistence and modulation of endogenous circatidal rhythmicity in *Lipophrys pholis* (Teleostei). *J Mar Biol Ass UK* 70:815–827.

Nursall, J. R. 1973. Some behavioral interactions of spottail shiner (*Notropis hudsonius*), yellow perch (*Perca flavescens*), and northern pike (*Esox lucius*). *J Fish Res Board Can* 30:1161–1178.

Nyberg, D. W. 1971. Prey capture in the largemouth bass. *Amer Midl Natur* 86:128–144.

O'Brien, L., J. Burnett, and R. K. Mayo. 1993. *Maturation of nineteen species of finfish off the northeast coast of the United States, 1985–1990*. Washington, DC: NOAA Tech Rept NMFS 113.

O'Brien, W. J. 1987. Planktivory by freshwater fish: Thrust and parry in the pelagia. In *Predation: Direct and indirect impacts on aquatic communities*, ed. W. C. Kerfoot and A. Sih, 3–16. Hanover, NH: University Press of New England.

O'Brien, W. J., B. Evans, and C. Luecke. 1985. Apparent size choice of zooplankton by planktivorous sunfish: Exceptions to the rule. *Env Biol Fish* 13:225–233.

Oelofsen, B. W., and K. Loock. 1981. A fossil cephalochordate from the early Permian Whitehill Formation of South Africa. *S Afr J Sci* 77:178–180.

Ogden, J. C. 1974. Grazing by the echinoid *Diadema antillarum* Philippi: Formation of halos around West Indian patch reefs. *Science* 182:715–716.

Ogden, J. C. 1977. Carbonate sediment production by parrotfish and sea urchins on Caribbean reefs. In *Reefs and related carbonates—ecology and sedimentology*, ed. S. H. Frost, M. P. Weiss, and J. B. Saunders, 281–288. Amer Ass Petrol Geol Stud Geol 4.

Ogura, M., M. Kato, N. Arai, T. Sasada, and Y. Sakaki. 1992. Magnetic particles in chum salmon (*Oncorhynchus keta*): Extraction and transmission electron microscopy. *Can J Zool* 70:874–877.

Ogutu-Ohwayo, R. 1990. The decline of the native fishes of lakes Victoria and Kyoga (East Africa) and the impact of introduced species, especially the Nile perch, *Lates niloticus*, and the Nile tilapia, *Oreochromis niloticus*. *Env Biol Fish* 27:81–96.

Okazaki, T. 1984. Genetic divergence and its zoogeographic implications in closely related species *Salmo gairdneri* and *Salmo mykiss*. *Japanese J Ichthyol* 31:297–311.

Olla, B. L., A. J. Bejda, and A. D. Martin. 1979. Seasonal dispersal and habitat selection of cunner, *Tautogolabrus adspersus*, and young tautog, *Tautoga onitis*, in Fire Island Inlet, Long Island, New York. *Fish Bull US* 77:255–261.

Olsen, P., and A. R. McCune. 1991. Morphology of the *Semionotus elegans* species group from the early Jurassic part of the Newark Supergroup of eastern North America with comments on

the Family Semionotidae (Neopterygii). *J Vert Paleo* 11:269–292.

Ono, R. D., J. D. Williams, and A. Wagner. 1983. *Vanishing fishes of North America*. Washington, DC: Stone Wall Press.

Osenberg, C. W., G. G. Mittelbach, and P. C. Wainwright. 1992. Two-stage life histories in fish: The interaction between juvenile competition and adult performance. *Ecology* 73:255–267.

Osenberg, C. W., E. E. Werner, G. G. Mittelbach, and D. J. Hall. 1988. Growth patterns in bluegill (*Lepomis macrochirus*) and pumpkinseed (*L. gibbosus*) sunfish: Environmental variation and the importance of ontogenetic niche shifts. *Can J Fish Aquat Sci* 45:17–26.

Osse, J. W. M., and M. Muller. 1980. A model of suction feeding in fishes with some implications for ventilation. In *Environmental physiology of fishes*, ed. M. A. Ali. New York: Plenum Press.

Overstrom, N. A. 1991. Estimated tooth replacement rate in captive sand tiger sharks (*Carcharias taurus* Rafinsesque, 1810). *Copeia* 1991:525–526.

Page, L. M. 1983. *Handbook of darters*. Neptune City, NJ: TFH Publ.

Paine, R. T., and A. R. Palmer. 1978. *Sicyases sanguineus*: A unique trophic generalist from the Chilean intertidal zone. *Copeia* 1978:75–81.

Pan Jiang. 1984. The phylogenetic position of the Eugaleaspida in China. *Proc Linn Soc NSW* 197:309–319.

Panella, G. 1971. Fish otoliths: Daily growth layers and periodical patterns. *Science* 173:1124–1127.

Parenti, L. R. 1984. Biogeography of the Andean killifish genus *Orestias* with comments on the species flock concept. In *Evolution of fish species flocks*, ed. A. E. Echelle and I. Kornfield, 85–92. Orono: University of Maine Press.

Parrish, J. K. 1988. Re-examining the selfish herd: Are central fish safer? *Anim Behav* 38:1048–1053.

Parrish, J. K. 1989a. Layering with depth in a heterospecific fish aggregation. *Env Biol Fish* 26:79–85.

Parrish, J. K. 1989b. Predation on a school of flat-iron herring, *Harengula thrissina*. *Copeia* 1989:1089–1091.

Parrish, J. K. 1992. Levels of diurnal predation on a school of flat-iron herring, *Harengula thrissina*. *Env Biol Fish* 34:257–263.

Parrish, J. K., and W. K. Kroen. 1988. Sloughed mucus and drag reduction in a school of Atlantic silversides, *Menidia menidia*. *Mar Biol* 97:165–169.

Parsons, G. R. 1990. Metabolism and swimming efficiency of the bonnethead shark *Sphyrna tiburo*. *Mar Biol* 104:363–367.

Partridge, B. L. 1982. Structure and function of fish schools. *Sci Amer* 245:114–123.

Partridge, B. L., J. Johansson, and J. Kalish. 1983. The structure of schools of giant bluefin tuna in Cape Cod Bay. *Env Biol Fish* 9:253–262.

Parzefall, J. 1993. Behavioural ecology of cave-dwelling fishes. In *The behaviour of teleost fishes*, 2d ed., ed. T. J. Pitcher, 573–606. London: Chapman and Hall.

Paszkowski, C. A., and B. L. Olla. 1985. Social interactions of coho salmon *Oncorhynhchus kisutch* smolts in seawater. *Can J Zool* 63:2401–2407.

Patterson, C. 1965. The phylogeny of the chimaeroids. *Philos Trans Roy Soc London B* 249:101–219.

Patterson, C. 1975. The braincase of pholidophorid and leptolepid fishes, with a review of the actinopterygian braincase. *Philos Trans Roy Soc London: Bio Sci* 269:275–579.

Patterson, C. 1982. Morphology and interrelationships of primitive actinopterygian fishes. *Amer Zool* 22:241–259.

Patterson, C., and G. D. Johnson. 1995. The intermuscular bones and ligaments of teleostean fishes. *Smithsonian Contrib Zool* 559:83.

Patterson, C., and A. E. Longbottom. 1989. An Eocene amiid fish from Mali, West Africa. *Copeia* 1989:827–836.

Patzner, R. A., and R. S. Santos. 1992. Field observations on the association between the clingfish *Diplecogaster bimaculata pectoralis*, Briggs, 1955 and different species of sea urchins at the Azores. *Z Fischk* 1:157–161.

Paxton, J. R., and W. N. Eschmeyer, eds. 1994. *Encyclopedia of fishes.* Sydney, Australia: University of New South Wales Press.

Peakall, D. 1992. *Animal biomarkers as pollution indicators.* New York: Chapman and Hall.

Pelster, B., and P. Scheid. 1992. Countercurrent concentration and gas secretion in the fish swim bladder. *Physiol Zool* 65:1–16.

Penney, R. K., and G. Goldspink. 1980. Temperature adaptation of sarcoplasmic reticulum of fish muscle. *J Therm Biol* 5:63–68.

Perra, P. 1992. By-catch reduction devices as a conservation measure. *Fisheries* 17(1):28–29.

Perrone, M. Jr., and T. M. Zaret. 1979. Parental care patterns of fishes. *Amer Natur* 113:351–361.

Petersen, C. W., R. R. Warner, S. Cohen, H. C. Hess, and A. T. Sewell. 1992. Fertilization success in a reef fish. *Ecology* 73:391–401.

Petrosky, B. R., and J. J. Magnuson. 1973. Behavioral responses of northern pike, yellow perch and bluegill to oxygen concentrations under simulated winterkill conditions. *Copeia* 1973:124–133.

Pfeiffer, W. 1982. Chemical signals in communication. In *Chemoreception in fishes*, ed. T. J. Hara, 307–326. Amsterdam: Elsevier.

Pfeiler, E. 1986. Towards an explanation of the developmental strategy in leptocephalus larvae of marine teleost fishes. *Env Biol Fish* 15:3–13.

Pietsch, T. W. 1974. Osteology and relationships of ceratioid anglerfishes of the family Oneirodidae, with a review of the genus *Oneirodes* Lutken. *Bull Natur Hist Mus Los Angeles Co* 18:1–112.

Pietsch, T. W. 1976. Dimorphism, parasitism and sex: Reproductive strategies among deepsea ceratioid anglerfishes. *Copeia* 1976:781–793.

Pietsch, T. W. 1978. The feeding mechanism of *Stylephorus chordatus* (Teleostei: Lampridiformes): Functional and ecological implications. *Copeia* 1978:255–262.

Pietsch, T. W., and D. B. Grobecker. 1978. The compleat angler: Aggressive mimicry in antennariid anglerfish. *Science* 201:369–370.

Pietsch, T. W., and D. B. Grobecker. 1987. *Frogfishes of the world: Systematics, zoogeography, and behavioral ecology.* Stanford: Stanford University Press.

Pister, E. P. 1981. The conservation of desert fishes. In *Fishes in North American deserts*, ed. R. J. Naiman and D. L. Soltz, 411–445. New York: John Wiley & Sons.

Pister, E. P. 1992. Ethical considerations in conservation of biodiversity. *Trans 57th N Amer Wildl Nat Res Conf* 355–364.

Pitcher, T. J. 1983. Heuristic definitions of shoaling behaviour. *Anim Behav* 31:611–613.

Pitcher, T. J., ed. 1993. *The behaviour of teleost fishes.* 2d ed. London: Chapman and Hall.

Pitcher, T. J., and P. J. B. Hart. 1982. *Fisheries ecology.* London: Croom Helm.

Pitcher, T. J., and J. K. Parrish. 1993. Functions of shoaling behaviour in teleosts. In *The behaviour of teleost fishes*, 2d ed., ed. T. J. Pitcher, 363–439. London: Chapman and Hall.

Pitcher, T. J., B. L. Partridge, and C. S. Wardle. 1976. A blind fish can school. *Science* 194:963–965.

Pitcher, T. J., and C. J. Wyche. 1983. Predator avoidance behaviour of sand-eel schools: Why schools seldom split. In *Predators and prey in fishes*, ed. D. L. G. Noakes, B. G. Lindquist, G. S. Helfman, and J. A. Ward, 193–204. The Hague: Dr. W. Junk.

Place, A. R., and D. R. Powers. 1979. Genetic variation and relative catalytic efficiencies: Lactate dehydrogenase B allozyme of *Fundulus heteroclitus*. *Proc Natl Acad Sci USA* 76:2354–2358.

Policansky, D. 1982a. Influence of age, size, and temperature on metamorphosis in the starry flounder, *Platichthys stellatus*. *Can J Fish Aquat Sci* 39:514–517.

Policansky, D. 1982b. The asymmetry of flounders. *Sci Amer* 246:116–122.

Policansky, D. 1982c. Sex change in plants and animals. *Ann Rev Ecol Syst* 13:471–495.

Policansky, D. 1993. Fishing as a cause of evolution in fishes. In *Evolution of exploited populations*, ed. R. Law and K. Stokes, 2–18. Berlin: Springer-Verlag.

Polloni, P., R. Haedrich, G. Rowe, and C. H. Clifford. 1979. The size-depth relationship in deep ocean animals. *Int Rev Ges Hydrobiol* 64:38–46.

Polovina, J. J., and S. Ralston, eds. 1987. *Tropical snappers and groupers, biology and fisheries management.* Boulder, CO: Westview Press.

Popper, A. N., and C. Platt. 1993. Inner ear and lateral line. In *The physiology of fishes*, ed. D. H. Evans, 99–136. Boca Raton, FL: CRC Press.

Posey, M. H., and W. G. Ambrose. 1994. Effects of proximity to an offshore hard-bottom reef on infaunal abundances. *Mar Biol* 118:745–753.

Poss, S. G., and B. B. Collette. 1995. Second survey of fish collections in the United States and Canada. *Copeia* 1995:48–70.

Potts, G. W. 1980. The predatory behaviour of *Caranx melampygus* (Pisces) in the channel environment of Aldabra Atoll (Indian Ocean). *J Zool Lond* 192:323–350.

Potts, G. W. 1983. The predatory tactics of *Caranx melampygus* and the response of its prey. In *Predators and prey in fishes. Developments in environ biol fish 2*, ed. D. L. G. Noakes, D. G. Lindquist, G. S. Helfman, and J. A. Ward, 181–191. The Hague: Dr. W. Junk.

Potts, G. W. 1984. Parental behaviour in temperate marine teleosts with special reference to the development of nest structures. In *Fish reproduction: Strategies and tactics*, ed. G. W. Potts and R. J. Wootton, 223–244. London: Academic Press.

Potts, G. W., and R. J. Wootton, eds. 1984. *Fish reproduction: Strategies and tactics.* London: Academic Press.

Pough, F. H., J. B. Heiser, and W. N. McFarland. 1989. *Vertebrate life.* 3d ed. New York: Macmillan.

Poulson, T. L. 1963. Cave adaptation in amblyopsid fishes. *Amer Midl Natur* 70:257–290.

Poulson, T. L., and W. B. White. 1969. The cave environment. *Science* 165:971–981.

Power, M. E. 1987. Predator avoidance by grazing fishes in temperate and tropical streams: Importance of stream depth and prey size. In *Predation: Direct and indirect impacts on aquatic communities*, ed. W. C. Kerfoot and A. Sih, 333–351. Hanover, NH: University Press of New England.

Power, M. E. 1990. Effects of fish in river food webs. *Science* 250:811–814.

Power, M. E., T. L. Dudley, and S. D. Cooper. 1989. Grazing catfish, fishing birds, and attached algae in a Panamanian stream. *Env Biol Fish* 26:285–294.

Pratt, H. L., and J. I. Castro. 1991. Shark reproduction: Parental investment and limited fisheries—an overview. In *Discovering sharks*, ed. S. H. Gruber, 56–60. Highlands, NJ: American Littoral Society, Spec Publ 14.

Pratt, H. L. Jr., S. H. Gruber, and T. Taniuchi, eds. 1990. *Elasmobranchs as living resources: Advances in the biology, ecology, systematics, and the status of fisheries.* Washington, DC: NOAA Tech Rept 90.

Pressley, P. H. 1981. Parental effort and the evolution of nest-guarding tactics in the threespine stickleback, *Gasterosteus aculeatus* L. *Evolution* 35:282–295.

Preston, J. L. 1978. Communication systems and social interactions in a goby-shrimp symbiosis. *Anim Behav* 26:791–802.

Price, D. J. 1984. Genetics of sex determination in fishes—a brief review. In *Fish reproduction: Strategies and tactics*, ed. G. W. Potts and R. J. Wootton, 77–89. London: Academic Press.

Primor, N., J. Parness, and E. Zlotkin. 1978. Pardaxin: The toxic factor from the skin secretion of the flatfish *Pardachirus marmoratus* (Soleidae). In *Toxins: Animal, plant and microbial*, ed. P. Rosenberg, 539–547. Oxford: Pergamon Press.

Pulliam, H. R., and T. Caraco. 1984. Living in groups: Is there an optimal group size? In *Behavioural ecology*, 2d ed., ed. J. R. Krebs and N. B. Davies, 122–142. Oxford: Blackwell Science.

Purcell, E. M. 1977. Life at low Reynolds numbers. *Am J Physics* 45:3–11.

Purnell, M. A. 1995. Microwear on conodont elements and macrophagy in the first vertebrates. *Nature* (London) 374:798–800.

Pusey, B. J. 1989. Aestivation in the teleost fish *Lepidogalaxias salamandroides* (Mees). *Comp Biochem Physiol* 92A:137–138.

Pusey, B. J. 1990. Seasonality, aestivation and the life history of the salamanderfish *Lepidogalaxias salamandroides* (Pisces: Lepidogalaxiidae). *Env Biol Fish* 29:15–26.

Pyle, R. L., and J. E. Randall. 1994. A review of hybridization in marine angelfishes (Perciformes: Pomacanthidae). *Env Biol Fish* 41:127–145.

Quattro, J., and R. C. Vrijenhoek. 1989. Fitness differences among remnant populations of the endangered Sonoran topminnow. *Science* 245:976–978.

Quinn, T. P. 1982. A model for salmon navigation on the high seas. In *Salmon and trout migratory behavior symposium*, ed. E. L. Brannon and E. O. Salo, 229–237. Seattle: School of Fisheries, University of Washington.

Quinn, T. P. 1984. Homing and straying in Pacific salmon. In *Mechanisms of migration in fishes*, ed. J. D. McCleave, G. P. Arnold, J. J. Dodson, and W. H. Neill, 357–362. New York: Plenum Press.

Quinn, T. P., and E. L. Brannon. 1982. The use of celestial and magnetic cues by orienting sockeye salmon smolts. *J Comp Physiol A* 147:547–552.

Quinn, T. P., and A. H. Dittman. 1990. Pacific salmon migrations and homing: Mechanisms and adaptive significance. *Trends Ecol Evol* 5:174–177.

Quinn, T. P., and W. C. Leggett. 1987. Perspectives on the marine migrations of diadromous fishes. In *Common strategies of anadromous and catadromous fishes*, ed. M. J. Dadswell, R. J. Klauda, C. M. Moffitt, R. L. Saunders, R. A. Rulifson, and J. E. Cooper. *Amer Fish Soc Symp* 1:377–388.

Rahel, F. J., J. D. Lyons, and P. A. Cochran. 1984. Stochastic or deterministic regulation of assemblage structure? It may depend on how the assemblage is defined. *Amer Natur* 124:582–589.

Ramanathan, V. 1988. The greenhouse theory of climate change: A test by an inadvertent global experiment. *Science* 240:293–299.

Randall, D. J. 1968. Functional morphology of the heart in fishes. *Amer Zool* 8:179–189.

Randall, D. J. 1970. The circulatory system. *Physiol Fish* 4:133–172.

Randall, J. E. 1958. Review of ciguatera tropical fish poisoning with a tentative explanation of its cause. *Bull Mar Sci* 8:235–267.

Randall, J. E. 1974. The effect of fishes on coral reefs. *Proc 2nd Internat Coral Reef Symp* 1:159–166.

Randall, J. E., and A. R. Emery. 1971. On the resemblance of the young of the fishes *Platax pinnatus* and *Plectorhynchus chaeto-*dontoides to flatworms and nudibranchs. *Zoologica* 1971:115–119.

Randall, J. E., and G. S. Helfman. 1972. *Diproctacanthus xanthurus*, a cleaner wrasse from the Palau Islands, with notes on other cleaning fishes. *Trop Fish Hobby* 197(11):87–95.

Randall, J. E., and R. H. Kuiter. 1989. The juvenile Indo-Pacific grouper *Anyperodon leucogrammicus*, a mimic of the wrasse *Halichoeres purpurescens* and allied species, with a review of the recent literature on mimicry in fishes. *Rev Fr Aquariol* 16:51–56.

Randall, J. E., and H. A. Randall. 1960. Examples of mimicry and protective resemblance in tropical marine fishes. *Bull Mar Sci Gulf Carib* 10:444–480.

Rasa, O. A. E. 1969. Territoriality and the establishment of dominance by means of visual cues in *Pomacentrus jenkinsi* (Pisces: Pomacentridae). *Z Tierpsychol* 26:825–845.

Raup, D. M. 1988. Diversity crises in the geological past. In *Biodiversity*, ed. E. O. Wilson and F. M. Peter, 51–57. Washington, DC: National Academy Press.

Raymond, H. L. 1988. Effects of hydroelectric development and fisheries enhancement on spring and summer chinook salmon and steelhead in the Columbia River basin. *N Amer J Fish Manage* 8:1–24.

Raymond, J. A. 1992. Glycerol is a colligative antifreeze in some northern fishes. *J Exp Zool* 262:347–352.

Raymond, J. A., P. Wilson, and A. L. DeVries. 1989. Inhibition of growth of nonbasal planes in ice by fish antifreezes. *Proc Natl Acad Sci USA* 86:881–885.

Reebs, S. 1992. Sleep, inactivity and circadian rhythms in fish. In *Rhythms in fishes*, ed. M. A. Ali, 127–135. New York: Plenum Press.

Reese, E. S. 1975. A comparative field study of the social behaviour and related ecology of reef fishes of the family Chaetodontidae. *Z Tierpsychol* 37:37–61.

Reese, E. S. 1981. Predation on corals by fishes of the family Chaetodontidae: Implications for conservation and management of coral reef ecosystems. *Bull Mar Sci* 31:594–604.

Reese, E. S. 1989. Orientation behavior of butterflyfishes (family Chaetodontidae) on coral reefs: spatial learning of route specific landmarks and cognitive maps. *Env Biol Fish* 24:79–80.

Reese, E. S., and F. J. Lighter. 1978. *Contrasts in behavior*. New York: Wiley-Interscience.

Regier, H. A., and J. D. Meisner. 1990. Anticipated effects of climate change on freshwater fishes and their habitat. *Fisheries* 15(6):10–15.

Reimchen, T. E. 1991. Evolutionary attributes of headfirst prey manipulation and swallowing in piscivores. *Can J Zool* 69:2912–2916.

Reinboth, R. 1975. *Intersexuality in the animal kingdom*. Berlin: Springer-Verlag.

Rensberger, B. 1992. The phantom fish-killer. *Washington Post National Weekly Edition*, 24–30 August, p. 38.

Reznick, D. N., and J. A. Endler. 1982. The impact of predation on life history evolution in Trinidadian guppies (*Poecilia reticulata*). *Evolution* 36:160–177.

Ribbink, A. J. 1987. African lakes and their fishes: Conservation scenarios and suggestions. *Env Biol Fish* 19:3–26.

Ribbink, A. J. 1991. Distribution and ecology of the cichlid fishes of the African Great Lakes. In *Cichlid fishes: Behaviour, ecology and evolution*, ed. M. H. A. Keenleyside, 36–59. London: Chapman and Hall.

Rice, C. D., and D. Schlenk. 1995. Immune function and cytochrome P4501A activity after acute exposure to 3,3′,4,4′,5-pentachlorobiphenyl (PCB 126) in channel catfish. *J Aquat Anim Health* 7:195–204.

Rich, P. V., and G. F. van Tets. 1985. Kadimakara, *extinct vertebrates of Australia*. Lilydale, Australia: Pioneer Design Studios.

Richards, W. J. 1976. Some comments on Balon's terminology of fish development intervals. *J Fish Res Board Can* 33:1253–1254.

Richards, W. J., and K. C. Lindeman. 1987. Recruitment dynamics of reef fishes: Planktonic processes, settlement and demersal ecologies, and fisheries analysis. *Bull Mar Sci* 41:392–410.

Richey, J. E., M. A. Perkins, and C. R. Goldman. 1975. Effects of kokanee salmon (*Oncorhynchus nerka*) decomposition on the ecology of a subalpine stream. *J Fish Res Board Can* 32:817–820.

Ricker, W. E. 1962. Regulation of the abundance of pink salmon stocks. In: *Symposium on pink salmon.* N. J. Wilimovsky, ed. 155–201. Vancouver, BC: Univ of British Columbia.

Ricker, W. E. 1975. Computation and interpretation of biological statistics of fish populations. *Bull Fish Res Board Can* 191:1–382.

Ricker, W. E. 1979. Growth rates and models. In *Fish physiology*, Vol. 8, ed. W. S. Hoar, D. J. Randall, and J. R. Brett, 677–743. London: Academic Press.

Riget, F. F., K. H. Nygaard, and B. Christensen. 1986. Population structure, ecological segregation and reproduction in a population of Arctic char (*Salvelinus alpinus*) from Lake Tasersvaq, Greenland. *Can J Fish Aquat Sci* 43:985–992.

Rimmer, D. W., and W. J. Wiebe. 1987. Fermentative microbial digestion in herbivorous fishes. *J Fish Biol* 31:229–236.

Rinne, J. N., and W. L. Minckley. 1991. *Native fishes of arid lands: A dwindling resource of the desert Southwest.* Fort Collins, CO: USDA Forest Service Gen Tech Rept RM-206.

Rivas, L. R. 1954. The pineal apparatus of tunas and related scombrid fishes as a possible light receptor controlling phototactic movements. *Bull Mar Sci Gulf Carib* 3:168–180.

Rivas, L. R. 1978. Preliminary models of annual life history cycles of the North Atlantic bluefin tuna. In *The physiological ecology of tunas*, ed. G. D. Sharp and A. E. Dizon, 369–393. New York: Academic Press.

Robbins, J. 1991. The real Jurassic Park. *Discover* 12(3):52–59.

Roberts, C. D. 1993. Comparative morphology of spined scales and their phylogenetic significance in the Teleostei. *Bull Mar Sci* 52:60–113.

Roberts, J. L. 1964. Metabolic responses of fresh-water sunfish to seasonal photoperiods and temperatures. *Helgolander wiss Meeresunters* 9:459–473.

Roberts, J. L. 1975a. Active branchial and ram gill ventilation in fishes. *Biol Bull* 148:85–105.

Roberts, J. L. 1975b. Respiratory adaptations in aquatic animals. In *Physiological adaptations to the environment*, ed. F. J. Vernberg, 395–414. New York: Intext Educational Publ.

Roberts, J. L. 1978. Ram gill ventilation in fish. In *The physiological ecology of tunas*, ed. G. D. Sharp and A. E. Dizon, 83–88. New York: Academic Press.

Roberts, L. 1991. Warm waters, bleached corals. *Science* 249:213.

Roberts, R. J. 1989. The pathophysiology and systematic pathology of teleosts. In *Fish pathology*, 2d ed., ed. R. J. Roberts, 56–134. Philadelphia: Bailliere Tindall.

Roberts, T. R. 1967. Tooth formation and replacement in characoid fishes. *Stanford Ichthyol Bull* 8:231–247.

Roberts, T. R. 1975. Geographic distribution of African freshwater fishes. *Zool J Linn Soc* 57:249–319.

Roberts, T. R. 1986. *Danionella translucida*, a new genus and species of cyprinid fish from Burma, one of the smallest living vertebrates. *Env Biol Fish* 16:231–241.

Roberts, T. S. 1975. Geographical distribution of African freshwater fishes. *Zool J Linnean Soc* 57:249–319.

Robertson, D. R. 1972. Social control of sex-reversal in a coral-reef fish. *Science* 117:1007–1009.

Robertson, D. R. 1982. Fish feces as fish food on a Pacific coral reef. *Mar Ecol Prog Ser* 7:253–265.

Robertson, D. R. 1987. Responses of two coral reef toadfishes (Batrachoididae) to the demise of their primary prey, the sea urchin *Diadema antillarum. Copeia* 1987:637–642.

Robertson, D. R. 1991. The role of adult biology in the timing of spawning of tropical reef fishes. In *The ecology of fishes on coral reefs*, ed. P. F. Sale, 356–386. San Diego, CA: Academic Press.

Robertson, D. R., C. W. Petersen, and J. D. Brawn. 1990. Lunar reproductive cycles of benthic-brooding fishes: Reflections of larval biology or adult biology? *Ecol Monogr* 60:311–329.

Robertson, D. R., H. P. A. Sweatman, E. A. Fletcher, and M. G. Cleland. 1976. Schooling as a mechanism for circumventing the territoriality of competitors. *Ecology* 57:1208–1220.

Robins, C. R., et al. 1991. *Common and scientific names of fishes from the United States and Canada.* 5th ed. Bethesda, MD: American Fisheries Society, Spec Publ 20.

Rogers, S. G., and M. J. Van Den Avyle. 1982. *Species profiles: Life histories and environmental requirements of coastal fishes and invertebrates (South Atlantic)—summer flounder.* US Fish Wildl Serv FWS/OBS-82/11.15. US Army Corps of Engineers TR EL-82-4:1–14.

Rombough, P. J., and D. Ure. 1991. Partitioning of oxygen uptake between cutaneous and branchial surfaces in larval and young juvenile chinook salmon *Oncorhynchus tshawytscha. Physiol Zool* 64:717–727.

Rome, L. C. 1990. Influences of temperature on muscle recruitment and muscle function in vivo. *Am J Physiol* 259:R210–R222.

Rome, L. C., D. A. Syme, S. Hollingworth, S. L. Lindstedt, and S. M. Baylor. 1996. The whistle and the rattle: the design of sound producing muscles. *Proc Natl Acad Sci USA* 93:8095–8100.

Romer, A. S. 1962. *The vertebrate body.* 3d ed. Philadelphia: W. B. Saunders.

Rosen, B. R. 1988. Progress, problems and patterns in the biogeography of reef corals and other tropical marine organisms. *Helgolander wiss Meeresunters* 42:269–301.

Rosen, D. E. 1978. Vicariant patterns and historical explanation in biogeography. *Syst Zool* 27:159–188.

Rosen, D. E., P. L. Forey, B. G. Gardiner, and C. Patterson. 1981. Lungfishes, tetrapods, paleontology, and plesiomorphy. *Bull Amer Mus Natur Hist* 167:163–275.

Rosen, R. A., and D. C. Hales. 1981. Feeding of the paddlefish, *Polyodon spathula. Copeia* 1981:441–455.

Rosenqvist, G. 1990. Male mate choice and female-female competition for mates in the pipefish *Nerophis ophidion. Anim Behav* 39:1110–1115.

Ross, M. R. 1983. The frequency of nest construction and satellite male behavior in the fallfish minnow. *Env Biol Fish* 9:65–67.

Ross, S. T. 1986. Resource partitioning in fish assemblages: A review of field studies. *Copeia* 1986:352–388.

Ross, S. T. 1991. Mechanisms structuring stream fish assemblages: Are there lessons from introduced species? *Env Biol Fish* 30:359–368.

Rothans, T. C., and A. C. Miller. 1991. A link between biologically imported particulate organic nutrients and the detritus food web in reef communities. *Mar Biol* 110:145–150.

Rothschild, B. J. 1986. *Dynamics of marine fish populations.* Cambridge: Harvard University Press.

Rubin, D. A. 1985. Effect of pH on sex ratio in cichlids and a poeciliid (Teleostei). *Copeia* 1985:233–235.

Russell, F. S. 1976. *The eggs and planktonic stages of British marine fishes.* London: Academic Press.

Russell, T. R. 1986. Biology and life history of the paddlefish—a review. In *The paddlefish: Status, management and propagation*, ed. J. G. Dillard, L. K. Graham, and T. R. Russell, 2–19. Bethesda, MD: North Central Division, American Fisheries Society, Spec Publ 7.

Sadovy, Y., and D. Y. Shapiro. 1987. Criteria for the diagnosis of hermaphroditism in fishes. *Copeia* 1987:136–156.

Safina, C., and J. Burger. 1985. Common tern foraging: Seasonal trends in prey fish densities and competition with bluefish. *Ecology* 66:1457–1463.

Sale, P. F. 1969. A suggested mechanism for habitat selection by juvenile manini *Acanthurus triostegus sandvicensis* Streets. *Behaviour* 35:27–44.

Sale, P. F. 1971. Extremely limited home range in a coral reef fish, *Dascyllus aruanus* (Pisces: Pomacentridae). *Copeia* 1971:324–327.

Sale, P. F. 1978. Coexistence of coral reef fishes—a lottery for living space. *Env Biol Fish* 3:85–102.

Sale, P. F. 1991a. Reef fish communities: Open nonequilibrial systems. In *The ecology of fishes on coral reefs*, ed. P. F. Sale, 564–598. San Diego, CA: Academic Press.

Sale, P. F., ed. 1991b. *The ecology of fishes on coral reefs*. San Diego, CA: Academic Press.

Sanderson, S. L., J. J. Cech Jr., and M. R. Patterson. 1991. Fluid dynamics in suspension-feeding blackfish. *Science* 251:1346–1348.

Sandlund, O. T., et al. 1988. Density, length distribution, and diet of age-0 arctic charr *Salvelinus alpinus* in the surf zone of Thingvallavatn, Iceland. *Env Biol Fish* 23:183–195.

Sandy, J. M., and J. H. S. Blaxter. 1980. A study of retinal development in larval herring and sole. *J Mar Biol Ass UK* 60:59–72.

Sano, M. N., M. Shimizu, and Y. Nose. 1984. Changes in structure of coral reef fish communities by destruction of hermatypic corals: Observational and experimental views. *Pacif Sci* 38:51–79.

Sargent, R. C., and M. R. Gross. 1994. Williams' principle: An explanation of parental care in teleost fishes. In *The behaviour of teleost fishes*, 2d ed., ed. T. J. Pitcher, 333–361. London: Chapman and Hall.

Satchell, G. H. 1991. *Physiology and form of fish circulation*. Cambridge: Cambridge University Press.

Saunders, M. W., and G. A. McFarlane. 1993. Age and length at maturity of the female spiny dogfish, *Squalus acanthias*, in the Strait of Georgia, British Columbia, Canada. *Env Biol Fish* 38:49–57.

Sayer, M. D., J. W. Treasurer, and M. J. Costello. 1996. *Wrasse biology and use in aquaculture*. London: Fishing News Books.

Sazima, I. 1977. Possible case of aggressive mimicry in a neotropical scale-eating fish. *Nature* 270:510–512.

Sazima, I. 1983. Scale-eating in characoids and other fishes. *Env Biol Fish* 9:87–101.

Sazima, I., and S. deA. Guimaraes. 1987. Scavenging on human corpses as a source for stories about man-eating piranhas. *Env Biol Fish* 20:75–77.

Sazima, I., and F. A. Machado. 1990. Underwater observations of piranhas in western Brazil. *Env Biol Fish* 28:17–31.

Schaeffer, B. 1967. Comments on elasmobranch evolution. In *Sharks, skates and rays*, P. W. Gilbert, R. F. Mathewson, and D. P. Rall (eds.), 3–36. Baltimore, MD: Johns Hopkins Press.

Schaeffer, B., and M. Williams. 1977. Relationships of fossil and living elasmobranchs. *Amer Zool* 17:293–302.

Scheich, H., and T. H. Bullock. 1974. The detection of electric fields from electric organs. In *Electroreceptors and other specialized receptors in lower vertebrates*, ed. A. Fessard, 201–256. New York: Springer-Verlag.

Schelske, C. L., and S. R. Carpenter. 1992. Lakes: Lake Michigan. In *Restoration of aquatic ecosystems: Science, technology, and public policy*, National Research Council (US), 380–392. Washington, DC: National Academy Press.

Schindler, D. E., J. F. Kitchell, X. He, S. R. Carpenter, J. R. Hodgson, and K. L. Cottingham. 1993. Food web structure and phosphorus cycling in lakes. *Trans Amer Fish Soc* 122:756–772.

Schlosser, I. J. 1982. Fish community structure and function along two habitat gradients in a headwater stream. *Ecol Monogr* 52:395–414.

Schlosser, I. J. 1985. Flow regime, juvenile abundance, and the assemblage structure of stream fishes. *Ecology* 66:1484–1490.

Schlosser, I. J. 1987. The role of predation in age- and size-related habitat use by stream fishes. *Ecology* 68:651–659.

Schlosser, I. J. 1991. Stream fish ecology: A landscape perspective. *BioScience* 41:704–712.

Schlupp, I., C. Marler, and M. J. Ryan. 1994. Benefit to sailfin mollies of mating with heterospecific females. *Science* 263:373–374.

Schmale, M. C. 1981. Sexual selection and reproductive success in males of the bicolor damselfish *Eupomacentrus partitus* (Pisces: Pomacentridae). *Anim Behav* 29:1172–1184.

Schmidt-Nielsen, K. 1983. *Scaling: Why is animal size so important*. Cambridge: Cambridge University Press.

Scholander, P. F., L. van Dam, J. W. Kanwisher, H. T. Hammel, and M. S. Gordon. 1957. Supercooling and osmoregulation in Arctic fish. *J Cell Comp Physiol* 49:5–24.

Schonewald-Cox, C. M., et al., eds. 1983. *Genetics and conservation*. Menlo Park, CA: Benjamin/Cummings.

Schreck, C. B., and P. B. Moyle, eds. 1990. *Methods for fish biology*. Bethesda, MD: American Fisheries Society.

Schultz, E. T. 1993. Sexual size dimorphism at birth in *Micrometrus minimus* (Embiotocidae): A prenatal cost of reproduction. *Copeia* 1993:456–463.

Schultz, R. J. 1971. Special adaptive problems associated with unisexual fish. *Amer Zool* 11:351–360.

Schultz, R. J. 1977. Evolution and ecology of unisexual fishes. In *Evolutionary biology*, Vol. 10, ed. M. K. Hecht, W. C. Steere, and B. Wallace, 277–331. New York: Plenum Press.

Schultze, H. P., and K. S. W. Campbell. 1987. Characterization of the *Dipnoi*, a monophyletic group. In *The biology and evolution of lungfishes*, ed. W. E. Bemis, W. W. Burggren, and N. E. Kemp, 25–38. New York: Alan R. Liss.

Schwartz, F. T. 1981. *World literature to fish hybrids with an analysis by family, species, and hybrid: Suppl 1*. Washington, DC: NOAA Tech Rept NMFS SSRF-750.

Schwassmann, H. O. 1971. Biological rhythms. In *Fish physiology*, Vol. 6, ed. W. S. Hoar and D. J. Randall, 371–428. New York: Academic Press.

Schwassmann, H. O. 1980. Biological rhythms: Their adaptive significance. In *Environmental physiology of fishes*, ed. M. A. Ali, 613–630. New York: Plenum Press.

Scoles, D. R., and J. E. Graves. 1993. Genetic analysis of the population structure of yellowfin tuna, *Thunnus albacares*, from the Pacific Ocean. *Fish Bull US* 91:690–698.

Scoppettone, G. G., and G. Vinyard. Life history and management of four endangered lacustrine suckers. In *Battle against extinction. Native fish management in the American West*, ed. W. L. Minckley and J. E. Deacon, 359–377. Tucson: University of Arizona Press.

Scott, H. W. 1973. New Conodontochordata from the Bear Gulch limestone (Namurian, Montana). *Publ Mus Mich St Univ, Paleontol Series* 1:85–99.

Scott, W. B., and E. J. Crossman. 1973. *Freshwater fishes of Canada*. Fisheries Research Board of Canada Bull 184.

Sedell, J. R., F. H. Everest, and F. J. Swanson. 1982. Fish habitat and streamside management: Past and present. *Proc Soc Amer For Ann Meet*.

Seeley, K. R., and B. A. Weeks-Perkins. 1991. Altered phagocytic activity of macrophages in oyster toadfish from a highly polluted subestuary. *J Aquat Anim Health* 3:224–227.

Seidensticker, E. P. 1987. Food selection of alligator gar and longnose gar in a Texas reservoir. *Proc Ann Conf Southeast Assoc Fish Wildl Agencies* 41:100–104.

Semon, R. 1898. Die Entwickelung der paarigen Flossen des *Ceratodus forsteri*. *Jen Denkschr 4, Semon Zool Forschungsreisen* 1:61–111.

Shafland, P. L., and W. M. Lewis. 1984. Terminology associated with introduced organisms. *Fisheries* 9(4):17–18.

Shapiro, D. Y. 1979. Social behavior, group structure, and the control of sex reversal in hermaphroditic fish. *Adv Stud Behav* 10:43–102.

Shapiro, D. Y. 1987. Differentiation and evolution of sex change in fishes. *BioScience* 37:490–497.

Shapiro, D. Y. 1991. Intraspecific variability in social systems of coral reef fishes. In *The ecology of fishes on coral reefs*, ed. P. F. Sale, 331–355. San Diego, CA: Academic Press.

Shapiro, D. Y., A. Marconato, and T. Yoshikawa. 1994. Sperm economy in a coral reef fish *Thalassoma bifasciatum*. *Ecology* 75:1334–1344.

Shapiro, D. Y., Y. Sadovy, and M. A. McGehee. 1993. Size, composition, and spatial structure of the annual spawning aggregation of the red hind, *Epinephelus guttatus* (Pisces: Serranidae). *Copeia* 1993:399–406.

Sharp, G. D. 1978. Behavioral and physiological properties of tunas and their effects on vulnerability to fishing gear. In *The physiological ecology of tunas*, ed. G. D. Sharp and A. E. Dizon, 397–449. New York: Academic Press.

Sharp, G. D., and A. E. Dizon, eds. 1978. *The physiological ecology of tunas*. New York: Academic Press.

Sharp, G. D., and S. W. Pirages. 1978. The distribution of red and white swimming muscles, their biochemistry, and the biochemical phylogeny of selected scombrid fishes. In *The physiological ecology of tunas*, ed. G. D. Sharp and A. E. Dizon, 41–78. New York: Academic Press.

Sharp, G. D., and S. Pirages. 1979. The distribution of red and white swimming muscles, their biochemistry, and the biochemical phylogeny of selected scombrid fishes. In *The physiological ecology of tunas*, ed. G. D. Sharp and A. E. Dizon, 41–78. New York: Academic Press.

Shaw, E. 1970. Schooling fishes: Critique and review. In *Development and evolution of behaviour*, ed. L. Aronson, E. Tobach, D. S. Lehrmann, and J. S. Rosenblatt, 452–480. San Francisco: W. H. Freeman.

Shaw, E. 1978. Schooling fishes. *Amer Sci* 66:166–175.

Shellis, R. P., and B. K. B. Berkowitz. 1976. Observations on the dental anatomy of piranhas (Characidae) with special reference to tooth structure. *J Zool* (London) 180:69–84.

Shelton, R. G. J. 1978. On the feeding of the hagfish *Myxine glutinosa* in the North Sea. *J Mar Biol Assoc UK* 58:81–86.

Shenker, J. M., and J. M. Dean. 1979. Utilization of an intertidal salt marsh creek by larval and juvenile fishes: Abundance, diversity and temporal variation. *Estuaries* 2:154–163.

Shenker, J. M., E. D. Maddox, E. Wishinski, and S. Pearl. 1993. Onshore transport of settlement-stage Nassau grouper (*Epinephelus striatus*) and other fishes in Exuma Sound, Bahamas. *Mar Ecol Prog Ser* 98:31–43.

Shepherd, A. R. D., R. M. Warwick, K. R. Clarke, and B. E. Brown. 1992. An analysis of fish community responses to coral mining in the Maldives. *Env Biol Fish* 33:367–380.

Shireman, J. V., D. E. Colle, and D. E. J. Canfield. 1986. Efficacy and cost of aquatic weed control in small ponds. *Water Res Bull* 22:43–48.

Shulman, M. J. 1985a. Recruitment of coral reef fishes: Effects of distribution of predators and shelter. *Ecology* 66:1056–1066.

Shulman, M. J. 1985b. Variability in recruitment of coral reef fishes. *J Exp Mar Biol Ecol* 89:205–219.

Shulman, M. J., and J. C. Ogden. 1987. What controls tropical reef fish populations: Recruitment or benthic mortality? An example in the Caribbean reef fish *Haemulon flavolineatum*. *Mar Ecol Prog Ser* 39:233–242.

Shulman, M. J., J. C. Ogden, J. P. Ebersole, W. N. McFarland, S. L. Miller, and N. G. Wolf. 1983. Priority effects in the recruitment of juvenile coral reef fishes. *Ecology* 64:1508–1513.

Shuttleworth, T. J., ed. 1988. *Physiology of elasmobranch fishes*. Berlin: Springer-Verlag.

Sidell, B. D. 1977. Turnover of cytochrome *c* in skeletal muscle of green sunfish (*Lepomis cyanellus* R.) during thermal acclimation. *J Exp Zool* 199:233–250.

Sidell, B. D., and T. S. Moerland. 1989. Effects of temperature on muscular function and locomotory performance in teleost fish. In *Advances in comparative and environmental physiology*, Vol. 5, 116–158. Berlin: Springer-Verlag.

Sih, A., P. Crowley, M. McPeek, J. Petranka, and K. Strohmeier. 1985. Predation, competition, and prey communities: A review of field experiments. *Ann Rev Ecol System* 16:269–311.

Simons, J. R. 1970. The direction of the thrust produced by the heterocercal tails of two dissimilar elasmobranchs: The Port Jackson shark, *Heterodontus portusjacksoni* (Meyer), and the piked dogfish, *Squalus megalops* (Macleay). *J Exp Biol* 52:95–107.

Simpson, B. R. C. 1979. The phenology of annual killifishes. *Symp Zool Soc Lond* 44:243–261.

Sinderman, C. J. 1990. *Principal diseases of marine fish and shellfish*. 2d ed. Vol. 1, *Diseases of marine fish*. San Diego, CA: Academic Press.

Skelton, P. H. 1987. *South African red data book—fishes*. South African National Scientific Programmes Rep 137.

Skulason, S., and, Smith, T. B. 1995. Resource polymorphisms in vertebrates. *Trends Ecol Evol* 10:366–370.

Skulason, S., S. S. Snorrason, D. Ota, and D. L. G. Noakes. 1993. Genetically based differences in foraging behaviour among sympatric morphs of arctic charr (Pisces: Salmonidae). *Anim Behav* 45:1179–1192.

Smatresk, N. J., and J. N. Cameron. 1982. Respiration and acid-base physiology of the spotted gar, a bimodal breather. I. Normal values, and the response to severe hypoxia. *J Exp Biol* 96:263–280.

Smith, B. R. 1971. Sea lampreys in the Great Lakes of North America. In *The biology of lampreys*, Vol. 1, ed. M. W. Hardisty and I. C. Potter, 207–247. New York: Academic Press.

Smith, B. R., et al. 1980. Proceedings of the 1979 Sea Lamprey International Symposium (SLIS). *Can J Fish Aquat Sci* 37:1585–2214.

Smith, C. L. 1975. The evolution of hermaphroditism in fishes. In *Intersexuality in the animal kingdom*, ed. R. Reinboth, 295–310. Berlin: Springer-Verlag.

Smith, C. L. 1978. Coral reef fish communities: A compromise view. *Env Biol Fish* 3:109–128.

Smith, C. L. 1988. Minnows first, then trout. *Fisheries* 13(4):4–8.

Smith, C. L., C. S. Rand, B. Schaeffer, and J. W. Atz. 1975. *Latimeria*, the living coelacanth, is ovoviviparous. *Science* 190:1105–1106.

Smith, C. L., and P. Reay. 1991. Cannibalism in teleost fish. *Rev Fish Biol Fish* 1:41–64.

Smith, G. R. 1981a. Late Cenozoic freshwater fishes of North America. *Ann Rev Ecol Syst* 12:163–193.

Smith, G. R. 1981b. Effects of habitat size on species richness and adult body sizes of desert fishes. In *Fishes in North American deserts*, ed. R. J. Naiman and D. L. Soltz, 125–171. New York: John Wiley & Sons.

Smith, G. R., and R. F. Stearley. 1989. The classification and scientific names of rainbow and cutthroat trouts. *Fisheries* 14(1):4–10.

Smith, G. R., and T. N. Todd. 1984. Evolution of species flocks of fishes in north temperate lakes. In *Evolution of fish species flocks*, ed. A. E. Echelle and I. Kornfield, 45–68. Orono: University of Maine Press.

Smith, J. B. 1990. From global to regional climate change: Relative knowns and unknowns about global warming. *Fisheries* 15(6):2–6.

Smith, J. B., and D. A. Tirpak, eds. 1989. *The potential effects of global climate change on the United States*. Appendix E, *Aquatic resources*. Washington, DC: US Environmental Protection Agency EPA-230-05-89-055.

Smith, J. L. B. 1956. *The search beneath the sea*. New York: Henry Holt.

Smith, M. P., D. E. G. Briggs, and R. J. Aldridge. 1987. A conodont animal from the lower Silurian of Wisconsin, USA, and the apparatus architecture of panderodontid conodonts. In *Palaeobiology of conodonts*, ed. R. J. Aldridge, 91–104. Chichester: Ellis Horwood.

Smith, P. J., R. I. C. C. Francis, and M. McVeagh. 1991. Loss of genetic diversity due to fishing pressure. *Fish Res* 10:309–316.

Smith, R. J. F. 1986. The evolution of chemical alarm signals in fishes. In *Chemical signals in vertebrates*. Vol. IV, *Ecology, evolution and comparative biology*, ed. D. Duvall, D. Muller-Schwarze, and R. M. Silverstein, 99–115. New York: Plenum Press.

Smith, R. J. F. 1992. Alarm signals in fishes. *Rev Fish Biol Fisheries* 2:33–63.

Smith, R. S., and D. L. Kramer. 1986. The effect of apparent predation risk on the respiratory behavior of the Florida gar (*Lepisosteus platyrhincus*). *Can J Zool* 64:2133–2136.

Smith, T. I. J. 1985. The fishery, biology, and management of Atlantic sturgeon, *Acipenser oxyrhynchus*, in North America. In *North American sturgeons: Biology and aquaculture potential. Developments in environmental biology of fishes 6*, ed. F. P. Binkowski and S. I. Doroshov, 61–72. Dordrecht: Dr. W. Junk.

Sneath, P. H. A., and R. R. Sokal. 1973. *Numerical taxonomy*. San Francisco: W. H. Freeman.

Sogard, S. M., and B. L. Olla. 1994. The potential for intracohort cannibalism in age-0 walleye pollock, *Theragra chalcogramma*, as determined under laboratory conditions. *Env Biol Fish* 39:183–190.

Sola, L., S. Cataudella, and E. Capanna. 1981. New developments in vertebrate cytotaxonomy. III. Karyology of bony fishes: A review. *Genetica* 54:285–328.

Soltz, D. L., and R. J. Naiman. 1981. Fishes in deserts: Symposium rationale. In *Fishes in North American deserts*, ed. R. J. Naiman and D. L. Soltz, 1–9. New York: John Wiley & Sons.

Somero, G. N., E. Dahlhoff, and A. Gibbs. 1991. Biochemical adaptations of deep-sea animals: Insights into biogeography and ecological energetics. In *Marine biology: Its accomplishments and future prospect*, ed. J. Mauchline and T. Nemoto, 39–57. Amsterdam: Elsevier.

Sorbini, L. 1975. Evoluzione e distribuzione del genere fossile *Eolates* e suoi rapporti con il genere attuale *Lates* (Pisces—Centropomidae). In Studi e Ricerche sui Giacimenti Terziari di Bolca. II. Museo Civico di Storia Naturale di Verona, Misc Paleontol, 1–54.

Soto, C. G., J. F. Leatherland, and D. L. G. Noakes. 1992. Gonadal histology in the self-fertilizing hermaphroditic fish *Rivulus marmoratus* (Pisces, Cyprinodontidae). *Can J Zool* 70:2338–2347.

SPA (Science and Policy Associates, Inc.). 1990. *Report on the workshop on effects of global climate change on freshwater ecosystems*. Washington, DC: SPA.

Sparholt, H. 1985. The population, survival, growth, reproduction and food of Arctic charr, *Salvelinus alpinus* (L.) in four unexploited lakes in Greenland. *J Fish Biol* 26:313–330.

Spieler, R. E. 1992. Feeding entrained circadian rhythms in fishes. In *Rhythms in fishes*, ed. M. A. Ali, 137–147. New York: Plenum Press.

Springer, V. G. 1982. Pacific Plate biogeography, with special reference to shorefishes. *Smithsonian Contrib Zool* 367:182.

Springer, V. G., and J. P. Gold. 1989. *Sharks in question*. Washington, DC: Smithsonian Institution Press.

Springer, V. G., and W. F. Smith-Vaniz. 1972. Mimetic relationships involving fishes of the family Blenniidae. *Smithson Contrib Zool* 112:1–36.

Stearley, R. F., and G. R. Smith. 1993. Phylogeny of the Pacific trouts and salmons (*Oncorhynchus*) and genera of the family Salmonidae. *Trans Amer Fish Soc* 122:1–33.

Stearns, S. C. 1992. *The evolution of life histories*. Oxford: Oxford University Press.

Steen, J. B., and T. Berg. 1966. The gills of two species of haemoglobin-free fishes compared to those of other teleosts—with a note of severe anaemia in an eel. *Comp Biochem Physiol* 18:517–526.

Stein, R. A. 1977. Selective predation, optimal foraging, and the predator-prey interaction between fish and crayfish. *Ecology* 58:1237–1253.

Stensio, E. A. 1963. The brain and cranial nerves in fossil, lower craniate vertebrates. Skr Norske, Vidlensk-Akad, Oslo, Marnuturv K.I., Ny-Serie 13:1–120.

Stepien, C. A., and R. H. Rosenblatt. 1991. Patterns of gene flow and genetic divergence in the northeastern Pacific Clinidae (Teleostei: Blenniodidei) based on allozyme and morphological data. *Copeia* 1991:873–896.

Stevens, E. D., H. M. Lam, and J. Kendall. 1974. Vascular anatomy of the countercurrent heat exchanger of skipjack tuna. *J Exp Biol* 61:145–153.

Stevens, E. D., and W. H. Neill. 1978. Body temperature relations of tuna, especially skipjack. In *Fish physiology*, Vol. 7, ed. W. S. Hoar and D. J. Randall, 315–424. New York: Academic Press.

Stevens, G. C. 1989. The latitudinal gradient in geographical range: How so many species coexist in the tropics. *Amer Natur* 133:240–256.

Stevens, J. D., ed. 1987. *Sharks*. New York: Facts on File Publishers.

Stevenson, D. K., and S. E. Campana, eds. 1992. Otolith microstructure examination and analysis. *Can Spec Pub Fish Aquat Sci* 117:1–126.

Stewart, D. J., J. F. Kitchell, and L. B. Crowder. 1981. Forage fishes and their salmonid predators in Lake Michigan. *Trans Amer Fish Soc* 110:751–763.

Stewart, P. A. 1978. Independent and dependent variables of acid-base control. *Respir Physiol* 33:9–26.

Stewart, P. A. 1983. Modern quantitative acid-base chemistry. *Can J Physiol Pharmacol* 61:1444–1461.

Stiassny, M. L. J., U. K. Schliewen, and W. J. Dominey. 1992. A new species flock of cichlid fishes from Lake Bermin, Cameroon with a description of eight new species of *Tilapia* (Labroidei: Cichlidae). *Ichthyol Explor Freshwater* 3:311–346.

Stobbs, R. 1988. Coelacanth mythology. *Ichthos: Newsl Soc Friends J L B Smith Inst Ichthyol* 2:18–19.

Stockner, J. G. 1987. Lake fertilization: The enrichment cycle and lake sockeye salmon (*Oncorhynchus nerka*) production. In *Sockeye salmon* (Oncorhynchus nerka) *population biology and future management*, ed. H. D. Smith, L. Margolis, and C. C. Wood. *Can Spec Publ Fish Aquat Sci* 96:198–215.

Stoner, A. W., and R. J. Livingston. 1984. Ontogenetic patterns in diet and feeding morphology in sympatric sparid fishes from seagrass meadows. *Copeia* 1984:174–187.

Stouder, D. J. 1987. Effects of a severe-weather disturbance on foraging patterns within a California surfperch guild. *J Exp Mar Biol Ecol* 114:73–84.

Stouder, D. J. 1990. Dietary fluctuations in stream fishes and the effects of benthic species interactions. Ph.D. dissertation, University of Georgia, Athens, GA.

Strahan, R. 1963. The behavior of *Myxine* and other myxinoids.

In *The biology of* Myxine, ed. A. Brodal and R. Fange, 22–32. Oslo: Scandinavian University Books.

Strange, E. M., P. B. Moyle, and T. C. Foin. 1993. Interactions between stochastic and deterministic processes in stream fish community assembly. *Env Biol Fish* 36:1–15.

Strauss, R. E., and C. E. Bond. 1990. Taxonomic methods: Morphology. In *Methods for fish biology*, ed. C. B. Schreck and P. B. Moyle, 109–140. Bethesda, MD: American Fisheries Society.

Stromberg, J.-O. 1991. Marine ecology of polar seas: A comparison Arctic/Antarctic. In *Marine biology: Its accomplishments and future prospect*, ed. J. Mauchline and T. Nemoto, 247–261. Amsterdam: Elsevier.

Strong, W. R., F. F. Snelson, and S. H. Gruber. 1990. Hammerhead shark predation on stingrays: An observation of prey handling by *Sphyrna mokarran*. *Copeia* 1990:836–840.

Sulak, K. J. 1975. Cleaning behaviour in the centrarchid fishes, *Lepomis macrochirus* and *Micropterus salmoides*. *Anim Behav* 23:331–334.

Suttkus, R. D. 1963. Order Lepisostei. In *Fishes of the western North Atlantic*. Part 3, *Soft-rayed bony fishes*, 61–88. New Haven, CT: Mem Sears Fndn Mar Res, No. 1, Part 1.

Swartzmann, G. L., and T. M. Zaret. 1983. Modeling fish species introduction and prey extermination: The invasion of *Cichla ocellaris* to Gatun Lake, Panama. In *Developments in environmental modeling: Analysis of ecological systems: State of the art in ecological modeling*, ed. W. K. Lauenroth, G. V. Skogerboe, and M. Flug, 361–371. New York: Elsevier Scientific Publ.

Sweatman, H. P. A. 1984. A field study of the predatory behaviour and feeding rate of a piscivorous coral reef fish, the lizardfish *Synodus englemani*. *Copeia* 1984:187–193.

Swofford, D. L. 1993. *PAUP: Phylogenetic analysis using parsimony, version 3.1.1*. Champaign: Illinois Natural History Survey.

Taborsky, M. 1984. Broodcare helpers in *Lamprologus brichardi*: Their costs and benefits. *Anim Behav* 32:1236–1252.

Tachibana, K., M. Sakaitanai, and K. Nakanishi. 1984. Pavoninins: Shark-repelling ichthyotoxins from the defense secretion of the Pacific sole. *Science* 226:703–705.

Tanaka, S. K. 1973. Suction feeding by the nurse shark. *Copeia* 1973:606–608.

Targett, T. E., K. E. Young, J. T. Konecki, and P. A. Grecay. 1987. Research on wintertime feeding in Antarctic fishes. *Antarctic J US* 22:211–213.

Tavolga, W. N., A. N. Popper, and R. R. Fay, eds. 1981. *Hearing and sound communication in fishes*. New York: Springer-Verlag.

Taylor, J. N., W. R. Courtenay Jr., and J. A. McCann. 1984. Known impacts of exotic fishes in the continental United States. In *Distribution, biology, and management of exotic fishes*, ed. W. R. Courtenay Jr. and J. R. Stauffer Jr., 322–373. Baltimore: Johns Hopkins University Press.

Taylor, L. R., L. J. V. Compagno, and P. J. Struhsaker. 1983. Megamouth—a new species, genus, and family of lamnoid shark (*Megachasma pelagios*, Family Megachasmidae) from the Hawaiian Islands. *Proc Calif Acad Sci* 43:87–110.

Taylor, M. H. 1984. Lunar synchronization of fish reproduction. *Trans Amer Fish Soc* 113:484–493.

Taylor, M. H. 1990. Estuarine and intertidal teleosts. In *Reproductive seasonality in teleosts: Environmental influences*, ed. A. D. Munro, A. P. Scott, and T. J. Lam, 109–124. Boca Raton, FL: CRC Press.

Tchernavin, V. V. 1953. *The feeding mechanisms of a deep sea fish* Chauliodus sloani *Schneider*. London: British Museum (Natural History).

Tesch, F.-W. 1977. *The eel*. Translated by J. Greenwood. New York: Chapman and Hall.

Thomas, P. 1988. Reproductive endocrine function in female Atlantic croaker exposed to pollutants. *Mar Environ Res* 24:179–183.

Thomas, P. 1990. Molecular and biochemical responses of fish to stressors and their potential use in environmental monitoring. *Amer Fish Soc Symp* 8:9–28.

Thomson, D. A., and C. E. Lehner. 1976. Resilience of a rocky intertidal fish community in a physically unstable environment. *J Exp Mar Biol Ecol* 22:1–29.

Thomson, K. S. 1967. Mechanisms of intracranial kinetics in fossil rhipidistian fishes (Crossopterygii) and their relatives. *J Linn Soc (Zool)* 46:223–253.

Thomson, K. S. 1969a. Gill and lung function in the evolution of the lungfishes (*Dipnoi*): An hypothesis. *Forma et Functio* 1:250–262.

Thomson, K. S. 1969b. The biology of the lobe-finned fishes. *Biol Rev* 44:91–154.

Thomson, K. S. 1976. On the heterocercal tail in sharks. *Paleobiology* 2:19–38.

Thomson, K. S. 1986. Marginalia—a fishy story. *Amer Sci* 74:169–171.

Thomson, K. S. 1990. The shape of a shark's tail. *Amer Sci* 78:499–501.

Thomson, K. S. 1991. *Living fossil. The story of the coelacanth*. New York: Norton.

Thorpe, J. E., ed. 1978. *Rhythmic activity of fishes*. London: Academic Press.

Thorpe, J. E., M. S. Miles, and D. S. Keay. 1984. Developmental rate, fecundity and egg size in Atlantic salmon, *Salmo salar* L. *Aquaculture* 43:289–305.

Thorson, T. B. 1971. Movement of bull sharks, *Carcharhinus leucas*, between Caribbean Sea and Lake Nicaragua demonstrated by tagging. *Copeia* 1971:336–338.

Thorson, T. B. 1972. The status of the bull shark, *Carcharhinus leucas*, in the Amazon River. *Copeia* 1972:601–605.

Thorson, T. B. 1982. Life history implications of a tagging study of the largetooth sawfish, *Pristis perotteti*, in the Lake Nicaragua-Rio San Juan system. *Env Biol Fish* 7:207–228.

Thorson, T. B. 1991. The unique roles of two liver products in suiting sharks to their environment. In *Discovering sharks*, ed. S. H. Gruber, 41–47. Highlands, NJ: American Littoral Society, Spec Publ 14.

Thorson, T. B., C. M. Cowan, and D. E. Watson. 1967. *Potamotrygon* spp.: Elasmobranchs with low urea content. *Science* 158:375–377.

Thorson, T. B., C. M. Cowan, and D. E. Watson. 1973. Body fluid solutes of juveniles and adults of the euryhaline bull shark *Carcharhinus leucas* from freshwater and saline environments. *Physiol Zool* 46:29–42.

Thorson, T. B., and E. J. Lacy Jr. 1982. Age, growth rate and longevity of *Carcharhinus leucas* estimated from tagging and vertebral rings. *Copeia* 1982:110–116.

Thresher, R. E. 1976. Field experiments on species recognition by the threespot damselfish, *Eupomacentrus planifrons* (Pisces: Pomacentridae). *Anim Behav* 24:562–569.

Thresher, R. E. 1977. Eye ornamentation of Caribbean reef fishes. *Z Tierpsychol* 43:152–158.

Thresher, R. E. 1978. Polymorphism, mimicry, and the evolution of the hamlets. *Bull Mar Sci* 28:345–353.

Thresher, R. E. 1980. *Reef fish: Behavior and ecology on the reef and in the aquarium*. St. Petersburg, FL: Palmetto Publ.

Thresher, R. E. 1984. *Reproduction in reef fishes*. Neptune City, NJ: TFH Publ.

Thuemler, T. F. 1985. The lake sturgeon, *Acipenser fulvescens*, in the Menominee River, Wisconsin-Michigan. In *North American sturgeons: Biology and aquaculture potential. Developments in environmental biology of fishes 6*, ed. F. P. Binkowski and S. I. Doroshov, 73–78. Dordrecht: Dr. W. Junk.

Todd, J. H., J. Atema, and J. E. Bardach. 1967. Chemical communication in social behavior of a fish, the yellow bullhead (*Ictalurus natalis*). *Science* 158:672–673.

Trautman, M. N. B. 1981. *The fishes of Ohio*. Rev. ed. Columbus: Ohio State University Press.

Tricas, T. C. 1982. Bioelectric-mediated predation by swell sharks, *Cephaloscyllium ventriosum*. *Copeia* 1982:948–952.

Tricas, T. C., and J. T. Hiramoto. 1989. Sexual differentiation, gonad development, and spawning seasonality of the Hawaiian butterflyfish, *Chaetodon multicinctus*. *Env Biol Fish* 25:111–124.

Tricas, T. C., and J. E. McCosker. 1984. Predatory behavior of the white shark (*Carcharodon carcharias*), with notes on its biology. *Proc Cal Acad Sci* 43:221–238.

Trippel, E. A., and M. J. Morgan. 1994. Sperm longevity in Atlantic cod (*Gadus morhua*). *Copeia* 1994:1025–1029.

Turingan, R. G., and P. C. Wainwright. 1993. Morphological and functional bases of durophagy in the queen triggerfish, *Balistes vetula* (Pisces, Tetraodontiformes). *J Morph* 215:101–118.

Turner, B. J., ed. 1984. *Evolutionary genetics of fishes*. New York: Plenum Press.

Turner, G. 1993. Teleost mating behaviour. In *The behaviour of teleost fishes*, 2d ed., ed. T. J. Pitcher, 307–331. London: Chapman and Hall.

Tyler, J. C. 1980. *Osteology, phylogeny, and higher classification of the fishes of the Order Plectognathi (Tetraodontiformes)*. Washington, DC: NOAA Tech Rept NMFS Circular 434.

Tyler, J. C., G. D. Johnson, I. Nakamura, and B. B. Collette. 1989. Morphology of *Luvarus imperialis* (Luvaridae), with a phylogenetic analysis of the Acanthuroidei (Pisces). *Smithsonian Contrib Zool* 485:78.

Tyus, H. M. 1985. Homing behavior noted for Colorado squawfish. *Copeia* 1985:213–215.

Ueber, E., and A. MacCall. 1990. The collapse of California's sardine fishery. In *Climate variability, climate change and fisheries*, ed. M. H. Glantz and L. E. Feingold, 17–23. Boulder, CO: Environmental and Societal Impacts Group, Nat Ctr Atmos Res.

Unger, L. M., and R. C. Sargent. 1988. Allopaternal care in the fathead minnow, *Pimephales promelas*: Females prefer males with eggs. *Behav Ecol Sociobiol* 23:27–32.

Upton, H. F. 1992. Biodiversity and conservation of the marine environment. *Fisheries* 17(3):20–25.

Urist, M. R. 1973. Testosterone-induced development of limb gills of the lungfish, *Lepidosiren paradoxa*. *Comp Biochem Physiol* 44A:131–135.

Utter, F., and N. Ryman. 1993. Genetic markers and mixed stock fisheries. *Fisheries* 18(8):11–21.

Valinsky, W., and L. Rigley. 1981. Function of sound production by the skunk loach, *Botia horae* (Pisces: Cobitidae). *Z Tierpsychol* 55:161–172.

Van Den Avyle, M. J., and J. Evans. 1990. Temperature selection of striped bass in a Gulf of Mexico coastal river system. *N Amer J Fish Manage* 10:58–66.

van den Berghe, E. P. 1988. Piracy as an alternative reproductive tactic for males. *Nature* 334:697–698.

van der Elst, R. P. 1979. A proliferation of small sharks in the shore-based natal sport fishery. *Env Biol Fish* 4:349–362.

van der Elst, R. P., and M. Roxburgh. 1981. Use of the bill during feeding in the black marlin (*Makaira indica*). *Copeia* 1981:215.

Vanderploeg, H. A., B. J. Eadie, J. R. Liebig, S. J. Tarapchak, and R. M. Glover. 1987. Contribution of calcite to the particle size spectrum of Lake Michigan seston and its interactions with the plankton. *Can J Fish Aquat Sci* 44:1898–1914.

Van Muiswinkel, W. B. 1992. Fish immunology and fish health. *Netherlands J Zool* 42:494–499.

Vannote, R. L., G. W. Minshall, K. W. Cummins, Z. J. R. Sedell, and C. E. Cushing. 1980. The river continuum concept. *Can J Fish Aquat Sci* 37:130–137.

Van Oosten, J. 1957. The skin and scales. *Physiol Fish* 1:207–244.

Van Vleet, E. S., S. Candileri, J. McNellie, S. B. Reinhardt, M. E. Conkright, and A. Zwissler. 1984. Neutral lipid components of eleven species of Caribbean sharks. *Comp Biochem Physiol* 79B:549–554.

Vari, R. P. 1979. Anatomy, relationships and classification of the families Citharinidae and Distichodontidae (Pisces, Characoidea). *Bull Brit Mus Nat Hist* 36:261–344.

Victor, B. C. 1991. Settlement strategies and biogeography of reef fishes. In *The ecology of fishes on coral reefs*, ed. P. F. Sale, 231–260. San Diego, CA: Academic Press.

Videler, J. J. 1993. *Fish swimming*. London: Chapman and Hall.

Vilches-Troya, J., R. F. Dunn, and D. P. O'Leary. 1984. Relationship of the vestibular hair cells to magnetic particles in the otolith of the guitarfish sacculus. *J Comp Neurol* 226:489–494.

Vladykov, V. D., and J. R. Greeley. 1963. Order Acipenseroidei. In *Fishes of the western North Atlantic; Soft-rayed bony fishes*, ed. Y. H. Olsen, 24–60. New Haven, CT: Mem Sears Fndn Mar Res, No. 1, Part 3.

Vladykov, V. D., and E. Kott. 1979. Satellite species among the Holarctic lampreys (Petromyzonidae). *Can J Zool* 57:860–867.

Vogel, S. 1981. *Life in moving fluids*. Boston: Willard Grant Press.

Vogel, S., and S. A. Wainwright. 1969. *A functional bestiary*. Reading, MA: Addison-Wesley.

von Westernhagen, H. 1988. Sublethal effects of pollutants on fish eggs and larvae. In *Fish physiology*, Vol. 11, *The physiology of developing fish*, Part A. Eggs and larvae, ed. W. S. Hoar and D. J. Randall, 253–346. San Diego, CA: Academic Press.

Voris, J. K., J. J. Voris, and L. B. Liat. 1978. The food and feeding behavior of a marine snake, *Enhydrina schistosa* (Hydrophiidae). *Copeia* 1978:134–146.

Voss, G. 1956. Solving life secrets of the sailfish. *Natl Geog Mag* 109:859–872.

Vrijenhoek, R. C. 1984. The evolution of clonal diversity in *Poeciliopsis*. In *Evolutionary genetics of fishes*, ed. B. J. Turner, 399–429. New York: Plenum Press.

Wainwright, P. C. 1987. Biomechanical limits to ecological performance: Mollusc crushing by the Caribbean hogfish, *Lachnolaimus maximus* (Labridae). *J Zool* (London) 213:283–297.

Wainwright, P. C. 1988. Morphology and ecology: Functional basis of feeding constraints in Caribbean labrid fishes. *Ecology* 69:635–645.

Wainwright, P. C., and B. A. Richard. 1995. Predicting patterns of prey use from morphology of fishes. *Env Biol Fish* 44:97–113.

Wainwright, P. C., R. G. Turingan, and E. L. Brainerd. 1995. Functional morphology of pufferfish inflation: Mechanism of the buccal pump. *Copeia* 1995:614–625.

Wainwright, S. A. 1983. To bend a fish. In *Fish biomechanics*, ed. P. W. Webb and D. Weihs, 68–91. New York: Praeger.

Wainwright, S. A. 1988. *Axis and circumference: The cylindrical shape of plants and animals*. Princeton, NJ: Princeton University Press.

Wainwright, S. A., N. D. Biggs, J. D. Currey, and J. M. Gosline. 1976. *Mechanical design in organisms*. New York: John Wiley & Sons.

Wainwright, S. A., F. Vosburgh, and J. H. Hebrank. 1978. Shark skin: Function in locomotion. *Science* 202:747–749.

Walker, M. M. 1984a. Learned magnetic field discrimination in yellowfin tuna, *Thunnus albacares*. *J Comp Physiol A* 155:673–679.

Walker, M. M. 1984b. Magnetite sensitivity and its possible physical basis in the yellowfin tuna, *Thunnus albacares*. In *Mechanisms of migration in fishes*, ed. J. D. McCleave, G. P. Arnold, J. J. Dodson, and W. H. Neill, 125–141. New York: Plenum Press.

Walker, M. M., J. L. Kirschvink, S.-B. R. Chang, and A. E. Dizon. 1984. A candidate magnetoreceptor organ in the yellowfin tuna, *Thunnus albacares. Science* 224:751–753.

Walker, W. F., and K. F. Liem. 1994. *Functional anatomy of the vertebrates: An evolutionary perspective.* 2d ed. Philadelphia: W. B. Saunders.

Wallace, A. R. 1860. On the zoological geography of the Malay Archipelago. *Proc Linnean Soc Lond* 4:172–184.

Wallace, A. R. 1876. *The geographical distribution of animals.* London: Macmillan.

Wallace, J. B., and A. C. Benke. 1984. Quantification of wood habitat in subtropical coastal plain streams. *Can J Fish Aquat Sci* 41:1643–1652.

Wallace, R. A., and K. Selman. 1981. Cellular and dynamic aspects of oocyte growth in teleosts. *Amer Zool* 21:325–343.

Wallin, J. E. 1989. Bluehead chub (*Nocomis leptocephalus*) nests used by yellowfin shiners (*Notropis lutipinnis*). *Copeia* 1989:1077–1080.

Wallin, J. E. 1992. The symbiotic nest association of yellowfin shiners, *Notropis lutipinnis*, and bluehead chubs, *Nocomis leptocephalus. Env Biol Fish* 33:287–292.

Walls, M., I. Kortelainen, and J. Sarvala. 1990. Prey responses to fish predation in freshwater communities. *Ann Zool Fennici* 27:183–199.

Walsh, W. J. 1983. Stability of a coral reef fish community following a catastrophic storm. *Coral Reefs* 2:49–63.

Walters, V. 1964. Order Giganturoidei. *Mem Sears Fnd Mar Res* 1(4):566–577.

Wankowski, J. W. J., and J. E. Thorpe. 1979. Spatial distribution and feeding in Atlantic salmon, *Salmo salar*, juveniles. *J Fish Biol* 14:239–247.

Waples, R. S. 1987. A multispecies approach to the analysis of gene flow in marine shore fishes. *Evolution* 41:385–400.

Ward, H. C., and E. M. Leonard. 1954. Order of appearance of scales in the black crappie, *Pomoxis nigromaculatus. Proc Okla Acad Sci* 33:138–140.

Ware, D. M. 1978. Bioenergetics of pelagic fish: Theoretical change in swimming speed and relation with body size. *J Fish Res Board Can* 35:220–228.

Warner, R. R. 1975. The adaptive significance of sequential hermaphroditism in animals. *Amer Natur* 109:61–82.

Warner, R. R. 1978. The evolution of hermaphroditism and unisexuality in aquatic and terrestrial vertebrates. In *Contrasts in behavior*, ed. E. S. Reese and F. J. Lighter, 77–101. New York: Wiley-Interscience.

Warner, R. R. 1982. Metamorphosis. *Science* 82 3:42–46.

Warner, R. R. 1988. Traditionality of mating site preferences in a coral reef fish. *Nature* (London) 335:719–721.

Warner, R. R. 1990. Resource assessment versus traditionality in mating site determination. *Amer Natur* 135:205–217.

Warner, R. R. 1991. The use of phenotypic plasticity in coral reef fishes as tests of theory in evolutionary ecology. In *The ecology of fishes on coral reefs*, ed. P. F. Sale, 387–398. San Diego, CA: Academic Press.

Warner, R. R., and R. K. Harlan. 1982. Sperm competition and sperm storage as determinants of sexual dimorphism in the dwarf surfperch, *Micrometrus minimus. Evolution* 36:44–55.

Warner, R. R., and S. G. Hoffman. 1980. Local population size as a determinant of mating system and sexual composition in two tropical marine fishes (*Thalassoma* spp.). *Evolution* 34:508–518.

Warner, R. R., D. R. Robertson, and E. G. Leigh Jr. 1975. Sex change and sexual selection. *Science* 190:633–638.

Watson, D. M. S. 1927. Reproduction of the coelacanth fish *Undina. Proc Zool Soc Lond* 1:453–457.

Weatherly, A. H., and H. S. Gill. 1987. *The biology of fish growth.* London: Academic Press.

Webb, P. W. 1982. Locomotor patterns in the evolution of actinopterygian fishes. *Amer Zool* 22:329–342.

Webb, P. W. 1984. Form and function in fish swimming. *Sci Amer* 251(1):72–84.

Webb, P. W. 1986. Locomotion and predator-prey relationships. In *Predator-prey relationships: Perspectives and approaches from the study of lower vertebrates*, ed. M. E. Feder and G. V. Lauder, 24–41. Chicago: University of Chicago Press.

Webb, P. W. 1993. Swimming. In *The physiology of fishes*, ed. D. H. Evans, 47–73. Boca Raton, FL: CRC Press.

Webb, P. W., and R. W. Blake. 1985. Swimming. In *Functional vertebrate morphology*, ed. M. Hildebrand, D. M. Bramble, K. F. Liem, and D. B. Wake, 110–128. Cambridge, MA: Belknap Press.

Webb, P. W., and R. S. Keyes. 1982. Swimming kinematics of sharks. *Fish Bull US* 80:803–812.

Webb, P. W., and J. M. Skadsen. 1980. Strike tactics of *Esox. Can J Zool* 58:1462–1469.

Webb, P. W., and D. Weihs. 1983. *Fish biomechanics.* New York: Praeger.

Wedemeyer, G. A., B. A. Barton, and D. J. McLeay. 1990. Stress and acclimation. In *Methods for fish biology*, ed. C. B. Schreck and P. B. Moyle, 451–489. Bethesda, MD: American Fisheries Society.

Weihs, D. 1975. Some hydrodynamical aspects of fish schooling. In *Swimming and flying in nature*, ed. W. Wu, C. J. Brokaw, and C. Brennen, 703–718. New York: Plenum Press.

Weihs, D., and H. G. Moser. 1981. Stalked eyes as an adaptation towards more efficient foraging in marine fish larvae. *Bull Mar Sci* 31:31–36.

Weinstein, M. P. 1979. Shallow marsh habitats as primary nurseries for fishes and shellfish, Cape Fear River, North Carolina. *Fish Bull US* 77:339–357.

Weis, J. S., and P. Weis. 1989. Effects of environmental pollutants on early fish development. *Rev Aquat Sci* 1:45–73.

Weisberg, S. B., R. Whalen, and V. A. Lotrich. 1981. Tidal and diurnal influence on food consumption of a saltmarsh killifish *Fundulus heteroclitus. Mar Biol* 61:243–246.

Weitzman, S. H. 1962. The osteology of *Brycon meeki*, a generalized characid fish, with an osteological definition of the family. *Stanford Ichthy Bull* 8:1–77.

Welcomme, R. L. 1984. International transfers of inland fish species. In *Distribution, biology, and management of exotic fishes*, ed. W. R. Courtenay Jr. and J. R. Stauffer Jr., 22–40. Baltimore: Johns Hopkins University Press.

Welcomme, R. L. 1985. *River fisheries.* FAO Tech Paper 262. Rome: FAO.

Wellington, G. M., and B. C. Victor. 1988. Variation in components of reproductive success in an undersaturated population of coral-reef damselfish: A field perspective. *Amer Natur* 131:588–601.

Wells, R. M. G., M. D. Ashby, S. J. Duncan, and J. A. MacDonald. 1980. Comparative study of the erythrocytes and haemoglogins in nototheniid fishes from Antarctica. *J Fish Biol* 17:517–527.

Wendelaar Bonga, S. E. W. 1993. Endocrinology. In *The physiology of fishes*, ed. D. H. Evans, 469–502. Boca Raton, FL: CRC Press.

Werner, E. E. 1984. The mechanisms of species interactions and community organization in fish. In *Ecological communities: Conceptual issues and the evidence*, ed. D. R. Strong, D. Simberloff, L. G. Abele, and A. B. Thistle, 360–382. Princeton, NJ: Princeton University Press.

Werner, E. E., and J. F. Gilliam. 1984. The ontogenetic niche and species interactions in size-structured populations. *Ann Rev Ecol Syst* 15:3939–425.

Werner, E. E., J. F. Gilliam, D. J. Hall, and G. G. Mittelbach. 1983.

An experimental test of the effect of predation risk on habitat use in fish. *Ecology* 64:1540–1548.

Werner, E. E., and D. J. Hall. 1974. Optimal foraging and the size selection of prey by the bluegill sunfish (*Lepomis macrochirus*). *Ecology* 55:1042–1052.

Werner, E. E., and D. J. Hall. 1979. Foraging efficiency and habitat switching in competing sunfish. *Ecology* 60:256–264.

Westby, G. W. 1979. Electrical communication and jamming avoidance between resting *Gymnotus carapo*. *Behav Ecol Sociobiol* 4:381–393.

Westneat, M. W., and P. C. Wainwright. 1989. Feeding mechanism of *Epibulus insidiator* (Labridae; Teleostei): Evolution of a novel functional system. *J Morph* 202:129–150.

Wheeler, A. 1975. *Fishes of the world*. New York: Macmillan.

Wheeler, A. 1985. *The world encyclopedia of fishes*. London: Macdonald.

Wheeler, A., and D. Sutcliffe, eds. 1991. The biology and conservation of rare fish. *J Fish Biol* 37 (suppl A):1–271.

Whitear, M. 1970. The skin surface of bony fishes. *J Zool* (London) 160:437–454.

Wildhaber, M. L., and L. B. Crowder. 1991. Mechanisms of patch choice by bluegills (*Lepomis macrochirus*) foraging in a variable environment. *Copeia* 1991:445–460.

Wiley, E. E. 1976. The phylogeny and biogeography of fossil and Recent gars (Actinopterygii: Lepisosteidae). *Univ Kansas Mus Natur Hist, Misc Publ* 64:1–111.

Wiley, M. L., and B. B. Collette. 1970. Breeding tubercles and contact organs in fishes: Their occurrence, structure, and significance. *Bull Amer Mus Natur Hist* 143:143–216.

Wilimovsky, N. J. 1956. *Protoscaphirhyncus squamosus*, a new sturgeon from the Upper Cretaceous of Montana. *J Paleo* 30:1205–1208.

Wilkens, H. 1988. Evolution and genetics of epigean and cave *Astyanax fasciatus* (Characidae, Pisces): Support of the neutral mutation theory. *Evol Bio* 23:271–367.

Wilkinson, C. R., and I. G. Macintyre, eds. 1992. The *Acanthaster* debate. *Coral Reefs* 11:51–122.

Williams, J. D., and G. H. Clemmer. 1991. *Scaphirhynchus suttkusi*, a new sturgeon (Pisces: Acipenseridae) from the Mobile Basin of Alabama and Mississippi. *Bull Alabama St Mus Natur Hist* 10:17–31.

Williams, J. E., et al. 1989. Fishes of North America, endangered, threatened, or of special concern; 1989. *Fisheries* 14(6):2–20.

Williams, T. 1992. The last bluefin hunt. *Audubon* 92(4):14–20.

Wilson, E. O. 1988. The current state of biological diversity. In *Biodiversity*, ed. E. O. Wilson and F. M. Peter, 3–18. Washington, DC: National Academy Press.

Wilson, M. V. H., and M. W. Caldwell. 1993. New Silurian and Devonian fork-tailed 'thelodonts' are jawless vertebrates with stomachs and deep bodies. *Nature* 361:442–444.

Winemiller, K. O. 1990a. Caudal eyespots as deterrents against fin predation in the neotropical cichlid *Astronotus ocellatus*. *Copeia* 1990:665–673.

Winemiller, K. O. 1990b. Spatial and temporal variation in tropical fish trophic networks. *Ecol Monogr* 60:331–367.

Winemiller, K. O., and K. A. Rose. 1992. Patterns of life-history diversification in North American fishes: Implications for population regulation. *Can J Fish Aquat Sci* 49:2196–2218.

Winn, H. E., and J. E. Bardach. 1959. Differential food selection by moray eels and a possible role of the mucous envelope of parrot fishes in reduction of predation. *Ecology* 40:296–298.

Winterbottom, R. 1974. A descriptive synonymy of the striated muscles of the Teleostei. *Proc Acad Nat Sci Philadelphia* 125:225–317.

Winterbottom, R., and A. R. Emery. 1981. A new genus and two new species of gobiid fishes (Perciformes) from the Chagos Archipelago, Central Indian Ocean. *Env Biol Fish* 6:139–149.

Winton, J. R., J. S. Rohovec, and J. L. Fryer. 1983. Bacterial and viral diseases of cultured salmonids in the Pacific Northwest. In *Bacterial and viral diseases of fish*, ed. J. H. Crosa, 1–20. Seattle: Washington Sea Grant Publication WSG-WO 83-1.

Wisby, W. J., and A. D. Hasler. 1954. The effect of olfactory occlusion on migrating silver salmon (*Oncorhynchus kisutch*). *J Fish Res Board Can* 1:472–478.

Wisner, R. L. 1958. Is the spear of istiophorid fishes used in feeding? *Pac Sci* 12:60–70.

Wisner, R. L., and C. B. McMillan. 1990. Three new species of hagfishes, Genus *Eptatretus* (Cyclostomata, Myxinidae), from the Pacific coast of North America, with new data on *E. deani* and *E. stoutii*. *Fish Bull US* 88:787–804.

Withers, P. C. 1992. *Comparative animal physiology*. New York: Saunders College Publishing.

Witte, F., et al. 1992. The destruction of an endemic species flock: Quantitative data on the decline of the haplochromine cichlids of Lake Victoria. *Env Biol Fish* 34:1–28.

Wittenberger, J. F. 1981. *Animal social behavior*. Boston: Duxbury Press.

Wolf, N. G. 1985. Odd fish abandon mixed-species groups when threatened. *Behav Ecol Sociobiol* 17:47–52.

Wolf, N. G., P. R. Swift, and F. G. Carey. 1988. Swimming muscle helps warm the brain of lamnid sharks. *J Comp Physiol* 157:709–715.

Wolke, R. E., R. A. Murchelano, C. D. Dickstein, and C. J. George. 1985. Preliminary evaluation of the use of macrophage aggregates (MA) as fish health monitors. *Bull Environ Contam Toxicol* 35:222–227.

Wood, C. W. 1991. Branchial ion and acid-base transfer in freshwater teleost fish: Environmental hyperoxia as a probe. *Physiol Zool* 64:68–102.

Wood, C. W. 1993. Ammonia and urea metabolism and excretion. In *The physiology of fishes*, ed. D. H. Evans, 379–426. Boca Raton, FL: CRC Press.

Woodley, J. D., et al. 1981. Hurricane Allen's impact on Jamaican coral reefs. *Science* 214:749–755.

Wooster, W. 1990. King crab deposed. In *Climate variability, climate change and fisheries*, ed. M. H. Glantz and L. E. Feingold, 13–17. Boulder, CO: Environmental and Societal Impacts Group, Nat Ctr Atmos Res.

Wootton, J. T., and M. P. Oemke. 1992. Latitudinal differences in fish community trophic structure, and the role of fish herbivory in a Costa Rican stream. *Env Biol Fish* 35:311–319.

Wootton, R. J. 1976. *The biology of the sticklebacks*. London: Academic Press.

Wootton, R. J. 1984. *A functional biology of sticklebacks*. London: Croom Helm.

Wootton, R. J. 1990. *Ecology of teleost fishes*. London: Chapman and Hall.

Wourms, J. P. 1972. The developmental biology of annual fishes. III. Pre embryonic and embryonic diapause of variable duration in the eggs of annual fishes. *J Exptal Zool* 182:389–414.

Wourms, J. P. 1981. Viviparity: The maternal-fetal relationship in fish. *Amer Zool* 21:473–515.

Wourms, J. P. 1988. The maternal-embryonic relationship in viviparous fishes. In *Fish physiology*. Vol. 11, *The physiology of developing fish*, Part B, *Viviparity and posthatching juveniles*, ed. W. S. Hoar and D. J. Randall, 1–134. San Diego, CA: Academic Press.

Wourms, J. P., and L. S. Demski. 1993a. The reproduction and development of sharks, skates, rays and ratfishes: Introduction, history, overview, and future prospects. *Env Biol Fish* 38:7–21.

Wourms, J. P., and L. S. Demski, eds. 1993b. The reproduction and development of sharks, skates, rays and ratfishes. *Env Biol Fish* 38:1–294.

Wourms, J. P., B. D. Grove, and J. Lombardi. 1988. The maternal-embryonic relationship in viviparous fishes. In *Fish physiology*, Vol. 11b, ed. W. S. Hoar and D. J. Randall, 1–134. San Diego, CA: Academic Press.

Wyanski, D. M., and T. E. Targett. 1981. Feeding biology of fishes in the endemic Antarctic Harpagiferidae. *Copeia* 1981:686–693.

Wydoski, R. S., and J. Hamill. 1991. Evolution of a cooperative recovery program for endangered fishes in the Upper Colorado River Basin. In *Battle against extinction: Native fish management in the American West*, ed. W. L. Minckley and J. E. Deacon, 123–135. Tucson: University of Arizona Press.

Yabe, M. 1985. Comparative osteology and myology of the superfamily Cottoidea (Pisces: Scorpaeniformes), and its phylogenetic classification. *Hokkaido Univ: Mem Fac Fish* 32(1):1–130.

Yant, P. R., J. R. Karr, and P. L. Angermeier. 1984. Stochasticity in stream fish communities: An alternative interpretation. *Amer Natur* 124:573–582.

Youson, J. H. 1988. First metamorphosis. In *Fish physiology*. Vol. 11, *The physiology of developing fish*. Part B, *Viviparity and posthatching juveniles*, ed. W. S. Hoar and D. J. Randall, 135–196. San Diego, CA: Academic Press.

Zachman, A., M. A. Ather, and J. Falcon. 1992. Melatonin and its effects in fishes: An overview. In *Rhythms in fishes*, ed. M. A. Ali, 149–165. New York: Plenum Press.

Zangerl, R. 1984. On the microscopic anatomy and possible function of the spine "brush" complex of *Stethacanthus* (Elasmobranchii: Symmoriida). *J Vert Paleont* 4:372–378.

Zaret, T. M. 1980. *Predation and freshwater communities*. New Haven, CT: Yale University Press.

Zaret, T. M., and R. T. Paine. 1973. Species introduction in a tropical lake. *Science* 182:449–455.

Zaret, T. W., and A. S. Rand. 1971. Competition in tropical stream fishes: Support for the competitive exclusion principle. *Ecology* 52:336–342.

Zelikoff, J. T. 1994. Immunological alterations as indicators of environmental metal exposure. In *Modulators of fish immune responses*. Vol. 1, *Models for environmental toxicology, biomarkers, immunostimulators*, ed. J. S. Stolen and T. C. Fletcher, 101–110. Fair Haven: SOS Publications.

Index

Note: Page numbers followed by f refer to illustrations; page numbers followed by t refer to tables.